D0772917

HANDBOOK OF INSTRUMENTATION AND CONTROLS

McGRAW-HILL HANDBOOKS

ABBOTT AND STETKA · National Electrical Code Handbook, 10th ed.
ALJIAN · Purchasing Handbook
AMERICAN INSTITUTE OF PHYSICS · American Institute of Physics Handbook
AMERICAN SOCIETY OF MECHANICAL ENGINEERS · ASME Handbooks:
Engineering Tables | Metals Engineering—Processes
Metals Engineering—Design | Metals Properties
AMERICAN SOCIETY OF TOOL AND MANUFACTURING ENGINEERS · Die Design Handbook
AMERICAN SOCIETY OF TOOL AND MANUFACTURING ENGINEERS · Handbook of Fixture Design
AMERICAN SOCIETY OF TOOL AND MANUFACTURING ENGINEERS · Tool Engineers Handbook, 2d ed.
BEEMAN · Industrial Power Systems Handbook
BERRY, BOLLAY, AND BEERS · Handbook of Meteorology
BLATZ · Radiation Hygiene Handbook
BRADY · Materials Handbook, 8th ed.
BURINGTON · Handbook of Mathematical Tables and Formulas, 3d ed.
BURINGTON AND MAY · Handbook of Probability and Statistics with Tables
CARROLL · Industrial Instrument Servicing Handbook
COCKRELL · Industrial Electronics Handbook
CONDON AND ODISHAW · Handbook of Physics
CONSIDINE · Process Instruments and Controls Handbook
CROCKER · Piping Handbook, 4th ed.
CROFT AND CARR · American Electricians' Handbook, 8th ed.
DAVIS · Handbook of Applied Hydraulics, 2d ed.
DUDLEY · Gear Handbook
ETHERINGTON · Nuclear Engineering Handbook
FACTORY MUTUAL ENGINEERING DIVISION · Handbook of Industrial Loss Prevention
FINK · Television Engineering Handbook
FLÜGGE · Handbook of Engineering Materials
FRICK · Petroleum Production Handbook, 2 vols.
GUTHRIE · Petroleum Products Handbook
HARRIS · Handbook of Noise Control
HARRIS AND CREDE · Shock and Vibration Handbook, 3 vols.
HENNEY · Radio Engineering Handbook, 5th ed.
HUNTER · Handbook of Semiconductor Electronics, 2d ed.
HUSKEY AND KORN · Computer Handbook
JASIK · Antenna Engineering Handbook
JURAN · Quality Control Handbook, 2d ed.
KALLEN · Handbook of Instrumentation and Controls
KETCHUM · Structural Engineers' Handbook, 3d ed.
KING · Handbook of Hydraulics, 4th ed.
KNOWLTON · Standard Handbook for Electrical Engineers, 9th ed.
KOELLE · Handbook of Astronautical Engineering
KORN AND KORN · Mathematical Handbook for Scientists and Engineers
KURTZ · The Lineman's Handbook, 3d ed.
LA LONDE AND JANES · Concrete Engineering Handbook
LANDEE, DAVIS, AND ALBRECHT · Electronic Designers' Handbook
LANGE · Handbook of Chemistry, 10th ed.
LAUGHNER AND HARGAN · Handbook of Fastening and Joining of Metal Parts
LE GRAND · The New American Machinist's Handbook
LIDDELL · Handbook of Nonferrous Metallurgy, 2d ed.
MAGILL, HOLDEN, AND ACKLEY · Air Pollution Handbook
MANAS · National Plumbing Code Handbook
MANTELL · Engineering Materials Handbook
MARKS AND BAUMEISTER · Mechanical Engineers' Handbook, 6th ed.
MARKUS · Handbook of Electronic Control Circuits
MARKUS AND ZELUFF · Handbook of Industrial Electronic Circuits
MARKUS AND ZELUFF · Handbook of Industrial Electronic Control Circuits
MAYNARD · Industrial Engineering Handbook
MERRITT · Building Construction Handbook
MOODY · Petroleum Exploration Handbook
MORROW · Maintenance Engineering Handbook
PERRY · Chemical Business Handbook
PERRY · Chemical Engineers' Handbook, 3d ed.
SHAND · Glass Engineering Handbook, 2d ed.
STANIAR · Plant Engineering Handbook, 2d ed.
STREETER · Handbook of Fluid Dynamics
STUBBS · Handbook of Heavy Construction
TERMAN · Radio Engineers' Handbook
TRUXAL · Control Engineers' Handbook
URQUHART · Civil Engineering Handbook, 4th ed.
WALKER · NAB Engineering Handbook, 5th ed.
WOODS · Highway Engineering Handbook
YODER, HENEMAN, TURNBULL, AND STONE · Handbook of Personnel Management and Labor Relations

HANDBOOK OF INSTRUMENTATION AND CONTROLS

A PRACTICAL DESIGN AND APPLICATIONS MANUAL FOR THE MECHANICAL SERVICES COVERING STEAM PLANTS, POWER PLANTS, HEATING SYSTEMS, AIR-CONDITIONING SYSTEMS, VENTILATION SYSTEMS, DIESEL PLANTS, REFRIGERATION, AND WATER TREATMENT

HOWARD P. KALLEN, B.M.E., M.M.E., P.E., Editor

Partner, Kallen & Lemelson, Consulting Engineers
Member, Instrument Society of America
American Society of Mechanical Engineers

FIRST EDITION

McGRAW-HILL BOOK COMPANY

NEW YORK TORONTO LONDON

1961

1982 Reissue

HANDBOOK OF INSTRUMENTATION AND CONTROLS

Library of Congress Catalog Card Number: 60-6886

11 12 13 14 15 FGBP 8 5 4 3 2 1

33235

PREFACE

As the building and plant services—power, heating, air conditioning, refrigeration, prime movers, and others—have become more complex in the years since World War II, the engineer has become increasingly aware of the importance of well-designed instrumentation and control systems to ensure proper operation of these building and plant services. Yet, as a practical matter, few engineers other than instrumentation specialists have the opportunity in their day-to-day activity to fully explore the many applications of instrumentation and control systems to their engineering problems.

This handbook has been prepared with the "nonspecialist" engineer in mind, although the specializing instrumentation engineer will find much material in each section of practical value to him. The material is presented in sufficient detail to provide a sound basis for system design, application, selection, and operation. It is intended to be a practical tool for all who are concerned with the mechanical services in institutional, commercial, and industrial buildings.

Thus, this book will be useful to engineers who design and apply mechanical systems to buildings, plant engineers, architects, consulting engineers, maintenance and operating engineers, power and utility engineers, and others in allied fields.

In its general arrangement the handbook gives emphasis to fundamentals as well as practical application. In this way it is felt that the material will be equally useful to the nonspecialist and specialist alike.

After a study of fundamentals in the second section, the reader is provided with both qualitative and quantitative data on pressure, temperature, flow liquid level, pH, and conductivity—the variables most frequently measured and controlled in mechanical services for buildings. These are then tied together in a detailed discussion of systems for boiler and power plants, heating plants, mechanical drives, air-conditioning ventilation, and refrigeration, with considerable space devoted to actual cases.

It is hoped that this handbook will contribute not only to a better understanding of the instruments and controls themselves, but to a fuller appreciation of what they can and cannot be expected to do when integrated with mechanical-service systems. In this way the hazard of misapplication may be avoided. It is quite true that no mechanical-service system can be better than the controls which cause it to function.

A handbook, by its very nature, is a compilation of the knowledge of many people, organizations, and companies. And, while individual credits are acknowledged throughout this handbook, I wish to express special appreciation to such individuals as A. C. Wenzel of Republic Flow Meters Co., R. E. Sprenkle of Bailey Meter Co., J. E. Haines of Minneapolis-Honeywell Regulator Co., and D. M. Considine of Hughes Aircraft Corp., and particularly to the Instrument Society of America and the American Society of Mechanical Engineers.

Finally, I wish to acknowledge the invaluable help of Mrs. Katherine Mangione and of my wife Claire in the preparation and proofreading of the manuscript and index.

Howard P. Kallen

CONTENTS

HANDBOOK OF INSTRUMENTATION AND CONTROLS

Section 1

INTRODUCTION

Today's advanced control techniques owe much to their early development in the mechanical services field. Closed-loop control systems were pioneered almost 40 years ago. Since that time, the mechanical services have been a proving ground for an almost endless number of developments that have pushed the controls art to new heights.

But this has been a two-way street; in the same period, controls had much to do with boosting electric generating efficiency from roughly 3½ lb of coal per kwhr to ¾ lb. And other services—air conditioning with its zone controls, electrical distribution's load-frequency control systems—have reaped the full measure from a dynamic controls technology. Yet, the ultimate in controls is nowhere in sight. We will witness many new and fundamental developments in the years ahead. And, no doubt, engineers will some day look upon today's control methods as relatively crude ancestors of "truly modern" control.

Primary Measurements. Pressure—force per unit area—is one of the most important of controlled variables. For the mechanical services, pressures range from less than an inch of mercury to several thousand pounds per square inch. And they must be measured accurately. Because of the broad range, numerous pressure-measuring elements have been developed. They vary from direct means such as spiral and bourdon gages for high pressures to inferential methods such as thermal (hot-wire) gages for high vacuums.

A relative newcomer to the field is the pressure transducer. It is based on electrical interpretation of pressure. A common design uses the strain-gage principle. The strain gage itself consists of a wire grid bonded to the surface of a small piece of Bakelite-impregnated paper. When the gage is cemented to the surface subjected to load, the wire grid stretches, changing its electrical resistance. By measuring resistance changes (with a potentiometer), we indirectly measure pressure. Transducers are compatible with the highest pressures the mechanical services field will be required to measure for years to come.

Temperatures that concern mechanical engineers are in the range of −200 to well over 1000°F. To meet the requirements of this wide spectrum, researchers constantly seek new principles of measurement, improving on existing methods.

Over certain temperature ranges, we have a wide selection of measuring elements. But range is only one of several key factors. Sensitivity, accuracy, response speed, cost, expected useful life, corrosion resistance—all need to be considered.

Briefly, present measuring practice breaks down like this: filled-system (vapor, gas, mercury) thermometers from −150 to 1000°F; resistance thermometers from −100 to 600°F; thermocouples from −300 to 2700°F; radiation (optical) pyrometers from 200 to 7000°F. We shall see more of radiation techniques for extremely high temperatures or in situations where it is impractical to contact the area being measured.

Liquid-, vapor-, and gas-flow measurement—of key importance to the mechanical services—has been the subject of several recent developments. But differential-pressure (orifice plate, flow nozzle, venturi tube) methods still are most frequently used. They are simple, easy to use, and readily adaptable to electric or pneumatic transmissions.

Area-type flow meters fit directly in the flow line and require no orifice plates or other primary elements. Pressure drop across the meter body is constant while flow area varies. For viscous fluids at least (oils, black liquor, etc.), we can expect stepped-up use of these meters.

Newest flow-sensing elements depend on electrical and magnetic techniques. One design employs a bucket-wheel rotor that spins freely between supports in the flow line. A powerful magnet inside the rotor produces an a-c signal with a frequency directly proportional to flow rate. These units can measure accurately continuous or intermittent flow of chemicals, liquefied gases, and solutions containing suspended particles at temperatures up to 1000°F. Range is from 0.08 to 5,200 gpm.

Recently developed methods of liquid-level measurement make use of radioactive materials. The gamma-ray system is based on the change in the number of gamma rays that can penetrate a layer of liquids. As layer thickness increases, the number decreases. Gamma-ray source is a minute quantity of radioactive material such as radium salts.

Controllers. A distinguishing mark of modern controls is their inability to tolerate time lags. Lacking "flywheel effect," they require faster, more accurate control response. To meet this need, more controllers will lean toward sophisticated control modes—combinations of proportional, automatic reset, and rate action.

On the question of pneumatic versus electrical and electronic controllers, we are likely to witness increasing use of all three types. This is largely because no one design will cover the full range of what is certain to be a fast-growing list of jobs.

Remote Transmission. Remote transmission of measurements—sending instrument signals over relatively long distances—has many advantages: safety, economy, convenience. Where high-pressure high-temperature fluids are present, remote transmission eliminates hazards by permitting control from a safe location. Thus, we shall see a strong trend continuing in this direction. Recent findings of an Edison Electric Institute questionnaire covering direct versus transmitted pressure for oil, hydrogen, high-pressure steam and water systems back up this thinking. Of 46 answering utility companies using some form of remote transmission (either electric or air or both), 63 per cent were on steam systems, 50 per cent on water systems, and 79 per cent on oil systems.

Telemetering, extension of the in-plant transmission idea, fills an important need in the far-flung power, pipeline, and water-supply systems of today. As these systems continue to grow, telemetering will take on even greater importance. Carrier current is the most common transmission medium on power systems, with private lines, leased lines, and microwave following in that order.

Systems Engineering. In the strict sense of the word, there is nothing new about *systems engineering* in the controls field. Instrument men have been engineering control systems for years.

Today, however, the term appears to be taking on an added meaning—complete *integration* of control-system design with the system it serves. The controls expert is called in and works closely with the equipment engineer from the early stages of design. The net result is a better control system, and equipment that is easier to control.

Scope. The scope of material covered in this handbook is intended to be of direct interest to the mechanical engineer engaged in the design, application, and operation of systems for mechanical services: steam plants, power plants, heating plants and heating systems, air-conditioning systems, diesel plants, and refrigeration systems. And, although the instrumentation and controls discussed have a broad application in other fields, notably the process industries, they are viewed here mainly as they apply to the mechanical services.

Instrumentation for certain mechanical services, because of its highly specialized nature, has been omitted. In this category is nuclear-reactor instrumentation. For an excellent treatment of this subject, the reader is referred to "Control of Nuclear Reactors and Power Plants" by M. A. Schultz, McGraw-Hill Book Company, Inc., 1955.

Credits. As editor of this handbook, I have received the most generous assistance and cooperation from a great number of U.S. instrumentation manufacturers, the Instrument Society of America, ASME, consulting engineering firms, and many users of mechanical-service instrumentation. While specific credits are made throughout the book, I wish to take this opportunity to express my deep appreciation to the many individuals and companies without whose cooperation preparation of this handbook would have been impossible.

Section 2

INSTRUMENTATION AND CONTROL FUNDAMENTALS

SYMBOLS AND NOTATIONS

In order to convey his ideas and information effectively and concisely, the controls engineer makes use of a system of symbols. They are, in effect, his "shorthand" and should be understood by the engineer who is responsible for design, selection, operation, or maintenance of control systems for the mechanical and electrical services.

A system of symbols has been standardized by the Instrument Society of America[1] and is given here in condensed form.

General. Instruments and instrumentation items are identified and represented by a system of letters and numbers, together with a number of simple basic pictorial symbols for illustrating the items on flow plans and other drawings.

Identifications. The identification shall consist of a combination of letters used to establish the *general identity* of the item with its purpose and functions. For some requirements this will be sufficient and complete; but usually it will be followed by a number that will serve to establish the *specific identity* of the item. The identifications shall be used for the complete designation of the item in written work, and in combination with the pictorial symbol for representation on flow plans or other drawings.

General Identifications. The general identifications shall consist of letters as listed in Table 2-1 used in combinations as shown in Table 2-2.

Table 2-1 covers the letters that may be employed, the definition or significance of each, and the permissible position or positions in which each may be used when combining.

Table 2-2 covers the permissible combinations of letters of identification and shows the significance of each such complete general identification.

In the use of the letters, or their combinations, the following rules and instructions apply:

1. All identifying letters shall in all cases be written in upper case. The

[1] *Tentative Recommended Practice, Instrumentation Flow Plan Symbols, ISA RP5.1,* Instrument Society of America.

only exceptions are the *optional* use of "d" and "r," and the use of "p" in the combined first letter "pH"; as per footnotes of Tables 2-1 and 2-2.

2. The *maximum* number of identifying letters in any combination shall be *three* (3). The only exception is in the use of "pH," or chemical symbols such as CO_2 where the self-defining pair is treated as a single letter.

Table 2-1. Letters of Identification

Upper-case letter	Definition, and permissible positions in any combination		
	First letter—process variable or actuation	Second letter—type reading or other function	Third letter—additional function
A	—	Alarm	Alarm
C	Conductivity	Control	Control
D	Density	—	—
E	—	Element (primary)	—
F	Flow	—	—
G	—	Glass (no measurement)	—
H	Hand (actuated)	—	—
I	—	Indicating	—
L	Level	—	—
M	Moisture	—	—
P	Pressure	—	—
R	—	Recording (recorder)	—
S	Speed	Safety	—
T	Temperature	—	—
V	Viscosity	—	Valve
W	Weight	Well	—

Note 1: When required the following may be used optionally as a first letter for other process variables:
(1) "A" may be used to cover all types of analyzing instruments.
(2) Readily recognized self-defining chemical symbols such as CO_2, O_2, etc., may be used for these specific analysis instruments.
(3) The self-defining symbol "pH" may be used for hydrogen-ion concentration.

Note 2: Although not a preferred procedure, when considered necessary it is permissible to insert a lower-case "r" after "F" to distinguish flow ratio. Likewise, lower-case "d" may be inserted after "T" or "P" to distinguish temperature difference or pressure difference.

3. A letter shall have only one definition or significance in its use as a "first" letter in any combination, to define the process variable.
4. A letter shall have only one definition or significance when used as *either* the "second" or the "third" letter in a combination, to define the type of device.
5. It is particularly important in writing the combinations of letters to adhere to the sequence of arrangement shown by Table 2-2.
6. No hyphens shall be used between letters or combinations of letters.

Table 2-2. Complete General Identifications
(Combinations of letters)

	First letter—process variable or actuation	Second and third letters—type of device												
		Controlling devices				Safety (relief) valves	Measuring devices		Glass devices for observation only (no measurement)	Alarm devices			Primary element	Wells
		Separate controllers			Self-actuated (integral) regulating valves									
		Recording	Indicating	Blind			Recording	Indicating		Recording	Indicating	Blind		
		-RC	-IC	-C	-CV	-SV	-R	I-	-G	-RA	-IA	-A	-E	-W
Temperature	T-	TRC	TIC	TC	TCV	TSV	TR	TI	///	TRA	TIA	TA	TE	TW
Flow	F-	FRC	FIC				FR	FI	FG	FRA	FIA		FE	///
Level	L-	LRC	LIC	LC	LCV		LR	LI	LG	LRA	LIA	LA		///
Pressure	P-	PRC	PIC	PC	PCV	PSV	PR	PI	///	PRA	PIA	PA	PE	///
Density	D-	DRC	DIC	DC			DR	DI	///	DRA	DIA			///
Hand	H-		HIC	HC	HCV		///	///	///	///	///		///	///
Moisture	M-	MRC	MIC	MC			MR	MI	///	MRA	MIA	MA	ME	///
Conductivity	C-	CRC	CIC				CR	CI	///	CRA	CIA	CA	CE	///
Speed	S-	SRC	SIC	SC	SCV	SSV	SR	SI		SRA	SIA	SA		///
Viscosity	V-	VRC	VIC				VR	VI	VG	VRA	VIA			///
Weight	W-	WRC	WIC				WR	WI		WRA	WIA		WE	///

Note: The optional additional process variables given in footnotes of Table 2-1, when used, shall be combined with second and third letters as per above.

Shaded spaces (///) indicate impossible combinations. Blank spaces indicate improbable combinations.

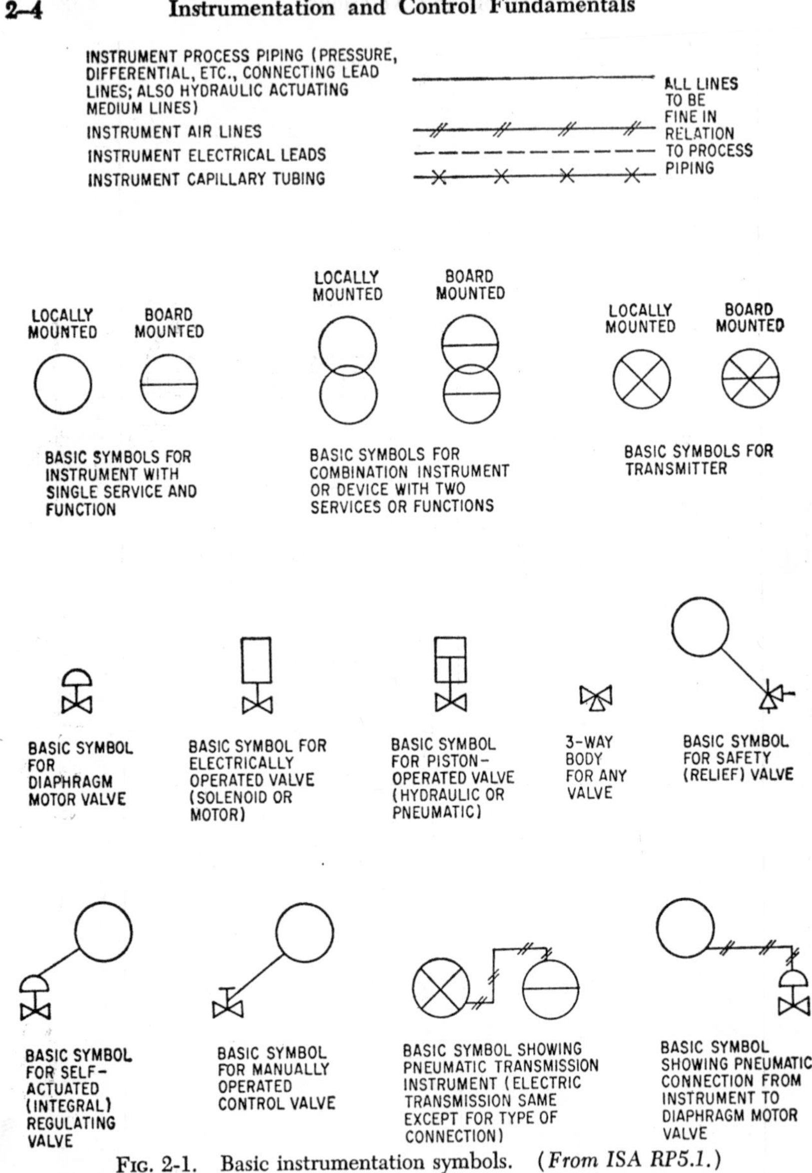

FIG. 2-1. Basic instrumentation symbols. (*From ISA RP5.1.*)

Specific Identifications. In most cases it will be necessary to supplement the general identification of an item by a numerical system, to establish its specific identity. Any system of item or serial numbers may be used, consistent with the requirements of the user. The numbers may pertain only to the same kind of item within one process unit; or may be a com-

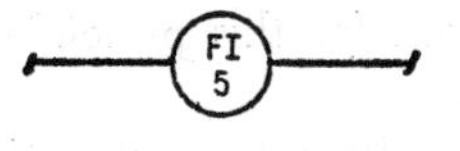

DISPLACEMENT-TYPE FLOW METER

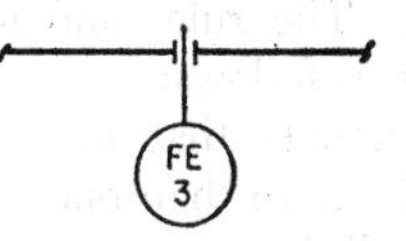

FLOW ELEMENT (PRIMARY) (WHEN NO MEASURING INSTRUMENT IS PROVIDED)

FLOW INDICATOR, DIFFERENTIAL TYPE, LOCALLY MOUNTED

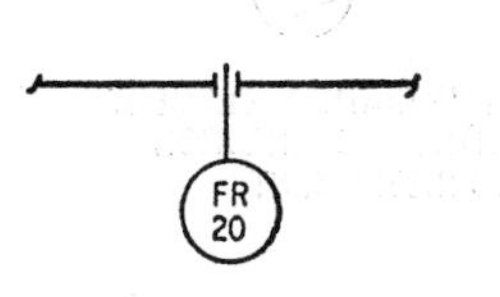

FLOW RECORDER, DIFFERENTIAL TYPE, MECHANICAL TRANSMISSION, LOCALLY MOUNTED

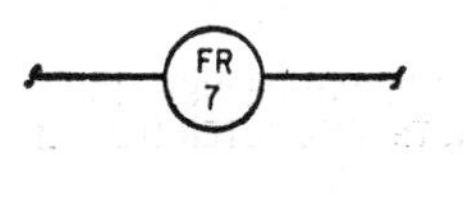

FLOW RECORDER, OF ROTAMETER OR OTHER IN-THE-LINE TYPE

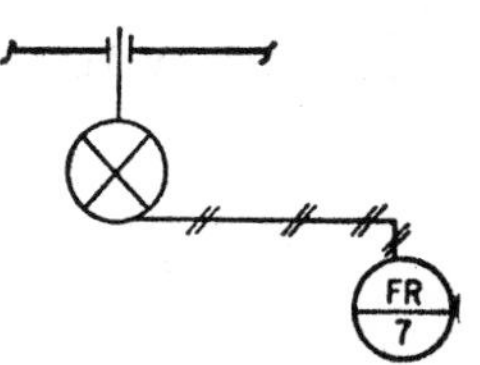

FLOW RECORDER, PNEUMATIC TRANSMISSION, TRANSMITTER LOCAL, RECEIVER MOUNTED ON BOARD

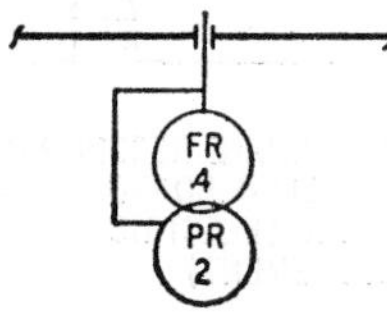

FLOW RECORDER, MECHANICAL TYPE, WITH DIRECT CONNECTED PRESSURE RECORDING PEN, LOCALLY MOUNTED (NOTE THAT IN LISTING SUCH A COMBINATION ITEM IN SPECIFICATIONS, ETC., IT WOULD BE WRITTEN AS FR-4 AND PR-2, THEREBY TREATING EACH ELEMENT AS SEPARATE ENTITY)

FLOW RECORDER WITH PRESSURE RECORDING PEN, BOTH ELEMENTS PNEUMATIC TRANSMISSION, TRANSMITTERS LOCAL, AND RECEIVER BOARD MOUNTED (RECEIVER SHOULD BE WRITTEN AS FR-5 AND PR-2, AND EACH TRANSMITTER IDENTIFIED BY ITS OWN ELEMENT)

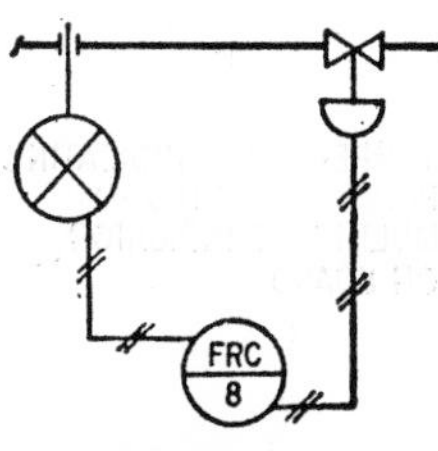

FLOW RECORDING CONTROLLER, PNEUMATIC TRANSMISSION WITH RECEIVER MOUNTED ON BOARD AND LOCAL TRANSMITTER

FIG. 2-2. Typical instrumentation symbols for flow. (*From ISA RP5.1.*)

plete serial-number system for a plant or an organization. In any case the series of consecutive numbers will be suitable for use with the general identifications.

When used in written work, the number shall be placed after the letters, and separated from them by a hyphen. For example, temperature-recording controller, item number one (1), is written as TRC-1.

Applying the Identifications. The identifications, wherever possible, shall be used to identify a complete instrumentation application with all its components, instead of independent identifications being assigned to the various pieces required. The rules and instructions for this principle of applying identifications are as follows:

1. For combination instruments that measure more than one kind of variable, or that provide more than one kind of function, each portion of the combination shall have its own identification. Components

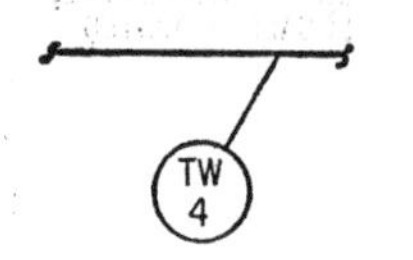

TEMPERATURE WELL

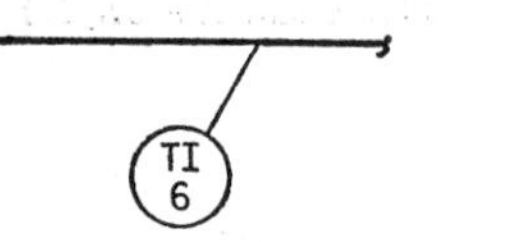

TEMPERATURE INDICATOR OR THERMOMETER (LOCAL)

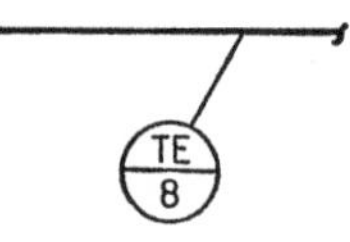

TEMPERATURE ELEMENT WITHOUT CONNECTION TO INSTRUMENT

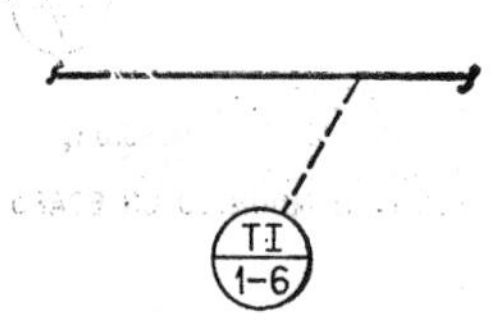

TEMPERATURE INDICATING POINT CONNECTED TO MULTIPOINT INDICATOR ON BOARD

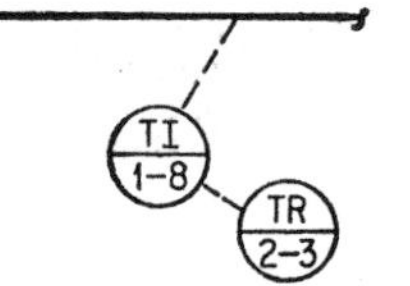

TEMPERATURE INDICATING AND RECORDING POINT CONNECTED TO MULTIPOINT INSTRUMENTS ON BOARD

TRC
1

TEMPERATURE RECORDING CONTROLLER, BOARD MOUNTED (ELECTRIC MEASUREMENT)

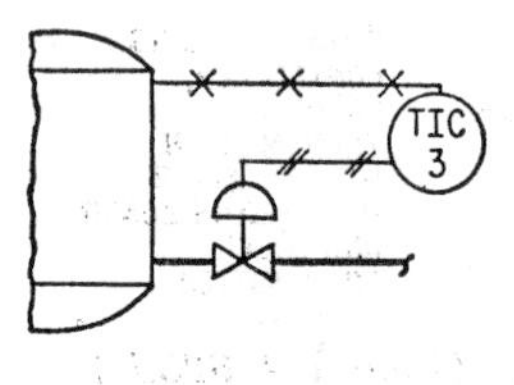

TEMPERATURE INDICATING CONTROLLER, FILLED SYSTEM TYPE, LOCALLY MOUNTED

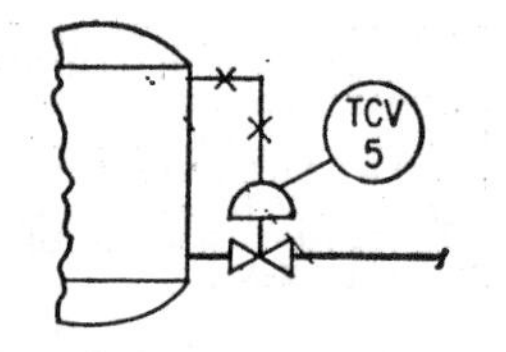

TEMPERATURE CONTROLLER OF SELF-ACTUATED TYPE

TRC
5
TR
2

TEMPERATURE RECORDING CONTROLLER AND TEMPERATURE RECORDER, COMBINED INSTRUMENT BOARD MOUNTED

FIG. 2-3. Typical instrumentation symbols for temperature. (*From ISA RP5.1.*)

pertaining to such a portion shall be identified accordingly, and the instrument or equipment common to both shall utilize *both* identifications. For example, a combination recorder for flow and pressure would be PR-1 and FR-5.

2. Multiple-pen or point instruments with all points of the same general

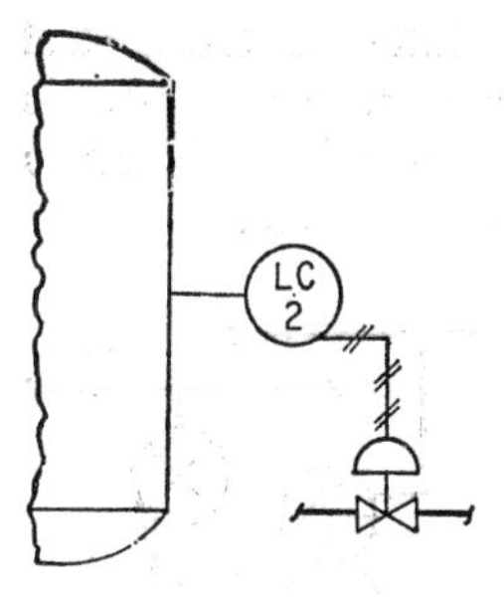

BLIND LEVEL CONTROLLER, INTERNAL TYPE

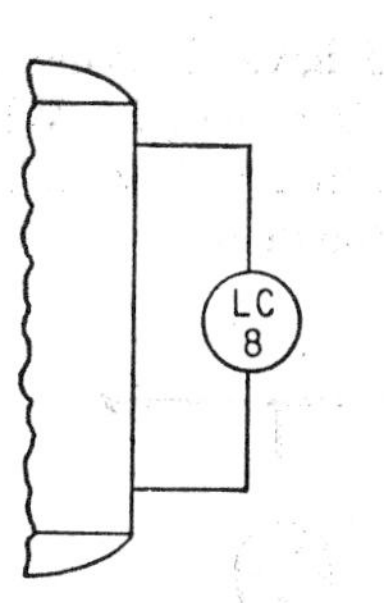

GAGE GLASS

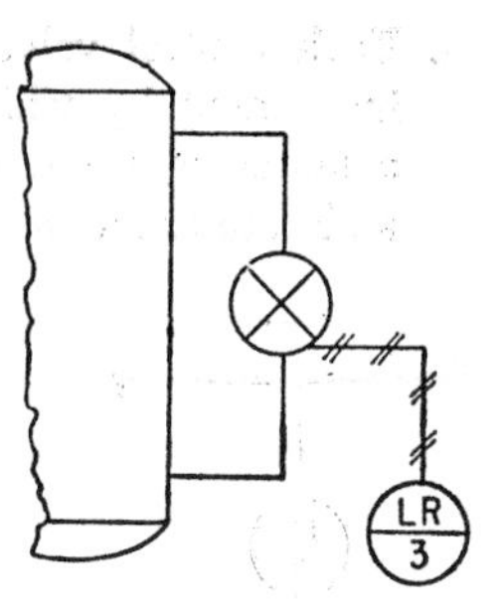

LEVEL RECORDER, PNEUMATIC TRANSMISSION, WITH BOARD MOUNTED RECEIVER. EXTERNAL TYPE TRANSMITTER

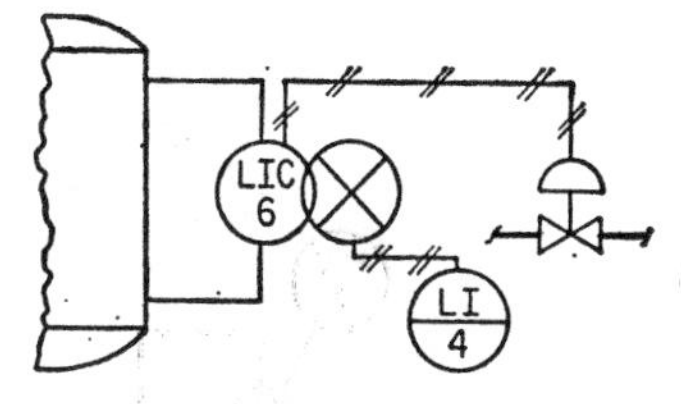

LEVEL INDICATING CONTROLLER AND TRANSMITTER COMBINED WITH BOARD MOUNTED LEVEL INDICATING RECEIVER

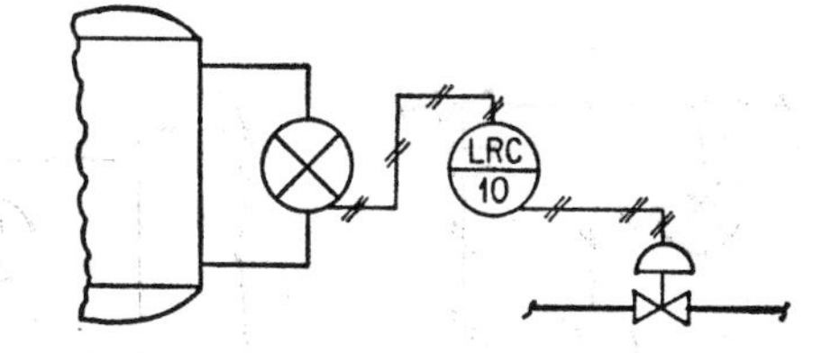

LEVEL RECORDING CONTROLLER, EXTERNAL TYPE, PNEUMATIC TRANSMISSION

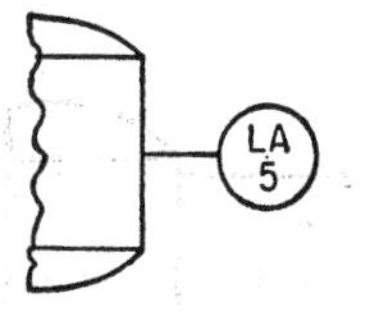

LEVEL ALARM, INTERNAL TYPE

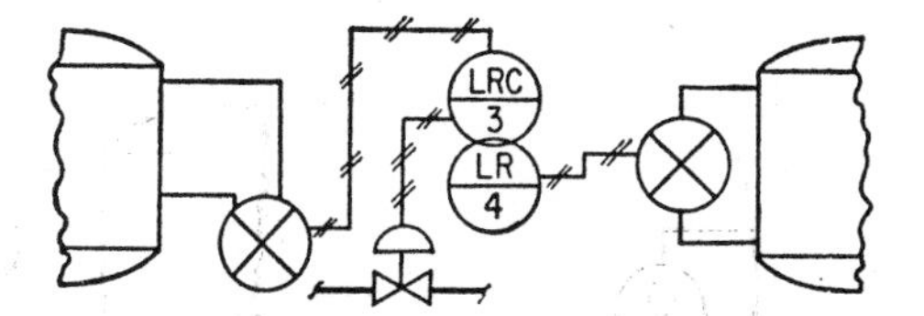

LEVEL RECORDING CONTROLLER AND LEVEL RECORDER, PNEUMATIC TRANSMISSION COMBINED RECEIVER BOARD MOUNTED

FIG. 2-4. Typical instrumentation symbols for level. (*From ISA RP5.1.*)

service, and all providing the same functions, shall have one identification. The separate elements and their components shall be identified by suffix numbers added to the number of the item: e.g., TR-1-1, TR-1-2, etc.

3. For remote-transmission instruments both the receiver and the transmitter shall have the same identifications, in agreement with the overall service and function of the item.

4. Each control valve shall have the same identification as the control instrument by which it is actuated. Where more than one valve is actuated by the same controller, they shall be identified by suffix letters added to the *number* of the item: e.g., TRC-1a, TRC-1b, etc.

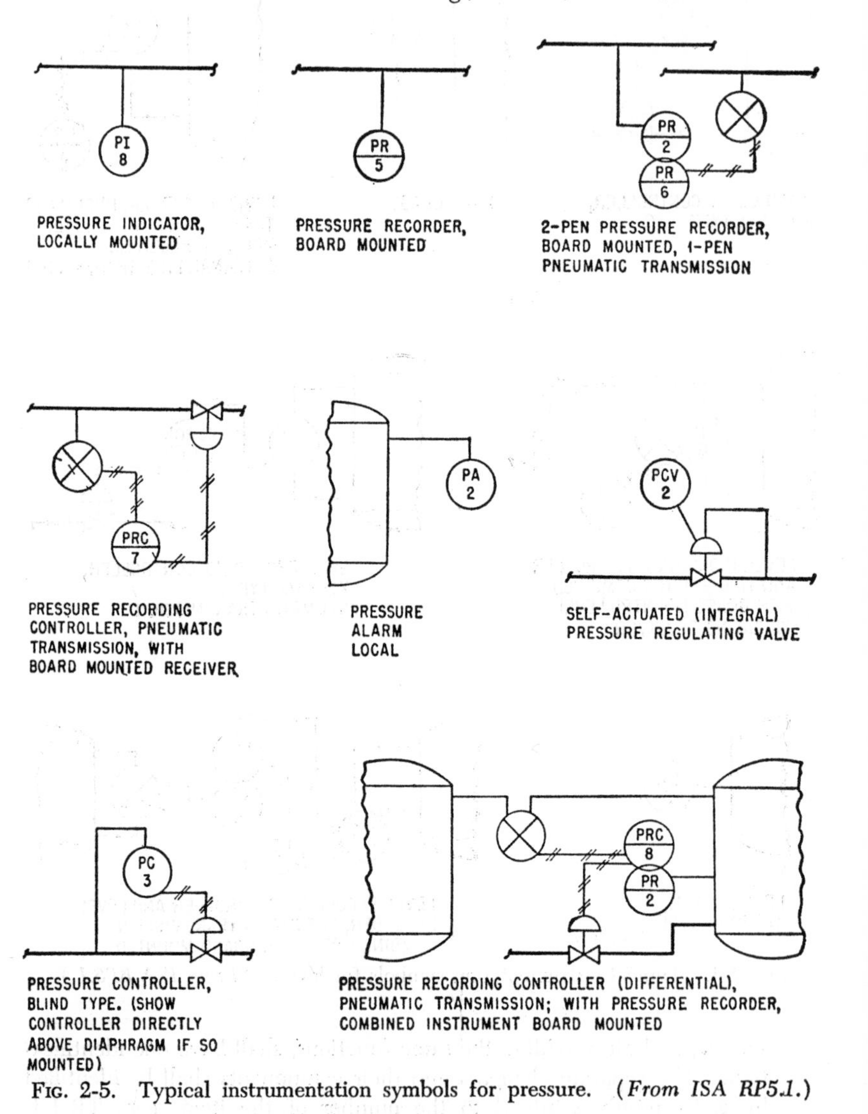

FIG. 2-5. Typical instrumentation symbols for pressure. (*From ISA RP5.1.*)

5. Where accessories such as valve positioners, air sets, switches, relays, etc., require identification, they shall be assigned the same identification as the instrument to which they connect or with which they are used.

6. Primary measuring elements shall be assigned the same identifications as the instruments to which they connect. Where an element does not connect to any instrument, such isolated item (only) shall be assigned separate primary-element identification. Where more than

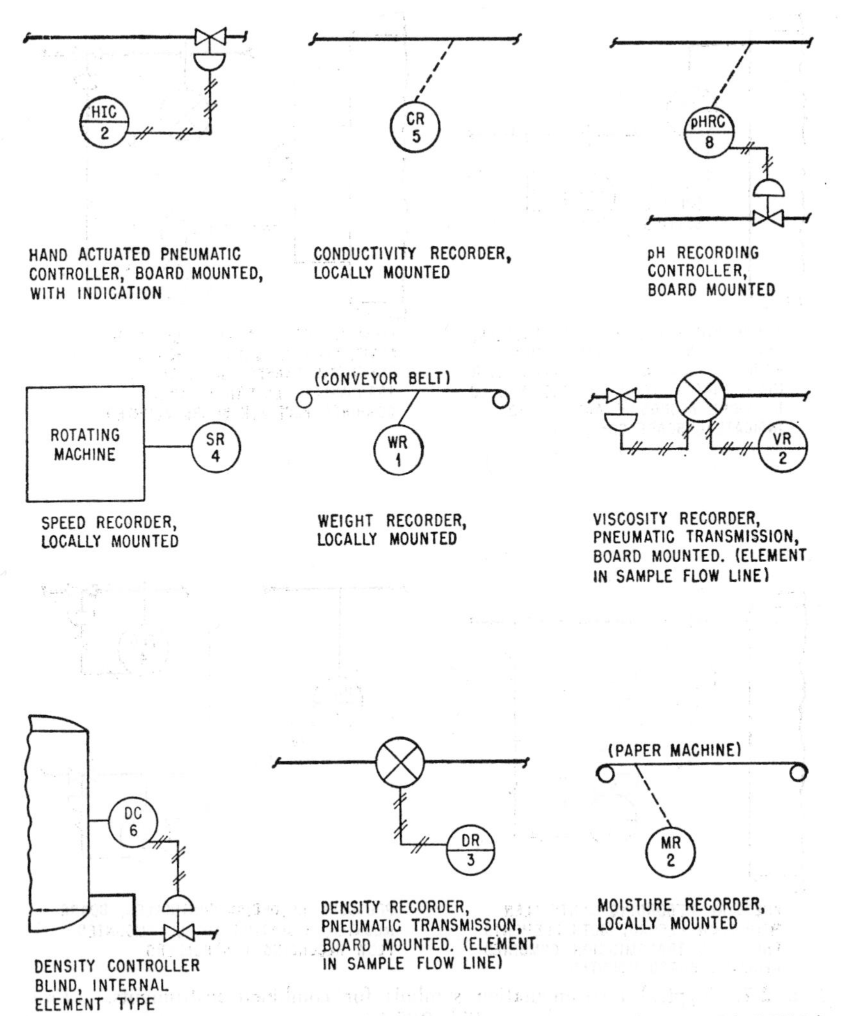

Fig. 2-6. Typical instrumentation symbols—miscellaneous. (*From ISA RP5.1.*)

one element connects to the same instrument they shall be identified by suffix numbers after the item number, in agreement with the point numbers of the instrument to which they connect.

Symbols. The symbols to be used to show instrumentation on flow plans and other drawings are illustrated in Figs. 2-1 to 2-7, inclusive. Figure

2-1 covers the basic pictorial symbols required. The remainder of the drawings show typical symbols, complete with identifications, covering the different process variables and types of equipment that are most likely to be encountered. All others shall be drawn by use of the basic pictorial

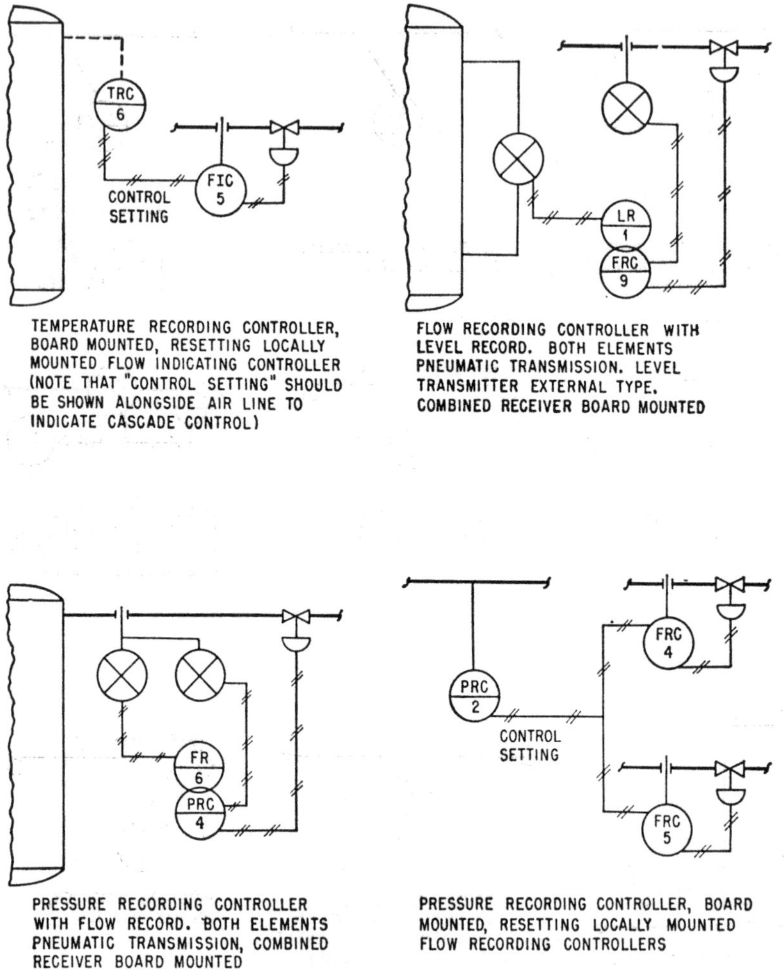

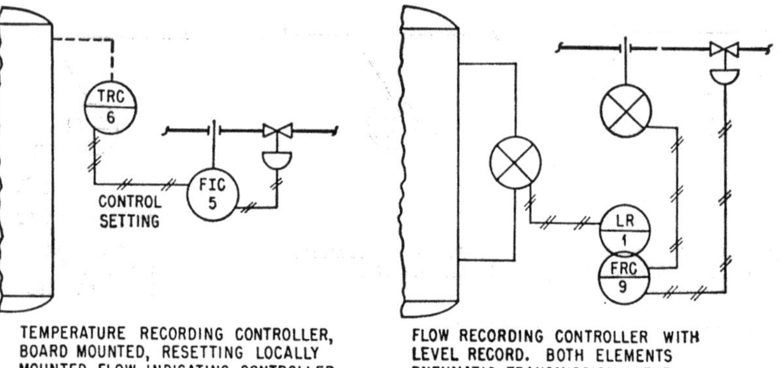

TRC
6
FIC
5
CONTROL
SETTING
TEMPERATURE RECORDING CONTROLLER, BOARD MOUNTED, RESETTING LOCALLY MOUNTED FLOW INDICATING CONTROLLER (NOTE THAT "CONTROL SETTING" SHOULD BE SHOWN ALONGSIDE AIR LINE TO INDICATE CASCADE CONTROL)
LR
1
FRC
9
FLOW RECORDING CONTROLLER WITH LEVEL RECORD. BOTH ELEMENTS PNEUMATIC TRANSMISSION. LEVEL TRANSMITTER EXTERNAL TYPE. COMBINED RECEIVER BOARD MOUNTED
FR
6
PRC
4
PRESSURE RECORDING CONTROLLER WITH FLOW RECORD. BOTH ELEMENTS PNEUMATIC TRANSMISSION, COMBINED RECEIVER BOARD MOUNTED
PRC
2
CONTROL
SETTING
FRC
4
FRC
5
PRESSURE RECORDING CONTROLLER, BOARD MOUNTED, RESETTING LOCALLY MOUNTED FLOW RECORDING CONTROLLERS

FIG. 2-7. Typical instrumentation symbols for combined instruments. (*From ISA RP5.1.*)

symbols of Fig. 2-1 with the proper identifications. The following notes pertain to the use of the symbols:

1. The circle, generally approximately $\frac{7}{16}$ in. in diameter, is used to depict instruments proper and most other instrumentation items. It is also used as a "flag" to enclose identifications and point out items such as valves which have their own pictorial symbol. Optionally

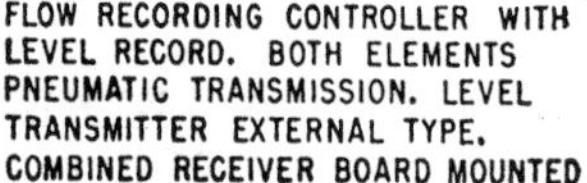

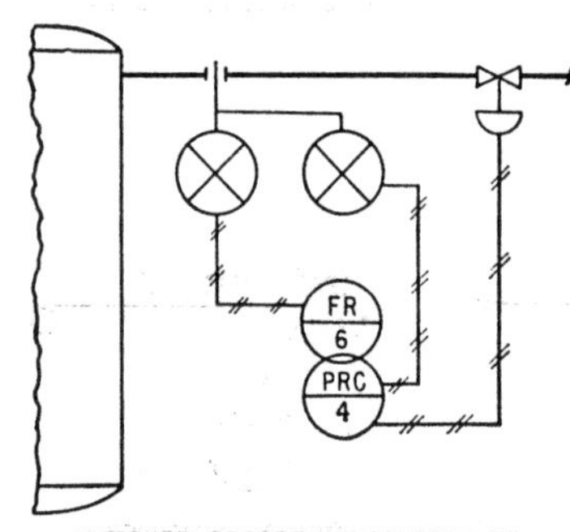

such items may have their identification written alongside the pictorial symbol, and the circle for "flagging" omitted.

2. It is generally unnecessary to repeat the identification for transmitter, control valve, primary element, etc., as they are determined by their connection to the instrument proper. Where such components are shown at a remote distance or on a separate sheet, a note of the identification may be added alongside the pictorial or connecting line symbol.
3. A brief explanatory notation may be added alongside a symbol if considered necessary to clarify the function or purpose of an item. A few such notes are easier to apply and use than a great variety of more complicated symbols.

AUTOMATIC-CONTROL TERMINOLOGY

Understanding the language of the power controls engineer is a necessary prerequisite to understanding control fundamentals. Standardization of control terminology has been difficult, largely because engineers often use several terms to refer to the same control phenomenon.

Recent publication of ASME Standard 105, "Automatic Control Terminology," by the Instruments and Regulators Division of the American Society of Mechanical Engineers, standardizes commonly used control terms and represents accepted practice. Terminology of this publication, as it applies to instrumentation and controls, is summarized in Table 2-3.

Table 2-3. Automatic-control Terminology*

An *automatic controller* is a device which measures the value of a variable quantity on condition and operates to correct or limit deviation of this measured value from a selected reference.

An automatic controller includes both the measuring means and the controlling means. True automatic-control systems always contain one or more feedback loops, at least one of which includes both the automatic controller and the process.

An *automatic-control system* is any operable arrangement of one or more automatic controllers connected in closed loops with one or more processes.

A *self-operated controller* is one in which all the energy to operate the final control element is derived from the controlled medium through the primary element.

This type of controller must have both self-operated measuring means and self-operated controlling means.

A *relay-operated controller* is one in which the energy transmitted through the primary element is either supplemented or amplified for operating the final control element by employing energy from another source.

This type of automatic controller may have a self-operated measuring means and a relay-operated controlling means, or a relay-operated measuring means and a self-operated controlling means, or a relay-operated measuring means and a relay-operated controlling means.

The *set point* is the position to which the control-point-setting mechanism is set.

Where the automatic controller possesses a set-point scale, the set point is the scale reading translated into units of the controlled variable. Where a setting scale is not provided, the set point is the position of the control-point-setting mechanism translated into units of the controlled variable.

* Abstracted from ASME Standard 105.

Table 2-3. Automatic-control Terminology (*Continued*)

In some types of automatic controllers, for example, those with the two-position differential gap, floating with neutral or proportional-position action, the set point is related to the position of a range of values of the controlled variable. The set point is often selected as the center of this range of values.

The set point may be varied manually or by automatic means, as in time-schedule or ratio control.

The *control point* is the value of controlled variable which, under any fixed set of conditions, the automatic controller operates to maintain.

In some types of automatic controllers, for example, those with two-position differential gap or floating with neutral controller action, the control point becomes a control range of values of the controlled variable rather than a single value.

In positioning-type controller action, the control point may lie anywhere within a predetermined range of values of the controlled variable. The control point may then differ from the set point by the amount of offset.

In floating controller action with zero neutral zone, the control point and the set point coincide.

Primary feedback is a signal which is related to the controlled variable and which is compared with the reference input to obtain the actuating signal.

The *controlling means* consists of those elements of an automatic controller which are involved in producing a corrective action.

A *motor operator* is a portion of the controlling means which applies power for operating the final control element.

The *final control element* is that portion of the controlling means which directly changes the value of the manipulated variable.

A *servomechanism* is an automatic-control system in which the controlled variable is mechanical position.

The *measuring means* consists of those elements of an automatic controller which are involved in ascertaining and communicating to the controlling means the value of the controlled variable.

The *primary element* is that portion of the measuring means which first either utilizes or transforms energy from the controlled medium to produce an effect which is a function of change in the value of the controlled variable.

The effect produced by the primary element may be a change of pressure, force, position, electrical potential, or resistance.

Neutral zone is a predetermined range of values of the controlled variable in which no change of position of the final control element occurs.

Neutral zone is commonly expressed in per cent of controller scale range. A neutral zone is employed in some types of floating controller action.

Sensitivity is the ratio of output response to a specified change in the input.

This term can be applied to any element in the control loop. For a measuring instrument the input is the measured variable; for an automatic controller it is the controlled variable.

For example, an automatic temperature controller having a net output pressure of 15 psi and a full-scale range of 0 to 150°F would have a sensitivity of 0.1 psi per degree F.

Where nonlinear relationships are involved, the point or points at which the sensitivity is given should be stated.

Resolution sensitivity is the minimum change in the measured variable which produces an effective response of the instrument or automatic controller.

Resolution sensitivity may be expressed in units of the measured variable or as a fraction or per cent of the full-scale value, or of the actual value.

Threshold sensitivity is the lowest level of the measured variable which produces effective response of the instrument or automatic controller.

Dead band is the range of values through which the input can be varied without initiating output response.

Differential gap, applying to two-position controller action, is the smallest range of values through which the controlled variable must pass in order to move the final control element from one to the other of its fixed positions.

Differential gap is commonly expressed in units of the controlled variable or per cent of controller scale range.

Table 2-3. Automatic-control Terminology (*Continued*)

Hysteresis is the difference between the increasing input value and the decreasing input value which effect the same output value. This term applies only where the output value is a continuous function of the input value.

Self-regulation is an inherent characteristic of the process which aids in limiting deviation of the controlled variable.

The *controlled variable* is that quantity or condition which is measured and controlled.

The *controlled medium* is that process energy or material in which a variable is controlled.

The *manipulated variable* is that quantity or condition which is varied by the automatic controller so as to affect the value of the controlled variable.

The *control agent* is that process energy or material of which the manipulated variable is a condition or characteristic.

Note that the *controlled variable* is a condition or characteristic of the *controlled medium*. For example, where temperature of water in a tank is automatically controlled, the controlled variable is temperature and the controlled medium is water.

Note that the *manipulated variable* is a condition or characteristic of the *control agent*. For example, when a final control element changes the fuel-gas flow to a burner, the manipulated variable is flow, and the control agent is fuel gas.

Error in measurement is the algebraic difference between a value which results from measurement and the corresponding true value.

A positive error denotes that the measured value is algebraically greater than the true value. Error is usually expressed in the units of the measured quantity, or as a fraction (or per cent) of the full-scale value, or of the actual value.

Accuracy in measurement is the degree of correctness with which a measuring means yields the "true" value referred to accepted engineering standards, such as the standard meter, gram, etc. It is assured that a true value always exists even though it may be impossible to determine.

Precision in measurement is the degree of reproducibility among several independent measurements of the same true value under specified conditions.

It is usually expressed in terms of deviation in measurement.

Deviation in measurement is a statistical number representing the randomness among independent measurements of the same true value.

Deviation is variously expressed as: (1) the difference between any measurement and the mean value of two or more; (2) the average of several independent variations from the mean (the average deviation); (3) the root-mean-square value computed from several individual deviations (this is the standard deviation); or (4) the ratio of (1) or (2) or (3) to the mean. In any case the number of observations should be stated.

Deviation is the difference between the actual value of the controlled variable and the value of the controlled variable corresponding with the set point.

Offset is the steady-state difference between the control point and the value of the controlled variable corresponding with the set point.

Offset is an inherent characteristic of positioning controller action.

Corrective action is the variation of the manipulated variable produced by the controlling means.

Cycling is a periodic change of the controlled variable. (Oscillation is a synonymous term.)

The *reference input* is the reference signal in an automatic controller.

The reference input is the output signal of the reference-input elements as determined by their response to the set-point input signal. It has the same units as the primary feedback.

FUNDAMENTALS OF MEASUREMENT AND CONTROL[2]

The fundamental concepts of measurement of system variables and automatic control form the foundation for the intelligent design and selection

[2] Minneapolis-Honeywell Regulator Co., Industrial Div.

of power control systems. The following paragraphs are applicable to control systems commonly encountered in the mechanical services, regardless of the variable being controlled, or the means of transmission.

Time Element in Measurement. Complete, immediate response to a change in a variable is an ideal which is not likely to be achieved in any physical system, including automatic control. The response may start immediately although it may take a long time to complete its effect. The responses of the measuring means, the controlling means, and the process determine the time element, or lag. Therefore, the *rate* of change of a variable is quite as important as the magnitude of change.

The time element, in a general sense, is called lag. *Lag* is the falling behind, or retardation, of one physical condition with respect to another

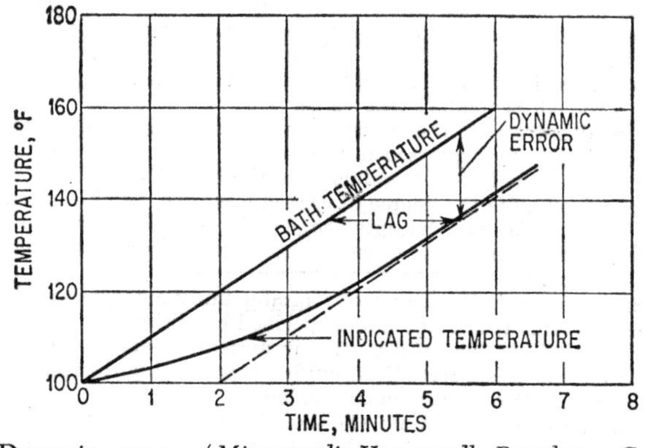

FIG. 2-8. Dynamic error. (*Minneapolis-Honeywell Regulator Co., Industrial Div.*)

physical condition to which it is related. For example, a temperature change at a thermometer bulb is not detected instantaneously. Heat must be transmitted through the bulb wall to the filling medium, and then the resulting change in pressure must be transmitted to the receiving spiral. So lag of the filled-system thermometer bulb involves heat transfer, the slight fluid flow of the fill material, and pressure transmission, in addition to the dynamics of the moving element (spiral). From the example just given, it is clear that the lag of a measuring means occurs in (1) the primary element, (2) the transmission system, and (3) the measuring element of the instrument. Each of these three must be considered in any problem of automatic control.

Dynamic Error and Lag. Usually the *error* and lag of an instrument as it measures a *gradual* change in a variable are important in automatic control.

Figure 2-8 shows what happens when a bath and an element both start at the same temperature, 100°F, and then the bath temperature is increased at a constant rate. After a short time, the measured temperature lags the bath temperature by a constant amount. This difference between the true and the measured temperature is called the *dynamic error*. It is simply the lag coefficient multiplied by the rate of change of the measured variable.

There is always dynamic error of measurement when the controlled variable is changing. In automatic control, this error is usually more important than any static error because operating conditions are seldom static, and the variable is usually fluctuating around the control point. In Fig. 2-9 you see the cycling of a bath temperature and how an instrument will indicate the cycle. The lag of the primary element may cause the cycle to be delayed, as well as reduced in amplitude. It is reduced in amplitude because the bath temperature reaches one extreme of its cycle swing and starts in the other direction before the measured value catches up with it.

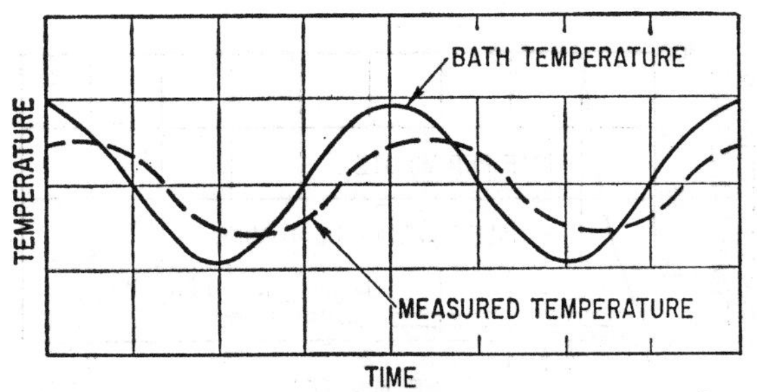

FIG. 2-9. Cycling of bath temperature. (*Minneapolis-Honeywell Regulator Co., Industrial Div.*)

The measured value meets the bath temperature going back in that direction and changes its direction before it reaches the cycle's extreme, and so it, too, cycles, but not so extensively.

The dynamic error is extremely important, for it may prevent the indicator or recorder from showing true conditions at the process. Obviously, for the best control results it is usually advisable that the measuring lag shall be as small as possible.

Static Error and Reproducibility in Controls. The factors of *static error* and *reproducibility* must be given proper consideration in any discussion of automatic control.

The *static error* of measurement is the deviation of the instrument reading from the true value of a static variable. Large static error is undesirable, but not necessarily detrimental to automatic control. This is particularly true where it is more important that the variable be held at a *constant* value than at an *exact* value. Static error will not interfere in this case.

Reproducibility is the degree of closeness with which the same value of a variable may be measured at different times. Reproducibility of a high order must be obtained if an exact quality of control is desired. For example, the calibration of a thermocouple at high temperature might drift because of contamination, and so the operating point would have to be shifted periodically to compensate for this drift.

Reproducibility is a time variable and not a static condition, and so it becomes more important than static error. Periodic checking and maintenance of an instrument are generally done to obtain reproducibility, rather than to determine the static error of indication.

Cause and Effect of Dead Zone. *Dead zone* is the largest range through which the controlled variable can change without the change being sensed by the controller. Among the causes of dead zone in such self-operated measuring means as the filled-system thermometer, flow meter, or pressure gage are the friction and lost motion of the moving parts of the mechanism. In a well-designed and well-maintained controller, the dead zone should be less than 0.2 per cent of scale. In a power-operated measuring means, such as the continuous-balance potentiometer, the dead zone can be reduced

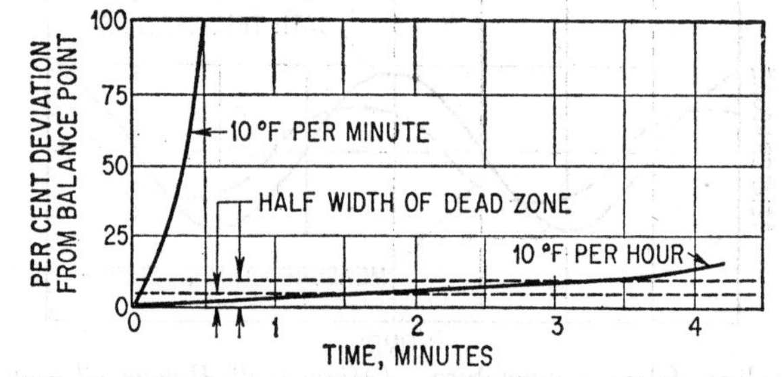

FIG. 2-10. Effect of dead zone. (*Minneapolis-Honeywell Regulator Co., Industrial Div.*)

to about 0.05 per cent of scale, depending upon the method of utilizing power to position the indicating element.

The effect of the dead zone is to create a delay in providing the initial impulse to the controller. Figure 2-10 illustrates this action. If the variable has previously balanced within the dead zone, the change in variable will not be detected by the controller until the controlled variable has reached either edge of the dead zone. A length of time, depending upon the rate of change of the variable, will have elapsed. This is sometimes a serious delay because the controller cannot make an early correction for the change. Also, the rate of variable change is not constant.

The length of time delay, or *dead time,* may be determined from Fig. 2-10. If the temperature is changing at a rate of 10°F per minute, then a dead zone of 0.1°F causes a dead time of 0.1 minute. If the temperature changes slowly, the dead zones shown cause dead times of 1.5 and 3.3 minutes. Therefore, the slower the variable changes, the more serious the measurement dead zone becomes.

Basic Characteristics. Every process, commercial, institutional, or industrial, exhibits two effects which must be taken into account when automatic-control equipment is being selected. These are (1) changes in the controlled variable due to altered conditions in the process, generally called *load changes*, and (2) a retardation or delay in the time it takes the process variable to reach a new value when load changes occur, called the *process lag*. This lag is caused by one or more of the process characteristics of *capacitance*, *resistance*, and *dead time*.

Process Load. Process load is defined as the total amount of control agent required by a process at any one time. In a heat exchanger, for example, where a fluid is continuously passing through tubes in a chamber through which steam (the control agent) flows, a certain quantity of steam is required to hold the temperature of the fluid at a given value when the fluid is flowing at a given rate. An increase in the fluid flow requires more steam, and thus constitutes a change in process load. A decrease in the inlet-fluid temperature with the original flow rate is also a load change.

Process load is directly related to that final control-element setting required to maintain the controlled variable at its set point. A load change causes deviation of the controlled variable from the desired (set-point) value and requires a new setting of the final control element. Both the magnitude and the rate of load change are prime factors in the application of automatic controllers.

Load Changes. The load changes in a process are not always easy to recognize. Some of the causes of process load changes are:

1. Greater or less demand for control agent by the controlled medium. In the heat-exchanger example cited above, an increase in the flow of the fluid or an increase in the temperature of the fluid constitutes a load change because the increased flow requires more steam, and, conversely, the increased temperature requires less steam.
2. A change in the heating value of fuel gas or in the pressure or quality of steam also causes a load change. If a process is operating at a certain temperature, and then suddenly for some reason the Btu content of the fuel gas drops, the burner output decreases and less heat is available for the process. Even though other conditions of the process are unchanged, more gas must be burned to maintain the same temperature. Changes in steam pressure similarly result in load changes.
3. Variations in the flow of control agent will also cause load changes. Varying pressure in a fuel-gas line will cause burner output to vary, for example.
4. Changes in ambient conditions often cause load changes, especially in outdoor installations where heat loss from radiation is significant.
5. The variable amount of heat generated or absorbed by a reaction (exothermic or endothermic process) will obviously effect load changes because, as the process itself generates or absorbs heat, more or less of the controlling agent is needed. The position of the final control element must be adjusted to meet the need.

The *capacitance* of a process is an extremely important factor in automatic control.

The type of *capacity* is determined by the energy or material contained in the process. The type of *capacitance* is determined both by the type of the quantity contained and the type of the reference variable. Several types of capacity and capacitance may exist together in one process.

A comparison between capacity and capacitance is illustrated in Fig. 2-11, showing a liquid stored in open vessels of equal volume. In this example, the capacities are equal (128 cu ft) but the capacitances are unequal. The liquid-volume capacity of an open vessel is the maximum quantity of liquid the vessel will hold without overflowing. The liquid-volume capacitance of an open vessel, with respect to the liquid head, is the change in liquid quantity per unit change in liquid head.

The capacitance of the slim vessel is the capacity (128 cu ft) divided by the reference variable, feet of level (8), or 16 cu ft of liquid per ft of level. The capacitance of the short vessel is the capacity (also 128 cu ft)

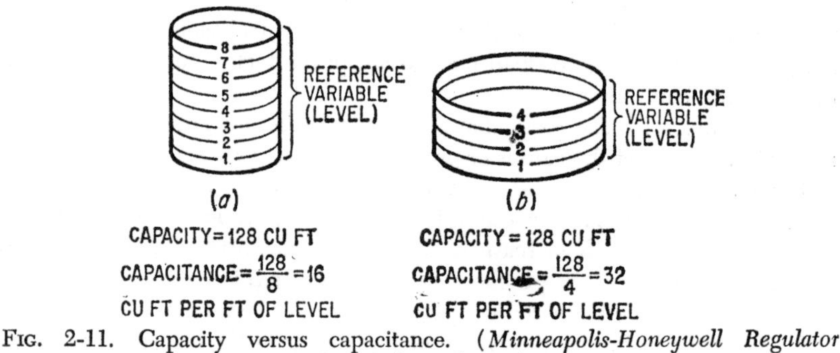

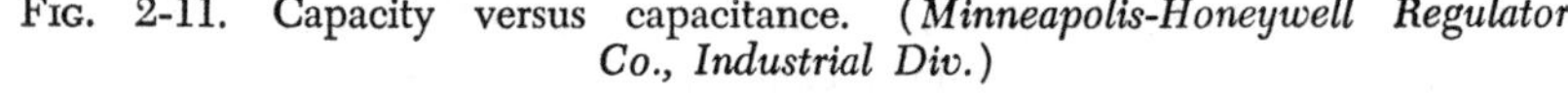

FIG. 2-11. Capacity versus capacitance. (*Minneapolis-Honeywell Regulator Co., Industrial Div.*)

divided by the reference variable, feet of level (4), or 32 cu ft of liquid per ft of level. So, whereas a change of 1 ft of liquid level in the tall tank changes the volume 16 cu ft, a change of 1 ft of liquid level in the short tank changes the volume 32 cu ft.

In a thermal process, if 200 Btu is required to raise the temperature of the controlled medium 10°F, then the capacitance would be 200 divided by 10, or 20 Btu/deg F. Equivalent expressions for this characteristic with other controlled variables are listed in Table 2-4 on page 2-33.

Capacitance and Rate of Reaction. In any process, a large capacitance relative to the flow of control agent may be favorable to automatic control. Its effect may be likened to a flywheel in a steam-driven electric generator whereby the power output is stabilized. The larger the wheel, the more it tends to operate at the same speed, regardless of changing conditions. Similarly, a relatively large process capacitance has a tendency to keep the controlled variable constant in spite of load changes. Large capacitance will make it easier to hold a variable at a certain value *but* more difficult to change to a new value.

As a comparison between large and small thermal capacitance, consider the difference in the ease of holding a constant temperature in Figs. 2-12 and 2-13. In the process of Fig. 2-12, heat is applied to a jacketed vessel containing a considerable volume of liquid, with the temperature being measured by a thermometer bulb.

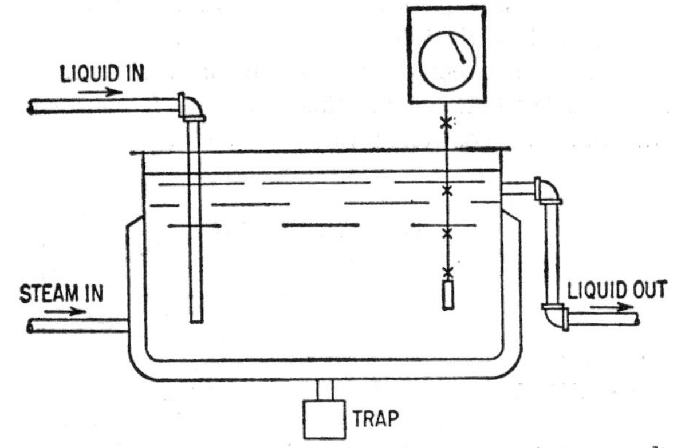

FIG. 2-12. Process with large thermal capacitance. (*Minneapolis-Honeywell Regulator Co., Industrial Div.*)

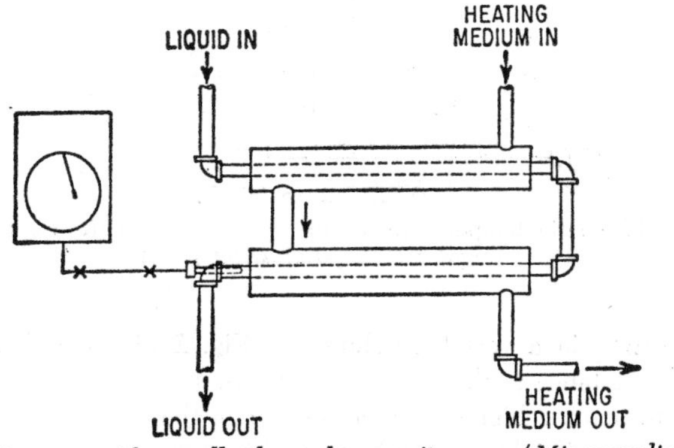

FIG. 2-13. Process with small thermal capacitance. (*Minneapolis-Honeywell Regulator Co., Industrial Div.*)

The mass of liquid exerts a stabilizing influence and tends to balance out changes which might be caused by variations in the rate of flow, minor variations in heat input, and sudden changes in ambient temperature.

Limited Thermal Capacitance. On the other hand, Fig. 2-13 illustrates a high-velocity heat exchanger with the thermometer bulb located in the exit liquid line. The rate of flow through this process is identical with that in the first process, but a comparatively small volume of liquid is flowing in the tubes at any one time. But, unlike the first process, the mass here is small, and so there is no stabilizing influence. The total volume of liquid in the exchanger at any moment is small in comparison with the rate of throughput, the heat-transfer area, and the heat supply. Slight variations in the rate of feed, or rate of heat supply, will be reflected almost immediately in fluctuations of the temperature of the liquid leaving the exchanger. If the process were manually regulated, these conditions would require continuous vigilance on the part of the operator in an almost impossible attempt to hold the temperature constant.

Although the over-all effect of large capacitance is generally favorable, it does introduce a lag between the time a change is made in the control agent and the time the controlled variable reflects the change. When a

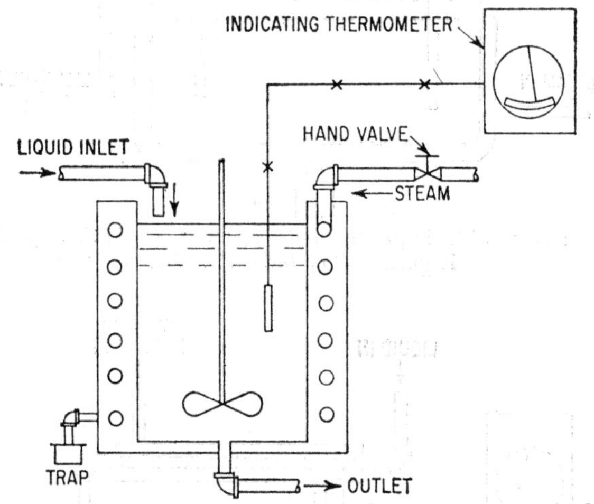

Fig. 2-14. Schematic temperature-control application. (*Minneapolis-Honeywell Regulator Co., Industrial Div.*)

liquid is heated in a vessel, as shown in Fig. 2-14, even though there is little or no resistance to the flow of heat to the liquid, it takes some time for the liquid to reach a higher temperature when the heat supply is increased. How much time it takes depends upon the capacitance of the liquid relative to the heat supply.

The time effect of capacitance is the determining factor in the reaction of the process. The corrective action brought about by an automatic controller to maintain the balance in the process depends upon the relative rate at which the process reacts (reaction rate). Therefore the first thing to consider in the analysis of a process is its capacitance.

Resistance. The second basic process characteristic is *resistance,* defined as the opposition to *flow.* When heat is being transferred by conduction through a solid, such as the wall of a vessel, a temperature drop occurs. This temperature drop depends upon the composition of the solid because some materials have better thermal conductivity (or less thermal resistance) than others. Thermal resistance is the change in temperature which occurs per unit rate of heat flow, with the unit rate of heat flow usually expressed in Btu per second.

Designers of heat exchangers are familiar with the fact that gas and liquid films on the tubes create a more serious resistance to heat flow than do the mere walls of the tubes themselves. These effects cannot be neglected in an analysis of the resistance characteristics of this type of equipment.

If a material is being heated in a process with high thermal resistance, it will take more control agent to effect a temperature change in the material than in a process with low thermal resistance. So the thermal resistance of this process will exert a strong influence upon the selection of the proper controller. However, the resistance characteristic exists in the measurement of most variables.

Transfer Lag. Many processes, particularly those involving temperature control, have more than one capacitance. If a flow of heat or other energy passes from one capacitance through a resistance to another capacitance, there is a *transfer lag.* The effect of this lag may be visualized by considering the system shown in Fig. 2-14. Steam coils are imbedded in a high-heat-capacity jacket surrounding a tank containing a well-stirred liquid which is to be manually controlled at a definite temperature. An indicating thermometer with its bulb in the liquid is used as the reference for adjustment of the hand valve that regulates steam flow.

As long as conditions in the system are constant, the rate of heat transfer to the liquid depends upon the resistance of the jacket and, of course, the temperature differential between the two. The steam flows at a rate sufficient to maintain a temperature somewhat greater than the liquid temperature, as determined by the temperature drop through the resistance to the liquid. If the flow of liquid into the kettle increases, a temperature drop is almost immediately indicated by the thermometer, and the valve must be opened to provide a greater heat-transfer rate—more Btu to increase the temperature. The thermal capacitance of the jacket, however, retards the building up of the higher temperature on the supply side. But the temperature will build up eventually, and so the effect of transfer lag is to appreciably retard the initial reaction of the process.

Effect of Transfer Lag. Transfer lag in temperature control is always unfavorable because it limits the rate at which heat input can be made to affect the controlled temperature. The result is a tendency to overshoot the set point, because the effect of the addition of heat is not immediately felt, and the controller calls for still more. From the above example, it is seen that a larger resistance increases the thermal head necessary for heat transfer. Furthermore, when load changes occur, a larger supply capacitance lengthens the time required to change the thermal head to a new value.

In heat exchangers, surfaces fouled by sludge or corrosion, as well as poor film coefficients due to insufficient fluid velocities, result in high resistances, and consequently large transfer lag.

Dead Time. Particularly in continuous processes, where it is necessary to transfer heat or other energy by means of a fluid flowing through some distance at a certain velocity, a third type of lag, *dead time,* often occurs. For example, in a tubular type of furnace commonly used in petroleum refining, oil is pumped at a constant rate through hundreds of feet of tubing which line the walls of the heater. It takes quite a long time for a given portion of the oil to pass through the heater.

If ideal conditions exist in which every variable except the inlet oil temperature is constant, it is evident that there will still be considerable lag in the recording of an inlet temperature change if the primary element is

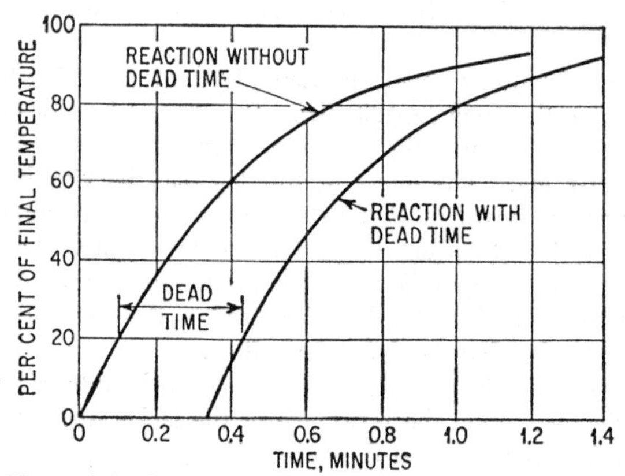

FIG. 2-15. Effect of dead time. (*Minneapolis-Honeywell Regulator Co., Industrial Div.*)

located in the outlet line. This lag is *dead time.* If the outlet temperature were to be controlled, it is obvious that any change in heat input in the heater would have this same dead time before it would be reflected at the primary element. Control action would be delayed that length of time.

Dead time is often created by installation of the control valve or measuring element at a distance from the process. For example, if a heated fluid flows from a heat exchanger at the rate of 200 fpm, and a thermometer bulb is located 50 ft from the outlet, then a dead time of 0.25 min, or 15 sec, will be introduced before any change in the process is detected by the temperature-measuring instrument.

In general, dead time introduces more difficulty in automatic control than does a lag at any other point in the controlled system. The effect of dead time is shown graphically in Fig. 2-15. It causes no change in the process reaction characteristic, but rather delays the reaction until a later time, and so there is a period during which the controller is helpless, since it cannot initiate a corrective action until a deviation occurs.

Table 2-4. Units of Measurement of Basic Characteristics*

Characteristics	Thermal	Pressure	Liquid level	Electrical
Capacity	Btu	Cu ft	Cu ft or lb	Coulomb
Potential	Degree	Psi	Ft	Volt
Capacitance	$\frac{\text{Btu}}{\text{Degree}}$	$\frac{\text{Cu ft}}{\text{Psi}}$	$\frac{\text{Cu ft}}{\text{Ft}}$ or $\frac{\text{lb}}{\text{ft}}$	$\frac{\text{Coulomb}}{\text{Volt}}$ = farad
Resistance	$\frac{\text{Degree}}{\text{Btu/sec}}$	$\frac{\text{Psi}}{\text{Cu ft/sec}}$	$\frac{\text{Ft}}{\text{Cu ft/sec}}$ or $\frac{\text{ft}}{\text{lb/sec}}$	$\frac{\text{Volt}}{\text{Coulomb/sec (Ampere)}}$ = ohm

* Courtesy of Minneapolis-Honeywell Regulator Co

Table 2-5. Summary of Characteristics*

Process characteristic	Temperature—usually 2 or more capacitances	Flow—usually 0 capacitance	Pressure—usually 1 capacitance	Liquid level—usually 1 capacitance
Capacitance	Medium or large	Negligible	Small or medium	Small or medium
Transfer lag	Medium or large	Small	Very small	Small or medium
Dead time	Small or large	Small	Small	Small

* Courtesy of Minneapolis-Honeywell Regulator Co.

Table 2-4 lists units of measurement. Table 2-5 summarizes the relative values of characteristics for process involving the most common variables. All these characteristics influence the selection of the mode of control.

MODES OF AUTOMATIC CONTROL[3]

The *mode of control* of an automatic controller is the manner in which the controller acts to restore the controlled variable to its desired value. Common modes of control are: (1) *two-position,* (2) *floating,* (3) *proportional,* (4) *proportional-plus-reset,* (5) *proportional plus rate,* and (6) *proportional-plus-reset plus rate.*

Two-position Control. In *two-position control,* the final control element, such as a valve, is quickly moved to one of two positions (high or low, open or closed) when the controlled variable deviates a predetermined amount from the set point of the controller. Sometimes these two positions are open and closed; sometimes high and low, in which case the mode may be referred to as "high-low" control.

The amount of deviation of the controlled variable necessary for control action in the two-position mode varies with the controller design, but in general is less than 2 per cent of the instrument range, and is usually not adjustable. This characteristic is known as the *neutral zone* of the con-

[3] Minneapolis-Honeywell Regulator Co., Industrial Div.

troller and is shown graphically in Fig. 2-16. In the illustration the valve is adjusted for a minimum opening of 20 per cent and for a maximum of 80 per cent. When the variable reaches the upper end of the neutral zone, the valve is moved quickly in the closing direction to its minimum position. The valve then remains in this position until the variable moves down scale to the lower end of the zone where the valve is actuated to open again.

The overlapping control action described above is usually desirable to prevent rapid cyclic operation of the final control element. Since the two valve positions must necessarily supply either too much or too little of the control agent, the controlled variable is inherently caused to cycle, although

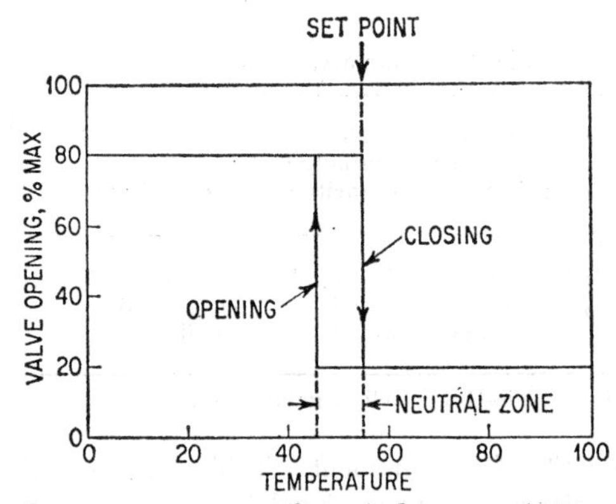

Fig. 2-16. Graphic representation of neutral zone. (*Minneapolis-Honeywell Regulator Co., Industrial Div.*)

under certain conditions the magnitude of its variation may be made extremely small.

Application of Two-position Control. Because of its simplicity, two-position control is very popular, and often adequate for process regulation. In general, it functions satisfactorily between the desired control limits if the service has the favorable characteristics of slow reaction rate and minimum transfer lag or dead time, and the valve travel is adjusted so that its movement permits an input just slightly above and slightly below requirements for normal operation.

Although two-position control will accommodate load changes to some extent, such changes must not be at a rapid rate. With changes in load, the controlled variable may deviate markedly from the average cycling before even cycling is finally reestablished. This deviation will depend upon the magnitude of the load change. Cycling at the new load will have a different average value—higher or lower, depending upon the direction of the load change.

An automatic gas hot-water heater is a good example of two-position control with load changes. When you draw off hot water to wash your hands, you will hear the burner go on and stay on for a few minutes. But, if your wife starts up the automatic washer, the burner may stay on for an hour. Then, if no further hot water is used subsequently, the burner will go on for only very short periods to make up for radiation losses. The inlet water temperature will also affect the length of time the burner must be on. If city water is at 70°F in the summer, it will take a shorter time to heat it to 140°F than in the winter when it is at 45°F.

Floating Control. The most common type of floating control which is not combined with another mode of control is known as *single-speed floating with neutral zone*—here simply called *floating control*—and is usually found only in electrically operated systems.

In floating control, the final control element is moved gradually at a constant rate toward either the open or the closed position, depending upon whether the controlled variable is above or below the neutral zone. While the controlled variable is within the neutral zone, the final control element remains in the last position attained at the time the variable entered the neutral zone. The final control element in floating control moves at a much slower rate than the final control element in two-position control because intermediate positioning is desired.

In floating control, there is no fixed number of positions for the final control element. The valve can assume any position between its two extremes as long as the controlled variable remains within the values corresponding to the neutral zone of the controller. Furthermore, when the controlled variable is outside the neutral zone of the controller, the final control element travels toward the corrective position until the value of the controlled variable is brought back into the neutral zone of the controller, or until the final control element reaches its extreme position.

Application of Floating Control. Single-speed floating control also inherently tends to produce cycling of the controlled variable. This cycling can be minimized by the timing of the actuator of the final control element. The actuator moves to its new position at a single predetermined rate, which must be timed to the natural cycle of the controlled system. If the actuator should move too slowly, the control system would not be able to keep pace with sudden changes, and, if the actuator should move too quickly, two-position control would result. Generally speaking, the actuator should move at a rate fast enough to keep pace with the most rapid load changes that can occur. The outstanding advantage of floating control is that gradual load changes can be counteracted by gradual shifting of the valve position.

Proportional Control. Proportional-position action, generally just called *proportional control;* and average-position action (proportional type), generally called *time-proportioning control,* are the two most widely used types of proportional control.

In proportional control, the final control element assumes a definite position for each value of the controlled variable. For instance, assume that a controller has a range of 0 to 100°F with a proportional band (which will be discussed later) of 100 per cent, and that the set point is 50°F. Assume also that at set point the valve is 50 per cent open.

Now, if the temperature of the controlled variable drops 5°, the valve will open 5 per cent wider so that it is 55 per cent open. By the same token, if the temperature increases 5°, the valve will close 5 per cent more, and so it is now 45 per cent open. So, for each change in the controlled variable, the valve changes its position by an amount which is proportional to the change of the variable.

In the *time-proportioning* control mode, provided by electric controllers only, it is the *off* time and *on* time, rather than the *position* of the final control element, which are proportioned to the value of the controlled variable. For each value of the controlled variable there is a definite on-off cycle. The length of the on plus off cycles is the same throughout the range of the controller, but the ratio of on time to off time within each cycle varies.

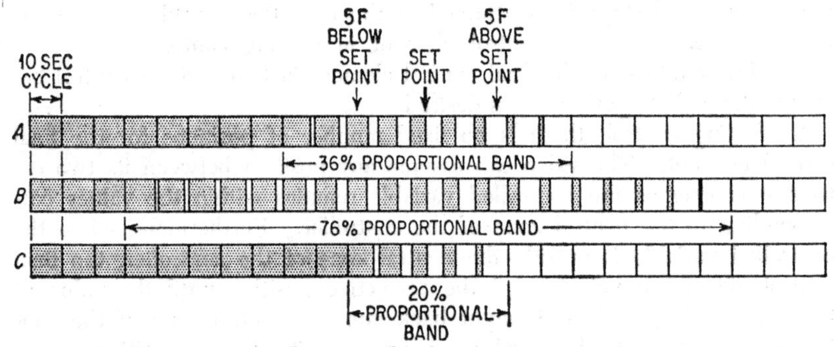

FIG. 2-17. On-time and off-time cycles for time-proportioning control. (*Hagan Chemicals & Controls, Inc.*)

Figure 2-17 shows these cycles and the relationship of on time to off time as the controlled variable deviates from the set point.

Time-proportioning control is an electric-control mode, well suited for application to electric furnaces. At set point, a time-proportioning controller with an 18-sec cycle time may cause the electric heaters to be on 9 sec and off 9 sec. If the temperature drops 5°, say, then the heaters may be on 11 sec and off 7 sec. On the other hand, if the temperature increases 2°, the heaters may be on 8 sec and off 10 sec. Notice that, regardless of the value of the variable, the total time cycle remains 18 sec; only the ratio of on time to off time within each cycle varies.

For purposes of flexibility in application, a calibrated adjustment called the *proportional band* is usually provided in both types of proportional controllers. The *proportional band* is the range of values of the controlled variable which corresponds to the full operating range of the final control element in proportional-position control, or to the percentage of on to off time in time-proportioning control. It is usually expressed as a percentage of the full-scale range of the controller, and the adjusting dial is calibrated as a percentage. For example, if a proportional temperature controller has a range of 0 to 1000°F, and acts to move a control valve through its full operating range as the recording pen moves from 500 to 700°F, the proportional-band adjustment would be $^{200}/_{1,000} \times 100$, or 20 per cent.

The selection of the size of the proportional band will significantly affect the amount of correction which the controller will call for upon a deviation from the set point. For instance, with an instrument range of 1 to 100°F and a 20 per cent proportional band, a change of 5°F causes the valve to move through 25 per cent of its full travel. With a 50 per cent proportional band, the same 5°F change causes valve travel of only 10 per cent, and a 100 per cent proportional band results in only 5 per cent valve travel for a 5°F change.

The proportional band of a time-proportioning controller is also expressed as a percentage of the full-scale range of the controller. Here, however, the proportional band refers to the number of time cycles between the high and low limits of the proportional band. For each value of the controlled variable, there is a corresponding time cycle which ratios the on time to off time.

The effect of the different proportional-band widths can be seen in the schematic diagram in Fig. 2-17. In all three cases, the controller time cycles are 10 sec long, and, when the controlled variable is at the set point, the process heating means is on 5 sec and off 5 sec. At 5° below set point, with the proportional band 36 per cent of full scale (line *A*), the process heating means is on 7 sec and off 3 sec. With a 76 per cent proportional band, still at 5°F below set point, the heating means is on only 6 sec and off 4 sec; and, when the proportional band is reduced to 20 per cent, the heating means is on 8⅓ sec and off 1⅔ sec. In every case you will notice that the on time plus off time equals 10 sec. The difference is that the proportional-band width adjusts the ratio of on time to off time within the cycle.

Manual Reset. The set point of the controller is at some intermediate value of the controlled variable within the proportional band, usually at or near the middle. With load changes, however, it is generally necessary to change the amount of valve opening in order to provide the proper normal rate of control-agent flow. This is accomplished by the *manual-reset* adjustment in the controller.

Effects of Adjustments. The effects of proportional-band and manual-reset adjustments are illustrated graphically in Fig. 2-18. All three curves show the same proportional band setting of 40 per cent. Curve *A* shows the set point at the middle of the band, in which case the control valve is half open. If, as the result of a load change, a lower flow of the control agent—such as a 25 per cent valve opening—will suffice to maintain the controlled variable at the set point, manual reset can shift the proportional band up scale as shown by curve *B*. Conversely, if a larger valve opening of, say, 75 per cent is required, manual reset can shift the band down scale, as shown by curve *C*.

By permitting intermediate positioning of the control valve, proportional control can regulate the control-agent flow to meet the process demand, as reflected by changes in the controlled variable. It initiates a corrective action almost immediately, at a rate and of a magnitude directly proportional to the deviation of the controlled variable. Since there is a fixed relationship between the controlled variable and the valve opening, however, proportional control cannot automatically accommodate load changes and still maintain the controlled variable at the set point.

For example, with reference again to the graph in Fig. 2-18, assume that the proportional band and manual reset are adjusted as shown for curve A,

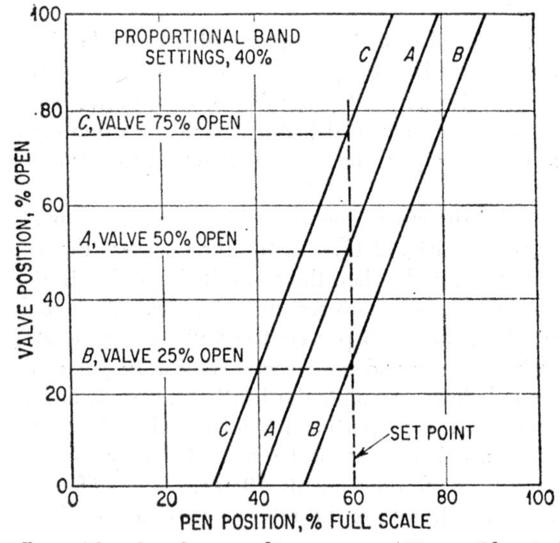

FIG. 2-18. Effect of band and reset adjustments. (*Hagan Chemicals & Controls, Inc.*)

which provides a 50 per cent valve opening when the pen is at the set point of 60 per cent of full scale. Now, suppose that there is a change in load in the process such that a 75 per cent valve opening is required to maintain the desired value of the controlled variable. The only way the controller can automatically provide this new valve opening—that is, without manual reset—is for the pen to be at 70 per cent of scale, or 10 per cent above the set point. This deviation is known as *offset*. Actually, where process load changes are small or infrequent, and some offset, or a temporarily larger offset, is permissible, proportional control is quite adequate.

In general, when proportional control is used, a larger process capacitance permits narrower proportional-band settings. Narrower bands result in faster corrective action and closer limits of control; hence slow process reaction rate is advantageous to proportional control. Greater transfer lag or dead time requires a wider proportional band and is therefore considered a limitation to the use of proportional control.

Another consideration in the application of the proportional mode is the existence of load changes in the presence of unfavorable process characteristics which require wider proportional bands. In these cases, the amount of offset which occurs with a load change varies directly with the band setting, and this offset may become excessive with the wider band.

Proportional-plus-reset Control. With proportional control, manual reset is necessary to provide for different valve openings to move the variable to the set point. This can be accomplished automatically to provide the *proportional-plus-reset* mode of control. The effect of automatic reset is illustrated graphically in Fig. 2-19. The controller has a 30 per cent

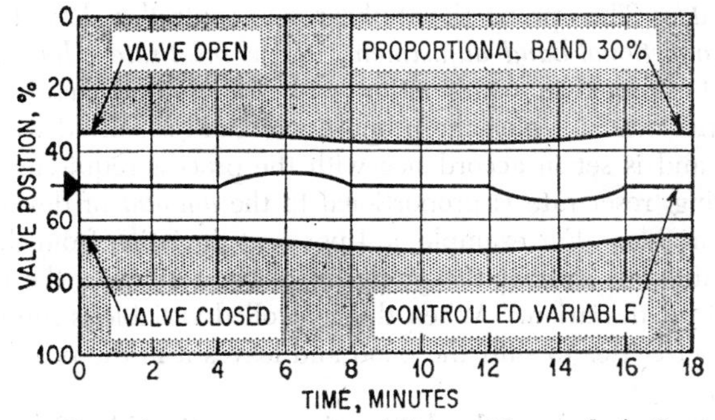

FIG. 2-19. Effect of automatic reset. (*Hagan Chemicals & Controls, Inc.*)

proportional band, and, at zero time with the temperature at the set point, provides a 50 per cent valve opening. At the time of 4 min, the temperature (not shown in the diagram) begins to drop below the set point because of an increase in process load.

Proportional action alone would cause the valve to assume a new position which is proportional to the deviation from set point. Remember that in proportional control there is one, and only one, position of the final control element for each value of the controlled variable within the proportional band. This new valve position—the only one the valve can assume for this deviation—is not sufficient to compensate for the increased load, and so the controlled variable will line out somewhat below the set point, and offset will result.

Automatic reset shifts the proportional band upward, however, and there is now a new position of the valve for this deviation from the set point, and the valve can open further. This increased valve opening permits sufficient flow of the control agent to bring the variable back to the set point and maintain it there as long as the load does not change again. If another load change occurs, automatic reset once more shifts the proportional band in the opposite direction of the load change, and permits the variable to return to set point.

In Fig. 2-19, the temperature has dropped sufficiently to cause the reset action to shift the proportional band upward by 10 per cent. With this reset proportional band, the valve is open 60 per cent to maintain the variable at set point with the increased process load. At 12 min, a load change of the same magnitude takes place in the opposite direction, and reset action shifts the proportional band downward so that a 50 per cent valve opening is adequate to maintain the variable at set point.

The proportional action functions with the reset action in producing a valve motion. For example, if the controlled variable suddenly deviated 10 per cent from the set point, proportional action would immediately cause the corresponding corrective movement of the control valve. Reset action would then function to return the variable to the set point because a new valve opening would be required to meet a sustained load change.

Adjustments. The proportional-plus-reset controller has two adjustments: a *proportional-band adjustment,* and a *reset-rate adjustment.* The reset-rate adjustment, in combination with the proportional-band setting, determines the rate at which the proportional band is shifted when a deviation occurs, and is set in accordance with the process requirements. With a given setting, reset rate is proportional to the *amount* of deviation of the controlled variable. For example, a 4 per cent deviation from the set point causes a continuous shift of the proportional band at twice the rate of that for a 2 per cent deviation. When the controlled variable is returned to the set point, reset ceases to shift the band and leaves it in the position attained at that time.

The reset response depends also on the proportional-band setting, being inversely proportional to the width of the proportional band.

Application. The proportional-plus-reset mode of control is the most generally useful of all modes, and its importance is constantly growing. It can handle most combinations of process characteristics satisfactorily.

Its primary advantage, of course, is the elimination of offset due to load changes, so that the controlled variable is maintained within close limits at all times. Small capacitances permit a higher reset rate, and enable the controller to counteract offset more quickly. Transfer lag and dead time in the controlled system, however, require a decreased reset rate in order to avoid excessive cycling of the controlled variable.

The limitation of proportional-plus-reset control lies in the large period of cycling and the attendant slow response when appreciable dead time is present in the controlled system. Transfer lag is not too serious a limiting factor so long as the measuring means is sensitive. If the measuring means is sluggish, transfer lag is converted to a dead time and the proportional band must be wider and the reset rate slower.

Rate Action. Processes with a large dead time (2 min or more) or a large transfer lag are sometimes difficult to control, even with the proportional-plus-reset mode. The proportional band must be set exceptionally wide and the reset rate unusually slow in order to avoid excessive cycling. Thus, when load changes occur, there is a rather wide deviation, and it takes a long time for the controlled variable to return to the set point. In such applications, which are most often found in temperature control, the addition of *rate action* to the proportional or proportional-plus-reset modes of control usually solves the control problem.

Basically, rate action provides an initially large corrective action when a deviation occurs, so that the final control element moves further at first than it normally would with proportional or proportional-plus-reset action. Then, having made this large initial change, the controller functions to remove this effect, leaving only the corrective action of the proportional or proportional-plus-reset responses to determine the position of the final control element. The result is an early, extra change in the control-agent

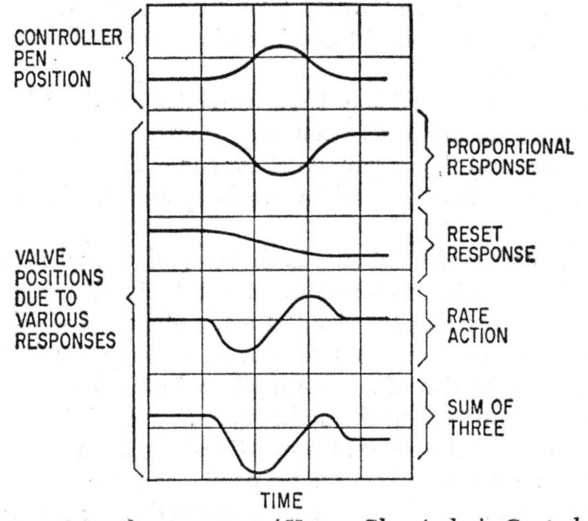

FIG. 2-20. Control responses. (*Hagan Chemicals & Controls, Inc.*)

supply, which tends to counteract the unfavorable effect of process lag. It is a temporary overcorrection proportional to the amount of deviation.

Advantages. Rate action is that in which there is a continuous relationship between the rate of change of the controlled variable and the position of the final control element. Thus, *proportional-plus-reset plus rate action* produces responses in proportion to (1) the deviation of the controlled variable (proportional action), (2) the amount and duration of the deviation (reset action), and (3) the rate of change of the controlled variable (rate action).

The individual control responses and the combined response of this mode are illustrated graphically in Fig. 2-20. Rate action occurs only when the controlled variable is changing, and is dependent on the proportional or proportional-plus-reset response of the controller for its functioning.

The most notable advantage of using rate action for counteracting the effect of dead time is the reduction of the period of cycling. Thus the controlled variable is stabilized at the set point more quickly when load changes occur. A rate-action adjustment is provided in controllers employing it so that the response can be adjusted in accordance with the amount of process lag. In pneumatic controllers, this mode can usually be added to the proportional-plus-reset controller.

Rate time—commonly expressed in minutes—is the time interval by which the rate action advances the effect of proportional action upon the final control element. It is determined by subtracting the time required for a certain movement of the final control element due to the combined effect of proportional-plus-rate action from the time required for the same movement due to the effect of proportional action alone. For example, if with proportional-plus-rate action a valve moves to a certain new position in 2 min, and this movement would take 4 min with proportional action alone, rate action has advanced the effect of proportional action by 2 min, and rate time is 2 min.

On process startup, the initially large correction provided by rate action brings the variable up to the set point very quickly, reduces overshoot, and cuts both the time and the magnitude of cycling. The variable lines out at the set point sooner than it would with the proportional-plus-reset mode alone.

Summary of Modes of Control. Each mode of control is applicable to processes having certain combinations of the basic characteristics. It is important to remember that the simplest mode of control which will do the job is the best one to use, both for reasons of economy and for best results. Frequently, the application of a too complicated control mode will result in poor control and contribute to the undesirable characteristics of the process.

If a controller has a 10 per cent proportional band, for example, and the variable deviates 5 per cent from the set point, the valve will be either fully open or fully closed. With such a large correction for deviation, even a load change will not allow offset to exist for any length of time before it is corrected. Therefore, with narrow proportional bands there is little need for reset.

Rate action is used to reduce long lags. In the measurement and control of rapidly changing variables, such as pressure and flow, the addition of rate action may actually increase upset instead of correcting it. In the measurement of flow with a *differential* meter—which responds very rapidly—pulsations in the flow from reciprocating pumps (actually "noise" and of no importance in measurement) will be picked up and transmitted to the controller. The rate unit in the controller will accelerate the corrective action in response to these pulsations and cause the final element to open and close too far and too rapidly—and to no purpose. In this instance, rate action in the system might actually cause it to react so that the final control element would be damaged.

Considerable experience is required in the selection of that mode best suited to the process. Table 2-6 summarizes the principal characteristics of the controlled system required for each mode discussed.

Table 2-6. Summary of Modes of Control Usage*

Mode of control	Process reaction rate	Transfer lag or dead time	Process load changes
Two-position	Fast	Slight	Small and slow
Proportional	Fast or medium	Small or moderate	Small
Proportional-plus-rate action	Slow or medium	†	Small
Proportional-plus-reset	†	Small or moderate	Slow
Proportional-plus-reset plus rate action	†	†	†

* Courtesy of Minneapolis-Honeywell Regulator Co.
† In general, any amount.

BLOCK-DIAGRAM ANALYSIS[4]

The study of control systems may be considerably simplified by the use of the block-diagram method. This technique is an organized system of thinking and enables one to outline the control problem.

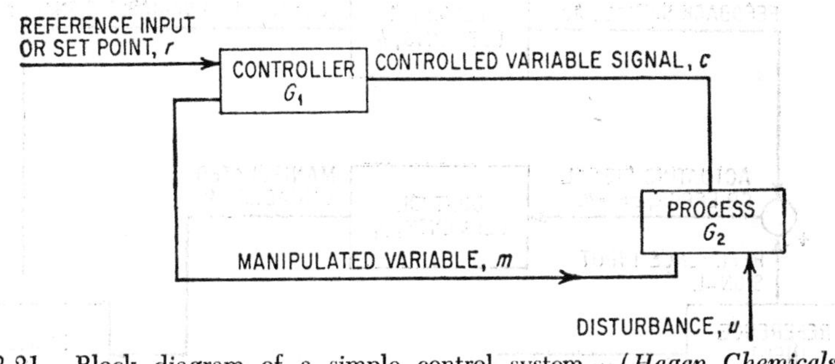

Fig. 2-21. Block diagram of a simple control system. (*Hagan Chemicals & Controls, Inc.*)

Figure 2-21 shows a control system reduced to its simplest elements—the process and the controller. It is noted that the controller has a reference input r, which is set to maintain the desired value of the controlled-variable signal c. By adjusting the manipulated variable m which acts on the process, the desired value of the controlled variable is maintained. Adjustments must be made upon the manipulated variable because the process is subject to a disturbance u, which acts to change the value of the controlled variable.

A simple illustration of the type of process shown in Fig. 2-21 might be a water tank (Fig. 2-22) in which it is desired to maintain a constant level. It could be assumed that the flow of water from the tank were variable and subject to the needs of the process and that the flow into the tank were controlled in order to maintain a constant level. In this case, the controlled

[4] Hagan Chemicals & Controls, Inc.

variable would be the water level; the reference input or set point would be the desired level as set on the controller by the float and valve adjustment; the manipulated variable would be the flow of water into the tank; and the disturbance would be changes in the rate of flow of water from the tank.

The simple block diagram on Fig. 2-21 may be extended and elaborated, if desired. In Fig. 2-23, the controller has been broken down into its various elements, namely, the reference-input elements, the control elements, and the feedback elements.

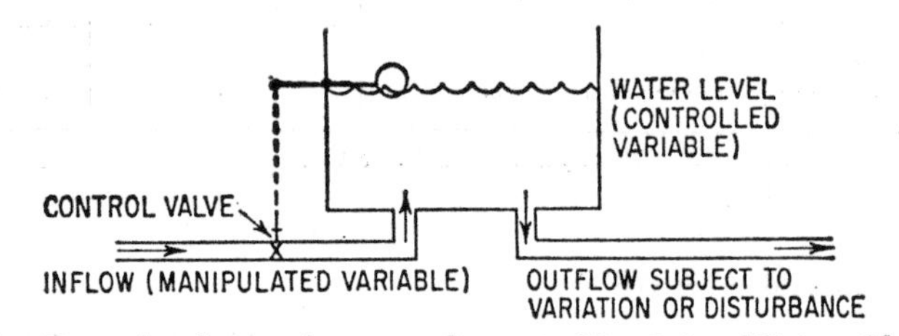

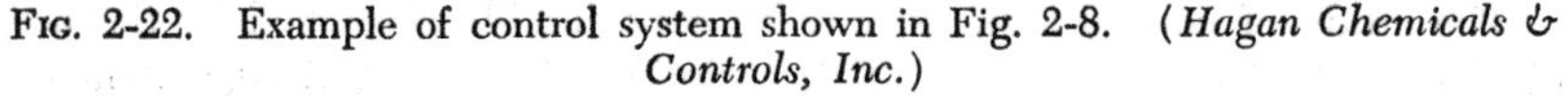

Fig. 2-22. Example of control system shown in Fig. 2-8. (*Hagan Chemicals & Controls, Inc.*)

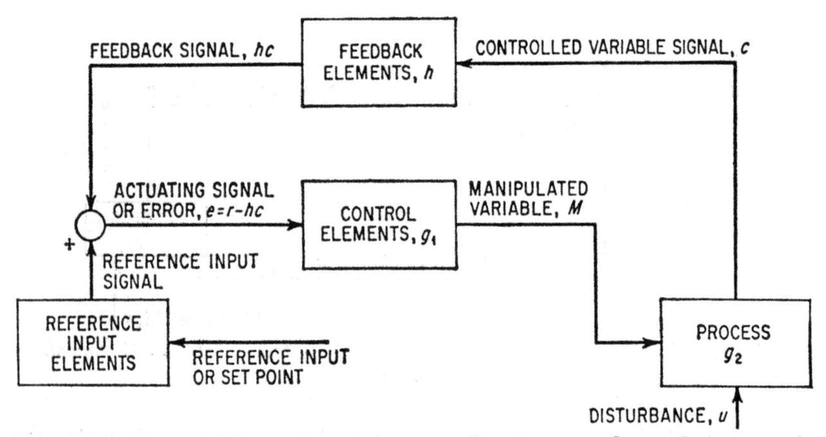

Fig. 2-23. Elements that make up the controller. (*Hagan Chemicals & Controls, Inc.*)

The reference-input portion of the circuit will usually receive a spring force or pneumatic-loading signal. The control elements will be actuated ordinarily by a pneumatic-loading signal to position a damper, valve, rheostat, or the like, usually with the aid of a power relay. The feedback elements will receive a signal of flow, pressure, temperature, or the like from the controlled variable. This will be converted into the same units as the reference-input signal (if they are not already comparable) before being subtracted from that signal to obtain the actuating signal of error.

The Effect of Feedback in a Controller. Feedback is a device or method used in a controller or control system in which the difference between the reference input and some function of the controlled variable is used to supply an actuating-error signal to the controller or control system.

In Fig. 2-24, K_1 represents the energy-modulating ability or transfer function of the controller, R is the reference input, and E is the actuating-error signal which is equal to R in the open-loop system. In addition to amplification or gain, the transfer function may introduce certain time-functional relationships. In this case, the symbol generally used would be G instead of K. However, to eliminate confusion, transfer functions will be denoted as K.

Negative Feedback. To demonstrate the application of feedback to the control system, let us substitute actual values for letters, and then make a

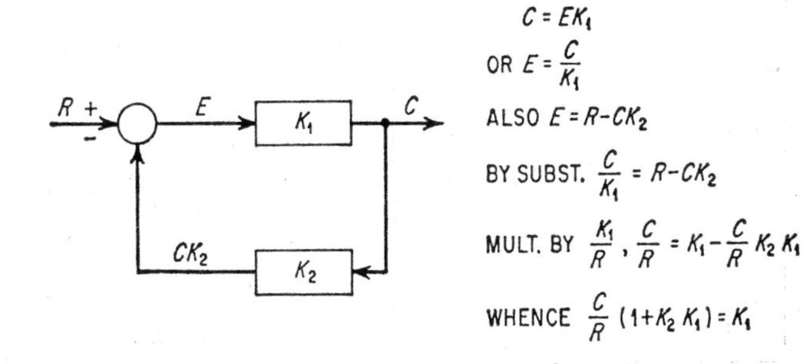

FIG. 2-24. Closed-loop control system. (*Hagan Chemicals & Controls, Inc.*)

plot of C vs. R for both open-loop and closed-loop control systems for a controller. Let K_1 have a value of 12.0 and K_2 have a value of 0.500; then C/R for the open loop is 12.0. For the closed loop of Fig. 2-24,

$$\frac{C}{R} = \frac{12.0}{1 + (12.0)(0.500)} = 1.71$$

It is evident that a considerable difference can exist in C/R values between open- and closed-loop systems. The actual effect on the controller is shown on Fig. 2-25. Notice how the looks of the open-loop curve have been straightened by the inclusion of negative feedback. This demonstrates that negative feedback tends to linearize the system. The proportional band has been increased seven times, thus increasing the stability but decreasing the sensitivity of the system. By reviewing the mathematical derivation of C/R values, it can be seen that the constants involved dictate the value of C/R. The amplification factor K_1 varies as the output of the controller. The amplification of the feedback circuit K_2 is varied by altering the proportional-band adjustment.

When the controlled process is operating under the desired conditions, no control system is really needed because theoretically the process could stay in equilibrium until acted on by some external force. It is these so-called "external" or "disturbing" forces that cause a process to deviate from the desired conditions. The greater the deviation from the desired condition, the more the control action is needed. If not enough stability is present, the system may hunt or oscillate. Therefore, the proportional band of the controller must be adjusted wide enough to give a stable system.

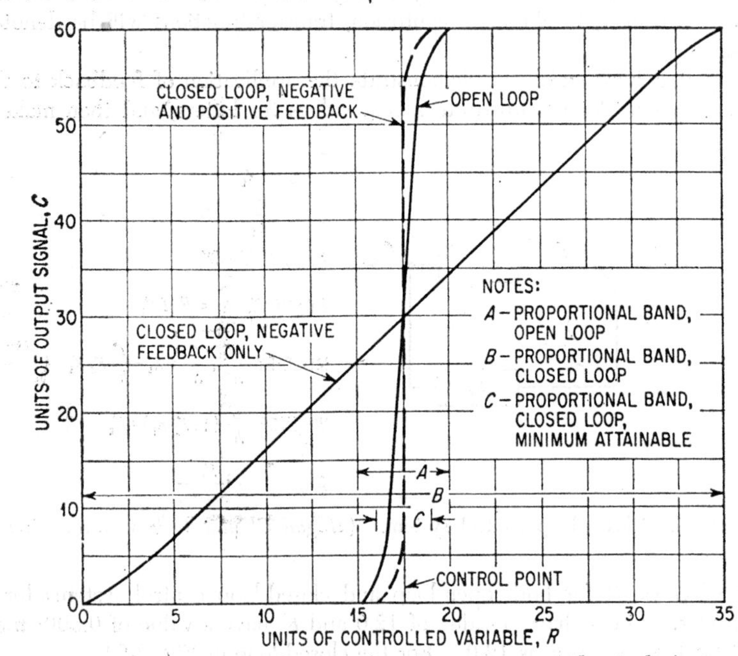

FIG. 2-25. Effect of feedback on a controller. (*Hagan Chemicals & Controls, Inc.*)

Combined Negative and Positive Feedback. In the previous paragraph it was deemed necessary to have proportional band of the proper amount in the system to make it stable. However, by the very definition of proportional band, there must be a change of controlled variable in order to have a change in output. In a great many cases it is desirable to have no change or at least a negligible change in the controlled variable for a change in output signal. This we can accomplish by introducing positive feedback in an amount sufficient to nullify the negative feedback and inherent system proportional band. This positive feedback is introduced with a time lag so that the proportional band will be reduced after it is no longer needed for stability.

For further clarification, we can substitute numbers for letters in Fig. 2-26 and make a plot of C vs. R for the closed loop with positive and negative feedback. Let K_1 have a value of 12.0, K_2 a value of 0.500, and K_3 a value of 0.535; then C/R for this system is

$$\frac{C}{R} = \frac{12.0}{1 + (12.0)(0.500) - (12.0)(0.535)}$$

The increase of C/R results in a decrease in the proportional band. The K_3 value we now see plays a very critical part in the controller.

In the preceding discussion and in Fig. 2-25, we have seen how positive feedback can be used to wipe out the proportional band due to negative feedback and that which is inherent from the control components, such as the gradient of a loading spring. If we do not provide a time lag in the introduction of positive feedback, we are worse off from the standpoint of stability than before, as the system will operate as an on-off control. By

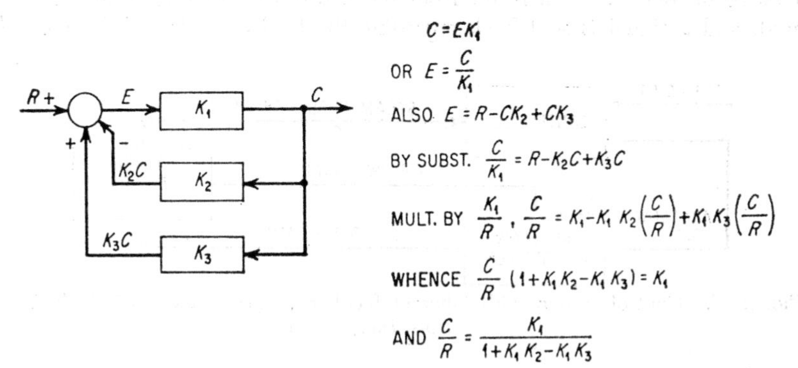

FIG. 2-26. System with both positive and negative feedback. (*Hagan Chemicals & Controls, Inc.*)

introducing a needle valve for example, and volume chamber between the negative- and positive-feedback elements (proportional-band adjustment and reset-adjustment bellows, respectively), we can adjust the system so that the positive feedback can reduce the proportional band to nearly zero. The time delay in the positive-feedback circuit must be adjusted to prevent instability.

Adjustment of a Controller to a Process. In adjusting, a controller process is first controlled by a machine with negative feedback only. The K_2 function is reduced until instability or hunting occurs. In the master sender, the procedure would be to close the needle valve on the reset adjustment (to remove positive feedback), and to move the proportional-band adjustment toward the main fulcrum (to reduce K_2) until hunting starts on load change. The proportional-band adjustment is then widened by moving this adjustment away from the fulcrum until hunting stops.

Next the K_3 value is adjusted until the C/R ratio of Fig. 2-26 is infinite, meaning that the proportional band is zero. With the master sender this is done by moving the reset-adjustment unit along the beam until the input-output is the vertical curve of Fig. 2-25. This is done with the reset-time-adjustment valve open and with the process being manually controlled.

With the controller again controlling the process, the valve for reset-time adjustment is set so that the reset will be as fast as possible without hunting or instability. The controller is now properly adjusted to that process.

It is seen from this discussion and a study of Fig. 2-25 that we have really set the controller to have a low gain (wide proportional band) at high frequencies and to have a high gain (narrow proportional band) at low frequencies. This is desirable in all controllers and will give as close control as possible with stability.

The Application of Feedback to a Control System. In applying feedback to the entire control system, it is observed that, in a feedback circuit, we originate a control signal which actuates some controlling device such as a valve or switch. The result from the controlled variable is then measured, and a signal is sent back regarding the final condition of the control

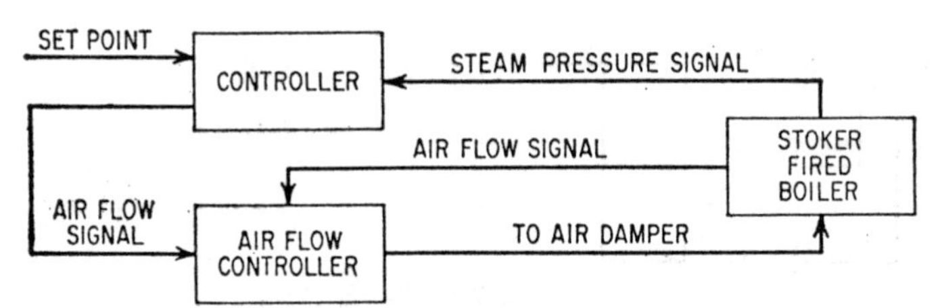

FIG. 2-27. Control system with inherent feedback. (*Hagan Chemicals & Controls, Inc.*)

variable. This signal is then compared with the input signal of requirement. If any difference exists between the requirement signal and that of the actual condition, the controller takes further action to reduce this difference to zero. This gives a positive and accurate means of controlling a process, since there is a continual check on what is occurring.

In an open circuit, there is no feedback of information regarding the effect of the input signal upon the controlled variable, so that we cannot expect the accuracy and positive control which is obtained with the feedback circuit. An example of an open system might be one in which we had a photoelectric cell designed to turn on the street lights in a portion of a city. If this cell were mounted on the top of a high building, exposed only to natural illumination, it might be desired to actuate the light switch when the illumination dropped to a certain degree of intensity. Once this cell had actuated the light switch, there would be no further check in the system regarding the intensity of illumination produced in the street.

A similar example of a feedback circuit might be a photoelectric cell designed to maintain a certain degree of illumination intensity inside a room or upon a machine. In this case, the cell would be actuated by the light emanating from a controlled source. It would make adjustments in voltage to the lamp, depending on the deviation of actual intensity from the set point.

Stiffening Control Systems. Most control systems have what might be called inherent feedback.

For example, in Fig. 2-27 consider the system without the air-flow feedback signal. In this case, assume that there is a drop in steam pressure which will cause the controller to send out an air-flow signal to the air-flow controller for more damper opening. If we assume that the damper regulates forced draft and that momentarily the wind-box pressure is low, it may be that insufficient air is supplied by this action to bring the steam pressure back to the normal condition. Eventually, this condition will be known at the pressure controller because the steam pressure will not rise to the set point. Depending on the boiler and furnace, the response of steam pressure to the air-flow signal may take considerable time. This time may be sufficient in some instances to produce a hunting action or an undesirable slowness of recovery.

In order to increase the accuracy of air supply, the air flow may be metered, and a signal sent back to the air-flow controller to compare with the input air-requirement signal. If there exists a deviation due to change in wind-box pressure or for any other reason, the difference between the two signals will be immediately reduced by controlling action. This method of speeding up corrective action within the control circuit is known as "stiffening" the system.

Section 3

MEASUREMENT OF PRESSURE AND DRAFT

Pressure, defined as a force per unit area, is one of the most important of the measured and controlled mechanical-service variables. An exceedingly wide range of pressure—all the way from high vacuum to pressure of 5,000 psi (pounds per square inch) or greater—must be measured and controlled accurately and reliably. Because of this large range, numerous pressure-measuring elements are required.

MEASUREMENT OF PRESSURE AND DRAFT

Bourdon Elements.[1] The bourdon-tube gage (Fig. 3-1) generally consists of a tube, oval in section, rolled into an arc of a circle, one end being attached to the socket and the other to a movable quadrant or "sector" which meshes with a pinion carrying the indicating pointer. The socket end of the tube is open to admit the fluid, and the other end is closed. The increase in pressure of the fluid in the tube tends to straighten out the tube, thus moving the tip. The movement of the tip is transferred to the sector, which turns the pinion and the pointer.

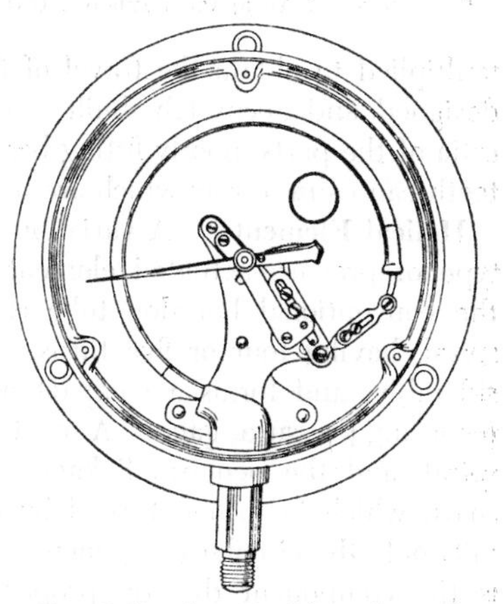

FIG. 3-1. A typical bourdon tube element. (*Crosby Valve & Gage Co.*)

The mechanical and mathematical principles which govern the tube are quite complex. The material of which the tube is made and heat treatment are likewise important. (See Table 3-1.)

The tip movement of a properly designed and accurately formed tube is a definite amount for each increment in pressure; therefore, the position of the pointer will indicate on the dial the pressure in the tube.

The tip movement is relatively quite small and must be considerably

[1] Crosby Valve & Gage Co.

Table 3-1. Typical Specifications of Bourdon Tubes of Different Materials*

Type of tube	Specifications	Range, lb
Phosphor bronze........	Forged brass socket, silver-soldered joints, ¼-in. connections, micrometer-pointer accuracy ½ of 1%, drawn tube	15–1,000
Alloy steel............	Forged steel socket, welded joints, ¼-in. connection to 1,000 lb, ½-in. connection for 1,000 lb and over, micrometer-pointer accuracy ½ of 1%:	
	Type 4130, drawn tube	15–5,000
	Bored tube	100–20,000
Stainless steel..........	Forged steel socket, welded joints, ¼-in. connection to 1,000 lb, ½-in. connection for 1,000 lb and over, micrometer-pointer accuracy ½ of 1%:	
	Type 316, drawn tube	15–2,000
	Type 403, drawn tube	3,000–15,000
	Type 403, bored tube	100–20,000
Beryllium copper.......	Forged stainless-steel socket, forged brass optional, silver-soldered joints, ¼-in. connection to 1,000 lb, ½-in. connection standard for 1,000 lb and over, micrometer-pointer accuracy ½ of 1%, drawn tube	5,000–20,000
K Monel.............	Forged Monel socket, welded joints, ¼-in. connection to 1,000 lb, ½-in. connection for 1,000 lb and over, micrometer-pointer accuracy ½ of 1%, drawn tube	15–5,000

* Courtesy of Acragage Division, International Register Co.

multiplied to obtain the travel of the pointer around the dial. A carefully designed and accurately built movement is a necessity. Friction and inertia of the parts, poorly fitting bearings, or improperly cut sector and pinion teeth can cause errors which are multiplied in the pressure indicated.

Helical Elements.[2] A variation of the simple bourdon type is the helical type of pressure-actuated element (Fig. 3-2). This element is similar to the conventional bourdon tube except that it is wound in the form of a spiral, having four or five turns. This increases the travel of the tip considerably, and forms a compact unit easily constructed and installed in a recording-pressure gage. A central shaft is usually installed within the spiral, and the pen-arm linkage is so arranged as to be driven from this shaft, which in turn is turned by the tip of the spiral. This design transmits only the circular component of the tip movement to the pen arm, which is the component directly proportional to the change in pressure. This type of element is also widely used for recording thermometers.

Spiral Elements.[3] The spiral type of measuring element, a second modification of the bourdon element, is used widely in industrial pressure gages. Briefly it is a thin-walled tube, which has been flattened on opposite sides to

[2] T. J. Rhodes, "Industrial Instruments for Measurement and Control," McGraw-Hill Book Company, Inc., New York, 1941.
[3] Minneapolis-Honeywell Regulator Co., Industrial Div.

produce an approximately elliptical cross section. The tube is then formed into a spiral. When pressure is applied to the open end, the tube tends to uncoil.

The spiral, illustrated in Fig. 3-3, provides large movement of the free end, which makes it ideal for use in a recorder. So the spiral is widely used in instruments measuring pressures from 10 to 4,000 psi. The spiral requires no sector and pinion to obtain sufficient pointer travel, but is connected directly to the pen or pointer shaft by a single link. This direct

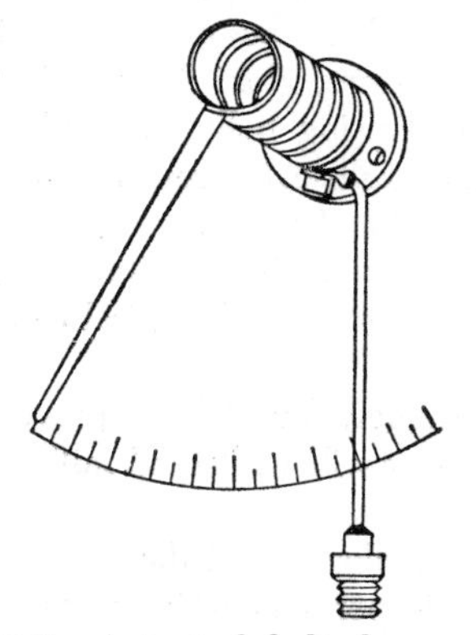

Fig. 3-2. A typical helical gage element (*Crosby Valve & Gage Co.*)

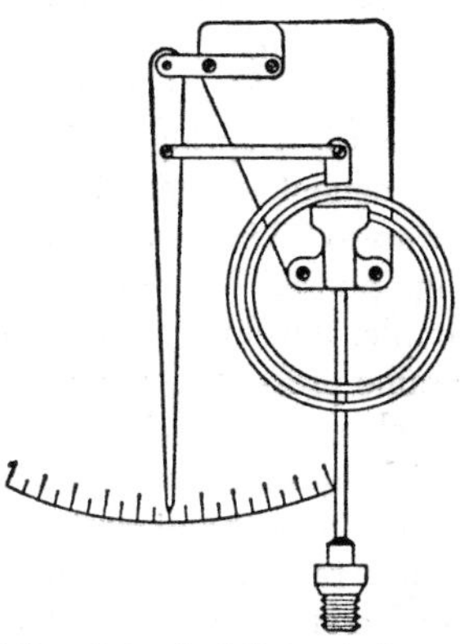

Fig. 3-3. A typical flat spiral gage element. (*Crosby Valve & Gage Co.*)

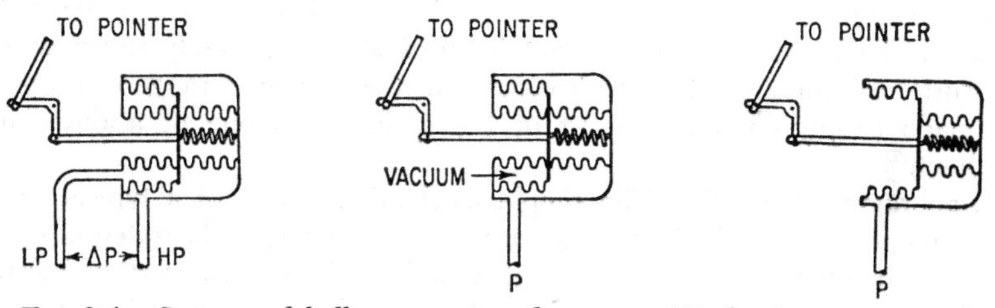

Fig. 3-4. Spring and bellows pressure elements. (*Taylor Instrument Co.*)

linkage reduces friction to a minimum, eliminates inertia or lost motion, and increases the accuracy as well as the response of measurement.

Spirals of bronze, steel, and stainless steel are usually available in ranges from 0–18 to 0–4,000 psi.

Spring and Bellows.[3] Pressures in what are termed the "intermediate" and "low" ranges cannot be measured satisfactorily with the spiral in industrial-type instruments. Spring-opposed bellows elements have been developed for these lower pressures. (See Fig. 3-4.)

A typical intermediate-range unit, with either brass or stainless-steel bellows, measures pressures with full-scale values between 100 in. of water and 40 psig (pounds per square inch gage), or vacuums with full-scale values

[3] Minneapolis-Honeywell Regulator Co., Industrial Div.

between 10 and 30 in. of mercury. It is made up of a metallic bellows enclosed in a shell, with the shell connected to the pressure source. Pressure, acting on the outside of the bellows, compresses the bellows, and moves its free end against the opposing force of the spring. A rod resting on the bellows transmits this motion through a linkage into linear pen or pointer movement.

For lower pressures and vacuums, a second type of spring-opposed element is employed. In this unit, the pressure is connected to the inside of a larger bellows. This creates an appreciably greater force per unit area, which acts against an opposing spring force. For the measurement of vacuums or combinations of vacuum and pressure, this element is also equipped with a spring which opposes the collapsing bellows. This low-pressure element is generally used for minimum full-scale pressures of 5 in. of water up to 90 in. of water and vacuums of 50 in. of water.

Spring-opposed bellows elements are very sensitive. The low-range type, for example, develops as much as 25 times the power of the spiral element for the same pressure change.

These elements are characterized by an extremely long life, as proved by tests which indicate that the bellows and springs will withstand millions of cycles of flexing without rupture.

Phosphor bronze is commonly used for the bellows, and the springs are made of carefully heat-treated metal which is permanent in its gradient (change in force per unit of compression).

The bellows gradient is small in comparison with the spring gradient, and has little effect on the calibration of the unit. The bellows, which does not generally have too linear a gradient in the first place, merely serves as a pressure enclosure. With such a construction a change in range can often be made simply by replacement of the spring with one of a different gradient.

Diaphragm Elements.[4] The simple mechanical action of a typical diaphragm-operated multipointer gage for pressure or draft is illustrated in Fig. 3-5. Such units consist essentially of a calibrated spring, a vertical lever, and a power diaphragm. The weight of the diaphragm is carried on pivots at the lower ends of the vertical lever and the calibrated spring.

Pressure applied on the side of the diaphragm which faces the calibrated spring causes movement of the system to the left, which motion is opposed by the flexing of the calibrated spring. This motion is transmitted by the vertical lever to the outside of the diaphragm unit where a drive link carries the motion to the indicating pointer.

Diaphragm units are used to measure draft, or pressure, or differential. When *draft* is being measured, the connecting tubing from the diaphragm-housing chamber containing the vertical lever is connected to the source of measurement, while the chamber on the spring side of the diaphragm housing is left open to the atmosphere. When *pressure* is being measured, the tubing from the chamber containing the calibrated spring is connected to

[4] Bailey Meter Co.

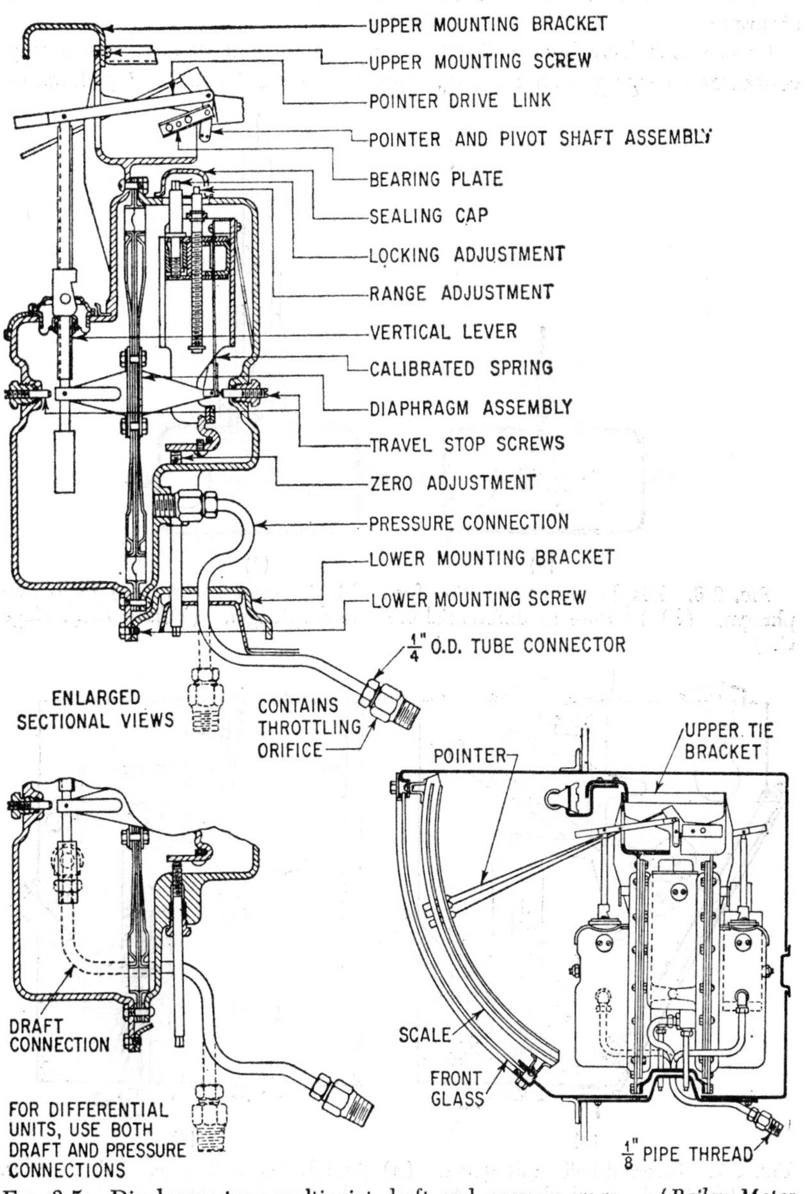

FIG. 3-5. Diaphragm-type multipoint draft and pressure gage. (*Bailey Meter Co.*)

the pressure source, while the chamber with the vertical lever is open to atmosphere.

Units which have been calibrated for pressure or draft have one tubing connector emerging from the gage case, whereas units measuring differen-

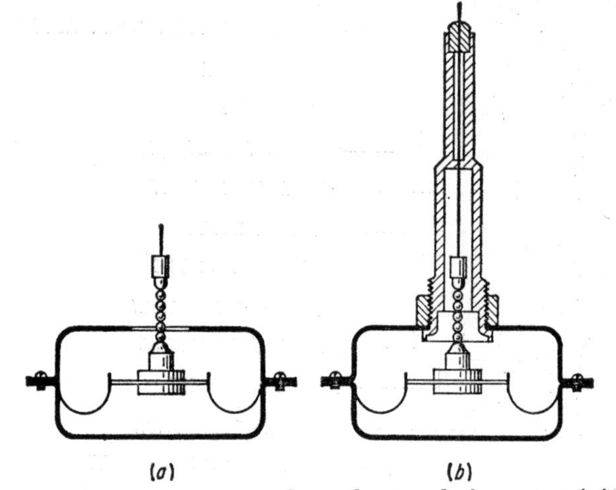

FIG. 3-6. Diaphragm construction for multipoint draft gages. (*a*) Draft diaphragm. (*b*) Pressure or differential-pressure diaphragm. (*Ellison Draft Gage Co.*)

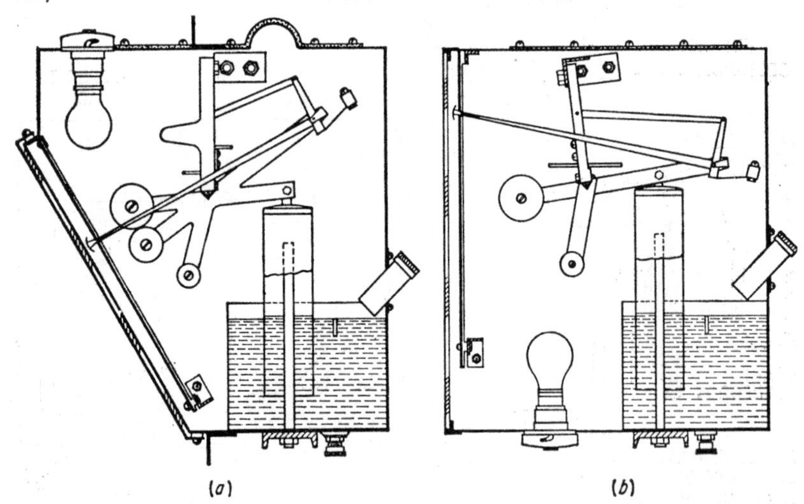

FIG. 3-7. Inverted-bell draft gages. (*a*) Straight-line bell gage. Tilted scales. (*b*) Straight-line bell gage. Straight scales. (*Ellison Draft Gage Co.*)

tial have two such connectors. The high-pressure side of a differential unit is connected to the point of measurement having the least draft (higher pressure), in case draft differentials are being measured by the unit.

Diaphragm units commonly withstand a maximum static pressure or draft of 150 in. of water without damage to the mechanism, but the maximum differential across the diaphragm generally should not exceed 50 to 60 in. of water. Another variation of diaphragm construction is shown in Fig. 3-6.

Inverted-bell Elements.[5] The inverted-bell type of pressure gage is confined to applications requiring a sensitive small-range pressure gage for relatively low pressures and for vacuums such as those in boilers. It is available in a single inverted bell (Fig. 3-7) for measuring static pressures (where the instrument is located at the point of measurement and compensation is unnecessary), or in a double-inverted-bell model for differential-pressure measurement and control where compensation is required.

The single-bell model consists of an inverted bell suspended in such a way that it is immersed in oil, and arranged so that the measured pressure raises the bell and moves a balance beam which has a counterweight on the opposite end. Movement of the balance beam is transmitted through a mechanical linkage to a pressure-indicating assembly.

The double-bell model incorporates two inverted bells partly immersed in oil which acts as a liquid seal, supported from the opposite ends of a balance beam and arranged so that a pressure can be introduced under each inverted bell.

In effect, this arrangement weighs the minutest difference in pressure between the two pressure lines. One of these lines may be exposed to the atmosphere, and the other may be connected, for example, to the interior of a furnace so that the instrument will indicate the pressure differential between the atmosphere and the interior of the furnace. As a result, the instrument is particularly adaptable to the measurement and control of furnace- and heater-draft conditions, particularly where atmospheric-pressure compensation is necessary.

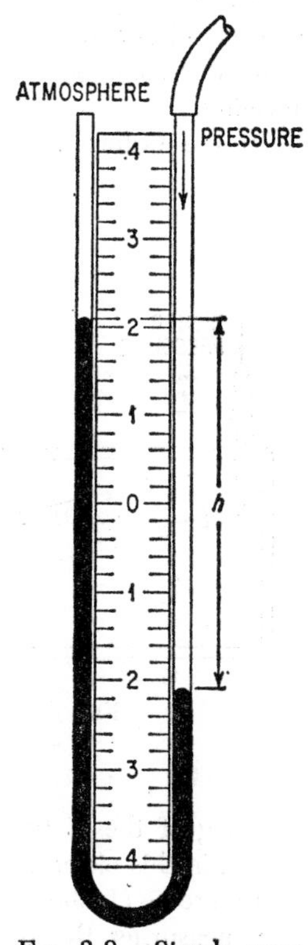

FIG. 3-8. Simple open manometer.

These types of pressure instruments are commonly sensitive to within ±0.0005 in. of water.

Manometer Elements.[6] The open vertical manometer or Uron tube (Fig. 3-8) indicates gage pressure or vacuum. The reading may be a static pressure or a total pressure. Since the instrument reading is a linear distance h (Fig. 3-8), the density of the liquid ρ must be known in order to determine the force per unit area, $P = \rho h$. The liquid density will vary

[5] Minneapolis-Honeywell Regulator Co., Industrial Div.

[6] C. F. Shoop and G. L. Tuve, "Mechanical Engineering Practice," McGraw-Hill Book Company, Inc., New York, 1956.

slightly with temperature. The manometer actually measures the difference between two pressures, and the true significance of this differential-pressure reading depends on the type of pressure openings and on the fluid in the connecting tubing (Fig. 3-11.)

In the inclined manometer (Fig. 3-9) one leg of the U tube is inclined at a known angle, so that the scale distance h is longer than h' ($h' = h \sin \alpha$). Usual multiplying factors are 10 to 1 or less, but 20 to 1 or even more can be obtained if the instrument is carefully adjusted and used. The two most common errors are faulty zero setting of the scale and inaccurate leveling.

In the cistern manometer of Fig. 3-10 a large supply chamber forms

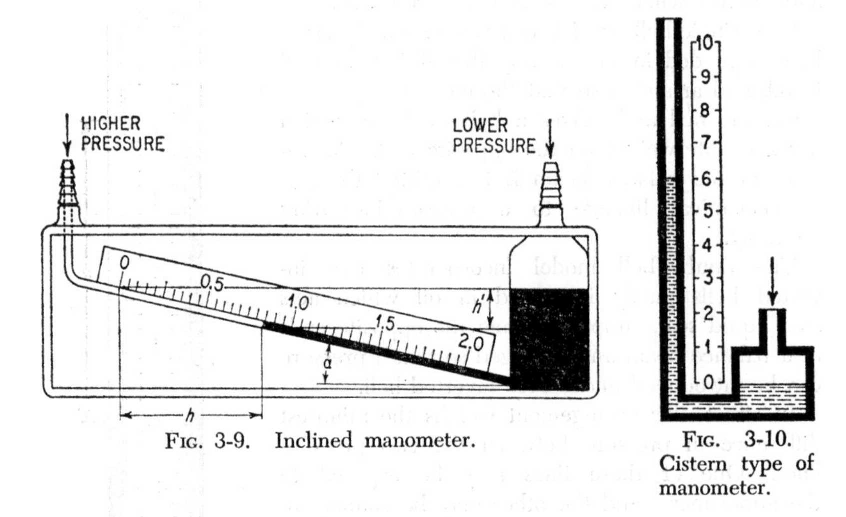

Fig. 3-9. Inclined manometer.

Fig. 3-10. Cistern type of manometer.

one leg of the U tube, and the zero of the scale is set at the level of the liquid in this cistern.

Figure 3-11 shows two ways of using a differential U-tube manometer for measuring the pressure drop across a metering orifice in a horizontal water pipeline. Manometer a is filled with mercury and is located below the pipe. The connecting tubes must be vented to be sure they are solidly filled with water. The true differential head at the pipe is then h' in. of mercury *less* h' in. of water. Manometer b is inverted, and the water rises from the pipe into each connecting tube and forms the indicating liquid in the manometer. Since the density of the air trapped in (or pumped into) the top of the manometer is small compared with the density of the water, the differential head h in inches of water may be regarded as that existing at the main pipeline.

The two-fluid manometer of Fig. 3-12 is useful for measuring air pressures beyond the range of a small water manometer. By using oil above mercury, the instrument may be calibrated in terms of the rise of the oil level in the small tube B. Many other types of two-fluid manometers

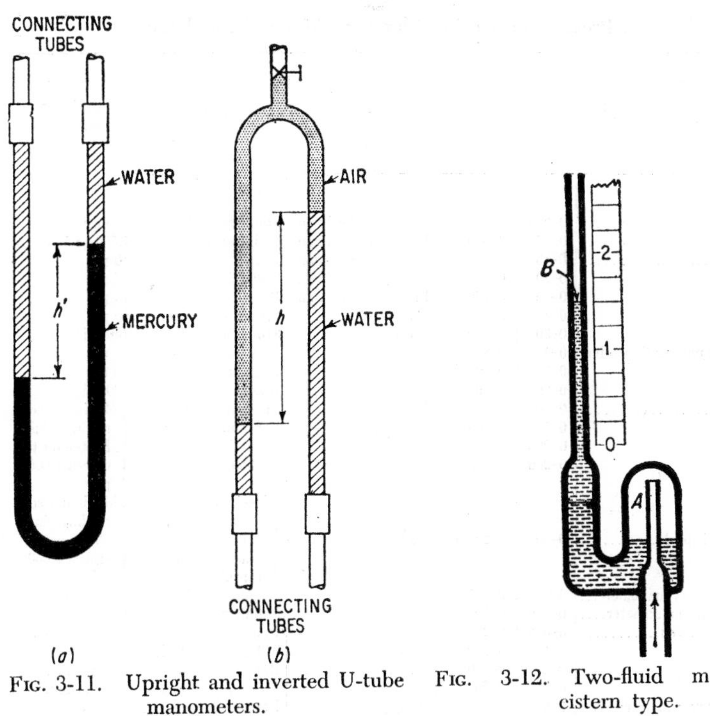

FIG. 3-11. Upright and inverted U-tube manometers.

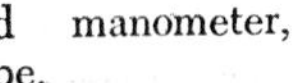

FIG. 3-12. Two-fluid manometer, cistern type.

Table 3-2. Typical Application of Pressure-measuring Instruments*

Element	Application	Minimum range	Maximum range
Extra-large spring and bellows	Pressure	0–5 in. water	0–90 in. water
	Vacuum	0–5 in. water vacuum	0–60 in. water
	Compound vacuum and pressure	5 in. water†	80 in. water†
Spring and bellows (brass)	Pressure	0–90 in. water	0–16.3 psig
	Vacuum	0–10 in. Hg vacuum	0–30 in. Hg vacuum
Spring and bellows (stainless steel)	Pressure	0–15 psig	0–40 psig
Spiral (bronze)	Compound vacuum and pressure	6.5 psig†	15 psig†
	Pressure	0–18 psig	0–400 psig
Spiral (steel)	Pressure	0–30 psig	0–4,000 psig
Spiral (stainless steel)	Pressure	0–50 psig	0–4,000 psig
Absolute-pressure bellows (brass)	Absolute pressure	0–100 mm Hg abs	0–50 in. Hg abs
Absolute-pressure bellows (stainless steel)	Absolute pressure	0–200 mm Hg abs	0–60 in. Hg abs

* Courtesy of Minneapolis-Honeywell Regulator Co., Industrial Div.
† Value represents span instead of range.

Table 3-3. Pressure-measuring Elements, Materials, and Ranges*

Application and measuring element	Range			Materials	Maximum pressure rating
	From	To	Units		
Gage pressure:					
Bell-type mercury manometer..	0–2.5	0–20	In. of water column	Stainless steel	500 psi
U-tube mercury manometer....	0–25	0–800	In. of water column	Stainless steel	1,500 or 5,000 psi
Open-frame bellows...........	0–10	0–40	In. of water column	Brass	10 psi
	0–20	0–120	In. of water column	316 stainless steel	15 psi
	0–40	0–120	In. of water column	Brass	15 psi
	0–4	0–30	Psi	316 stainless steel	50 psi
	0–30	0–100	Psi	316 stainless steel	150 psi
Bellows-type pneumatic receiver	0–4	0–8	Psi	Brass and aluminum	20 psi
	0–8	0–20	Psi	Brass and aluminum	30 psi
	0–20	0–80	Psi	Brass and aluminum	100 psi
Helical......................	0–80	0–600	Psi	Phosphor bronze	Maximum scale reading plus 50%—not to exceed 10,000 psi
	0–500	0–10,000	Psi	Beryllium copper	
	0–200	0–5,000	Psi	316 stainless steel	
	0–200	0–10,000	Psi	Carbon steel	
Liquid-filled helical†..........	0–80	0–2,000	Psi	Steel, 304 or 316 stainless steel, Monel, nickel, Hastelloy, or tantalum	Maximum scale reading plus 50%—not to exceed 2,000 psig
Bell-type mercury manometer..	0–0.2	0–1.5	In. Hg	Stainless steel	Full vacuum
U-tube mercury manometer....	0–2.0	0–30	In. Hg	Stainless steel	Full vacuum
Open-frame bellows...........	0–0.8	0–2.9	In. Hg	Brass	20 in. Hg vacuum
	0–1.5	0–9.0	In. Hg	316 stainless steel	Full vacuum
	0–2.9	0–9.0	In. Hg	Brass	Full vacuum
	0–8	0–30	In. Hg	316 stainless steel	Full vacuum
Bellows-type pneumatic receiver	0–8	0–30	In. Hg	Brass and aluminum	Full vacuum
Differential pressure:					
Bell-type mercury manometer..	0–2.5	0–20	In. of water column	Stainless steel	500 psi
U-tube mercury manometer....	0–25	0–800	In. of water column	Stainless steel	1,500 or 5,000 psi
Absolute pressure:					
Bell-type mercury manometer..	0–5	0–37	Mm Hg abs	Stainless steel	500 psi
U-type mercury manometer‡...	0–1	0–30	Psia	Stainless steel	1,500 or 5,000 psi
Opposed bellows..............	0–4	0–30	Psia	Brass	40 psia
	0–8	0–30	Psia	316 stainless steel	40 psia

* Courtesy of Fischer & Porter Co.

† Liquid-filled helical element is recommended for slurry service, highly corrosive service (chlorine, acids, fluorides, bromine, iodine, etc.), dirty or freezing fluids, viscous service, paper-stock service, condensation in gas or vapor service, etc. Materials listed are diaphragm seal options.

‡ To obtain absolute pressure, a high-vacuum pump connected to the low-pressure side of the manometer must keep reference pressure at zero.

have been used, but capillary and meniscus errors are common, and, unless the densities of the two fluids are widely different, the line of separation between them is likely to be indefinite.

Tables 3-2 and 3-3 list applications for the different elements.

MANOMETER TABLES[7]

Density of Mercury (Table 3-4). The values of the density of mercury given here have international acceptance. At 0°C, the density is generally considered accurate to one part in 100,000 for natural mercury.

[7] Condensed from *Recommended Practice, Manometer Tables, ISA RP2.1,* Instrument Society of America.

Table 3-4. Density of Mercury*

Temperature °C	Temperature °F	Density, g/cu cm
−20	−4	13.6446
−10	14	13.6198
0	32	13.5951
5	41	13.5827
10	50	13.5704
15	59	13.5581
20	68	13.5458
25	77	13.5336
30	86	13.5214
35	95	13.5091
40	104	13.4969
45	113	13.4847
50	122	13.4725
60	140	13.4482
70	158	13.4240
80	176	13.3998
90	194	13.3755
100	212	13.3515

Temperature, °F	Density, lb/cu in.	Density, g/cu cm
0	0.49275	13.6391
10	0.49225	13.6253
20	0.49175	13.6116
32	0.491157	13.5951
40	0.49076	13.5841
50	0.49026	13.5704
60	0.48977	13.5568
70	0.48928	13.5431
80	0.48879	13.5295
90	0.48830	13.5159
100	0.48780	13.5023
110	0.48732	13.4888
120	0.48683	13.4753
130	0.48634	13.4617
140	0.48585	13.4482
150	0.48536	13.4347
160	0.48488	13.4213
170	0.48439	13.4079
180	0.48391	13.3944
190	0.48342	13.3809
200	0.48293	13.3675

* At 0°C (32°F) = 13.5951 g/cu cm
= 0.491157 lb wt/cu in.
= 848.719 lb wt/cu ft

At other temperatures the density can be computed from the following:

$$V = V_0(1 + 0.0001818t) \quad \text{where } t \text{ is in deg C} \qquad (1)$$

$$= V[1 + 0.0001010(t - 32)] \quad \text{where } t \text{ is in deg F} \qquad (2)$$

$$D = \frac{1}{V} \qquad (3)$$

where V = specific volume at temperature t
V_0 = specific volume at 0°C or 32°F in the same unit as V
D = mass per unit volume

Density of mercury from "Smithsonian Physical Tables." These values agree with those computed from formulas (1) and (3) above.

Table 3-5. Density of Water*

(Distilled and free from air)

Temperature, °C	Density, lb/cu in.	Density, g/cu cm
0	0.0361218	0.999841
3.98	0.0361265	0.999973
5	0.0361262	0.999965
10	0.0361167	0.999701
15	0.0360951	0.999102
20	0.0360627	0.998207
25	0.0360209	0.997048
30	0.0359704	0.995651
35	0.0359121	0.994037
40	0.0358465	0.992221
45	0.035774	0.99021
50	0.035695	0.98804
60	0.035522	0.98324

Temperature, °F	Density, lb/cu in.	Density, g/cu cm
35	0.0361232	0.999882
40	0.0361265	0.999971
45	0.0361236	0.999892
50	0.0361167	0.999701
55	0.0361060	0.999406
60	0.0360919	0.999015
65	0.0360746	0.998536
68	0.0360627	0.998207
70	0.0360542	0.997971
75	0.0360309	0.997327
80	0.0360050	0.996608
85	0.0359764	0.995819
90	0.0359454	0.994960
95	0.0359121	0.994037
100	0.0358764	0.993051
105	0.0358387	0.992006
110	0.035799	0.99090
115	0.035757	0.98974
120	0.035713	0.98854
130	0.035622	0.98600
140	0.035522	0.98324
150	0.035414	0.98025
200	0.034792	0.96304

* From *J. Research NBS*, vol. 18, *Research Paper 971*, February, 1937, and from Table 93 in N. E. Dorsey, "Water-substance," 1940.

The density at 0°C is the standard used internationally in defining pressure in terms of the height of a mercury column.

Density of Water (Table 3-5). The values given here differ slightly from those given by various experimenters. The data for temperatures above 40°C (104°F) are less reliable than those for lower temperatures, which is indicated by dropping one digit from the values. The density is affected somewhat by the content of absorbed air. Also, the dissolved salts will affect the density, which effect is avoided by the use of distilled water.

Table 3-6. Gravity Corrections at Sea Level at Various Latitudes for Manometers Filled with Any Liquid

(Corrections in the same unit as the height of the column)

Latitude, degrees	Gravity, cm/sec^2	Height of liquid column in any unit				
		20	40	60	80	100
0	978.039	−0.054	−0.107	−0.161	−0.214	−0.268
10	978.195	−0.050	−0.101	−0.151	−0.201	−0.252
20	978.641	−0.041	−0.083	−0.124	−0.165	−0.206
25	978.960	−0.035	−0.070	−0.104	−0.139	−0.174
30	979.329	−0.027	−0.054	−0.082	−0.110	−0.137
35	979.737	−0.019	−0.038	−0.057	−0.076	−0.095
40	980.171	−0.010	−0.020	−0.030	−0.040	−0.050
45	980.621	−0.001	−0.002	−0.003	−0.004	−0.005
—	980.665	0	0	0	0	0
50	981.071	+0.008	+0.017	+0.025	+0.033	+0.041
55	981.507	+0.017	+0.034	+0.052	+0.069	+0.086
60	981.918	+0.026	+0.051	+0.077	+0.102	+0.128
65	982.288	+0.033	+0.066	+0.099	+0.132	+0.166
70	982.608	+0.040	+0.079	+0.119	+0.159	+0.198
80	983.059	+0.049	+0.098	+0.146	+0.195	+0.244
90	983.217	+0.052	+0.104	+0.156	+0.208	+0.260

The unit height of a water column at either 15°C (60°F) or 68°F is used to define a pressure. The Instrument Society of America has adopted as a recommended practice 68°F (20°C) as the definition.

Gravity Corrections at Sea Level at Various Latitudes or at Various Values of Gravity (Tables 3-7 and 3-8). In Table 3-7 are given the gravity corrections at sea level for a column of any liquid in any unit of height, millimeters, centimeters, inches, or feet, against latitude. The values of gravity corresponding to latitude are from the formulas and tables given in the "Smithsonian Physical Tables and Meteorological Tables."

Table 3-8 presents the gravity corrections, again for any liquid or in any unit of height, against evenly divided values of gravity. This table is convenient to use if the value of gravity is known.

When Table 3-6 is used, an additional correction is required for the

Table 3-7. Gravity Corrections for Manometers Filled with Any Liquid

(Corrections in the same unit as the height of the column)

Gravity, cm/sec²	Height of liquid column in any unit				
	20	40	60	80	100
978.0	−0.054	−0.109	−0.162	−0.217	−0.272
978.5	−0.044	−0.088	−0.132	−0.177	−0.221
979.0	−0.034	−0.068	−0.102	−0.136	−0.170
979.5	−0.024	−0.048	−0.071	−0.095	−0.119
980.0	−0.014	−0.027	−0.041	−0.054	−0.068
980.5	−0.003	−0.007	−0.010	−0.013	−0.017
980.665	0	0	0	0	0
981.0	+0.007	+0.014	+0.021	+0.027	+0.034
981.5	+0.017	+0.034	+0.051	+0.068	+0.085
982.0	+0.027	+0.054	+0.082	+0.109	+0.136
982.5	+0.037	+0.075	+0.112	+0.150	+0.187
983.0	+0.048	+0.095	+0.143	+0.191	+0.238
983.5	+0.058	+0.116	+0.174	+0.231	+0.289

elevation of the station above sea level. The altitude correction to be applied to the value of gravity, which is, strictly speaking, only for free air, and applies only approximately to large elevations found in mountainous areas, is as follows:

$$\text{Correction (cm/sec}^2) = -94 \times 10^{-6}h$$

where h is the elevation in feet of the manometer above sea level.

As an example of the use of the tables, assume a liquid column height of 72 in. (after application of scale correction, before or after application of the temperature correction) at latitude 38° and elevation above sea level of 150 ft.

From Table 3-6 the value of gravity at sea level at 38° latitude is 979.997 cm/sec².

The altitude correction is $-94 \times 150 \times 10^{-6}$ or −0.014.

The value of gravity is 979.983 cm/sec².

By interpolation in Table 3-7, the gravity correction is −0.048 in.

The column height corrected for gravity is 72 − 0.048 or 71.952 in.

Neglecting the altitude correction and using Table 3-6 to obtain the correction, there is obtained 0.049 in. Correcting gravity for the altitude effect introduces neglible error in this case.

Conversion Factors for Various Pressure Units (Table 3-8). The number of digits given here is greater than required for most manometer use. The factors were used insofar as applicable in preparing the other tables.

The factors apply only to manometer readings after all corrections have been made to obtain the pressures in the units stated.

Table 3-8. Conversion Factors for Various Pressure Units

(Equivalent value in various units)

Pressure unit value	Mm Hg, 0°C	In. Hg, 0°C	Millibars	Psi	Psf	Oz/sq in.	G/sq cm	Cm water, 60°F	In. water, 60°F	Cm water, 20°C	In. water, 20°C	Cm water, 25°C	In. water, 25°C	Atmosphere
Mm Hg	1	0.03937	1.3332	0.019337	2.7845	0.30939	1.3595	1.3609	0.53577	1.3620	0.53620	1.3635	0.53682	1.31579×10^{-3}
In. Hg	25.400	1	33.864	0.49116	70.727	7.8585	34.532	34.566	13.609	34.594	13.620	34.634	13.635	0.0334211
Millibars	0.75006	0.029530	1	0.014504	2.0886	0.23206	1.0197	1.0207	0.40186	1.0215	0.40218	1.0227	0.40265	9.86923×10^{-4}
Psi	51.715	2.0360	68.947	1	144	16	70.307	70.376	27.707	70.433	27.729	70.515	27.762	0.0680457
Psf	0.35913	0.014139	0.47880	0.0069444	1	0.11111	0.48824	0.48872	0.19241	0.48912	0.19257	0.48969	0.19279	4.72540×10^{-4}
Oz/sq in.	3.2322	0.12725	4.3092	0.0625	9	1	4.3942	4.3985	1.7317	4.4021	1.7331	4.4072	1.7351	4.25286×10^{-3}
G/sq cm	0.73556	0.028959	0.98066	0.014223	2.0482	0.22757	1	1.0010	0.39409	1.0018	0.39441	1.0030	0.39487	9.67841×10^{-4}
Cm water, 60°F	0.73483	0.028930	0.97970	0.014209	2.0461	0.22735	0.99901	1	0.3937	1.0008	0.39402	1.0020	0.39448	9.66887×10^{-4}
In. water, 60°F	1.8665	0.073483	2.4884	0.036092	5.1972	0.57747	2.5375	2.5400	1	2.5421	1.0008	2.5450	1.0020	2.45590×10^{-3}
Cm water, 20°C	0.73424	0.028907	0.97891	0.014198	2.0445	0.22717	0.99821	0.99919	0.39338	1	0.3937	1.0012	0.39416	9.66105×10^{-4}
In. water, 20°C	1.8650	0.073424	2.4864	0.036063	5.1930	0.57700	2.5355	2.5380	0.99919	2.5400	1	2.5430	1.0012	2.45392×10^{-3}
Cm water, 25°C	0.73339	0.028873	0.97777	0.014181	2.0421	0.22690	0.99705	0.99803	0.39292	0.99884	0.39324	1	0.3937	9.64984×10^{-4}
In. water, 25°C	1.8628	0.073339	2.4835	0.036021	5.1870	0.57633	2.5325	2.5350	0.99803	2.5371	0.99884	2.5400	1	2.45106×10^{-3}
Atmosphere	760	29.9212	1,013.25	14.6960	2,116.22	235.136	1,033.23	1,034.25	407.183	1,035.08	407.512	1,036.29	407.986	1

1 millibar = 1,000 dynes/sq cm.
1 kg/sq meter = 0.1 g/sq cm.
1 kg/sq cm = 1,000 g/sq cm.
1 g/sq meter = 0.0001 g/sq cm.

PRESSURE INSTRUMENTS

PRIMARY SPECIFICATION SHEET

SHEET No. ____
TAG No. ____
DATE ____
REVISED ____
BY ____

GENERAL

1 DESCRIPTION RECORDER ☐ INDICATOR ☐ BLIND ☐
CONTROLLER ☐ TRANSMITTER ☐
2 CASE RECTANGULAR ☐ CIRCULAR ☐
OTHER ____
3 CASE COLOR BLACK ☐ OTHER ____
4 MOUNTING FLUSH ☐ SURFACE ☐ YOKE ☐
5 NO. PTS.--RECORDING ____ INDICATING ____
6 CHART TYPE 12" CIRC. ☐ OTHER ____
7 CHART RANGE ____ NUMBER ____
8 SCALE RANGE ____ TYPE ____
9 CHART DRIVE SPRING ☐ ELECTRIC ☐ PNEUM. ☐
10 CHART SPEED ____ WIND ____
11 V ____ C ____ EX.PRF. ☐ AIR PRESS. ____
12 OTHER ____

TRANSMITTER

13 TYPE PNEUMATIC ☐ ELECTRIC ☐
14 OUTPUT 3-15 PSI ☐ OTHERS ____
15 RECEIVERS ON SHEET NO. ____

CONTROL

16 TYPE PNEUMATIC ☐ ELECTRIC ☐
OTHER ____
17 PROP ____ % AUTO-RESET ☐ RATE ACTION ☐ ON-OFF ☐
OTHER ____
18 OUTPUT 3-15 PSI ☐ OTHER ____
19 ON MEASUREMENT INCREASE:
OUTPUT: INCREASES ☐ DECREASES ☐

AUTO MANUAL SWITCH

20 NO. POSITIONS ____ EXTERNAL ☐ INTERNAL ☐
INTERNAL ☐

SETPOINT ADJUSTMENTS

21 MANUAL INTERNAL ☐ EXTERNAL ☐
22 AUTO-SET PNEUMATIC ☐ ELECTRIC ☐
23 BAND FIXED ☐ ADJUSTABLE ☐
24 OTHERS ____

NOTE:

PRESSURE ELEMENT

25 SPIRAL ☐ BELLOWS ☐ BOURDON ☐
DIAPHRAGM ☐ HELICAL ☐
OTHER ____

MATERIAL

26 BRONZE ☐ STAINLESS ☐ STEEL ☐
OTHER ____
27 ABSOLUTE PRESS. COMPENSATION ____
28 STATIC HEAD COMPENSATION ____
HEAD ____
29 RANGE ____
PSIG ☐ IN.HG.VAC. ☐ PSIA ☐
OTHER ____
30 CONNECTION-NPT 1/4" ☐ 1/2" ☐
BACK ☐ BOTTOM ☐ OTHER ____

ACCESSORIES

31 FILTER & REGULATOR ____
32 AIR SUPPLY GAGE ____
33 LOCAL INDICATOR ____
34 CHARTS & INKSET ____
35 MOUNTING YOKE ____
36 PULSATION DAMPENER ____
37 SYPHON ____
38 ALARM SWITCH ____
HERMETICALLY SEALED ☐ E.P. ☐ G.P. ☐

OPERATING CONDITIONS

PRESSURE, NORMAL ____ MAX. ____
TEMPERATURE NORMAL ____ MAX. ____
FLUID ____
SEAL FLUID ____ S.G. @ 60°F ____

FIG. 3-13*a*. Pressure instruments primary specification sheet. (*Instrument Society of America.*)

Properties of Manometer Liquids (Table 3-9). Water is probably the most widely used manometric liquid, being not only low in cost, readily available, and stable in nature, but also having a low factor of thermal expansion. Its major disadvantages are its limited range of utility—usually from 0.05 to 2.0 psi, and only 40 to 100°F—and its corrosive effect on ferrous metals. One of the most desirable methods of increasing its visibility in glass tubing is by the addition of a few drops of fluorescein solution; coloring by inks or dyes is not satisfactory.

Table 3-9. Properties of Manometric Liquids

Liquid	Specific gravity, 20:20	Action with water vapor	Vapor pressure at 68°F	Coefficient of thermal expansion			Melting point, °F	Boiling point, °F	Flash point, °F
				per °F $\times 10^6$	per °C $\times 10^6$	Range, °F			
			Mm Hg						
1. Ethyl alcohol, C_2H_6O	0.7939	Absorbs	43.9	600	1080	50–86	−179	173	55
2. Kerosine, 41 API at 60°F	0.8200 60:60	Negligible	....	480	864	30–100	−20	300+	120
3. Ellison gage oil	0.8340 60:60	Negligible	....	466	839	30–100	....	300+	140
4. Benzene (benzol), C_6H_6	0.8794	Negligible	74.7	687	1237	68	42	176	12
5. Butyl Cellosolve $C_6H_{14}O_2$ (ethylene glycol monobutyl ether)	0.9019	Absorbs	0.85	...	......	....	−100	340	165
6. Water	1.000	—	17.5	115	207	68	32	212	Nonflammable
7. Alcohol glycol	1.000	Absorbs	....	427	769	30–100	−60	173	70
8. Carbitol, $C_6H_{14}O_3$ (Diethylene glycol monoethyl ether)	1.024–30	Absorbs	....	...	......	....	−76	202	210
9. *n*-Butyl phthalate, $C_{16}H_{22}O_4$	1.0477	Negligible	10^{-4}	433	780	....	−31	644	340
10. Ethylene glycol (glycol), $C_2H_6O_2$	1.1155 20:4	Absorbs slowly	0.09	354	638	68	+0.8	387	241
11. Halowax oil	1.19–1.25	—	0.3–50 C	367	660	....	−24– −42	.....	203
12. Glycerin (glycerol), $C_3H_8O_3$	1.260 20:4	Absorbs	Low	281	505	68	64	554	320
13. *o*-Dibromobenzene, $C_6H_4Br_2$	1.956 20:4	Negligible	....	432	778	30–100	35.2	430	150+
14. 1,1-Dibromoethane, $C_2H_4Br_2$	2.089 20:4	Negligible	34.7	532	958	30–100	40	230	75+
15. Acetylene tetrabromide (tetrabromoethane), $C_2H_2Br_4$	2.964 20:4	Absorbs slightly	....	370	660	....	−4	.....	Nonflammable
16. Mercury	13.570	Negligible	0.0012	101	181.8	−20 to 250	−38	679	Nonflammable

Mercury is the second most widely used manometric liquid, being quite stable and reasonably available, immiscible with other liquids, and having a low thermal expansion and low vapor pressure. It can be used over a wide range of temperature conditions—from −35 to over 250°F, and normally in the pressure range from 0.2 to 20–30 psi. Many metals, such as copper, tin, silver, and zinc, are soluble in mercury, so that instrument parts in contact with it should be glass, carbon steel, stainless steel, or good-grade iron.

PRESSURE-INSTRUMENT SPECIFICATION

As a guide to engineers who specify and select pressure instrumentation, the Instrument Society of America has developed recommended practice specification forms that may be used as a guide.[8] (See Fig. 3-13.)

PRESSURE INSTRUMENTS

SECONDARY SPECIFICATION SHEET

SHEET NO.
TAG NO.
DATE
REVISED
BY

REV	QUAN.	TAG NO.	RANGE	PRESS. ELEMENT	ELEMENT MAT'L.	CONNECTION	MEAS. INC. OUTPUT	SERVICE	NOTES

FIG. 3-13*b*. Pressure instruments secondary specification sheet. (*Instrument Society of America.*)

The purpose of this recommended practice is to provide uniformity in instrument specifications, in both content and form. The complexities of modern instrumentation make necessary some type of specification form to transmit details to all interested parties. It is felt that certain advantages will be gained if these forms are generally used throughout the industry. Some of the advantages of industry-wide use of these forms are:

1. Assures accuracy and completeness in specification of instruments.
2. Promotes uniformity in terminology.[9]

[8] *Tentative Recommended Practice, ISA RP20.1,* Instrument Society of America.

[9] Wherever applicable, the terminology used is that published in ASME Industrial Instruments and Regulators Division, Standard 105, "Automatic Control Terminology" (copyrighted 1954), and/or that developed by Scientific Apparatus Manufacturer's Association (SAMA), Recorder Controller Section, in their various instrument standards or tentative standards.

PRESSURE INSTRUMENTS

Specification Sheet Instructions

Prefix numbers designate line number on corresponding specification sheet.

1) Description - Check one or combination of two or more. Example: Recorder Controller.

2) Case - Check one.
Describe other type or special consideration to type checked on Line 2.

3) Check black or write in color required.

4) Mounting - Check one.

5) Write in number of Recording Pens and/or Indicating Pointers exclusive of Index Pointer.

6) Check 12" size, or write in size required.

*7) Write in chart range and number.

*8) Write in scale range and type.

*9) Check type.

10) Write revolutions per day, and days spring wound drive to run on complete wind.

11) Write in voltage and cycles, check if explosion proof required when chart drive is electric, air supply pressure if pneumatic.

12) Write in any special type or deviation from standard drives on Line 11.

13) Check one.

14) If pneumatic transmission, check if output is 3-15 PSI. Write in if other than 3-15 PSI, or if electric, write in electric characteristics.

15) Write in Specification Sheet No. or numbers on which Receiver Instruments appear. (For cross reference).

16) Check one or write in other type.

17) Write in proportional band, percent of scale or chart range desired. Check if auto-reset and/or rate action is desired.

18) Check if 3-15 PSI pneumatic or write in controller output.

*19) When form is used to specify a single instrument, check either Increase (Inc.) or Decrease (Dec.). When form is used to specify multiple instruments of the same general description, write in abbreviation (Inc.) or (Dec.) in column provided on Secondary Sheet.

20) Check one and write in number of positions.

21) If manual, check location.

22) If automatically set - check operating medium.

23) Check if index set travel is fixed in proportion to the impulse, or is adjustable.

24) Write in any other, such as linkage or cam, or deviation from Lines 21, 22, 23.

*25) Check type - Bourdon tube, helix, bellows, receiver.

*26) Check material desired.

27) Check if absolute pressure compensation is required.

28) Check if required. State amount of compensation required.

*29) Range - Calibration range of element. Check or write in range units.

*30) Pipe connections are female type. Check or write in pipe size.

31) Write in if individual units or combination type desired, or if it is to be purchased separately.

32) Write in "Furnish" if for instrument without integral gages. If "By Others", show Specification Sheet No. and Item No.

33) Specify size dial and scale for receiver type gauge for local indication when using blind type transmitter. If "By Others", write Specification Sheet No.

**When more than one instrument is specified, show data on Secondary Sheet, leaving lines unmarked on Base Sheet.*

FIG. 3-13c. Pressure instruments specification sheet instructions. (*Instrument Society of America.*)

3. Details can be communicated without transformation or misunderstanding.
4. Facilitates quoting, purchasing, receiving, accounting, and ordering procedures by uniform transmittal of information.
5. Provides a permanent record and means for checking the installation.
6. Facilitates training of personnel associated with any phase of instrumentation.
7. Improves efficiency from the initial step to the final installation.

PRESSURE INSTRUMENT

SPECIFICATION SHEET INSTRUCTIONS, CONT'D.

34) Write in number of days supply required.

35) Write in Yes or No if for type where yoke is optional.

36) State type required. Fixed, adjustable, etc.

37) State diaphragm and bowl material. State thread size. If prefilled type, give length of capillary and capillary material. State ambient temperature limits.

38) Write in location, rating in amps., volts & cycles for contacts. Check housing type.

Operating Conditions - When form is used for Single Instruments only, Operating Conditions are to be listed on Secondary Sheet.

Secondary Sheet - For listing Multiple Instruments.

List all instruments of the same type, specified on Primary Sheet, with variations as shown. Variations not provided for to be noted on Primary Sheet as Note (1), etc., and entered on Secondary Sheet where applicable.

Rev. -.Revision Number
Quan. - Show Quantity Required.
Tag No. - Identification of Item Number
Range - See Instructions #29, 7, 8
Press. Element - See Instruction #25
Element Mat'l. - See Instruction #26
Meas. Inc. Output - See Instruction #19
Connection - See Instruction #30
Service - Fill in Service
Note - Note number appearing on Primary Sheet.

FIG. 3-13*c*. (*Continued*)

The make-up of the forms is such that variations as required by the individual user can be made without materially affecting the basic form. For example, space is provided at the top for serial number, company or plant number, model number, revisions, etc. If the form is used to specify a single instrument, space can also be used for maintenance records, service changes, shop instructions, etc.

To aid in interpreting the meaning of each item to be checked on the primary specification sheets, an instruction form is provided. This is called

PRESSURE GAGES

SPECIFICATION SHEET

SHEET NO. ______
TAG NO. ______
DATE ______
REVISED ______
BY ______

GENERAL SPECIFICATION

1 TYPE INDICATING ☐ RECEIVER ☐
OTHER ______
2 MOUNTING SURFACE ☐ LOCAL ☐ FLUSH ☐
3 DIAL DIAMETER ______
4 DIAL COLOR BLACK ☐ WHITE ☐
5 CASE MAT'L CAST IRON ☐ ALUMINUM ☐ PHENOL ☐
OTHER ______
6 RING TYPE SCREWED ☐ HINGED ☐ SLIP ☐
OTHER ______
7 MANUFACTURER'S MODEL NO. ______

8 PRESS ELEMENT BOURDON ☐ BELLOWS ☐
OTHER ______
9 ELEMENT MAT'L BRONZE ☐ STEEL ☐
TYPE ___ STNL.STL ☐ OTHER ______
10 SOCKET MAT'L BRONZE ☐ STEEL ☐
TYPE ___ STNL.STL ☐ OTHER ______
11 CONNECTION-NPT 1/4" ☐ 1/2" ☐
BOTTOM ☐ BACK ☐
12 MOVEMENT BRONZE ☐ STNL.STL ☐
NYLON ☐ OTHER ______

REV	QUAN	TAG NO.	RANGE		OPER. PRESS	SERVICE	ACCESSORIES	NOTES
			TUBE	DIAL				

FIG. 3-13*d*. Pressure gages specification sheet. (*Instrument Society of America.*)

"Specification Sheet Instructions." Each line is numbered and keyed by the same number as on the primary sheet. If deviations from the ISA forms are made by users, care should be taken not to lose the advantage of the keyed-number system.

CALIBRATION OF PRESSURE INSTRUMENTS

A gage[10] is always calibrated to some standard of pressure, and many different types are available, depending on the pressure and accuracy required.

For accurate measurements of low pressures, that is below 15 psi, manometers provide a primary standard of pressure measurement. Their principle is based on the visible measurement of pressure head created by such liquids as mercury, oil, or water. (See "Manometer Elements," page 3-7.)

For pressures above 15 psi, various forms of test pumps and dead-weight testers are available. A test pump usually consists of an adjustable piston in a closed cylinder, to which are attached a test gage and the gage to be calibrated (Fig. 3-14). In this case the accuracy of calibration obtained depends entirely on the accuracy of the test gage used, and therefore is not considered as a primary standard of pressure measurement.

Pressure produced by a definite weight acting on a piston and cylinder of definite area is an accurate primary source of pressure measurement. This is the principle of the dead-weight tester (Fig. 3-15). The weights

[10] Crosby Valve & Gage Co.

PRESSURE GAGES

SPECIFICATION SHEET INSTRUCTIONS

Prefix numbers designate line numbers on corresponding specification sheet.

1) Check or write in general type.

2) Mounting - Check one.

3) Write in nominal gage size.

4) Check one.

5) Check one or write in other material desired.

6) Check one or write in other type.

7) If desired, write in manufacturer's model number or style.

8) Check one or write in other type.

9) Check one or write in other material desired.

10) Check one or write in other material desired.

11) Check socket connection size desired, also whether bottom or back connected.

12) Check one or write in other.

Rev. - Revision Number
Quan. - Show quantity required.
Tag No. - Identification of item no.
Range - Show dial range if standard indicating gage. If receiver gage, show both tube and dial ranges.
Oper. Press. - If desired, show normal operating pressure.
Service - Fill in service.
Accessories - Write in accessories desired.

Notes: Note number or letter of note appearing in space provided at the bottom of the sheet for notes.

Fig. 3-13*e*. Pressure gages specification sheet instructions. (*Instrument Society of America.*)

in this case are applied to the piston by a sensitive scale-balance multiplying-lever system. Thus, by this method, a reasonable size weight can be used to exert a very high pressure.

Operating a Dead-weight Pressure-gage Tester (Fig. 3-15). Place tester in level position on bench, so that the upright cylinder will be exactly vertical and the weight piston will work smoothly in it without friction.

Before using the tester, be sure that piston and cylinder are thoroughly clean and entirely free from foreign matter of any sort. The best results

DIFFERENTIAL PRESSURE INSTRUMENTS

PRIMARY SPECIFICATION SHEET

SHEET NO. ____
TAG NO. ____
DATE ____
REVISED ____
BY ____

GENERAL

1 DESCRIPTION RECORDER ☐ INDICATOR ☐ BLIND ☐
CONTROLLER ☐ TRANSMITTER ☐
2 CASE RECTANGULAR ☐ CIRCULAR ☐
OTHERS ____
3 CASE COLOR BLACK ☐ OTHER ☐
4 MOUNTING FLUSH ☐ SURFACE ☐ YOKE ☐
5 NO. PTS.-RECORDING ____ INDICATING ____
6 CHART TYPE 12" CIRC. ☐ OTHER ____
7 CHART RANGE ____ NUMBER ____
8 SCALE RANGE ____ TYPE ____
9 CHART DRIVE SPRING ☐ ELECTRIC ☐ PNEUM. ☐
10 CHART SPEED ____ WIND ____
11 V ____ C ____ E.P. ☐ AIR PRESS. ____
12 OTHER ____

TRANSMITTER

13 TYPE PNEUMATIC ☐ ELECTRIC ☐
14 OUTPUT 3-15 PSI ☐ OTHER ☐
15 RECEIVERS ON SHEET NO. ____

CONTROL

16 TYPE PNEUMATIC ☐ ELECTRIC ☐
OTHER ____
17 PROP. ____ % AUTO-RESET ☐ RATE-ACTION ☐ ON-OFF ☐
OTHER ____
18 OUTPUT 3-15 PSI ☐ OTHER ____
19 ON MEASUREMENT INCREASE
OUTPUT: INCREASES ☐ DECREASES ☐

AUTO-MANUAL SWITCH

20 NO. POSITIONS EXTERNAL ☐ INTERNAL ☐
INTEGRAL ☐

SETPOINT ADJUSTMENTS

21 MANUAL INTERNAL ☐ EXTERNAL ☐
22 AUTO-SET PNEUMATIC ☐ ELECTRIC ☐
23 BAND FIXED ☐ ADJUSTABLE ☐
24 OTHER

STATIC PRESSURE OR RECEIVER ELEMENT

25 TYPE ____ MATERIAL ____
26 RANGE ____
27 FOR OTHER ELEMENTS SEE TAG NO. ____
ON SHEET NO. ____

DIFFERENTIAL UNIT

28 FLOW ☐ LEVEL ☐ PRESSURE ☐
29 MERCURY ☐ BELLOWS ☐ DIAPHRAGM ☐
30 OTHERS ____
31 MAT'L BODY ____ DIAPHRAGM OR BELLOWS ____
32 BODY RATING PSIG ____ @ 60°F
33 DIFFERENTIAL RANGE ____
34 DIFFERENTIAL CONN. 1/4" ☐ 1/2" ☐ OTHER ____

ACCESSORIES

35 FILTER & REGULATOR ____
36 AIR SUPPLY GAGE ____
37 LOCAL INDICATOR ____
38 CHARTS & INKSET ____
39 MOUNTING YOKE ____
40 PULSATION DAMPENER ____
41 MERCURY ____
42 LUBRICATOR & ISO. VALVE ____
43 INTEGRATOR ____
44 PRIMARY MEASURING ELEMENT ____
TYPE ____
45 ALARM SWITCH ____
HERMETICALLY SEALED ☐ E.P. ☐ G.P. ☐

SERVICE CONDITIONS

FLUID ____ FLOW UNITS ____
FLOW NORMAL ____ FULL SCALE ____
@ 60°F & ____ PSIA
OPERATING PRESS. PSIG ____ TEMP. ____ °F
SPEC. GRAV. @ 60°F ____ & ____ PSIA
SPEC. GRAV. @ F.T. ____ & ____ PSIA
VAPOR GAS MOL. WT. ____
VISCOSITY @ ____ °F ____
SUPER-COMPRESSIBILITY FACTOR @ O.P. ____
SEAL FLUID OR PURGE ____
SPEC. GRAV. @ 60°F ____
PIPE I.D. ____ TYPE TAPS ____
FLANGE RATING & FACING ____
ORIFICE BORE NEAR. 1/8" ☐ EVEN MULTPL'R ☐
PLATE MAT'L ____ ACTUAL BORE ____
CHART MULTIPLIER ____
FLOW EQUALS ____ X DIFF. X ____

Fig. 3-13*f*. Differential-pressure instruments primary specification sheet. (*Instrument Society of America.*)

are obtained by using only high-grade lubricating oil as furnished with the tester.

When ready to fill tester with oil, close the two valves (3 and 4) on gage-connection arm by turning clockwise. The cap over the cylinder having been removed, the handwheel should be screwed into the oil reservoir as far as it will go. Oil can now be slowly poured into the cylinder, and the handwheel gradually unscrewed until the instrument is completely filled. Next, slightly open valve 4 until oil shows at gage connection, and then close.

The gage to be tested can now be applied, as shown in Fig. 3-15, and the No. 4 valve opened. The weight piston with tray may then be inserted in the cylinder, making the tester complete and ready for use.

To ensure accuracy of readings, the piston should be revolved slowly during testing so as to minimize any friction in the cylinder. Also keep the piston about ⅜ or ½ in. above the bottom during testing by gradually screwing in the handwheel. Do not force the weight piston up too high.

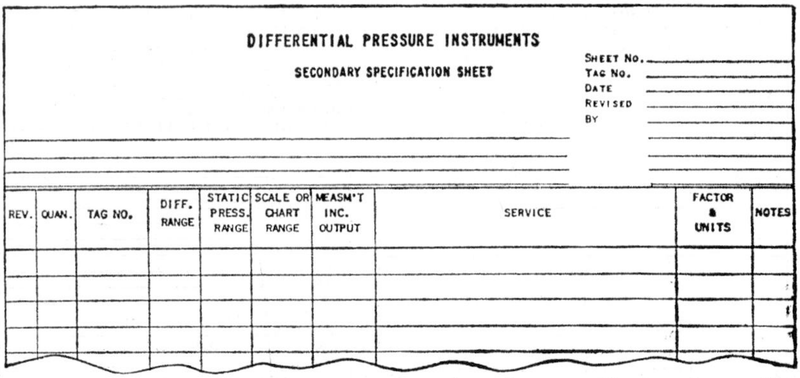

DIFFERENTIAL PRESSURE INSTRUMENTS

SECONDARY SPECIFICATION SHEET

SHEET NO.
TAG NO.
DATE
REVISED
BY

REV.	QUAN.	TAG NO.	DIFF. RANGE	STATIC PRESS. RANGE	SCALE OR CHART RANGE	MEASM'T INC. OUTPUT	SERVICE	FACTOR & UNITS	NOTES

FIG. 3-13g. Differential-pressure instruments secondary specification sheet. (*Instrument Society of America.*)

When testing at low pressures, the combined large and small areas of the piston should be used. This is accomplished by closing No. 1 valve on the side of the vertical cylinder and opening No. 2 valve on the opposite side. For testing at higher pressures no additional weights are required. It is only necessary to reverse this adjustment of the cylinder valves, open No. 1 and close No. 2, which makes effective only the small-area piston,

Table 3-10. Pressures Exerted on Gage

(In pounds)

	Single or combined	Small
Plunger and weight holder	5	20
¼-lb weight	1¼	5
½-lb weight	2½	10
1-lb weight	5	20
2-lb weight	10	40
4-lb weight	20	80

which exerts a pressure four times that of the combined areas. This applies to the weight holder as well as to each of the weights, increasing the testing capacity to full maximum. Before making this change, it is advisable to remove all pressure in the tester by unscrewing the handwheel.

The piston and weight holder exert a pressure on the gage of exactly

DIFFERENTIAL PRESSURE INSTRUMENTS

Specification Sheet Instructions

Prefix numbers designate line number on corresponding specification sheet.

1) Description - Check one or combination of two or more.
Example: Recorder Controller

2) Case - Check one.
Describe other type or special consideration to type checked on Line 2.

3) Check black or write in color required

4) Mounting - Check one.

5) Write in number of Recording Pens and/or Indicating Pointers, exclusive of Index Pointer.

6) Check 12" size, or write in size required.

*7) Write in chart range and number.

*8) Write in scale range and type.

9) Check type.

10) Write revolutions per day and days spring wound drive to run on complete wind.

11) Write in voltage and cycles, check if explosion proof required when chart drive is electric, air supply pressure if pneumatic.

12) Write in any special type or deviation from standard drives on Line 11.

13) Check one.

14) If pneumatic transmission, check if output is 3 - 15 PSI. Write in if other than 3 - 15 PSI, or if electric. Write in electric characteristics.

15) Write in Specification Sheet No. or numbers on which Receiver Instruments appear. (For cross references).

16) Check one or write in other type.

17) Write in proportional band, percent of scale or chart range desired. Check if auto-reset and/or rate action is desired.

18) Check if 3-15 PSI pneumatic or write in controller output.

*19) When form is used to specify a single instrument, check either Increase (Inc.) or Decrease (Dec.). When form is used to specify multiple instruments of the same general description, write in abbreviation (Inc.) or (Dec.) in column provided on Secondary Sheet.

20) Check one and write in number of positions.

21) If manual, check location.

22) If automatically set - check operating medium.

23) Check if index set travel is fixed in proportion to the impulse, or is adjustable.

24) Write in any other, such as linkage or cam, or deviation from Lines 21, 22 or 23.

*25) Write in type - bourdon tube, helix, bellows, receiver. Specify material.

*26) Range - Calib. range of element.

27) Where element cannot be completely specified on this sheet, but to be located in the case of instrument covered on this sheet, write in Tag No. of element and Sheet No. on which item is specified.

28) Check application.

29) Check type differential unit.

30) Write in, if not included on Line 29, or write in special to one checked on Line 29.

31) Write in body material, required types available - Cast Iron, Steel or Stainless Steel (Stnl. Stl.). Write in bellows or diaphragm material.

32) Write in body rating - check manufacturer for rating available.

33) Write in differential range required if sheet used for single instrument. Write "As Listed" if for multiple instruments and listed on secondary sheet.

***When more than one instrument is specified, show data on secondary sheet, leaving lines unmarked on base sheet.**

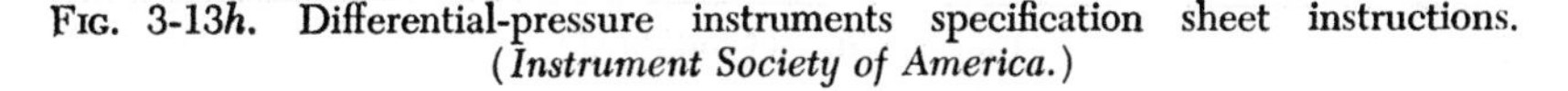

Fig. 3-13*h*. Differential-pressure instruments specification sheet instructions. (*Instrument Society of America.*)

5 psi when both large and small areas are used, but, when the smaller single area is used, this increases to 20 psi. Every weight is marked with the number of pounds pressure per square inch it will exert on the gage.

The pressures exerted on the gage are given in Table 3-10, page 3–24.

When each test is completed, close No. 4 valve. This shuts off connection to the gage, from which the oil may be drained into an oil can by placing the can under the No. 3 valve and opening the valve.

After the gage has been removed and another one applied, the No. 3 valve should again be closed and the No. 4 valve opened before starting

DIFFERENTIAL PRESSURE INSTRUMENT

SPECIFICATION SHEET INSTRUCTIONS, CONT'D.

34) Check connection size required.

35) Write in if individual units or combination type desired, or if to be purchased separately.

36) Write in "Furnished", if for instrument without integral gages. If "by other", show Specification Sheet No. and Item No.

37) Specify size dial and scale for Receiver Type Gage for Local Indication when using Blind Type Transmitter. If by others, write Specification Sheet No.

38) Write in number of days supply required.

39) Write in Yes or No if for type where yoke is optional.

40) If for mercury, write in "Standard", if for bellows type, write in "Yes" or "No". Not Pneumatic Type.

41) For mercury manometers only. Write in "Furnish" or "By Others".

42) For grease packed differential shaft only. Specify if required.

43) Write in if to be furnished with instrument.

44) Specify Type - (Orifice Plate) (Pitot) (Flow Nozzle) (Venturi Section).

45) Write in location, rating in amps, volts & cycles for contacts.
Check housing type.

Service Conditions - When form is used for Multiple Instruments, Operating Conditions are to be listed on another Secondary Sheet.

Secondary Sheet - For Listing Multiple Instruments.

List all instruments of the same type, specified on Primary Sheet, with variations as shown. Variations not provided for to be noted on Primary Sheet as Note (1), etc., and entered on Secondary Sheet where applicable.

Rev. - Revision Number
Quan. - Show quantity required.
Tag No. - Identification of item number.
Diff. Range - Manometer or Differential Range in inches of water (Dry Calibration).
Static Press. Range - See instruction #26.
Scale or Chart Range - See instructions #7, #8.
Measm't. Inc. Output - See instruction #19.
Service - Fill in Service or Location.
Factor & Units . Multiplying Factor for Flow and Units of Measurement.
Notes - Note number appearing on Primary Sheet.

FIG. 3-13*h*. (*Continued*)

second test. Never take off weights or remove piston without first completely unscrewing the handwheel and removing all pressure from tester.

When tester is not in use, oil may be left in it, but the piston should be taken out and carefully cleaned before being put away; likewise the cap should always be screwed on top of the cylinder to prevent dirt from entering the cylinder. Opening both valves 3 and 4 will permit draining of all oil from the machine.

Operating a Pneumatic-type Pressure-instrument Tester (Fig. 3-16).[11] Essentially a weight-controlled relay, the calibrator operates on the force-

[11] Republic Flow Meters Co.

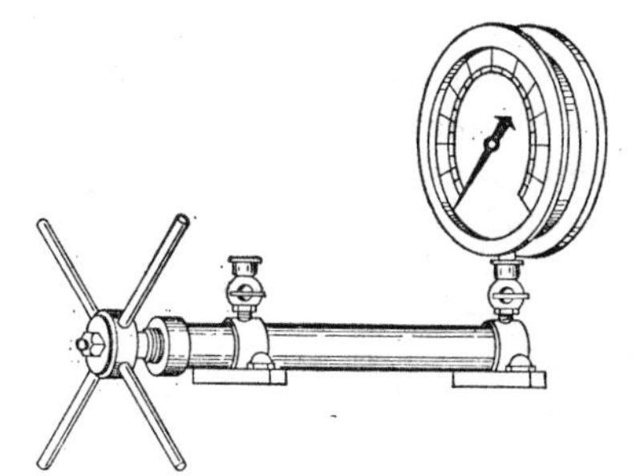

FIG. 3-14. Test pump. (*Crosby Valve & Gage Co.*)

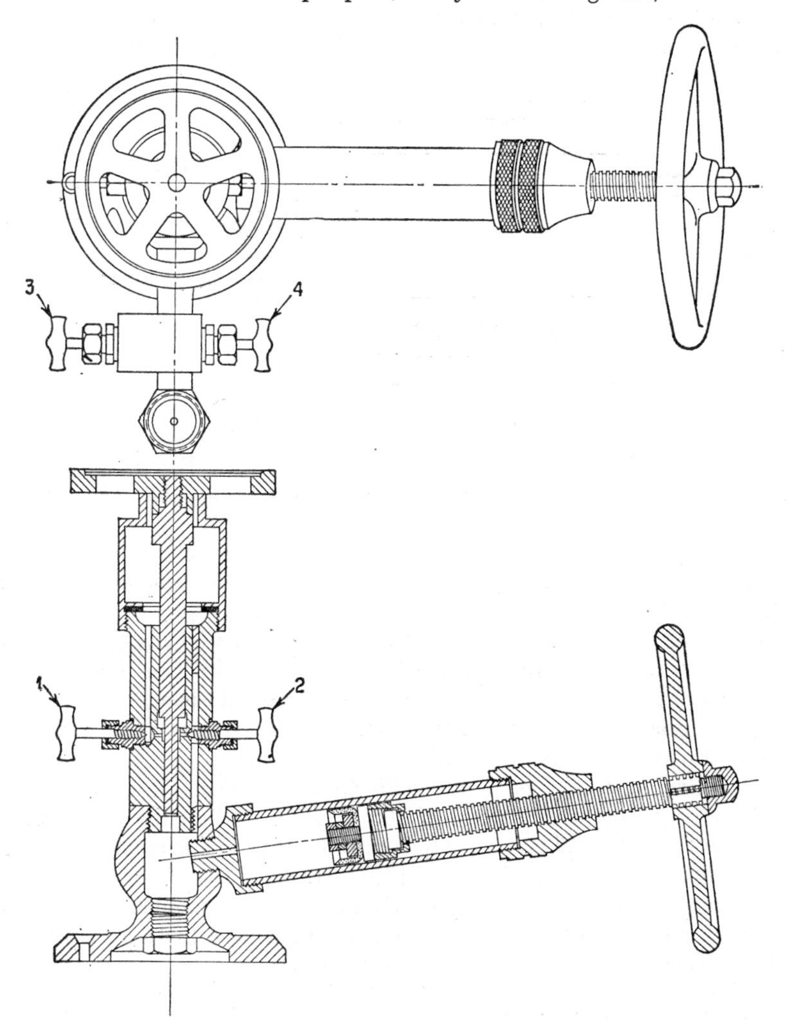

FIG. 3-15. Deadweight gage tester. (*Crosby Valve & Gage Co.*)

balance principle. Weights on the weight beam are counterbalanced by air pressure under a reaction bellows. The more weight on the weight beam, the more will be the pressure under the bellows. Thus the bellows pressure can be made to correspond to accurate weights. This pressure is conducted to instruments to be calibrated.

Connections. Air pressure from a 35-psi source should be supplied through an air-set or pressure regulator to the calibrator. Output from the calibrator should be applied to both the instrument to be adjusted and to a manometer. The manometer is used only for the original calibration and can be removed after this is completed.

Testing in Steps. Assume, for example, that a gage or recorder is to be tested in steps of 1 in. Hg. Adjust the calibrator by rotating the weight-hanger ring inward or outward on the weight beam so that 500 g of weight

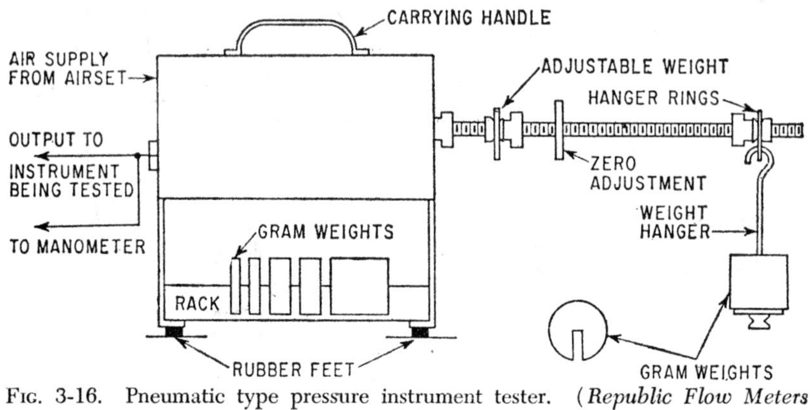

Fig. 3-16. Pneumatic type pressure instrument tester. (*Republic Flow Meters Co.*)

will produce an output of 10 in. Hg. Fifty-gram increments will then deliver 1-in. Hg. steps.

Calibrating in Pounds per Square Inch Using Gram Weights. A number of instruments with a 15-psi range are to be tested. One pound (per square inch) is equal to 2.04 in. Hg. Adjust the weight-hanger ring so that 60 g will produce an output of 2.04 in. Hg. Sixty-gram steps will then give even pound outputs; 30 g will produce ½ lb, etc.

Calibrating in Inches of Water with Gram Weights. It is desired to check a transmitter with a range of 100 in. H_2O. Adjust the weigh-beam hanger so that 20 g will give an output of 10 in. H_2O on a manometer. One gram weights will give an increment of ½ in. H_2O, and so on.

Checking Calibration of Evenly Graduated Scale. A number of receivers used with pneumatic transmitters having an output of 3 to 15 psi are to be tested. They are used with standard 0–100 charts.

In this case it would be convenient to check 10 per cent steps so that the standard chart would be direct-reading. Since 24.5 in. Hg is equal

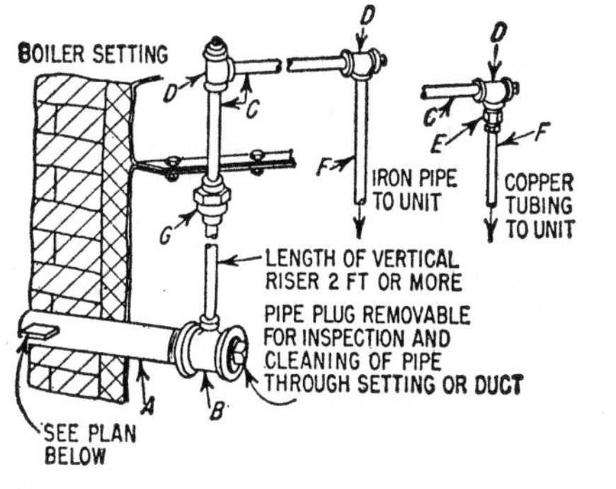

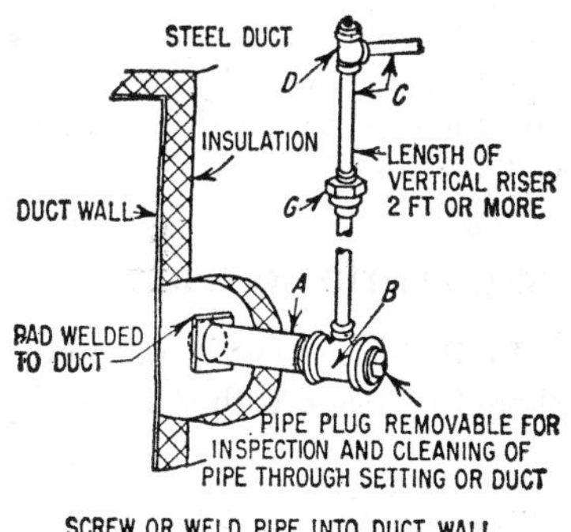

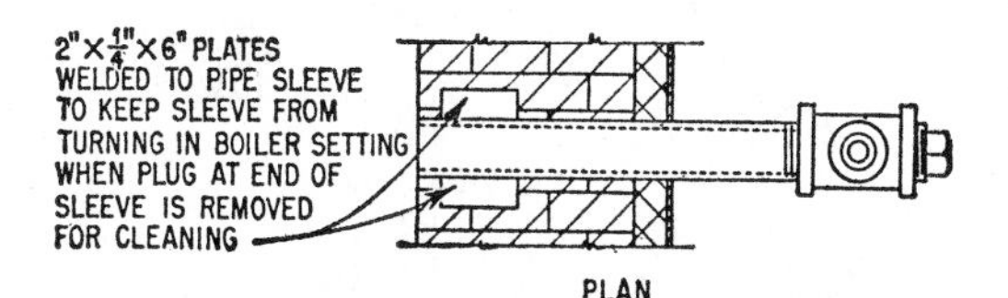

THE PIPING FROM THESE CONNECTIONS SHOULD RUN AS DIRECT AS POSSIBLE TO METER, GAGE OR CONTROLLER AND MUST HAVE NO BRANCH CONNECTIONS TO OTHER EQUIPMENT

(DIMENSIONS IN TABLE ARE IN INCHES)

TO	CONNECTING PIPING TO UNIT	BOILER SETTING		STEEL DUCT		SIZE IRON PIPE C	SIZE IRON TEE D	IRON PIPE F TO UNIT	COPPER TUBE F TO UNIT	SIZE IRON UNION G
		SLEEVE A	TEE B	SLEEVE A	TEE B					
MULTIPOINTER	COPPER TUBING	2	2×2×$\frac{3}{8}$	1	1×1×$\frac{3}{8}$	$\frac{3}{8}$	$\frac{3}{8}$	—	$\frac{1}{2}$ O.D.	$\frac{3}{8}$
	IRON PIPE	2	2×2×$\frac{3}{8}$	1	1×1×$\frac{3}{8}$	$\frac{3}{8}$	$\frac{3}{8}$	$\frac{3}{8}$	—	$\frac{3}{8}$
BOILER METER	COPPER TUBING	2	2×2×$\frac{1}{2}$	2	2×2×$\frac{1}{2}$	$\frac{1}{2}$	$\frac{1}{2}$	—	$\frac{5}{8}$ O.D.	$\frac{1}{2}$
	IRON PIPE	2	2×2×$\frac{1}{2}$	2	2×2×$\frac{1}{2}$	$\frac{1}{2}$	$\frac{1}{2}$	$\frac{1}{2}$	—	$\frac{1}{2}$
FURNACE DRAFT CONTROLLER OR DRIVE	COPPER TUBING	2	2×2×$\frac{3}{4}$	2	2×2×$\frac{3}{4}$	$\frac{3}{4}$	$\frac{3}{4}$	—	$\frac{3}{4}$ O.D.	$\frac{3}{4}$
	IRON PIPE	2	2×2×$\frac{3}{4}$	2	2×2×$\frac{3}{4}$	$\frac{3}{4}$	$\frac{3}{4}$	$\frac{3}{4}$	—	$\frac{3}{4}$
RATIO CONTROLLER	COPPER TUBING	2	2×2×$\frac{3}{8}$	1	1×1×$\frac{3}{8}$	$\frac{3}{8}$	$\frac{3}{8}$	—	$\frac{1}{2}$ O.D.	$\frac{3}{8}$
	IRON PIPE	2	2×2×$\frac{3}{8}$	1	1×1×$\frac{3}{8}$	$\frac{3}{8}$	$\frac{3}{8}$	$\frac{3}{8}$	—	$\frac{3}{8}$

NOTE: FOR LENGTHS OF F EXCEEDING 75 FT, USE NEXT LARGER SIZE OF PIPE OR TUBING THAN SPECIFIED IN TABLE

FIG. 3-17. Draft connections to boiler settings and ducts. (*Bailey Meter Co.*)

to 12 psi, the range of the receivers, the calibrator weigh-beam hanger should be set so that 60 would produce 2.45 in. Hg output. This will then give 10 per cent steps on the chart for each 60 g applied.

Adjusting Minimum and Maximum Readings. The calibrator is suitable for setting minimum and maximum readings of instruments on a production-line basis. The calibrator weight-hanger ring can be set so that one weight will produce a desired minimum output pressure and a second weight will produce the desired maximum output pressure. By interchanging the weights, both outputs can be obtained immediately. Checking each setting with a manometer once the calibrator is adjusted and weights are selected is not necessary.

INSTALLATION OF PRESSURE INSTRUMENTS

In Boilers.[12] Draft, pressure, or differential connections to a controller should be made with ¾-in. steel pipe and screwed fittings.

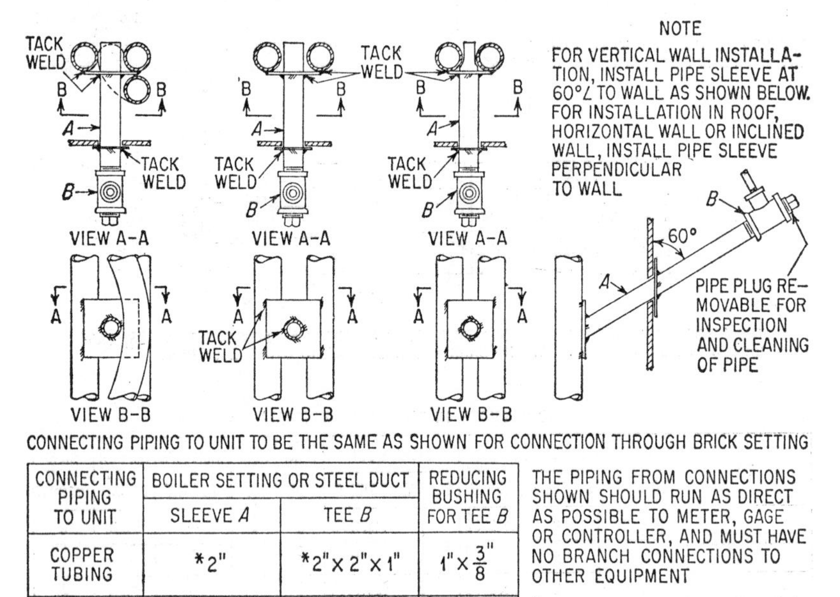

CONNECTING PIPING TO UNIT	BOILER SETTING OR STEEL DUCT SLEEVE *A*	BOILER SETTING OR STEEL DUCT TEE *B*	REDUCING BUSHING FOR TEE *B*
COPPER TUBING	*2"	*2"x 2"x 1"	1"x 3/8"
IRON PIPE	*2"	*2"x 2"x 1"	1"x 3/8"

THE PIPING FROM CONNECTIONS SHOWN SHOULD RUN AS DIRECT AS POSSIBLE TO METER, GAGE OR CONTROLLER, AND MUST HAVE NO BRANCH CONNECTIONS TO OTHER EQUIPMENT

*FOR STEEL DUCT ONLY, IF DESIRED, USE 1" SLEEVE *A* AND 1"x 1"x 1" TEE *B*

FIG. 3-18. Draft connections to water-wall setting. (*Bailey Meter Co.*)

The most satisfactory method of making the connection to a boiler setting, wind box, air duct, etc., is by means of a length of 2-in. steel pipe tightly cemented into the brickwork or welded into the steel casing, as shown in Figs. 3-17 and 3-18. Connecting nipples must stop flush with the inside surface of the wall through which they pass, and must also be smooth and free from burrs. The arrangement of pipe and vertical riser shown will minimize the accumulation of dust particles or moisture in the connecting piping.

For all changes in the direction of the piping, it is advisable to use tees

[12] Bailey Meter Co.

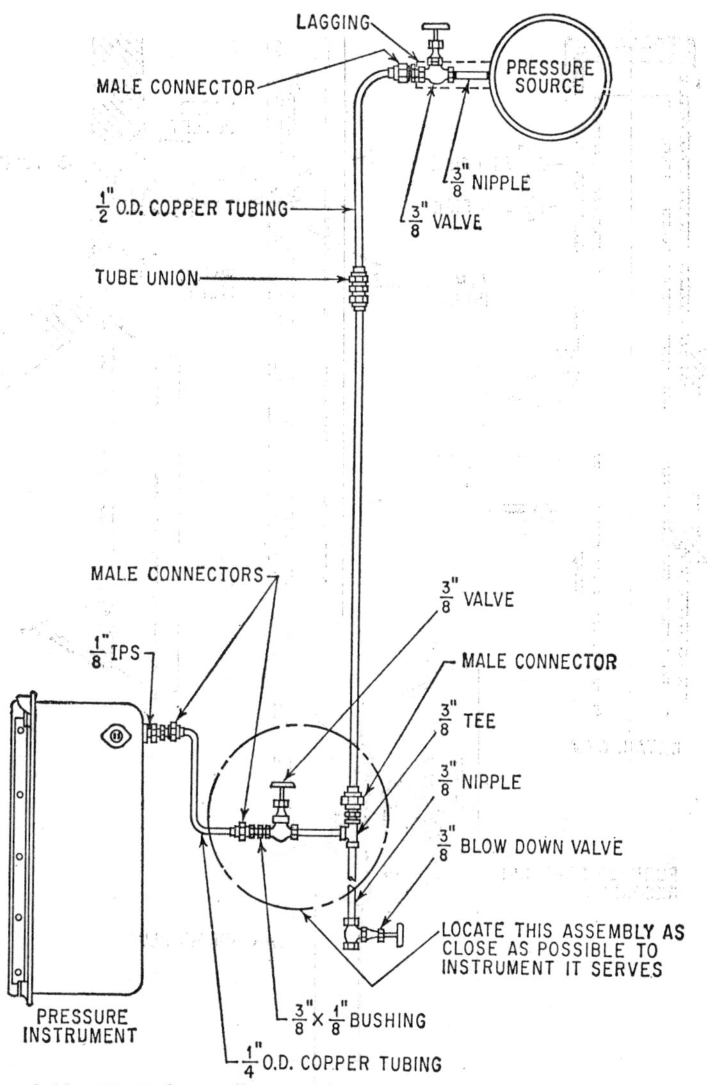

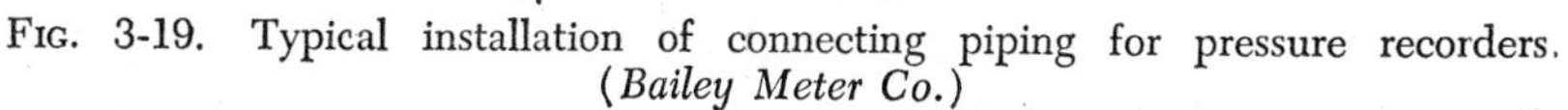
FIG. 3-19. Typical installation of connecting piping for pressure recorders. (*Bailey Meter Co.*)

instead of elbow connectors, to facilitate periodic cleaning out of any accumulation of dust, soot, etc.

Valves may be installed between the controller and the point of measurement for convenient shutoff.

For air and gas installations, the connecting piping should be installed in such a manner as to avoid the trapping of water or other liquids. If it is not possible to install the piping so that liquids will drain back through

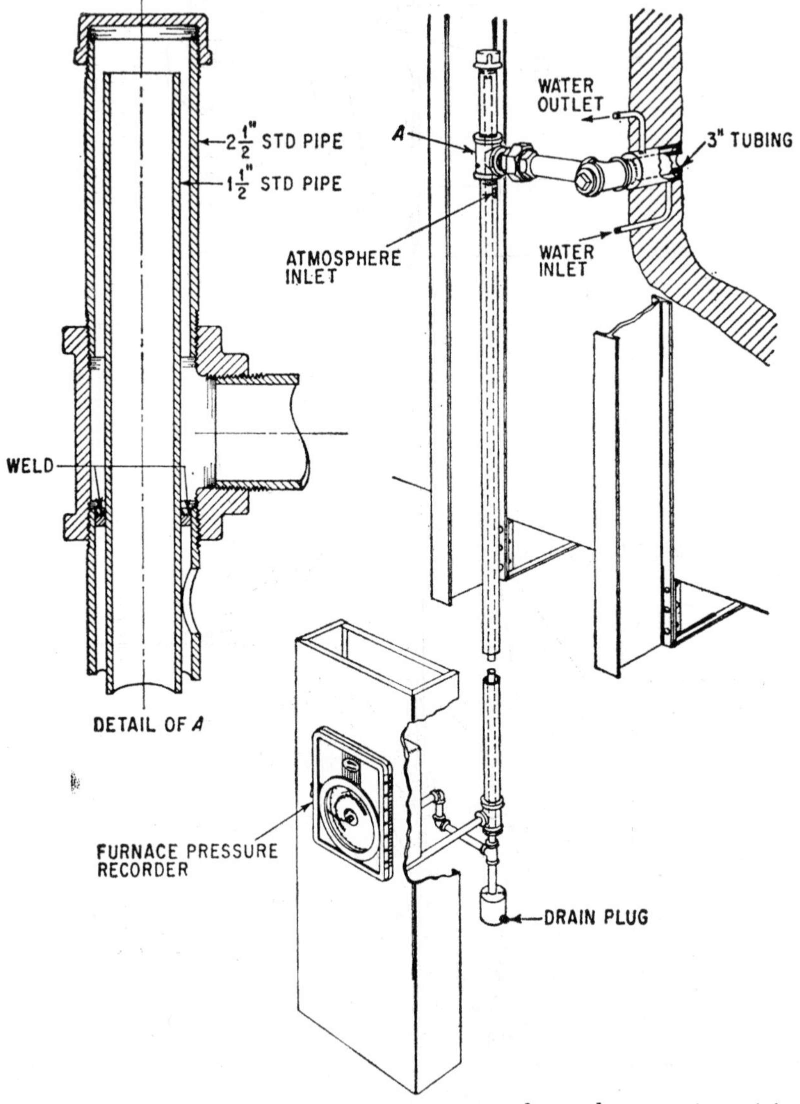

FIG. 3-20. Installation of furnace-pressure recorder and connecting piping. (*Bailey Meter Co.*)

the pressure-source connection, drip pockets should be installed at any low points with provisions for periodic draining. When the unit is used with steam or liquids, the piping should be installed in such a manner as to avoid trapping of air. If this is not possible, air vents should be provided wherever air or other gases might be trapped.

Figures 3-19 and 3-20 show typical pressure-instrument hookups.

Section 4

TEMPERATURE MEASUREMENT

Most temperature measurements made in building services and industry in general are in the range of −150 to +3000°F. Progress in science and engineering, however, is continually widening this range, and instrument engineers are constantly seeking new principles of measurement and improving existing methods to satisfy industry's ever-increasing demands.

To meet the requirements of this wide spectrum of temperature measurements, several physical principles have become well established as the bases of temperature-measuring instruments. The practical ranges of these instruments are shown graphically in Fig. 4-1.

Over certain bands of the temperature spectrum there is a rather wide selection of available instruments. The range of an instrument, however, is only one of several pertinent factors that must be considered when a solution to a temperature-measuring problem is sought. Sensitivity, accuracy, speed of response, expected useful life, cost, availability of various modes of automatic control, and resistance to corrosion, vibration, and other conditions of institutional and industrial applications are a few of the factors which require attention in selecting temperature equipment.

TEMPERATURE-MEASURING THEORY[1]

Response Characteristics. If the bare bulb of a filled-system thermometer—usually a cylindrical element filled with a liquid or gas which expands or contracts upon temperature changes—is suddenly immersed in an agitated tank maintained at 270°F, the thermometer pen rises as indicated in curve *A* of Fig. 4-2. The curve is exponential, or nonlinear, and it approaches the tank temperature gradually. At a lower tank temperature of 170°F, the same procedure yields a curve similar to curve *B* of Fig. 4-2. Any thermometer bulb attains a given percentage of the total change in a given time, irrespective of the magnitude of the change in bath temperature. The bulb that responds to 95 per cent of a 100°F temperature change in one minute will also respond to 95 per cent of a 1000°F change in one minute.

[1] Minneapolis-Honeywell Regulator Co., Industrial Div.

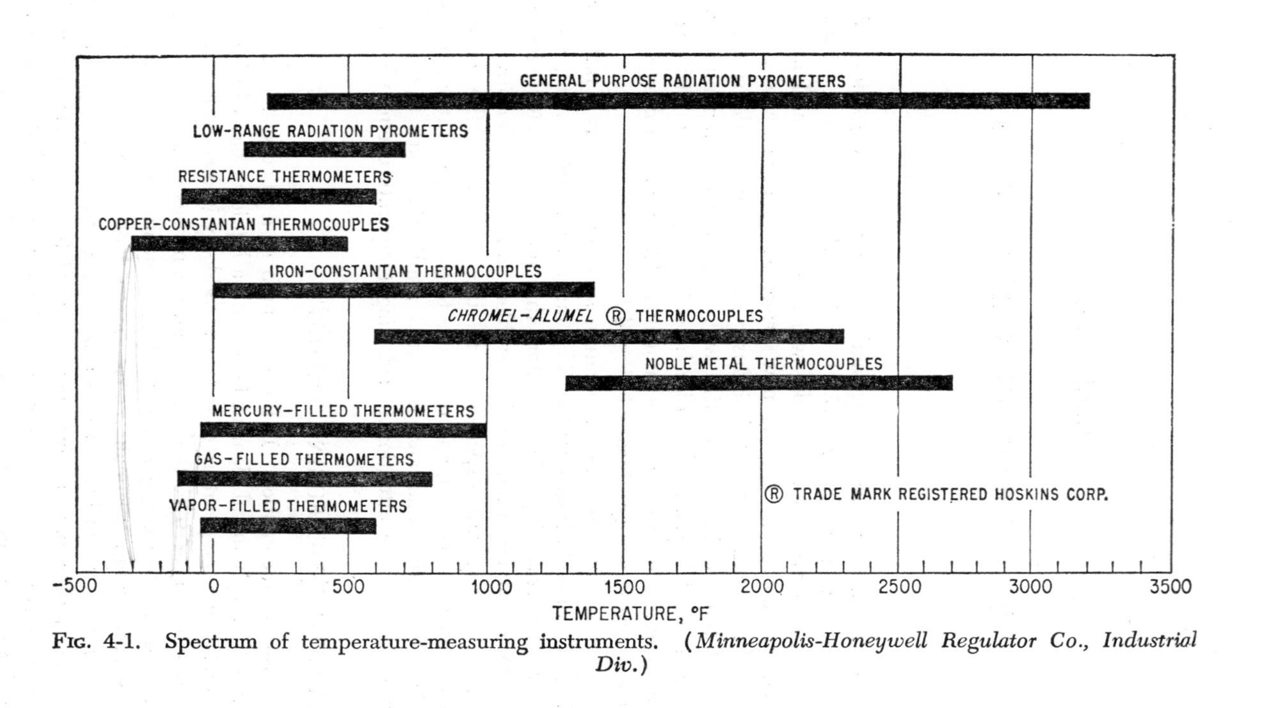

FIG. 4-1. Spectrum of temperature-measuring instruments. (*Minneapolis-Honeywell Regulator Co., Industrial Div.*)

And so, because of the practical difficulties encountered, it has been determined mathematically that the *lag coefficient* for a bare primary element is the time required to reach 63.2 per cent of the total change. For thermocouples in wells, for instance, other percentages have been established. In Fig. 4-2, the lag coefficient is 0.1 min, shown by the broken lines.

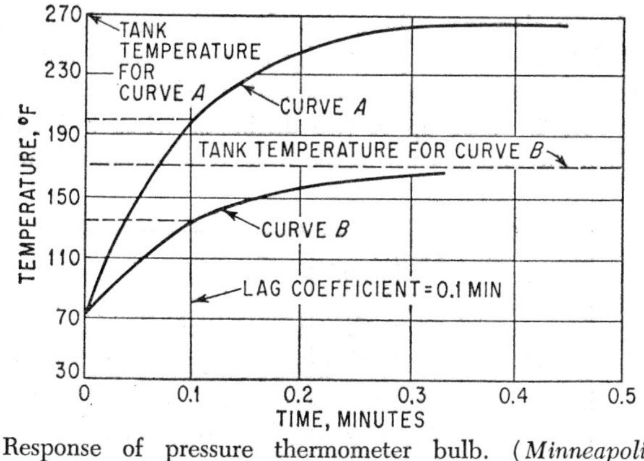

FIG. 4-2. Response of pressure thermometer bulb. (*Minneapolis-Honeywell Regulator Co.*)

Factors Influencing Response. The factors which influence the responsiveness of a thermometer bulb or thermocouple are:

1. Thermal capacity of element
2. Thermal conductivity of element
3. Surface area per unit mass of element
4. Film coefficients of heat transfer
5. Mass velocity of fluid surrounding element
6. Thermal capacity of fluid surrounding element
7. Thermal conductivity of fluid surrounding element

The material of the element of protecting tube, such as stainless steel in a thermometer bulb, determines the rate of transfer to the inside of the element. An element having a larger mass with high thermal capacity and low thermal conductivity requires a longer time to reach 99+ per cent of the final temperature than one with smaller mass and lower thermal capacity. On the other hand, a larger surface area per unit of mass increases the rate of heat transfer.

The movement of fluid surrounding the element is also a very important factor in speed of response. If the fluid is moving very slowly, a film of cooled fluid will build up around the element. The thermal capacity and conductivity of the fluid determine the amount of heat made available for transfer to the element. Liquids have a higher heat capacity and conductivity than air, and so a thermometer bulb or a thermocouple in a stirred liquid will respond much faster than one in air or in other gases.

Transmission Lags. The transmission lag of a filled-system thermometer depends on the length and internal diameter of the capillary tubing and the volume of the receiving element. Mercury- or liquid-filled thermometers have a transmission lag which is usually negligible because a very small volume of liquid must pass through the capillary on a change in temperature. Gas- and vapor-actuated thermometers have a small but noticeable transmission lag because of the compressibility of the filling medium.

Electric primary elements—thermocouples, resistance-thermometer bulbs, and radiation detectors—have practically no transmission lag because electric current flows through a wire at the speed of light.

A thermocouple is made up of two dissimilar wires welded together at one end. When the temperature at the welded or measuring junction varies

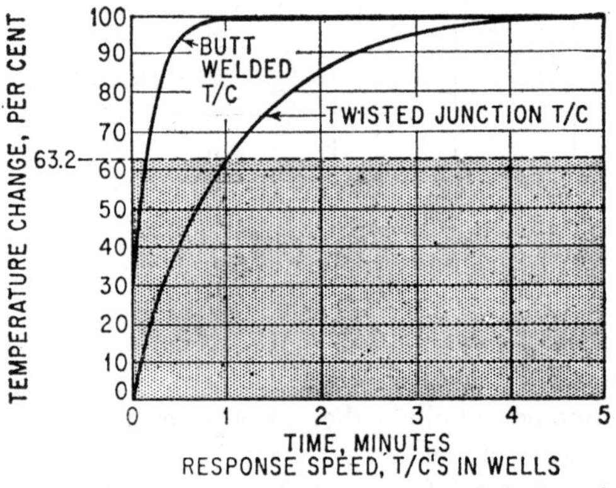

FIG. 4-3. Response of thermocouples in unagitated liquid. (*Minneapolis-Honeywell Regulator Co.*)

from that at the unwelded, or reference junction, an emf which varies with this temperature difference is generated in the circuit.

A resistance element operates on the principle that the electrical resistance of certain kinds of wire (nickel, platinum) changes with temperature. Usually a resistance-thermometer bulb is made up of a coil of wire wound on a core and installed in a rigid capsule.

The radiation element employs a thermopile (a number of small thermocouples in series) upon which the radiation from the heated body is focused by a lens.

The speed of response of the thermocouple has been increased with the development of the butt-welded thermocouple. This butt-welded construction gives a significantly faster response than the twisted thermocouple because the mass of the hot or measuring junction is considerably reduced, and the insulator is isolated by a loop so that it does not carry heat away from this measuring junction. The relative responses of butt-welded and twisted thermocouples are shown in Fig. 4-3.

The radiation type of primary element detects temperature by receiving radiant energy from the heated source. The radiation detector is not immersed in the heated source, as is a thermometer bulb or a thermocouple. Figure 4-4 shows how rapidly the radiation unit responds. There are two reasons why it is so fast: (1) radiant heat transfer is practically instantaneous, and (2) the mass of the thermopile in the radiation unit is very small.

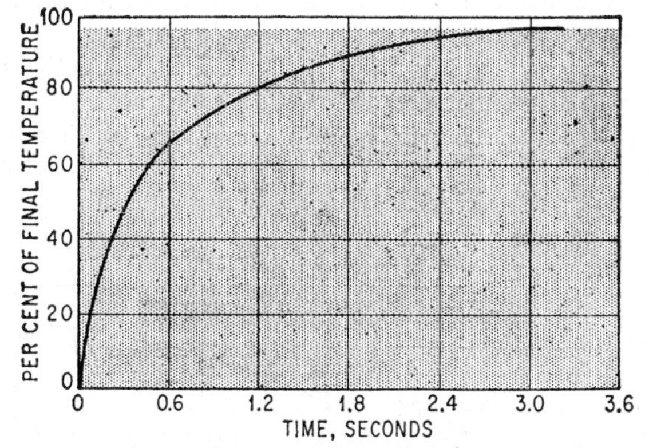

FIG. 4-4. Response of a radiation unit, without target tube. (*Minneapolis-Honeywell Regulator Co.*)

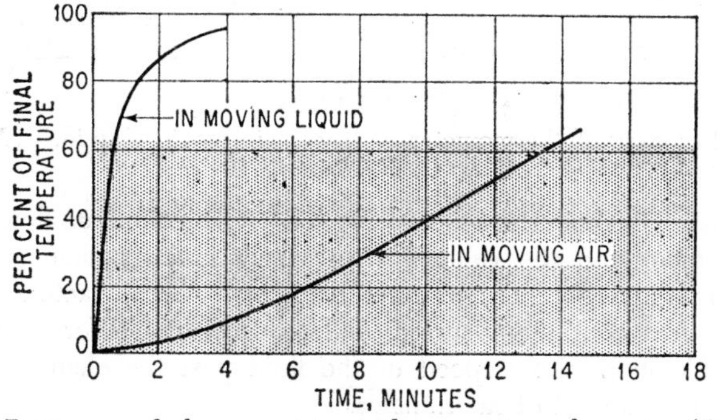

FIG. 4-5. Response of bare resistance-thermometer element. (*Minneapolis-Honeywell Regulator Co.*)

High-speed resistance-thermometer bulbs have been developed which compare favorably with thermocouples. The response curve for a resistance-thermometer element is shown in Fig. 4-5. Notice how much faster the bulb responds in moving liquid than it does in moving air.

Protecting Wells and Speed of Response. Often it is not feasible to expose a bare thermocouple or a resistance or filled-system thermometer bulb to the medium whose temperature is being measured. Corrosion and oxidation are accelerated at higher temperatures and pressures, so it is often necessary to protect the primary element with a suitable well.

Although a well protects the primary element, it will also slow its response. Figure 4-6 shows the response curve for a thermometer bulb with a metal protecting well. In this case, the addition of the well causes the lag coefficient to increase from 0.10 to 1.66 min, an increase of about 16 times.

Effect of Air Space. Air space between a bulb or a thermocouple and its protecting well causes additional lag in response of the primary element. A dead air space insulates and nullifies the advantages gained by selecting materials of low thermal mass and high thermal conductivity. In high-speed resistance-thermometer bulbs, for example, heat transfer from the outside of the well to the winding is promoted by making the clearance between the well and the protecting tube and between the protecting tube and winding as small as possible. Grounding the tip of a thermocouple to the well for good thermal contact also helps to speed response.

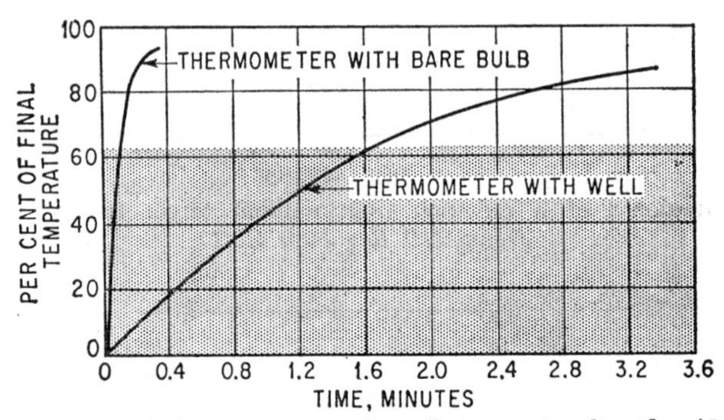

FIG. 4-6. Response of thermometer with well, in moving liquid. (*Minneapolis-Honeywell Regulator Co.*)

Variation of Temperature-measuring Lag. The kind of fluid surrounding the element and the velocity with which it flows have an important influence on the lag of temperature elements. These factors govern the rate at which a continuous supply of heat is available for transfer to the measuring element. Low speed of the fluid past the element greatly increases the resistance to heat transfer.

The effect of the velocity of water flowing past a thermometer bulb is considerable. The lag coefficient, shown in Fig. 4-7, rapidly decreases from 0.10 min at a flow of 2.0 fpm to 0.05 min at a flow of 20.0 fpm. In order to maintain the lag coefficient at a reasonable minimum, a liquid should flow past the bulb at a speed of at least 60 fpm. If the thermometer bulb or well has rough or uneven surfaces, a flow of 120 fpm would be desirable in order to carry away the insulating film around the bulb.

The thermal capacity and conductivity of air are low, so that a temperature element has a much higher lag coefficient in air than in a moving liquid. Figure 4-5 shows the difference between the two. Air should circulate around a bulb at a rate of at least 400 fpm in order to reduce the lag coefficient to a minimum.

Radiation is especially important in the measurement of high temperatures in furnaces. At lower temperatures, radiation is not so effective as it is at higher temperatures. Figure 4-8 shows the response curve for a bare thermocouple suddenly inserted into a furnace being held at a series of different temperatures. The lag coefficient decreases from 5.7 min at 500°F to 1.75 min at 1700°F.

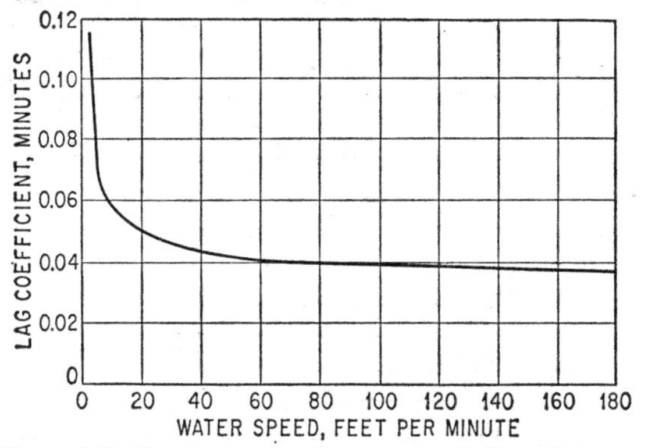

Fig. 4-7. Effect of fluid velocity on thermometer bulb. (*Minneapolis-Honeywell Regulator Co.*)

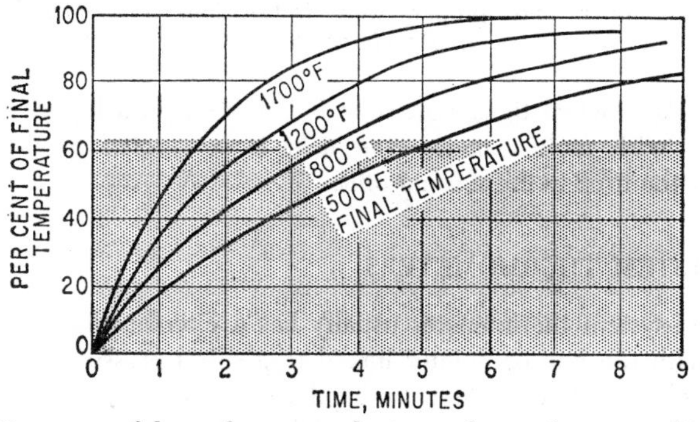

Fig. 4-8. Response of bare thermocouple inserted into furnace. (*Minneapolis-Honeywell Regulator Co.*)

Insufficient depth of immersion of a thermocouple or thermometer well, as illustrated by Fig. 4-9, causes inaccuracy and additional response lag. In this case, a low-resistance path for heat flow is formed along the well toward the outside wall. The heat lost by conduction cannot transfer to the thermocouple, so that the measured temperature is materially lower than the actual temperature of the fluid. Protecting wells with high thermal conductivity generally must be deeply immersed, with proper insulation at the head of the assembly, in order to obtain a low lag coefficient.

The radiation type of primary element does not depend on contact between bodies, and its response is not affected directly by velocity and type of liquid. However, a smoky atmosphere between the measured object and the radiation element absorbs radiant energy and causes inaccuracy. When a closed-end target tube is used, it must be immersed deeply enough into the material being measured so that the end of the tube receives direct

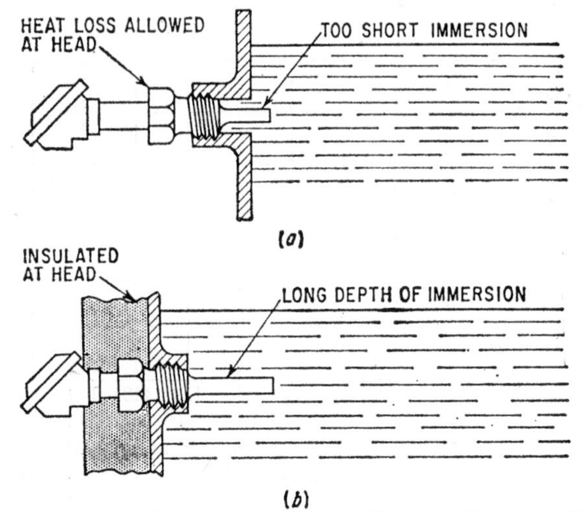

FIG. 4-9. Improper and proper well installation. (*Minneapolis-Honeywell Regulator Co.*)

radiation. And, when the target tube is immersed in a liquid—as is sometimes done with molten metal—the heat is *conducted* to the tube and *radiated* by the tube to the sensing unit.

FILLED-SYSTEM THERMOMETERS

The filled-system thermometer usually has a circular chart or an indicating scale with the pen or pointer linked to a spiral pressure element. The measuring element is connected by capillary tubing to a bulb located some distance away in the process. A fluid fill in the system creates an internal pressure, which moves the instrument pen or pointer in a definite relationship to the temperature change of the process medium.

Filled-system thermometers are generally classified as vapor-filled, gas-filled, or mercury-filled, depending on the fluid in the sealed measuring system. Table 4-1 lists typical characteristics.

Classification and Description.[2] Four fundamental physical principles are utilized in the filled-system thermometer: the differential expansion of two metals, change in volume of a liquid, change in saturated vapor pres-

[2] L. E. Smith, The Bristol Co., "Operation of the Filled System Thermometer," *Refrig. Engin.*, November, 1954.

Table 4-1. Typical Characteristics of Liquid-, Vapor-, and Gas-filled Systems*

	Class IA, liquid expansion, fully compensated	Class IB, liquid expansion, case-compensated	Class II, vapor pressure, uniform scale	Class II, vapor pressure, nonuniform scale	Class IIIB, gas pressure, case-compensated
Bourdon spring type	5-turn helical	5-turn helical	5-turn helical	5-turn helical	5-turn helical
Bourdon material	Beryllium copper	Beryllium copper	Beryllium copper	Beryllium copper	Beryllium copper
Tubing lengths	Up to 150 ft	Up to 40 ft	Up to 150 ft	Up to 150 ft	Up to 200 ft
Typical sensitive bulb size	3/8 × 3 in.	3/8 × 3 in.	9/16 × 4 in.	9/16 × 4 in.	7/8 × 6 in.
Filling fluid	Hydrocarbon	Hydrocarbon	Volatile liquid	Volatile liquid	Nitrogen
Temperature limits	−300 to +500°F	−300 to +500°F	−50 to 450°F	−50 to 450°F	−400 to +1000°F
Type scale	Uniform	Uniform	Uniform	Increasing increments	Uniform
Minimum span	40°F	40°F	100°F	100°F	200°F
Maximum span	500°F	500°F	150°F	150°F	1000°F
Case compensation	Complete	Complete	None required		Complete
Tubing compensation	Complete	Partial	None required		Partial
Elevation error	None	None	Can correct	Can correct	None
Cross ambient error	None	None	Yes	Yes	None
Barometric error	None	None	Very slight	Very slight	Negligible
Overrange available	Yes	Yes	Small	Small	Yes
Underrange available	Yes	Yes	Yes	Yes	Yes
Torque (output at pen spindle)	12–20 gr × cm/%	25–30 gr × cm/%	12–20 gr × cm/%	12–20 gr × cm/%	10–20 gr × cm/%
Linearity (maximum deviation from)	0.5% span	0.5% span	0.5% span	10–20%	0.5% span
Sensitiveness	0.2% span	0.2% span	0.2% span	0.2% span	0.2% span
Hysteresis	0.25% span	0.25% span	0.4% span	0.4% span	0.4% span
Repeatability	0.4% span	0.4% span	0.4% span	0.4% span	0.4% span
Accuracy (factory calibration)	0.5% span	0.5% span	0.5% span	0.5% span	0.5% span
Guaranteed accuracy, 1% span:					
Bare bulb, agitated water	4–8 sec	4–8 sec	2–5 sec	2–5 sec	5–10 sec
Bare bulb, agitated air	20–40 sec	20–40 sec	10–25 sec	10–25 sec	25–50 sec
Well, agitated water	12–24 sec	12–24 sec	6–15 sec	6–15 sec	15–30 sec
Well, agitated air	60–120 sec	60–120 sec	30–75 sec	30–75 sec	75–150 sec
Capillary tubing materials	Copper or 321 stainless steel				
Protective armor materials	Bronze, 302 stainless steel, lead, or polyvinyl chloride-covered bronze				
Sensitive bulbs	Copper, lead, Everdur, or 316 stainless steel				
Bushings	304 stainless steel, brass, or 316 stainless steel				
Wells	304 stainless steel, hardware bronze, 316 stainless steel, Hastelloys, nickel, Monel, glass, tantalum, lead, or silver				

* Courtesy of Fischer & Porter Co.

sure of a volatile liquid, and change in pressure of a constant volume of gas (as affected by a change in temperature).

The first of these principles, the differential expansion of two metals, is used as a means of compensation for temperature changes in the measuring element, as will be brought out later. A simple bimetallic strip, composed of a layer of high-expanding material such as brass, and a layer of low-expanding material such as invar, will deflect when its temperature is changed, and, if the strip is coiled in the arc of a circle, its angular motion in degrees equals $C(TL/t)$ where C is the constant of the strip per degrees Fahrenheit, T is the temperature change of the strip in degrees Fahrenheit, L is the length in inches, and t is the thickness in inches. Incidentally, this bimetallic strip is used as a self-contained thermometer element, which is helically coiled and directly connected to the pen. This system, illustrated

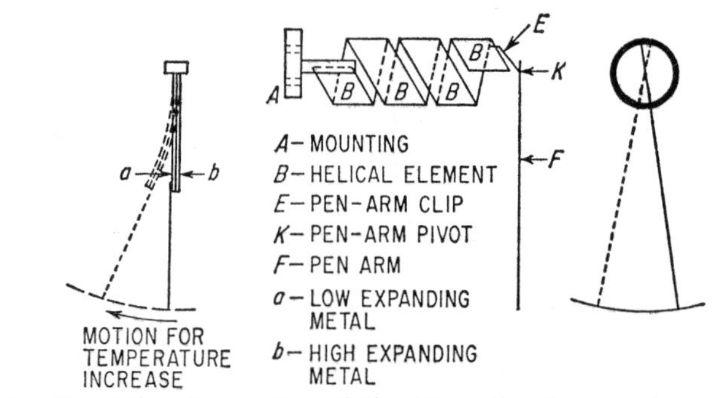

FIG. 4-10. Principle of operation of the bimetallic thermometer. (*The Bristol Co.*)

in Fig. 4-10, has its limitations, but is useful for measuring temperature at the instrument.

There are four general classes of filled thermal systems in general use:

Class I—liquid-filled thermal systems:

1. With full compensation—class IA
2. With compensating means within the case only—class IB

Class II—vapor-pressure thermal systems:

1. Designed to operate with the measured temperature above the temperature of the rest of the thermal system—class IIA
2. Designed to operate with the measured temperature below the temperature of the rest of the thermal system—class IIB
3. Designed to operate with the measured temperature above and below the temperature of the rest of the thermal system—class IIC
4. Designed to operate with the bulb temperature above, below, and at the temperature of the thermal system—class IID

Class III—gas-filled thermal systems:

1. With a second thermal system minus the bulb or equivalent means of compensation—class IIIA

2. With compensating means within the case only—class IIIB

Class V—mercury-filled thermal systems:

1. With full compensation, the compensating means being a second thermal system minus the bulb or equivalent means of compensation —class VA
2. With compensating means within the case only—class VB

Class I—Liquid-filled System. This class of thermal system is based on the principle that liquids expand as the temperature increases, and contract as the temperature decreases. This system (Fig. 4-11) is filled completely with liquid *J* at a predetermined pressure after evacuation of the system. As the temperature of bulb *L* increases, the liquid *J* in bulb *L* expands. The weakest part of the system must give to allow for this expansion, and so the measuring element *B* uncoils. This moves the pen arm *F*. In this

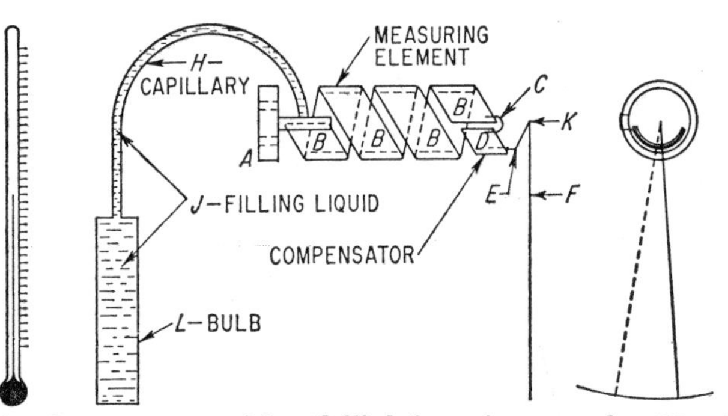

FIG. 4-11. Case-compensated liquid-filled thermal system, class IB. *J*, liquids; *A*, mounting; *B*, hollow helical tube; *C*, compensator clip; *D*, compensator; *E*, pen-arm clip; *F*, pen arm; *K*, pen-arm pivot; *H*, tubing; *L*, bulb. (*The Bristol Co.*)

system, the liquid in the capillary *H* and in the element *B* will also expand with temperature increase; hence they act as bulbs also. In the simple system (class IB), to prevent any effect from the liquid expansion in the element *B*, a bimetallic compensator *D* is interposed between the end of the spring and the pen arm by clip *C* and clip *E*. This coiled bimetallic strip will rotate the pen arm in the direction opposite to the element rotation when the temperature varies—the length is adjusted to cause no movement of the pen for a given temperature change of the element. The error in the capillary due to temperature changes is not compensated, but the error is kept to a minimum by using a small bore capillary and only a short length of capillary tubing. Ten feet of capillary is usually the recommended maximum length. Under conditions where the ambient temperature (temperature around the capillary) may vary only slightly (for instance, in an air-conditioned room), a longer length of capillary may be permissible. The amount of immersion of the bulb for this type is critical.

The filling liquid used must possess certain characteristics—it must not freeze at the minimum temperature to which any part of the system will be subject; it must have a vapor pressure at any temperature of the bulb relatively low compared to pressure in the system; it must have a linear expansion characteristic over the bulb range; it must remain stable over the lifetime of the system; and it should have a low viscosity rating. Instrument manufacturers use many different liquids for filling—ethyl alcohol for the lower temperatures, metaxylene for the middle-temperature ranges, and tetrahydronaphthalene (tetralin) for the higher ranges are some of the satisfactory liquids. Temperature ranges between −125 and 600°F are available. In all ranges, a uniform chart is possible—that is, equal increments of temperature give equal angular distances on the chart. The bulb volume depends on the temperature range, the filling liquid used, and the volumetric change of the element required for full scale.

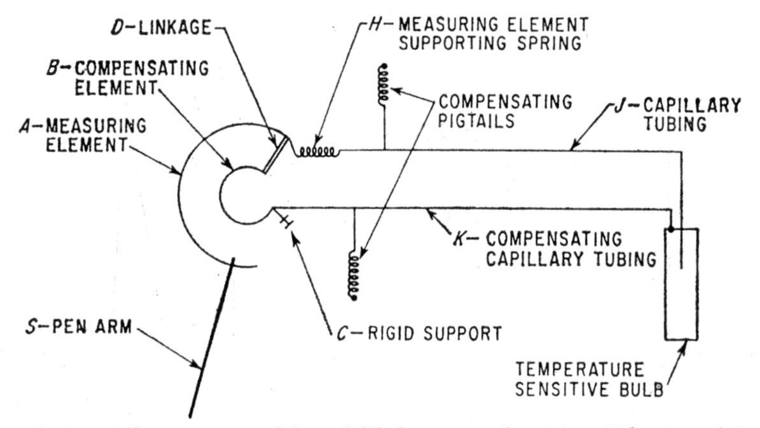

Fig. 4-12. Fully compensated liquid-filled system, class IA. (*The Bristol Co.*)

The spring element itself may be a C, a helical, or a spiral spring generally with a flat cross-sectional area which reduces the volume to a minimum—although the volume is small if not compensated by a bimetallic strip, errors would amount to at least 5 per cent for a 25°F ambient change. This element is a very precise measuring device. It takes only about 0.10 cc change in the element volume to deflect the pen through the 100 per cent full chart range. In order to be accurate to 1 per cent of the full range, the element must be consistent and respond to a volumetric charge of 0.0010 cc. It can be seen that there can be no internal change in the volume of the complete system of this magnitude—hence the importance of proper design of the bulb, capillary, and measuring element, and of necessary precautions in all welded, brazed, and soldered joints.

Class IA system (Fig. 4-12) is the fully compensated liquid-filled system. The measuring device consists of two identical elements, wound in opposite directions. One is known as the compensating element, and the other as the measuring element. The compensating element *B* is rigidly fastened to base C at its lower end, with the freely deflected end connected to the bottom of the main element A by the linkage system D shown. The upper end of the main element is connected to the pen arm S.

The main element, which is driven by the compensating element, is floating—that is, it is connected to its capillary by a flexible coil of capillary H. The capillary J from the main element runs as usual to the sensitive bulb.

A second capillary K, the same size and length as the main capillary, runs from the compensating element alongside the main capillary and terminates at the top of the bulb. Thus, if the capillary and element volumes were identical and volumetric-deflection characteristics of the elements were identical, the system would be automatically compensated. However, variations are present, and a compensation procedure must be used by proper adjustments.

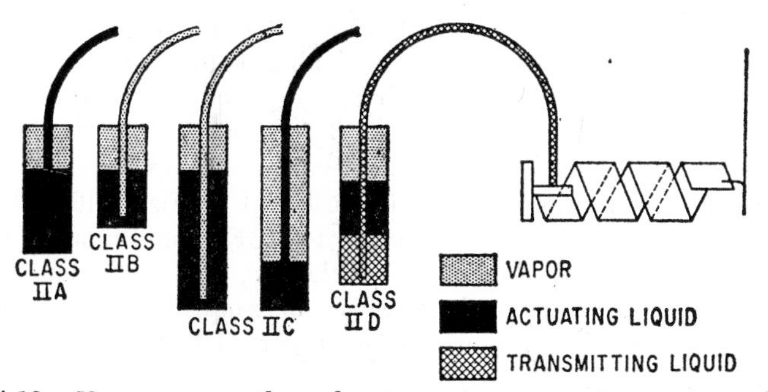

FIG. 4-13. Vapor-pressure thermal systems: to measure temperature when the bulb temperature is above (class IIA), below (class IIB), above and below (class IIC), and above, below, and at (class IID) the temperature of the rest of the thermal system. (*The Bristol Co.*)

The maximum length of tubing is dependent on the ambient temperature to which the capillary will be subjected. If the capillary is to be installed indoors, where temperature changes may vary only 30 to 40°F from season to season, a much longer length is permissible than if the capillary is exposed to outside temperature, where ambient temperatures may easily vary three times the above range. The measuring-element deflection is the limiting factor.

Class II—Vapor-pressure System. The vapor-pressure thermal system (class II) is based on the physical characteristic that all enclosed liquids at a given temperature will create a definite vapor pressure if the liquid only partially occupies the enclosed space. This vapor pressure will increase with temperature.

Four types (Fig. 4-13) of vapor-pressure systems are used in industry, the type selected depending mainly on the proposed application. The first type (class IIA) is designed to operate with measured temperature (temperature of the bulb) always above the temperature of the rest of the system. This type can use a relatively small bulb. The system must be filled with the proper amount of actuating liquid so that the dividing surface between the liquid and the vapor is within the bulb. The temperature indicated is the temperature at this interface. Temperature from ambient to 600°F can be measured.

The second type of vapor-pressure system (class IIB) is used when the bulb temperature is always lower than the remaining parts of the filled system. The amount of filling liquid is the main difference between this and class IIA. Only a small amount of the volatile liquid is necessary, as shown in Fig. 4-13, to retain the dividing line between the liquid and vapor in the bulb. The bulb in this case can be smaller than in class IIA. Systems can be designed to read temperatures to at least −60°F.

In many cases, it is necessary to read temperature below as well as above room temperature. The system can easily be designed to take care of this need. One way to do this is to use a larger-volume bulb so that, when the bulb is colder than the capillary and element, the liquid and vapor dividing line is still in the bulb, and, when the bulb is hotter than the rest of the system, there will remain in the bulb sufficient liquid to produce an interface in the bulb between the liquid and the vapor. The bulb volume must therefore be greater than the combined volume of capillary and element. This type is called class IIC and is illustrated in Fig. 4-13. With this system there will be a region of about 5 to 10°F around room or ambient temperature where the indication is not definite, and, as the ambient temperature may vary greatly, a large region of temperature cannot be indicated. Systems of this type are designed to read temperatures from 0 to 250°F approximately.

The fourth type of vapor-pressure system (class IID) is designed to operate with the bulb temperature above, below, and at the temperature of the rest of the thermal system. In this type the system is filled so that a nonvolatile liquid (called the transmitting liquid), which does not mix with the regular actuating liquid, occupies the element, the capillary, and a portion of the bulb. In the bulb is the actuating liquid with a vapor space. The bulb is provided with a special trap which reduces the possibility of the actuating liquid's escaping up the capillary. Irrespective of the ambient temperature or the bulb temperature, the actuating liquid is always in the bulb, and the true temperature of the bulb is indicated, resulting in accurate readings. The transmitting liquid can be a nonfreezing solution such as water and glycerin or water and ethylene glycol, while the actuating liquid can be methyl ether, propane, butane, or ethyl ether, all of which are practically insoluble in the water solutions. This type of class II system is not recommended over 200 or below −30°F.

Elevation errors are present in the vapor-pressure system which are not present in the class I type. This is due to the head of liquid between the element and the bulb which, if the measuring element or case is installed above the bulb, will act against the vapor pressure in the bulb and cause a lower pen reading. If the element is below the bulb, the pressure head will add to the vapor pressure and cause a higher reading. The pen can be set up or set back, respectively, to these cases at any position if the system has been calibrated with the bulb at the same level as the element. If vapor is in the capillary and elements, as in class IIB, no elevation error is present. Class IIC cannot be corrected for elevation for all operating temperatures of the bulb. Ambient temperatures of the capillary and measuring element have no effect on the vapor pressure of the system for a pure filling liquid. Any length of capillary can be used.

Vapor-pressure thermal systems produce a nonlinear output. It is now possible to obtain a usable linear output by a simple linkage system, which is so designed that the motion of the element is amplified in decreasing degree as the pressure in the measuring element increases. This device allows the use of uniform charts with a system that is inherently cheaper to produce, and with over-all errors less than the comparative types of the other classes of the systems. The recommended span for this linear-conversion type is from a minimum of 50 to a maximum of 150°F. The conversion linkage can be used with any of the four vapor-pressure types of systems at any temperature level desired from −50 to 600°F, using the recommended spans of 50 to 150°F.

Class III—Gas-filled Systems. The gas-filled thermal system (class IIIA, Fig. 4-14) is a thermal system filled with a gas and operating on the principle of pressure change with temperature change. This system utilizes the basic principle of the universal gas law of $PV = RT$, where P is absolute pressure, V is volume, T is absolute temperature, and R is the universal gas constant. The system is evacuated and then filled with nitrogen under a certain required pressure. It can be supplied from −150 to 1000°F (it is possible, with limitations, to extend this range); spans from 100 to 1000°F are available, although the shorter ranges are limited and not recommended.

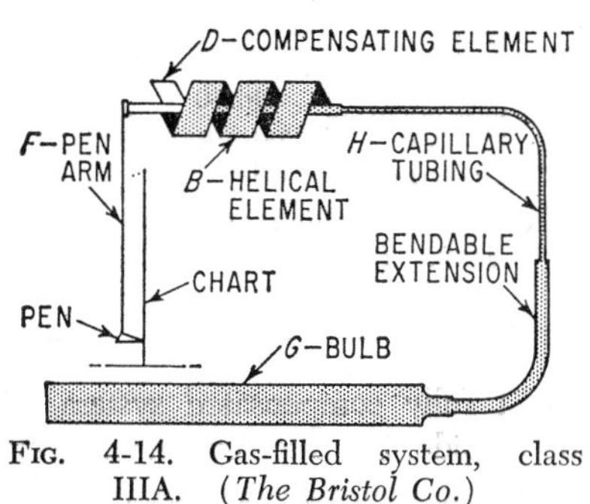

FIG. 4-14. Gas-filled system, class IIIA. (*The Bristol Co.*)

Ambient temperature errors are present for the capillary and the element as for class I systems, because the gas density in these parts will also change with temperature. The errors can be reduced considerably, however, by using a large bulb. The larger the bulb, the smaller the error.

The error caused by ambient temperature around the element is due to two effects: First, the most obvious one, is that due to the volume of gas in the element. As this volume can be only 1 per cent of the volume of the bulb, the error may become insignificant. The second effect due to the change in elastic modulus of the element with temperature change becomes more important at the operating pressures required. The greatest part of these two errors can be taken care of by a bimetallic compensator (Fig. 4-14). Theoretically, this compensation is only perfect for one temperature of the bulb (that is, for one pressure in the system). The greater the pressure in the system, the longer is the bimetal required for compensation.

The capillary error will depend on the volume of the capillary and the temperature range. To keep the error to within ½ per cent of full scale for a 50 per cent ambient temperature change, for instance, only 20 ft of capillary can be used for a 100°F range, whereas, for a 1000°F interval, 200 ft may be used for one design where the bulb volume is 50 cc. The ambient error will be proportional to ambient temperature changes and to the length of capillary.

For long lengths of capillary, where the ambient temperature errors may become large or where a smaller bulb is desired, it is possible to use a small-bulb gas-filled system (class IIIB). This system is a duplicate of the

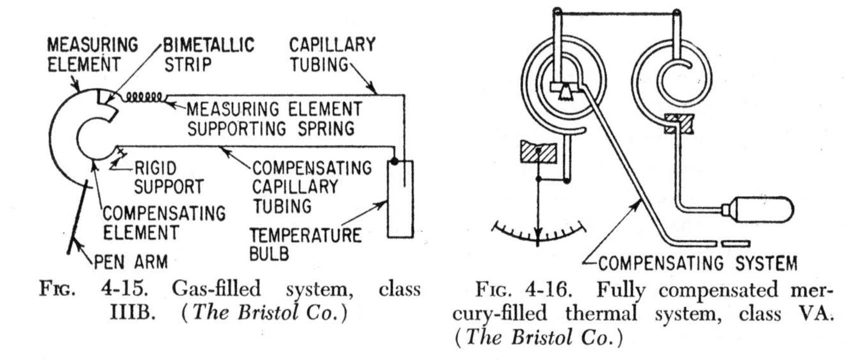

FIG. 4-15. Gas-filled system, class IIIB. (*The Bristol Co.*)

FIG. 4-16. Fully compensated mercury-filled thermal system, class VA. (*The Bristol Co.*)

class I fully compensated system with two elements and two capillaries side by side, but the elements are compensated by a bimetallic compensator on the compensating element. This is not a perfect compensated system, but is satisfactory for industrial work. It is possible to use a smaller bulb than that used for class IIIA. It is especially useful where the capillary is long and accuracy at one temperature is desired. It is a low-torque system and should be used with discretion. A schematic diagram is shown in Fig. 4-15.

Class V—Mercury-filled Systems. The final class of filled thermal systems is a mercury-filled type (class V). This system is completely filled with mercury or mercury-thallium eutectic amalgam operating on the principle of liquid expansion. With full compensation (class VA), the compensating means is either a second thermal system minus the bulb, as shown in Fig. 4-16 (similar to class IA), or a compensated capillary system. The former is similar to that described for the liquid-filled thermal system. The latter consists of a wire of invar or other low-expanding material inserted in steel or stainless-steel capillary. It is possible to proportion the volumes of the wire and the capillary so that, with ambient changes around the capillary, the volumetric expansion of the mercury will be equal to the differential volume change between the capillary and the wire. The element is compensated by a bimetal.

With the case-compensation type (class VB, Fig. 4-17) the compensating means is within the case itself. This is similar to liquid-filled systems, case-compensated (class IB), where the element is compensated by a bimetallic compensator. The capillary is not compensated, and so there is some error due to ambient. Only short lengths of capillary are recommended for this type.

Mercury-filled thermal systems can be used from −80 to 1000°F with a uniform chart or scale. All parts of a mercury system must be made of a ferrous material, and all joints must be welded. Bulbs are larger than those encountered with the organic liquid-filled system for the same temperature

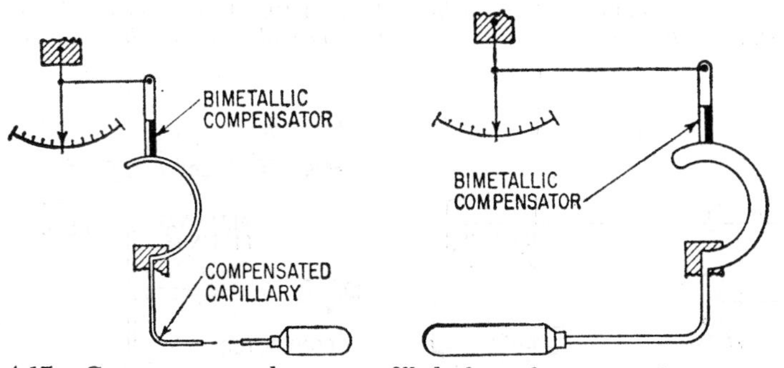

FIG. 4-17. Case-compensated mercury-filled thermal system, class VB. (*The Bristol Co.*)

spans. In many applications, the use of mercury in a filled system is prohibited because of the detrimental effect of mercury on the particular operation.

Bulbs.[3] Vapor-filled and gas-filled thermal systems are generally available with copper, steel, or stainless-steel bulbs. Where required, other materials, such as Monel, can be used. Lead or plastic coatings can be applied for corrosion resistance, and abrasion-resistant coatings can be used if necessary. Mercury-filled thermal systems have stainless-steel bulbs, but when necessary the bulb can be coated with lead, plastic, and other materials. The tubing that protects the capillary is known as flexible tubing, and is normally the bronze-armored type, but is also available in other materials. Protective tubing can also be of some smooth material, such as lead or stainless steel, and is then known as bendable tubing.

Plain Bulbs. These are especially useful where portability is a factor, and where the medium to be measured is not under pressure. A plain bulb with bendable extension neck can be used to measure the temperatures

[3] Minneapolis-Honeywell Regulator Co., Industrial Div.

of several adjacent vessels. And the fact that the bulb can be easily installed and removed is often advantageous. For more permanent installation, the bulb shown in Fig. 4-18 is also available with a flange on the extension.

Union-connected Bulbs. This type, shown in Fig. 4-19, is provided with fittings for installation in vessels under pressure where plant safety requirements do not specify the use of a well or socket. Fittings may be permanently fixed or adjustable. These bulbs are often installed in pipeline tees. If the opening must be closed when the bulb is removed, a plug and chain can be supplied.

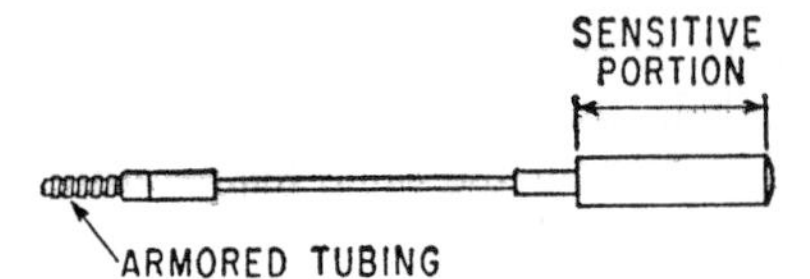

Fig. 4-18. Plain bulb. (*Minneapolis-Honeywell Regulator Co.*)

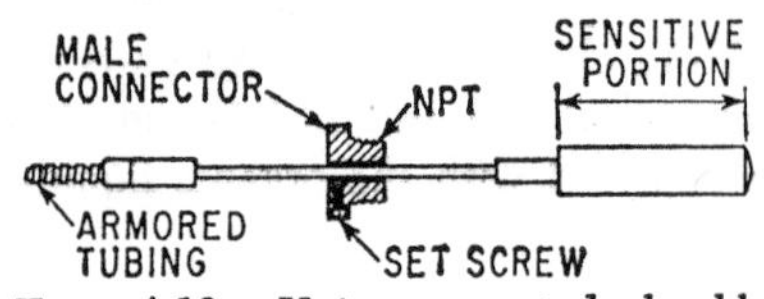

Fig. 4-19. Union-connected bulb. (*Minneapolis-Honeywell Regulator Co.*)

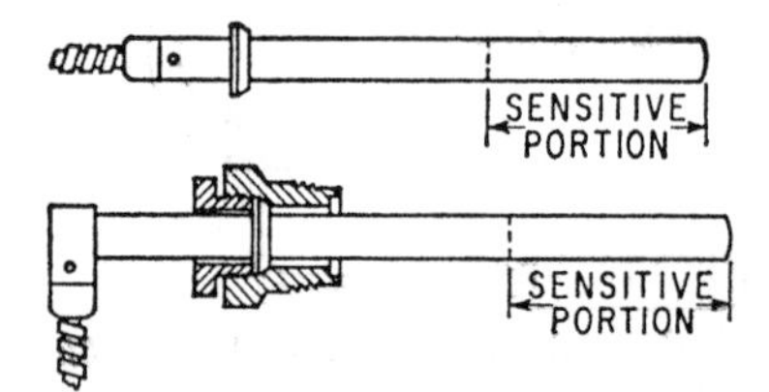

Fig. 4-20. Plain rigid-extension bulb (top) and rigid-angle-extension bulb with optional union connection and bushing. (*Minneapolis-Honeywell Regulator Co.*)

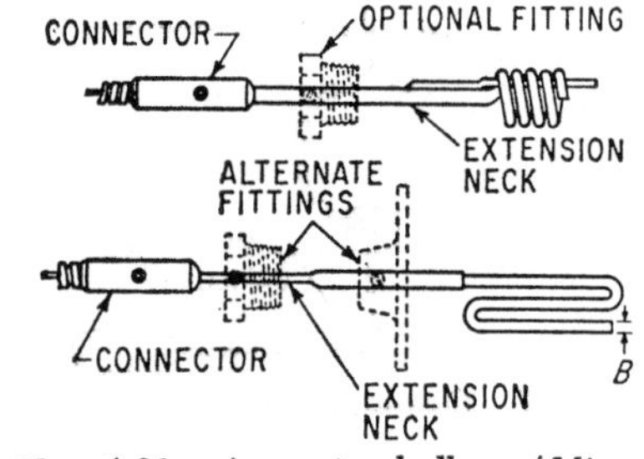

Fig. 4-21. Averaging bulbs. (*Minneapolis-Honeywell Regulator Co.*)

Rigid-extension Bulbs. These, both plain and union-connected, are available in both vapor- and mercury-filled systems with rigid extensions, in either the angle or the straight type, as illustrated in Fig. 4-20. This bulb is especially well suited for use with a rotary joint for insertion in the bearing of a rotating cylinder whose internal temperature is to be measured. They are also useful in vessels with agitators or flow lines, where turbulence might damage an unprotected plain bulb, but where it is not desirable to use a well.

Averaging Bulbs. Because of the large surface area exposed by a long, small-diameter active portion, the averaging bulbs (Fig. 4-21) have a very quick response and are especially effective for measuring the temperature

Table 4-2. Typical Bulb Specifications.†

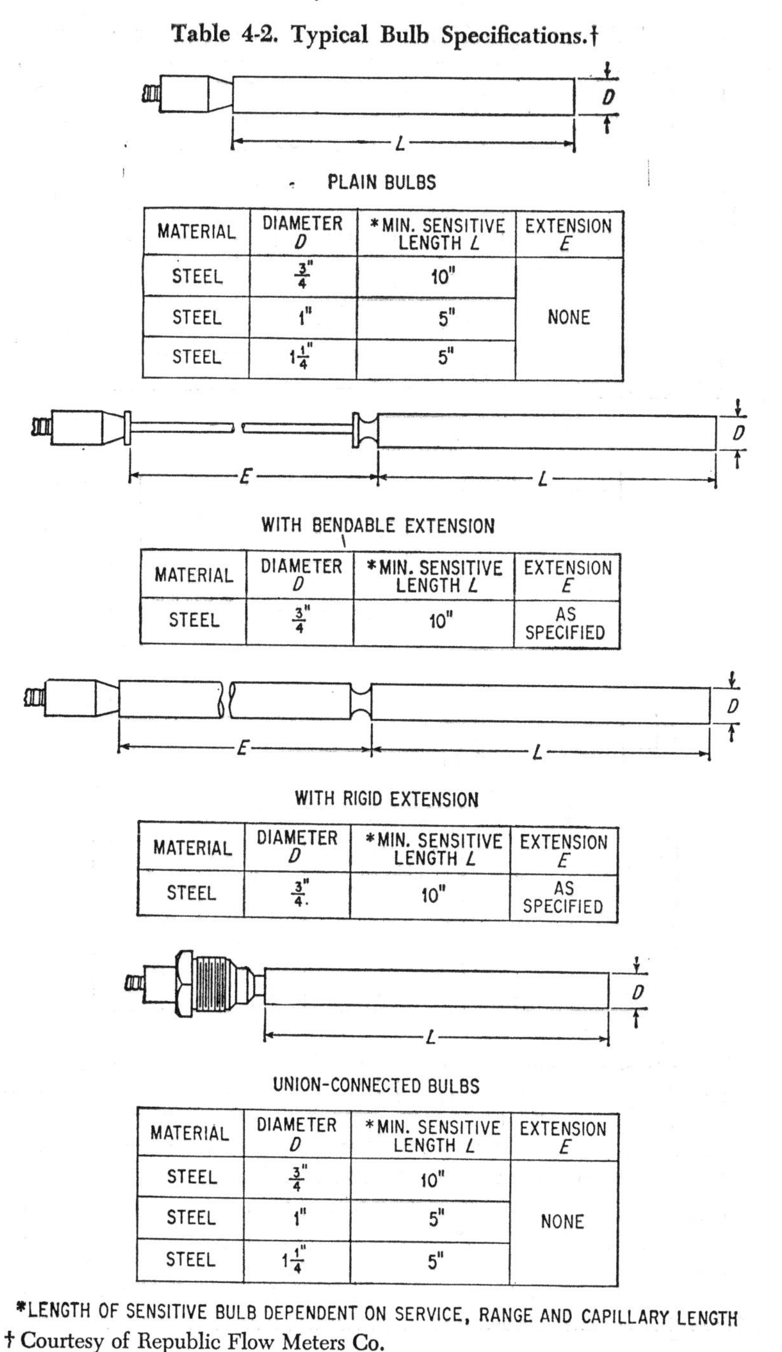

PLAIN BULBS

MATERIAL	DIAMETER D	*MIN. SENSITIVE LENGTH L	EXTENSION E
STEEL	$\frac{3}{4}$"	10"	NONE
STEEL	1"	5"	
STEEL	$1\frac{1}{4}$"	5"	

WITH BENDABLE EXTENSION

MATERIAL	DIAMETER D	*MIN. SENSITIVE LENGTH L	EXTENSION E
STEEL	$\frac{3}{4}$"	10"	AS SPECIFIED

WITH RIGID EXTENSION

MATERIAL	DIAMETER D	*MIN. SENSITIVE LENGTH L	EXTENSION E
STEEL	$\frac{3}{4}$"	10"	AS SPECIFIED

UNION-CONNECTED BULBS

MATERIAL	DIAMETER D	*MIN. SENSITIVE LENGTH L	EXTENSION E
STEEL	$\frac{3}{4}$"	10"	NONE
STEEL	1"	5"	
STEEL	$1\frac{1}{4}$"	5"	

*LENGTH OF SENSITIVE BULB DEPENDENT ON SERVICE, RANGE AND CAPILLARY LENGTH

† Courtesy of Republic Flow Meters Co.

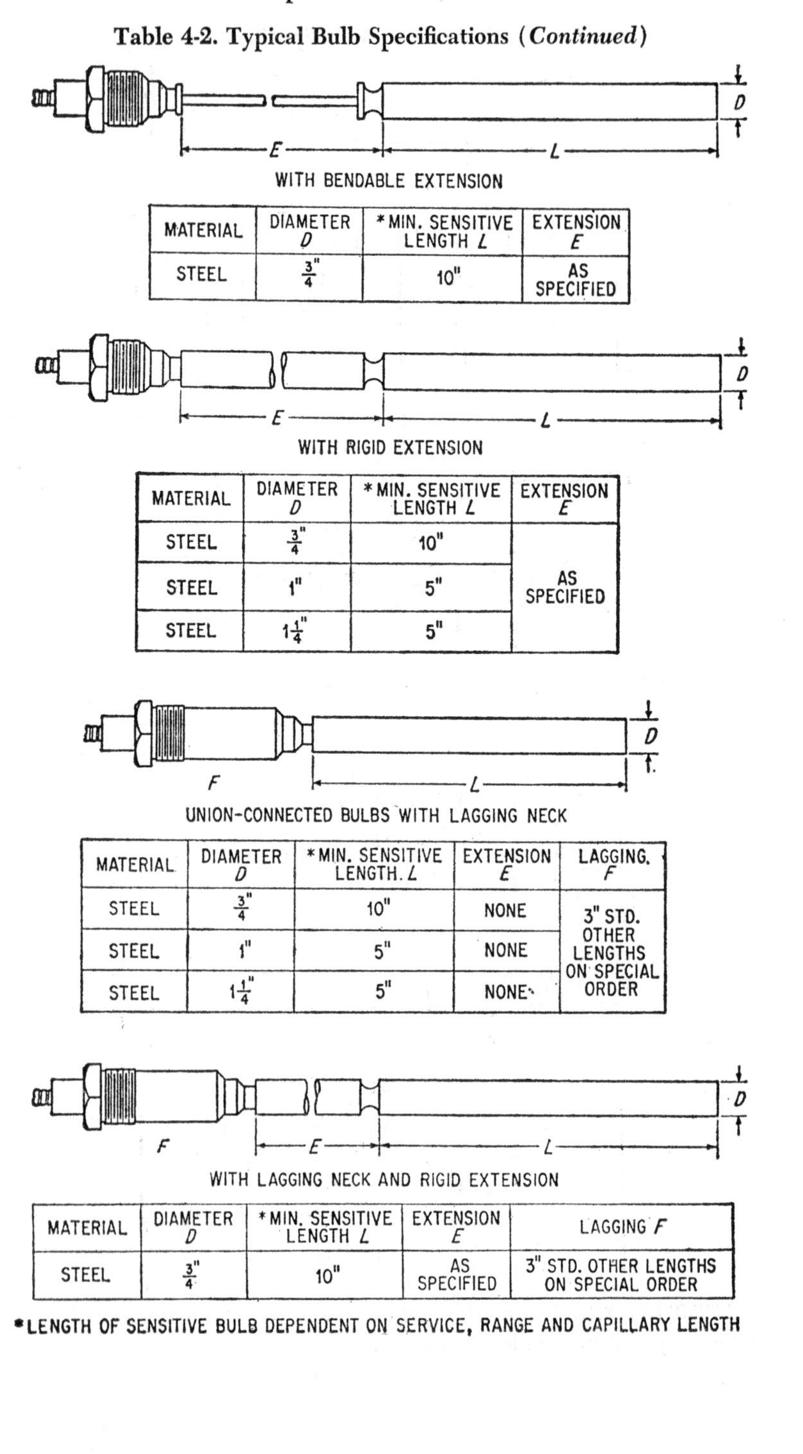

Table 4-2. Typical Bulb Specifications (*Continued*)

WITH BENDABLE EXTENSION

MATERIAL	DIAMETER D	*MIN. SENSITIVE LENGTH L	EXTENSION E
STEEL	$\frac{3}{4}$"	10"	AS SPECIFIED

WITH RIGID EXTENSION

MATERIAL	DIAMETER D	*MIN. SENSITIVE LENGTH L	EXTENSION E
STEEL	$\frac{3}{4}$"	10"	AS SPECIFIED
STEEL	1"	5"	
STEEL	$1\frac{1}{4}$"	5"	

UNION-CONNECTED BULBS WITH LAGGING NECK

MATERIAL	DIAMETER D	*MIN. SENSITIVE LENGTH L	EXTENSION E	LAGGING F
STEEL	$\frac{3}{4}$"	10"	NONE	3" STD. OTHER LENGTHS ON SPECIAL ORDER
STEEL	1"	5"	NONE	
STEEL	$1\frac{1}{4}$"	5"	NONE	

WITH LAGGING NECK AND RIGID EXTENSION

MATERIAL	DIAMETER D	*MIN. SENSITIVE LENGTH L	EXTENSION E	LAGGING F
STEEL	$\frac{3}{4}$"	10"	AS SPECIFIED	3" STD. OTHER LENGTHS ON SPECIAL ORDER

*LENGTH OF SENSITIVE BULB DEPENDENT ON SERVICE, RANGE AND CAPILLARY LENGTH of air or other gases, as well as the temperature of slowly moving liquids, where heat-transfer conditions are not favorable.

The term *averaging* is used because the bulbs measure the average temperature over their active portion. The coiled type is ideally suited for duct-temperature measurement, whereas the long straight type is available in lengths up to 25 ft and can be threaded back and forth across an oven, furnace chamber, or duct coil.

Table 4-2. Typical Bulb Specifications (*Continued*)

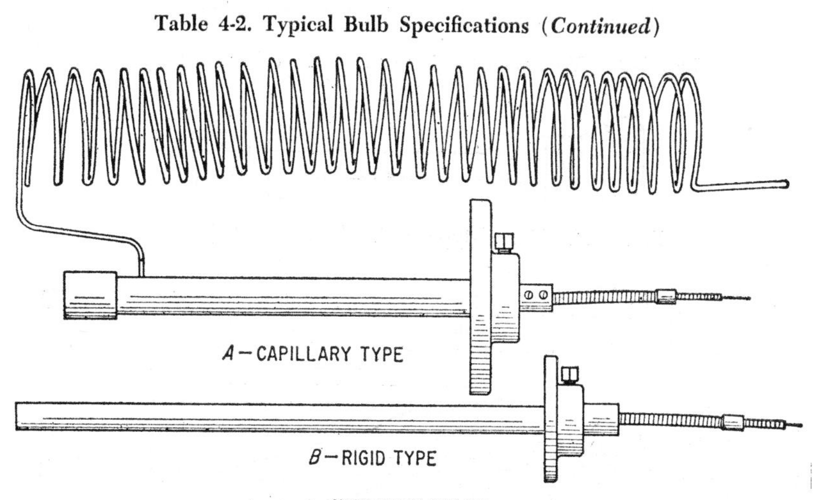

AVERAGING BULBS

ASSEMBLY	MATERIAL	COIL DIAMETER	LENGTH COILED	LENGTH UNCOILED	ACCESSORIES FURNISHED
A	STEEL	$3\frac{3}{4}''$	6"	25'	FLANGE AND $\frac{3}{4}''$ COUPLING
A	COPPER	$3\frac{3}{4}''$	6"	25'	
A	STN. STEEL	$3\frac{3}{4}''$	6"	25'	
A	STEEL	$3\frac{3}{4}''$	10"	40'	

ASSEMBLY	MATERIAL	DIAMETER	LENGTH	ACCESSORIES FURNISHED
B	STEEL	$\frac{3}{4}''$	42"	PROTECTING TUBE AND FLANGE

Table 4-2 lists typical bulb specifications.

Separable Wells. Wells for thermal bulbs are used for protection against corrosion, abrasion, erosion, impact, and high pressure. They are also required for use where it is necessary to remove the bulb without interrupting operations. Wells are furnished in type 316 stainless steel, steel, or brass, with a wide variety of other materials for special applications. Figure 4-22 illustrates a standard plain well, with length as required. Figure 4-23 illustrates a well with a 3-in. lagging extension for added convenience in

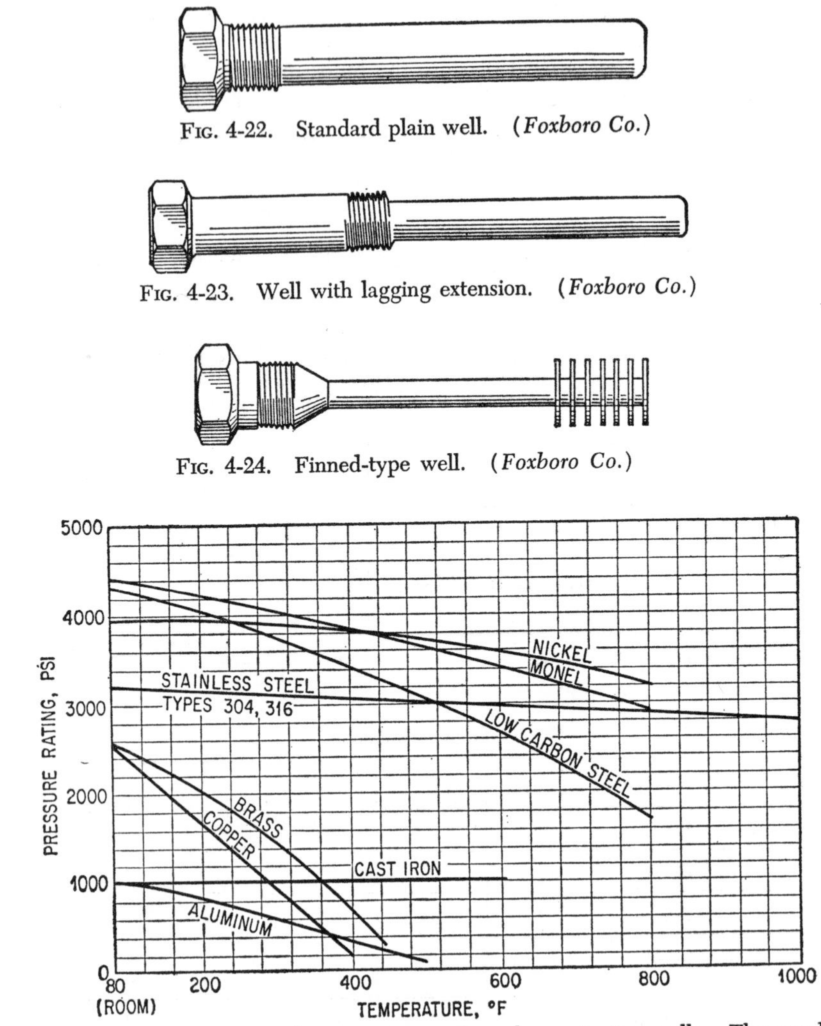

FIG. 4-22. Standard plain well. (*Foxboro Co.*)

FIG. 4-23. Well with lagging extension. (*Foxboro Co.*)

FIG. 4-24. Finned-type well. (*Foxboro Co.*)

FIG. 4-25. Pressure and temperature ratings for protective wells. The graph above gives pressure-temperature ratings of variously constructed wells. Ratings are based on a static condition of the fluids into which wells are immersed. (*Taylor Instrument Co.*)

installing in lagged lines or vessels. Other sockets such as the aluminum-finned type shown in Fig. 4-24 are for special applications. Figure 4-25 plots pressure and temperature ratings of wells.

Tubing. Connecting capillary tubing is required on all filled-type thermal systems. The capillary tubing connects the sensitive bulb and the receiving element, thereby forming an integral system. The length of capillary tubing required depends on the distance between the sensitive bulb and the exhibiting instrument. Protective armor is supplied over the capillary tubing to provide mechanical strength, to ensure against crimping capillary bore and to provide additional corrosion protection. Materials for capillary tubing and protective armor are selected on the basis of corrosion resistance, appearance, installation, protection required, etc. (Table 4-3).

Table 4-3. Typical Capillary Tubing Materials.*

	CAPILLARY TUBING		PROTECTIVE ARMOR	
	MATERIAL	DIAM	MATERIAL	DIAM
	COPPER	1"/16	BRONZE	5"/16
	COPPER	1"/16	302 STAINLESS STEEL	5"/16
	COPPER	1"/16	POLYVINYL CHLORIDE COVERED BRONZE	7"/16
	COPPER	1"/16	LEAD	5"/16
	321 STAINLESS STEEL	1"/16	BRONZE	5"/16
	321 STAINLESS STEEL	1"/16	302 STAINLESS STEEL	5"/16
	321 STAINLESS STEEL	1"/16	POLYVINYL CHLORIDE COVERED BRONZE	7"/16
	321 STAINLESS STEEL	1"/16	LEAD	5"/16
	* STAINLESS STEEL	3"/32	NONE REQUIRED	

* Courtesy of Fischer & Porter Co.

Installation. Bare sensitive bulbs should be used whenever possible. They permit faster response and are not so expensive as sensitive bulbs installed in wells. Always locate the bulb where temperature is representative and where velocity is adequate for proper heat transfer. Avoid conduction errors by immersion of sensitive bulb plus part of the bulb extension. Use capillary bulbs or averaging bulbs for gas or vapor service when forced circulation or velocity is low. Radiation effects should be considered when installing sensitive bulb on the side of oven, duct, etc. Proper shielding is necessary for accurate temperature measurement. Figures 4-26 to 4-29 show typical installations.

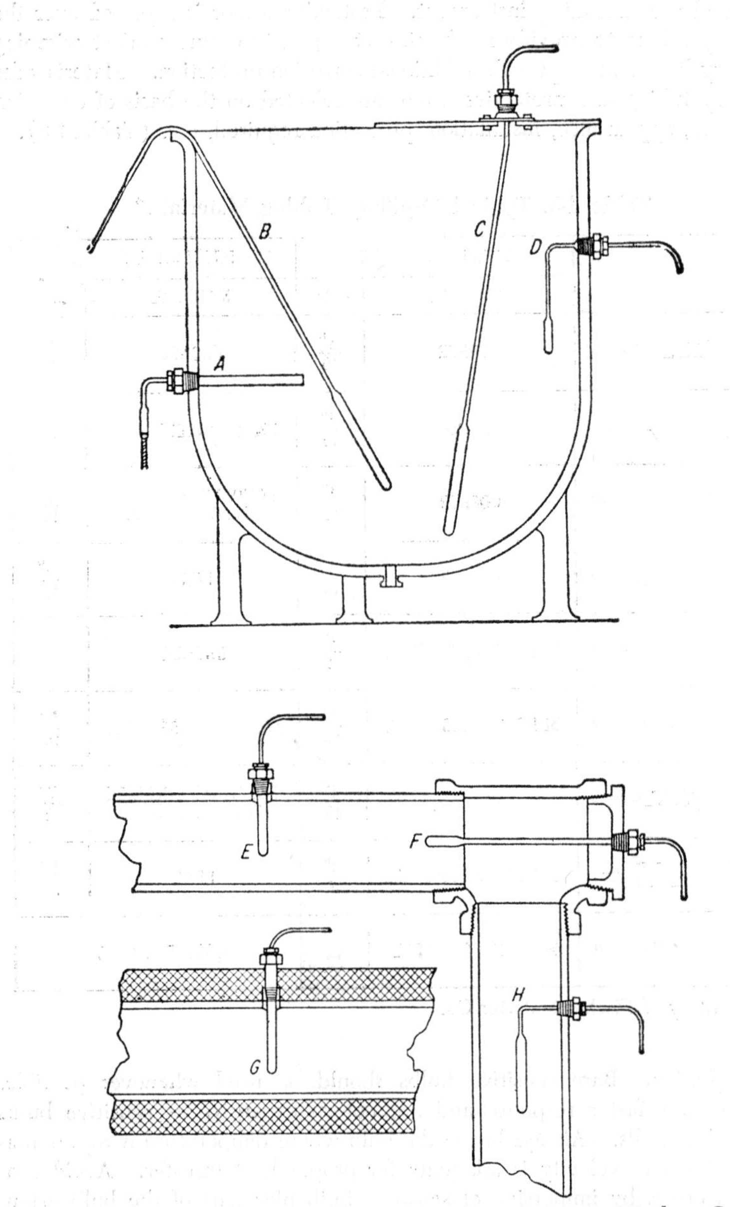

FIG. 4-26. Typical bulb installation in vessels and pipes. (*Foxboro Co.*)

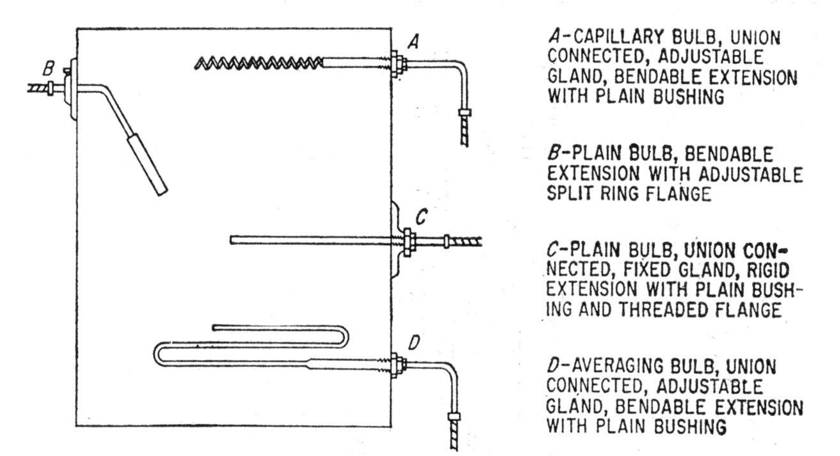

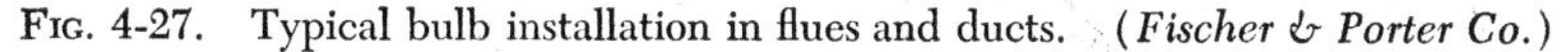

FIG. 4-27. Typical bulb installation in flues and ducts. (*Fischer & Porter Co.*)

BIMETAL THERMOMETERS

Bimetal thermometers, widely used in buildings today because they combine simple rugged construction with good accuracy, are based on the principle of differential thermal expansion of two dissimilar metals.

Bimetals.[4] The bimetals are composed of two or more layers of metallic alloys having different coefficients of thermal expansion and usually different physical characteristics. The alloys are special in nature and are selected

Table 4-4. Coefficient of Thermal Expansion of Metals

Metal	Temperature, °C	Coefficient, $\times 10^{-4}$
Iron	40	0.1210
Brass	20	0.189
Copper	−191 to +16	0.1409
Zinc	10 to 100	0.2628
Tungsten	20 to 100	0.0336
Tin	18 to 100	0.2692
Glass	0 to 100	0.0833
Nickel	40	0.1279
Invar	20	0.009
Steel	40	0.1322

for their physical properties over a broad range of temperatures. Among the properties of importance in determining the suitability of an alloy for bimetal use are coefficient of expansion (Table 4-4), modulus of elasticity, elastic limit after cold rolling, electrical conductivity, ductility, metallurgical stability, and strength at various temperatures. Once selected, the alloys for a particular bimetal must be laminated to produce a precision thermo-

[4] W. M. Chace Co.

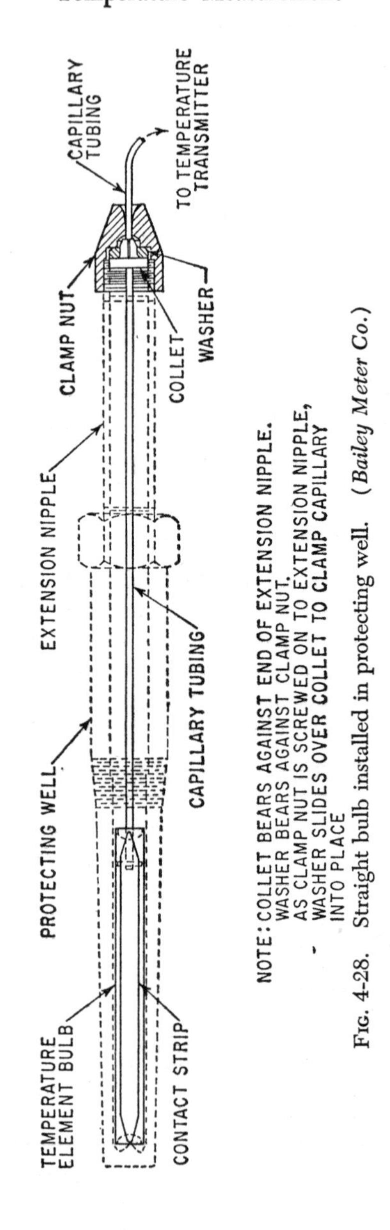

Fig. 4-28. Straight bulb installed in protecting well. (*Bailey Meter Co.*)

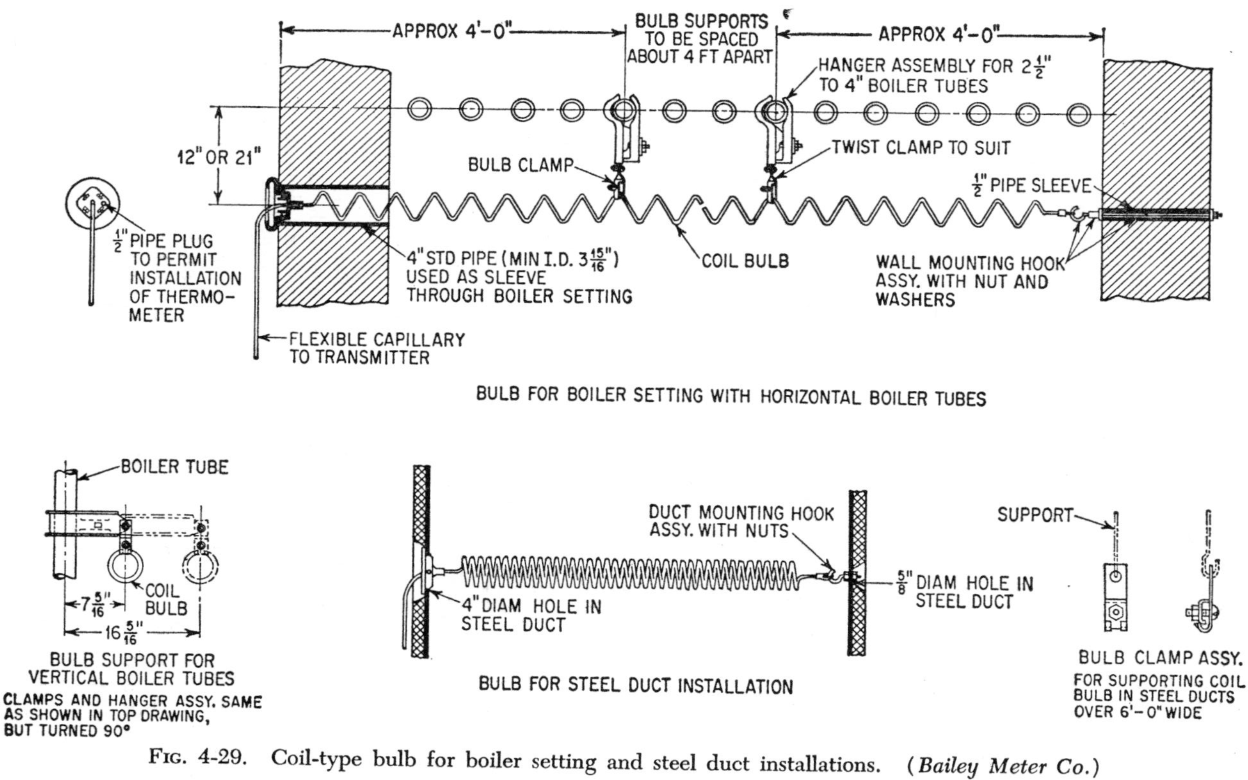

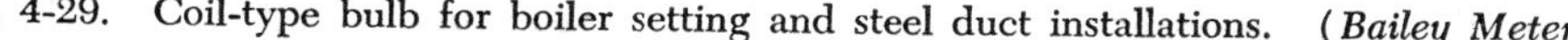
FIG. 4-29. Coil-type bulb for boiler setting and steel duct installations. (*Bailey Meter Co.*)

static element. This lamination is accomplished by a high-temperature weld process, which involves the use of no intermediate bonding material. The direct weld thus allows utilization of the full physical properties of the special alloys over the maximum possible temperature range.

The fundamental property of all thermostatic bimetals is the ability to bend or change curvature with temperature change. The temperature change may result from any type of heating (or cooling), such as radiation, convection, or conduction. The bimetal can also be heated by the I^2R effect developed when the element carries an electrical current.

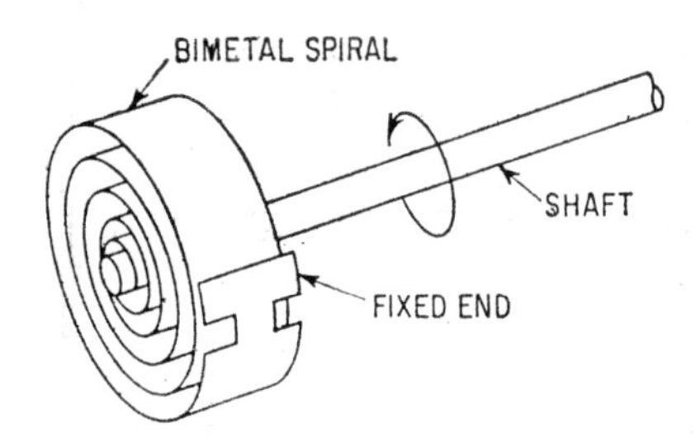

(*a*)

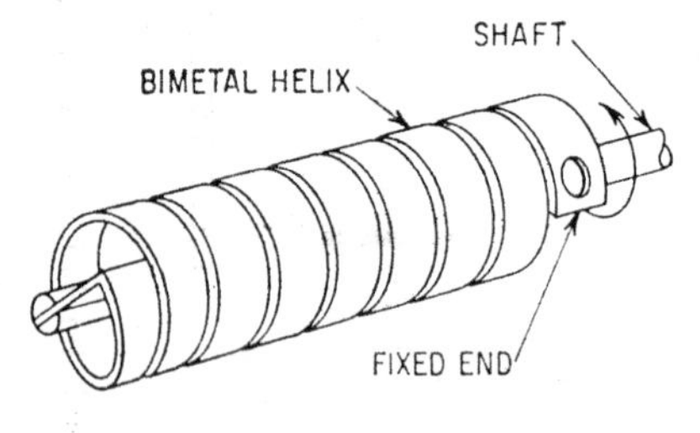

(*b*)

FIG. 4-30. Elements used in bimetal thermometers. (*W. M. Chace Co.*)

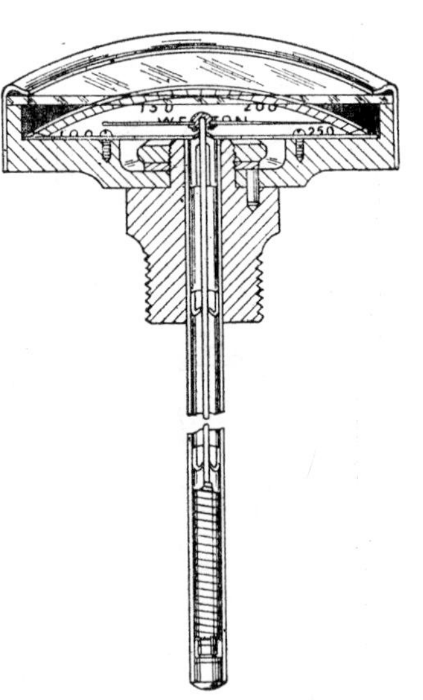

FIG. 4-31. Sectional view of bimetal thermometer using multiple-helix element. (*Weston Instrument Div. of Daystrom.*)

When the thermostatic bimetal is restrained from bending as it is subjected to a temperature change, a force is developed. The force is equal in magnitude to that which would occur when a similar beam is bent or curved through a distance corresponding to the unallowed deflection. A bimetal is a constant-moment device; that is, since the difference in coefficients of expansion, and therefore the tendency to bend, is constant along the length of the element, the bending moment is also constant.

Figure 4-30 shows typical elements used in bimetal thermometers. Figure 4-31 shows the construction of a bimetal thermometer using a continuous multiple-helix element. In each case movement of shaft actuates a needle that indicates on a dial face. Generally, bimetal dial thermometers are guaranteed to be accurate to within 1 per cent of the range over the entire temperature scale.

INSTITUTIONAL AND INDUSTRIAL LIQUID-IN-GLASS THERMOMETERS

The liquid-in-glass thermometer (Figs. 4-32 and 4-33), and in particular the mercury-in-glass thermometer, is the most frequently used of all tem-

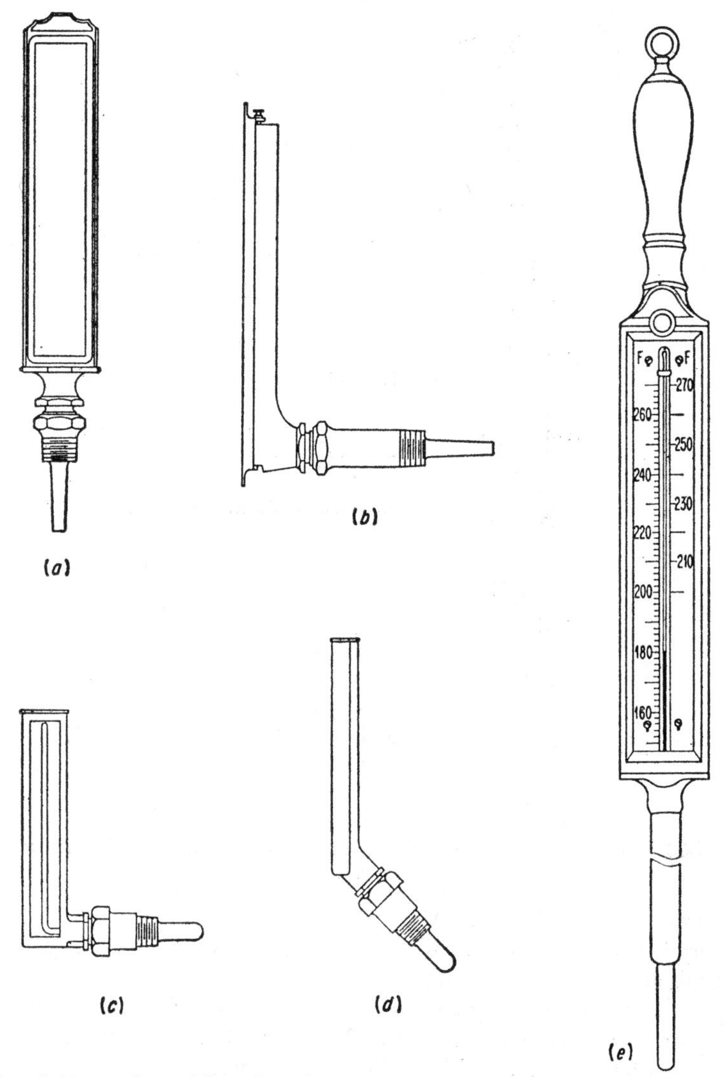

Fig. 4-32. Industrial liquid-in-glass thermometers. (*a*) Straight form without extension neck; (*b*) regular angle-form glass; (*c*) side-angle form; (*d*) 135° angle form; (*e*) hand type. (*Weston Instrument Div. of Daystrom.*)

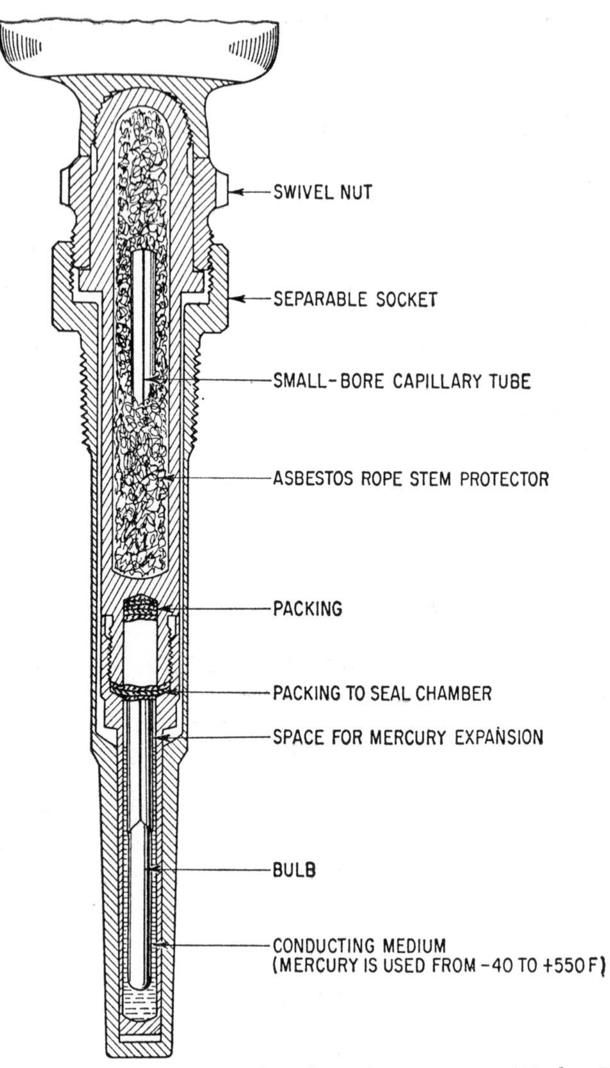

FIG. 4-33. Section through liquid-in-glass thermometer. (*Taylor Instrument Co.*)

perature-measuring devices. Most important in the building field is the mercury-in-glass thermometer with a range of −38 to +950°F. Suitable materials are used for other ranges.

Space above the mercury column in the bore of the thermometer is filled with a dry inert gas under pressure. This prevents the mercury from boiling at high temperatures and helps to eliminate mercury separations.

For temperatures below the freezing point of mercury (−38°F), spirit-filled tubes must be used. When the thermometer is to be used for indicating the temperature of air or gases, it should be ordered with a bare bulb (without the metallic bulb chamber) and with a suitable guard.

To prevent frost creeping up the scale case of the thermometer, for refrigeration applications, a ring of low-conductivity material is often interposed between the scale case and the stem.

THERMOCOUPLE PYROMETERS

The principle of thermoelectricity, discovered by Seebeck in 1821, is the basis for one of the most commonly used temperature-sensing devices—the *thermocouple*. When two dissimilar metals are welded together at one end and this junction is heated, a voltage is developed on the free ends. In modern practice, the two free ends are connected to a millivoltmeter or potentiometer which measures the emf created and indicates or records this emf in terms of temperature. Instruments of this type are known as *thermocouple pyrometers*. In addition to indicators and recorders, these instruments are available with electrical, electronic, mechanical, and pneumatic components to effect the various modes of automatic control.

Peltier and Thomson Effects.[5] The electromotive force developed in a thermoelectric circuit is ascribed to two phenomena, one known as the Peltier effect, and the other as the Thomson effect. The Peltier effect governs the emf resulting solely from the *contact* of two dissimilar metals. (However, its magnitude varies with the temperature at the point of contact.) The emf resulting from the less predominant Thomson effect is that produced by a temperature gradient along a single wire.

Since there are two points of contact, and two wires with temperature gradient, it follows that there are two Peltier emf's and two Thomson emf's. The total emf acting in the circuit is the result of all four, with polarity being determined by the particular materials used and by the relationship of the temperatures at the two junctions. This emf can be measured by connecting a potentiometer, or other instrument for measuring emf, into the circuit at any point.

Commercial thermocouples generate on the order of 20 to 50 mv through the range of their ordinary operating temperatures (Fig. 4-34).

Since the materials used for commercial thermocouples are selected so that the Thomson effect can be disregarded, the total emf becomes the sum of the two generated by the Peltier effect. If the temperature at one junction (reference junction) is kept constant, or if its emf generations are compensated for, the effective emf of the thermocouple becomes that generated by changes in temperature at the uncompensated or measuring junction. This emf is used to measure a temperature change.

There are two laws of thermoelectricity which become important for they govern both thermocouple theory and practice. They are known as the law of intermediate temperatures and the law of intermediate metals.

[5] B. B. Ritchey and C. H. Vogelsang, Current from Temperature, *Instrumentation*, vol. 9, no. 5.

Law of Intermediate Temperatures. In the most common installations, it is not practical to maintain the reference junction of a thermocouple at a constant temperature. So, some means must be provided to bring the emf developed under existing conditions at the reference junction to a value equal to that which would be generated with the reference junction maintained at a standard temperature, usually 32°F (cold-junction compensation).

The law of intermediate temperatures provides a means for relating the emf generated by a thermocouple under ordinary conditions to a standardized constant temperature. In effect, the law states that the sum of the emf's generated by two thermocouples—one with its junctions at 32°F and some reference temperature, the other with its junctions at the same reference temperature and the measured temperature—is equivalent to that produced by a *single* thermocouple with its junctions at 32°F and the measured temperature.

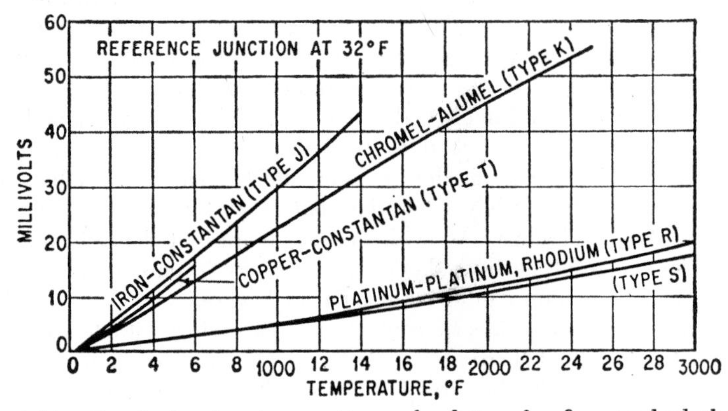

FIG. 4-34. Graph showing temperature-emf relation for five standard thermocouple wire types. (*Minneapolis-Honeywell Regulator Co., Industrial Div.*)

This is represented in Fig. 4-35, where the measured temperature is 700°F. Therefore, by adding an emf equal to that produced by thermocouple *A* in Fig. 4-35 with its junctions at 32°F and the reference temperature equal to that of thermocouple *B*, a total emf equivalent to that generated by the hypothetical thermocouple *C* results. In most pyrometers, this is done by a temperature-sensitive resistor, which measures the variations in reference-junction temperature caused by ambient conditions, and automatically provides the necessary emf by means of a voltage drop produced across it. Thus, the instrument calibration becomes independent of reference-temperature variations.

Law of Intermediate Metals. When thermocouples are used, it is generally necessary to introduce additional metals into the circuit. This happens when an instrument is used to measure the emf, and when the junction is soldered or welded.

It would seem that the introduction of additional metals would modify the emf developed by the thermocouple and destroy its calibration. However, the law of intermediate metals states that the introduction of a third metal into the circuit will have no effect on the emf generated, so long as the junctions of the third metal with the other two are at the same temperature.

Any number of different metals can be introduced, providing all the junctions are at the same temperature. Thus, in Fig. 4-36, the circuits shown all generate the same emf, even though the second and third circuit diagrams show materials *C, D, E,* and *F* inserted between *A* and *B.* In practice, this means that, if the temperature inside an instrument containing a number of different metals in the thermoelectric circuit is maintained

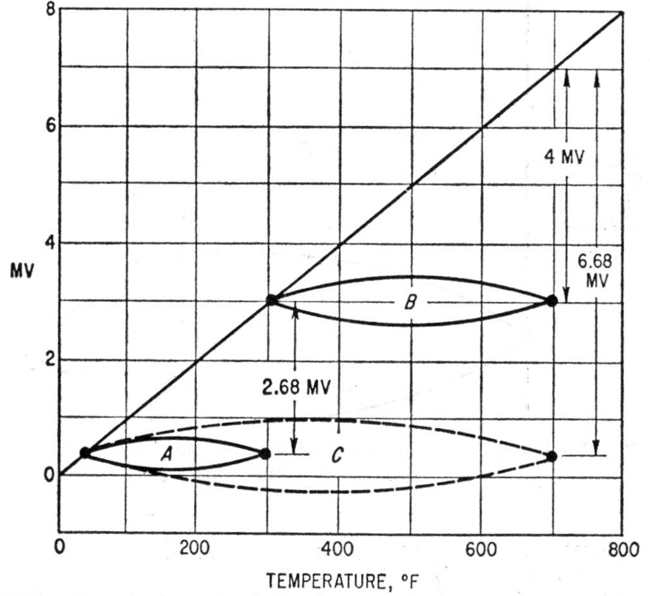

FIG. 4-35. According to the law of intermediate temperatures, the emf of thermocouple *A* plus the emf of thermocouple *B* is equal to the emf of thermocouple *C* above.

uniform, the net emf generated by the thermocouple itself will be unaffected.

An elementary thermoelectric pyrometer system is shown in Fig. 4-37. The instrument is usually located away from the point at which the temperature is measured. Since the temperature-sensing resistor for maintaining a constant reference-junction emf can be most conveniently located in the instrument as a part of its circuit, it is necessary to locate the reference junction itself in the instrument. Therefore, the thermoelectric circuit must be extended from the measuring junction, at the point where the temperature measurement is desired, to the reference junction in the instrument. This is done through the use of extension wires as explained below.

Because the thermocouple assembly is exposed to elevated temperatures and adverse atmospheric conditions, it is an expendable unit and must be replaced periodically. To do this, a terminal head is supplied, to which wires known as extension wires are connected. Since these wires are in the thermoelectric circuit, they must be made of the same material as the

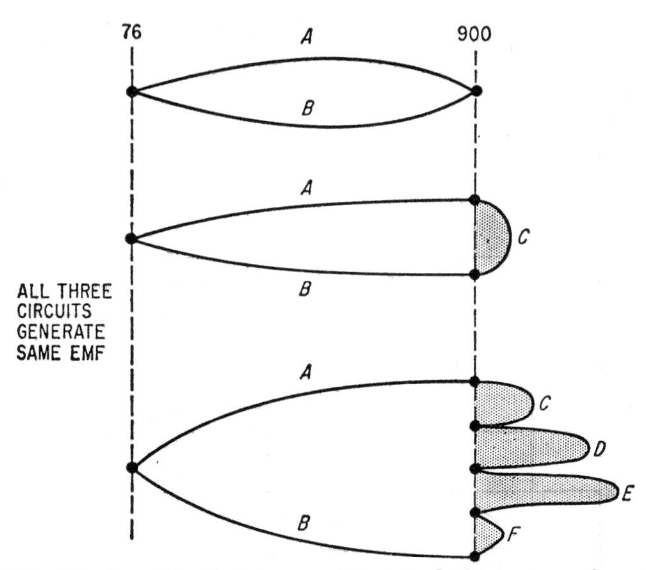

Fig. 4-36. No harmful effect is caused by introducing any number of metals at a thermocouple junction so long as all connections are at the same temperature.

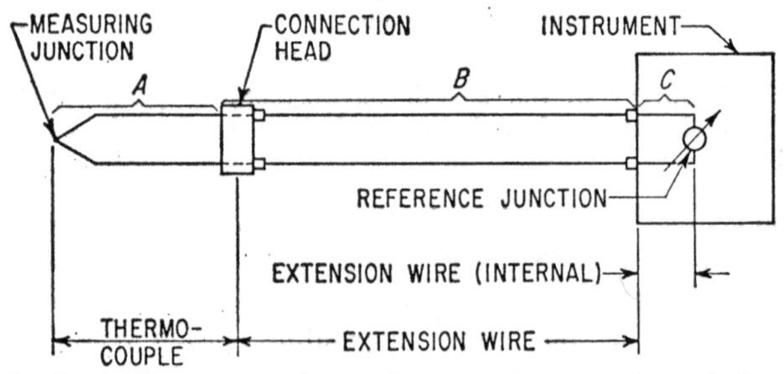

Fig. 4-37. In reality, every thermoelectric system consists of three separate thermocouples: *A*, the thermocouple proper; *B*, the external lead wire; and *C*, the internal lead wire.

thermocouple assembly (or materials having essentially the same temperature-emf curve).

Thermocouple Materials. Several combinations of two-metal conductors serve well as thermocouples. These combinations of wire must possess reasonably linear temperature-emf relationships, they must develop an emf per degree of temperature change that is detectable with standard measuring equipment, and in many applications they must be physically able to withstand sustained high temperatures, rapid temperature changes, and the effects of corrosive atmospheres. Since these requirements are quite severe, no one combination of wires serves satisfactorily for all conditions.

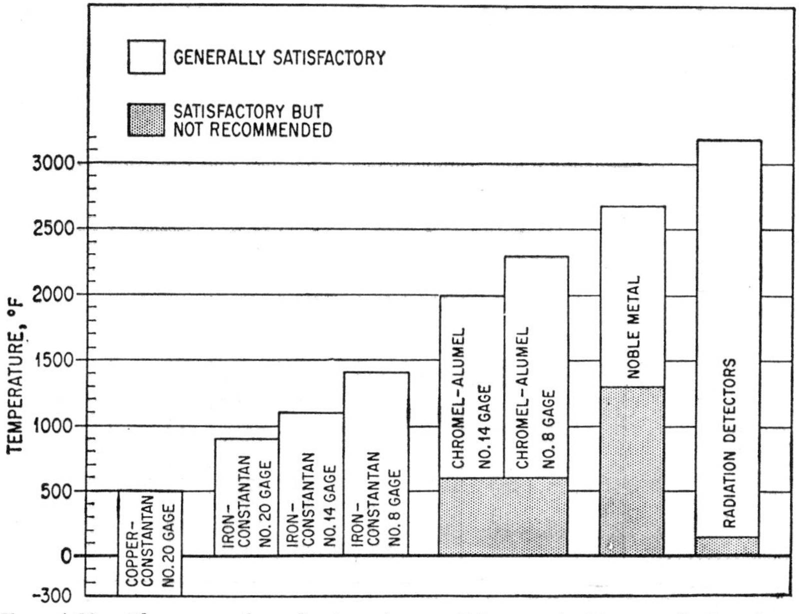

Fig. 4-38. Thermocouple selection chart. (*Minneapolis-Honeywell Regulator Co.*)

Throughout the years, designers have standardized on few wire combinations, the temperature limitations of which are shown in Fig. 4-38 and in Table 4-5.

Table 4-5. Recommended Upper Temperature Limits for Protected Thermocouples*

(In degrees Fahrenheit; for various wire sizes, AWG)

Thermocouple type	No. 8, 0.128 in.	No. 14, 0.064 in.	No. 20, 0.032 in.	No. 24, 0.020 in.	No. 28, 0.013 in.
Copper-constantan	—	700	500	400	400
Iron-constantan	1400	1100	900	700	700
Chromel-Alumel	2300	2000	1800	1600	1600
Platinum rhodium-platinum	—	—	—	2700	—

* Courtesy of Instrument Society of America.

Copper-Constantan (type T). The superiority of copper-constantan over other types of thermocouples for measurement of relatively low temperatures, especially subzero temperatures, is well established. Copper-constantan thermocouples stand up well against corrosion, are reproducible to a high degree of precision, and are generally preferred for ranges between −300 and +600°F.

Iron-Constantan (type J). These are suitable for use in reducing atmospheres where there is a deficiency of free oxygen. Above 1000°F the rate of oxidation of the iron wire increases rapidly and the use of no. 8 gage wire is suggested. Number 8 gage iron–constantan-protected thermocouples are generally considered satisfactory and economical up to 1600°F.

Table 4-6. Thermocouple and Extension-wire Chemical and Electrical Data*

Type of wire	Chemical composition—per cent weight							Melting point, °F	Resistance, ohms/circular mil-foot	Temperature coefficient of resistance/°F
	Al	Cr	Cu	Fe	Mn	Ni	Lb/cu in.			
Alumel.......	2			0.5	2.5	94	0.316	2600	177	0.0008
Chromel......		10				90	0.316	2642	425	0.00018
Constantan...			55			45	0.321	2400	294	0.000011
Copper.......			100				0.322	1982	10.371	0.00218
Iron..........				100			0.284	2795	60.14	0.00278
KA_2S.........		19		71	1	9	0.286	2575	438	0.00052
Manganin....			85		12	3	0.303	1868	290	0.000083
Nickel........						100	0.321	2646	60	0.00278

* Minneapolis-Honeywell Regulator Co.

Chromel-Alumel[6] (type K). Type K was developed especially for use in oxidizing atmospheres where there is an excess of free oxygen. Under these conditions, Chromel-Alumel will generally give better service life than iron-constantan, particularly at higher temperatures. Complete absence of free oxygen (a reducing atmosphere) has a tendency to alter the thermoelectric characteristics of these wires, causing a loss of accuracy. Number 8 gage Chromel–Alumel-protected thermocouples are generally considered satisfactory and economical up to 2100°F.

Platinum–Platinum Rhodium (types R and S). These thermocouples are used for the higher-temperature ranges. They are adversely affected by atmospheres containing reducing gases, and so should always be protected with impervious *sillimanite* tubes when used above 1000°F in the presence of such gases.

Table 4-6 lists chemical and electrical data. Table 4-7 gives ISA recommended symbols. Tables 4-8 and 4-9 list limits of error of thermocouples and extension wires.

[6] Registered trademark of the Hoskins Mfg. Co.

Table 4-7. Recommended Symbols for Types of Thermocouple Wire*

Thermocouple			Thermocouple symbols		
Combination	Positive wire	Negative wire	Combination	Positive	Negative
Iron-constantan (type J)	Iron	Constantan	J	JP	JN
Iron-constantan (type Y)	Iron	Constantan	Y	YP	YN
Chromel-Alumel	Chromel	Alumel	K	KP	KN
Platinum, 10% rhodium-platinum	Platinum, 10% rhodium	Platinum	S	SP	SN
Platinum, 13% rhodium-platinum	Platinum, 13% rhodium	Platinum	R	RP	RN
Copper-constantan	Copper	Constantan	T	TP	TN

* Courtesy of Instrument Society of America.

Table 4-8. Limits of Error* of Thermocouples and Extension Wires for Standard Wire Sizes†

Thermocouples				Extension wires		
Type	Temperature range, °F	Limits of error		Temperature range, °F	Limits of error, °F	
		Standard	Special		Standard	Special
Copper-constantan	−300 to −75 −150 to −75 −75 to +200 200 to 700	— ±2% ±1½°F ±¾%	±1% ±1% ±¾°F ±⅜%	−75 to +200	±1½	±¾
Iron-constantan	0 to 530 530 to 1400	±4°F ±¾%	±2°F ±⅜%	0 to 400	±4	±2
Chromel-Alumel	0 to 530 530 to 2300	±4°F ±¾%	—	0 to 400	±6	—
Platinum rhodium-platinum	0 to 1000 1000 to 2700	±5°F ±0.5%	—	See Table 4-9.		

* The limit of error of a thermocouple (or of thermocouple extension wire) is the maximum deviation in degrees from the standard temperature-emf values for the type of thermocouple in question when the reference-junction temperature of the thermocouple is at a known reference temperature and the measuring junction is at the temperature to be measured.
† Courtesy of Instrument Society of America.

Connection Heads. The purpose of the thermocouple connection or terminal head is to provide facilities for making positive electrical connections between thermocouple and extension wires and to provide a means of attachment for a protecting tube and extension-wire conduit. The head contains a terminal block for all electrical connections. Typical heads include a general-purpose head for most installations, a screw-cover head for applications which must be completely weatherproof, and other connection means.

Selecting Thermocouple-wire Size. There is no general rule that can be applied to the selection of the thermocouple-wire size. Where sensitivity is desired, the smaller-gage wires are suggested. Where longer life is wanted and for the higher temperatures, heavier thermocouple wires (no. 8 gage) are suggested. Table 4-10 gives ISA limits for thermocouples.

Table 4-9. Limits of Error* of Alternate Extension Wires for Standard Wire Sizes†

Type of thermocouple	Type of extension wire	Temperature range, °F	Limits of error, °F
Chromel-Alumel	Copper-constantan	75 to 200	±6
Chromel-Alumel	Iron-alloy	75 to 400	±6
Platinum rhodium–platinum	Copper-alloy	75 to 400	±12

* The limit of error of a thermocouple (or of thermocouple extension wire) is the maximum deviation in degrees from the standard temperature-emf values for the type of thermocouple in question when the reference-junction temperature of the thermocouple is at a known reference temperature and the measuring junction is at the temperature to be measured.

† Courtesy of Instrument Society of America.

Table 4-10. Recommended Wire Sizes for Thermocouples and Extension Wires*

Thermocouples:	
Iron-constantan	8, 14, 20, 24, and 28 AWG†
Chromel-Alumel	8, 14, 20, 24, and 28 AWG†
Copper-constantan	14, 20, 24, and 28 AWG†
Platinum rhodium–platinum	24 AWG† only
Extension wires	14 and 16 AWG†‡

* Courtesy of Instrument Society of America.

† American Wire Gage, also known as Brown & Sharpe (B&S).

‡ These sizes apply to all types of extension wires.

Selecting the Thermocouple Length. Thermocouples must always be of sufficient length to minimize the effect of conduction from the hot end of the thermocouple along the elements and the protecting tube. Insufficient insertion causes low readings. Though no general rule can hold for all conditions, it is suggested that, if possible, the thermocouple be inserted for a minimum distance the equivalent of four times the outside diameter of the protecting tube or well. If the thermocouple must pass through the walls of a muffle furnace or a water-cooled wall, the insertion should be increased up to 10 times the outside tube diameter. It should extend a sufficient distance outside so that the connection-head temperature does not exceed 400°F.

Selecting the Proper Protecting Tube or Well. Although maximum sensitivity is obtained by using a thermocouple without outer protection,

Table 4-11. Metal Protecting Tubes*

Material	Composition of material	Maximum temperature, °F	
		Oxidizing atmosphere	Reducing atmosphere
Carbon steel	Carbon steel	1000	1000
Cast iron	Cast iron	1300	1600
Inconel	14% chromium, 80% nickel	2200	2200
Cast T	15% chromium, 35% nickel	2200	2200
Nickel	Nickel	1800	1800
Resisteat	28% chromium iron	1800	1800
316 stainless	18% chromium, 8% nickel molybdenum	1800	1800
304 stainless	18% chromium, 8% nickel	1800	1800
Wrought iron	Wrought iron	1200	1300
Seamless steel	Steel	1000	1000

* Courtesy of Minneapolis-Honeywell Regulator Co.
Note: For temperatures in excess of 2100°F continuous duty, a ceramic protecting tube is recommended.

Table 4-12. Ceramic Protecting Tubes*

Material	Recommended maximum temperature, °F	Remarks
Sillramic	3000	Excellent mechanical strength and resistance to thermal shock. For oxidizing or reducing atmosphere
Fused silica	2300	Withstands severe thermal shock
Firebrick	2650	Secondary protection for sillimanite tubes
Durax (silicon carbide)	3000	Secondary protection where rapid temperature changes and mechanical shock are likely to be encountered
Silica	2900	Used in glass-tank crowns for secondary protection
Mullite	3000	For secondary protection. Same characteristics as sillimanite
Vycor	1800	Exceptionally stable and resistant to thermal shock

* Courtesy of Minneapolis-Honeywell Regulator Co.

most applications require the use of a protecting tube or well to protect against corrosion or mechanical injury.

Tables 4-11 and 4-12 show tube materials suggested for specific applications. These suggestions must be general and apply to average conditions.

Extension Wire. Extension wire theoretically extends the thermocouple to the reference junction in the instrument. This wire is generally furnished in the form of a matched pair of conductors having insulation designed to meet the service needs of a particular application. Table 4-13 lists ISA recommended symbols for extension wire.

The simplest procedure is to use for extension wire the same types of wire that the thermocouple itself is made of. However, in installations with noble-metal couples where several hundred feet of extension wire must be used, or where numerous couples are employed, the cost of such a procedure may become too expensive. For such cases, alternative, lower-cost materials with similar characteristics at lower temperatures are available. Similarly, for Chromel-Alumel thermocouples, either Chromel-Alumel or iron-Cupronel extension wire is available.

Table 4-13. Recommended Symbols for Types of Extension Wire*

Thermocouple	Extension wire			Extension-wire symbols		
	Combination	Positive wire	Negative wire	Combination	Positive	Negative
Iron-constantan type J†	Iron-constantan	Iron	Constantan	JX	JPX	JNX
Iron-constantan type Y†	Iron-constantan	Iron	Constantan	YX	YPX	YNX
Chromel-Alumel type K	Chromel-Alumel	Chromel	Alumel	KX	KPX	KNX
Chromel-Alumel	Iron-alloy	Iron	Alloy	WX	WPX	WNX
Chromel-Alumel	Copper-constantan	Copper	Constantan	VX	VPX	VNX
Platinum, 10 or 13% rhodium-platinum type S or R	Copper-alloy	Copper	Alloy	SX	SPX	SNX
Copper-constantan type T	Copper-constantan	Copper	Constantan	TX	TPX	TNX

* Courtesy of Instrument Society of America.

† Iron-constantan type J is the calibration most widely used in industry today and is published in Pyrometric Practice, *NBS Tech. Paper 170*, p. 306, Table IV, column L, Feb. 16, 1921, and in "International Critical Tables," vol. I, McGraw-Hill Book Co., Inc., New York, 1926.

Iron-constantan type Y is the calibration indicated in *NBS Research Paper 1080*, Tables 9 and 10, March, 1938.

Thermocouple Installation. Select the location and depth of insertion in the vessel with care, to avoid stagnant areas of the measured fluid which do not have a representative temperature.

Avoid direct flame impingement on the protecting tube because it will materially shorten tube life, and temperature readings will not be representative.

If you make your own thermocouples, clean the ends of the thermocouple wire well before fastening them to the terminal head, and be sure they are inserted with proper polarity as identified on the terminal block.

When measuring high temperatures, install the thermocouple vertically wherever possible to prevent sagging of the tube (Fig. 4-39).

Extension-wire Installation. Use only the correct type of extension wire for a given type of thermocouple (Table 4-14).

Observe the color coding of the individual wires, and connect the negative-wire to the negative-wire terminal at both the thermocouple connection head and the instrument.

Install wires in conduit wherever possible, and ground the conduit well to prevent leakage from power installations or lighting circuits.

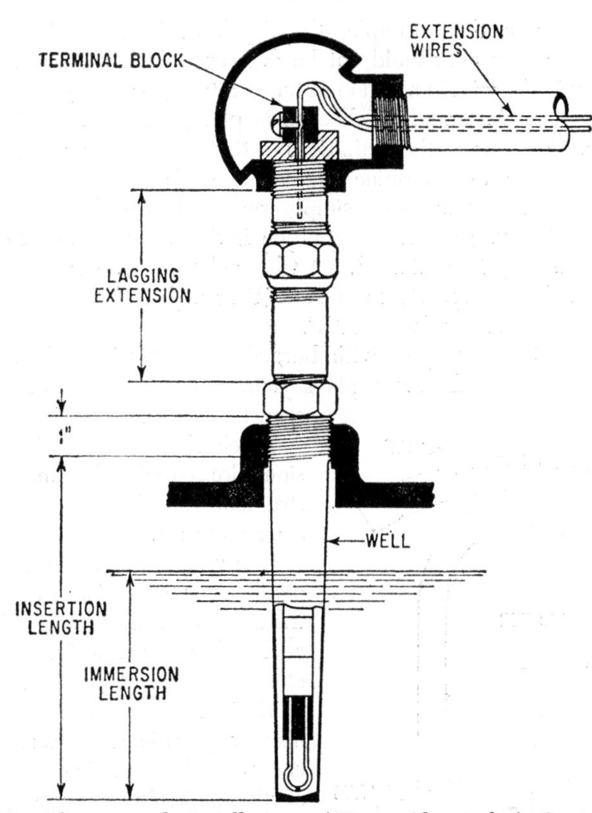

FIG. 4-39. Thermocouple installation. (*Hagan Chemicals & Controls Corp.*)

Keep the wires at least a foot away from any a-c line. If this precaution is not taken, induced alternating current may affect the pyrometer readings. Never run other electrical wires in the same conduit with extension wires.

General Maintenance. Thermocouples should be checked regularly at a rate determined by experience. Once a month is usually sufficient for base-metal thermocouples.

Table 4-14. Recommended Types of Extension Wire*

Thermocouple type	*Recommended type of extension wires*
Iron-constantan type J	Iron-constantan type JX
Iron-constantan type Y	Iron-constantan type YX
Copper-constantan type T	Copper-constantan type TX
Chromel-Alumel type K	Chromel-Alumel type KX†
	Iron-alloy type WX‡
	Copper-constantan type VX§
Platinum rhodium-platinum (type S)	Copper-alloy type SX
Platinum rhodium-platinum (type R)	Copper-alloy type SX

* Courtesy of Instrument Society of America.
† For maximum accuracy.
‡ For use up to 400°F (more economical but introduces some error).
§ Should not be used where head temperature exceeds 200°F.

In reinserting a thermocouple, the depth of insertion should not be changed and, above all, should not be decreased (errors are caused by inhomogeneity in the wire in a region of temperature gradients).

The thermocouple should be checked in place, if possible—not disturbed in any way. Because of the limitations of this procedure, many plants have conducted surveys to determine the average life of their thermocouples in various applications, and they simply replace the couples after the established proper interval; optimum operation is thus assured, and the problem of periodic checking is eliminated. A Chromel-Alumel thermocouple should not be exposed to temperatures of 1600°F or higher if it is to be used for accurate measurements below 1000°F.

Do not use thermocouples with burned-out protecting tubes. Replace the tubes before the thermocouple is damaged, or the thermocouple will soon be useless.

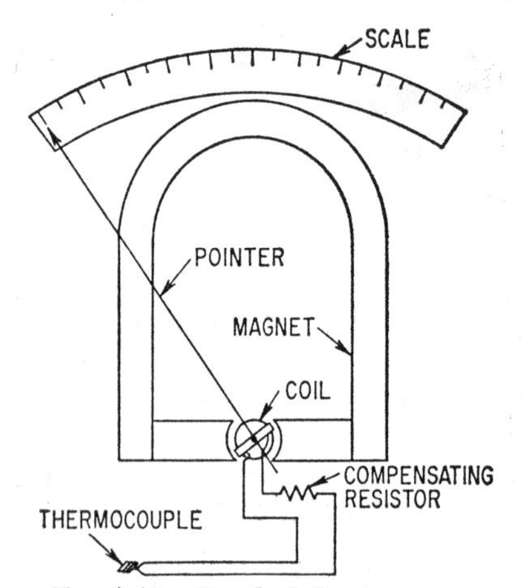

Fig. 4-40. Simple deflection pyrometer. (*Minneapolis-Honeywell Regulator Co.*)

To check for grounds in the extension-wire circuit, disconnect the instrument and thermocouple, and use a magneto set.

If pyrometers are connected in parallel for operation from a single thermocouple, the circuit should be analyzed for possible feedback effects of one instrument on the other.

Millivoltmeter Pyrometers. For many applications, the sensitivity, accuracy, and automatic-control features of the millivoltmeter are quite suitable. Indicating-controlling millivoltmeters are used frequently as excess-temperature alarms and cutoffs on nearly all types of heating equipment.

The basic components of a millivoltmeter pyrometer, sometimes referred to as a simple deflection-type pyrometer, are illustrated schematically in Fig. 4-40. The instrument is essentially a d'Arsonval galvanometer. A magnetic field, set up by the permanent magnet and pole pieces, surrounds a coil which is suspended by pivots and jeweled bearings. An indicating pointer is attached to this coil. Electrical current from the thermocouple passes through the coil and sets up an opposing magnetic field. This action causes the coil to turn and the pointer to move across the scale. To retard its movement and to return it to zero when no thermocouple current is passing through the coil, the coil and pointer deflect against hairsprings. A bimetal spiral attached to the hairspring mounting provides reference-junction compensation.

Potentiometer Pyrometers. A common type of potentiometer pyrometer is illustrated schematically in Fig. 4-41. This principle of potentiometric measurement has brought about higher-speed operation, greater sensitivity and accuracy of measurement, and better reproducibility. Consequently, plant operations can be performed under closer supervision and control.

The unbalance between the d-c millivoltage developed by the thermocouple and that from a standard voltage source is changed to an a-c voltage of proportional magnitude in a converter-transformer. The converter is essentially a flat metal reed oscillating between two contacts connected to the opposite ends of the primary winding on an input transformer. The unbalanced d-c voltage is impressed across the converter and the center tap of the primary winding on the input transformer. As the reed moves from one contact to the other, any unbalanced d-c voltage will cause direct current to flow first in one direction through one-half the primary winding, and then in the opposite direction through the other half. This action generates an alternating flux in the input transformer core, which

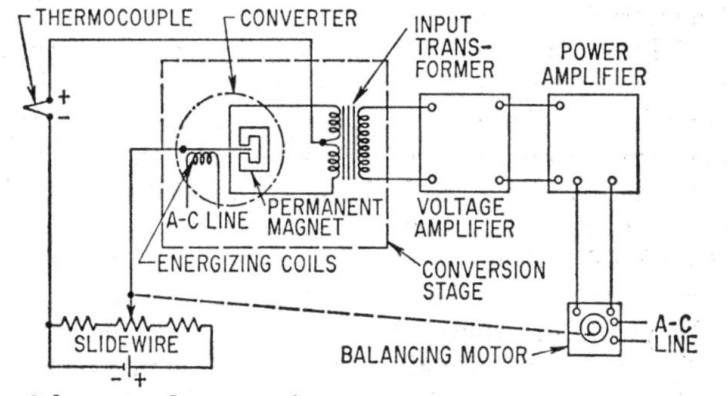

FIG. 4-41. Schematic diagram of potentiometer pyrometer system. (*Minneapolis-Honeywell Regulator Co.*)

in turn induces an alternating voltage in the transformer secondary winding.

The action of the converter is related to the a-c supply voltage by the energizing coil, which is excited by the a-c supply voltage through a step-down transformer. The reed is polarized by a permanent magnet and, therefore, is actuated by the energizing coil to oscillate in synchronism with the a-c supply voltage. This reed closes one contact to the transformer for one-half the supply-voltage cycle and the other contact for the other half, making one complete oscillation for each cycle. The direct current flowing in each half of the transformer primary winding, therefore, will create an alternating current in the transformer secondary winding of the same frequency as the supply current. This alternating current is amplified so that it will actuate a two-field balancing motor.

If there is an unbalance between the slidewire voltage and the thermocouple voltage, the balancing motor functions to rebalance the two voltages by moving the contactor on the slidewire, at the same time moving the instrument pen or pointer to a new temperature value. The direction in which the motor rotates, determined by a definite phase relationship, depends, of course, on whether the measured variable is higher or lower than the value indicated by the instrument. When no unbalance exists, there is no movement of instrument pen or pointer.

SUPERHEATER-TUBE-TEMPERATURE MEASUREMENTS[7]

The primary value of superheater-tube-temperature measurements is the knowledge gained of temperature distribution across the boiler. This information serves as a check on circulation through the tubes so that safe metal temperature will not be exceeded. Measurements may be taken by thermocouples installed on individual tubes near the superheater outlet header. These measurements not only assist in determining and testing the design of the unit and the proper location of soot blowers, but are also important operating guides in the use of blower equipment, and in determining the most suitable method of firing to maintain proper temperature distribution across the boiler.

Additional benefits may be obtained from these measurements. During startup, tube temperatures indicate when water in pendent-type superheater or reheater tubes has boiled out. There has been at least one instance where tube-temperature measurements have indicated a flow restriction in one of the tubes. On some boiler designs, tube-temperature measurements indicate excessive carry-over of water from the drums or excessive moisture introduced in desuperheating.

If measurements are only for test purposes of short duration, protection need not be provided. However, it has been found desirable on many units recently installed or currently being designed to provide tube-temperature-measuring equipment which can be used during operation of the boiler and which, therefore, must have long life.

The only satisfactory means of protecting the thermocouples from the hot products of combustion is to seal the couple from these gases. The couple must be protected not only within the combustion chamber, but also in the chamber between the furnace roof and boiler casing, if there is any possibility that products of combustion may enter this section. This would be expected on a positive-pressure boiler. In addition, products of combustion may enter this chamber even on a negative-pressure boiler. The furnace roof is not tight, and a flow of combustion gases will take place in this chamber between any two openings where a pressure differential exists.

If the tube-temperature measurements at the reheater outlet or superheater outlet are made in a section of the boiler-roof vestibule in which no products of combustion will be present, protection from the atmosphere is not necessary, and unprotected couples may be used. Figure 4-42 illustrates one method of attaching a suitable couple to a tube where unprotected couples will be satisfactory. The thermocouple is made up of 14-gage Chromel and Alumel thermocouple wire in two-hole porcelain

[7] T. W. Jenkins, Jr., Leeds & Northrup Co.

insulators 1 in. long. Note that the measuring junction is welded in a flattened section of steel tubing, which, in turn, is welded to the superheater tube. The welding ensures good thermal contact between the thermocouple and the tube. To minimize conduction errors, it is good practice to fasten the couple to the tube with asbestos tape reinforced with iron wire or Nichrome wire for 12 to 18 in. from the measuring junction. It is expected that the atmosphere will consist of reasonably still air and that the air temperature will approach that of the tubes, thus making further precautions for minimizing conduction errors unnecessary. Mechanical

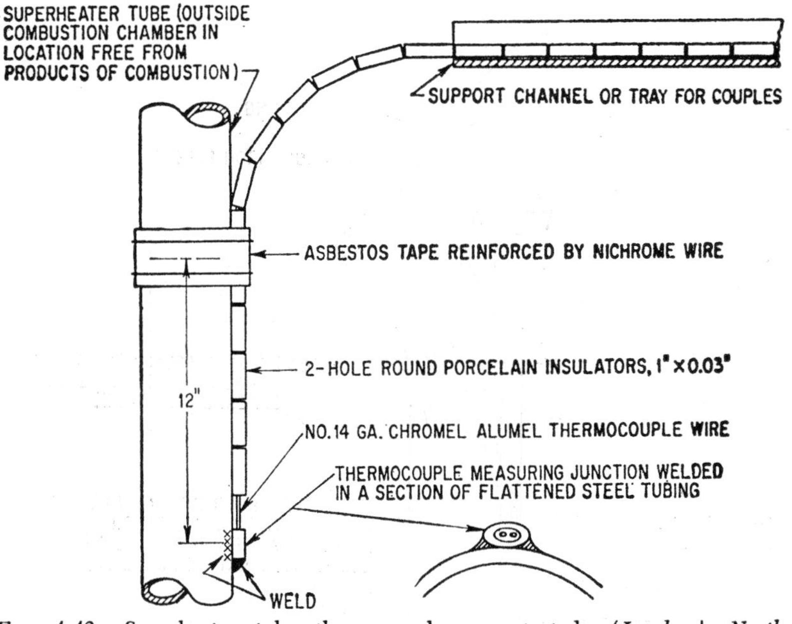

FIG. 4-42. Superheater tube thermocouple, unprotected. (*Leeds & Northrup Co.*)

support of the thermocouples is required. Fourteen-gage couples are used as a compromise between the small size to minimize conduction errors and a large size to provide satisfactory mechanical strength. The thermocouples may be supported individually, or a number of couples may be placed in a supporting channel or tray and brought through the boiler casing to suitable terminals. If no sharp bends are to be made in the thermocouple (less than a 15-in. radius), the 1-in.-long two-hole insulators will be suitable for the entire length of the thermocouple. However, if short-radius bends are to be made, fish-spine insulators should be used on each thermocouple lead where the bend occurs.

Twenty-gage glass-insulated wire has been used for this type of installation. Where temperatures will exceed 1000°F, the glass-insulated 20-gage wire will not have so long a life expectancy as the thermocouple illustrated in Fig. 4-42.

Chromel-Alumel thermocouples are preferred for this application over iron-constantan couples. For a given wire size, Chromel-Alumel couples can withstand a higher temperature than iron-constantan. In general for 14-gage protected couples, the maximum temperature for Chromel-Alumel is 2000°F, and for iron-constantan only 1100°F. For a 14-gage unprotected couple, the maximum for Chromel-Alumel is 1700°F and for iron-constantan approximately 900°F.

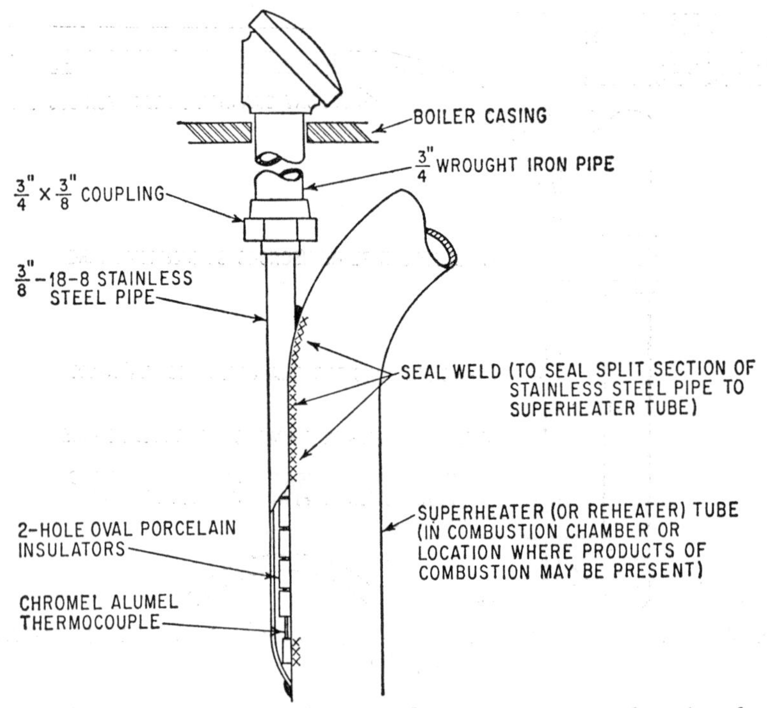

Fig. 4-43. Superheater tube thermocouple with protection tube. (*Leeds & Northrup Co.*)

This method of measurement is not suitable if products of combustion are present in the atmosphere. Contamination of the couple will occur, causing a change in calibration. Under these conditions, protection should be provided similar to that required for couples measuring tube temperatures within the combustion chamber of the furnace.

Figure 4-43 illustrates one method of making a measurement under these conditions. The thermocouple without the protection tube is basically the same as that illustrated in Fig. 4-42. However, protection has been added to seal out the products of combustion. Note that the protection tube has been ground off on one side and welded to the superheater (or reheater) tube. This provides good thermal contact between the protection tube and the superheater tube, thus bringing the protection-tube temperature and thermocouple-lead temperature close to the superheater-tube temperature. This minimizes the conduction error. The mass of the protection tube is relatively small so that the temperature of the superheater tube is not appreciably changed by the addition of this couple. The protection tube is ⅜-in. 18-8 stainless-steel pipe where it is exposed to the flow of the hot products of combustion. However, in the roof vestibule, the atmosphere temperature is considerably lower and ¾-in. wrought-iron pipe is suitable. The basic purpose of the protection tube is to protect the thermocouple from contamination by the hot products of combustion. The protection will be no better than the seal provided by the seal weld. In

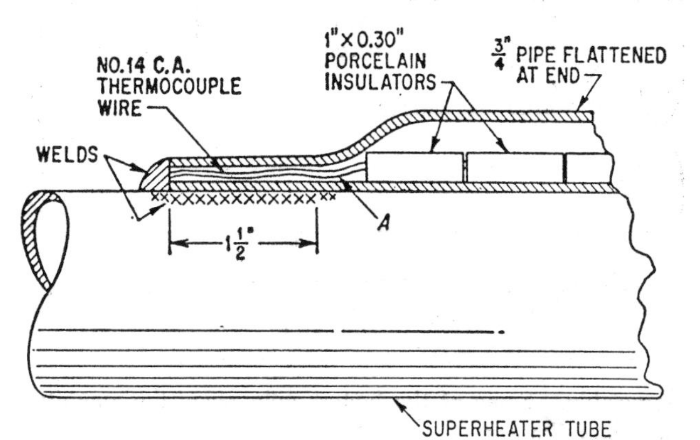

FIG. 4-44. Protected superheater tube thermocouple, alternate arrangement. (*Leeds & Northrup Co.*)

general, it is most suitable to grind the protection tube in the field to suit the contour of the superheater or reheater tube.

Figures 4-44 and 4-45 illustrate alternative methods which have been used for making this measurement. Each of these two methods enables the use of a simpler weld to seal out the hot products of combustion. However, each has the limitation that it does not provide such good thermal contact between the protection tube and the superheater-tube element as the arrangement shown in Fig. 4-43. It is expected that the conduction errors will be greater. Most of the heat picked up from the hot gases by the thermocouple protection tube is conducted to the superheater element in the vicinity of the measuring junction of the thermocouple, thus raising the temperature at this point. In the arrangement shown in Fig. 4-44, the actual measuring junction is at point *A*. The conduction error could be reduced by the use of a ⅜-in. pipe protection tube instead of ¾-in., and by extending the weld of the protection tube to the superheater further along the tube. Although the thermocouple protection tube and the superheater element may apparently be in close contact, actually the contact is only a series of point contacts, or line contacts, and is not efficient from a heat-transfer standpoint. Welding one to another provides good thermal contact.

The conduction errors cannot be accurately estimated and may or may

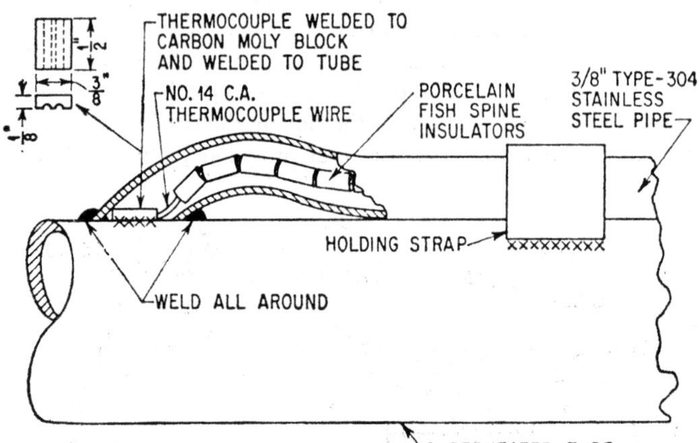

FIG. 4-45. Protected superheater tube thermocouple, alternate arrangement. (*Leeds & Northrup Co.*)

not be appreciable. Some of the factors influencing the magnitude of this error include the following:

1. The temperature difference between the hot products of combustion and superheater element
2. The mass of the protection tube relative to that of the superheater element
3. The rate of heat transfer from the hot gases to the thermocouple protection tube
4. The velocity of the hot gases
5. The velocity of the steam flow through the superheater element

In the face of so many variables, it is felt that steps which can be readily taken to minimize the conduction errors are justified.

One of the major limitations of this type of measuring element is the difficulty of replacing the element. This limitation does not apply to the unprotected thermocouples illustrated in Fig. 4-42. Of course, where replacement presents a problem, the obvious answer is to design the thermocouple and its installation so that reasonable life of the thermocouple may be expected and replacement becomes unnecessary.

In this connection, protection from contamination by the hot products of combustion has been discussed. In addition, the thermocouples should be protected from excessively high temperatures which would exist if the thermocouple and its protection tube extended for any appreciable distance through the hot products of combustion.

A 14-gage Chromel-Alumel couple would not be expected to last long at temperatures greater than about 2000°F. Accordingly, it is good practice to keep the thermocouple and its protection tube adjacent to, and in contact with, the superheater-tube element where this thermocouple is brought

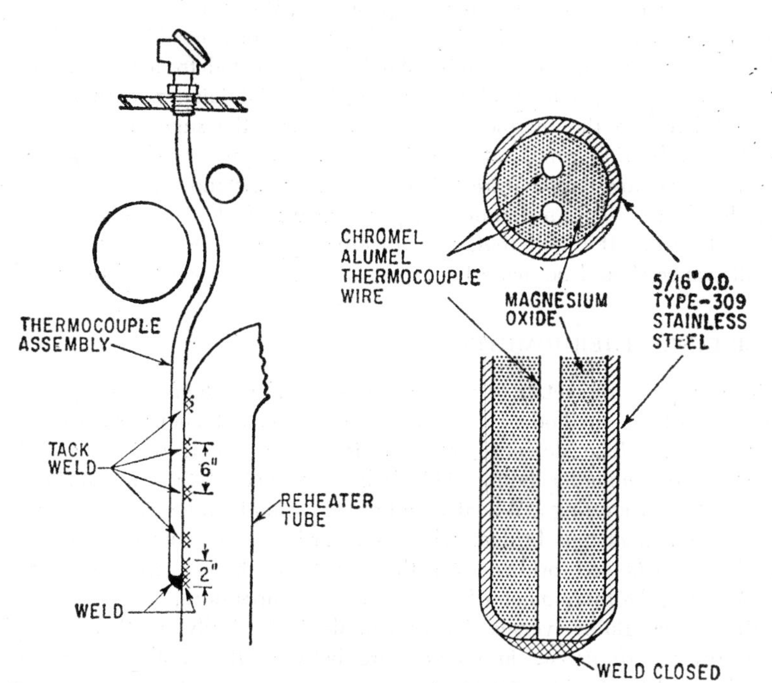

FIG. 4-46. Superheater tube thermocouple, swaged type. (*Leeds & Northrup Co.*)

through the hot products of combustion. This will provide a "cooling" of the thermocouple by the superheater tube.

Figure 4-46 illustrates a swaged type of thermocouple. The particular assembly shown consists of a Chromel-Alumel thermocouple encased in magnesium oxide enclosed in a 5⁄16-in. OD type 309 stainless-steel sheath. The measuring junction is welded to the closed end of the sheath. This swaged type of construction is similar to that employed in the electric-heating elements of domestic electric stoves. This type of construction provides a thermocouple element that is sufficiently flexible so that it may be bent around obstructions. It should be noted that the element is welded to the superheater tube for a distance of approximately 2 in. from the measuring junction, and tack-welded at approximately 6-in. intervals along the rest of the superheater tube where the thermocouple element is maintained in close proximity with the tube to prevent overheating from the hot gases. The field welding has no bearing on the efficiency of the seal of the thermocouple against the effects of the hot products of combustion. Thus, the technique of the field welding should not affect thermocouple life.

Swaged-type thermocouples can be provided in diameters larger and smaller than the 5/16-in. OD shown. However, this size is felt to be a reasonable compromise, considering conduction errors, flexibility, and long life. The thermocouple assembly should be protected from the high temperatures of the products of combustion by maintaining reasonably close contact between the thermocouple assembly and the superheater element in the combustion zone. It is estimated that approximately 1500°F will be the maximum temperature this couple can withstand with a reasonable length of life. This maximum temperature would depend on the size of element used. It is felt that, as with other types of elements, some slack or allowance should be made for expansion between supports of the couple.

RESISTANCE THERMOMETERS

The property of metals to increase in electrical resistance as temperature rises provides a method of temperature measurement known as *resistance thermometry*. The detecting element is a wire-wound resistor called a *resistance-thermometer* bulb. This bulb is connected to a measuring instrument incorporating a Wheatstone bridge. Simple indicating instruments employ a *deflectional* bridge, whereas recorders and controllers generally employ a *balanced* bridge. All these instruments interpret changes in resistance at the thermometer bulb in terms of temperature.

Resistance Bulbs versus Thermocouples.[8] Basically a thermocouple measures the *difference* in temperature between the "hot" or measuring junction and the "cold" or reference junction. For this difference to represent an accurate measure of the temperature at the measuring junction, the reference junction must be maintained at a constant temperature. This presents a problem because the reference junction is actually every electrical connection in the instrument-measuring circuit, and these connections are seldom at the same temperature owing to temperature variations within the instrument, and seldom remain at constant temperature because of ambient temperature changes. The usual method of solving this problem in commercial instruments is to group all the reference junctions as close together as possible in an enclosure protected against drafts and to install within this enclosure a temperature-sensitive element adjusted to compensate automatically for the effects of ambient temperature.

The accuracy of this method of maintaining constant reference-junction temperature is limited by the faithfulness with which the compensating

[8] Foxboro Co., Inc.

element can reproduce the temperature of each and every junction (electrical connection) within the enclosure. Although theoretically such compensation may be perfect, in actual practice it is not possible to maintain all the junctions at precisely the same temperature, especially under conditions of changing ambient temperature. For this reason, commercial

Table 4-15. Resistance versus Temperature of Various Metals

Metal	Resistivity, microhm-cm	Relative resistance R_t/R_0 at 0°C											
		−200	−100	0	100	200	300	400	500	600	700	800	900
Alumel....	28.1	.....	.....	1.000	1.239	1.428	1.537	1.637	1.726	1.814	1.899	1.982	2.066
Copper....	1.56	0.117	0.557	1.000	1.431	1.862	2.299	2.747	3.210	3.695	4.208	4.752	5.334
Iron......	8.57	.....	.....	1.000	1.650	2.464	3.485	4.716	6.162	7.839	9.790	12.009	12.790
Nickel....	6.38	.....	.....	1.000	1.663	2.501	3.611	4.847	5.398	5.882	6.327	6.751	7.156
Platinum..	9.83	0.177	0.599	1.000	1.392	1.773	2.142	2.499	2.844	3.178	3.500	3.810	4.109
Silver.....	1.50	0.176	0.596	1.000	1.408	1.827	2.256	2.698	3.150	3.616	4.094	4.586	5.091

thermocouple instruments can be subject to errors of as much as 2°F or more, depending on the amount and speed of ambient temperature changes. Errors of this magnitude are unimportant—in fact are not readable—in the higher-temperature ranges, but, in the lower-temperature ranges, they may become a serious percentage of the total scale reading.

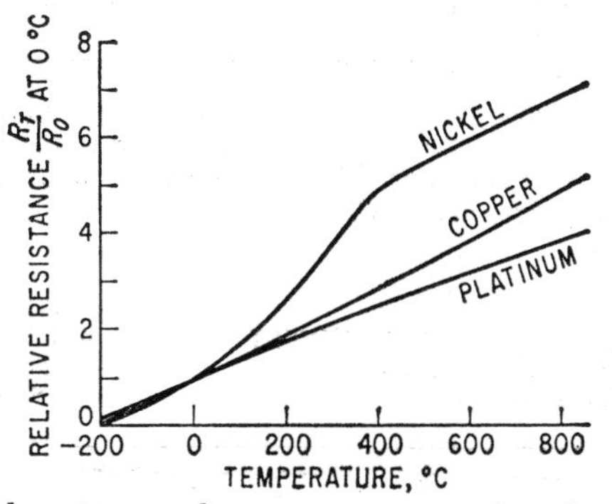

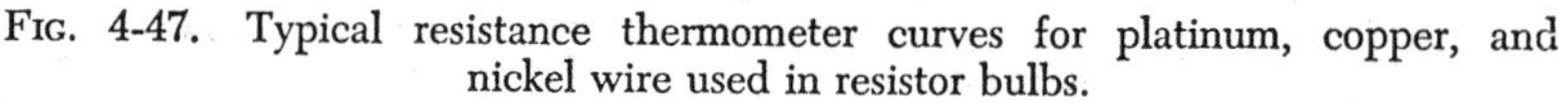
FIG. 4-47. Typical resistance thermometer curves for platinum, copper, and nickel wire used in resistor bulbs.

A resistance bulb, basically, measures temperature *directly* in that the resistance of the coil of wire within the bulb is a direct function of its temperature (Table 4-15). The accuracy of this measurement is entirely unaffected by the ambient temperature to which the instrument is exposed; consequently no compensation is needed.

Resistance bulbs also have the advantage of greater sensitivity because the change of resistance per degree Fahrenheit in a resistance bulb is a much larger quantity—hence is more easily measured—than the microscopic change of voltage per degree Fahrenheit in a thermocouple.

Resistance bulbs have a maximum temperature limit, however, above which they cannot be used. Also when resistance bulbs and thermocouples

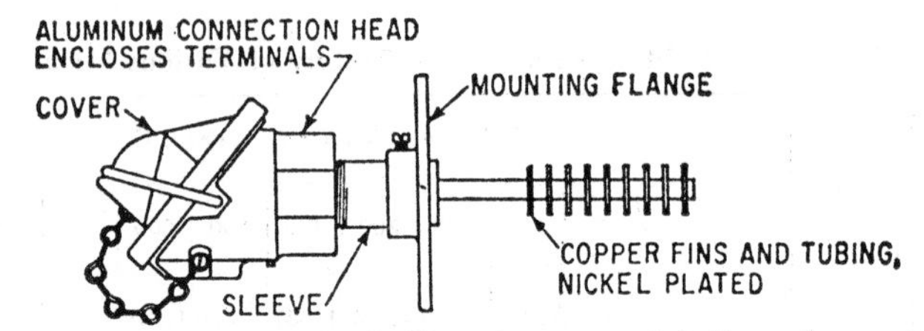

FIG. 4-48. Duct-type sensing bulb. This type of bulb is designed for temperature measurements in ducts and pipes where protection from moisture and fumes is needed. The unit is complete with weatherproof head and "fins." (*Wheelco Instruments Div., Barber Coleman Co.*)

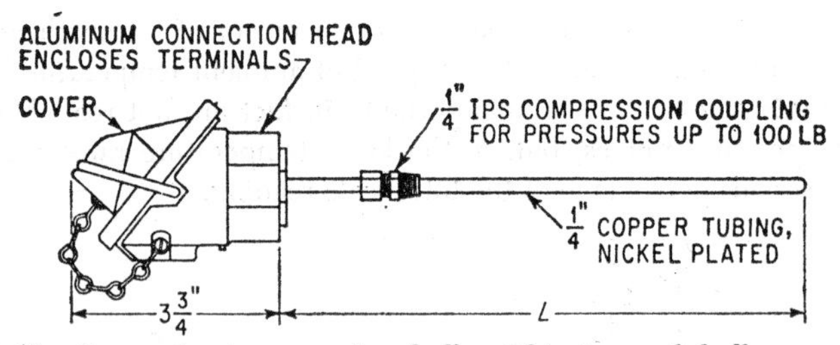

FIG. 4-49. Immersion-type sensing bulb. This type of bulb may be used for pressures up to approximately 100 lb. Adjustable compression coupling may be located where desired. For higher pressures, wells are recommended. (*Wheelco Instruments Div., Barber Coleman Co.*)

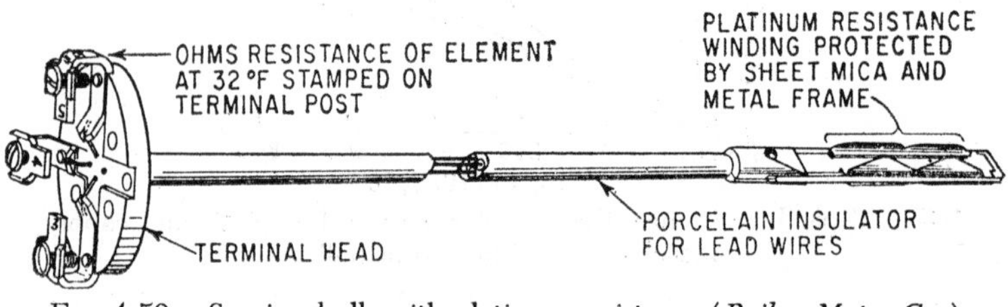

FIG. 4-50. Sensing bulb with platinum resistor. (*Bailey Meter Co.*)

are compared for use in a protective well, conventional resistance bulbs are at a disadvantage because they are larger (hence more difficult to install) and somewhat slower in response. For these reasons, resistance bulbs should always be used, in preference to thermocouples, up to the temperature limit of resistance-bulb operation. Above this limit, in the higher-temperature ranges, thermocouples are the better means of temperature measurement.

Bulbs and Wells. Resistance bulbs and wells are available in a variety of sizes, shapes, materials, insulations, and winding types. Resistance characteristics for three most common resistance-thermometer materials are shown in Fig. 4-47. Nickel, despite some limitations, is most frequently used. As shown by the curve (Fig. 4-47), nickel's temperature-resistance characteristics are nonlinear.

Notwithstanding cost, platinum is more suitable than either nickel or copper for resistance-thermometer applications. But its use is generally confined to jobs that cannot be properly handled by resistance nickel or copper.

A number of winding shapes are commercially available that can be expected to give reliable performance: woven wire-mesh cloth, strips of thin nickel foil, helical wire coil mounted in a thin-wall insulated metal or ceramic tube, two narrow mica strips that support a winding of straight bare wire, spaced winding of bare wire over a round insulating arbor. Construction of sensing bulbs is shown in Figs. 4-48 through 4-51.

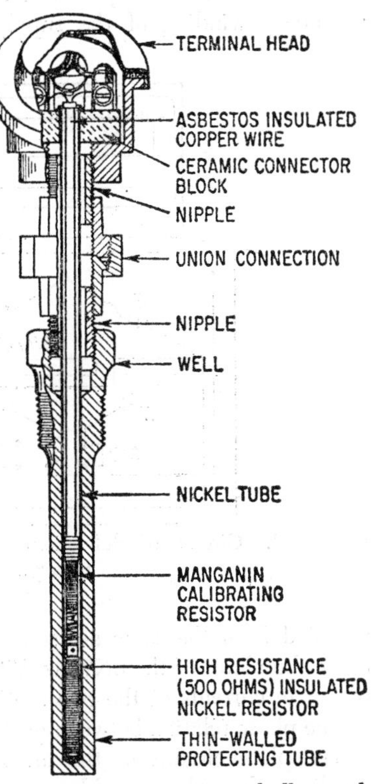

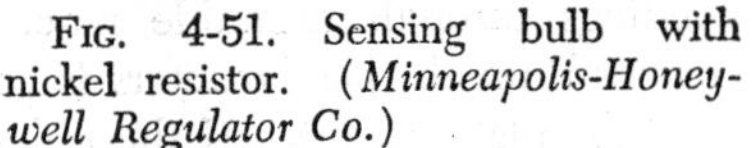
FIG. 4-51. Sensing bulb with nickel resistor. (*Minneapolis-Honeywell Regulator Co.*)

Measuring Circuits. Deflectional Wheatstone Bridge.[9] In the circuit of the deflectional-type indicator (Fig. 4-52), points A, B, and C represent the terminals of the bulb located at the point of temperature measurement. D and E are ratio arms of equal resistance. P is a fixed resistance equal to the resistance of the bulb corresponding to the highest-temperature reading on the indicator scale. X is a fixed resistance equal to the resistance of the bulb corresponding to the lowest-temperature reading on the scale. BAT is the battery which energizes the circuit, and RH is the rheostat for adjusting the battery current to the correct value. When the battery current is adjusted, the switch is thrown to the side marked STD to include resistor X in the bridge circuit. The rheostat is then adjusted until the indicator deflects to the lowest-temperature reading on the scale. After the current is adjusted,

[9] Minneapolis-Honeywell Regulator Co.

the switch is thrown to the BULB position. The galvanometer pointer deflects to full scale on open circuit.

Balanced Bridge.[10] Figure 4-53 shows the complete schematic circuit. The unbalanced voltage of the bridge is proportional to the magnitude of the change in resistance of *T* produced by a change in temperature. Its phase relative to the supply voltage depends on the direction of unbalance. This bridge output signal is amplified by a two-stage resistance-coupled amplifier having a voltage gain of about 2,500. The amplified signal is applied to the grids of the 6N7 double-triode motor-control tube. The voltage applied to the plates of this tube is obtained from a center-tapped secondary winding of the supply transformer whose primary winding is

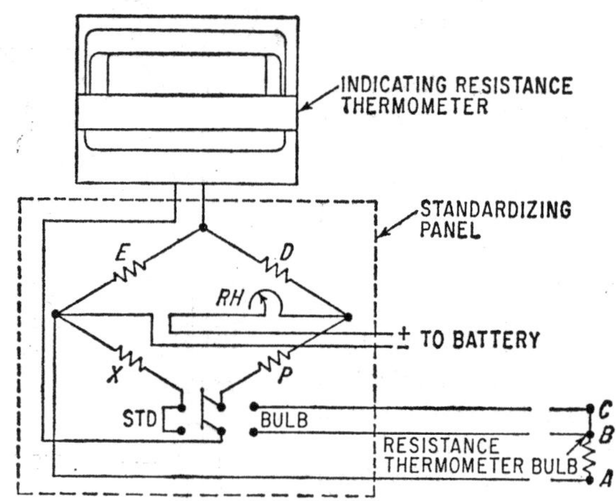

FIG. 4-52. Circuit of deflectional-type resistance thermometer. (*Minneapolis-Honeywell Regulator Co.*)

supplied from the same a-c source as the supply to the measuring circuit. Since the plates of the 6N7 are 180° out of phase and the grids are in phase, the plate currents of the two sections of this tube selectively are controlled by the phase relation between grid and plate voltages.

The plate circuits of the motor-control tube are coupled to the motor windings through saturable reactors *A* and *B*. The flow of alternating current through these reactors is controlled by the direct current flowing in the tube plate circuit. If the unbalance of the measuring circuit is such as to give an amplifier voltage in phase with the plate of triode *A*, this section will pass more current and triode *B* less current than at balance. Consequently, motor current will flow through reactor *A*, directly through winding W_A, and through the motor capacitor and winding W_B. This phase relationship of winding currents will cause the motor to rotate in a direction to rebalance the measuring circuit. By the same analysis, unbalance in the

[10] Bailey Meter Co.

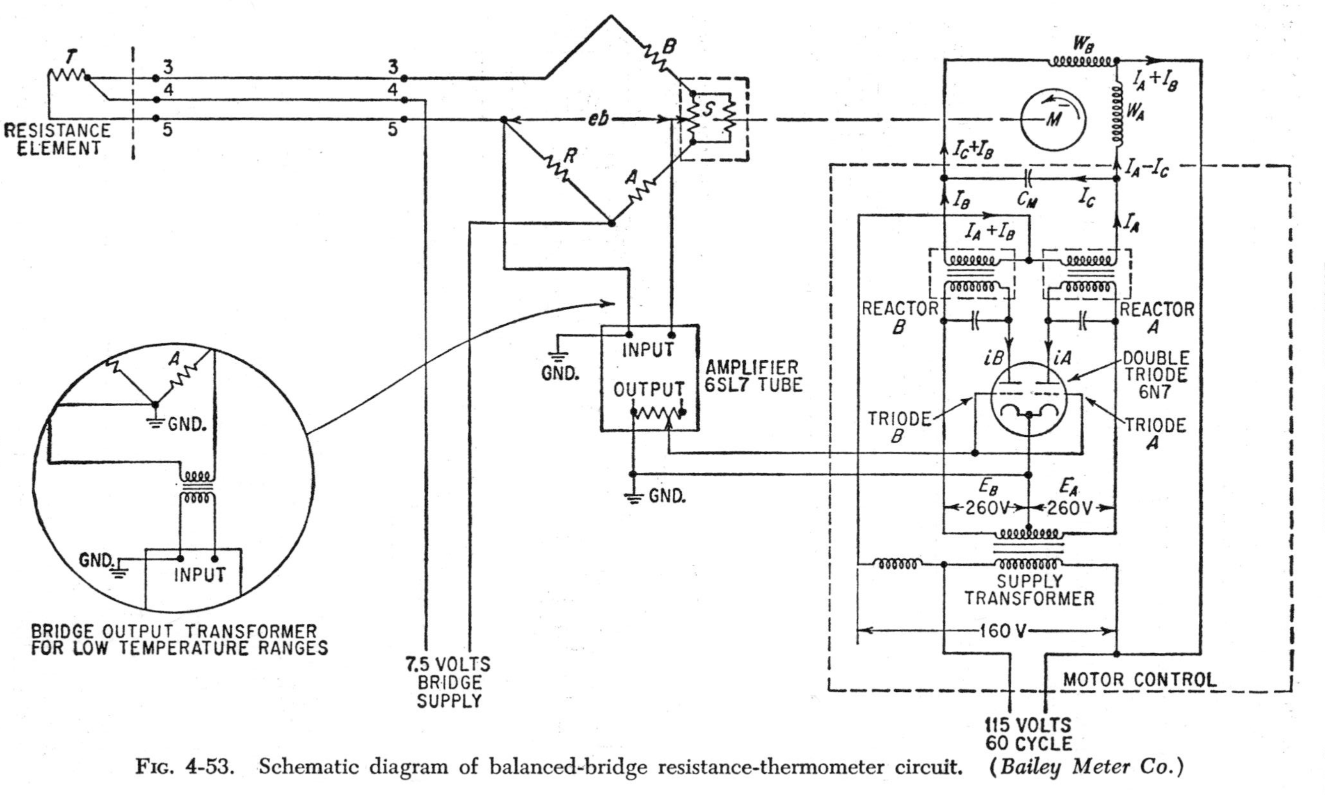

FIG. 4-53. Schematic diagram of balanced-bridge resistance-thermometer circuit. (*Bailey Meter Co.*)

opposite direction will produce an amplified voltage in phase with triode B and a resulting motor rotation opposite to what it was before, but again in a direction to restore balance.

RADIATION PYROMETERS

The industrial radiation pyrometer is a practical application of the Stefan-Boltzmann law of radiant energy, which states that the intensity of radiant energy emitted from the surface of a body increases proportionately to the

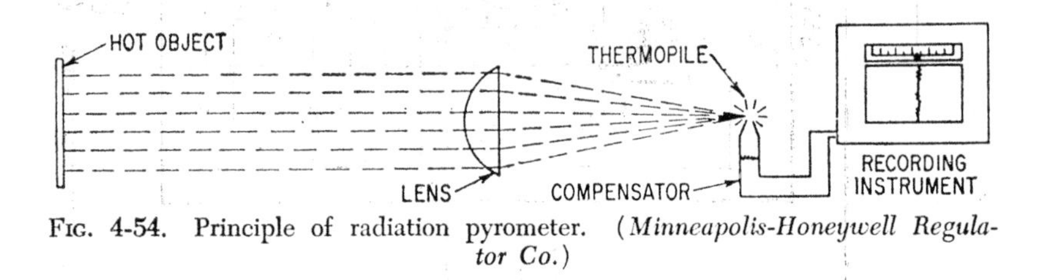

FIG. 4-54. Principle of radiation pyrometer. (*Minneapolis-Honeywell Regulator Co.*)

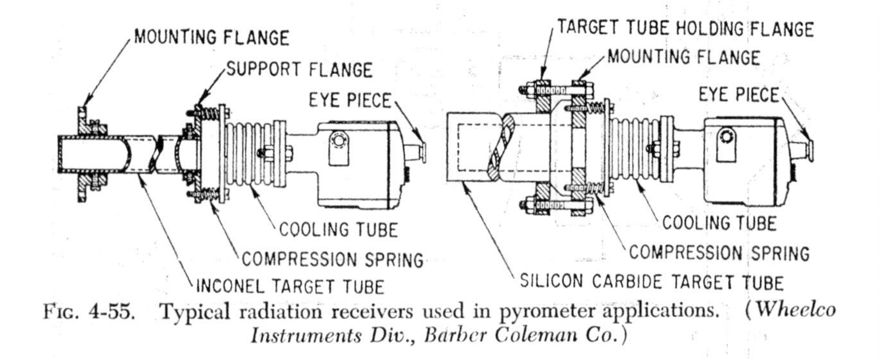

FIG. 4-55. Typical radiation receivers used in pyrometer applications. (*Wheelco Instruments Div., Barber Coleman Co.*)

fourth power of the absolute temperature of the body. As illustrated in Fig. 4-54, energy from the target—a portion of the object whose temperature is being measured—is focused on a thermopile (a number of small thermocouples connected in series) by the pyrometer lens. This thermopile generates an emf proportional to the amount of energy falling upon it; so the emf is also proportional to the temperature of the target. This emf is fed to a millivoltmeter or potentiometer which indicates, records, and controls temperatures in the same manner as the conventional thermocouple-type pyrometer.

Applications. For a number of common temperature-measuring problems, the radiation pyrometer possesses many advantages over the thermocouple pyrometer. The radiation pyrometer is ideally suited for the following types of measurement:

1. Where very high temperatures are involved, temperatures definitely beyond the practical range for thermocouple measurements
2. Where furnace atmospheres are detrimental to thermocouples and cause erratic measurement and short life (Fig. 4-56)
3. Where, for other reasons, it is impractical to contact the material whose temperature is to be measured

The radiation pyrometer excels in these applications because it measures without coming into physical contact with the temperature source. The

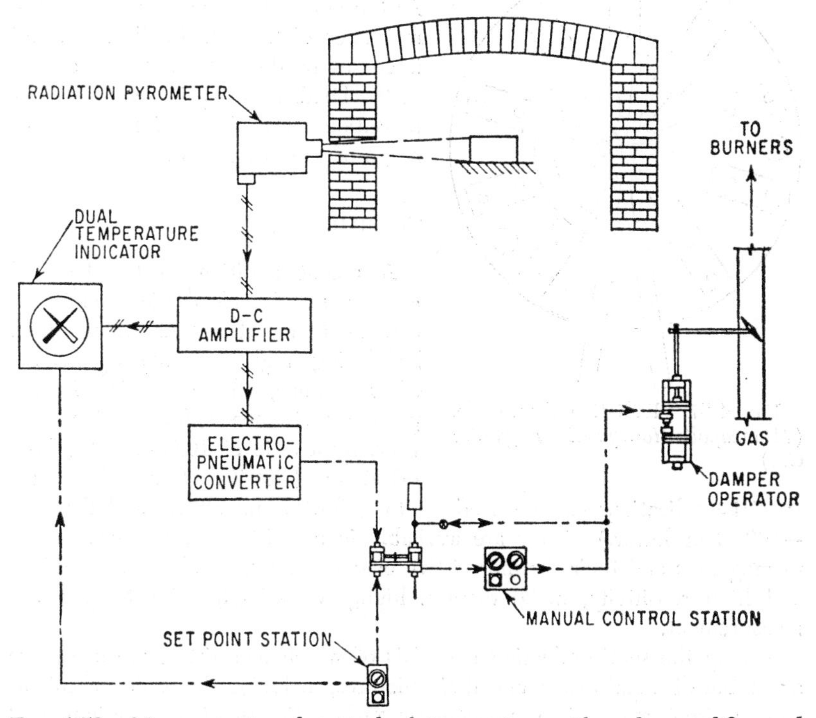

FIG. 4-56. Measurement and control of temperature with a d-c amplifier and radiation pyrometer. (*Hagan Chemicals & Controls, Inc.*)

measuring unit can be sighted directly on an object considerably distant from the unit. Where it is necessary to protect the lens and measuring unit from direct impingement of flames, a closed-end target tube can be used.

Where ambient temperatures are relatively high, the pyrometer is commercially available with an air-cooled or water-cooled fitting, which prevents conduction of heat from the furnace walls and, where an open-end sighting tube is used, provides means for purging with compressed air to prevent dirt or smoke from obscuring the lens. A safety shutter can be included to protect the radiation-measuring element from flame damage.

Thermopile. The temperature-sensing element in one commonly used pyrometer is a thermopile (Fig. 4-57), which is a group of very small thermocouples connected in series much like the cells of a storage battery, so that the emf output of the thermocouples is additive. The tiny thermocouple junctions, about pin-point size, are flattened and blackened so that they absorb all the radiant energy reaching the thermopile. To compensate for the effects of varying ambient temperatures upon the thermopile, a nickel resistance spool provides a variable shunt across the emf produced. As the ambient temperature varies, the resistance of the nickel coil varies, and increases or decreases the emf output of the head, resulting in accurate compensation over the entire measuring range of the instrument.

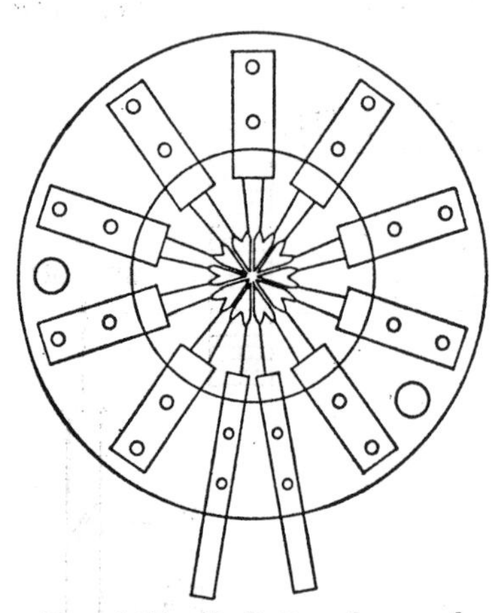

FIG. 4-57. Radiation thermopile. (*Minneapolis-Honeywell Regulator Co.*)

THERMISTORS

A recent addition to the field of components for electronic circuitry is ceramic temperature-sensitive resistors. Exhibiting high negative coefficients of resistance, these semiconductors possess resistance values which may vary, by a ratio of 10,000,000:1, from −100 to +450°C. (Thermistors for special applications can be used up to +500°C and above, and down to −180°C or lower.) They are available in a resistance range from ohms to megohms and their thermoresistive characteristics, coupled with stability and high sensitivity, make them a highly versatile tool for temperature measurement.

Among the semiconducting materials of which thermistors are made are a number of metal oxides and their mixtures, including the oxides of cobalt, copper, iron, magnesium, manganese, nickel, tin, titanium, uranium, and zinc. The oxides, usually compressed into the desired shape from powders, are heat-treated to recrystallize them, resulting in a dense ceramic body. Electric contact is made by various means—wires imbedded before firing the material, plating, or metal-ceramic coatings baked on.

Forms in general use are beads as small as 0.015 in. in diameter, disks ranging from 0.2 to 1.0 in. in diameter, and rods from 0.03 to 0.25 in. in diameter and up to 2 in. in length. Flakes a few microns thick are employed as infrared-radiation detectors or bolometers. Various methods of mounting are used. Beads are suspended from wire leads or imbedded in probes; disks are mounted in spring-loaded stacks with or without heat-dissipating fins; other disks and rods are pigtail-mounted. Beads and small disks may be covered with a thin adherent coat of glass to reduce composition changes of the thermistor at high temperature. They may be mounted by their leads in evacuated or gas-filled bulbs to minimize or control the degree of thermal coupling to the surroundings. Some types have associated heaters to control the thermistor resistance for various purposes (Fig. 4-58).

Figure 4-59 shows how the logarithm of the specific resistance varies with temperature for three typical thermistor materials and for a metal. The specific resistance of the thermistor represented by curve 1 decreases by a

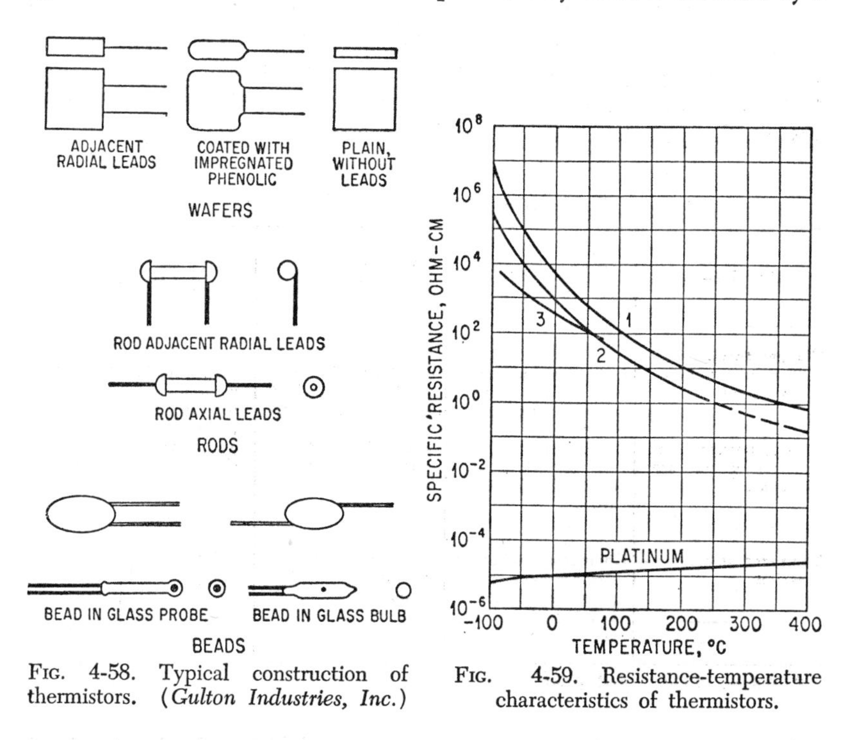

FIG. 4-58. Typical construction of thermistors. (*Gulton Industries, Inc.*)

FIG. 4-59. Resistance-temperature characteristics of thermistors.

factor of 50 as the temperature is increased from 0 to 100°C. Over this same temperature range, the resistivity of a typical metal will increase by a factor of 1.4.

Terminal Resistance. Thermistors range in terminal resistance at room temperature from about 1 ohm to the order of 10^8 ohms, depending on composition, shape, and size. Within a given type, they commonly vary 10 to 30 per cent in resistance from the nominal at reference temperature. Some types can be adjusted mechanically by grinding off part of the contact area to bring resistance values within closer limits.

Temperature Coefficient. All practical types of thermistors have negative temperature coefficients of resistance, values ranging from 1 to 5 per cent per degree centigrade near room temperature. These values increase with lowered temperature, doubling at about −60°C. They are halved at approximately 150°C. Within a given type, individual thermistors may have a temperature coefficient varying from the nominal about 0.5 per cent for compositions best in this respect, or as much as 5 per cent where composition and process are not optimized for this factor.

The application of thermistors to the measurement of temperature (Table 4-16) follows the usual principles of resistance thermometry. Conventional bridge or other resistance-measuring circuits are commonly employed. The high-temperature coefficient of thermistors results in their having greater available sensitivity as temperature-sensing elements than metal resistance thermometers or common thermocouples. To illustrate, the use of thermistors permits the adaptation of conventional temperature recorders to the measurement of 1°C spans, which is not feasible with ordinary resistance-thermometer or thermocouple elements.

Table 4-16. Thermistor Applications*

If problem is	Using a thermistor will provide	Because of these characteristics	Type recommended	For the approximate range
Temperature sensing:				
Temperature measurement	Quick response, small size, simple related circuitry, remote location	Resistance of thermistor is dependent on ambient temperature	Bead, bead in glass probe, bead in metal probe	−60°C to +400°C
Temperature alarm and temperature control	Large resistance change with slight change in temperature	Change in resistance used to activate relay	Bead, bead in glass probe	−60°C to +400°C
Temperature compensation	Compensation in transistor circuits, compensation for materials with positive coefficients	Exhibits negative temperature coefficient	Wafer	−60°C to +150°C

* Courtesy of Gulton Industries, Inc.

The application of thermistors to temperature control is closely related to their use as temperature-measuring devices. The high sensitivity of thermistors to small temperature changes results in greater temperature-control precision than is available from conventional primary elements. The size and shape of the sensitive element are dictated by several factors such as the space available, the required speed of response to temperature changes, and the amount of power which must be dissipated in the element by the control circuit to permit the arrangement to perform its function. If the amount of power dissipated in the sensing element is great enough to raise its temperature considerably above that of the surroundings, its temperature will become dependent upon the thermal conductivity of the surroundings, and may vary with the kind of mounting, or with motion of liquid or gaseous media in which it may be immersed. This property has been utilized in flow meters and devices for measuring heat conductivity of liquids and solids.

TEMPERATURE MEASUREMENT BY COLOR[11]

The color of a hot body is a function of its temperature, and is fairly independent of emissivity. Everyone who deals with high temperatures has used the color of a glowing workpiece as a crude indication of temperature. Some work has been done along these lines but, so far, instruments based on this principle have not found wide application.

Many chemical compounds undergo molecular rearrangement or decomposition at elevated temperatures. Some of these chemical transformations are accompanied by a marked change in color, which may be used as an indication of an attained temperature level. Temperature-indicating colors incorporate some of the more dramatic color-changing mechanisms into different temperature-sensitive paints. Some of these show a single color change at a predetermined temperature level. Most of the paints, however, exhibit several successive color transformations as the temperature increases. Although these color changes are time-dependent in that the duration of heating will to some degree determine the temperature at which they occur, nevertheless they provide a useful means of investigating temperature distribution in many applications. These heat-sensitive paints are intended for monitoring the safe upper-temperature limits of heat processes, such as the operation of electrical transformers, steel tanks, etc. They have good weathering properties and good adhesion at high temperatures.

TEMPERATURE MEASUREMENT BY CHEMICAL SIGNAL[12]

For the measurement of surface temperatures, chemical signals can be deposited directly on the surfaces where temperatures have to be measured. These are obtainable commercially as temperature-indicating crayons, temperature-indicating liquids, and temperature-indicating pellets.

In use the workpiece is stoked with the proper temperature-indicating crayon from time to time during the heating process. Below the temperature rating of the crayon, this will leave a dry mark. When its stated temperature is reached or exceeded, the crayon will leave a liquid smear.

When the heating cycle is of short duration, it is sufficient to make a mark on the workpiece with the proper crayon before heating begins. When the workpiece reaches the stated temperature, the crayon mark will liquefy, and on subsequent cooling will solidify as a translucent smear which is readily distinguishable from the original crayonlike mark.

Temperature-indicating lacquer consists of materials very similar in nature to the crayons, suspended in an inert volatile nonflammable liquid.

[11] G. M. Wolten, Tempil Corp.
[12] *Ibid.*

TEMPERATURE INSTRUMENTS (FILLED SYSTEM)

SPECIFICATION SHEET

SHEET NO. ______
TAG NO. ______
DATE ______
REVISED ______
BY ______

GENERAL

1	DESCRIPTION	RECORDER ☐ CONTROLLER ☐ INDICATOR ☐ TRANSMITTER ☐ BLIND ☐
2	CASE	RECTANGULAR ☐ CIRCULAR ☐ OTHER ______
3	CASE COLOR	BLACK ☐ OTHER ______
4	MOUNTING	FLUSH ☐ SURFACE ☐ YOKE ☐
5	NO. PTS.-- RECORDING ______	INDICATING ______
6	CHART SIZE	12" CIRC. ☐ OTHER ______
7	CHART RANGE ______	NUMBER ______
8	SCALE RANGE ______	TYPE ______
9	CHART DRIVE	SPRING ☐ ELECTRIC ☐ PNEUM. ☐
10	CHART SPEED ______	WIND ______
11	V ______ C ______ EX.PRF. ☐	AIR PRESS. ______
12	OTHER ______	

TRANSMITTER

13	TYPE	PNEUMATIC ☐ ELECTRIC ☐
14	OUTPUT	3-15 PSI ☐ OTHER ______
15	RECEIVERS ON SHEET NO. ______	

CONTROL

16	TYPE	PNEUMATIC ☐ ELECTRIC ☐ OTHER ______
17	PROP. ______ %	AUTO-RESET ☐ RATE-ACTION ☐ ON-OFF ☐ OTHER ______
18	OUTPUT	3-15 PSI ☐ OTHER ______
19	ON MEASUREMENT INCREASE	OUTPUT: INCREASES ☐ DECREASES ☐
20	ELECTRIC SWITCH TYPE - ON MEASUREMENT INCREASE	CONTACTS OPEN ☐ CLOSE ☐
21	CONTACT RATING AMPS. ______	VOLTS ______

AUTO-MANUAL SWITCH

22	NO. POSITIONS ______	EXTERNAL ☐ INTERNAL ☐ INTEGRAL ☐

SETPOINT ADJUSTMENTS

23	MANUAL	INTERNAL ☐ EXTERNAL ☐
24	AUTO-SET	PNEUMATIC ☐ ELECTRIC ☐
25	BAND	FIXED ☐ ADJUSTABLE ☐
26	OTHER ______	

NOTE:

THERMAL ELEMENT

27	CLASS	IA ☐ IB ☐ IIA ☐ IIB ☐ IIC ☐ IID ☐ IIIA ☐ IIIB ☐ VA ☐ VB ☐
28	RANGE ______	OVERRANGE PROTECTION

BULB

29	PLAIN ☐	UNION CONN. ☐ SANITARY ☐
30	EXTENSION	RIGID ☐ ANGLE ☐ BENDABLE ☐ OTHER ______
31	INSERTION LENGTH, INCHES ______	
32	MATERIAL	316SS ☐ OTHER ______
33	BULB IS ______ FEET ABOVE INSTRUMENT CASE	______ FEET BELOW INSTRUMENT CASE

BULB CONNECTIONS

34	FLANGE ☐	THREADED ☐ CLAMP ☐ OTHER ______
35	BUSHING	3/4" ☐ 1" ☐
	WELL	3/4" ☐ 1" ☐ EXT. ☐
36	MATERIAL	316SS ☐ OTHER ______
37	SANITARY	3A ☐ OTHER ______

CAPILLARY TUBING

38	LENGTH ______	
39	TYPE	ARMORED ☐ PLAIN ☐
40	MATERIAL	CAPILLARY: STNL. STL. ☐ COPPER ☐ OTHER ______ ARMOR: STNL. STEEL ☐ BRONZE ☐ PLASTIC OVER BRO. ☐ LEAD ☐ OTHER ______
41	CONNECTION AT CASE	BACK ☐ BOTTOM ☐

ACCESSORIES

42	FILTER & REGULATOR ______
43	AIR SUPPLY GAGE ______
44	LOCAL INDICATOR ______
45	CHARTS & INKSET ______
46	MOUNTING YOKE ______
47	PORTABLE CASE FEATURES ______
48	MOUNTING ACCESSORIES FOR WET & DRY BULBS ______
49	ALARM SWITCH ______
	HERMETICALLY SEALED ☐ E.P. ☐ G.P. ☐

FIG. 4-60*a*. Temperature instruments—filled system. (*Instrument Society of America.*)

It is especially useful for applications to smooth surfaces such as polished metal. Applied by dipping, spraying, or brushing, it dries quickly to a dull and opaque coating. On heating, the dried lacquer melts as soon as its temperature rating is reached. On cooling, the melted lacquer material solidifies with a characteristic glossy, crystalline, or transparent appearance distinctively different from the original coating. Unlike crayons, the lacquer cannot be deposited on a surface which is already hot, but must be applied before heating begins.

Temperature-indicating pellets are intended for use in connection with heating of large units, operations inside a furnace, and in general where observation must be made at some distance. They must be placed on the workpiece before heating begins. After melting at its stated temperature rating, the liquid residue will survive a long period of heating. Pellets provide an attained-temperature signal suitable for subsequent examination.

TEMPERATURE INSTRUMENT SPECIFICATION

To promote uniformity in specifying temperature instrumentation, the Instrument Society of America has prepared recommended specification forms[13] that may be used as a standard. These forms may be modified to suit special needs.

Figure 4-60 is a summary of these typical temperature instrument specification forms.

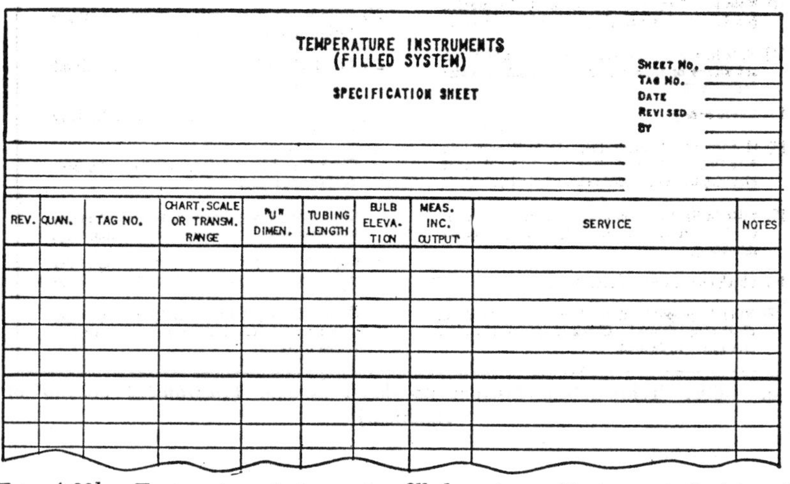

TEMPERATURE INSTRUMENTS
(FILLED SYSTEM)

SPECIFICATION SHEET

SHEET NO.
TAG NO.
DATE
REVISED
BY

REV.	QUAN.	TAG NO.	CHART, SCALE OR TRANSM. RANGE	"U" DIMEN.	TUBING LENGTH	BULB ELEVATION	MEAS. INC. OUTPUT	SERVICE	NOTES

FIG. 4-60*b*. Temperature instruments—filled system. (*Instrument Society of America.*)

[13] *Tentative Recommended Practice, ISA RP20.1*, Instrument Society of America.

TEMPERATURE INSTRUMENTS

(FILLED SYSTEM)

Specification Sheet Instructions

Prefix numbers designate line number on corresponding specification sheet.
Refer to Scientific Apparatus Manufacturer's Association (SAMA) tentative standard on filled system thermometers.

1) Description - Check one, or combination of two or more.
Example: Recorder Controller.

2) Case - check one. Describe other type, or special consideration to type checked on Line 2.

3) Check black or write in color required.

4) Mounting - Check one. If stem type is wanted, specify on Line 12.

5) Write in number of Recording Pens and/or Indicating Pointers, exclusive of Index Pointer.

6) Check 12" size, or write in size required.

*7) Write in chart range and number.

*8) Use manufacturer's standard scale range for indicators, showing concentric or eccentric size, and dial color under "Type".

9) Check type.

10) Write revolutions per day and days spring wound drive to run on complete wind.

11) Write in voltage and cycles, check if explosion proof required, air supply pressure if pneumatic.

12) Write in any special type or deviation from standard drives of Line 11. Also, special indicator features.

13) Check one.

14) If pneumatic transmission, check if output is 3-15 PSI. Write in if other than 3-15 PSI, or if electric, write in electric characteristics.

15) Write in Specification Sheet No. or numbers on which Receiver Instruments appear. (For cross reference).

16) Check one or write in other type.

17) Write in proportional band, percent of scale or chart range desired. Check if auto-reset and/or rate action is desired.

18) Check if 3-15 PSI pneumatic or write in controller output.

*19) When form is used to specify a single instrument, check either Increase (Inc.) or Decrease (Dec.). When form is used to specify multiple instruments of the same general description, write in abbreviation (Inc.) or (Dec.) or (Open) or (Close) for electric contacts in column provided on Secondary Sheet.

20) If electric contact controller, check contact action desired.

21) This is electrical rating of contact or relay supplied in electric contact controller.

22) Check one and write in number of positions.

23) If manual, check location.

24) If automatically set - check operating medium.

25) Check if index set travel is fixed in proportion to the impulse, or is adjustable.

26) Write in any other, such as linkage or cam, or deviation from Line 23, 24 or 25.

27) Filled Thermal System Instruments are classified as follows:

Class IA: Liquid filled, uniform scale, fully compensated.

Class IB: Liquid filled, uniform scale, case compensated only.

Class IIA: Vapor Pressure, increasing scale, with measured temp. above case & tubing temp.

Class IIB: Vapor Pressure, increasing scale, with measured temp. below case & tubing temp.

Class IIC: Vapor Pressure, increasing scale, with measured temp. above and below case & tubing temp.

FIG. 4-60c. Temperature instruments—filled system. (*Instrument Society of America.*)

TEMPERATURE INSTRUMENTS

(FILLED SYSTEM)

Specification Sheet Instructions

Class IID: Vapor Pressure, increasing above, at, and below case & tubing temp.

Class IIIA: Gas filled, uniform scale, fully compensated.

Class IIIB: Gas filled, uniform scale, case compensated only.

Class VA: Mercury filled, uniform scale, fully compensated.

Class VB: Mercury filled, uniform scale, case compensated only.

28) Range should match manufacturer's standard Over-range protection to be manufacturer's standard.

29) (a) A "Plain" bulb consists of sensitive portion, extension and connector, to which the capillary is sealed. Used with open vessels.

(b) A "Union Connection" bulb consists of sensitive portion, extension, connector to which the capillary is sealed, a fixed or movable seating part on the extension and a jam nut for attachment to a bushing or well.

(c) Sanitary bulbs usually conform to N.D.A. code numbered fittings.

30) The bulb extension may be bendable, rigid, or rigid angle type.

Write in other manufacturer's types such as finned, capillary, flush, etc., under "OTHER".

*31) Insertion length is the bulb length below the threaded portion of the bushing or well on a union bulb, or the length below the connector on a plain bulb. (Same as well length).

32) Check type 316 stainless steel or write in manufacturer's other standard materials.

*33) Applies to Class II Thermal System bulbs only.

34) For use on ovens, etc., where vessel material cannot be drilled and tapped for bulb mounting.

35) Sizes other than those shown are special. Check if manufacturer's standard well extension required.

36) Write in manufacturer's other Standard bulb materials.

37) Specify N.D.A. Code number, or other arrangement.

*38) Specify length - five feet is manufacturer's standard base length except on blind transmitters. Check with manufacturer for standard lengths for blind transmitters.

39) Armor is used over small diameter capillary. "Plain" is larger diameter capillary without armor.

40) Check capillary or armor material or write in special capillary, armor or sheath material.

41) As required for flush or surface mounting.

42) Write in if individual units or combination type desired, or if to be purchased separately.

43) Write in "Furnish", if for instrument without integral gages. If "By Others", show Specification Sheet No. and Item No.

44) Specify dial size and scale for Receiver Type Gage for local indication when using Blind Type Transmitter, if "By Others", write Specification Sheet No.

45) Write in number of days supply required.

46) Write in "Yes" or "No" if for type where yoke is optional.

47) Specify leatherette case or portable handle and legs.

48) Write in "Yes" or "No".

49) Write in location, rating in amps., volts and cycles for contacts. Check housing type.

**When more than one instrument is specified, show data on Secondary Sheet, leaving lines unmarked on Base Sheet.*

FIG. 4-60d. Temperature instruments—filled system. (*Instrument Society of America.*)

TEMPERATURE INSTRUMENTS

(FILLED SYSTEM)

Specification Sheet Instructions

Secondary Sheet - For Listing Multiple Instruments

List all instruments of the same type, specified on Primary Sheet, with variations as shown. Variations not provided for to be noted on Primary Sheet as Note (1) etc., and entered on Secondary Sheet where applicable.

Rev. - Revision number
Quan. - Show quantity required
Tag No. - Identification of Item Number
Chart, Scale or Trans. Range - Check with manufacturer for Standard Charts, Scales and Transmitter Ranges.
Bulb Insertion Length - See Instruction #31
Bulb Elevation - See Instruction #33
Tubing Length - See Instruction #38
Meas. Inc. Output - See Instruction #19
Service - Fill in service or location
Notes - Note number appearing on Primary Sheet

FIG. 4-60*e*. Temperature instruments—filled system. (*Instrument Society of America.*)

TEMPERATURE INSTRUMENTS
(POTENTIOMETER PYROMETER & RESISTANCE)

SPECIFICATION SHEET

SHEET NO. ______
TAG NO. ______
DATE ______
REVISED ______
BY ______

GENERAL

1 DESCRIPTION RECORDER ☐ INDICATOR ☐ BLIND ☐ CONTROLLER ☐ TRANSMITTER ☐
2 CASE RECTANGULAR ☐ CIRCULAR ☐ OTHER NO'S. ______
3 CASE COLOR BLACK ☐ OTHER ______
4 MOUNTING FLUSH ☐ SURFACE ☐ YOKE ☐
5 NO. PTS.--RECORDING ______ INDICATING ______
6 CHART TYPE ______" STRIP ☐ 12" CIRC. ☐ OTHER ______
7 CHART RANGE & NO. ______
8 SCALE RANGE & TYPE ______
9 REV. PER DAY ______ OR INCHES PER HOUR ______
10 PEN OR POINTER SPEED ______ SECONDS FULL SCALE TRAVEL
11 PRINTING SPEED ______ SECONDS PER POINT
12 BALANCING ELECTRONIC ☐ MECHANICAL ☐ MANUAL ☐ OTHER ______
13 V ______ CYCLES ______

TRANSMITTER

14 TYPE PNEUMATIC ☐ ELECTRIC ☐
15 OUTPUT 3-15 PSI ☐ OTHER ______
16 RECEIVERS ON SHEET NO. ______

CONTROL

17 TYPE PNEUMATIC ☐ ELECTRIC ☐ OTHER ______
18 PROP. ______ % AUTO-RESET ☐ RATE ACTION ☐ ON-OFF ☐ OTHER ______
19 OUTPUT 3-15 PSI ☐ OTHER ______
20 ON MEASUREMENT INCREASE OUTPUT: INCREASES ☐ DECREASES ☐
21 ELECTRIC SWITCH TYPE: ON MEASUREMENT INCREASE CONTACTS OPEN ☐ CLOSE ☐
22 CONTACT RATING AMPS. ______ VOLTS ______

AUTO-MANUAL SWITCH

23 NO. POSITIONS ______ EXTERNAL ☐ INTERNAL ☐ INTEGRAL ☐

SETPOINT ADJUSTMENTS

24 MANUAL INTERNAL ☐ EXTERNAL ☐
25 AUTO-SET PNEUMATIC ☐ ELECTRIC ☐
26 BAND FIXED ☐ ADJUSTABLE ☐
27 OTHER ______

MEASUREMENT

28 THERMOCOUPLE TYPE ☐ MATERIAL J(IC) ☐ K(CA) ☐ T(CC) ☐ OTHER ______ REFERENCE JUNCTION COMPENSATION ☐ AUTOMATIC STANDARDIZATION ☐
29 RADIATION TYPE ☐
30 RESISTANCE BULB TYPE ☐ MATERIAL NICKEL ☐ PLATINUM ☐ OTHER
31 OTHER ______
32 RANGE ______ TO ______ °F ☐ °C ☐
33 OTHER ______

ACCESSORIES

34 FILTER & REGULATOR ______
35 AIR SUPPLY GAGE ______
36 CHARTS & INKSET ______
37 DRY CELL ______
38 ALARM CONTACTS ______
39 OTHER ______

NOTE:

FIG. 4-60*f*. Temperature instruments—potentiometer pyrometer and resistance. (*Instrument Society of America.*)

TEMPERATURE INSTRUMENTS
(POTENTIOMETER PYROMETER & RESISTANCE)

SPECIFICATION SHEET

SHEET NO. ____
TAG NO. ____
DATE ____
REVISED ____
BY ____

REV.	QUAN.	TAG NO.	RANGE	TYPE THMCPL. OR RESIST. BULB	SCALE & CHART	NO. POINTS	MEAS. INC. OUTPUT	SERVICE	NOTES

FIG. 4-60*g*. Temperature instruments—potentiometer pyrometer and resistance. (*Instrument Society of America.*)

TEMPERATURE INSTRUMENTS

(POTENTIOMETER PYROMETER AND RESISTANCE)

Specification Sheet Instructions

Prefix numbers designate line number on corresponding specification sheet.

1) Description - Check one or combination of two or more.
Example: Recorder Controller

2) Case - Check One.
Describe other type or special consideration to type checked on Line 2.

3) Check black or write in color required.

4) Mounting - Check One.

*5) Write in number of Recording Pens and/or Indicating Pointers, exclusive of Index Pointer. For multipoint instruments, write in number of recording and/or indicating points.

6) Specify width of chart and check if strip chart. Check if 12" circular chart. Write in if other.

*7) Write in chart range and number.

*8) Write in scale range and type.

9) Write in number of revolutions per day if circular chart or inches per hour if strip chart.

10) Specify pen or pointer speed in seconds for full scale travel.

11) Specify printing speed in seconds per point.

12) Check type of balancing. Describe other type or special consideration to type checked on Line 12.

13) Specify voltage and frequency of power supply.

14) Check One.

15) If pneumatic transmission, check if putout is 3-15 PSI. Write in if other than 3-15 PSI, or if electric characteristics.

16) Write in Specification Sheet No. or numbers on which Receiver Instruments appear. (For cross reference).

17) Check one or write in other type.

18) Write in proportional band, percent of scale or chart range desired. Check if auto-reset and/or rate action is desired.

19) Check if 3-15 PSI pneumatic or write in controller output.

*20) When form is used to specify a single instrument, check either Increase (Inc.) or Decrease (Dec.). When form is used to specify multiple instruments of the same general description, write in abbreviation (Inc.) or (Dec.) or (Open) or (Close) for electric contacts in column provided on Secondary Sheet.

*21) If electric contact controller, check contact action required.

22) This is electrical rating of contact or relay supplied in an electric contact controller.

23) Check one and write in number of positions.

24) If manual, check location.

25) If automatically set - check operating medium.

26) Check if index set travel is fixed in proportion to the impulse, or is adjustable.

27) Write in any other, such as linkage or cam, or deviation from Lines 24, 25 or 26.

*28) Check if instrument to be calibrated for thermocouple measurement. Also check thermocouple material or write in after other.

29) Check if instrument to be calibrated for a radiation pyrometer.

30) Check if instrument to be calibrated as resistance thermometer and check type of bulb or write in after other.

****When more than one instrument is specified, show data on secondary sheet, leaving lines unmarked on base sheet.***

FIG. 4-60*h*. Temperature instruments—potentiometer pyrometer and resistance. (*Instrument Society of America.*)

TEMPERATURE INSTRUMENTS

(POTENTIOMETER PYROMETER AND RESISTANCE)

Specification Sheet Instructions

31) Specify if instrument to be calibrated other than thermocouple, radiation or resistance thermometer. Example: Direct Current Millivolt Recorder.

*32) Fill in range of temperature calibration.

*33) Fill in range if other than temperature. Example: 10 to 50 millivolts D.C.

34) Write in if individual units or combination type desired, or if to be purchased separately.

35) Write in "Furnish", if for instrument without integral gages. If "By Others", show Specification Sheet No. and Item No.

36) Write in number of days supply required.

37) Write in if dry cell required.

38) Write in if alarm contacts desired including contact action, rating and range, over which contact to be adjustable.

39) Write in any other optional features desired. Example: Internal Illumination, etc.

Secondary Sheet - For listing multiple instruments.

List all instruments of the same type, specified on Primary Sheet with variations as shown. Variations not provided for to be noted on Primary Sheet etc., and entered on Secondary Sheet where applicable.

Rev. - Revision Number
Quan. - Show Quantity required.
Tag No. - Identification of Item Number
Range - Show range of Instrument - See Instructions #32, #33.
Type Thmcple. or Res. Bulb - Show type of thermocouple or resistance bulb for which Instrument is to be calibrated.
Scale and/or Chart - Check with manufacturer for standard charts and scales. See Instructions #7, #8.
No. Points - Show number of records or number of indicating points. See Instruction #5.
Meas. Inc. Output - See Instruction #20.
Service - Fill in Service or location
Note - Note number appearing on Primary Sheet.

**When more than one instrument is specified, show data on Secondary Sheet, leaving lines unmarked on Base Sheet.*

Fig. 4-60i. Temperature instruments—potentiometer pyrometer and resistance. (*Instrument Society of America.*)

THERMOCOUPLES

SPECIFICATION SHEET

SHEET NO. ____
TAG NO. ____
DATE ____
REVISED ____
BY ____

GENERAL

1 DESCRIPTION ELEMENT ONLY ☐ COMPLETE ASSEMBLY ☐
ASSEMBLY LESS WELL ☐ OTHER ____
2 MOUNTING SCREWED ☐ FLANGED ☐
OTHER ____
3 PROTECTING WELL OR TUBE DRILLED BAR STOCK ☐
CLOSED END TUBE ☐ OTHER
4 NIPPLE & UNION LENGTH "A" ____

HEAD

5 COVER SCREWED ☐ CAP TYPE ☐
OTHER ____
6 CONN. SIZE-INS SPT CONDUIT ____ TUBE ____
7 TERM. BLOCK SINGLE ☐ DUPLEX ☐
OTHER ____

PROTECTING WELL OR TUBE

8 MATERIAL 304SS ☐ 316SS ☐
OTHER ____
9 CONSTRUCTION TAPERED ☐ STRAIGHT ☐
OTHER ____
10 WELL DIMENSIONS MFGR STD ☐ O.D. ____" I.D. ____"
11 MALE THREAD SIZE 1"SPT ☐ OTHER ____
12 LAG EXTENSION "T" NONE ☐ SHOWN BELOW ☐
3" ☐
13 FLANGE SIZE ____ RATING & TYPE ____
MATERIAL ____
14
15
16

REV	QUAN	TAG NO.	FIG NO.	CALIB.	GAUGE	ELEM. LENGTH	"U" DIMEN.	"T" DIMEN.	SERVICE	NOTES

FIG. 4-60j. Thermocouples. (*Instrument Society of America.*)

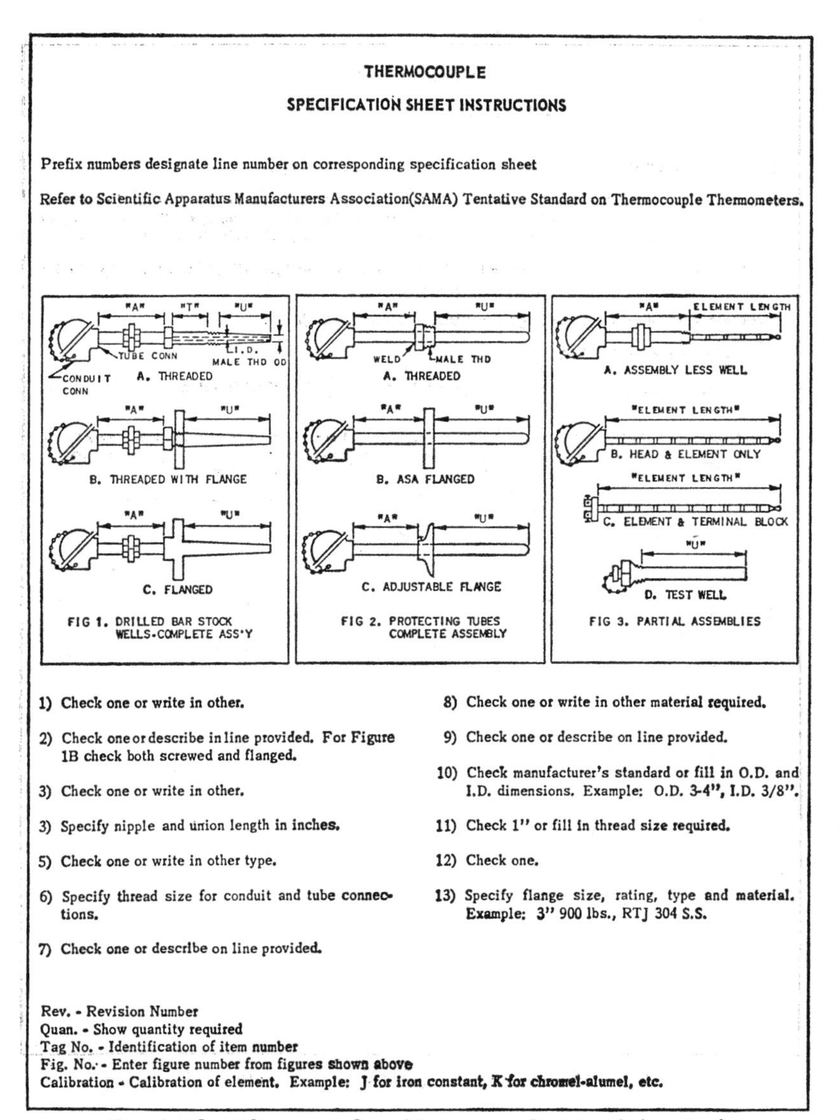

THERMOCOUPLE

SPECIFICATION SHEET INSTRUCTIONS

Prefix numbers designate line number on corresponding specification sheet

Refer to Scientific Apparatus Manufacturers Association(SAMA) Tentative Standard on Thermocouple Thermometers.

1) Check one or write in other.

2) Check one or describe in line provided. For Figure 1B check both screwed and flanged.

3) Check one or write in other.

3) Specify nipple and union length in inches.

5) Check one or write in other type.

6) Specify thread size for conduit and tube connections.

7) Check one or describe on line provided.

8) Check one or write in other material required.

9) Check one or describe on line provided.

10) Check manufacturer's standard or fill in O.D. and I.D. dimensions. Example: O.D. 3-4", I.D. 3/8".

11) Check 1" or fill in thread size required.

12) Check one.

13) Specify flange size, rating, type and material. Example: 3" 900 lbs., RTJ 304 S.S.

Rev. - Revision Number
Quan. - Show quantity required
Tag No. - Identification of item number
Fig. No. - Enter figure number from figures shown above
Calibration - Calibration of element. Example: J for iron constant, K for chromel-alumel, etc.

FIG. 4-60*k*. Thermocouple. (*Instrument Society of America.*)

THERMOCOUPLE

SPECIFICATION SHEET INSTRUCTIONS

Gage - Specify B & S wire gauge size of element required. Example: 14 gauge.

Element Length - Element length is the length from the terminal block to the free end of the element. If a complete assembly is ordered, this column not required. To figure element length, add "U" dimension plus "T" dimension plus nipple extension plus 2-½" (constant). Show dimension to nearest ¼".

"U" Dimen. - "U" dimension is length below thread to the free end of a well protecting tube. Length of well below thread or flange. (See SAMA Tentative Standard).

"T" Dimen. - "T" dimension is lag length (see SAMA Tentative Standard).

Service - Fill in service if desired.

Notes: Note number or letter appearing in space provided at the bottom of the sheet for notes.

Secondary Sheet - For additional thermocouple with the same general specifications as shown on the Primary Specification Sheet. All columns are the same as the Primary Specification Sheet. A number or letter appearing in notes column refer to notes on Primary Specification Sheet.

FIG. 4-60*l*. Thermocouple. (*Instrument Society of America.*)

INDICATING BIMETAL THERMOMETERS

SPECIFICATION SHEET

SHEET NO. ________
TAG NO. ________
DATE ________
REVISED ________
BY ________

GENERAL

1 BULB TYPE THREADED ☐ PLAIN ☐ UNION ☐
OTHER ________
2 CASE MATERIAL MFGR STD ☐ OTHER ________
3 DIAL SIZE ________ COLOR ________
4 FORM STRAIGHT ☐ ANGLE ☐
FIXED ☐ ADJUSTABLE ☐
5 BULB OR UNION THREAD SIZE ________ 1/2" ☐ 3/4" ☐
6 MFGR MODEL NO. ________

WELL

7 MATERIAL 304SS ☐ 316SS ☐
OTHER ________
8 CONSTRUCTION BUILT-UP ☐ DRILLED BAR STOCK
OTHER ________
9 THREAD SIZE NPT MALE 1" ☐ OTHER ________
FEMALE 1/2" ☐ OTHER ________
10 LAG EXTENSION "T" NONE SHOWN BELOW ☐
3"
11 MFGR. MODEL NO. ________

REV	QUAN	TAG NO.	RANGE	OPER TEMP.	"U" DIMEN.	"T" DIMEN.	SERVICE	NOTES

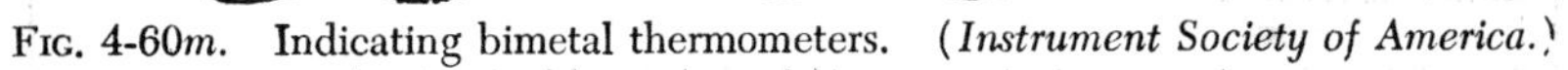

Fig. 4-60*m*. Indicating bimetal thermometers. (*Instrument Society of America.*)

INDICATING BI-METAL THERMOMETERS

SPECIFICATION SHEET INSTRUCTIONS

Prefix numbers designate line number on corresponding specification sheet.

Refer to Scientific Apparatus Manufacturers Association (SAMA) Tentative Standard on Bi-Metallic Thermometers.

1) Bulb Type - Check one or write in other type.

2) Case Material - Check manufacturer's standard, or write in other type.

3) Write in dial size and color.

4) Check whether straight or angle form. Straight form is with plane of the dial parallel to the axis of the bulb. Angle form is with dial at right angles to the axis of the bulb. For the straight form only, check whether axis of bulb is fixed or adjustable with respect to dial.

5) Write in bulb or union thread size, if other than 1/2" or 3/4" NPT is required.

6) Write in manufacturer's model number for bulb, if desired.

7) Check one or write in other material.

8) Check one or write in other construction.

9) Check one or write in other size for both male (outside) or female (inside) thread.

10) Check one. 3" is SAMA preferred.

11) Write in manufacturer's model number for well, if desired.

Rev. - Revision Number
Quan. - Show quantity desired
Tag No. - Identification of item number
Range - Check with manufacturer for standard ranges
Oper. Temp. - If desired, show normal operating temperature
"U" Dimen. - "U" dimension is bulb insertion length. With well "U" dimension is length below thread. (See SAMA Tentative Standard).
"T" Dimen. - "T" dimension is lag length. (See SAMA Tentative Standard).
Service - Fill in service

Notes: Note number or letter appearing in space provided at the bottom of the sheet for notes.

FIG. 4-60n. Indicating bimetal thermometers. (*Instrument Society of America.*)

Section 5

MEASUREMENT OF FLOW

Peak operating efficiency of entire systems demands the maintenance of maximum operating economies. Such flows as steam, air, water, oil, and gas must, therefore, be accurately metered from the standpoint of both total flow and final distribution.

PRIMARY ELEMENTS FOR DIFFERENTIAL-PRESSURE FLOW METERS[1]

Pressure differentials are produced by various types of primary elements located in the line so that the flowing fluid passes through them. Their

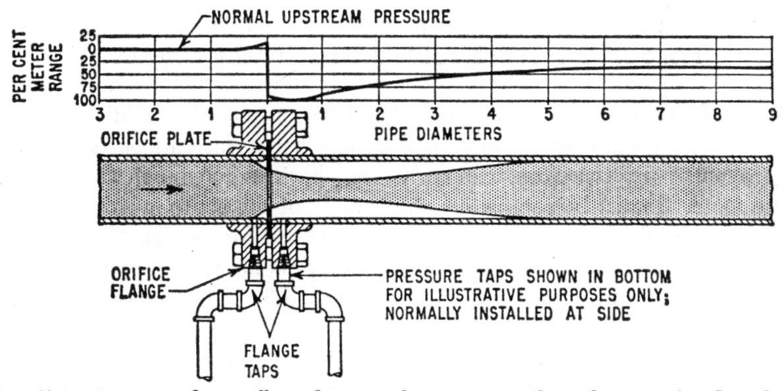

FIG. 5-1. Pressure-drop effect due to placing an orifice plate in the flow line. (*Taylor Instrument Co.*)

function is to increase the fluid velocity temporarily by decreasing the cross-sectional area over a short length so that a static pressure decrease will occur. Rate of flow can then be determined from the pressure drop. Various types of these elements are used.

The Orifice Plate (Fig. 5-1). This is the most simple and flexible. In its most common form, it is merely a circular hole in a thin flat plate, which is clamped between the flanges at a joint in the pipeline so that the

[1] Taylor Instrument Co.

hole is correctly located within the pipe. It can be readily rebored or replaced to accommodate flow-capacity changes. It can be used for measuring forward or reverse flows. It is the least expensive and is as accurate as the venturi or the flow nozzle. Generally, it is furnished so that the hole is concentric with the pipe, but it is also available in eccentric form when wet gases or liquids, with high gas content, are metered (Fig. 5-2). It is also provided in the segmental form when solids are present in the flowing stream. A calibrated orifice section should be used in pipe sizes of less than 2 in. if accuracies of better than 10 per cent are required.

The importance of a sharp leading edge on the orifice cannot be overestimated. Though it is true that the orifice will produce differential pressure with its edge rounded, the significance of the instrument's reading is severely reduced. Tests have clearly shown that a slight rounding can immediately introduce an error of 2 to 10 per cent of maximum flow-meter capacity. It is of interest to note that the corrosion rate increases as the

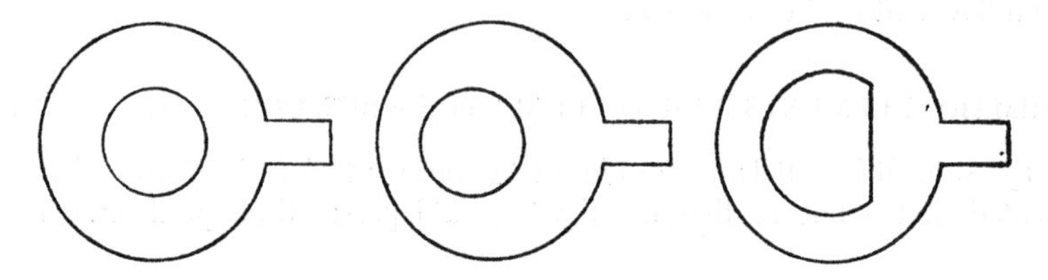

FIG. 5-2. Typical orifice-plate designs. (*Bailey Meter Co.*)

degree of sharpness increases, and therefore, even under mildly corrosive conditions, the sharp orifice edge will be the first surface attacked by corrosive agents. It, therefore, cannot be emphasized too much that orifice-plate material must be completely stable and not subject to even the mildest attack by the fluid if accuracy is to be maintained for any length of time.

The Flow Nozzle. This is supported in a pipe between standard flanges, and its general appearance can be seen in Fig. 5-3. The rounded approach has a curvature equivalent to the quadrant of an ellipse, and the distance between the front face and the nozzle tip is usually one-half the pipe diameter whenever the throat ratio of the nozzle lies in the limits of 0.4 to 0.8. When throat ratios of less than 0.4 are used, there is a curvature change, and the over-all length of the nozzle proper is materially increased. These are known as long-radius nozzles. The flow nozzle has a coefficient which lies between the venturi tube and the orifice plate. The rounded upstream approach contributes to its efficiency, but the absence of a recovery cone renders it impossible to capitalize on this efficiency. Therefore the metering loss is more than the venturi tube, and furthermore it is a function of the throat ratio. Figure 5-4 shows the pressure recovery as a function of nozzle-throat diameter.

The flow nozzle lends itself very well to the measurement of wet gases such as saturated steam with condensate in suspension. If the steam is dry or possesses superheat, then the nozzle is not necessary unless other conditions require it. Little droplets carried in suspension in a gas stream can exert a considerable erosive effect, and the curved surface on the nozzle face guards this device against such action, thus contributing to long, useful life. Ordinarily, as throat ratios increase, the quality of performance of primary elements decreases, but with a nozzle this effect does not appear until throat ratios exceed 0.8. Because of this property, plus the fact that

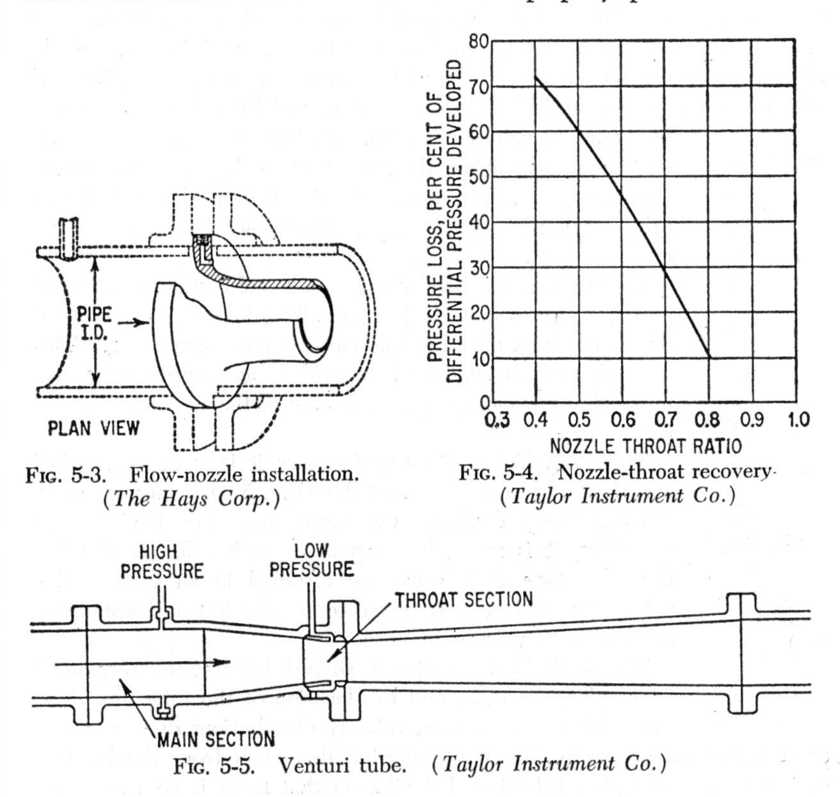

FIG. 5-3. Flow-nozzle installation. (*The Hays Corp.*)

FIG. 5-4. Nozzle-throat recovery. (*Taylor Instrument Co.*)

FIG. 5-5. Venturi tube. (*Taylor Instrument Co.*)

it is more efficient than an orifice plate, the flow nozzle can be used to meter high-velocity fluids. This condition is often encountered when plant capacities are increased with no attendant changes in the piping proper. Also, in those instances where high-temperature steam is being produced, there is a tendency to hold the pipe size down to an absolute minimum because of the high accelerating costs of high-temperature piping as the pipe size is increased. Here again the nozzle is a useful tool in measuring steam flow for it can handle the resultant high-velocity high-temperature steam flow.

The Venturi Tube. The important feature associated with the venturi tube (Fig. 5-5) is its pressure-recovery characteristics. For this reason it is recommended where metering conditions require extremely low pressure loss. Low-pressure gas lines and water mains through which widely varying quantities will flow probably represent the greatest number of applications of the venturi tube. Because of its streamlined approach and exit, use of the venturi tube is often considered when the flow of liquids with solids in suspension must be measured. When it is kept in mind that the fluid friction factor along a side wall of a pipe varies as the fifth power of the diameter, it is seen that, whenever a finite length exists between two pressure taps, the differential developed is the sum of the energy-transfer effect and the frictional effect. In a venturi tube, there is an appreciable distance between the taps and a persistent decrease in diameter, both of which can exert a serious influence on the differential developed whenever fluid-friction values are high. In the case where solids are carried in suspension, the fluid friction depends on the concentration of solids, and the venturi-tube differential will exhibit a sensitivity to the percentage of solids carried along in the stream.

Venturi tubes generally are constructed with a system of pressure taps, which project radially into the pipe and feed into a common chamber known as a piezometer ring. This multiple-tap arrangement provides an opportunity for measuring an average pressure around the circumference of the stream.

FIG. 5-6. Pitot tube. (*Taylor Instrument Co.*)

The Pitot Tube. In certain cases it is more economical and desirable to use a pitot tube (Fig. 5-6) in preference to other primary devices. On large line sizes having high velocities, it is much less expensive to install. It also has the advantage that it can be installed in an existing line where it would be impractical to cut the line and install any other type of primary device.

The fluid to be metered should be free of suspended matter which might plug the tube, and the location of the tube should be where there is no possibility of uneven velocity distributions due to fittings or disturbances upstream for about 20 pipe diameters from the location of the tube. The pitot tube is not flexible in that there is no way of increasing differential (as can be done with other types of primary devices, i.e., changing orifice or throat size). Consequently, it can be used only between the limits of (1) a minimum velocity of the flowing medium sufficient to give a good readable differential on the instrument, and (2) a maximum velocity governed by the physical strength of the tube itself, which is supported at one end only.

The Pitot-Venturi Flow Element. This is used to measure the velocity of either liquid or gaseous mediums. Its design (Fig. 5-7) is a combination of the pitot tube and the venturi element, and it is so arranged that the differential produced is greater than that developed by a pitot tube alone. For velocities in excess of 1,000 fpm, the differential developed by this element is approximately 10 times that produced by the pitot tube. An exact relationship between flow and differential is obtained through accurate positioning of the calibrating ring[2] which is seen on the external surface of the main body of the element.

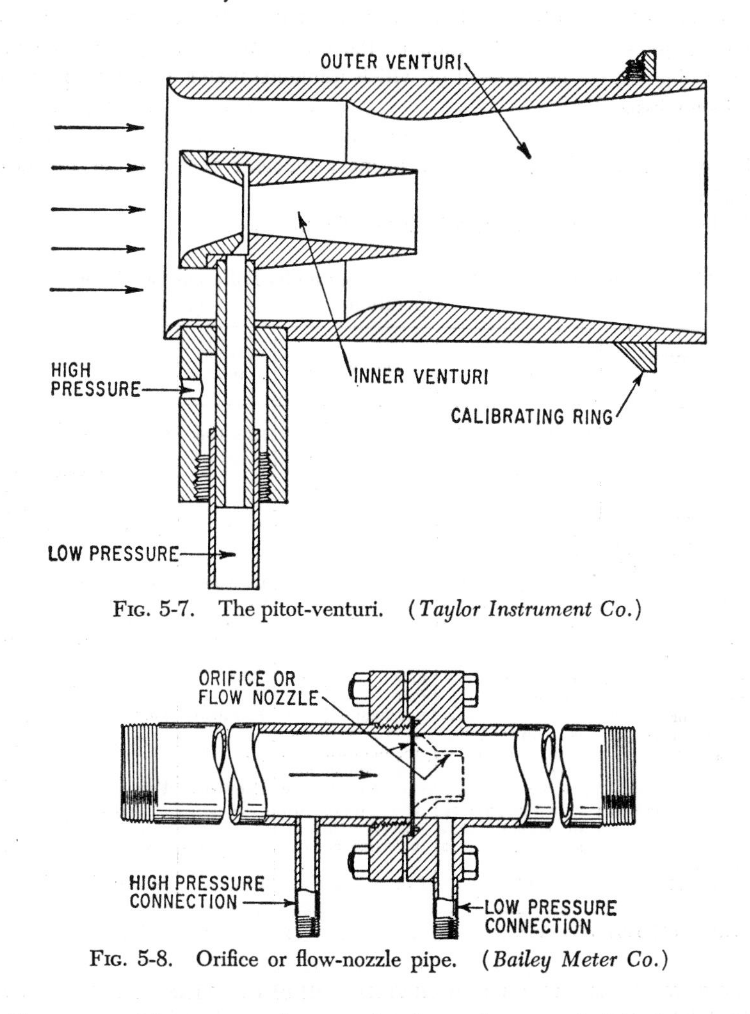

FIG. 5-7. The pitot-venturi. (*Taylor Instrument Co.*)

FIG. 5-8. Orifice or flow-nozzle pipe. (*Bailey Meter Co.*)

No measurable pressure loss results from the use of this device, and, because of its compactness, it can be located where space is at a premium. When the element is located at the average velocity position, it can then be used as a flow meter.

Orifice and Flow-nozzle Pipes. For line sizes of 2½-in. diameter and less, the orifice or flow-nozzle pipe can be used (Fig. 5-8). It consists of

[2] Patent No. 2,441,042, Taylor Instrument Co.

two sections of steel tubing and contains a pair of special flanges which hold the orifice or flow nozzle in place.

Use of the orifice or flow-nozzle pipe eliminates the difficulty of drilling and tapping small-size pipes for nipples. It also removes any possibility of errors due to eccentricity of the pipe, inaccurate measurement of the internal diameter, improper location of the meter pressure tap, locating the primary element off center—all of which influence meter accuracy, particularly in small-size lines.

Future changes in capacity of the meter can be effected by replacing the orifice or flow nozzle with a new primary element designed for the new conditions.

Table 5-1 compares various primary devices.

Table 5-1. Comparison of Primary Devices under Different Conditions of Operation*

Application considerations	Concentric orifice	Flow nozzle	Venturi tube	Pitot tube
Accuracy	Excellent	Good	Excellent	Good for one-point measurement. For measurement of total flow, a velocity traverse is necessary for good accuracy
Suitability for liquids containing solids in suspension	Poor	Good	Excellent	Unsatisfactory
Pressure recovery	Poor	Poor	Good	Excellent
Suitability for viscous flows	Fair	Good	Good	Requires a velocity traverse of the conduit
Low cost in large sizes	Good	Fair	Poor	Excellent
Low cost in small sizes	Excellent	Fair	Poor	Good
Suitability for liquids containing traces of vapors	Excellent—if flow is in upward direction through a vertical conduit	Excellent—if flow is in upward direction through a vertical conduit	Excellent	Fair
Suitability for gases containing traces of condensate	Excellent—if flow is in a downward direction through a vertical conduit	Excellent—if flow is in a downward direction through a vertical conduit	Excellent	Fair
Ease of changing capacity	Excellent	Fair	Poor	Impossible

* Courtesy of Foxboro Co., Inc.

SELECTING THE DIAMETER RATIO, DIFFERENTIAL RANGE, AND PIPING LAYOUT[3]

To measure flow accurately, the primary element must be located in the piping layout so that its operation follows well-defined principles which have been established.

Certain minimum requirements for straight run of pipe, both preceding and following the element, must be satisfied in order to avoid the effects of

[3] Penn Instruments Div., Burgess-Manning Co.

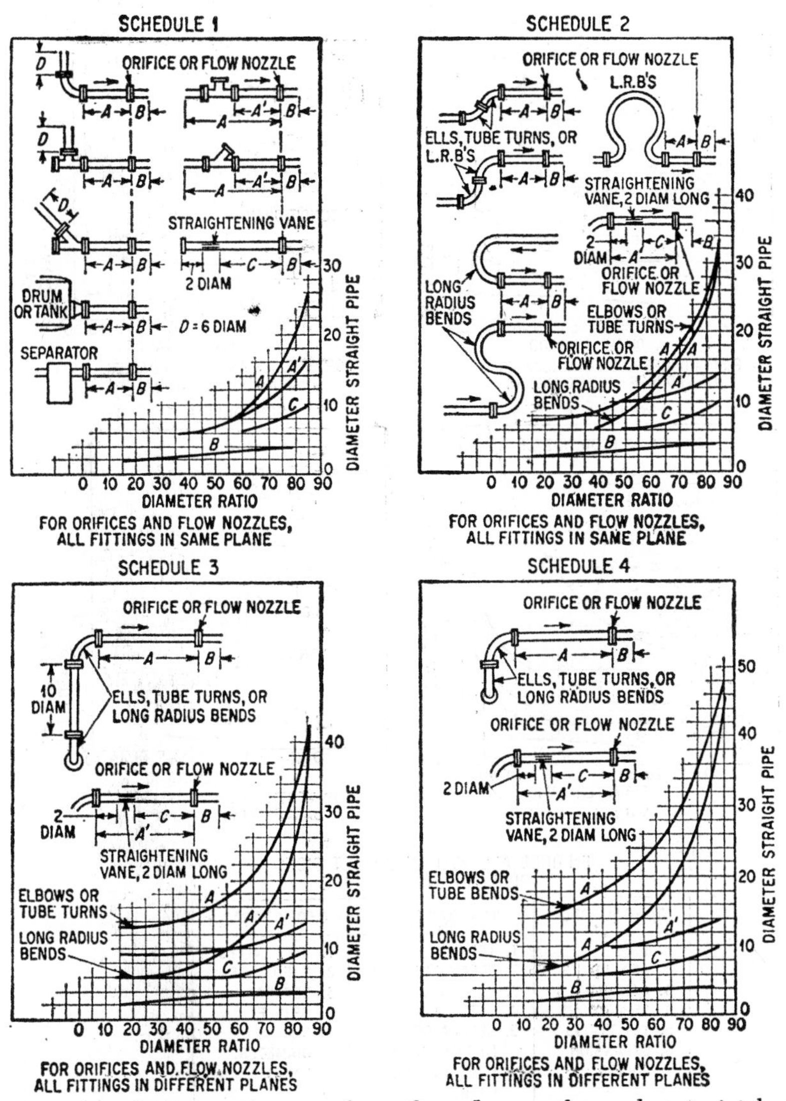

Fig. 5-9. Piping requirements for orifices, flow nozzles, and venturi tubes. (1944 *ASME Paper, Piping Arrangements for Acceptable Flow-meter Accuracy, by R. E. Sprenkle, director of education, Bailey Meter Co.*)

swirls, crosscurrents, and large eddies in the flowing fluid. Straightening vanes minimize these length requirements.

Two considerations affect the length of straight piping: (1) the character of the piping before the straight run, and (2) the diameter ratio of the primary elements (ratio of primary element diameter to inside pipe dia-

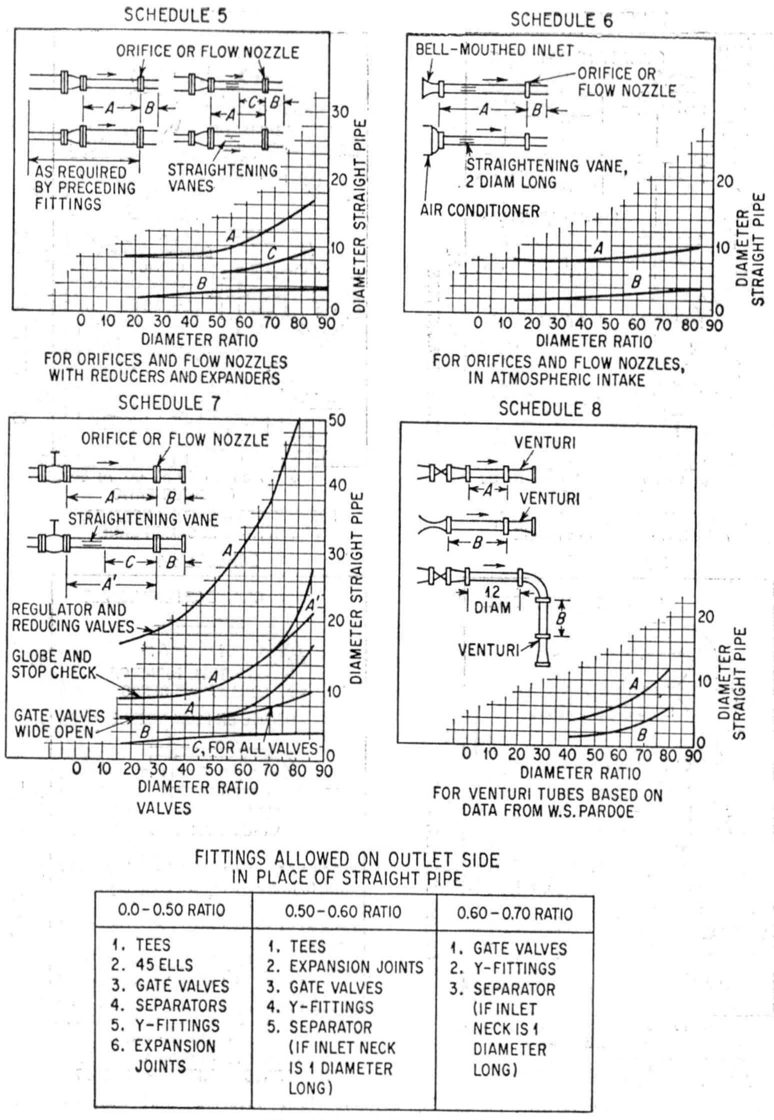

FITTINGS ALLOWED ON OUTLET SIDE IN PLACE OF STRAIGHT PIPE

0.0 - 0.50 RATIO	0.50 - 0.60 RATIO	0.60 - 0.70 RATIO
1. TEES 2. 45 ELLS 3. GATE VALVES 4. SEPARATORS 5. Y-FITTINGS 6. EXPANSION JOINTS	1. TEES 2. EXPANSION JOINTS 3. GATE VALVES 4. Y-FITTINGS 5. SEPARATOR (IF INLET NECK IS 1 DIAMETER LONG)	1. GATE VALVES 2. Y-FITTINGS 3. SEPARATOR (IF INLET NECK IS 1 DIAMETER LONG)

FIG. 5-9. (*Continued*)

meter). The need for straight pipe increases (*a*) as the piping preceding the straight run becomes more complicated, and (*b*) as the orifice bore becomes larger with respect to the inside diameter of the pipe.

Regarding the first consideration, the effect of certain of the more common types of installation on metering accuracy has been studied, and the result of these studies has been condensed into Figs. 5-9 to 5-12.

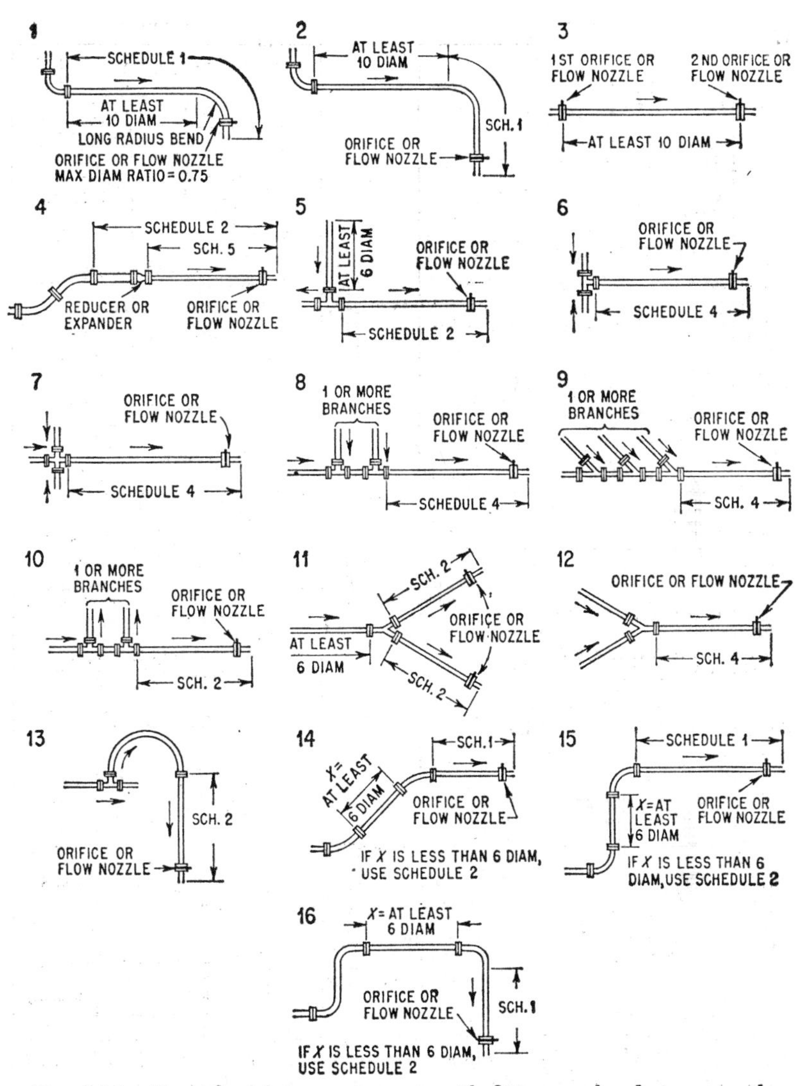

FIG. 5-10. Typical piping arrangements with fittings or bends in same plane. (1944 *ASME Paper, Piping Arrangements for Acceptable Flow-meter Accuracy, by R. E. Sprenkle, director of education, Bailey Meter Co.*)

As for the second consideration, the diameter ratio, we can set up a quick method for its approximate determination, thereby also settling the differential range.

For these purposes, Tables 5-2 to 5-19 give all the necessary information. Their use is best illustrated by an example. Suppose we wish to measure

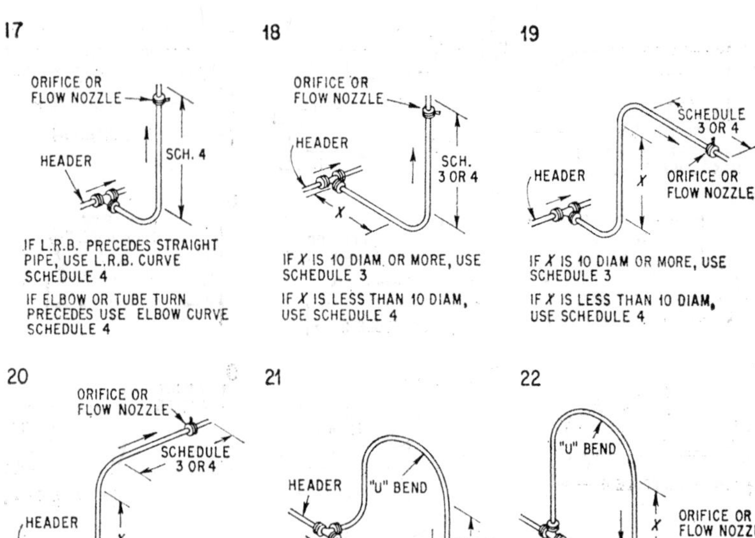

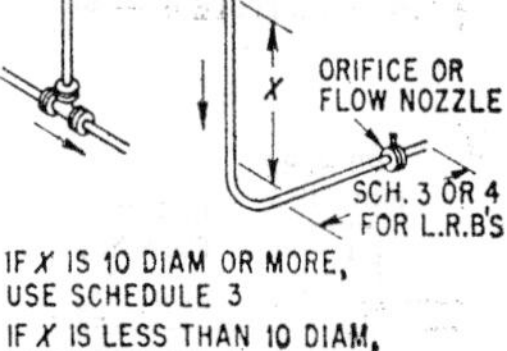

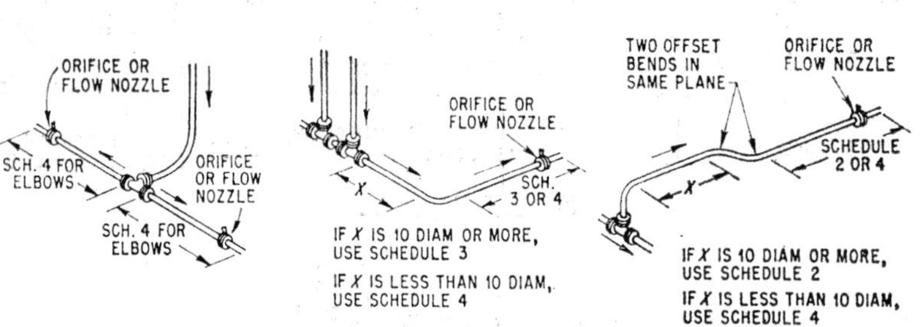

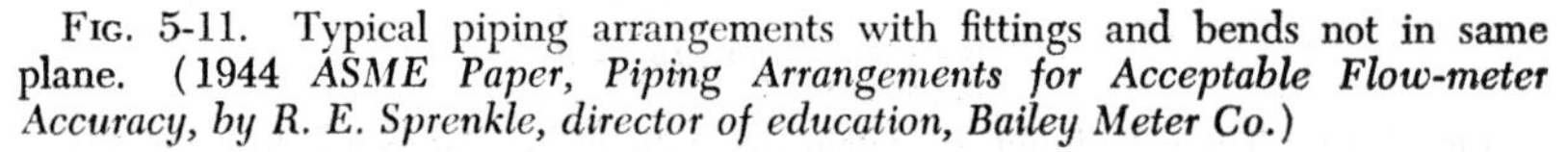

Fig. 5-11. Typical piping arrangements with fittings and bends not in same plane. (1944 *ASME Paper, Piping Arrangements for Acceptable Flow-meter Accuracy, by R. E. Sprenkle, director of education, Bailey Meter Co.*)

30,000 lb/hr of saturated steam at 500 psi in a 6-in. extra-heavy pipe with a single elbow preceding the 5-ft-long meter run. From Table 5-7 we find that the maximum capacity with an orifice plate is 37,500 lb/hr, using a 75.36-in. differential meter in a 6-in. extra-heavy pipe, with the largest recommended ratio of orifice bore to inside pipe diameter, which is 0.70d/D. All capacities in Table 5-7 are based on this maximum d/D of 0.70 for orifice plates and 0.75 flow nozzles, and on the use of a meter with a 75.36-in. water differential range.

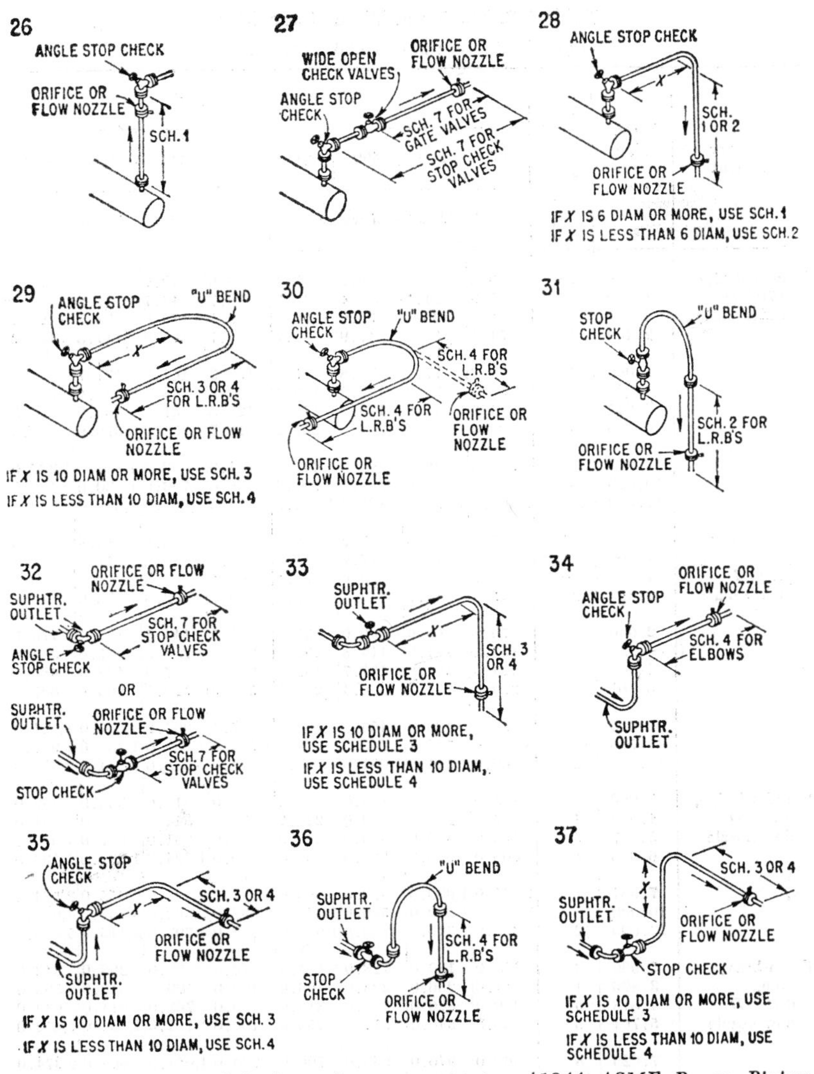

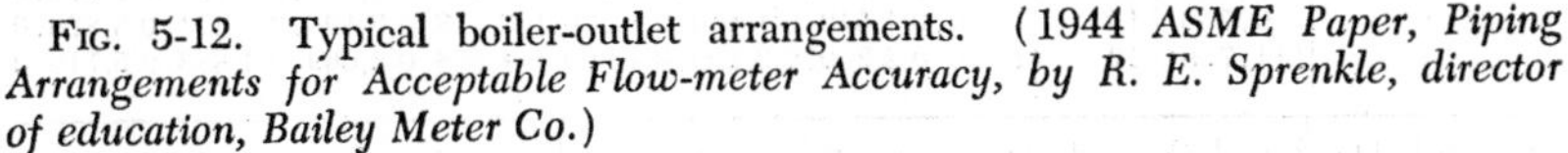
FIG. 5-12. Typical boiler-outlet arrangements. (1944 *ASME Paper, Piping Arrangements for Acceptable Flow-meter Accuracy, by R. E. Sprenkle, director of education, Bailey Meter Co.*)

Table 5-6 is a listing of most commercial meter ranges available.

Since the 75.36-in. range is the basis for Table 5-2, it has a capacity factor of 1.0, and the entire family covers a capacity span from 0.5 to 3.0.

A tabulation is then made up of the five possible ranges and their maximum capacity. From Table 5-7 the maximum capacity is obtained for the 75.36-in. range, which is 37,500 lb of steam per hr. This capacity is mul-

Table 5-2. Maximum Water Capacities

(With differential of 75.36 in. water)

Primary element	Inside pipe diameter	Nominal pipe size, in.	Water temperature, °F: 60	200	400	600	60	200	400	600
			1,000 lb/hr of water				Gal/min of water			
Standard pipe, 0.70d/D orifice plate	1.049	1*	9.3	9.1	8.7	7.7	18.6	19.0	20.1	23.2
	1.610	1½*	21.9	21.5	20.4	18.2	43.7	44.7	47.4	54.6
	2.067	2*	36.1	35.5	33.6	30.0	72.1	73.6	78.2	90.0
	2.469	2½*	51.4	50.5	47.9	42.7	102.7	104.9	111.3	128.3
	3.068	3	79.2	77.9	73.8	65.8	158.3	161.7	171.7	197.8
	4.026	4	136.4	134.2	127.2	113.4	272.6	278.4	295.6	340.6
	5.047	5	214.3	210.9	199.8	178.1	428.5	437.5	464.6	535.3
	6.065	6	309.5	304.5	288.6	257.2	618.7	631.9	670.9	773.0
	7.981	8	535.5	526.8	499.3	445.1	1,071.0	1,093.0	1,161.0	1,337.0
	10.020	10	844.2	830.6	787.1	701.6	1,688.0	1,723.0	1,830.0	2,108.0
	12.000	12	1,209.6	1,190.0	1,128.0	1,005.0	2,418.0	2,469.0	2,622.0	3,021.0
Extra-heavy pipe, 0.70d/D orifice plate	0.957	1*	7.7	7.6	7.2	6.4	15.4	15.8	16.8	19.3
	1.500	1½*	19.0	18.7	17.7	15.8	38.0	38.8	41.2	47.4
	1.939	2*	31.7	31.2	29.6	26.4	63.4	64.8	68.8	79.2
	2.323	2½*	45.5	44.7	42.4	37.8	90.9	92.8	98.6	113.6
	2.900	3	70.8	69.6	66.0	58.8	141.5	144.5	153.4	176.7
	3.826	4	123.2	121.2	114.8	102.4	246.2	251.5	267.0	307.6
	4.813	5	194.9	191.8	181.7	162.0	389.7	397.9	422.5	486.8
	5.761	6	279.3	274.8	260.4	232.1	558.3	570.1	605.4	697.5
	7.625	8	488.8	480.9	455.7	406.2	977.1	997.9	1,060.0	1,221.0
	9.750	10	799.4	786.4	745.2	664.3	1,598.0	1,632.0	1,733.0	1,996.0
	11.750	12	1,160.0	1,141.0	1,081.0	963.8	2,318.0	2,368.0	2,514.0	2,896.0
Standard pipe, 0.75d/D flow nozzle	3.068	3	155.0	151.0	143.0	127.0	310.0	313.0	333.0	382.0
	4.026	4	267.0	260.0	250.0	220.0	534.0	544.0	581.0	661.0
	5.047	5	420.0	410.0	390.0	350.0	840.0	851.0	907.0	1,052.0
	6.065	6	600.0	590.0	560.0	500.0	1,200.0	1,224.0	1,302.0	1,503.0
	7.981	8	1,050.0	1,030.0	970.0	860.0	2,099.0	2,137.0	2,255.0	2,584.0
	10.020	10	1,650.0	1,620.0	1,530.0	1,360.0	3,298.0	3,362.0	3,557.0	4,087.0
	12.000	12	2,350.0	2,310.0	2,200.0	1,950.0	4,698.0	4,793.0	5,115.0	5,860.0
Extra-heavy pipe, 0.75d/D flow nozzle	2.900	3	138.0	136.0	128.0	114.0	276.0	282.0	298.0	343.0
	3.826	4	241.0	236.0	223.0	198.0	482.0	490.0	518.0	595.0
	4.813	5	380.0	370.0	350.0	310.0	760.0	768.0	814.0	932.0
	5.761	6	550.0	540.0	510.0	450.0	1,100.0	1,121.0	1,186.0	1,352.0
	7.625	8	960.0	940.0	890.0	790.0	1,920.0	1,951.0	2,069.0	2,374.0
	9.750	10	1,560.0	1,540.0	1,450.0	1,290.0	3,118.0	3,196.0	3,371.0	3,876.0
	11.750	12	2,290.0	2,250.0	2,100.0	1,900.0	4,578.0	4,669.0	4,883.0	5,710.0

* Sizes 1, 1½, 2, and 2½ in. to be used in orifice pipe assemblies only.

1. Locate capacity in table above.
2. If line size is greater than 12 in., multiply capacity of 12-in. standard pipe listing by square of actual inside diameter in inches divided by 144.
3. If temperature is not covered by table, multiply capacity under 60°F listing by correction factor from Table 5-3.
4. If liquid is other than water, multiply capacity by specific-gravity correction factor from Table 5-4.

Table 5-3. Temperature Correction Factors Based on Water at 60°F
(Includes area-expansion changes for stainless-steel primary elements)

°F	Factor	°F	Factor	°F	Factor	°F	Factor	°F	Factor
				Weight Units					
50	1.0002	165	0.9900	265	0.9712	370	0.9432	475	0.9043
60	1.0000	170	0.9892	270	0.9701	375	0.9414	480	0.9022
70	0.9997	175	0.9885	275	0.9688	380	0.9398	485	0.9002
75	0.9994	180	0.9876	280	0.9678	385	0.9382	490	0.8979
80	0.9992	185	0.9866	285	0.9666	390	0.9365	495	0.8956
85	0.9990	190	0.9860	290	0.9653	395	0.9348	500	0.8932
90	0.9986	195	0.9851	295	0.9642	400	0.9342	505	0.8908
95	0.9982	200	0.9842	300	0.9630	405	0.9316	510	0.8882
100	0.9978	205	0.9830	305	0.9616	410	0.9298	515	0.8854
105	0.9973	210	0.9824	310	0.9603	415	0.9281	520	0.8831
110	0.9968	212	0.9819	315	0.9690	420	0.9263	525	0.8805
115	0.9962	215	0.9814	320	0.9577	425	0.9256	530	0.8776
120	0.9957	220	0.9806	325	0.9564	430	0.9228	535	0.8760
125	0.9952	225	0.9795	330	0.9550	435	0.9208	540	0.8718
130	0.9946	230	0.9784	335	0.9536	440	0.9188	545	0.8686
135	0.9941	235	0.9777	340	0.9521	445	0.9169	550	0.8657
140	0.9935	240	0.9766	345	0.9508	450	0.9150	555	0.8625
145	0.9929	245	0.9755	350	0.9493	455	0.9128	560	0.8594
150	0.9923	250	0.9746	355	0.9479	460	0.9108	565	0.8564
155	0.9916	255	0.9734	360	0.9464	465	0.9086	570	0.8534
160	0.9908	260	0.9723	365	0.9448	470	0.9064	575	0.8501
				Volume Units					
50	0.9950	165	1.0140	265	1.0377	370	1.0729	475	1.1236
60	1.0000	170	1.0150	270	1.0391	375	1.0750	480	1.1264
70	1.0007	175	1.0159	275	1.0406	380	1.0770	485	1.1273
75	1.0012	180	1.0169	280	1.0420	385	1.0790	490	1.1325
80	1.0016	185	1.0179	285	1.0436	390	1.0814	495	1.1357
85	1.0022	190	1.0191	290	1.0451	395	1.0834	500	1.1389
90	1.0027	195	1.0202	295	1.0466	400	1.0855	505	1.1422
95	1.0032	200	1.0213	300	1.0481	405	1.0876	510	1.1456
100	1.0038	205	1.0228	305	1.0496	410	1.0900	515	1.1491
105	1.0045	210	1.0236	310	1.0513	415	1.0922	520	1.1527
110	1.0050	212	1.0240	315	1.0529	420	1.0945	525	1.1563
115	1.0057	215	1.0249	320	1.0544	425	1.0967	530	1.1603
120	1.0065	220	1.0261	325	1.0561	430	1.0992	535	1.1643
125	1.0072	225	1.0273	330	1.0580	435	1.1016	540	1.1686
130	1.0080	230	1.0286	335	1.0598	440	1.1043	545	1.1731
135	1.0088	235	1.0297	340	1.0617	445	1.1070	550	1.1775
140	1.0096	240	1.0309	345	1.0634	450	1.1097	555	1.1820
145	1.0104	245	1.0323	350	1.0652	455	1.1124	560	1.1865
150	1.0112	250	1.0336	355	1.0670	460	1.1152	565	1.1907
155	1.0121	255	1.0349	360	1.0690	465	1.1180	570	1.1951
160	1.0130	260	1.0363	365	1.0708	470	1.1208	575	1.2000

tiplied by the capacity factors of 1.25, 1.50, 1.75, and 2.00 given in the second column, or the maximum capacity at all other standard differentials.

For a differential range of, in. of water	Maximum capacity from Table 5-7, lb./hr	% of maximum capacity	Orifice ratio from Table 5-5	Straight piping, diameters, from Figs. 5-9 to 5-12	
				Upstream	Downstream
75.36	37,500	80	0.640	10.1	3.6
117.75	46,875	64	0.582	8.4	3.4
169.56	56,250	53.3	0.536	7.6	3.2
230.79	65,625	45.7	0.499	7.1	3.1
301.44	75,000	40.0	0.469	6.8	3.0

Table 5-4. Specific-gravity Correction Factors for Liquids

Specific gravity	Factor	Specific gravity	Factor	Specific gravity	Factor	Specific gravity	Factor
			Weight Units				
0.50	0.7071	1.15	1.0724	1.80	1.3416	2.45	1.5653
0.55	0.7416	1.20	1.0954	1.85	1.3602	2.50	1.5811
0.60	0.7746	1.25	1.1180	1.90	1.3784	2.55	1.5969
0.65	0.8062	1.30	1.1402	1.95	1.3964	2.60	1.6125
0.70	0.8366	1.35	1.1619	2.00	1.4142	2.65	1.6279
0.75	0.8660	1.40	1.1832	2.05	1.4318	2.70	1.6432
0.80	0.8944	1.45	1.2042	2.10	1.4491	2.75	1.6583
0.85	0.9219	1.50	1.2247	2.15	1.4663	2.80	1.6733
0.90	0.9487	1.55	1.2450	2.20	1.4832	2.85	1.6882
0.95	0.9747	1.60	1.2650	2.25	1.5000	2.90	1.7029
1.00	1.0000	1.65	1.2845	2.30	1.5166	2.95	1.7176
1.05	1.0247	1.70	1.3038	2.35	1.5330	3.00	1.7321
1.10	1.0488	1.75	1.3229	2.40	1.5492		
			Volume Units				
0.50	1.4142	1.15	0.9325	1.80	0.7454	2.45	0.6389
0.55	1.3483	1.20	0.9129	1.85	0.7351	2.50	0.6325
0.60	1.2911	1.25	0.8944	1.90	0.7255	2.55	0.6262
0.65	1.2404	1.30	0.8771	1.95	0.7162	2.60	0.6202
0.70	1.1953	1.35	0.8607	2.00	0.7071	2.65	0.6143
0.75	1.1547	1.40	0.8452	2.05	0.6957	2.70	0.6086
0.80	1.1181	1.45	0.8304	2.10	0.6901	2.75	0.6030
0.85	1.0847	1.50	0.8165	2.15	0.6820	2.80	0.5976
0.90	1.0541	1.55	0.8032	2.20	0.6742	2.85	0.5924
0.95	1.0259	1.60	0.7905	2.25	0.6667	2.90	0.5872
1.00	1.0000	1.65	0.7785	2.30	0.6594	2.95	0.5822
1.05	0.9759	1.70	0.7670	2.35	0.6523	3.00	0.5774
1.10	0.9534	1.75	0.7559	2.40	0.6455		

Table 5-5. d/D Diameter Ratios

% of maximum capacity	Orifice plate	Flow nozzle	% of maximum capacity	Orifice plate	Flow nozzle
10	0.235	0.263	56	0.548	0.600
12	0.258	0.288	58	0.557	0.608
14	0.280	0.313	60	0.565	0.617
16	0.298	0.333	62	0.574	0.625
18	0.318	0.352	64	0.582	0.634
20	0.335	0.370	66	0.589	0.641
22	0.350	0.381	68	0.598	0.649
24	0.367	0.403	70	0.605	0.657
26	0.381	0.419	72	0.612	0.664
28	0.395	0.434	74	0.619	0.670
30	0.408	0.448	76	0.627	0.678
32	0.422	0.463	78	0.634	0.685
34	0.435	0.475	80	0.640	0.692
36	0.443	0.483	82	0.647	0.698
38	0.458	0.502	84	0.654	0.704
40	0.469	0.514	86	0.659	0.711
42	0.480	0.527	88	0.669	0.710
44	0.491	0.538	90	0.673	0.724
46	0.500	0.548	92	0.678	0.729
48	0.510	0.560	94	0.684	0.735
50	0.520	0.570	96	0.689	0.740
52	0.530	0.580	98	0.694	0.745
54	0.539	0.590	100	0.700	0.750

Table 5-6. Commercial Meter Ranges

Differential range			Differential range		
In. Hg	In. of water (wet)	Capacity factor	In. Hg	In. of water (wet)	Capacity factor
.....	1.85	0.157	6.000	75.36	1.000
.....	2.32	0.175	6.370	80.12	1.031
.....	3.71	0.222	7.375	92.58	1.108
.....	4.63	0.248	8.048	103.31	1.171
.....	5.56	0.272	8.714	112.09	1.219
.....	7.41	0.314	9.375	117.75	1.250
.....	9.26	0.351	9.552	120.00	1.262
1.054	13.25	0.419	10.099	127.02	1.298
1.475	18.51	0.496	11.062	138.94	1.358
1.500	18.84	0.500	12.574	161.42	1.462
1.600	20.12	0.513	13.500	169.56	1.500
1.840	23.15	0.554	14.328	180.00	1.545
2.229	28.00	0.609	14.335	183.65	1.561
2.286	29.41	0.625	14.749	185.26	1.568
2.536	31.90	0.651	16.000	201.23	1.634
2.584	32.42	0.656	16.875	212.00	1.677
3.375	42.39	0.750	18.375	230.79	1.750
3.569	45.91	0.781	22.399	286.96	1.951
3.687	46.29	0.784	24.000	301.44	2.000
4.019	50.55	0.819	25.394	319.03	2.058
4.218	53.00	0.838	29.529	370.52	2.217
5.150	66.29	0.937	37.500	471.00	2.500
5.577	71.74	0.976	54.000	678.24	3.000

Table 5-7. Maximum Steam Capacities

(With differential of 75.36 in. water; expressed in 1,000 lb/hr of steam)

Primary element	Inside pipe diameter	Nominal pipe size, in.	Steam pressures, psig															
			100	150	200	250	300	350	400	450	500	600	700	800	900	1,000	1,500	2,000
Standard pipe, 0.70d/D orifice plate	1.049	1*	0.60	0.71	0.81	0.89	0.97	1.04	1.12	1.18	1.25	1.37	1.48	1.59	1.67	1.77	2.27	2.75
	1.610	1½*	1.4	1.7	1.9	2.1	2.3	2.46	2.63	2.78	2.94	3.22	3.48	3.74	3.93	4.17	5.34	6.48
	2.067	2*	2.3	2.8	3.1	3.5	3.8	4.05	4.34	4.58	4.84	5.30	5.74	6.16	6.48	6.87	8.79	10.68
	2.469	2½*	3.3	3.9	4.5	4.9	5.4	5.78	6.18	6.52	6.89	7.56	8.18	8.77	9.24	9.79	12.55	15.23
	3.068	3	5.1	6.1	6.9	7.6	8.3	8.91	9.53	10.1	10.6	11.7	12.6	13.5	14.2	15.1	19.38	23.52
	4.026	4	8.8	10.4	11.9	13.1	14.3	15.3	16.4	17.3	18.3	20.1	21.7	23.3	24.5	26.0	33.37	40.50
	5.047	5	13.8	16.4	18.6	20.6	22.5	24.1	25.8	27.2	28.8	31.5	34.1	36.6	38.5	40.8	52.45	63.65
	6.065	6	19.9	23.7	26.9	29.8	32.5	34.8	37.2	39.3	41.5	45.5	49.3	52.9	55.7	59.0	75.74	91.92
	7.981	8	34.4	41.0	46.6	51.6	56.2	60.2	64.4	68.0	71.9	78.8	85.3	91.5	96.3	102.	131.2	159.2
	10.020	10	54.3	64.6	73.4	81.3	88.5	94.9	102.	107.	113.	124.	134.	144.	152.	161.	206.7	250.9
	12.000	12	77.8	92.6	105.	117.	127.	136.	146.	154.	162.	178.	193.	207.	218.	231.	293.5	356.1
Extra-heavy pipe, 0.70d/D orifice plate	0.957	1	0.50	0.59	0.67	0.74	0.81	0.87	0.93	0.98	1.04	1.14	1.23	1.32	1.39	1.47	1.89	2.29
	1.500	1½*	1.22	1.45	1.65	1.82	1.99	2.13	2.28	2.41	2.55	2.79	3.02	3.24	3.41	3.62	4.63	5.62
	1.939	2*	2.04	2.43	2.76	3.06	3.33	3.57	3.82	4.03	4.26	4.67	5.05	5.42	5.71	6.05	7.74	9.40
	2.323	2½*	2.92	3.48	3.96	4.38	4.77	5.11	5.47	5.77	6.10	6.69	7.24	7.77	8.18	8.67	11.11	13.48
	2.900	3	4.55	5.41	6.16	6.82	7.42	7.96	8.51	8.98	9.50	10.4	11.3	12.1	12.7	13.5	17.32	21.02
	3.826	4	7.92	9.43	10.7	11.9	12.9	13.9	14.8	15.6	16.5	18.1	19.6	21.0	22.2	23.5	30.14	36.58
	4.813	5	12.5	14.9	17.0	18.8	20.4	21.9	23.5	24.7	26.2	28.7	31.0	33.3	35.1	37.1	47.70	57.89
	5.761	6	18.0	21.4	24.3	26.9	29.3	31.4	33.6	35.4	37.5	41.0	44.5	47.7	50.2	53.2	68.35	82.95
	7.625	8	31.4	37.4	42.5	47.1	51.3	55.0	58.8	62.0	65.6	71.9	77.8	83.5	87.9	93.1	119.7	145.3
	9.750	10	51.4	61.2	69.5	77.0	83.8	89.9	96.2	101.	107.	118.	127.	137.	144.	152.	188.3	228.6
	11.750	12	74.6	88.7	101.	111.	122.	130.	140.	147.	156.	171.	185.	198.	209.	221.	266.5	323.4
Standard pipe, 0.75d/D flow nozzle	3.068	3	9.8	11.8	13.4	14.9	16.3	17.5	18.6	19.6	20.6	22.5	24.4	26.3	27.5	29.2	37.8	45.9
	4.026	4	17.1	20.	23.	25.	28.	30.	32.	34.	36.	39.	42.	45.	48.	51.	65.2	79.1
	5.047	5	27.	32.	36.	40.	43.	47.	51.	54.	57.	63.	67.	71.	76.	80.	102.	124.
	6.065	6	39.	46.	53.	58.	63.	68.	72.	77.	81.	89.	96.	103.	110.	116.	148.	179.
	7.981	8	67.	82.	92.	101.	109.	117.	125.	133.	140.	153.	165.	178.	188.	200.	256.	311.
	10.020	10	105.	126.	144.	160.	174.	186.	197.	208.	219.	240.	260.	280.	295.	313.	403.	489.
	12.000	12	151.	183.	210.	230.	250.	270.	280.	300.	320.	350.	380.	400.	424.	450.	572.	695.

Table 5-7. Maximum Steam Capacities (*Continued*)

Primary element	Inside pipe diameter	Nominal pipe size, in.	Steam pressures, psig															
			100	150	200	250	300	350	400	450	500	600	700	800	900	1,000	1,500	2,000
Extra-heavy pipe, 0.75d/D flow nozzle	2.900	3	8.8	10.6	12.1	13.3	14.5	15.7	16.7	17.7	18.7	20.5	22.	23.	25.	26.	33.8	41.0
	3.826	4	15.3	18.5	20.8	23.	25.	27.	29.	31.	32.	36.	39.	41.	43.	46.	58.8	71.4
	4.813	5	24.	29.	33.	36.	39.	43.	46.	49.	51.	57.	61.	65.	67.	72.	93.1	113.
	5.761	6	35.	41.	47.	53.	57.	61.	65.	69.	73.	80.	86.	92.	98.	104.	133.	162.
	7.625	8	61.	72.	82.	92.	100.	108.	115.	122.	128.	141.	153.	163.	171.	182.	233.	283.
	9.750	10	100.	119.	135.	151.	165.	177.	188.	200.	210.	230.	250.	267.	281.	298.	367.	446.
	11.750	12	145.	172.	196.	220.	240.	260.	270.	290.	310.	330.	360.	390.	407.	432.	520.	631.

* Sizes 1, 1½, 2, and 2½ in. to be used in orifice pipe assemblies only.

1. Locate capacity in table above.
2. If line size is greater than 12 in., multiply capacity of 12-in. standard pipe listing by square of actual inside diameter in inches divided by 144.
3. If pressure is not covered by capacity table, multiply capacity under 100 psig listing by correction factor from Table 5-9.
4. If superheat or moisture is involved, multiply capacity by correction factor from Table 5-10 or 5-11.
5. Use Table 5-8 to determine superheat by subtracting equivalent temperature from total steam temperature.

Table 5-8. Pressure-Temperature Equivalents for Steam
(Based on 14.7 psia)

Psig	°F	Psig	°F	Psig	°F	Psig	°F	Psig	°F
0	212.01	65	311.77	360	438.24	690	503.84	1,000	546.36
1	215.33	70	316.00	370	440.78	700	505.42	1,050	552.26
2	218.50	75	320.03	380	443.28	710	506.99	1,100	557.94
3	221.52	80	323.89	390	445.74	720	508.53	1,150	563.44
4	224.39	85	327.59	400	448.14	730	509.96	1,200	568.74
5	227.14	90	331.14	410	450.50	740	511.56	1,250	573.90
6	229.79	95	334.57	420	452.81	750	513.05	1,300	578.94
7	232.32	100	337.87	430	455.08	760	514.55	1,350	583.78
8	234.76	110	344.15	440	457.32	770	516.00	1,400	588.45
9	237.12	120	350.04	450	459.52	780	517.45	1,450	593.02
10	239.39	130	355.59	460	461.69	790	518.90	1,500	597.48
11	241.60	140	360.85	470	463.82	800	520.32	1,550	601.82
12	243.73	150	365.85	480	465.91	810	521.75	1,600	606.11
13	245.79	160	370.61	490	467.97	820	523.13	1,650	610.24
14	247.80	170	375.17	500	470.00	830	524.52	1,700	614.30
15	249.75	180	379.54	510	472.00	840	525.90	1,750	618.22
16	251.65	190	383.74	520	473.98	850	527.26	1,800	622.11
17	253.50	200	387.77	530	475.92	860	528.62	1,850	625.90
18	255.30	210	391.67	540	477.84	870	529.95	1,900	629.64
19	257.06	220	395.43	550	479.83	880	531.27	1,950	633.24
20	258.27	230	399.07	560	481.60	890	532.59	2,000	636.83
21	260.45	240	402.59	570	483.44	900	533.90		
22	262.08	250	406.01	580	485.26	910	535.20		
23	263.68	260	409.33	590	487.05	920	536.43		
24	265.25	270	412.55	600	488.82	930	537.74		
25	266.79	280	415.69	610	490.57	940	539.01		
30	274.02	290	418.76	620	492.30	950	540.22		
35	280.63	300	421.74	630	494.00	960	541.49		
40	286.72	310	424.65	640	495.69	970	542.73		
45	292.38	320	427.49	650	497.36	980	543.96		
50	297.66	330	430.26	660	499.01	990	545.17		
55	302.63	340	432.98	670	500.64				
60	307.32	350	435.64	680	502.25				

Since the actual capacity is to be 30,000, this value is divided by these maximum capacities to obtain the third column "% of maximum capacity." From these values and by using Table 5-5, the approximate orifice ratio listed in the fourth column is determined.

Using Figs. 5-9 to 5-12, the data can be completed by listing in columns 5 and 6 the upstream and downstream straight-pipe requirements.

Using any one of the five ranges, we can have our meter read 30,000 lb/hr flow rate at its highest graduation, but to do so the orifice bore must be made larger as the differential range becomes smaller. However, as the orifice bore becomes larger (large d/D ratio) we need a longer run of straight pipe. Because we have only 5 ft of straight pipe available, the only acceptable solution is the highest differential range 301.44 in.

Table 5-9. Pressure Correction Factor for Steam

Psig	Factor	Psig	Factor	Psig	Factor	Psig	Factor	Psig	Factor
1	0.3929	55	0.7902	205	1.3646	410	1.8861	1,000	2.9787
2	0.4044	60	0.8164	210	1.3795	420	1.9096	1,050	3.0618
3	0.4155	65	0.8418	215	1.3943	430	1.9317	1,100	3.1420
4	0.4264	70	0.8663	220	1.4089	440	1.9536	1,150	3.2223
5	0.4369	75	0.8901	225	1.4234	450	1.9734	1,200	3.3021
6	0.4472	80	0.9132	230	1.4377	460	1.9966	1,250	3.3812
7	0.4572	85	0.9357	235	1.4519	470	2.0178	1,300	3.4604
8	0.4669	90	0.9576	240	1.4660	480	2.0389	1,350	3.5386
9	0.4764	95	0.9791	245	1.4799	490	2.0596	1,400	3.6168
10	0.4857	100	1.0000	250	1.4937	500	2.0780	1,450	3.6959
11	0.4949	105	1.0204	255	1.5074	510	2.1006	1,500	3.7958
12	0.5036	110	1.0404	260	1.5210	520	2.1209	1,550	3.8504
13	0.5125	115	1.0601	265	1.5344	530	2.1413	1,600	3.9525
14	0.5211	120	1.0794	270	1.5477	540	2.1613	1,650	4.0168
15	0.5295	125	1.0982	275	1.5610	550	2.1789	1,700	4.0839
16	0.5378	130	1.1168	280	1.5741	560	2.2014	1,750	4.1673
17	0.5466	135	1.1350	285	1.5871	570	2.2211	1,800	4.2436
18	0.5539	140	1.1530	290	1.5999	580	2.2411	1,850	4.3286
19	0.5618	145	1.1707	295	1.6129	590	2.2606	1,900	4.4062
20	0.5696	150	1.1881	300	1.6255	600	2.2765	1,950	4.4926
21	0.5779	155	1.2051	310	1.6507	650	2.3713	2,000	4.5704
22	0.5847	160	1.2220	320	1.6755	700	2.4634		
23	0.5921	165	1.2387	330	1.7001	750	2.5534		
24	0.5995	170	1.2551	340	1.7243	800	2.6414		
25	0.6067	175	1.2713	350	1.7482	850	2.7278		
30	0.6414	180	1.2874	360	1.7718	900	2.7960		
35	0.6741	185	1.3032	370	1.7952	950	2.8966		
40	0.7051	190	1.3189	380	1.8183	1,000	2.9787		
45	0.7347	195	1.3342	390	1.8403				
50	0.7630	200	1.3495	400	1.8637				

All this is based on using an orifice, whereas by using a nozzle much greater capacity could be secured for each of these differential ranges. Therefore, if with an orifice the highest differential range is still insufficient, the more expensive flow nozzle will be useful.

The correction factor tables are useful, both with meters already installed but operating under conditions other than those for which the meter was calibrated, and with the capacity tables.

Correction for Steam Pressure. Assume that the steam capacity of the meter is to be determined for 90 psi, which is not listed in Table 5-7. Table 5-9 lists correction factors based on 100 lb as unity, showing a factor of 0.9576 for 90 psi. Therefore, the capacities listed in Table 5-7 under 100 psi can be multiplied by the factor 0.9576 to obtain the correct capacity for 90 psi.

Correction for Steam Pressure Different from That for Which Meter Was Calibrated. Assume that the steam meter had been calibrated for a capacity of 500,000 lb/hr at 400 psi but is operated at 450 psi. From

Table 5-10. Superheat Correction Factors for Steam

°F	Pounds per square inch															
	0	100	200	300	400	500	600	800	1,000	1,200	1,500	1,600	1,700	1,800	1,900	2,000
10	0.992	0.992	0.992	0.991	0.990	0.989	0.988	0.986	0.984	0.982	0.978	0.976	0.975	0.972	0.968	0.967
20	0.984	0.984	0.983	0.982	0.981	0.979	0.977	0.974	0.971	0.967	0.957	0.956	0.953	0.949	0.944	0.941
30	0.976	0.976	0.975	0.974	0.972	0.970	0.968	0.963	0.958	0.953	0.941	0.939	0.935	0.930	0.925	0.920
40	0.969	0.969	0.968	0.966	0.964	0.961	0.958	0.952	0.947	0.941	0.927	0.924	0.920	0.914	0.907	0.902
50	0.962	0.962	0.960	0.958	0.956	0.953	0.950	0.942	0.935	0.928	0.914	0.911	0.906	0.900	0.893	0.887
60	0.955	0.955	0.953	0.951	0.948	0.945	0.941	0.934	0.926	0.918	0.902	0.898	0.893	0.887	0.879	0.873
70	0.949	0.949	0.947	0.944	0.941	0.937	0.933	0.925	0.917	0.909	0.893	0.887	0.882	0.875	0.867	0.861
80	0.942	0.942	0.940	0.937	0.934	0.930	0.925	0.917	0.909	0.901	0.883	0.877	0.871	0.864	0.856	0.849
90	0.936	0.936	0.934	0.931	0.927	0.922	0.918	0.909	0.900	0.891	0.873	0.867	0.861	0.851	0.846	0.839
100	0.930	0.930	0.927	0.924	0.919	0.915	0.911	0.902	0.892	0.882	0.864	0.858	0.852	0.845	0.837	0.830
110	0.924	0.924	0.921	0.917	0.913	0.909	0.904	0.895	0.885	0.875	0.856	0.849	0.844	0.834	0.828	0.821
120	0.919	0.919	0.916	0.912	0.908	0.903	0.898	0.888	0.878	0.868	0.848	0.841	0.835	0.828	0.820	0.812
130	0.913	0.913	0.910	0.906	0.901	0.896	0.891	0.881	0.871	0.861	0.841	0.834	0.828	0.818	0.812	0.804
140	0.907	0.907	0.904	0.900	0.895	0.890	0.885	0.875	0.865	0.854	0.834	0.828	0.821	0.813	0.805	0.797
150	0.900	0.901	0.898	0.894	0.889	0.885	0.880	0.869	0.859	0.848	0.827	0.821	0.814	0.806	0.798	0.790
160	0.894	0.897	0.894	0.890	0.885	0.880	0.874	0.863	0.853	0.842	0.821	0.814	0.807	0.800	0.791	0.783
170	0.888	0.891	0.888	0.884	0.879	0.874	0.869	0.858	0.847	0.835	0.814	0.808	0.801	0.791	0.785	0.777
180	0.883	0.886	0.883	0.879	0.874	0.869	0.864	0.853	0.842	0.830	0.808	0.802	0.795	0.787	0.779	0.771
190	0.878	0.882	0.878	0.874	0.869	0.864	0.859	0.848	0.836	0.824	0.803	0.797	0.789	0.780	0.773	0.765
200	0.873	0.877	0.873	0.869	0.864	0.859	0.854	0.843	0.831	0.819	0.798	0.791	0.784	0.776	0.767	0.759
210	0.868	0.872	0.869	0.864	0.859	0.854	0.849	0.838	0.826	0.813	0.793	0.786	0.778	0.769	0.762	0.754
220	0.863	0.868	0.864	0.860	0.855	0.850	0.844	0.833	0.821	0.809	0.788	0.781	0.773	0.765	0.757	0.749
230	0.858	0.863	0.860	0.855	0.850	0.845	0.839	0.828	0.816	0.804	0.783	0.776	0.766	0.759	0.752	0.744
240	0.853	0.859	0.855	0.851	0.846	0.840	0.834	0.823	0.811	0.799	0.778	0.771	0.764	0.756	0.747	0.739
250	0.848	0.854	0.851	0.847	0.842	0.836	0.830	0.819	0.807	0.794	0.774	0.766	0.760	0.750	0.743	0.732

Table 5-10. Superheat Correction Factors for Steam (*Continued*)

°F	Pounds per square inch															
	0	100	200	300	400	500	600	800	1,000	1,200	1,500	1,600	1,700	1,800	1,900	2,000
260	0.844	0.850	0.847	0.843	0.838	0.832	0.826	0.814	0.802	0.789	0.769	0.762	0.755	0.746	0.738	0.730
270	0.839	0.846	0.843	0.839	0.834	0.828	0.822	0.810	0.798	0.786	0.765	0.758	0.750	0.741	0.734	0.726
280	0.835	0.842	0.839	0.834	0.829	0.824	0.818	0.806	0.794	0.781	0.760	0.753	0.746	0.738	0.730	0.721
290	0.830	0.837	0.834	0.830	0.825	0.820	0.814	0.802	0.790	0.777	0.756	0.749	0.742	0.734	0.726	0.717
300	0.826	0.834	0.831	0.826	0.821	0.815	0.809	0.798	0.786	0.774	0.753	0.745	0.738	0.730	0.721	0.713
350	0.805	0.814	0.811	0.807	0.802	0.797	0.791	0.780	0.768	0.755	0.733	0.726	0.719	0.711	0.703	0.695
400	0.786	0.796	0.793	0.790	0.785	0.780	0.774	0.763	0.751	0.738	0.716	0.709	0.702	0.694	0.686	0.678
450	0.768	0.778	0.776	0.773	0.769	0.764	0.759	0.747	0.735	0.721	0.702	0.695	0.687	0.679	0.671	0.668
500	0.751	0.763	0.761	0.758	0.754	0.749	0.744	0.733	0.720	0.708	0.689	0.682	0.673	0.665	0.659	0.649
600	0.721	0.736	0.733	0.730	0.726	0.720	0.716	0.705	0.695	0.682	0.664	0.657	0.648	0.645	0.633	0.625
700	0.695	0.709	0.708	0.705	0.701	0.696	0.692	0.681	0.672	0.660	0.641	0.635	0.626	0.619	0.611	0.604
800	0.671	0.686	0.686	0.683	0.679	0.675	0.670	0.660	0.651	0.639	0.621	0.615	0.606	0.600	0.592	0.585
900	0.649	0.665	0.666	0.663	0.660	0.656	0.651	0.641	0.632	0.620	0.603	0.596	0.589	0.582	0.575	0.568
1000	0.629	0.646	0.647	0.645	0.641	0.638	0.633	0.624	0.614	0.603	0.586	—	—	—	—	—
1100	0.611	0.629	0.630	0.628	0.625	0.621	0.617	—	—	—	—	—	—	—	—	—

Table 5-11. Moisture Correction Factors for Steam

Quality, %	Factor	Quality, %	Factor
0	1.000	6	1.033
1	1.006	7	1.038
2	1.011	8	1.043
3	1.016	9	1.049
4	1.022	10	1.055
5	1.027		

Table 5-9, the factor for 400 psi is found to be 1.8637, and for 450 psi the factor is 1.9734.

Divide:

$$\frac{\text{Actual pressure factor}}{\text{Calibrated pressure factor}} = \frac{1.9734}{1.8637} \quad \text{or } 1.0589$$

Therefore, the meter capacity or readings must be multiplied by 1.0589 to be correct for the difference in pressure.

Correction for Steam, Temperature, or Moisture. For corrections for differences in temperature, moisture content, and specific gravity, follow the same procedure, always dividing the factor for actual conditions by the

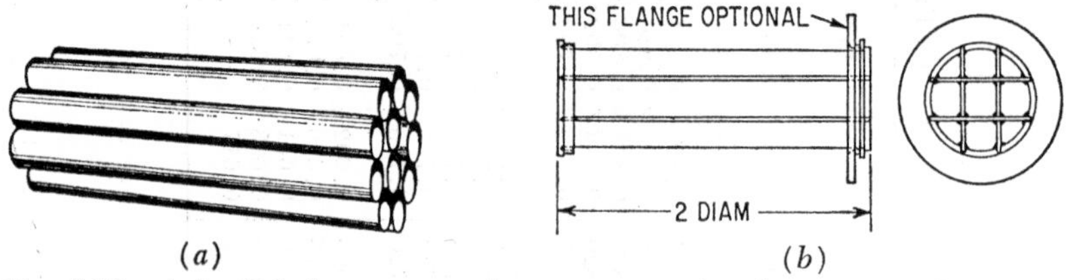

Fig. 5-13. (*a*) Tubular-type straightening vane. (*Foxboro Co.*) (*b*) "Egg-crate-type" straightening vane. (*Bailey Meter Co.*)

factor for the calibrated conditions, and multiply the meter readings by the resulting factor.

Water-meter Correction Factors. For liquids, different correction factors must be used when the maximum capacity is expressed in volumetric units such as gallons, cubic feet, or liters, as opposed to those used when the maximum capacity is expressed in weight units such as pounds, tons, or kilograms. Therefore, two sets of tables are furnished, each for temperature and specific-gravity corrections—one set for volume and one set for weights. The use of the correction factor tables is exactly the same as for steam.

Straightening Vanes. If the flow within 15 pipe diameters upstream from the pressure tap contains swirls, cross currents, or large eddies, or is otherwise turbulent, then accurate measurement is impossible unless straightening vanes (Fig. 5-13) are installed above the orifice.

Table 5-12. Maximum Air or Gas Capacities

(With differential of 81.36 in. of water; expressed in 1,000 cfm at 60°F)

Primary element	Inside pipe diameter	Nominal pipe size, in.	Air pressure, psig													
			5	10	15	20	25	50	75	100	125	150	175	200	225	250
Standard pipe, 0.70d/D orifice plate	1.049	1	0.0841	0.095	0.104	0.112	0.118	0.150	0.177	0.200	0.221	0.240	0.257	0.274	0.289	0.304
	1.610	1½	0.198	0.223	0.245	0.265	0.278	0.355	0.418	0.473	0.522	0.566	0.608	0.647	0.683	0.718
	2.067	2	0.327	0.367	0.403	0.436	0.457	0.583	0.686	0.776	0.856	0.930	0.998	1.062	1.122	1.179
	2.469	2½	0.466	0.523	0.574	0.622	0.656	0.837	0.985	1.114	1.229	1.335	1.433	1.524	1.610	1.692
	3.068	3	0.718	0.806	0.886	0.958	1.010	1.289	1.518	1.716	1.894	2.056	2.207	2.348	2.480	2.607
	4.026	4	1.235	1.387	1.525	1.650	1.738	2.219	2.613	2.954	3.260	3.539	3.799	4.042	4.270	4.487
	5.047	5	1.940	2.180	2.395	2.593	2.743	3.501	4.122	4.661	5.144	5.585	5.994	6.378	6.737	7.080
	6.065	6	2.804	3.149	3.460	3.743	3.896	4.973	5.856	6.621	7.307	7.933	8.515	9.060	9.571	10.06
	7.981	8	4.856	5.455	5.992	6.485	6.844	8.735	10.29	11.63	12.83	13.94	14.96	15.91	16.81	17.67
	10.020	10	7.657	8.590	9.445	10.22	10.78	13.76	16.20	18.32	20.22	21.95	23.56	25.07	26.48	27.83
	12.000	12	10.98	12.33	13.55	14.65	14.91	19.03	22.40	25.33	27.95	30.35	32.57	34.66	36.61	38.48
Standard pipe, 0.75d/D flow nozzle	3.068	3	1.435	1.612	1.772	1.916	2.020	2.578	3.035	3.432	3.788	4.112	4.414	4.696	4.961	5.213
	4.026	4	1.470	2.774	3.050	3.300	3.476	4.437	5.224	5.907	6.519	7.078	7.596	8.083	8.539	8.973
	5.047	5	3.880	4.360	4.790	5.186	5.487	7.003	8.245	9.323	10.29	11.17	11.99	12.76	13.48	14.16
	6.065	6	5.608	6.298	6.920	7.486	7.792	9.945	11.71	13.24	14.61	15.86	17.03	18.12	19.14	20.11
	7.981	8	9.712	10.91	11.98	12.97	13.69	17.48	20.58	23.27	25.68	27.88	29.93	31.84	33.64	35.35
	10.020	10	15.31	17.18	18.89	20.44	21.56	27.51	32.40	36.63	40.42	43.89	47.11	50.12	52.95	55.64
	12.000	12	21.96	24.66	27.10	29.30	29.81	38.05	44.80	50.66	55.91	60.70	65.15	69.32	73.23	76.95

1. Locate capacity in table above.
2. If line size is greater than 12 in., multiply capacity of 12-in. standard pipe listing by square of actual inside diameter in inches divided by 144.
3. If pressure is not covered by capacity table, multiply capacity under 100 psig listing by correction factor from Table 5-14.
4. If temperature is other than 60°F, multiply capacity by correction factor from Table 5-18.
5. If gas is other than air, multiply capacity by correction factor from Table 5-19.

Table 5-13. Commercial Meter Ranges

Differential range		Capacity factor	Differential range		Capacity factor
In. Hg	In. of water (dry)		In. Hg	In. of water (dry)	
.....	1.00	0.111	6.000	81.36	1.000
.....	2.00	0.157	6.370	86.48	1.031
.....	2.50	0.175	7.375	100.00	1.108
.....	4.00	0.222	8.048	109.45	1.171
.....	5.00	0.248	8.714	118.51	1.219
.....	6.00	0.272	9.375	127.13	1.250
.....	8.00	0.314	9.552	129.53	1.262
.....	10.00	0.351	10.099	137.11	1.298
1.054	14.30	0.419	11.062	150.00	1.358
1.475	20.00	0.496	12.574	171.00	1.462
1.500	20.34	0.500	13.500	183.06	1.500
1.600	21.72	0.513	14.328	194.30	1.545
1.840	25.00	0.554	14.335	194.95	1.561
2.229	30.23	0.609	14.749	200.00	1.568
2.286	31.09	0.625	16.000	217.23	1.634
2.536	34.43	0.651	16.875	228.85	1.677
2.584	35.00	0.656	18.375	249.17	1.750
3.375	45.77	0.750	22.399	304.63	1.951
3.569	48.54	0.781	24.000	325.44	2.000
3.687	50.00	0.784	25.394	344.39	2.058
4.019	54.57	0.819	29.529	400.00	2.217
4.218	57.21	0.838	37.500	508.50	2.500
5.150	70.04	0.937	54.000	732.24	3.000
5.577	75.85	0.976	55.302	750.00	3.036

Table 5-14. Gas-pressure Correction Factors

(Based on 30 in. of mercury)

Psig	Factor	Psig	Factor	Psig	Factor	Psig	Factor
1	0.3704	27	0.6031	65	0.8337	200	1.3683
2	0.3820	28	0.6103	70	0.8594	205	1.3839
3	0.3932	29	0.6175	75	0.8844	210	1.3995
4	0.4041	30	0.6244	80	0.9086	215	1.4150
5	0.4148	31	0.6314	85	0.9324	220	1.4303
6	0.4251	32	0.6383	90	0.9554	225	1.4455
7	0.4353	33	0.6450	95	0.9780	230	1.4605
8	0.4452	34	0.6517	100	1.0000	235	1.4754
9	0.4548	35	0.6584	105	1.0215	240	1.4900
10	0.4643	36	0.6651	110	1.0427	245	1.5046
11	0.4736	37	0.6715	115	1.0633	250	1.5190
12	0.4827	38	0.6780	120	1.0837	255	1.5333
13	0.4917	39	0.6844	125	1.1036	260	1.5474
14	0.5004	40	0.6907	130	1.1232	265	1.5615
15	0.5091	41	0.6970	135	1.1424	270	1.5753
16	0.5176	42	0.7032	140	1.1614	275	1.5891
17	0.5260	43	0.7094	145	1.1799	280	1.6027
18	0.5342	44	0.7155	150	1.1982	285	1.6164
19	0.5422	45	0.7215	155	1.2163	290	1.6297
20	0.5502	46	0.7276	160	1.2341	295	1.6431
21	0.5581	47	0.7335	165	1.2516	300	1.6562
22	0.5659	48	0.7394	170	1.2689	305	1.6695
23	0.5735	49	0.7454	175	1.2860	310	1.6823
24	0.5810	50	0.7511	180	1.3028	315	1.6954
25	0.5885	55	0.7796	185	1.3194	320	1.7085
26	0.5958	60	0.8071	190	1.3359	325	1.7209

Table 5-15. Maximum Air or Gas Capacities for Low Pressures

(With differential of 4.0 in. of water; expressed in cubic feet per minute at 60°F)

Primary element	Nominal pipe size, in.	Inside pipe diameter	Air pressure, inches of mercury, absolute												
			30	35	40	45	50	55	60	65	70	75	80	85	90
Standard pipe, 0.70d/D orifice plate	1	1.049	15.9	17.2	18.4	19.5	20.5	21.5	22.5	23.4	24.3	25.1	26.0	26.8	27.5
	1½	1.610	37.6	40.6	43.4	46.0	48.5	50.9	53.2	55.3	57.4	59.4	61.4	63.3	65.1
	2	2.067	61.7	66.6	71.2	75.6	79.7	83.5	87.3	90.8	94.2	97.6	100.8	103.9	106.9
	2½	2.469	88.5	95.6	102.2	108.4	114.3	119.8	125.2	130.3	135.2	139.9	144.5	149.0	153.3
	3	3.068	136.4	147.3	157.5	167.0	176.1	184.7	192.9	200.8	208.4	215.7	222.7	229.6	236.2
	4	4.026	234.8	253.6	271.1	287.6	303.1	317.9	332.1	345.6	358.7	371.2	383.4	395.2	406.7
	5	5.047	370.4	400.1	427.7	453.6	478.2	501.5	523.8	545.2	565.8	585.6	604.9	623.5	641.5
	6	6.065	526.2	568.3	607.6	644.4	679.3	712.5	744.2	774.5	803.8	832.0	859.3	885.7	911.4
	8	7.981	924.3	998.3	1067	1132	1193	1252	1307	1360	1412	1461	1509	1556	1601
	10	10.020	1456	1573	1681	1783	1880	1971	2059	2143	2224	2302	2378	2451	2522
	12	12.000	2013	2174	2324	2465	2599	2726	2847	2963	3075	3183	3287	3388	3487

1. Locate capacity in table above.
2. If line size is greater than 12 in., multiply capacity of 12-in. standard pipe listing by square of actual inside diameter in inches divided by 144.
3. If pressure is not covered by capacity table, multiply capacity under 30 inches of mercury listing by correction factor from Table 5-16.
4. If temperature is other than 60°F, multiply capacity by correction factor from Table 5-17.
5. If gas is other than air, multiply capacity by specific-gravity correction factor from Table 5-18.

The greatest possibility of excessive turbulence occurs when the pipeline upstream from the orifice has bends in two planes, or contains a regulating valve or similar flow-disturbing apparatus. Under these conditions vanes are often essential to accurate measurement. They may be omitted if there are long straight runs of pipe upstream from the orifice.

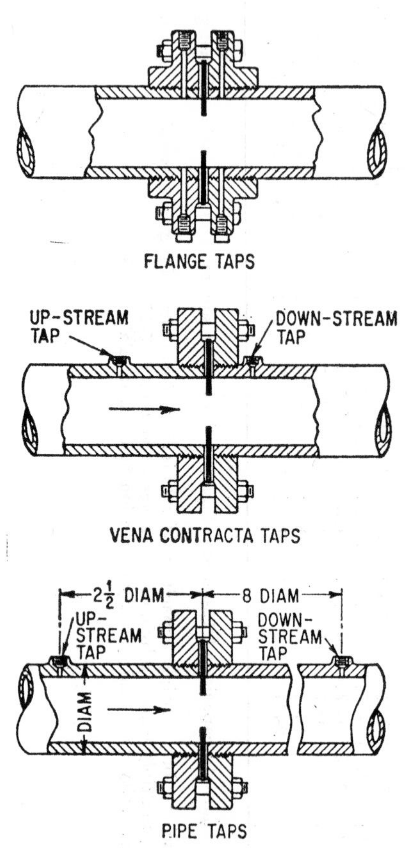

FIG. 5-14. Locations of orifice taps. (*Minneapolis-Honeywell Regulator Co.*)

Location of Taps.[4] *Flange Taps.* These are the most universally used and are the only taps concerning the use of which satisfactory agreement exists between the ASME and the AGA. The center of the upstream tap is located 1 in. from the upstream face of the orifice plate, and the center of the downstream tap is placed 1 in. from the downstream face of the orifice plate (Fig. 5-14).

Table 5-16. Meter Capacity Factors
(Based on 4 in. of water differential)

Inches	Factor	Inches	Factor
1.00	0.500	11	1.658
2.00	0.707	12	1.782
2.25	0.750	13	1.803
3	0.866	14	1.870
4	1.000	15	1.937
5	1.118	16	2.000
6	1.225	17	2.062
7	1.323	18	2.121
8	1.414	19	2.179
9	1.500	20	2.236
10	1.581		

Vena Contracta Taps. This type is employed mostly in large-size pipes where the use of a flange union is impractical, or in any size pipes where requirement for special material rules out the use of a flange union. The center of the upstream tap is located one internal pipe diameter from the upstream face of the orifice plate; the location of the center of the downstream tap, which is the point of minimum pressure or highest velocity, is a function of orifice ratio, and may be determined from the curve shown in Fig. 5-15. This point of minimum pressure and highest local velocity is called the point of vena contracta.

[4] Fischer & Porter Co.

Table 5-17. Gas-pressure Correction Factors
(In inches of mercury absolute)

In. Hg	Factor	In. Hg	Factor	In. Hg	Factor	In. Hg	Factor
29	0.9832	45	1.2247	61	1.4259	77	1.6021
30	1.0000	46	1.2382	62	1.4376	78	1.6124
31	1.0165	47	1.2516	63	1.4491	79	1.6227
32	1.0328	48	1.2649	64	1.4606	80	1.6330
33	1.0488	49	1.2780	65	1.4719	81	1.6431
34	1.0646	50	1.2910	66	1.4832	82	1.6532
35	1.0801	51	1.3038	67	1.4944	83	1.6633
36	1.0954	52	1.3165	68	1.5055	84	1.6733
37	1.1105	53	1.3291	69	1.5165	85	1.6832
38	1.1254	54	1.3416	70	1.5275	86	1.6931
39	1.1401	55	1.3540	71	1.5384	87	1.7029
40	1.1547	56	1.3662	72	1.5492	88	1.7127
41	1.1690	57	1.3784	73	1.5599	89	1.7224
42	1.1832	58	1.3904	74	1.5705	90	1.7320
43	1.1972	59	1.4023	75	1.5811		
44	1.2110	60	1.4142	76	1.5916		

Table 5-18. Gas-temperature Correction Factors

°F	Factor	°F	Factor	°F	Factor	°F	Factor
0	1.0633	110	0.9552	250	0.8557	550	0.7175
10	1.0518	120	0.9468	260	0.8498	575	0.7087
20	1.0408	130	0.9388	270	0.8439	600	0.7003
30	1.0302	140	0.9309	280	0.8382	625	0.6922
40	1.0199	150	0.9233	290	0.8326	650	0.6843
50	1.0098	160	0.9158	300	0.8271	700	0.6694
60	1.0000	170	0.9085	325	0.8138	750	0.6554
65	0.9953	175	0.9049	350	0.8011	800	0.6423
70	0.9905	180	0.9013	375	0.7891	850	0.6299
75	0.9859	190	0.8944	400	0.7775	900	0.6182
80	0.9813	200	0.8876	425	0.7664	950	0.6071
85	0.9768	210	0.8809	450	0.7558	1000	0.5967
90	0.9724	220	0.8744	475	0.7457		
95	0.9679	230	0.8680	500	0.7359		
100	0.9636	240	0.8619	525	0.7265		

Table 5-19A. Gas Specific-gravity Factors
(Density of air at 30 in. Hg and 60°F = 0.076561 pcf)

Specific gravity	Factor	Specific gravity	Factor	Specific gravity	Factor	Specific gravity	Factor
0.30	1.8257	0.70	1.1953	1.10	0.9535	1.50	0.8165
0.35	1.6903	0.75	1.1547	1.15	0.9325	1.55	0.8032
0.40	1.5811	0.80	1.1180	1.20	0.9129	1.60	0.7906
0.45	1.4907	0.85	1.0847	1.25	0.8944	1.65	0.7785
0.50	1.4142	0.90	1.0541	1.30	0.8770	1.70	0.7669
0.55	1.3484	0.95	1.0260	1.35	0.8607	1.75	0.7559
0.60	1.2910	1.00	1.0000	1.40	0.8452		
0.65	1.2403	1.05	0.9759	1.45	0.8305		

Table 5-19B. Pressure Equivalents

Psi	In. Hg	Ft water	In. water
14.7	29.92	33.91	406.9
1.0	2.036	2.307	27.68
0.4913	1.0	1.1322	13.59
0.433	0.8819	1.0	12.00
0.03611	0.0735	0.0833	1.0

Pipe Taps. Use of these taps permits direct determination of actual permanent pressure loss, thus giving a head differential less than that for flange taps at a given rate of flow. The center of the upstream tap must be 2½ nominal pipe diameters from the orifice, while the center of the downstream tap must be 8 nominal pipe diameters below the orifice.

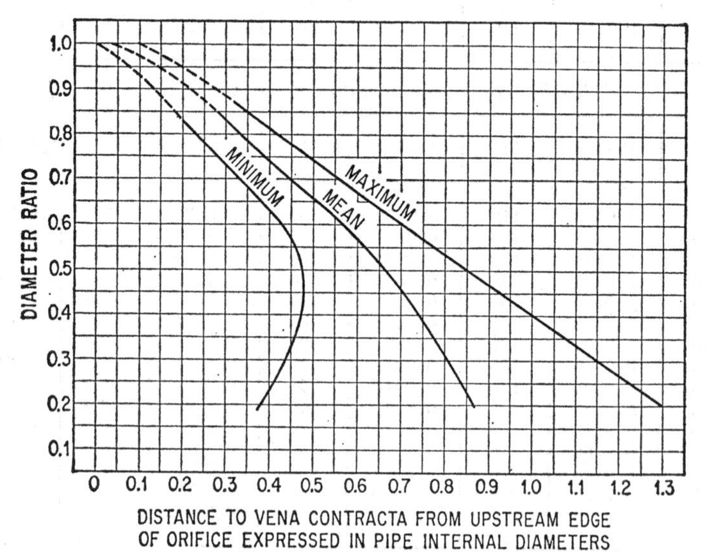

Fig. 5-15. Distance to vena contracta from upstream edge of orifice expressed in pipe internal diameters. (*ASME Fluid Meters Report, 4th ed.,* 1937, *p.* 35.)

MECHANICAL AND ELECTRIC METER BODIES

Since the differential pressure across the orifice or other primary element is proportional to the square of the fluid velocity, it may be used as an accurate index of the rate of flow. The meter body measures this differential, and translates it into units of flow for indicating, recording, and integrating.

Every meter body is designed to operate between zero differential and a maximum differential, so that, for a given installation, the primary element is designed with the proper throat size to produce this maximum differential when the rate of flow is at the desired maximum.

The Mercury-float-type Manometer (Fig. 5-16). This operates on the basic principles of the ordinary U tube. Any change in differential pressure causes the mercury to move up or down. The higher of the two pressures

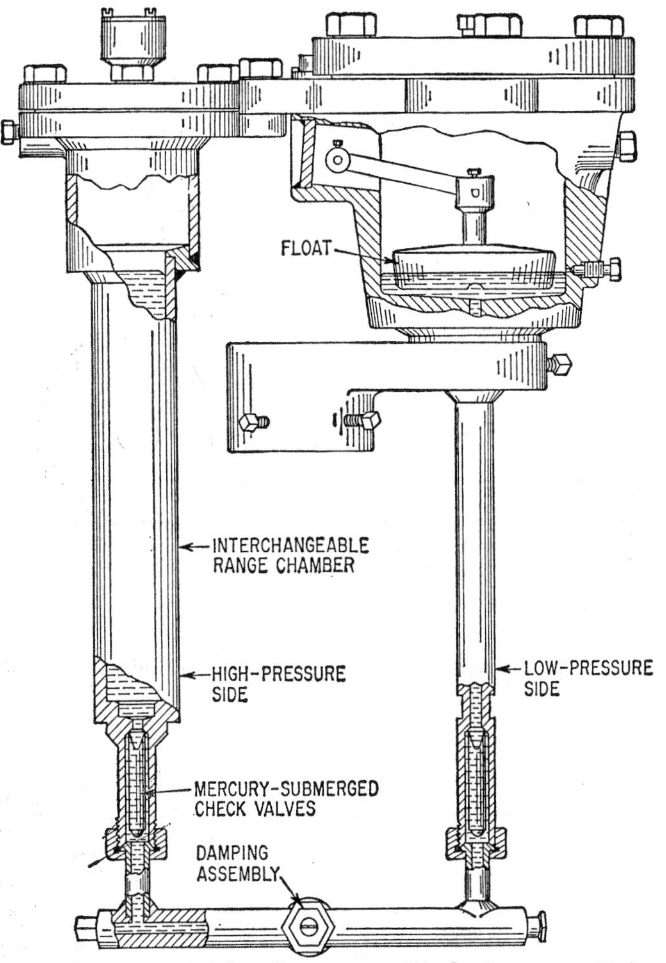

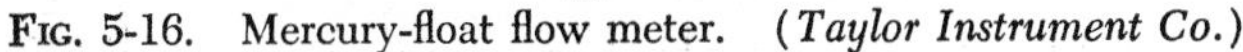

FIG. 5-16. Mercury-float flow meter. (*Taylor Instrument Co.*)

being measured is piped to the high-pressure chamber. The low-pressure chamber houses the float mechanism, and this actuates the instrument pen or pointer. As the mercury moves up in the low-pressure leg, it raises the float, which moves upward for full-scale travel of the pen or pointer. The distance between the mercury levels in the two legs determines how far the

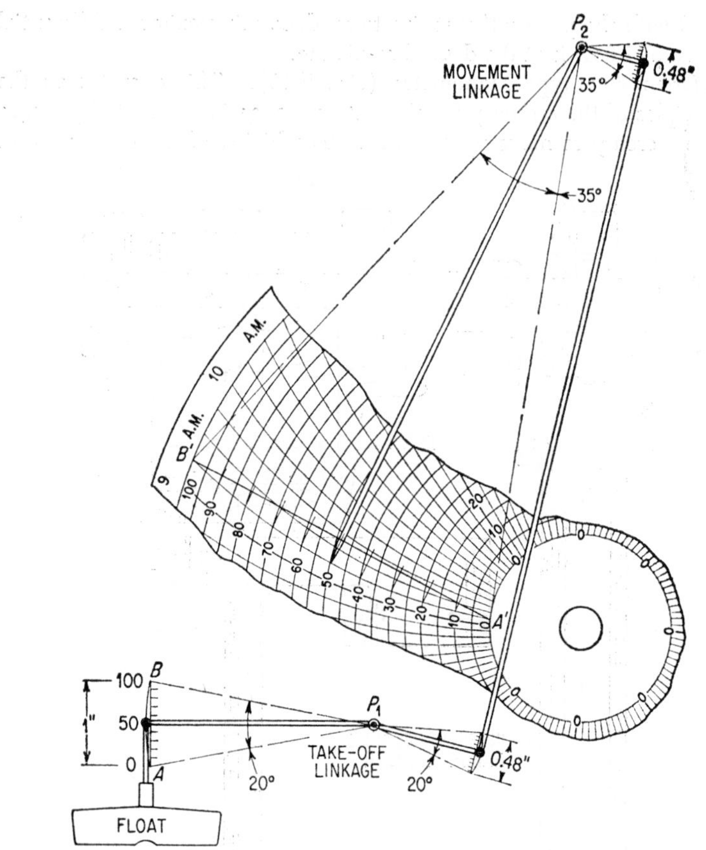

FIG. 5-17. Conveying float motion to recording pen. (*Taylor Instrument Co.*)

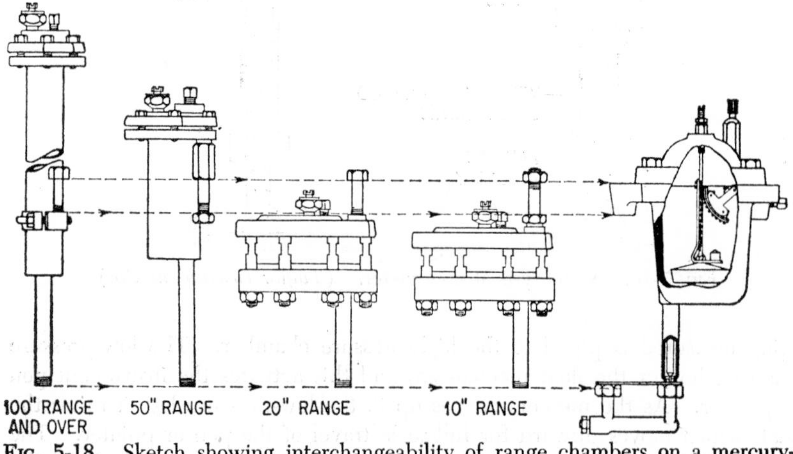

FIG. 5-18. Sketch showing interchangeability of range chambers on a mercury-float meter body. (*Foxboro Co.*)

float moves, and, consequently, this motion is a measure of the differential pressure across the mercury legs.

Figure 5-17 shows how float motion is conveyed to the recorder pen.

Interchangeable Range Chambers. This convenient interchangeability (Fig. 5-18) is an important feature, especially if the meter is apt to be moved from one application to another, or if there are seasonal differences

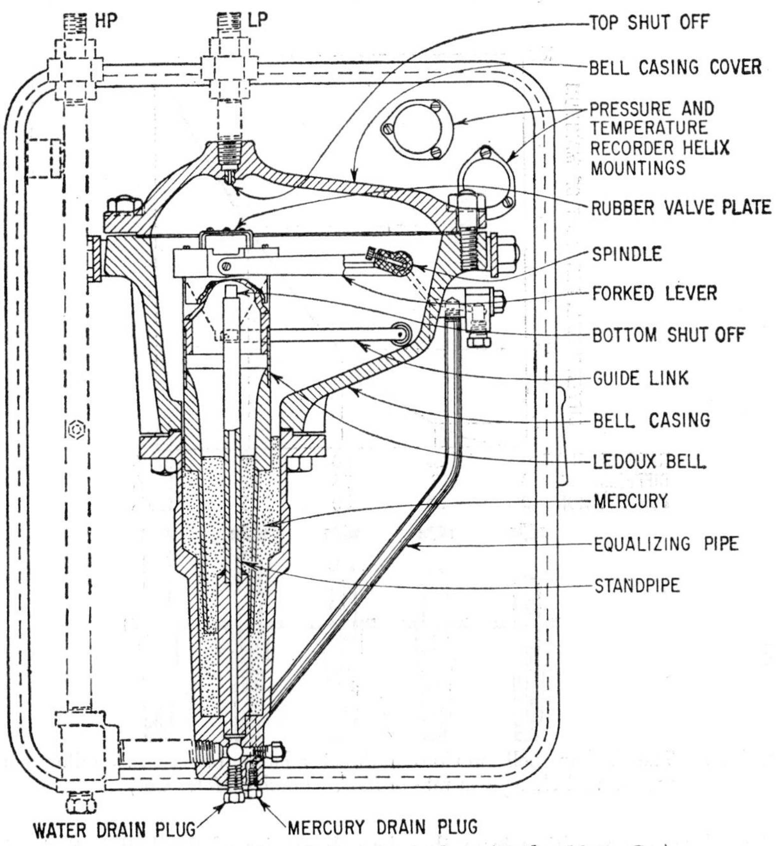

FIG. 5-19. Ledoux-bell meter body. (*Bailey Meter Co.*)

in the normal rate of flow. For example, a winter load of steam in a system might require a meter with a 100-in. chamber, whereas a 20-in. chamber would be more suitable for the summer load. The same meter can be quickly adapted to either condition as satisfactorily as though a second meter had been bought for the purpose.

Ledoux-bell Meter Body. The pressure existing at the inlet side of the primary element is applied to the interior of the mercury-sealed Ledoux bell (Fig. 5-19), and the lower pressure at the outlet side is applied to the exterior of the bell.

Changes in the differential pressure move the bell up and down like a frictionless piston, as illustrated in the schematic drawing of Fig. 5-20. As the bell rises, its walls emerge from the mercury, changing the buoyant force, and thereby counterbalancing the upward force of the differential pressure. Because of the internal shape, vertical movements of the bell cause the movements of the direct-connected recording pen to be directly proportional to the changes in rate of flow.

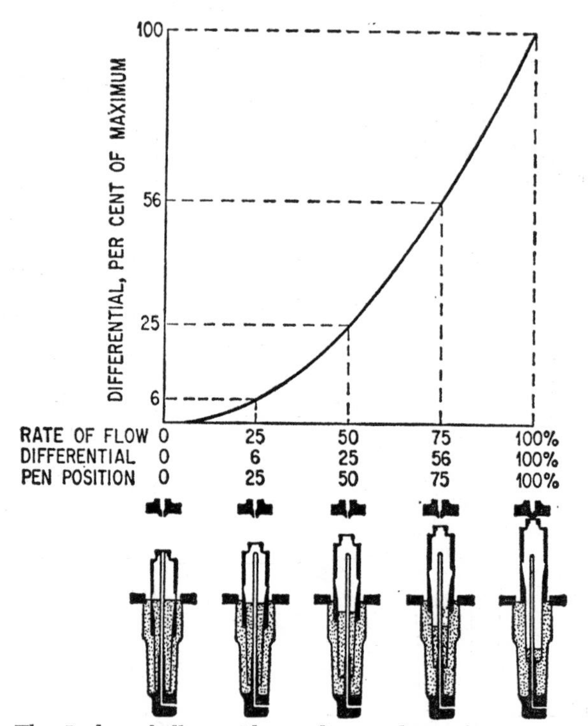

Fig. 5-20. The Ledoux bell provides a direct relation between recording-pen motion and changes in rate of flow. (*Bailey Meter Co.*)

An advantage resulting from the shaped interior of the bell is that, at low rates of flow where differential pressures are small, a much greater area is subjected to these differentials, to provide ample power for operation of the meter.

The bell casing itself serves as the mercury reservoir, thus ensuring permanent calibration since mercury cannot be spilled out.

A guide link keeps the Ledoux bell from rubbing against the standpipe or bell casing, even when the meter is not level as often happens in marine service.

The maximum capacity of the meter may be altered without the primary element being changed when conditions make it desirable, by replacing the

Ledoux-bell assembly with another of different maximum differential and changing the lower part of the bell casing.

Ring-balance Meter Body.[5] The ring-balance flow meter is a radial-torque meter which uses a hollow ring body to convert the differential generated by a differential medium, or by a difference in static pressure, into a rotation which is transmitted to the recorder or indicator.

Figure 5-21 shows the ring assembly in a typical flow-measuring system; *a* indicates the position of the ring body at zero flow condition.

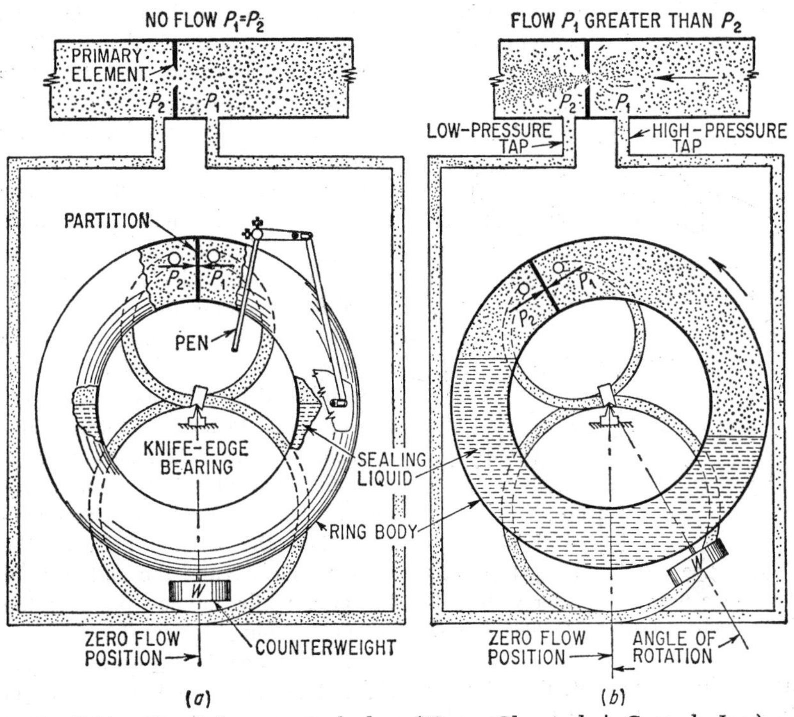

Fig. 5-21. Ring-balance meter body. (*Hagan Chemicals & Controls, Inc.*)

The ring assembly is mounted on a knife-edge bearing, which permits rotation about the axis of the ring. The ring is divided into two pressure compartments by the baffle at the top, and by the sealing liquid which fills the lower part of the ring. The two-ring compartments thus formed are connected to the differential-pressure pipes by means of flexible tubing, to permit the ring to rotate freely under the action of the difference in pressure in the compartments. These may be the S-shaped, self-compensating tubes shown in Fig. 5-21 for relatively higher static pressures, or parallel tubes for the lower-pressure applications.

[5] Hagan Chemicals & Controls, Inc.

The ring rotation is transmitted to the recording pen or indicating pointer by a linkage. When this meter is used to register units of flow, a square-root chart or scale is employed.

Ring torque is a function of the differential pressure acting on the baffle. This torque is resisted by an external calibration weight, rigidly attached to the bottom of the ring body. The meter is in equilibrium when the ring torque is balanced by this counterweight.

The mercury or other sealing fluid exerts no force tending to rotate the ring, and acts only as a seal for the differential pressure in the two compartments. This is true because in the circular ring body all hydrostatic forces resulting from the deflection of the sealing liquid are directed normally to the containing circle, and therefore pass radially through the exact center of rotation without producing meter torque.

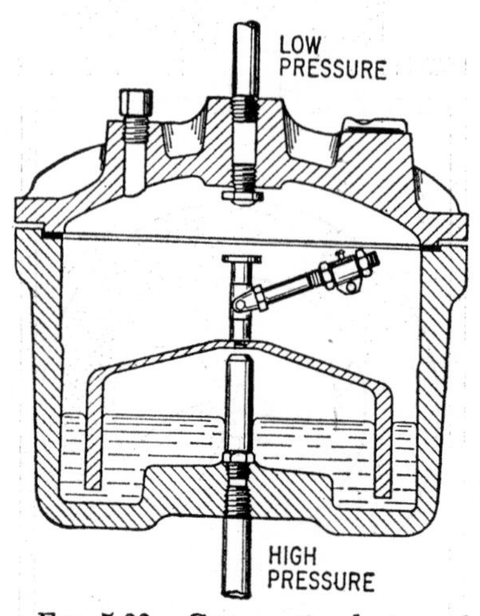

FIG. 5-22. Cross-sectional view of low-range flow manometer with bell-type float. (*Taylor Instrument Co.*)

Low-range Meter Body.[6] The measurement of small differentials such as 1, 2½, and 5 in. of water requires a method which will produce more float motion than that which would result from using a U-type mercury-float meter body. This is accomplished by allowing an inverted cup or "bell" to float on the mercury surface. Figure 5-22 shows a cross-sectional view of this type of instrument. The high-pressure tap connects into the volume between the mercury surface and the underside of the cup, while the low-pressure tap enters the volume external to the cup. As a difference in pressure appears across the instrument, the effective weight of the cup is reduced, which in turn causes it to elevate and reduce its submergence. As the thickness of the side wall of the cup is diminished, the vertical travel for a given differential change will increase. The variation in range, therefore, is accomplished by altering the thickness of the cup's side walls.

The bell-type manometer can be used for either gas or liquid service. When it is used for liquid service, the low-pressure chamber can be vented through the low-pressure taps. To vent air from underneath the bell, the vent screw located at the base of the manometer is opened until liquid issues from it. The bell-type manometer is used to measure flows where pressure loss is of great importance.

Aneroid Meter Body. In one design the balancing agent is a torque tube, actuated by a strong but sensitive metal bellows. Other designs use

[6] Taylor Instrument Co.

a stuffing box and a pressure-tight bearing to transmit the bellows motion. This establishes the range of the manometer, and is also the means of connection between the manometer and the instrument on which it is installed. Figure 5-23 pictures a manometer cut away to show internal structure.

The bellows is anchored and sealed on the downstream (left) side of the manometer, so that reduced-pressure influence is from within, while upstream pressure is applied to the outside of the bellows. The right end of the bellows is free and will move linearly, away from the high-pressure

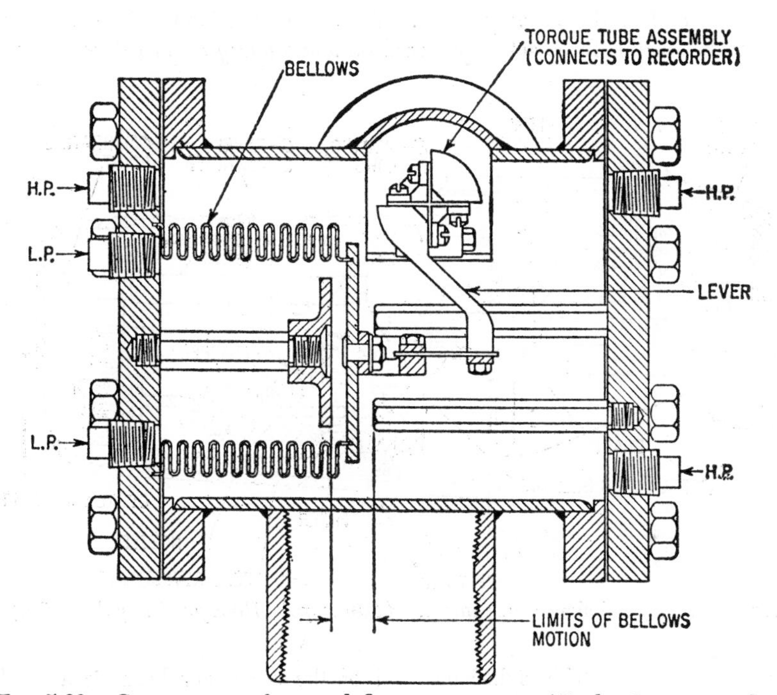

FIG. 5-23. Cross section of aneroid flow manometer. (*Taylor Instrument Co.*)

side of the manometer. Bellows sensitivity is such that extremely small changes in differential will cause further movement of the bellows.

Force-balance Flow Transmitter. This transmitter (Fig. 5-24) is a mercuryless type of meter body designed to meet requirements for control within narrow limits and for easy installation and maintenance.

The unit operates on the force-balance principle, in which the force of differential pressure across a primary element is balanced by the force of air pressure acting in a weigh-beam system. The air pressure required to balance this force thereby becomes a measure of the differential pressure or flow, and is transmitted by a single air line to a pneumatic receiving instrument.

The schematic diagram[7] of Fig. 5-24 shows how a typical unit operates. The *low-* and *high-pressure chambers A* and *B* are separated by a flexible *diaphragm,* which is connected to one end of a *primary beam D.* The beam extends through a *sealing bellows E,* which confines the measured fluid within the high-pressure chamber. Near the bellows is the fulcrum for the primary beam. Toward the opposite end of this beam is a bevel-edged *rider G,* which is movable, to provide the range-changing feature, but which can be considered a fixed knife-edge in the operation of the unit.

Differential pressure in the meter body creates an upward force on the left-hand end of the primary beam. Because of the intermediate fulcrum on the beam, this force acts downward on a *secondary beam F* through the

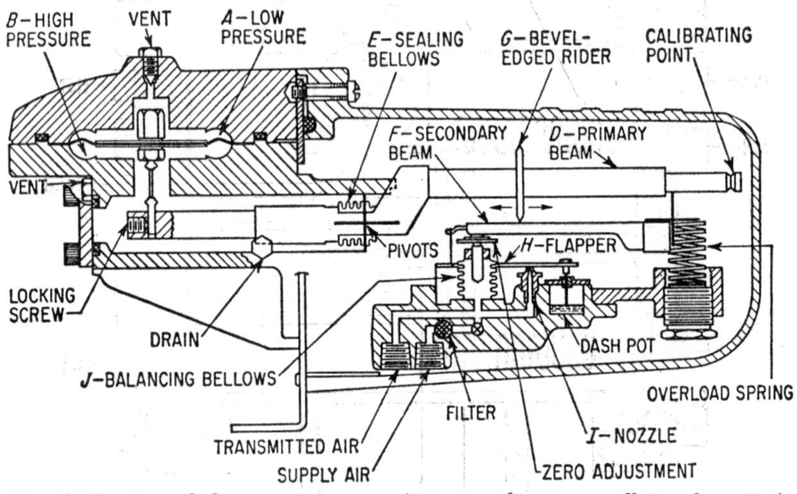

FIG. 5-24. Force-balance-type meter. (*Minneapolis-Honeywell Regulator Co.*)

bevel-edge rider. The right-hand end of the secondary beam is a fixed fulcrum point, and the other end is linked to a *flapper H.* So downward force on the secondary beam moves the flapper toward a *nozzle I.*

Air pressure from the supply is bled through a restriction into the force-balance system, which comprises the flapper and nozzle and a *balancing bellows J.* Thus, when the nozzle is covered, air pressure builds up in the balancing bellows, and an upward force is created in opposition to the downward force due to the differential pressure. In this manner, output air pressure from the converter is directly proportional to the measured differential pressure, and can be translated directly into terms of flow or differential-pressure readings by the bellows of a pneumatic receiver.

Electric Meter Bodies. These units are based on either inductance or resistance principles. They are suitable for remote transmission, but their

[7] Minneapolis-Honeywell Regulator Co.

use may be limited in "hazard" areas, because of explosion-proof requirements.

In one inductance design[8] (Fig. 5-25), the pressures transmitted to the bottom of the high- and low-pressure housings act on the floating end of the bellows to move them until the spring effect of the unit balances the difference in the applied pressures. Since the movement of the bellows is proportional to the differential pressure acting on it, the transformer core attached to the bellows moves in direct relation to the differential pressure. The movement of the core with respect to the three fixed coils of the differential transformer produces an electrical output.

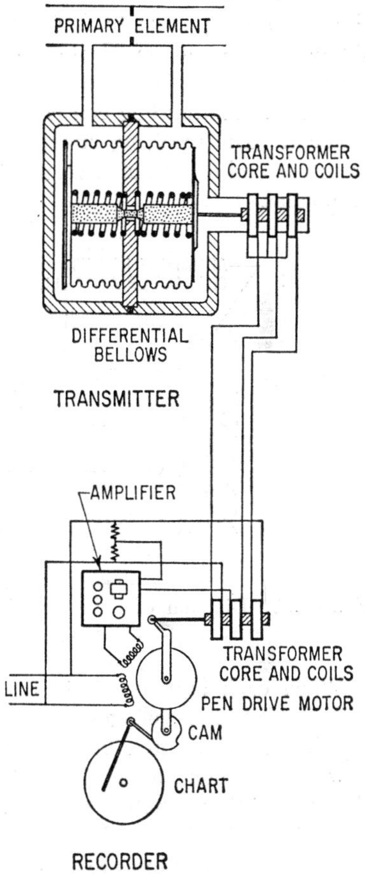

Differential transformers (one at the transmitter and one at the receiver) each consist of three coils—two primary coils with a single secondary coil mounted between them—and a core. The two primary coils of each transformer are connected so that their magnetic fluxes are opposed, and a zero voltage is induced in the secondary coil when the core is centered. All primary coils are connected in series so that the same current flows through each.

When the core is "out of center," it alters the coupling between the primary and secondary coils and results in the induction of a voltage into the secondary coil. The magnitude of the voltage is a function of the amount of displacement of the core from the geometric center of the coils. The phase of the voltage is determined by the direction of movement of the core from the center of the coils.

Fig. 5-25. Electric flow transmitter with electronic recorder. (*The Hays Corp.*)

The measuring element of the transmitter is mechanically linked to the core of the differential transformer of the transmitter. When the core is moved with respect to the three fixed coils of the differential transformer, an output voltage is produced in the center coil. This voltage is transmitted on three wires to the receiver. At the receiver, a similar differential transformer produces a voltage, which is a function of the indicator or pen

[8] The Hays Corp.

position. An electronic amplifier detects the difference between the two voltages and acts upon a motor, which moves the core of the receiver unit until the voltage difference becomes zero. The indicator or pen then reflects the position of the measuring element of the transmitter.

Still another electric meter body operating on the inductance principle[9] utilizes a characterized bell and balanced bridge (Fig. 5-26). Essentially, the system comprises a transmitter coil in the meter body, which is connected by three wires to an identical coil in the receiving instrument. Within each coil is a magnetic armature of the same size and weight. The armature in the transmitting coil is attached through a rod to the flow-sensitive element. The receiving armature is suspended within the instrument coil from one end of a counterweighted rocker arm linked to the instrument pen or pointer.

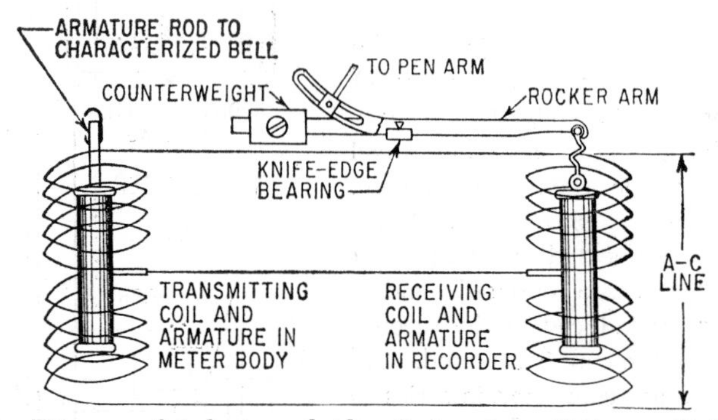

Fig. 5-26. Diagram of inductance-bridge flow meter. (*Minneapolis-Honeywell Regulator Co.*)

When each armature is in the same relative position in its respective coil, the ratios of voltages across the two sections of the divided coils are equal. A change in differential pressure, resulting from a change in the flow measurement, alters the position of the characterized bell, and moves the transmitting armature to a new position. This momentarily unbalances the bridge circuit, thereby producing unequal voltages and currents, and causing a magnetic force to act upon the receiving armature.

As a result, the receiving armature moves to a new position corresponding to that of the transmitting armature. At this point, the voltage ratios again become equal across the two sections of the divided coils, and the armature movement ceases. Thus, the instrument pointer or pen is driven by the armature movement to a calibrated position that is indicative of the new flow condition.

Electrical-resistance Meter Body. In one design[10] (Fig. 5-27), the upstream pressure is admitted at (3). The downstream, or low, pressure is

[9] Minneapolis-Honeywell Regulator Co.
[10] Republic Flow Meters Co.

admitted at (4). The high-pressure connection leads to the bottom of the oil seal, from which the pressure is transmitted, through the medium of the oil, to the mercury in the high-pressure chamber, which is part of the high-pressure leg of the mercury U tube. The low-pressure connection leads to the bottom of the oil chamber, which contains the low-pressure leg of the mercury U tube, the contact chamber, and the scale. In operation,

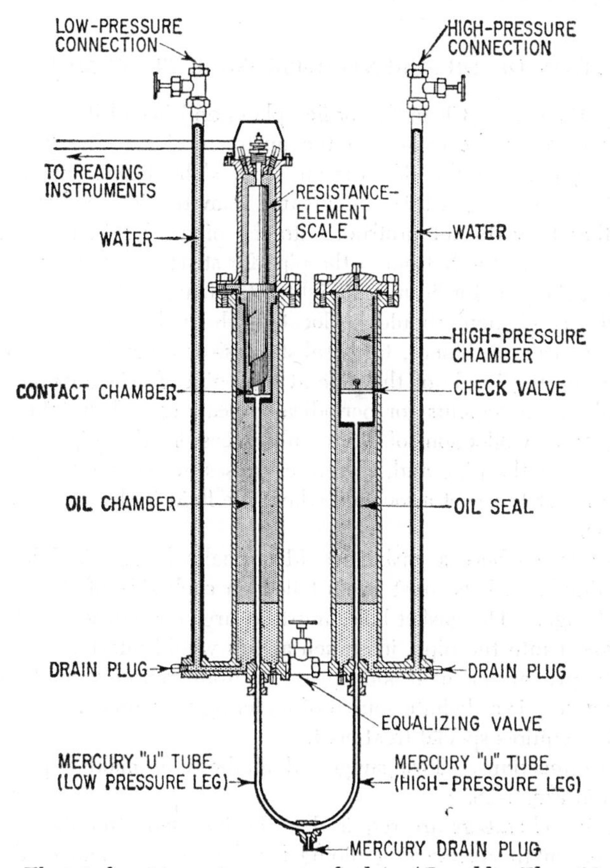

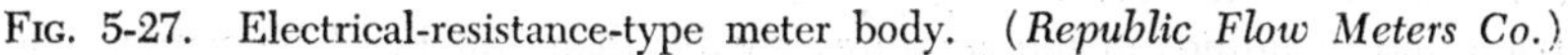
FIG. 5-27. Electrical-resistance-type meter body. (*Republic Flow Meters Co.*)

the oil seal and the oil chamber are completely filled with oil and water, as shown. The oil seal prevents scale, dirt, oxide, or other foreign matter from entering the meter body. Some forms of gases require a variation from this method of sealing.

To balance the differential head, the mercury level is depressed in the high-pressure chamber of the meter body, and raised in the low-pressure, or contact, chamber. As the mercury rises in the contact chamber, it engages the resistance element, or scale, and progressively reduces the resistance in the electrical circuit, thereby increasing the current to the reading instruments. The resistances are so divided, by the scale, that the conductance of the circuit corresponding to any given contact is directly proportional to the flow of fluid in the pipe. The indicator, recorder, and integrator, while measuring the conductance regulated by the flow, actually register the rate of flow through the pipe, directly in pounds, gallons, or cubic feet per hour or per day.

INSTALLATION OF DIFFERENTIAL-PRESSURE FLOW METERS

Primary Element. Clean the orifice plate, and install it with the square and sharp edge of the orifice on the upstream side. The beveled edge, if any, should be on the downstream side. The inside diameter of the gaskets should be at least 1/16 in. greater than the inside diameter of the pipe, so that they do not protrude into the pipe. Drain holes, if any, in orifice plates are at the bottom of the pipe for steam or gas installations, and at the top of the pipe for liquids. Tighten the flange bolts evenly.

The primary element should be located where the fluid is cleanest. In pipes where scale, sediment, tarry substances, or other form of matter will accumulate on the inside of the pipe at the point of orifice or nozzle installation, make arrangements for periodically cleaning and draining out these accumulations. Collection of such substances in the pipe changes the effective area of the pipe and, therefore, causes inaccuracy in the metering. It is always best to select a location where the fluid is clean and, thus, avoid this trouble.

In every case where a gasket would normally be required between the flanges, suitable gaskets must be installed on each side of the orifice plate or nozzle flange. The gasket holes must be large enough so that the gaskets do not project into the pipe, inasmuch as this would interfere with the flow through the primary element and cause inaccurate metering conditions.

Pulsating flow is a definite source of metering inaccuracy. Such a condition usually requires special treatment.

Meter Connections. Two suggested methods of making pressure taps are shown in Fig. 5-28.

Condensing chambers are required, to be installed after the primary-element valves, in the connecting piping for the measurement of the flow of steam and of liquids at temperatures greater than 250°F. The condensing chambers have two main functions, each of equal importance to accurate metering. One of these is to cool and/or condense the fluid at the primary element. The other is to provide adequate volume for displacement of fluid in the meter pressure casing during changes in measurement, without noticeably changing the physical head imposed on the meter by the vertical distance between the meter and the primary element.

The condensing chambers must be installed and maintained level lengthwise and with each other. This is important in that the physical head, due to the connecting piping, on the meter must be equal in both lines. If

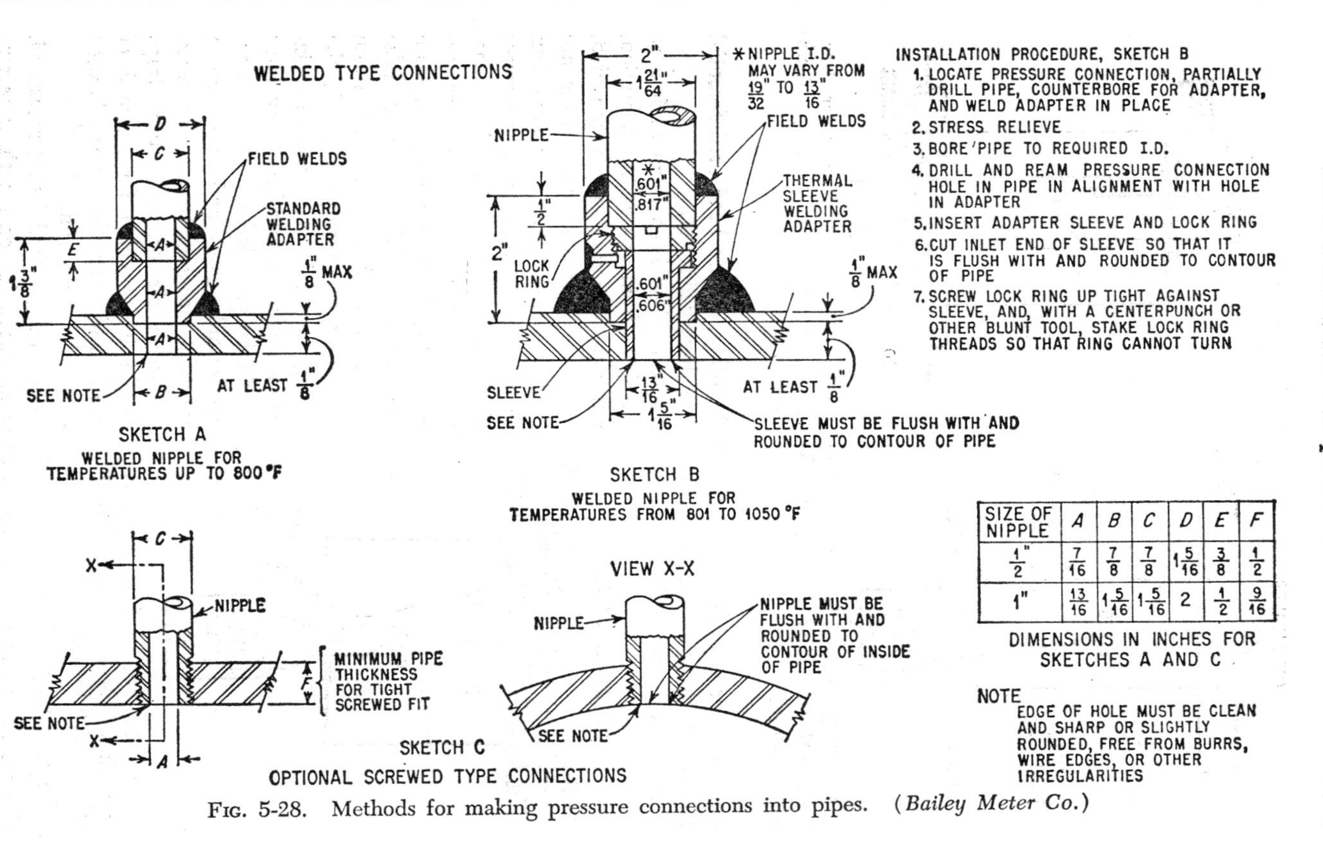

SIZE OF NIPPLE	A	B	C	D	E	F
$\frac{1}{2}$"	$\frac{7}{16}$	$\frac{7}{8}$	$\frac{7}{8}$	$1\frac{5}{16}$	$\frac{3}{8}$	$\frac{1}{2}$
1"	$\frac{13}{16}$	$1\frac{5}{16}$	$1\frac{5}{16}$	2	$\frac{1}{2}$	$\frac{9}{16}$

FIG. 5-28. Methods for making pressure connections into pipes. (*Bailey Meter Co.*)

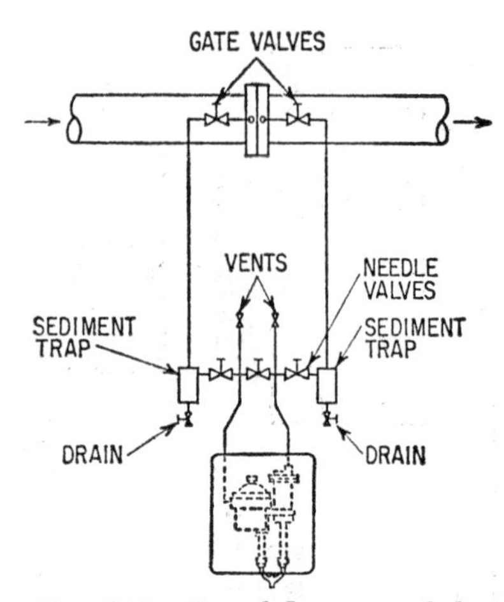

Fig. 5-29. Liquid flow, meter below line. (*Mason-Neilan Div. of Worthington Corp.*)

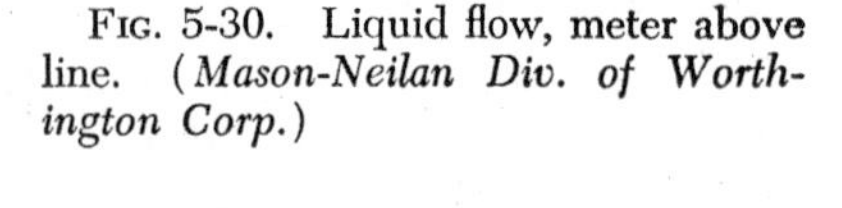

Fig. 5-30. Liquid flow, meter above line. (*Mason-Neilan Div. of Worthington Corp.*)

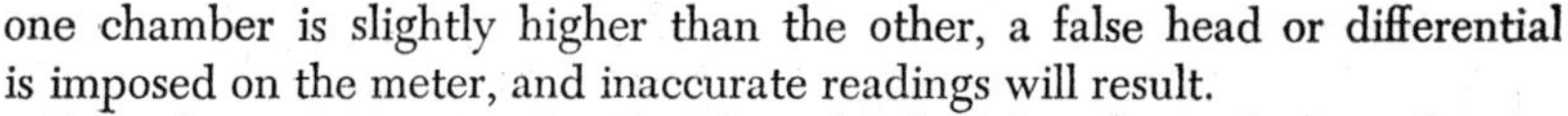

one chamber is slightly higher than the other, a false head or differential is imposed on the meter, and inaccurate readings will result.

In order to maintain the chambers level and to keep them as free as possible from excessive vibration, adequate supports should be provided.

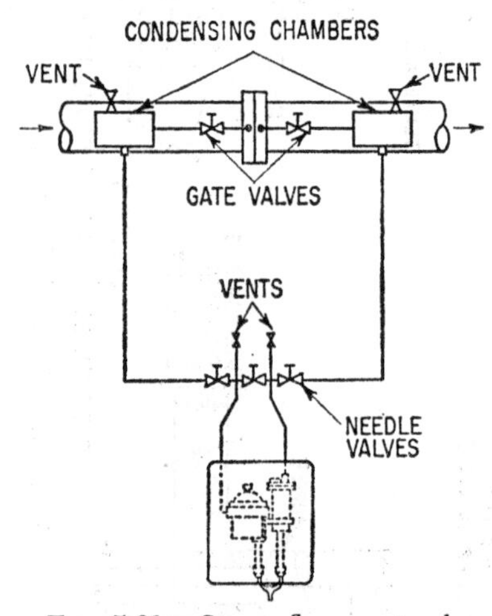

Fig. 5-31. Steam flow, meter below line. (*Mason-Neilan Div. of Worthington Corp.*)

When differential-type meters are used to measure flow, it is important that only the differential pressures across the orifice plate be transmitted to the meter. Piping arrangements should be designed to eliminate the possibility of measuring false differentials. For liquid and steam (or other condensable vapors), the piping should be arranged to maintain a uniform liquid head, and traps should be provided for venting noncondensable gases and for draining sediment accumulations. For moisture-laden air or wet gases, traps should be provided for removal of condensate, unless the meter is located above the meter run.

Piping should resist the corrosive action of the fluid being measured. Copper tubing of proper size, ⅜ or ½ in. depending on length, is recommended for air, noncorrosive gas, light oils, brine, and steam, as long as the pressure and temperature ratings of the tubing are not exceeded.

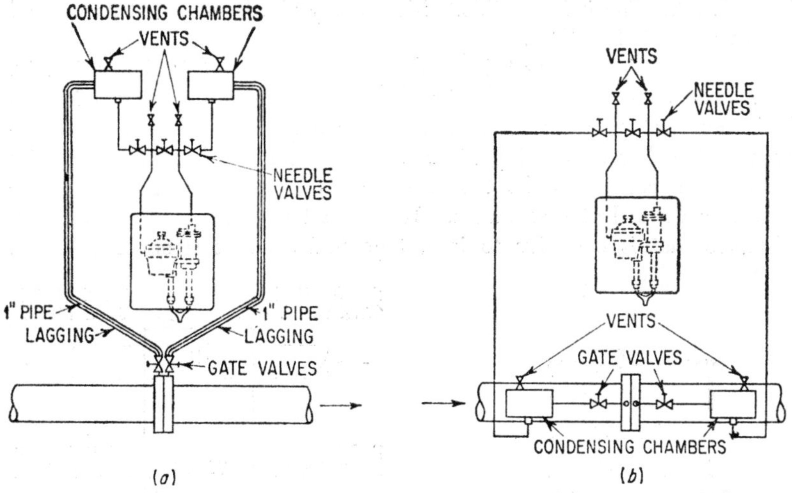

FIG. 5-32. Steam flow, meter above line. (*Mason-Neilan Div. of Worthington Corp.*)

Blow out all pipe, tubing, and fittings with compressed air before connecting them to the controller. If thread compound is used, apply above the second or third male thread in moderate amounts.

Under the usual conditions of liquid flow, the meter may be mounted above or below the line or on a vertical line, as shown in Figs. 5-29 and 5-30. Whenever the flow of a corrosive liquid, highly volatile liquid, tar, sulfur-content oil, liquids with solids in suspension, or a highly viscous liquid is being measured, it should be prevented from entering the connecting lines and meter chambers by means of sediment traps or a liquid or air purge. With a purge system the meter may be mounted above or below the line.

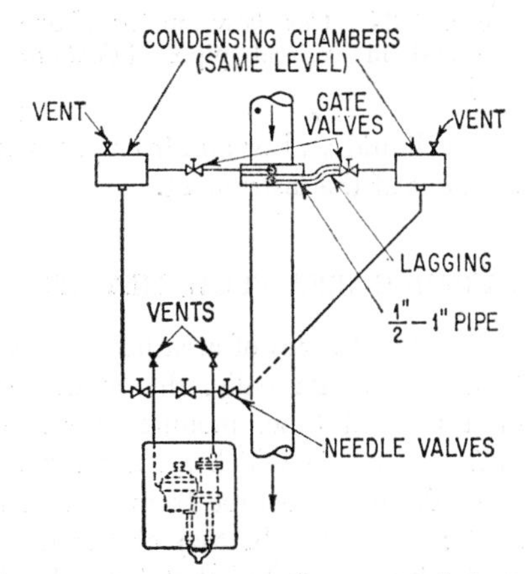

FIG. 5-33. Steam flow, meter below orifice in vertical run. (*Mason-Neilan Div. of Worthington Corp.*)

The measurement of steam or other condensable vapor requires the use of condensing chambers (Figs. 5-31 and 5-32) to condense the vapor and provide two equal liquid heads on the meter chambers. Whenever possible, the meter should be installed below the level of the steam pipeline as shown in Fig. 5-31. Figure 5-32 shows a method of installing the meter above the line but is recommended only where installation below the line is impossible. Meters may also be mounted on vertical pipe as shown in Fig. 5-33.

A common piping layout for a dry-gas measurement is shown in Fig. 5-34. The meter may also be installed below the line as shown in Fig. 5-35. For wet-gas measurement, the meter may be installed above or below the line, but condensing pipes must be installed above the line with pressure connections to the top of the pipe. Measurement may also be made in a vertical pipe as shown. If the gas being measured contains any substance that is corrosive to iron, type 304 stainless steel, or mercury,

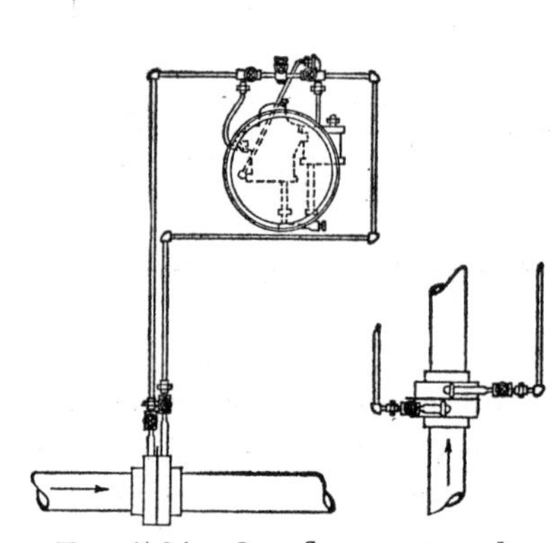

Fig. 5-34. Gas flow, meter above line and in vertical pipe. (*Foxboro Co.*)

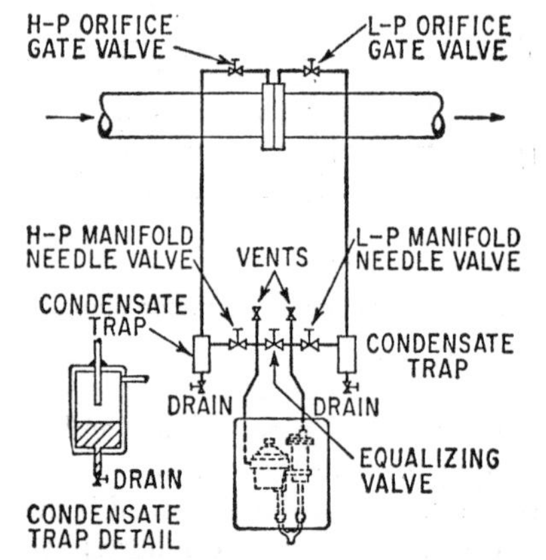

Fig. 5-35. Gas flow, meter below line. (*Mason-Neilan Div. of Worthington Corp.*)

then a liquid seal or an air or gas purge must be used to keep the corrosive fluid out of the instrument.

VARIABLE-AREA FLOW METERS

Two basic types of area flow meter are (1) the rotameter and (2) piston-type meter. Basically, the rotameter type of variable-area meter consists of a tapered tube, mounted vertically with the smaller end of the tube facing down, in which a metering float is located (Fig. 5-36). The fluid flows through the tube from bottom to top. When no fluid is flowing, the float rests at the bottom of the tapered tube, and its maximum diameter is usually so selected that it blocks the small end of the tube almost completely. When flow commences in the pipeline and the fluid reaches the float, the buoyant effect of the fluid lightens the float. However, as the float has a greater density than the fluid, the buoyant effect is not sufficient to lift the float. Therefore, the flow passage remains blocked, and fluid pressure starts increasing. When the upward fluid pressure plus the fluid buoyant effect exceeds the downward pressure due to the weight of the float, then the float rises and truly "floats" within the fluid stream.

With upward movement of the float toward the larger end of the tapered tube, an annular passage is opened between the inner wall of the glass tube and the periphery of the float, and through this opening the fluid passes. The float continues to rise until the annular passage is large enough to handle all the fluid coming through the pipe. Concurrently, the fluid velocity pressure drops until it, plus the fluid buoying effect, exactly equals the float weight. The float then comes to rest in dynamic equilibrium.

Any further increase in flow rate causes the float to rise higher in the tube, and a decrease causes it to sink to a lower level. Every float position corresponds to one particular flow rate and to no other. It is necessary merely to provide a reading or calibration scale on the outside of the tube, and flow rate can be determined by direct observation of the metering float (Fig. 5-37).

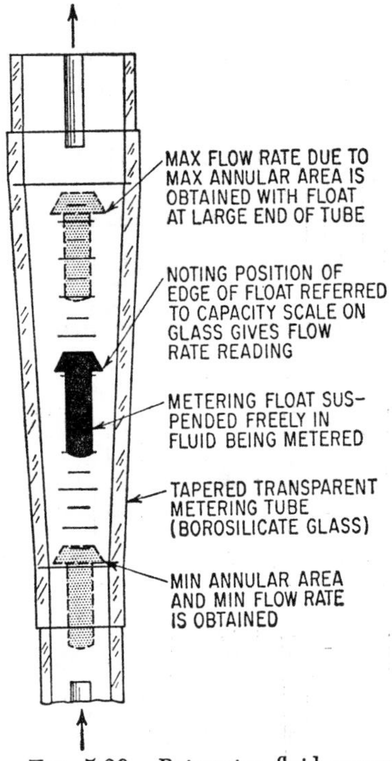

FIG. 5-36. Rotameter fluid passes through this annular opening between periphery of float head and inside diameter of tapered tube. Flow rate varies directly as area of annular opening varies. (*Fischer & Porter Co.*)

Both[11] the differential-head meter and the variable-area meter utilize the principle of the conservation of energy as expressed for flow in piping systems by the Bernoulli theorem: A restriction placed in the piping system increases the velocity of the fluid stream as it passes the restriction, and causes a corresponding pressure reduction, the value of which can be calculated by the Torricelli equation. As applied to both the variable-head and the variable-area types of flow-rate meters, the Torricelli equation, in its simplest form, reads

$$Q = CA\sqrt{2gh}$$

where Q = rate at which fluid is flowing, cfs
C = flow coefficient or efficiency with which pressure has been converted to velocity
A = free area of opening through restricted passage, sq ft
g = acceleration of gravity, 32.2 ft/sec^2
h = pressure reduction in the flowing stream created by restriction, ft of flowing fluid

[11] Fischer & Porter Co.

We note that the flow rate is directly proportional to the area of the restriction and to the square root of the pressure reduction:

$$Q \propto A$$
$$Q \propto \sqrt{h}$$

The area available for flow past the float is inferred from the position that the float seeks in the tapered tube, and the flow-area value A is then sub-

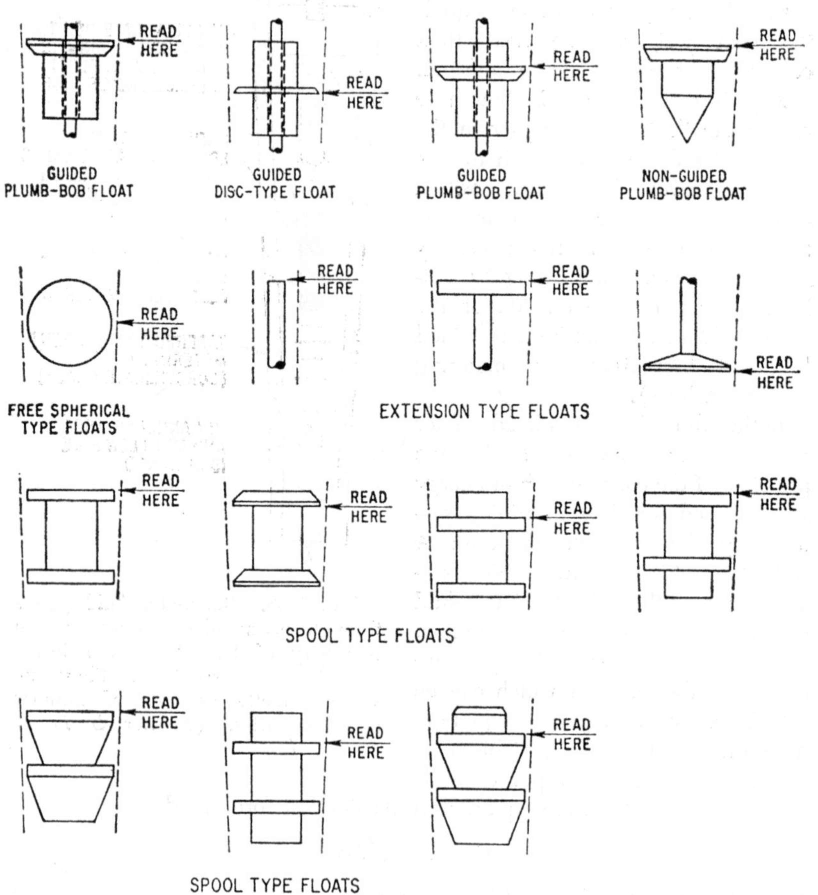

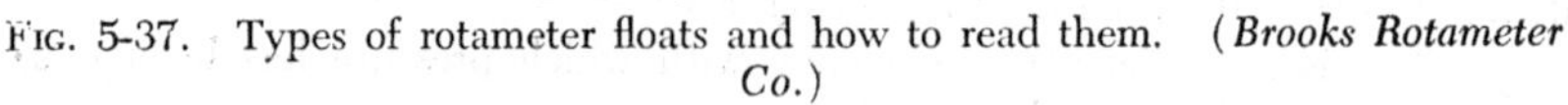

FIG. 5-37. Types of rotameter floats and how to read them. (*Brooks Rotameter Co.*)

stituted in the Torricelli equation. The variable-head meter utilizes a fixed restriction in the fluid stream with a manometer, to permit observation of the differential created by the various flow rates passing this restriction.

While area in the variable-area (constant-head) meter and head in the variable-head (constant-area) meter are usually considered the only variables, other influences may also affect the accuracy of flow measurement. These additional variables are:

1. Viscosity
2. Pulsation
3. Turbulence (approach piping conditions)
4. Sharpness and dimensions of metering element
5. Installation and maintenance attention

The efficiency with which pressure is converted into velocity, as expressed by the coefficient C, is primarily affected by viscosity. The value of coefficient C will be maximum if velocity distribution is uniform in the flow stream when it reaches the restriction. If the flow stream is not uniform, the coefficient C will be reduced. However, changes in viscosity affect the coefficient insofar as they alter the ratio of inertial forces of resistance to viscous forces of resistance. This ratio is the Reynolds number expressed as fluid density × fluid velocity × diameter of flow passage/fluid viscosity:

$$\frac{\rho V D}{\mu}$$

Pulsation will cause vertical oscillation of the float in the variable-area meter, and of the mercury column or other exhibiting devices in the variable-head meter. The mean reading of the oscillation will never represent true flow in the variable-head meter, and frequently does not represent true flow in the variable-area meter.

If the dimensional size of the restriction is not maintained at its original value, the calculations by Torricelli's equation will not be accurate. Also the value of C changes markedly with the degree of sharpness of the restriction edge. It decreases as the edge becomes sharper and increases as the edge is rounded. A jagged restriction edge causes the value of C to be erratic, and accurate calculations are impossible.

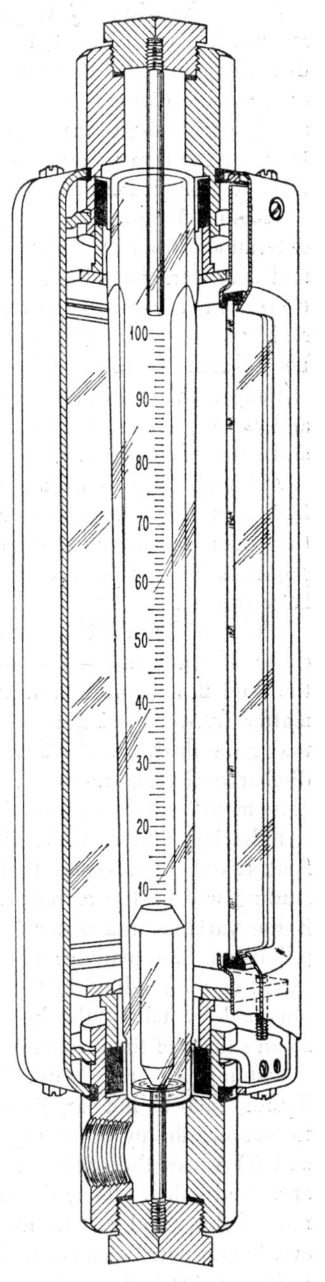

Fig. 5-38. Construction of float-type area meter. (*Fischer & Porter Co.*)

Scale Linearity. Q is proportional to A in the first power; so their relationship is linear. A, in turn, is proportional to the metering tube inside diameter squared (at the elevation to the float metering edge) minus the float outside diameter squared. The plotting of float travel versus Q yields an almost straight line. (This slight curvature can be corrected readily in either the transmitter or the remote exhibiting instrument.) Flow ranges of 10 to 1 or greater are handled easily.

Pressure Drop. Because h is a constant in each instrument, it can be determined from the float weight. By the proper choice of float weight and diameter, the value of h may be preselected, and the metering pressure drop, which is constant over the entire range of the meter, can be made as low as desired. (Metering pressure drop as low as 0.1 in. of water has been obtained with specially fabricated floats.)

Range. Variable-area meters are particularly suitable for small flows and have been used to measure accurately less than 0.1 cc/min of liquid and 1.0 cc/min of gas.

Although there is no theoretical limitation to the size in which the variable-area meter may be constructed, the bulk and weight will increase as the square of the pipe diameter. Therefore, cost, weight, and handling problems become considerations when variable-area meters are used for large pipe sizes.

Effect of Dirt. The variable-area meter tends to be self-cleaning because of the annular flow-passage arrangement. The scouring action of the fluid flow against both the float and the tube wall discourages foreign matter from adhering to either surface. A large obstruction may lodge under the float, but the float can rise beyond the top of the metering tube, discharge the obstruction, and fall back into its proper position. Where glass metering tubes are used, visual inspection can be made effortlessly.

Pulsating Flow. Under ideal conditions, when the oscillations of the float exactly correspond to the amplitude of flow pulsation, the average reading will be the correct reading. However, in a pulsating-flow stream, if the variable-area meter is provided with a dashpot to dampen out the pulsations, the readings obtained will be erroneous to exactly the same extent as the readings of the variable-head meter, because the flow readings obtained by taking the linear average of the pulsations will be the root mean square of the actual flow.

Piston-type Area Meters.[12] Designed especially for metering viscous liquids, such as hot tar, Bunker C oil, and black liquor, which are difficult to measure in any other way, the piston-area-type flow meter (Fig. 5-39, *A* and *B*) varies the orifice area while the differential pressure is maintained at a fixed value. Thus the area of the orifice is directly proportional to the rate of flow, and can be used as the basis for flow measurement with an evenly graduated chart or scale.

Area-meter body readings are transmitted electrically to an evenly graduated indicating or recording instrument. The armature in the transmitting

[12] Minneapolis-Honeywell Regulator Co.

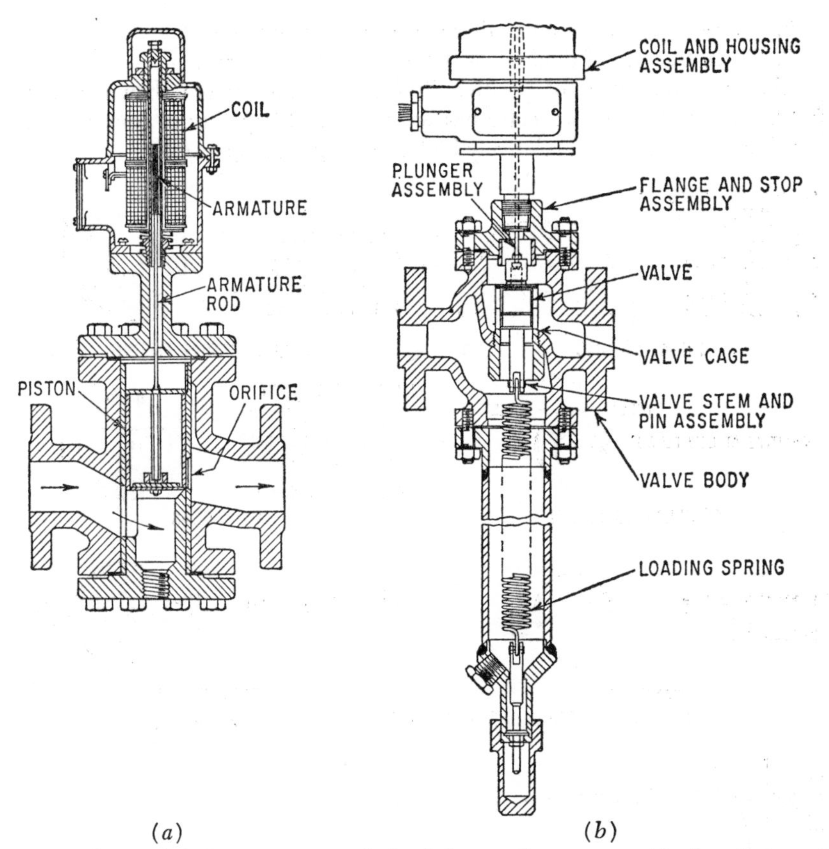

FIG. 5-39. (*a*) Piston-type weight-loaded area flow meter with electric transmission. (*Minneapolis-Honeywell Regulator Co.*) (*b*) Piston-type spring-loaded area flow meter with electric transmission. (*Bailey Meter Co.*)

coil is connected to a movable piston; so changes in piston position are instantly shown by the instrument as changes in flow rate.

POSITIVE-DISPLACEMENT FLOW METERS[13]

The positive-displacement meter is a flow-sensing instrument. It responds to variations in flow rate and delivers a corresponding mechanical signal; the rotation of its spindle.

This mechanical output signal can be coupled to a variety of devices.

These devices fall into several classes:

1. Those that totalize and register quantity of fluid through the meter

[13] Abstracted from R. W. Henke, Badger Meter Mfg. Co., Positive Displacement Flow Meters, *Control Eng.* (McGraw-Hill publication), May, 1955.

2. Those that control some other piece of equipment
3. Those that are of themselves the control or the process

The available types of positive-displacement meters are shown in Fig 5-40. The nutating-disk meter has widest use. Simplicity of construction, sturdiness, long life, and relatively high accuracy are responsible.

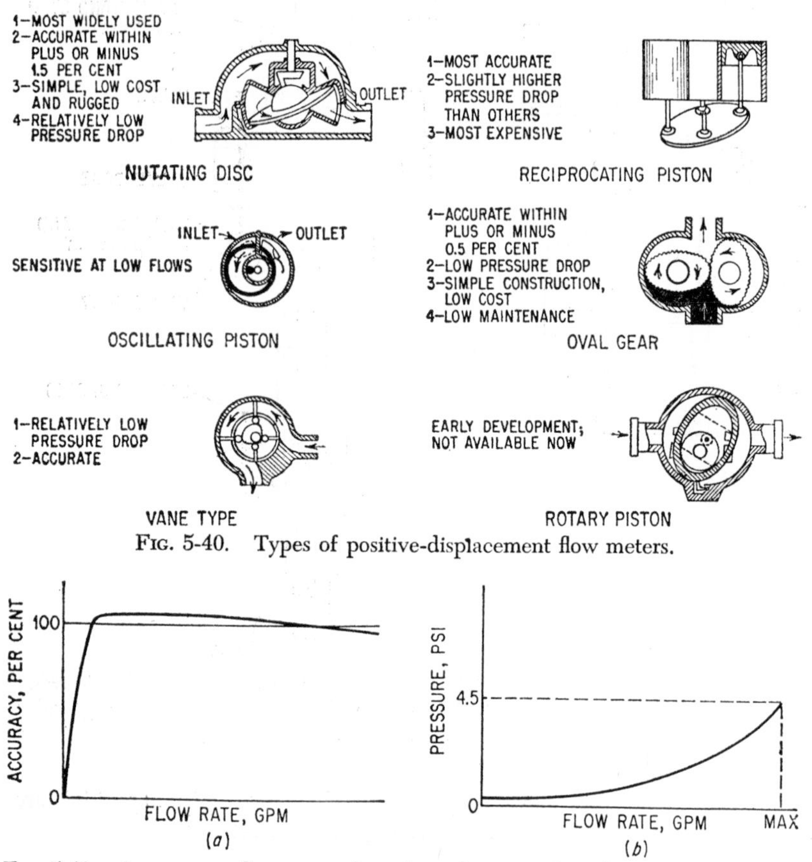

Fig. 5-40. Types of positive-displacement flow meters.

Fig. 5-41. Accuracy and pressure drop through meter plotted against rate of flow for a typical nutating-disk positive-displacement meter.

Figure 5-41*a* shows a typical accuracy curve for a nutating-disk meter. Note that the error over the sensible flow range of 0.5 gpm to the maximum rated capacity of about 40 gpm is within plus or minus 1.5 per cent. A typical pressure-loss curve for the same meter (Fig. 5-41*b*) indicates a pressure loss of about 4.5 psi at the maximum flow rate.

These curves point out two important characteristics of positive-displacement meters: high accuracy of volumetric registration and relatively low-pressure drops through the meter over extremely wide flow ranges.

The oscillating-piston meter is the next most popular type of positive-displacement meter. The characteristics of this meter parallel closely those of the nutating disk. But the oscillating piston has greater sensitivity than the disk in the lower flow ranges. This advantage is offset to some extent by the more critical machining requirements and increased sensitivity to damage or fouling by foreign matter.

The rotary-piston meter was one of the first types of positive-displacement meters and has been largely replaced by the other types.

Vane-type meters are widely used in applications that most need their high capacity and can best stand their higher cost. Figure 5-42 shows pressure-drop curves for various sizes of rotary-van-type meters. Accuracy can be held high enough to meet U.S. Bureau of Standards requirements.

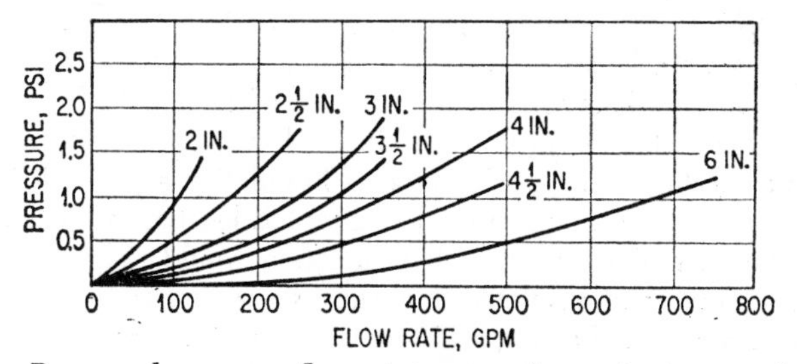

FIG. 5-42. Pressure-drop-versus-flow-rate curves for various sizes of vane-type positive-displacement meters.

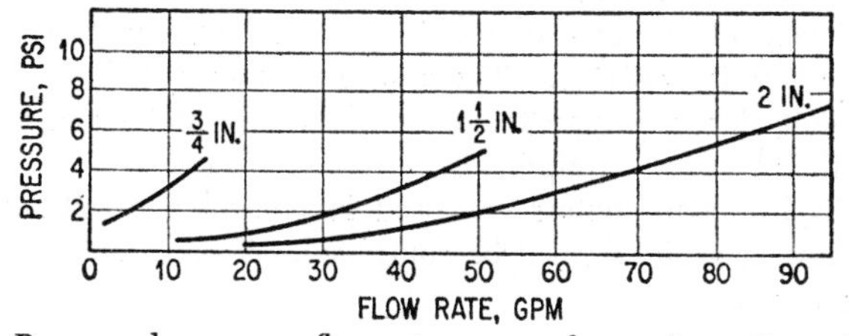

FIG. 5-43. Pressure-drop-versus-flow-rate curves for various sizes of reciprocating-piston meters.

Reciprocating-piston meters are the most accurate of all the positive-displacement meters and are also the most expensive. Figure 5-43 shows a typical pressure-flow curve for this type of meter.

In addition to the basic units, many accessory devices are used in conjunction with positive-displacement meters. These may register the total flow through the meter, permit a preset quantity of liquid to flow through the meter, or control some other flow in accordance with the flow through the meter.

Positive-displacement meters equipped with registers are used for data-gathering purposes. These data can be used to

1. Account for total product
2. Account for constituents used

3. Account for waste of material
4. Check the delivery of materials

This information can be collected in two ways, depending on the scope of the operation, on what level the data are collected, and for what purpose. For independent installations, any of the direct-reading registers can be used. To obtain the data from these units, the operator must go to the spot where the meter is installed. This type of unit is particularly adapted to single installations, mobile units, or those in the vicinity of an operator's station.

The estimated cost of such an installation ranges from $30 to $1,500, dpending on the size, type, and application of the meter. Error varies from 0.5 to 1.5 per cent, depending on the type of meter. Available flow ranges vary from 0.25 to 26 gpm in one meter up to 5 to 3,000 or 4,000 gpm, depending again on the size and type of meter.

For large installations where the number of meters to be read or the distance between stations prohibits individual readings, the remote-indicating unit is advantageous. In this type of installation a transmitting head is mounted on the meter, and the totalizing unit is located in a central control station. The cost of a single such installation would range from $150 to $3,000. Accuracy and available flow ranges are similar to those above.

Even with remote-indicating registers it may be impractical to try to keep a record of all the meter registrations in a large installation by visual examination and manual recording of data. In these applications, recording mechanisms can be substituted for the registers. Not only is the operator relieved of the task of reading the meters, but also a permanent record of the flow is available. The cost of such an installation ranges from $500 to $5,000, depending on the size and type of meter and the type and capacity of the recording equipment.

Positive-displacement meters are used in control applications when the volumetric flow of a liquid must be controlled. Typical applications are:

1. Water treatment
2. pH control
3. Injection of corrosion inhibitors

The two meter-controlled systems described below are typical examples showing the application of positive-displacement meters in chemical-feeding equipment.

In these applications it is necessary to feed or blend one or two liquids in definite proportion to another liquid. The sensing device on the meters shown in Figs. 5-44 and 5-45 is a pneumatic pilot valve that applies high and low pressure alternately across the driving piston of the chemical-feeder pumps. The frequency of alternation is directly proportional to the revolutions of the control meter's spindle. Thus, as the flow rate in the metered line varies, the chemical feeder is speeded up or slowed down, thus maintaining constant ratio.

Figure 5-44 shows a positive-displacement meter as the control instrument in a flow-responsive boiler-feedwater system. The signal from the transmitting head actuates the chemical pump to maintain the ratio of treating agent to feedwater at a predetermined value. The meter equipment alone for such an installation ranges in cost from $150 to $1,000, depending on the size and materials of construction of the meter.

Positive-displacement meters are often used to control the proportioning of water-softening chemicals in the "lime soda" or "hot process phosphate" processes. Meters are similarly applied to control the "hot zeolite" process. A typical installation is shown in Fig. 5-45. The water-softening chemicals are introduced by the chemical feeders, and the meter controls the rate of feeding to maintain a predetermined ratio of treatment to water.

When a positive-displacement meter is used in flow-control instrumentation (Fig. 5-46), the spindle is connected to one side of a differential, which is driven from the other side by a component rate control and integrator.

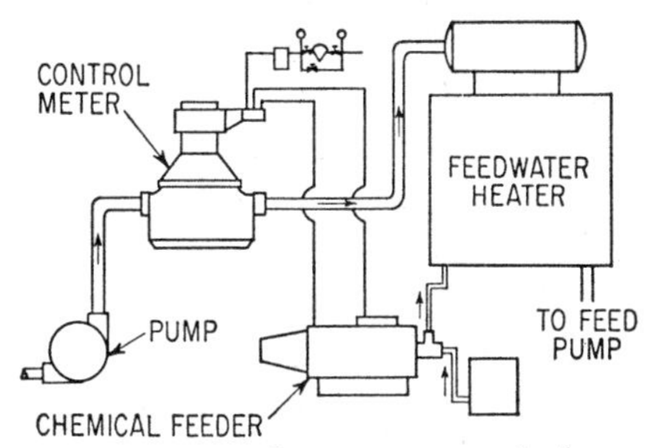

Fig. 5-44. Flow-responsive boiler-feedwater system in which meter maintains correct proportion of treating agent to feedwater. (*B-I-F Industries, Inc.*)

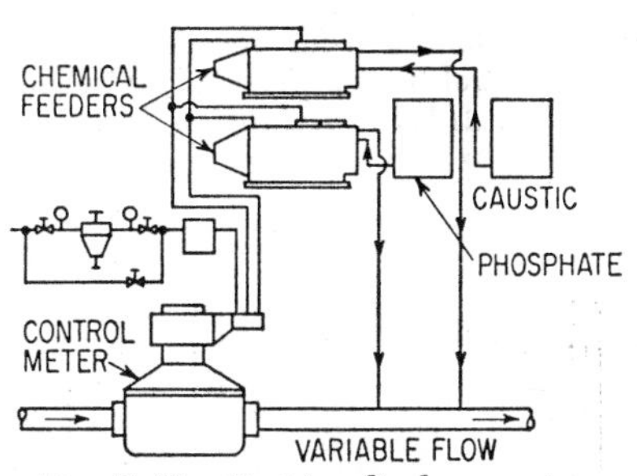

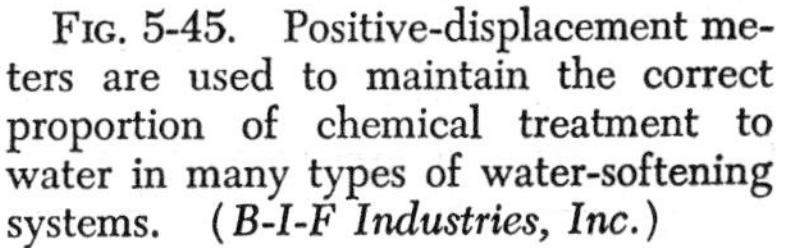

Fig. 5-45. Positive-displacement meters are used to maintain the correct proportion of chemical treatment to water in many types of water-softening systems. (*B-I-F Industries, Inc.*)

If the meter and component integrator speeds are equal, the differential connection to the pneumatic controller stands still, and the valve position remains fixed. If the speeds differ, the valve varies the flow in the proper direction to bring the meter spindle speed back into correspondence with the integrator speed. A master rate integrator and rate control maintains over-all control of many units.

This type of control can be stacked for blending several liquids in predetermined proportions—for example, blending oils. In this application, varying the setting of the master rate control adjusts the rate of flow of the final product, while the setting of the component control devices adjusts the proportion of the various constituents that make up the final product. The cost of a single bank of this equipment will range from $1,000 to $4,000. A typical system used for automatic flow-responsive sampling is shown in Fig. 5-47.

In this installation, the positive-displacement meter controls a pumping unit that extracts sample material from the delivery line and pumps it into

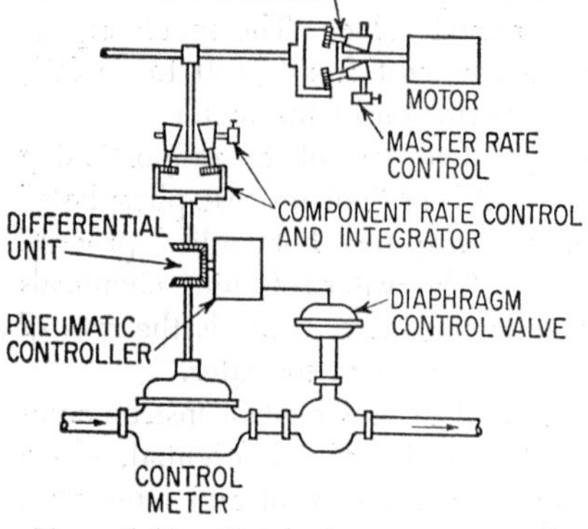

FIG. 5-46. Total flow-rate control and blending system. One master control, stacked with several component control units, permits controlled blending. (*B-I-F Industries, Inc.*)

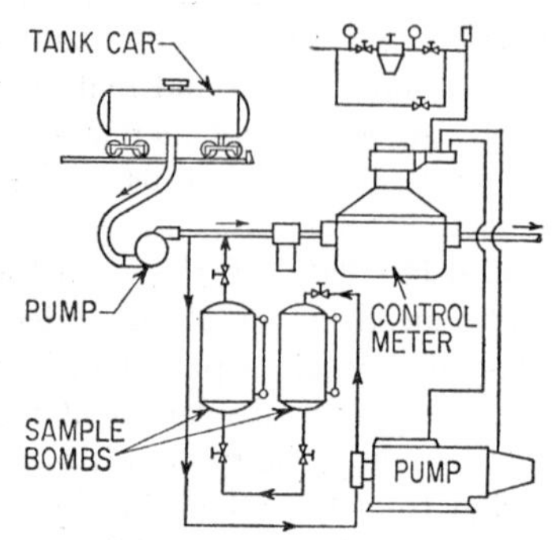

FIG. 5-47. System for extracting a sample proportional to total flow in delivery or production line. (*B-I-F Industries, Inc.*)

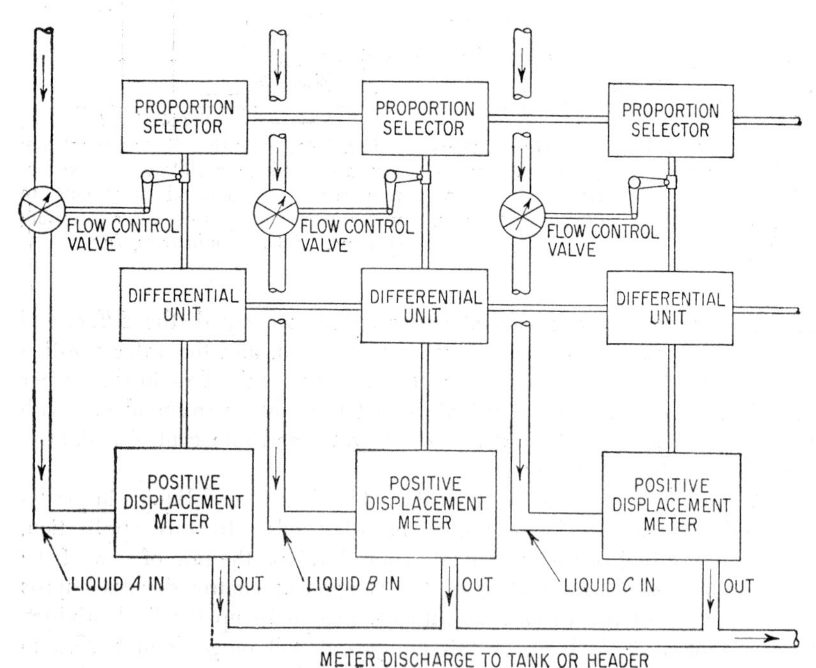

FIG. 5-48. Schematic diagram of commercially available automatic blending system. Cost per bank runs about the same as for the system shown in Fig. 5-46. (*Bowser, Inc.*)

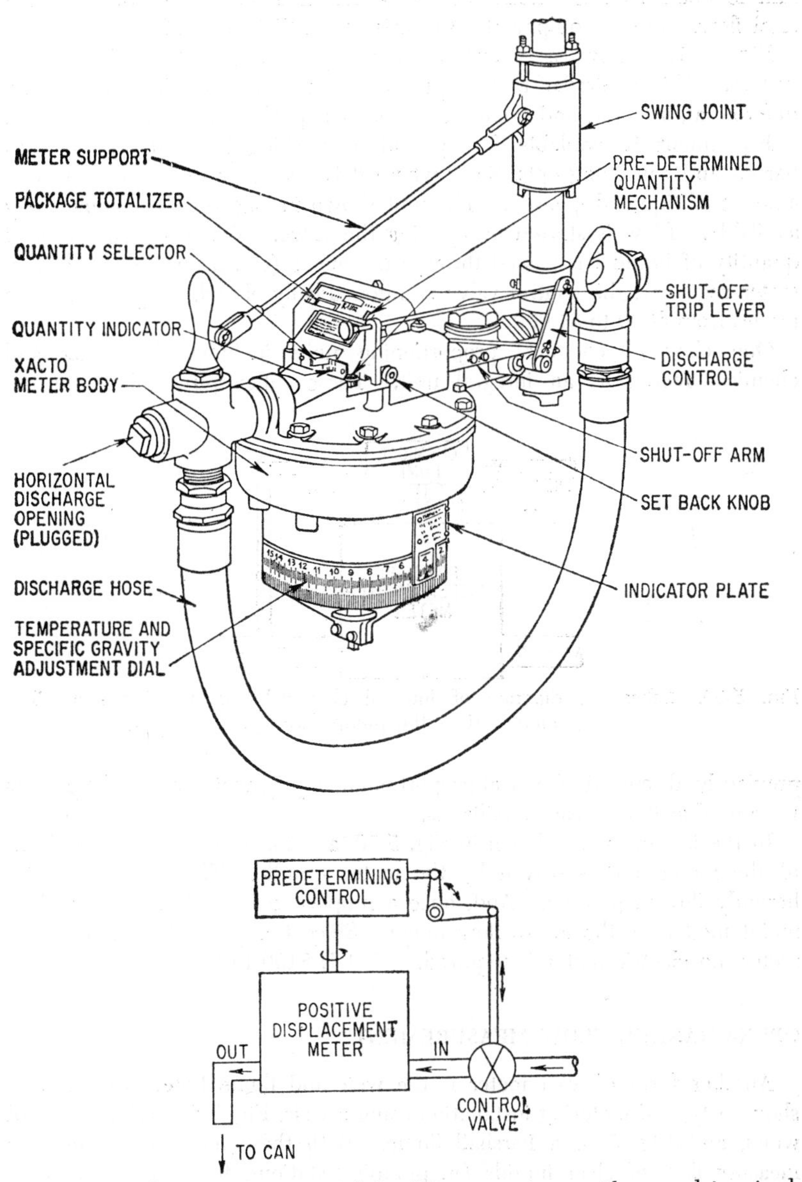

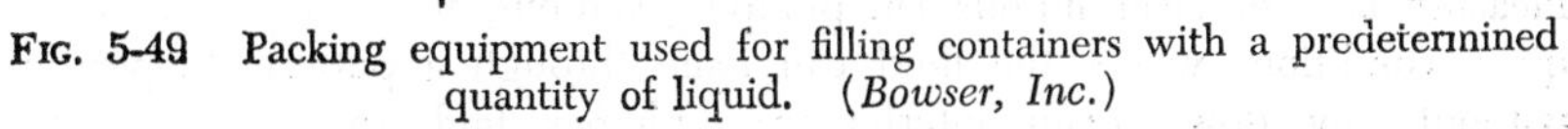

Fig. 5-49 Packing equipment used for filling containers with a predetermined quantity of liquid. (*Bowser, Inc.*)

sample bombs. The amount of the sample is always proportional to the total flow. Meter equipment cost varies from $150 to $1,000.

Figure 5-48 shows a typical commercially available automatic-blending system. This is similar to the system shown in Fig. 5-46. Each meter, differential, selector, and related equipment is packaged as an integral unit.

Equipment is available for specialized packing applications, such as barrel- and can-filling operations, where it is necessary to control the exact amount of liquid dispensed. The unit shown in Fig. 5-49 is commercially available. Flow is started by opening the valve. When a predetermined quantity of liquid has passed through the meter (as set on the preselecting meter register), the control valve is automatically closed. Equipment cost ranges from $300 to $1,000.

One class of chemical proportioning units includes the meter and chemical feeder as an integral unit. These differ physically from the

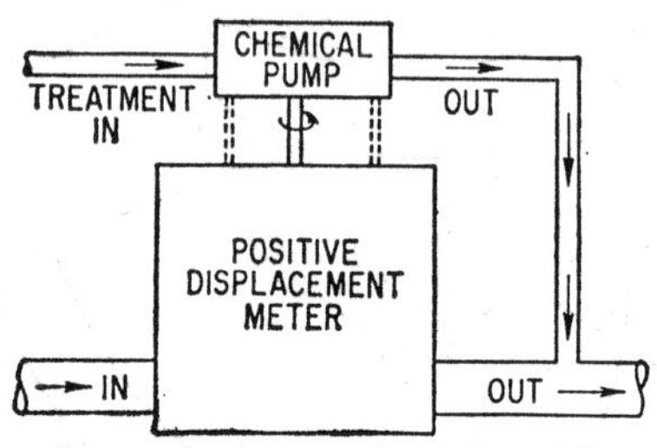

Fig. 5-50. Schematic diagram of integral chemical proportioning unit. The meter drives the pump directly.

previously discussed chemical-proportioning equipments where the meters were used as flow-sensing auxiliaries.

In the integral units shown in Fig. 5-50, a pump is mounted on the head of the meter and is driven by the meter spindle. Thus the unit is inherently flow-responsive. And, once a proportion is established, it will be maintained over the entire flow range. Since the pump is driven by the meter, no electric motor is required. Cost is $150 to $300.

OPEN-CHANNEL FLOW MEASUREMENT[14]

Another form of area meter is the weir and flume type. Figure 5-51 shows a typical metering installation using a weir, Fig. 5-52 various types of weirs, and Fig. 5-53 a Parshall flume. Both these methods are used to measure flow of clear liquids (in pumping stations, for example) and require conditions where the fluid can issue through a patterned opening without any downstream interference. Flumes find application where liquids are carrying solids in suspension (sewage and industrial wastes) and

[14] Abstracted from H. W. Stoll, Taylor Instrument Co., "The Use of Primary Elements in Measurement of Flow."

in locations where sudden changes in elevation are not available, such as might be encountered in measuring the flow of water through an irrigation canal. A convenient method of determining the flow-stream area through the weir or flume is to measure the height of fluid above a specific reference level. In a weir, this reference level is its base, or sill as it is often called. In flumes the reference level is measured by means of a stilling well on the side one-third of the distance from the point of first convergence to the throat. This height can be found either through the use of a graduated scale or float mechanism or by measurement of hydrostatic pressure.

The open-channel flow meter measures the flow of liquids like sewage, by measuring the liquid head created by a differential device such as a weir or flume installed in the channel. The transmitter for the meter is placed

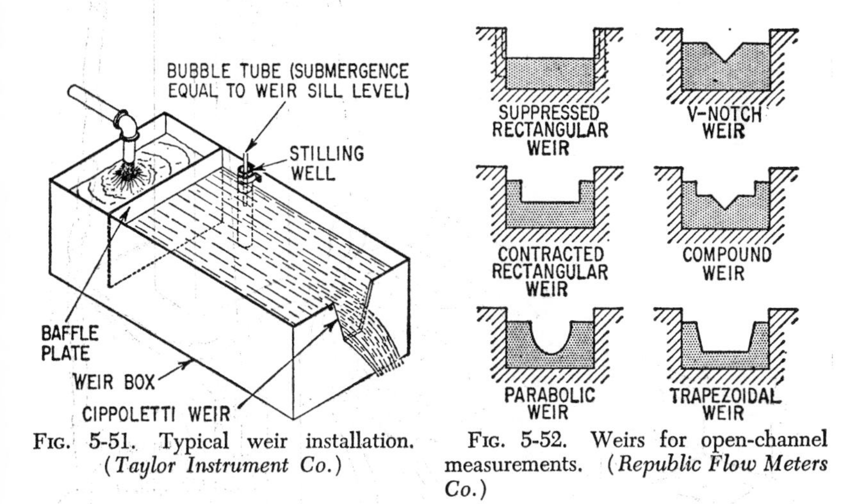

FIG. 5-51. Typical weir installation. (*Taylor Instrument Co.*)

FIG. 5-52. Weirs for open-channel measurements. (*Republic Flow Meters Co.*)

directly over a still well connected to the channel. Suspended below the transmitter is a float, which follows the movements of liquid in the still well. If the still well is correctly located in relation to the weir or flume, the level of liquid in the still well will be proportional to the rate of flow. Thus the float will rise and fall as flow through the weir or flume increases and decreases.

In one design[15] using electric transmission (Fig. 5-54) the float motion operates the pulley mechanism to raise and lower a resistance scale in a chamber partially filled with mercury, and, also, to move a pointer on a mechanical indicator. As the scale moves down, for example, contact rods on the scale progressively touch the mercury, short-circuiting the resistance in the scale. The lengths of the contact rods are arranged to fit the flow characteristic curve of the type of differential device used. In this manner, resistance in the scale is made to vary inversely with flow. Electrical-

[15] Republic Flow Meters Co.

reading instruments connected to this resistance circuit, which is in effect an electrical transmitter, will be actuated by any change in resistance and will read directly in units of flow.

A second electric type[16] (Fig. 5-55) contains a pair of inductance coils with a movable magnetic core located within them; the electric receiver

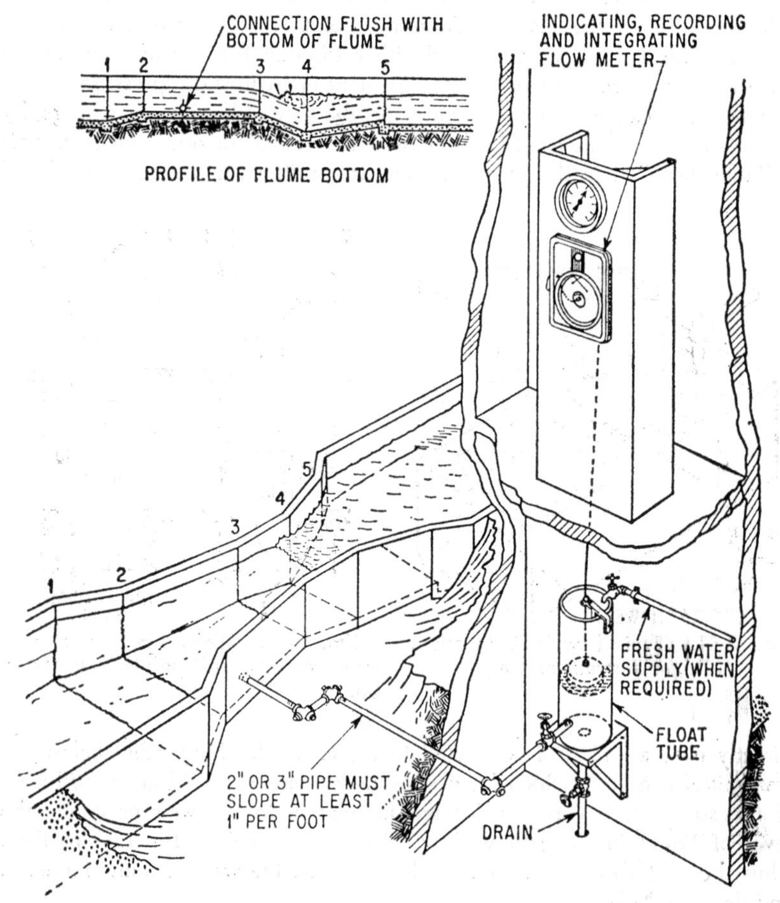

FIG. 5-53. Typical installation of flow meter with Parshall venturi flume. (*Bailey Meter Co.*)

(indicating, recording, and/or totalizing) likewise contains a pair of inductance coils and a core.

The transmitter and receiver are connected by three wires forming a null-balance inductance bridge similar to a Wheatstone bridge. Alternating current is connected to the receiver only. A sensitive galvanometer in the receiver has its connection across two of the wires joining the transmitter

[16] Penn Instruments Div., Burgess-Manning Co.

and receiver. As long as the galvanometer boom remains in the center position, it indicates null or zero voltage; it shows that the electric bridge is in balance and that no change of position of the core in the transmitter has occurred.

A change in flow through a flume or weir moves the float vertically, thus repositioning the core in the transmitter coils. This core movement un-

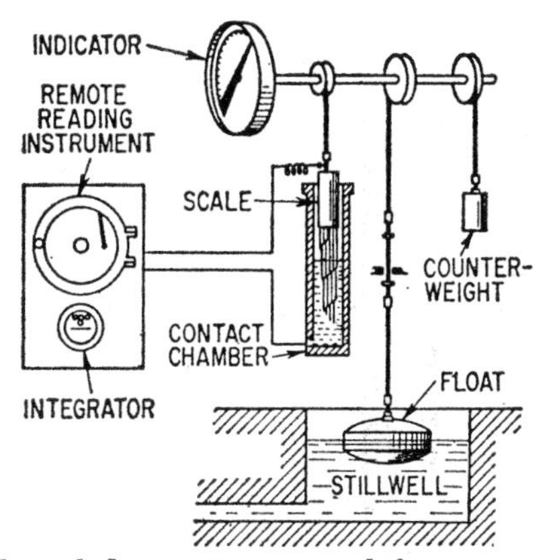

FIG. 5.54. Open-channel flow meter arranged for remote electric transmission of measurements. (*Republic Flow Meters Co.*)

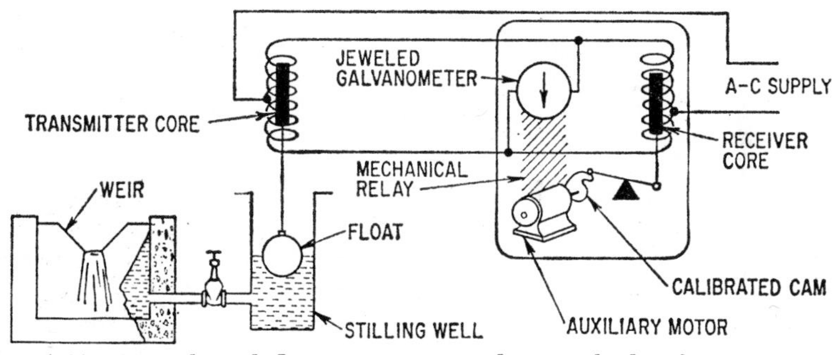

FIG. 5-55. Open-channel flow meter using inductance bridge for transmission. (*Penn Instruments Div., Burgess-Manning Co.*)

balances the inductances of the transmitter and receiver coils, causing a deflection of the galvanometer boom. The indicator pointer, pen arm, and integrator will show the change in rate of flow because the servomotor moves these parts simultaneously as it moves the receiver core to a position corresponding to the new position of the transmitter core. When this is accomplished, the galvanometer returns to its null (zero) balance.

Other designs featuring direct-reading mechanical flow meters and pneumatic transmitters for remote reading are commercially available.

Section 6

MEASUREMENT AND CONTROL OF LIQUID LEVEL

FUNDAMENTAL PRINCIPLES[1]

Liquid level, like most variables, can be measured directly or inferred. The methods discussed in this section can be classified as follows:

Direct methods:
- Ball float
- Conductivity electrode

Inferential methods:
- Pressure gage
- Diaphragm box
- Purge systems
- Differential-pressure converter
- Mercury manometer
- Nonindicating pressure controller
- Displacement transmitter

Each method of measuring level discussed is based on one of the following physical principles: hydrostatic head, float movement, displacement, electrical conductivity.

Hydrostatic Head. Many methods are based on the principle of measuring the hydrostatic head. This head is the weight of liquid above a reference or datum line. At any point its force is exerted equally in all directions and is independent of the volume of liquid involved or the shape of the vessel. Head is often expressed in terms of pressure or level height. Measurement of pressure due to liquid head can be translated to level height above the datum line as follows,

$$H = \frac{P}{D}$$

where H = height of level
P = pressure due to hydrostatic head
D = density of liquid

[1] Minneapolis-Honeywell Regulator Co.

Table 6-1. Typical Liquid-level Instruments*

Systems	Applications	Components	Principles of operation	Limitations	Advantages
Pressure gage	Open vessels	Pressure element. Standard instrument components	Static pressure connected to spiral or spring and bellows element	Instruments must be located near vessel and near minimum level of liquid. Sealing equipment needed for corrosive or viscous liquids	Simple and inexpensive for clear liquids
Diaphragm box	Open vessels	Flexible diaphragm in box connected to pressure gage or mercury manometer	Box at minimum level inside or outside vessel. Static pressure transmitted through diaphragm, which seals gage line from liquid	Length of tubing to instrument should not be greater than 250 ft. Closed-type box system needs sealing equipment for corrosive or viscous liquids	Recording instrument need not be located at minimum tank level
Purge or bubbler	Open or closed vessels. Depends on instrument used with purge system	Pressure-measuring instrument. Compressed-air supply. Pneumatic accessories	Supply air pressure balances pressure due to liquid head	Limited only by range of instrument used and supply line pressure	Ideal for corrosive fluids, also viscous liquids and liquids with suspended solids. Pressure gages with air purge need not be installed at minimum level. Mechanical manometers with purges do not need dropped-range tubes
Differential-pressure transmitter	Open or closed vessels. Pneumatic transmission of variations in level to other instruments	Chamber with diaphragm. Pivoted beam. Feedback bellows. Pneumatic system	Differential-pressure change moves diaphragm and beam. Feedback bellows from pneumatic system balances beam.	Ambient temperatures 32 to 225°F. Upper limit of range is 200 in. Seals or purges needed for certain few liquids	Small and light. Easy installation and maintenance. Range continuously adjustable in field. Calibration checked by set of weights in field. Fast response
Mercury manometer	Open or closed vessels. Interface level	Mechanical manometer uses dropped-range tube. Standard instrument components	Pressure applied to legs or U tube containing mercury. Float transmits motion	Sealing equipment needed for corrosive and viscous liquids, also for liquids with excessive dirt or solids	Uses flow-meter components. Works at high pressures

Nonindicating pressure controller	Open or closed vessels. Control only—no indication, record, or transmission	Bellows. Electric switches or slide-wires, or pneumatic control system	Pressure due to liquid head operates bellows. Transmits motion to control components	Cannot use fluids chemically injurious to bellows. Not suitable where high degree of accuracy and sensitivity is needed	Inexpensive. Small size saves space and makes installation easier
Ball float	Open or closed vessels. Control only—no indication, record, or transmission	Internal float, or external float with cage. Mechanical or pneumatic connection to valve	Float moves up or down with level. Transmits motion	Narrow range. Cannot be used with corrosive liquids or liquids which would deposit solids on ball. Accuracy and speed of response limited by mechanical linkage and stuffing box	Simple and inexpensive. Suitable for viscous liquids
Remote-reading tape and float	Open or closed vessels. Remote indication only	Float. Perforated steel tape. Sprocket-wheel assembly. Transmitter. Electronic precision indicator	Motion of float changed to variable emf. Transmitted to remote indicator	Cannot be used with liquids that would deposit solids on float	One instrument provides indications transmitted from many remotely located tanks. Current flows only while operator is actually gaging at receiver
Displacement transmitter	Open or closed vessels. Pneumatic transmission of variations in level to other instruments	Cylindrical float. Sealing bellows. Balance beam. Feedback bellows. Pneumatic system	Change in level varies buoyant force, opposed by pneumatic feedback bellows	Cannot be used with liquids that would corrode displacer or deposit solids on it. However, various trims are available to combat corrosion	Sensitive to small-level changes and specific-gravity variation. Requires no stuffing box. Remarkably friction-free
Conductivity electrode	Open or closed vessels. Interface level. Control at set point or limit points—no indication, record, or transmission	Electrode(s). Detecting relay	Conducting liquid completes circuit with electrode at predetermined level to initiate control action	Liquid must not be chemically injurious to stainless-steel electrodes. Not suited for viscous electrically conductive liquids that tend to build up deposit and short-circuit electrodes	Simple and inexpensive. Flexible. Range unlimited for high-low control

* Courtesy of Minneapolis-Honeywell Regulator Co.

Temperature changes sufficient to alter the liquid density appreciably will affect the accuracy of measurement and may be particularly significant where automatic control is employed.

If a pressure greater than atmospheric is imposed on the surface of the liquid in a closed vessel, this pressure adds to the pressure due to hydrostatic head and requires the use of instruments that measure differential pressure.

The added pressure in the vapor space above the liquid is imposed on both the high- and low-pressure connections of the differential-pressure instrument and mechanically canceled out.

Float Movement. A float moves up or down with changes in liquid level, this movement being translated by various means into control action (usually positioning of a control valve on the inflow or outflow line to the vessel).

Displacement. Archimedes' principle states that the resultant pressure of a fluid on a body immersed in it acts vertically upward through the center of gravity of the displaced fluid and is equal to the weight of the fluid displaced. This resultant upward force exerted by the fluid on the body is called *buoyancy.*

Measurement of the buoyant force exerted upon a partially submerged element at various degrees of submergence enables a scale to be calibrated in terms of liquid level.

Electrical Conductivity. The fact that certain liquids conduct electricity, while air and certain other liquids—in a relative sense—do not, is used to distinguish between the presence and absence of liquid and to provide on-off control at predetermined levels.

Table 6-1 compares different types of liquid-level instruments.

BALL-FLOAT DEVICES[2]

Direct-operated Float Valves. The simplest of level controllers is generally known as a float-operated lever valve of a type as shown in Fig. 6-1. Here a float, attached to a lever mechanism, operates a valve member through mechanical levers of either the first- or second-class leverage type.

Ordinarily this type of level controller is used to maintain a level in a tank by allowing liquid to be fed into the tank at the same rate in which it is being fed into the vessel. This application may be on a water-storage tank on top of a building, hotel, or laundry, or at the bottom of a cooling tower where the valve adds make-up water to take care of evaporative conditions. The use of this float valve is limited to open-tank work and is not for pressure tanks. The valve is a double-ported balanced valve because pressure of the supply liquid as furnished to the valve is balanced, thus holding unbalanced pressure forces at a minimum. Otherwise the float would not have sufficient power to operate the valve. However, a

[2] Fisher Governor Co.

slight amount of leakage may always be anticipated in this type of float valve since a double-ported valve does not provide tight shutoff.

Another factor which must be watched in this type of unit is the effect that may result from flow velocity and velocity unbalance as a result of the liquid flow across the valve seats. The only available power for operating the valve comes from the ball float (see Fig. 6-2), which, in order to have power available in both directions, should be weighted, or heavy enough to float in the liquid at the halfway point. (See Fig. 6-3.) Most common

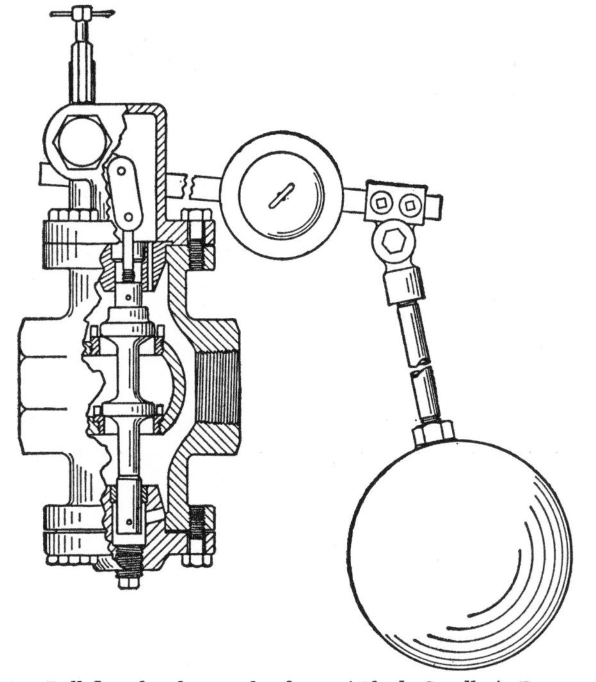

Fig. 6-1. Ball-float level control valve. (*Black, Sivalls & Bryson, Inc.*

float sizes are 8 and 10 in., with an occasional 12-in. float for larger valves such as 4- and 6-in. sizes. Larger floats are not practical because of cost and space limitation. The power potentialities of a 10-in.-diameter float may be analyzed as follows:

Area at section through center is 78 sq in.

Assume level rises or falls 1 in. above or below center line of float.

Volume of segment of 1 in. × 78 sq in. is 78 cu in.

Weight of water/cu in. is 0.036 lb.

Weight of 78 cu in. of water is 2.81 lb.

Power available for operation of valve is the equivalent of the liquid displacement—for 1 in. of level change around the float—the available power would be 2.81 lb.

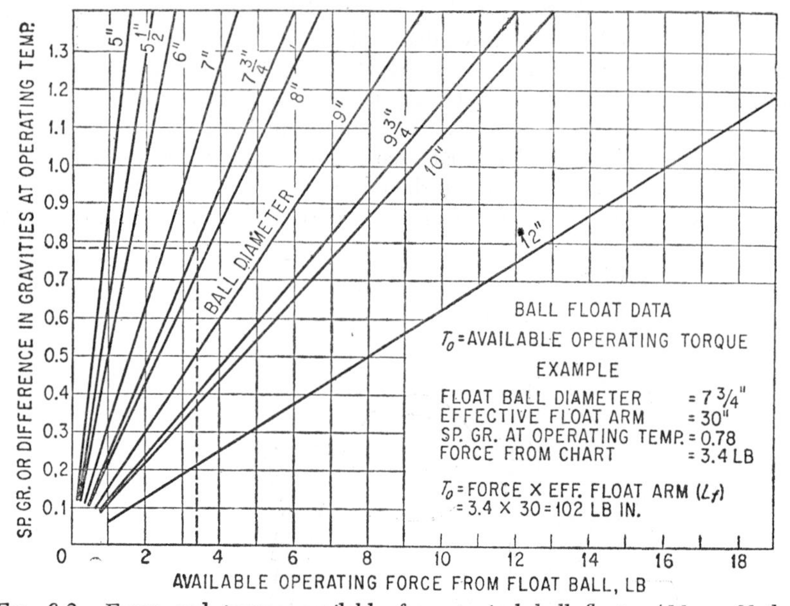

FIG. 6-2. Force and torque available from typical ball floats. (*Mason-Neilan Div. of Worthington Corp.*)

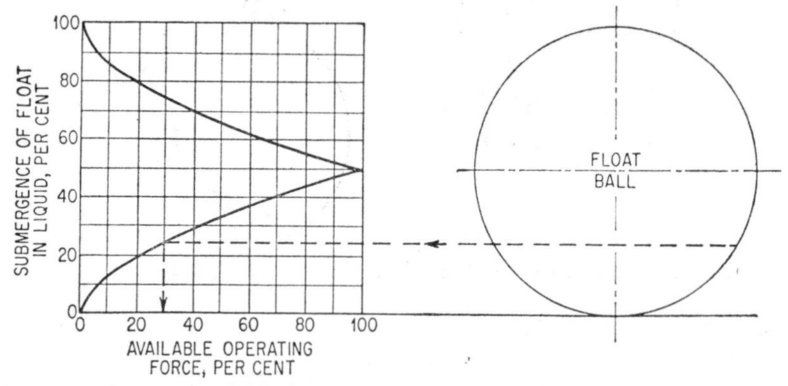

FIG. 6-3. Greatest available force is realized with float submerged 50 per cent. (*Mason-Neilan Div. of Worthington Corp.*)

This power or force is transmitted to the inner-valve mechanism through a simple system of levers. The ratio of this leverage is a comparison of the amount of level change that is used to get complete valve travel. If the amount of valve travel needed is 1 in., and the level is allowed to change 12 in., the ratio of lever system is 12 to 1. With a force of 2.81 lb from the float, multiplied 12 times, the power or force transmitted to the valve stem is 2.81 × 12 = 33.7 lb.

The above assumes a level change on the float of 1 in. If there were no friction forces and no velocity forces, the float would merely ride the level and no displacement of the float would be necessary. But those forces are present and must be overpowered. Yet it is desirable to use only a small segment of the float through the center where the smallest amount of level change produces the largest amount of power.

The above power is ample for operating small valves up to 2 in. with pressure drop across the valve up to several hundred pounds, but in sizes 6 in. and larger the drop across the valve should not exceed 10 or 15 lb. Otherwise the velocity forces can become greater than the potential float power, and the valve will go out of control. In fact, the inner-valve position may try to control the float position, instead of the float having dominant power over the valve.

These forces become greater as the valve size increases, varying approximately equivalent to the capacity relationship of valves. Thus, the larger the valve, the larger the float needed to overpower these forces, and to make sure that the valve position is commanded by the float.

To help minimize these forces and to ensure good throttling of the valve, some form of a throttling inner-valve design is recommended.

In summary, these direct-operated float-controlled valves have wide application, but also have limitations in size and pressure drops, since the size of a float must be restricted by physical considerations, and thus it becomes impossible to produce, in all instances, sufficient power to assure stable control of the valve position.

Internal Pilot-operated Float and Lever Valves. To overcome some of the difficulties of power requirements of the double-ported valve, the internal pilot-actuated lever or float valve is applied.

Here, the float or direct-actuated level operator actuates both the valve stem and a pilot valve of small size, which requires only a relatively small amount of power.

This type of float-operated valve is generally single-seated, making it particularly desirable for installations such as on the head or roof tank on top of a building or hotel, where it is necessary to have tight shutoff at times of no consumption of water, as at night or on holidays. The leakage due to the use of a double-ported valve might cause the tank to overflow onto the roof or flood a building. Typical operation of this valve (see Fig. 6-4) is as follows: Inlet pressure or supply comes in over the valve disk and aids in holding the valve closed. The composition seat (soft disk) ensures a tight shutoff condition. The supply pressure equalizes across the operating piston through the orifice in the piston. When the pilot valve is closed, the pressure across the pistons is equalized, and the supply pressure over the valve disk holds the valve tightly closed. When the float or float controller causes the valve to open, the valve stem moves upward, and the pilot valve is opened. The area of the piston is larger than the valve area, and so there is now a preponderance of force pushing upward, a force greater than that holding the valve on its seat. As a result, the valve-and-piston assembly moves upward, opening the valve to a distance determined by stem movement, since, as soon as the pilot orifice begins to close, while the piston moves upward and the stem begins closing the pilot orifice, the valve comes to a balance and opens no farther. If the float device now closes the pilot valve, the pressure across the piston equalizes, and the valve is closed by the action of supply pressure across the main disk.

One of the disadvantages of these piston-operated or balanced valves is that they will seize or stick if the liquid is dirty. Foreign materials such as sand, scale, and dust will settle behind the piston and render the valve inoperative. To prevent this, the piston is loosely fitted, and a synthetic-rubber cup takes the place of the regular clearance, acting like piston rings

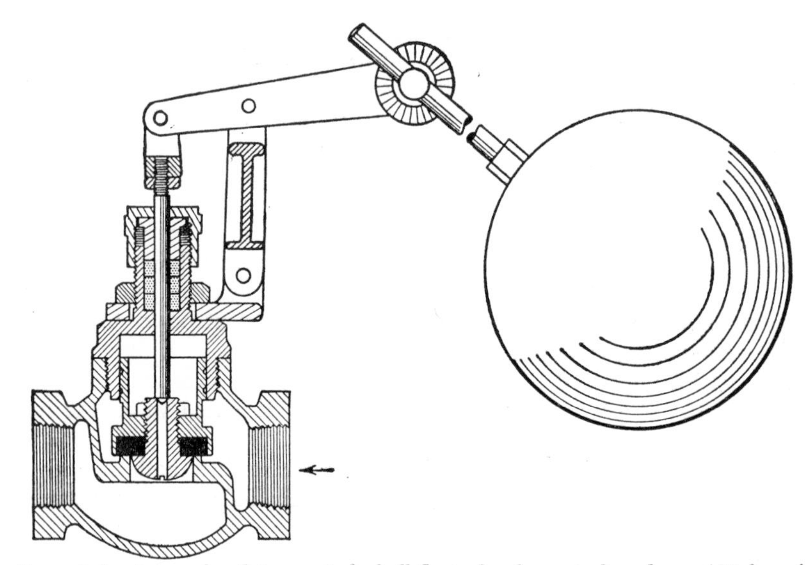

Fig. 6-4. Internal pilot-operated ball-float level control valve. (*Kieley & Mueller, Inc.*)

or packing in a pump. Sand, scale, etc., works its way past or around this cup, instead of sticking to the piston and making it inoperative.

Tank-level Controls. This section covers devices used for level control and measurement where the tank is closed or partially closed.

Tank Stuffing Boxes. One of the most common types is a tank stuffing box as illustrated in Fig. 6-5.

The valve is operated directly by the float motion, as shown in Fig. 6-6. Limitations as discussed under float valves are applicable to this type of controller.

Float Cages. Where it is neither practical nor desirable to have the float in the vessel, the float cage shown in Fig. 6-7 is applied to the tank. Float cages are generally manufactured in four ratings of 125-lb iron, 250-lb iron, 600-lb steel, and 1,500-lb steel.

The float cage is mounted on the tank, in which the level is to be maintained, with one connection from the bottom of the cage into the liquid in the vessel, and the other from the top of the cage into the vapor space above the level. As the level in the vessel changes, that same change is transmitted into the float cage, since the two levels must equalize. Any change in level results in changing the float position, since the float rides

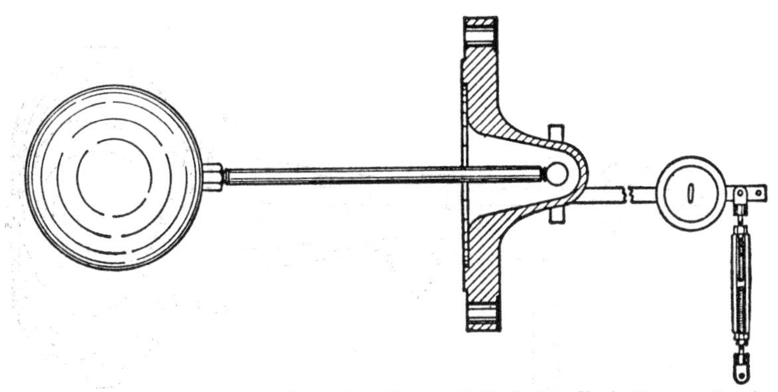

FIG. 6-5. Float flange with stuffing box. (*Black, Sivalls & Bryson, Inc.*)

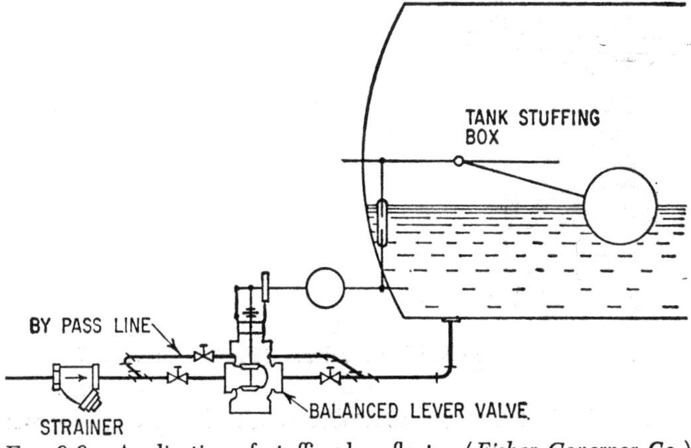

FIG. 6-6. Application of stuffing-box float. (*Fisher Governor Co.*)

on the level in the cage. This motion is transmitted through a rotary shaft to an outside lever, where its motion is converted to work, such as operating a control valve.

The equalizing connections are generally 1½- or 2-in. screwed connections, as these sizes are large enough to ensure satisfactory level equalization. Where high pressures or high temperatures are present, the connections to the cages are flanged, in order to procure the best possible protection against leakage, or for safety reasons. In some cases, the liquid-equalizing connection may be made larger in order to secure rapid equalizing of the level between float cage and vessel. An example is the condenser hotwell application on large power-plant surface condensers. Here, because a small-size liquid reservoir handles large quantities of condensate, the level may change very rapidly, and so a 3-in. connection is used. This is necessary to prevent the level in the cage from lagging behind the level in the hotwell and setting the system into a continuous cycling or hunting

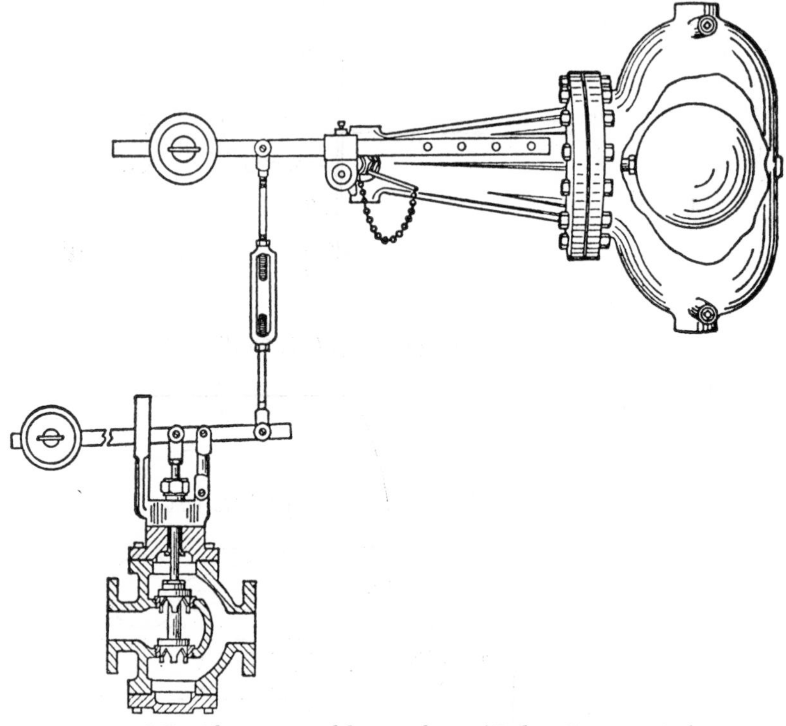

FIG. 6-7. Float cage and lever valve. (*Fisher Governor Co.*)

action, due to the fact that the controller is out of step or phase with the actual level in the hotwell.

The float-cage units are used to operate lever valves in direct-connected assemblies as illustrated in Fig. 6-7. Here the valve should be close to or adjacent to the cage, since connections with pulleys, cables, and long rods do not prove entirely satisfactory. It must be remembered that the only power available for operating these valves comes from the float ball; therefore, the limitation generally applicable to direct-operated float valves will apply.

These float-cage types are also used to translate level changes to operating switches to start motors, ring bells, sound alarms, or remotely signal and indicate level positions. Figure 6-8 shows an assembly of float cage and pilot with an explosion-proof switch, for operating signal alarms.

External-pilot-operated Level Control. A pilot-operated liquid-level controller may be generally defined as a control device or mechanism in which the primary change of a liquid level is transmitted through the action of a float or other means to an amplifying relay requiring but small motive power. The amplifying relay in turn sends an energizing medium to actuate the final element of control, usually a control valve.

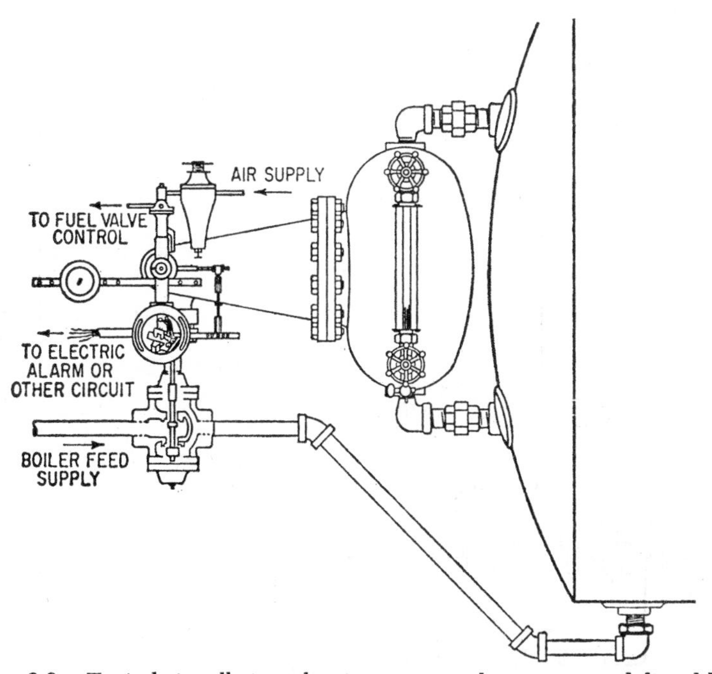

FIG. 6-8. Typical installation showing use as direct-connected liquid-level controller, with pilot controlling fuel valve and mercury switch operating at an indicated level to make or break circuit to an electric alarm, pump, or blower. (*Black, Sivalls & Bryson, Inc.*)

The advantages of a pilot-operated control mechanism are:

1. It is not necessary to place the control valve adjacent to the float-control device. The liquid-level controller or measuring unit may be installed on a vessel at a remote distance from the control valve. Though distances up to 200 or 300 ft are common, distances up to 1,000 ft or more may be used. And, if an electrical circuit is applied for indication or control function, the distance between the control member and the control valve is no longer limited. An example of a remote application would be maintenance of a liquid level in a storage tank on top of a building, by controlling the speed of a pumpout

pump located in the basement of the building. The float-cage member may be placed on the tank, the valve in the steam line to the pump, and the operating-medium supply pressure brought to the valve by means of proper tubing and fittings. Thus, a change in level within the vessel causes a change in the pressure supplied to the control valve, actuating it so that the pump speed maintains the level at a predetermined point by either increasing or decreasing the rate of pump discharge.

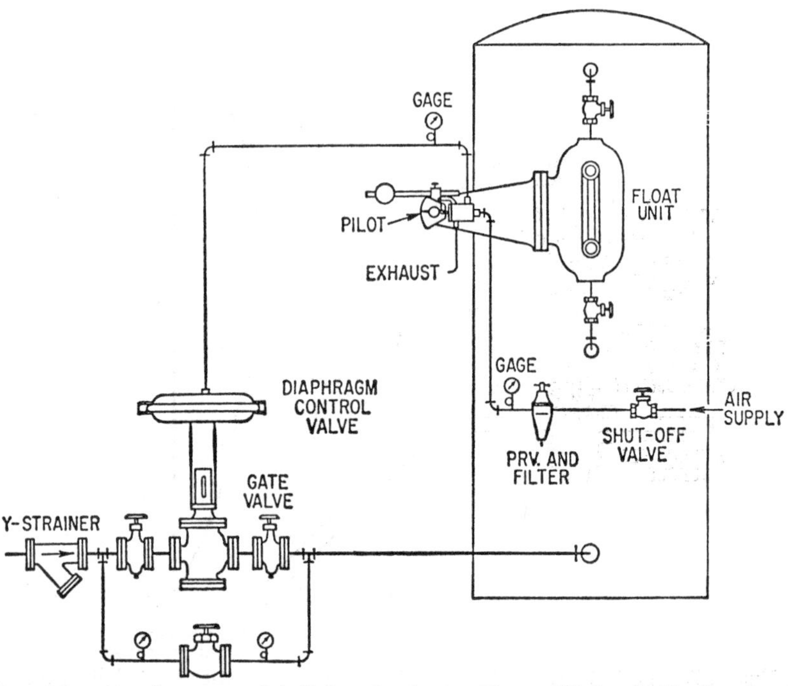

Fig. 6-9. Air-pilot-actuated ball-float level controller. (*Kieley & Mueller, Inc.*)

2. It is necessary to have only small increments of power output from the float device in order to operate the pilot to transmit an output pressure change. The amplifying effect of the pilot or relay allows for considerable power development at the primary control valve, ensuring its positive placement at a given position, and accomplishing a more satisfactory resulting control.

In the application of a pilot-operated liquid-level controller of the ball-float type, the air or hydraulic pressure is transmitted from the pilot valve to a diaphragm-actuated control valve. In general, a range of pressure changes to the control valve is 3 to 15 lb in the most common type of application. These pressure changes, acting on a large diaphragm and suitable spring combination, provide a positive movement of the valve for every pressure condition sent out by the pilot member. Large diaphragms and heavy springs ensure a very stable condition of the valve, regardless of velocity and hydraulic forces that might be caused by velocity-flow effects through the valve. Furthermore, no effect of these forces existent within the valve structure is transmitted back to the liquid-level controller to disturb its proper position, since there is no direct mechanical hookup between the control valve and the float member of the liquid-level controller. (See Fig. 6-9.)

DISPLACEMENT-TYPE LEVEL CONTROLS

Principles. When a body is immersed or partly immersed in any liquid, it loses weight equal to the weight of the liquid displaced.

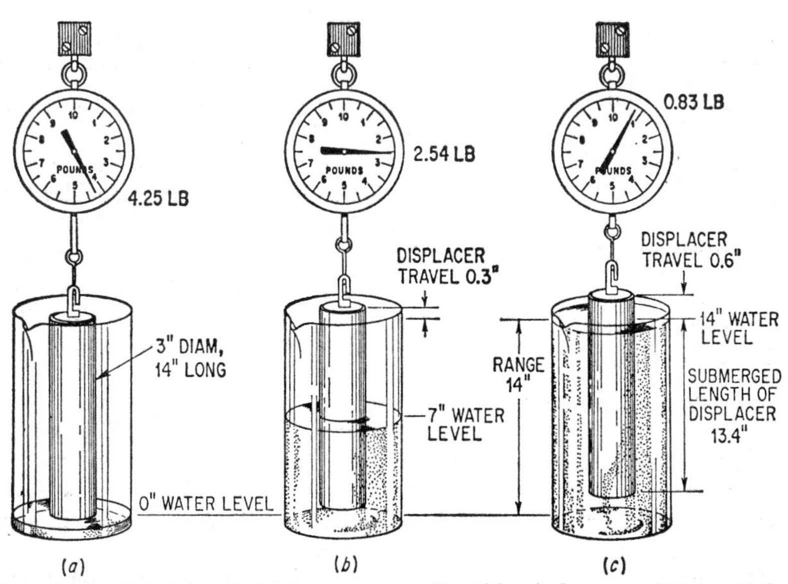

FIG. 6-10. Principle of displacement-type liquid-level devices. (*Mason-Neilan Div. of Worthington Corp.*)

Figure 6-10 diagrammatically illustrates the method of using the displacement of liquid for level measurement. The displacer shown is the size used in the 14-in. range. It is 3 in. in diameter and 14 in. long and weighs 4.25 lb.

In *a* the displacer is suspended in a glass jar by a spring scale having a range of 0 to 10 lb. The water level is just even with the bottom of the displacer. Consequently, the spring scale is supporting the full weight of the displacer as indicated on the dial at 4.25 lb.

With a 7-in. water level as shown in *b*, the displacer is partly immersed. The loss in displacer weight is equal to the weight of the water displaced, and the net weight of the displacer now becomes 2.54 lb.

In *c* the water level is 14 in., and the net weight supported by the spring scale is 0.83 lb. This increase in water level from zero to 14 in. has decreased to net displacer weight from 4.25 to 0.83 lb, a net change in weight of 3.42 lb.

Note also that, as the net weight of the displacer has been decreased, the spring scale has lifted the displacer an amount directly proportional to the increase in water level.

The amount of water displaced by the 14-in. change in level is equal to the cross-sectional area of the displacer multiplied by the submerged length (14 in. minus the displacer travel of 0.6 in.). This represents a volumetric displacement of 95 cu in. The weight of 95 cu in. of water is 3.42 lb.

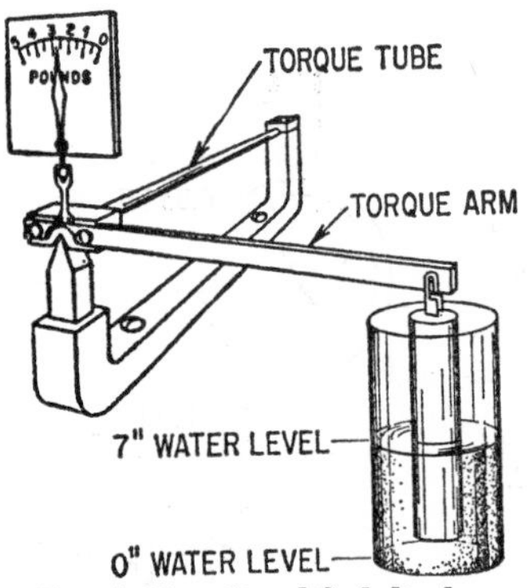

FIG. 6-11. Simplified displacement indicator. (*Mason-Neilan Div. of Worthington Corp.*)

The travel of the displacer for a 14-in. level change can be varied within limits by changing the range of the spring scale. If a 50-lb scale is used, the travel is materially reduced. Conversely, the use of a 5-lb scale approximately doubles the travel. The change in travel will not be quite directly proportional to the spring-scale range, since the volumetric displacement varies with the scale range. For practical purposes this error may be disregarded.

Using an accurate spring scale, the scale dial could be calibrated in terms of level, providing a simple and accurate level indicator for liquids of known constant specific gravity. Level ranges up to 30 or 40 ft are practical. Such a measuring device could be adapted for indication, recording, and control so long as the vessel was open to the atmosphere.

The principle of liquid-level measurement in a closed vessel under pressure is shown in Fig. 6-11. The torque tube and torque arm have replaced the tension spring in the spring scale (Fig. 6-10). The torque tube has been selected with respect to diameter, wall thickness, and length to give the desired spring rate for a torque arm length of 8 in. Since the change in the displacer position is very slight in comparison with the corresponding change in liquid level, it is evident that the range of measurement of liquid-level-control instruments is greatly increased by the use of this displacement principle. See Fig. 6-12 for torque tube construction.

Operation of Pilot. Referring to Fig. 6-13, the operating medium of air or gas at 20 psi pressure is supplied from the auxiliary filter-regulator to orifice *J*, into relay diaphragm chamber *L*, through small tubing *D* inside the bourdon tube, and to nozzle *A*. The nozzle *A*, when not restricted by flapper *B*, is large enough to bleed off all the air coming through orifice *J*, and the pressure will be zero between the orifice and the nozzle. When the nozzle is restricted by the flapper because of a rise in liquid level, pressure is built up in the system between *A* and *J*. Thus any change in liquid level results in a change in pressure in the chamber *L*.

The intermittent-bleed pneumatic-type relay wastes air or gas only when pressure to the diaphragm-control valve is being reduced. The double

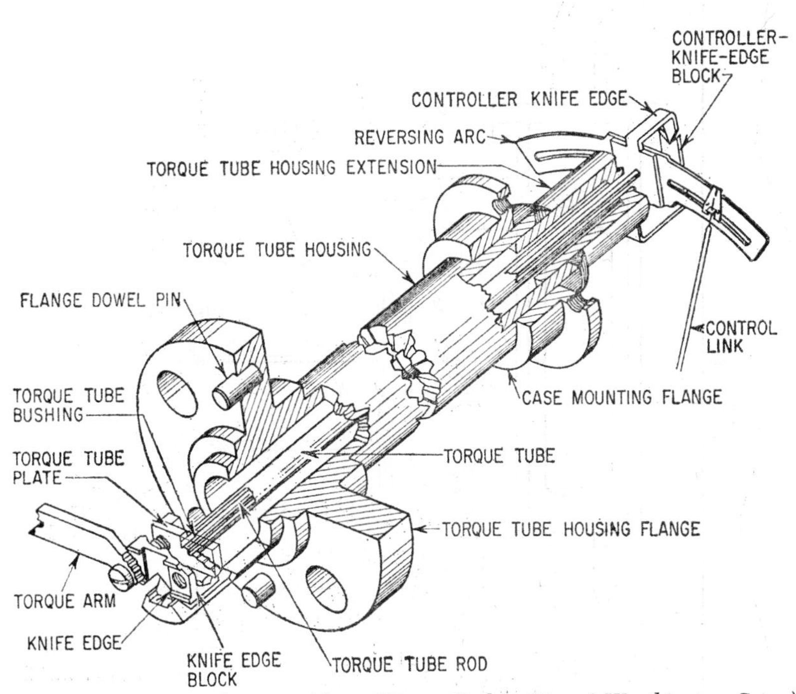

FIG. 6-12. Torque tube assembly. (*Mason-Neilan Div. of Worthington Corp.*)

diaphragm assembly with exhaust ports between diaphragms *M* and *P* is free-floating and always pressure-balanced. If there is an increase in pressure in chamber *L*, the diaphragm assembly is pushed downward, and the inlet valve *O* is pushed open. This allows supply pressure to come into chamber *N* until it pushes the relay-diaphragm assembly back into its original position and the inlet valve *O* is closed again. A decrease in pressure in chamber *L* will cause the diaphragm assembly to move upward and to open exhaust valve *K*, allowing pressure under small diaphragm *P* to bleed out until the diaphragm assembly again returns to its original position and exhaust valve *K* is closed.

The ratio of the two diaphragm areas in the relay is 3:1, or such that a

5-lb change on large diaphragm *M* results in a 15-lb change in pressure to the diaphragm-control valve.

The third part of the pilot assembly is the proportional-band-adjustment mechanism, which consists of a three-way valve assembly *H* in a branch from the diaphragm-control-valve supply line to compensating bourdon

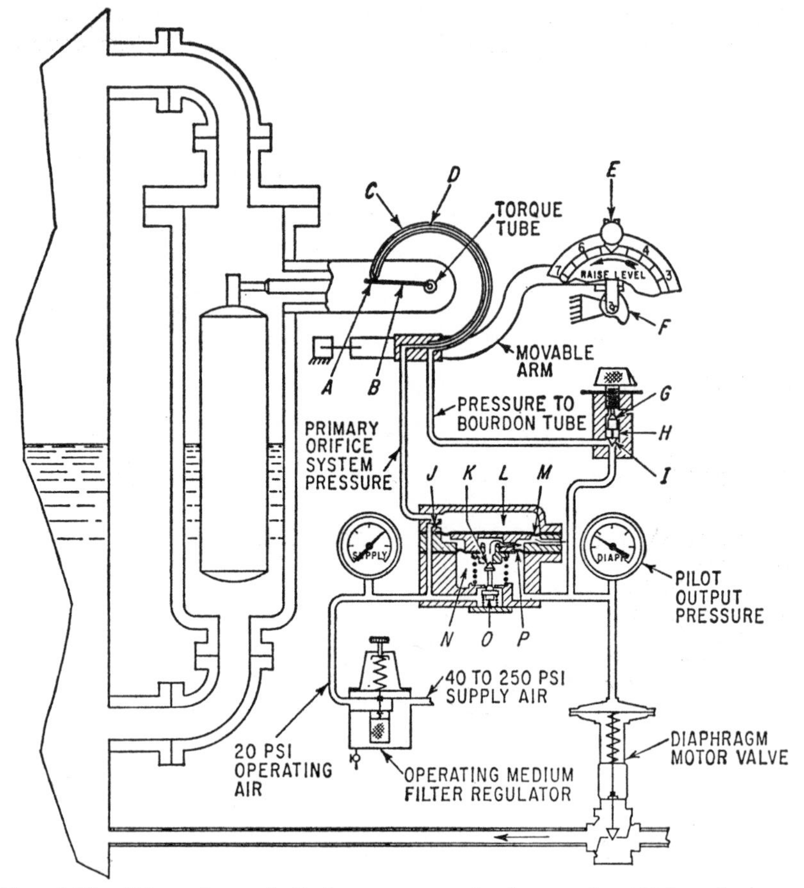

FIG. 6-13. Pilot relay and displacement-type level control. (*Fisher Governor Co.*)

tube *C*. The three-way valve *H* is manually positioned between the inlet port *I*, and the exhaust port *G*. When the valve is seated against exhaust port *G*, all the diaphragm pressure is transmitted to bourdon tube *C*. This causes the bourdon tube to "back away," and the flapper has to move a relatively large distance to close the nozzle. On the other hand, if the valve is seated against the inlet port *I*, no pressure is transmitted to the bourdon tube, with the result that a very small flapper movement is all that is necessary to close the nozzle. Intermediate positions of the valve, of course, result in intermediate pressure to the bourdon tube.

Assuming that the pilot and diaphragm-control valve are both direct-acting as shown in the schematic drawing, the operating cycle of the complete level controller is explained as follows:

Consider the level in the vessel at a point midway on the float, and the pilot adjusted to give 9 lb on the diaphragm of the control valve. Inlet flow to the vessel equals outlet flow. Now, if there is a decrease in outlet

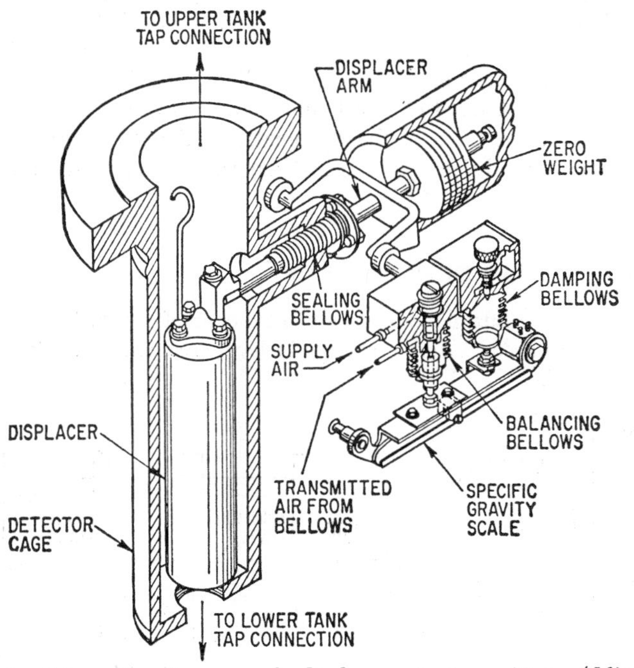

FIG. **6-14.** Schematic diagram of displacement transmitter. (*Minneapolis-Honeywell Regulator Co.*)

flow, the level in the vessel and in the float cage will rise. The float then will rise, causing flapper *B* to rise toward nozzle *A*. This will build up pressure in the relay chamber *L*, and the relay-diaphragm assembly will move downward, opening relay supply valve *O*. Operating medium then flows into chamber *N* until the relay-diaphragm assembly is pushed back into its original position, and valve *O* is closed again. The pressure in chamber *N* is transmitted to the diaphragm of the control valve, causing it to move toward its seat.

At the same time, the pressure in bourdon tube *C* is being increased through the three-way valve assembly *H*, which causes nozzle *A* to move away from the flapper, thus stopping the pressure build-up in chamber *L*. The unit is again in equilibrium with the level at a higher point, and the diaphragm-control pressure is increased to partially close the control valve so that inlet flow again equals outlet flow. If an increase in outflow takes place, the reverse of the above cycle will occur, with a decrease in liquid level causing an increase in control-valve opening.

Displacement Transmitter. The variable-displacement-type liquid-level detector and force-balance transmitter (Fig. 6-14) balances the buoyant force upon a cylindrical metal float in the liquid against the force produced by a pneumatic-feedback bellows. The pressure in the bellows is used for remote transmission, and may also be used for pneumatic control.

The system acts like an automatically balanced weigh beam. On one end are the zero balancing weights. On the other end is a combination of three forces: the weight of the displacer cylinder itself at zero liquid level, the upward buoyant force of any liquid above zero level, and the exactly balancing downward force of the pneumatic-feedback bellows. Since the upward force instantly produces the downward force, there is very little motion.

HYDROSTATIC-HEAD DEVICES

One of the most flexible and convenient means of measuring liquid level, especially where a considerable change in level is encountered, is the static-pressure method.

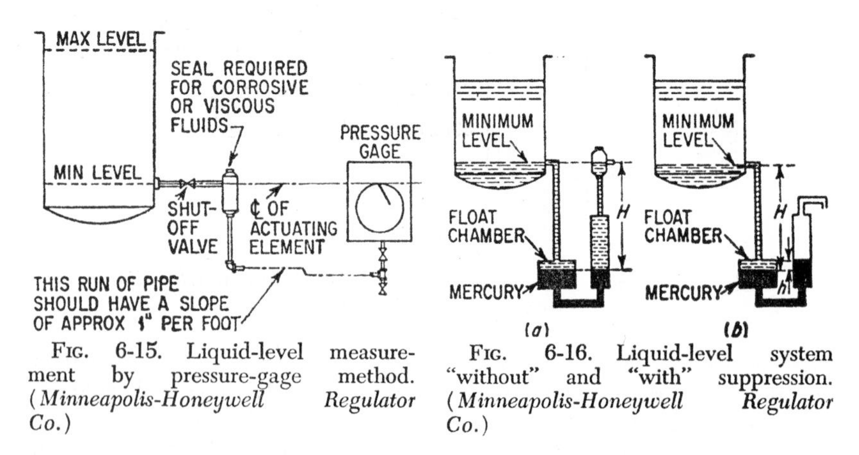

Fig. 6-15. Liquid-level measurement by pressure-gage method. (*Minneapolis-Honeywell Regulator Co.*)

Fig. 6-16. Liquid-level system "without" and "with" suppression. (*Minneapolis-Honeywell Regulator Co.*)

This method is based on the fact that the static pressure exerted by any liquid is directly proportional to the height of the liquid above the point of measurement, irrespective of volume. This relationship may be expressed mathematically by the formula $P = hgw$, where P is the pressure head, h the height of liquid, g the specific gravity of the liquid, and w the weight of a unit volume of water. Thus, any instrument which measures pressure can be calibrated in terms of the height of a given liquid and used to measure liquid level.

In open vessels, various methods utilize the static-pressure principle as a basis for liquid-level measurement.

Pressure-gage Method. The simplest method is a properly calibrated *pressure gage* with a pressure tap located at the line of minimum level in the vessel (Fig. 6-15). If the center line of the gage is not at the same elevation as the pressure tap, the head effect in the gage line is compensated for by recalibration of the instrument. A seal is needed for liquids which contain entrained solids or have a corrosive effect on the gage element.

Manometer Method. Liquid level in open vessels can also be measured with the mercury-manometer flow meter—either mechanical or electric. The minimum level tap is connected to the high-pressure (float chamber) side of the meter body with the range-tube side open to the atmosphere—as illustrated in the two schematic diagrams for systems without suppression *a* and with suppression *b*, both shown in Fig. 6-16. System *a* has a reservoir and connecting line to the range tube containing the vessel liquid with a head *H* equal to that on the float chamber, thus permitting the float to rise to its equalized position for minimum level.

System *b* does not have a reservoir. It depends on additional mercury (*h* in the diagram) in the range tube to suppress (compensate for) the liquid head in the float chamber and connecting line. Therefore, when the liquid is at a minimum level in the tank, the differential pressure is zero, even though the connecting line and high-pressure chamber are filled with liquid. This system eliminates extra piping—important when the meter is considerably below the vessel or where the liquid may freeze in the outer leg.

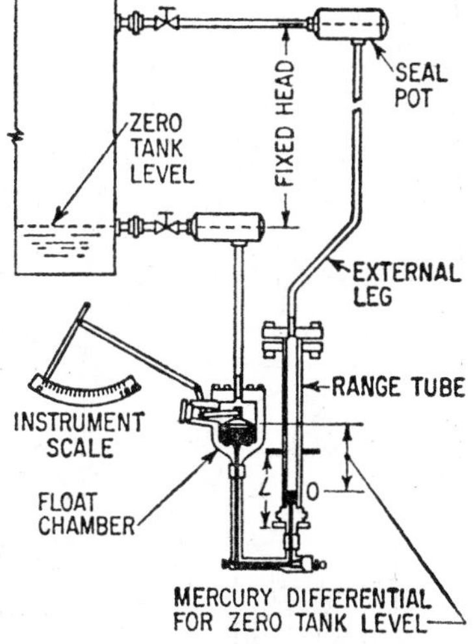

FIG. 6-17. Liquid-level measurement in closed vessels. (*Minneapolis-Honeywell Regulator Co.*)

The measurement of liquid level in closed vessels or pressure vessels can also be accomplished by the static-pressure method. In closed vessels, however, the pressure existing above the liquid is additive to that exerted by the liquid head and must therefore be compensated for. An instrument for this purpose is the mercury manometer, which can be installed to measure a differential pressure which changes only as the level varies.

With the mechanical meter body, the range-tube side is connected near the top of the vessel, and the float-chamber side near the bottom, as illustrated in Fig. 6-17. A fixed pressure equal to the pressure in the vessel, plus the constant head of liquid in the vertical connecting line, is thus exerted on the range-tube side; while a variable pressure, dependent on the liquid level in the vessel, is exerted on the float-chamber side. The pressure in the vessel acting on both sides of the meter body is seen to be canceled out.

Since, at minimum level, the static pressure exerted by the head of liquid on the range-tube side is a maximum—the reverse of normal conditions for manometer readings—use is made of what is termed a "dropped" range-tube meter, which permits the pen or pointer to assume its normal zero

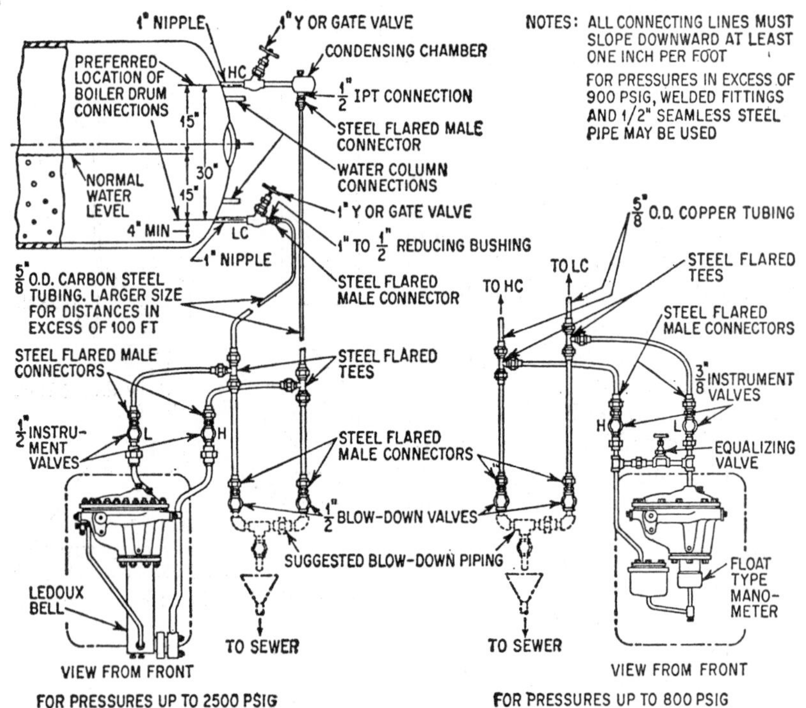

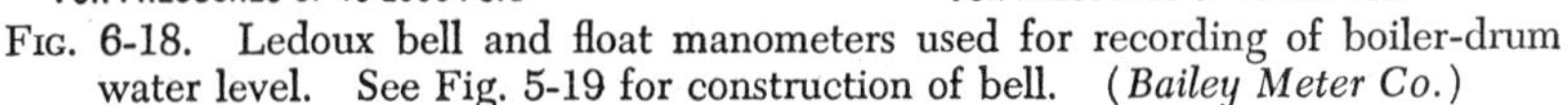

Fig. 6-18. Ledoux bell and float manometers used for recording of boiler-drum water level. See Fig. 5-19 for construction of bell. (*Bailey Meter Co.*)

position, and to travel from left to right on the chart or scale toward maximum as the level increases. With the electric meter body, it is merely necessary to connect the meter body with its float-chamber side to the top of the closed vessel and then to reverse the electrical connections to the recorder or indicator for this normal reading.

Figures 6-18 and 6-19 show manometers for general application. They are used principally for boiler level and for noncorrosive liquids at temperatures above atmospheric.

Vapor condenses in the condensing chamber, maintaining a constant

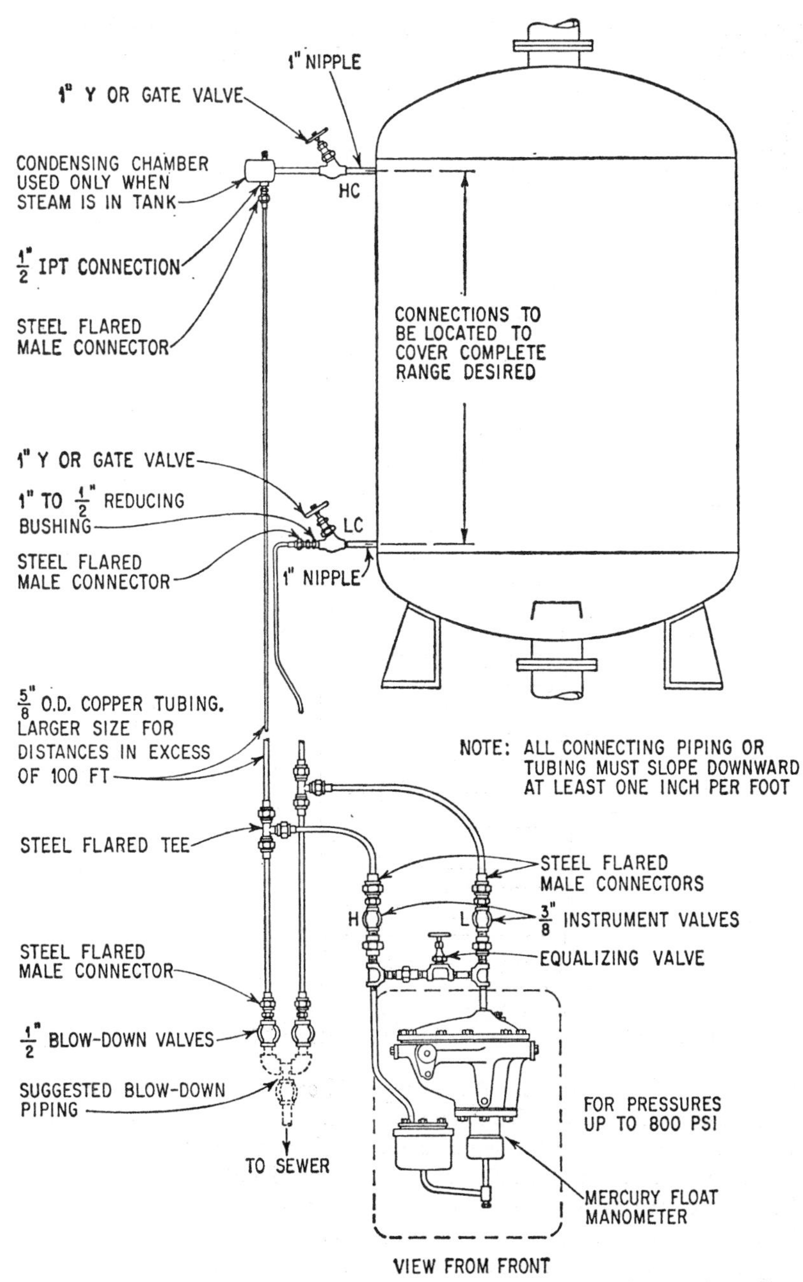

FIG. 6-19. Mercury manometer used for recording of tank, heater, hotwell, or reservoir level. (*Bailey Meter Co.*)

liquid head on one side of the instrument. Pressure on the other side, connected to the lower tap, varies with the level of liquid. The pressure difference is read in terms of liquid level at the instrument.

These units can be equipped with transmitters for remote reading or control of liquid level.

Diaphragm Type. Essentially this is the type of control device in which the head of the liquid is transmitted to the diaphragm, which is opposed

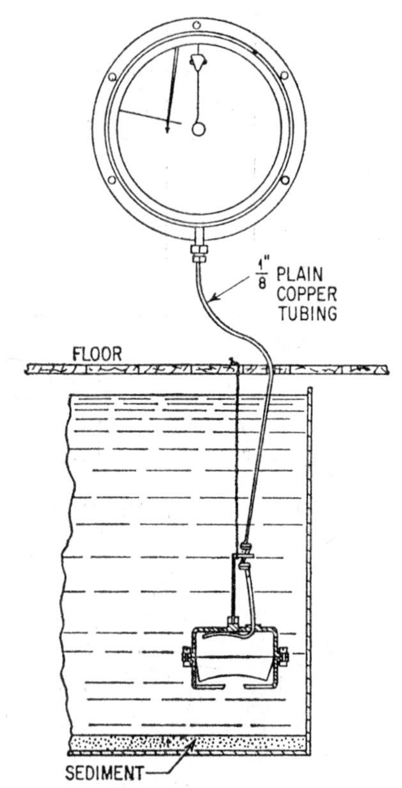

Fig. 6-20. Diaphragm-box level indicator. (*Foxboro Co.*)

by spring, weights, or levers. Any change of head in the system, of course, changes the position of the diaphragm member.

Such construction may involve a diaphragm box located in an open tank with the head of the water being applied against the diaphragm box (Fig. 6-20), or it may take the form of a diaphragm-actuated valve member, where the head of the system is impressed on the diaphragm. The commonly called "altitude valve," usually applied in waterworks systems, is an illustration of this type of unit.

A typical example is shown in Fig. 6-21. A control valve, which regulates the rate of input to an altitude tank, is located in the main feed line. The head from the overhead tank is brought to the diaphragm of the small pilot valve located on the upper portion of the main valve. Hydrostatic power to operate the valve is taken from the inlet side of the valve. The pilot valve consists of a diaphragm, which is spring-opposed, and a small needle-point inner valve. The operating-medium supply is brought through a strainer and an auxiliary regulator, reduced to a pressure of approximately 20 lb, which is sufficient to close the main valve.

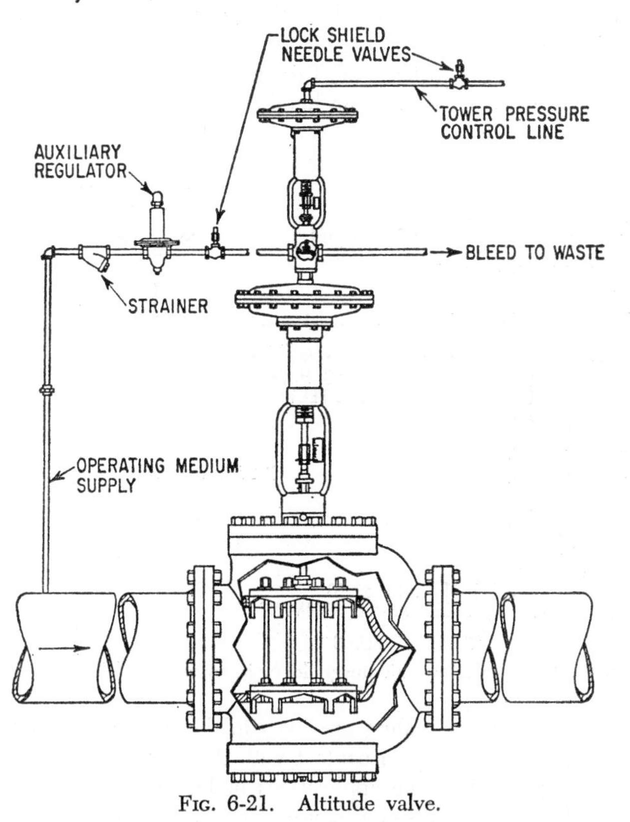

FIG. 6-21. Altitude valve.

When the level in the overhead receiver is low and, resultantly, the pressure head transmitted is low, the pilot valve is open, allowing the pressure from the main-valve diaphragm to bleed the waste. As the level in the vessel increases in height, the head also increases and the pilot valve closes. The operation-medium supply, coming through the small auxiliary regulator and passing through a restricted-orifice needle valve, then builds up on the main-valve diaphragm so that the main valve closes. In the intermediate position, the pilot valve would not completely close. Its ratio of opening, as compared to the orifice in the needle valve, allows the pressure on the main-valve diaphragm to be proportioned anywhere from 3 to 15 lb. This causes the main valve to assume proportioning or throttling action, feeding liquid into the tank at the same rate at which it is being withdrawn, thus maintaining the level at a predetermined point.

An example of a diaphragm-box unit for measuring liquids in open vessels is shown in Fig. 6-20. In this system a hermetically sealed box member with a small rubber diaphragm is placed in the vessel. As the head changes, it compresses the diaphragm, and the entrapped air is transmitted back to an air-actuated instrument, of either the recording or the indicating-controller type. If it is a controller, the operating medium is supplied from this instrument to a diaphragm-control valve to maintain level height within the vessel. Variations of this construction are applicable for specialized operation where, for the handling of sump water, slime, etc., it

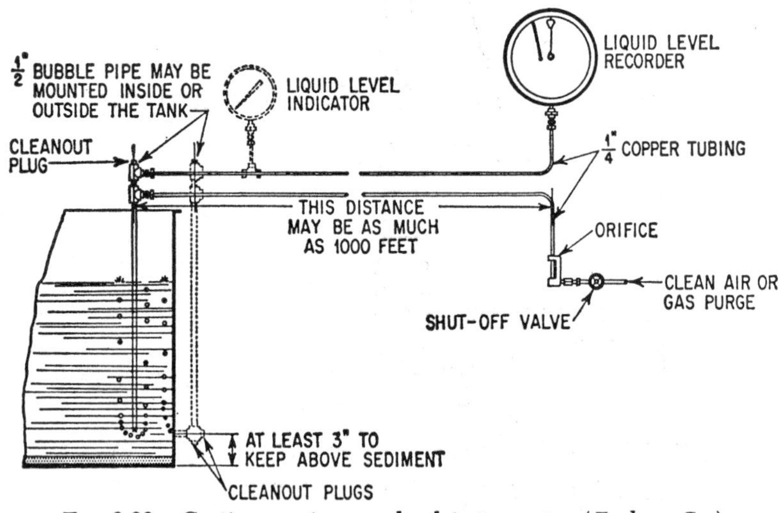

Fig. 6-22. Continuous air-purge level instrument. (*Foxboro Co.*)

may be necessary to put sealing members in the system so that the liquid itself does not come directly in contact with the diaphragm member.

Air-bubble Type. In some applications of liquid control it may not be possible, for physical reasons or because of the nature of the material being handled, to use a float and float-cage member or a diaphragm-head member. On such applications in an open-tank arrangement, the air-bubble method, as shown in Fig. 6-22, may be applicable.

In this type of application a pipe is lowered into the vessel, and a constant supply of air is fed into the pipe through a small dampening orifice. This supply of air pressure must be higher than the head of the liquid, so that the pressure within the pipe will build up until air escapes out of the bottom. The pressure of the air in the pipe is equal to the head of the liquid. This pressure, in turn, is transmitted to a pilot-actuated instrument, either indicating-recording, or nonindicating. As the level changes within the vessel, the air-bubble head pressure will change. A higher level will require slightly greater air pressure before the air bubbles out again. A lower level will require less air pressure. This pressure through the control instrument is translated into operating-medium pressure change to the diaphragm-control valve, through which the level in the vessel is being controlled. It is a simple and efficient method, but it cannot be used satisfactorily in vessels under pressure, because the slightest variations of pressure within the vessel would in turn cause a change in the control valve, without any resultant level change having required such a valve-position change. It may also contaminate the vapors in a pressure vessel, and for that reason may not be applicable.

ELECTRIC- AND PNEUMATIC-LEVEL TRANSMISSION

Remote-transmission systems, both pneumatic and electric, are used to send level measurements over relatively long distances. Reasons for the use of remote transmission are safety, economy, and convenience: safety,

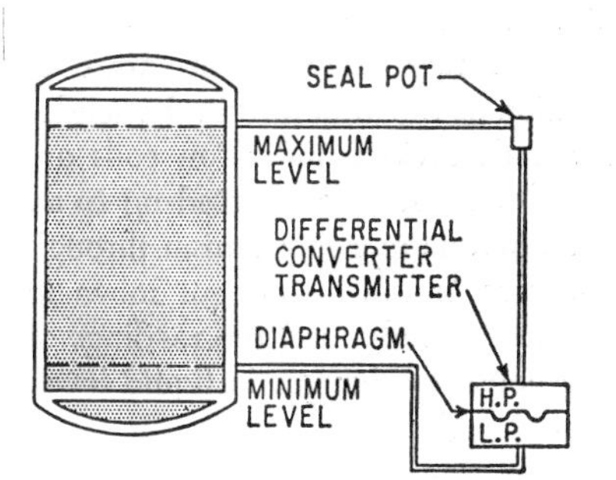

Fig. 6-23. Differential-pressure converter system for closed vessels.

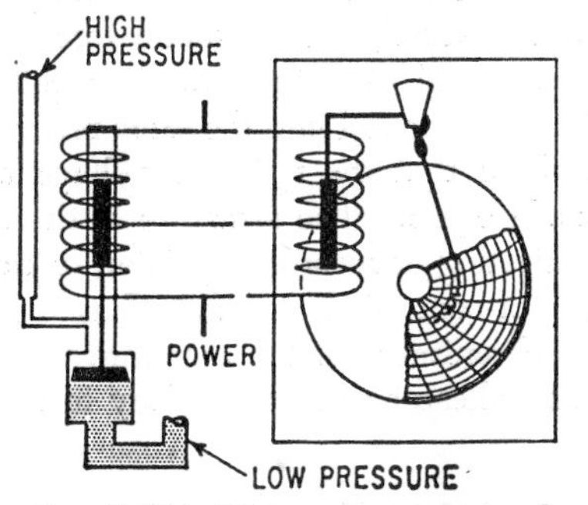

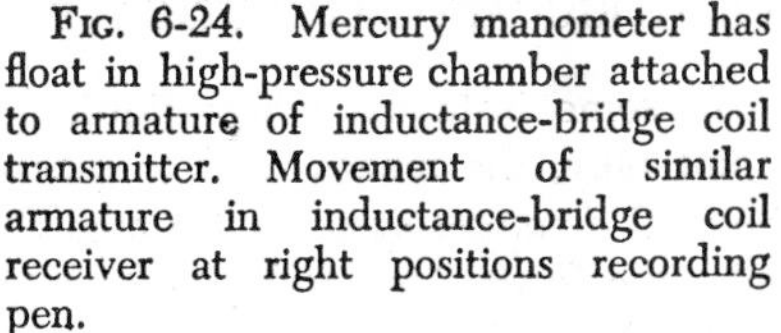

Fig. 6-24. Mercury manometer has float in high-pressure chamber attached to armature of inductance-bridge coil transmitter. Movement of similar armature in inductance-bridge coil receiver at right positions recording pen.

because corrosive or flammable liquids or vapors and high pressures are kept out of the control room, and the operator can observe and control the level at a safe distance; economy, because long lengths of connecting piping to the instrument are eliminated and replaced by a single length of small-diameter copper tubing or wire; and convenience, because many instruments can be grouped in a central location.

Pneumatic Remote Transmission. This is used to transmit liquid-level readings from the point of measurement to recording instruments at distances as great as 600 ft. A single receiving instrument can be used to record measurements from several transmitters.

When liquid level in closed vessels is to be transmitted, the constant head in the outer leg of the tank (Fig. 6-23) is applied to the top of the diaphragm in the transmitter. A constant-level pot is used to maintain a constant head on the high-pressure (top) side of the diaphragm. This is necessary because, if the outer leg were not kept completely full, the evaporation and condensation of tank contents would result in the accumulation of some liquid in the outer leg. Since the level of this liquid would not be constant, accurate measurement would be impossible.

The variable head in the tank is applied to the low-pressure (bottom) side of the diaphragm. Thus the differential pressure is at a maximum when the tank is empty, and at zero when the tank is full.

Electrical Type. In the differential type of unit, "electrical" means that the electrical type of transmitters may be used to transmit level position through electrical circuits and then may be transferred over to controlling operations (Fig. 6-24). The differential manometer is connected into the vessel in which the level is to be measured and controlled. The position of the level head is transmitted to the manometer, and the float, floating on the mercury, changes the position of the armature in the divided induction coil. As the armature is moved up and down through varying level heights in the vessel, the magnetic forces as reflected in the induction coils are transmitted over to a receiver coil. These forces draw the armature of the receiver coil into the same relative position as the armature in the sender coil, so that the ratios of voltage across both divided induction coils are made equal. The armature in the receiver coil can be connected up to an indicating or recording pointer, or can operate a flapper mechanism through an air-controlled circuit. Thus the armatures in the instrument move in synchronism with the armature within the meter body, and the readings of the indicator or recorder vary in direct response to changes in liquid level.

ELECTRIC- AND MAGNETIC-TYPE LEVEL-LIMIT DEVICES

In some instances where liquid level is to be controlled at one specific point only, the means used to measure this level can be relatively simple. Several common systems are available for such applications, two of which take the form of nonindicating controllers, namely, (1) *the electrode type,* and (2) *the float type.*

The Electrode Method. This method uses the electrical conductivity of the solution for level measurement. A typical application of this system is illustrated in Fig. 6-25. Low-voltage current flows between the tank wall and the electrodes as long as they are covered by the liquid. When the level drops below the lower electrode, the interruption of the circuit deenergizes a relay contact, which causes liquid to be added to the tank until the level rises to the upper electrode. This energizes a second relay contact, which, in turn, closes the first contact, stopping the flow to the tank. The two-electrode system is used to prevent instability of control in turbulent liquids. The electrodes are usually adjusted so that one is about three-quarters of an inch above the other.

The electrode liquid-level-measuring system can, of course, be used only with solutions which contain electrolytes and which will not be ignited by any arcing of the current—such as a size solution for cloth. The system has some flexibility since the electrodes can easily be raised or lowered to adjust the set point of the level.

Float Type. The operating principle behind magnetic liquid-level controls involves three factors: (1) an Alnico permanent magnet, (2) its magnetic field of force, and (3) a stainless-steel magnetic piston (attached to a float rod), which moves freely inside the nonmagnetic enclosing tube.

Figure 6-26 illustrates a switch mechanism as used for low-level cutoff. The magnet and mercury-to-mercury contact switch are assembled to a

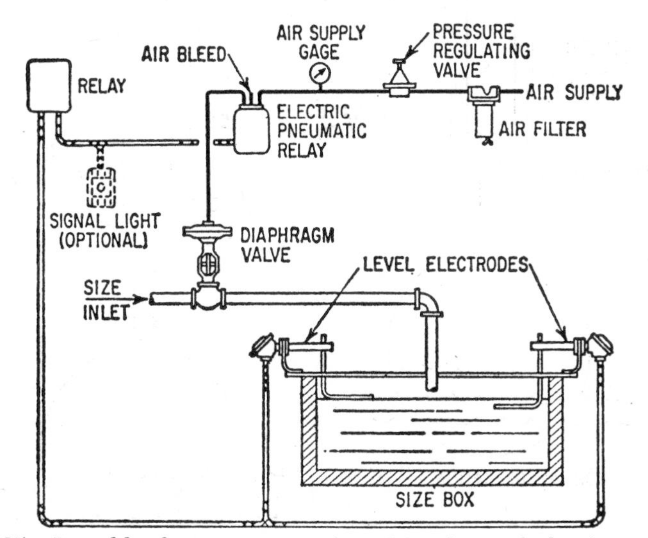

FIG. 6-25. Liquid-level measurement by electrode method. (*Minneapolis-Honeywell Regulator Co.*)

swinging relay arm which operates on frictionless socket-type stainless-steel pivots.

In the normal level position (at left) the magnetic piston (3), buoyed up by the float riding the liquid level, is within the magnetic field (2) of the permanent Alnico magnet (1). The magnet, on its swinging arm, is thus held against the nonmagnetic enclosing tube (5) in the "swing-in" position, which maintains a "closed" circuit between the common and right-hand legs of the SPDT (single-pole double-throw) mercury switch and an "open" circuit between the common and left-hand switch leg.

Should the liquid level recede, the magnetic piston (3) will be drawn down slowly by the float. At the predetermined "low"-level position (at right), the magnetic piston is drawn below the magnetic field (2). This releases the magnet (1) which swings (gravity assisted by tension spring)

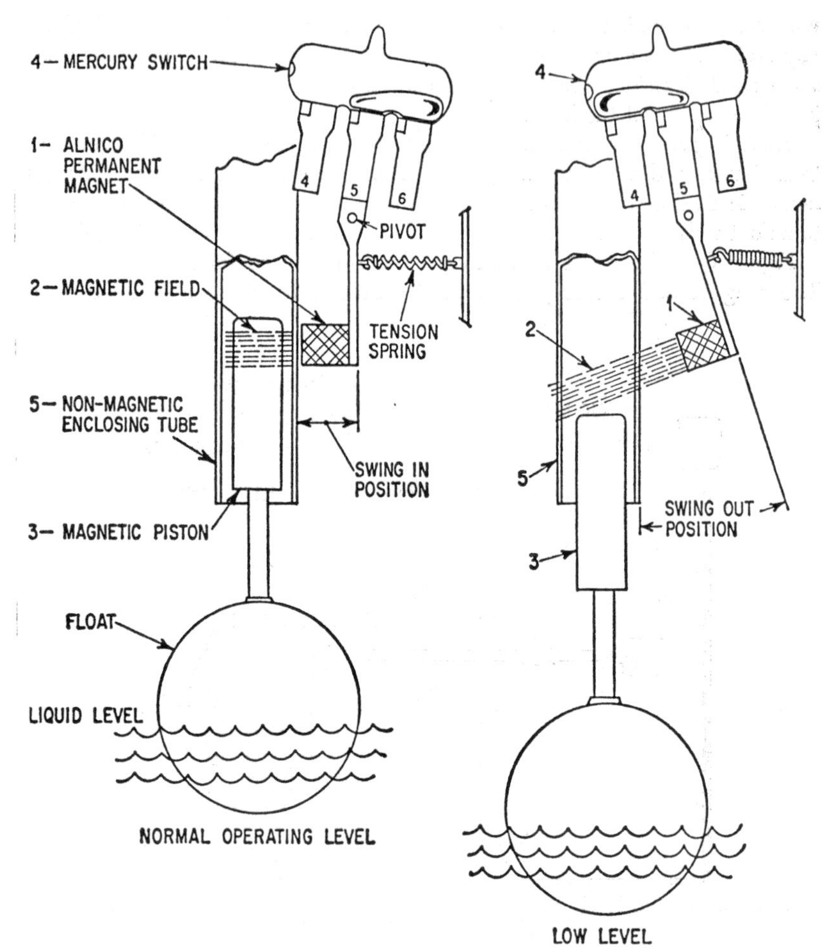

FIG. 6-26. Electric-magnetic float-type limit switch. (*Magnetrol, Inc.*)

away from the enclosing tube (5) to "break" the common-to-right leg circuit of the switch and "make" the common-to-left leg circuit.

As the liquid is replaced in the float chamber, the float is raised and again brings the magnetic piston (3) into the magnetic field (2), attracting the magnet (1), and reversing the switch action to normal level.

BOILER-WATER-LEVEL INDICATORS[3]

Gage glasses and remote indicators measure changes in level of liquids of constant density. In steam-boiler application, however, the density of the water varies with its temperature, which, in turn, depends on the boiler

[3] W. J. Kinderman, Yarnall Waring Co., *Power Eng.*

pressure. As the operating pressures are raised, the temperature of the water in the boiler increases, and its density is reduced. These changes affect the accuracy of water-level readings.

Tests show that the average temperature of the water in a boiler gage glass is always lower than the steam temperature, and that the water level indicated by the gage is lower than the actual level in the boiler drum.

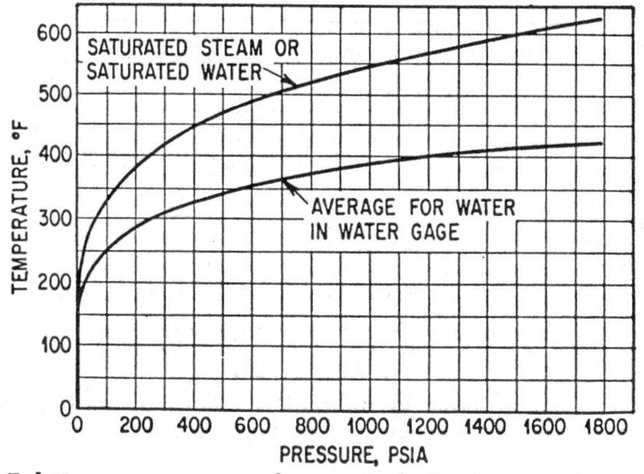

FIG. 6-27. Relative temperatures of water in boiler drum and in water gage at various operating pressure, found in average installations.

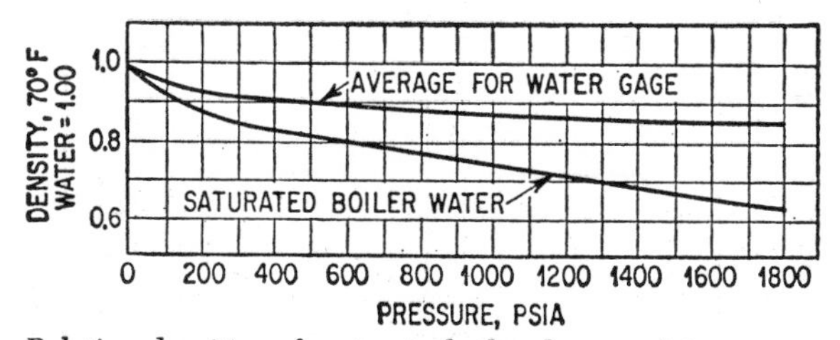

FIG. 6-28. Relative densities of water in boiler drum and in water gage at various operating pressures, resulting from variations of Fig. 6-27.

These differences may vary with the design of gage and arrangement of steam and water connections, but the temperature conditions to be expected in an average installation may be approximated, and are shown in Fig. 6-27. The higher the boiler pressure, the higher the boiler temperature and the greater the drop in temperature in the water gage.

Difference in temperature of boiler drum and water gage (Fig. 6-27) produces corresponding differences in the density of the water at these two locations. These variations are shown in Fig. 6-28; the higher the pressure, the lower the density and the greater the difference in density of the water in the drum and that in the gage.

Water-level gage-glass reading is the accepted standard of boiler-water-level reference, and other water-level-indicating equipment should be calibrated to operate according to this standard. In many cases, the difference in water level between the gage glass and the steam drum, caused by the variations in density mentioned above, is compensated by lowering the gage with respect to the boiler drum an amount equal to the difference.

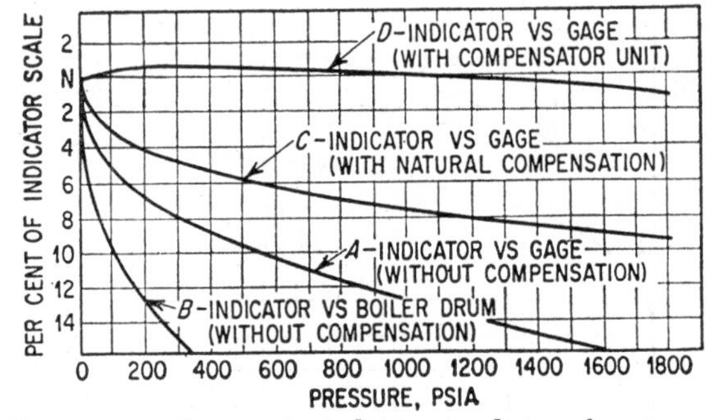

Fig. 6-29. Percentage of error in indicator readings of water-gage levels at various operating pressures and corresponding temperatures.

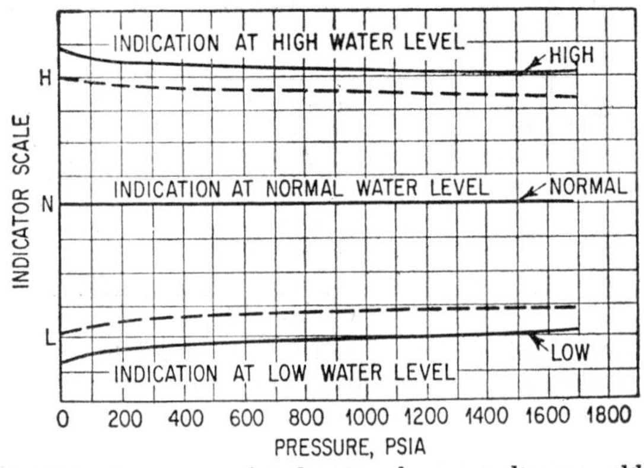

Fig. 6-30. Variation in range of indication for an indicator calibrated to full accuracy at 1,600 psi. Dotted lines show cold-water calibration.

Water-level indicators are normally calibrated with respect to reference water-glass readings at room temperature, with attention to accuracy at the normal level. On low-pressure applications, such direct calibration may be sufficiently accurate; but, on higher-pressure installations, allowance for density correction corresponding to operating conditions becomes desirable, and instrument calibration to such conditions is made.

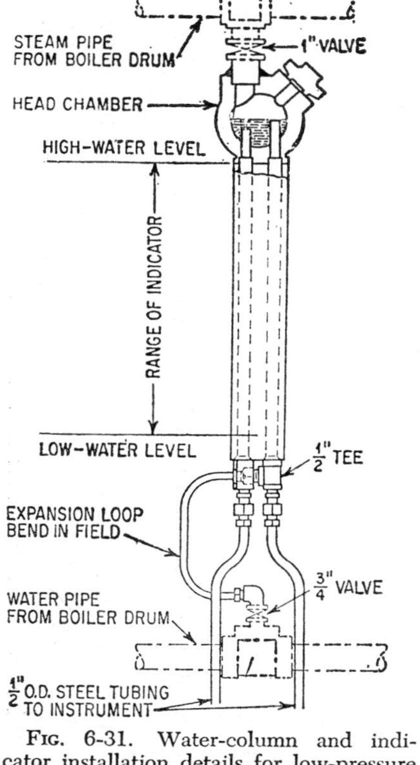

Fig. 6-31. Water-column and indicator installation details for low-pressure and high-pressure units. Details of a typical temperature-compensating unit. (*Yarnall Waring Co.*)

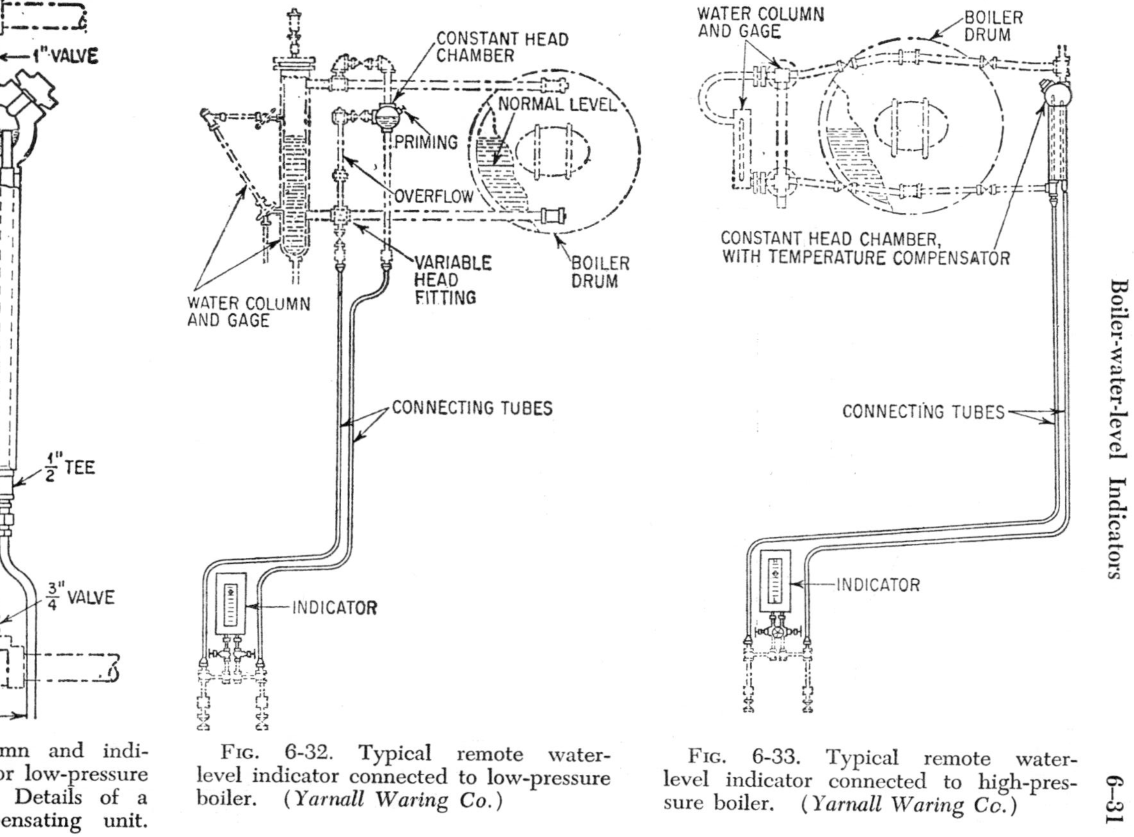

Fig. 6-32. Typical remote water-level indicator connected to low-pressure boiler. (*Yarnall Waring Co.*)

Fig. 6-33. Typical remote water-level indicator connected to high-pressure boiler. (*Yarnall Waring Co.*)

Remote water-level indicators operating on the differential-pressure principle are usually equipped with connecting tubes leading to the spaces in the boiler drum above and below water level. One of these tubes has a fixed static head; the other a head which corresponds to changing water

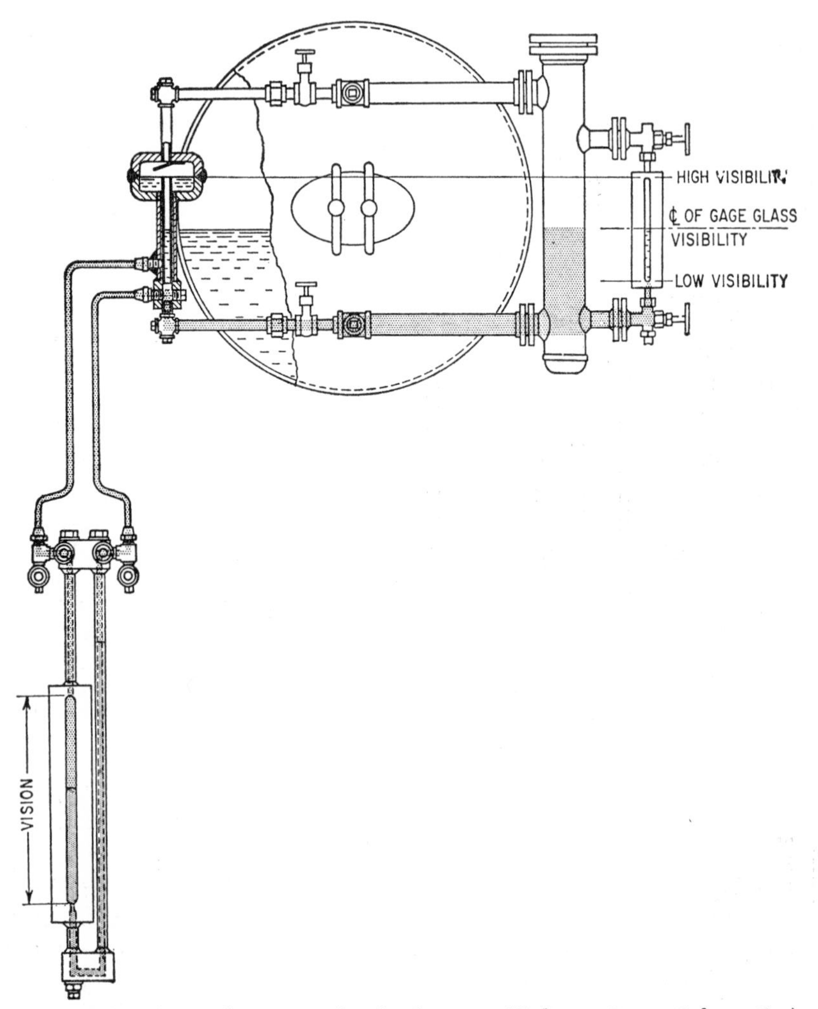

Fig. 6-34. Remote drum water-level indicator. (*Reliance Gauge Column Co.*)

level. The differential pressure operates the indicating element to register all fluctuations of the water level in the boiler drum overhead. Figures 6-32, 6-33, and 6-34 show how a typical remote water-level indicator is connected to the boiler fittings for both low pressure and high pressure.

Variations in water density in boiler drum, overhead gage glass, and remote indicator installed on the instrument panel or other eye-level location produce errors of indication that increase as boiler pressures are raised. The amount of this error in the reading of the average remote indicator with respect to the gage-glass reading at normal level, and without compensation of any kind, is shown in curve *A* of Fig. 6-29. The corresponding error of indication with respect to the actual level in the boiler is shown by curve *B*. These curves indicate percentages of error that seriously affect the accuracy of the indicator reading, especially at high pressures. Some form of compensation, therefore, is required.

Under ordinary operating conditions, the indicator constant head fitting on the upper part of the boiler drum is filled with steam, which heats the water in its connecting tube, and the resulting increase in temperature of the water in this tube lowers its density and decreases the fixed head against

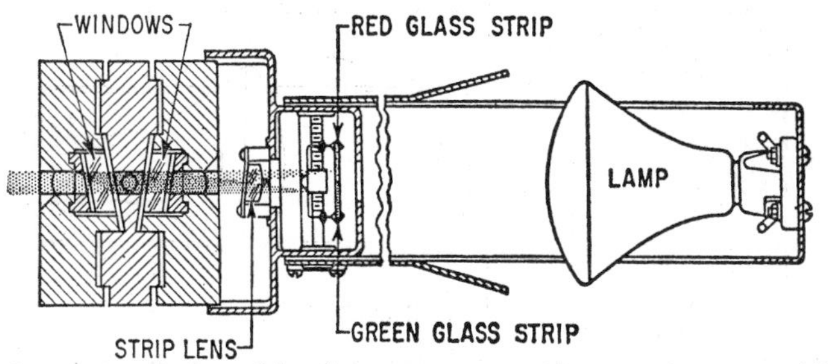

FIG. 6-35. Sectional plan of bicolor water gage with steam in gage. (*Diamond Power Specialty Corp.*)

which the varying head of the second connecting tube must operate. This change in the differential pressure at the indicator produces a higher reading, which is closer to the actual water levels as shown in curve *C* of Fig. 6-29, and up to 700 psi boiler pressure the error is not over 7 per cent, which is not considered excessive.

At higher boiler pressures, however, the compensation produced by normal heating of the constant-head connection is not sufficient for required accuracy between readings of the indicator at eye level and the water gage overhead. For pressure above 700 psi, some form of special temperature compensation is needed; Fig. 6-31 shows a typical temperature-compensating unit for these conditions. Steam and hot condensate, overflowing into the right-hand variable head tube, heat the left-hand constant-head tube by radiation, thereby duplicating actual normal level-gage-glass conditions at these higher pressures. The resultant proportional adjustment of the temperature and the density of the water in both tubes ensures true remote indication of the water level in the overhead gage within the narrow limits shown by curve *D* of Fig. 6-29.

Water-level readings at the remote indicator are corrected to be effective for gage-glass readings at normal water level. In a typical remote water-level indicator, the range of pointer travel with respect to the travel of the water level in the overhead gage is reduced somewhat at high boiler pressures. As shown on dotted line curves in Fig. 6-30, the indication at high-water level is lowered and at low-water level is raised. These variations, however, are well within accepted limits for good operation, especially in view of variations found in individual water-glass readings, errors in water-gage conditions following blowdown, or other change from normal boiler operation. Remote-indicator readings, therefore, may be considered a reliable check on the actual water level in the boiler.

Since full accuracy of readings under operating conditions is more important than that at low pressure and temperature, it is desirable to calibrate

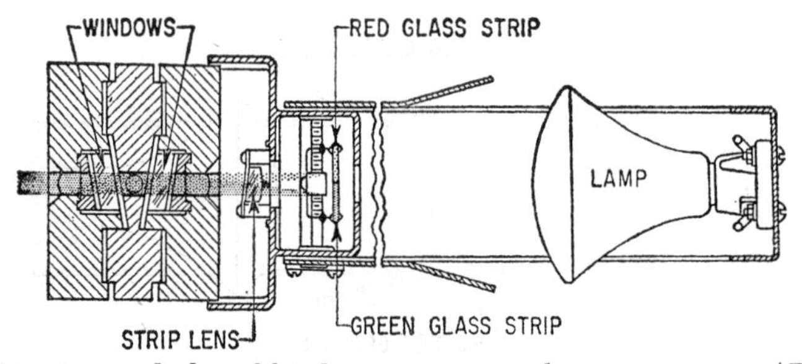

Fig. 6-36. Sectional plan of bicolor water gage with water in gage. (*Diamond Power Specialty Corp.*)

to operating conditions, and thus introduce departure from this calibration at lower pressures. Full-line curves in Fig. 6-30 illustrate this condition for a water pressure of 1,600 psi. A similar adjustment for density may be made for any desired operating condition.

Bicolor Gages. Operation of the bicolor water gage is based on the optical principle that the refraction (bending) of a ray of light differs as it passes obliquely through different media. When a light beam passes obliquely through a column of steam, the amount of bending to which it is subjected is not the same as when it passes through a similar column of water.

Figures 6-35 and 6-36 show a bicolor water gage in diagrammatic sectional plan. Parallel strips of red and green glass (in which the colors are permanently fused) are placed between the illuminator lamps and the strip lens that is adjacent to the gage. The windows of the gage, protected on the water sides by sheet mica, are set at angles to each other that utilize to best advantage the difference in refraction between water and steam.

Beams of red and green light are projected through the entire length of the strip lens and strike the face of the gage window at different angles, each color beam being refracted, as shown in the sketches.

When steam occupies the space between the windows, the index of refraction is such that the green light beam is bent out of the field of vision. The red beam, however, is bent so that it emerges from the gage into the line of vision of the observer, who then sees a "red glass."

When water occupies the space between the gage windows, the index of refraction is such that the red beam of light is bent out of the field of

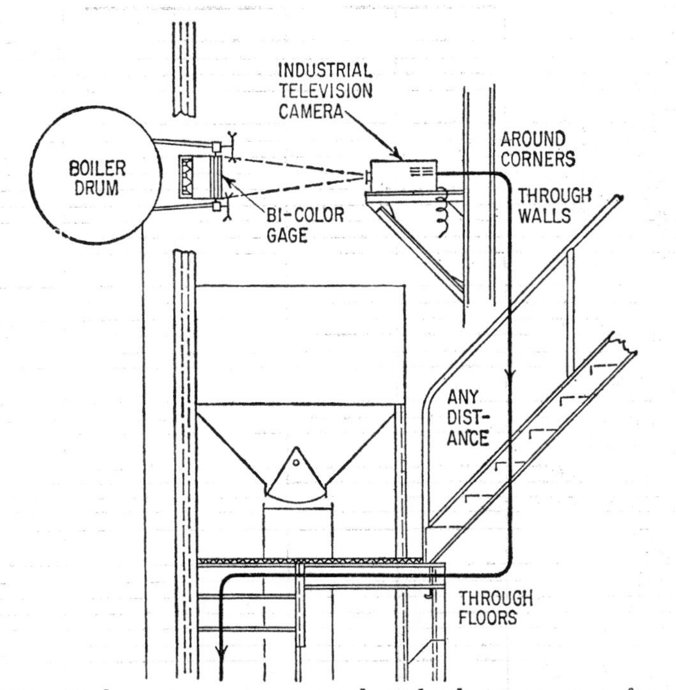

FIG. 6-37. Bicolor water gage using industrial television camera for remote reading. (*Diamond Power Specialty Corp.*)

vision. At the same time the green beam of light is bent so that it emerges from the gage, and the observer then sees a "green glass."

Figure 6-37 shows a bicolor gage used in conjunction with an industrial television camera that sends a "picture" of drum-level conditions to a screen on a centrally located panel board.

LEVEL INSTRUMENT SPECIFICATIONS

Liquid-level-instrument specification forms developed by the Instrument Society of America may be used as a standard, or may be modified to suit special needs (Fig. 6-38).

LEVEL INSTRUMENTS

SPECIFICATION SHEET

SHEET NO.
TAG NO.
DATE
REVISED
BY

	GENERAL					
1	TYPE					
2						
3	TAG NO.					
4	VESSEL OR EQUIPMENT NO.					
	BODY					
5	MATERIAL					
6	TOP CONN. LOCATION					
7	BTM CONN. LOCATION					
8	CONN. - SIZE					
9	CONN. SCREWED OR FLANGED					
10	CASE MOUNTING					
11	FLANGE ORIENTATION					
12	ROTATABLE HEAD					
13						
	FLOAT OR DISPLACER					
14	DIAMETER OR LENGTH					
15	EXTENSION					
16	MATERIAL					
17	TORQUE TUBE MATERIAL					
18	AIR FIN					
19						
	TRANSMITTER					
20	TYPE					
21	OUTPUT					
22	RECEIVERS ON SHEET NO.					
	CONTROL					
23	TYPE					
24	PROPORTIONAL-% / RESET					
25	OUTPUT					
26	ON LEVEL INCREASE: OUTPUT					
27						
	ACCESSORIES					
28	FILTER & REGULATOR					
29	GAGE GLASS CONNECTIONS					
30	GAGE GLASS					
31	PURGE CONNECTION					
32	ELECTRIC SWITCH					
33						
34						
	SERVICE CONDITIONS					
35	UPPER LIQUID					
36	LOWER LIQUID					
37	SP. GR. UPPER / LOWER					
38	PRESS MAX / NORM					
39	TEMP MAX / NORM					
40						
41						

NOTES:

FIG. 6-38*a*. Level instruments specification sheet. (*Instrument Society of America.*)

LEVEL INSTRUMENTS
SPECIFICATION SHEET INSTRUCTIONS

Prefix numbers designate line number on corresponding specification sheet.

1) & 2) Specify whether float or displacement type, whether external or internal on displacement, and whether kidney, flange or shaft on float type.

3) Identification of item number.

4) Vessel or equipment number on which instrument is mounted.

5) Specify material of chamber and flange, such as bronze, cast iron, steel, etc.

6) On displacement type, specify location of top connection.

7) On displacement type, specify location of bottom connection.

8) Specify connection size such as 1-1/2 or 2" on displacement type and kidney float type, or flange size on flange and shaft float types.

9) Specify whether screwed or flanged, and pressure rating. Also specify facing if flanged.

10) On displacement types only, specify "left hand" or "right hand" case mounting. When facing case, for left hand mounting, case is on left side of chamber. For right hand mounting, case is on the right of chamber.

11) Specific orientation of flange connection when one or both connections are at side. Standard orientation is with flange opposite torque arm.

12) On displacement types, where there are one or more side connections, specify if case is to be rotatable with respect to flange orientation.

13) For additional specifications not covered by lines 5 through 11.

14) Specify ball float diameter or displacer length. On displacement types, displacer length is same as range.

15) Specify float extension rod length.

16) Specify float material, including float extension material, such as: 316 stainless, monel, etc.

17) Specify torque tube material, such as: Inconel, 316 stainless, etc.

18) Specify whether air-fin extension on torque tube is required.

19) For additional specifications not covered by lines 13 through 17.

20) Specify whether pneumatic, electric, or other.

21) If pneumatic transmission, specify if output is 3-15 PSI, or other. If electric, specify characteristics.

22) Write in specification sheet no. on which receiver instrument appears.

23) Specify whether pneumatic, electric, or other.

24) Specify proportional band. If reset required, specify reset rate in repeats per minute.

25) If pneumatic controller, specify if output is 3-15 PSI, or other. If electric, specify characteristics.

26) Write in "increases" or "decreases".

27) For additional specifications not covered by lines 23 through 27.

28) Write in if individual units or combination type desired, or if to be purchased separately.

29) Specify size of connections, if required.

30) Specify type of gage glass or model number, if required.

31) Specify size of purge connections and location, if required.

32) Specify type of electric switch, if required. Write in rating in amps., volts and cycles for contacts. Write in housing type.

33) & 34) For additional specifications not covered by lines 29 through 32.

FIG. 6-38b. Level instruments specification sheet instructions. (*Instrument Society of America.*)

LEVEL INSTRUMENTS
SPECIFICATION SHEET INSTRUCTIONS

35) Show type of upper liquid if on interface service only.

36) Show type of lower liquid for all services.

37) Write in specific gravity of upper liquid (if interface) and power liquid.

38) Write in maximum and normal pressures.

39) Write in maximum and normal temperatures.

40) For additional specifications not covered by
& lines 35 through 39.
41)

Fig. 6-38c. Level instruments specification sheet instructions. (*Instrument Society of America.*)

Section 7

FINAL CONTROL ELEMENTS

It is the function of the final control element, and of the power unit that operates it, to carry out faithfully the commands given to it by the control circuit. Sluggish response by the element to system commands can be as harmful to close regulation as incorrect measurement or excessive transmission lags. Thus the final control element is a major link in the control loop.

The control valve is the final control element most commonly used in institutional and industrial-service control systems. But dampers and damper drivers are also important for control of air and gas flow through ducts, fans, blowers, etc. The controlled-volume or proportioning pump, commonly used for metering chemical solutions, may also be considered as a form of final control element.

Regardless of whether the final control element is a control valve or a control damper, it will consist of a device that varies fluid flow by varying pressure drop. In effect, the device is a continually varying orifice. The second major component of the final control element is the operator or power unit. The most common methods of operator actuation are pneumatic and electric, but hydraulic operators also find important applications in institutional and industrial building services.

Because of the great range of applications for final control elements, the engineer finds a great many different designs available to choose from. They vary from simple on-off types to special valve bodies for high-accuracy proportional control.

Among the important topics to be discussed in this section are valve and operator construction, material selection, sizing the control valve, valve-piping arrangements, and special applications.

The selection of the proper type of valve for a specific job must be determined by the character of control required, and the conditions under which it must operate; not by the general type of application. Each type of valve has desirable characteristics, but no single type possesses every feature.

Valve Action. Whether the apparatus is best protected by having the controlling medium turned on or off in case of air-supply failure determines the choice of valve action.

Air-to-close (Direct-acting). This type of valve is closed by air pressure and opened by the diaphragm-motor spring. If the air supply fails, such a valve opens.

Air-to-open (*Reverse-acting*). Opened by air pressure and closed by the motor spring, this type of valve closes in case of air failure.

Various valve-disk and valve-seat constructions are available in both air-to-close and air-to-open types. In general, however, these valves fall in two main service classifications:

For short time lags and large capacities, simpler control systems are adequate, permitting the use of a high-sensitivity controller and a quick-opening *disk* valve. Since, for this type of service, the character of flow is not too important, the size and material of the valve are the principal considerations.

On the other hand, as the time lag increases and the capacity decreases, the control problem becomes more difficult and requires the more complicated types of controllers and valves. The larger the time lag, the smaller and more precise must be the changes in the rate of flow of the controlling medium for a given change in temperature, pressure, rate of flow, or liquid level to assure throttling control. These applications require *throttling*-type valves—with definite disk travel-flow characteristics and the ability to throttle anywhere within the maximum and minimum flow requirements.

VALVE BODIES AND INNER VALVES

Single-seated Valves. The quick-opening *disk* type of valve (Figs. 7-1 and 7-2) is used principally on simple, high-sensitivity, small-time-lag appli-

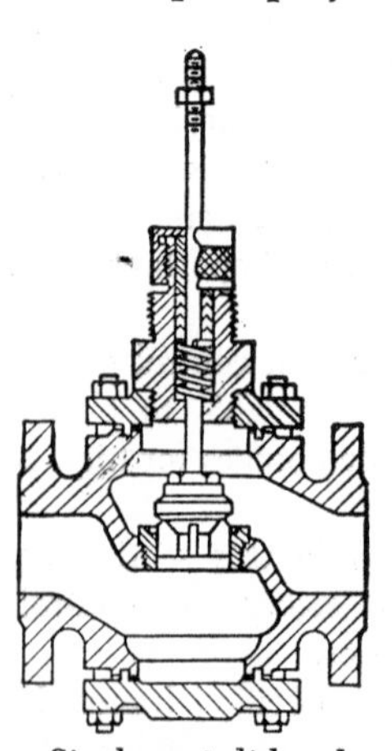

FIG. 7-1. Single-seat disk valve, air-to-close type. (*Taylor Instrument Co.*)

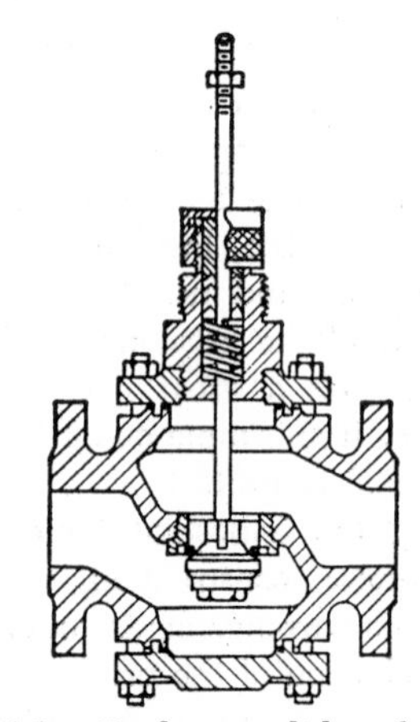

FIG. 7-2. Single-seat disk valve, air-to-open type. (*Taylor Instrument Co.*)

cations and for open-and-shut service; or, with stable load conditions and constant line pressures, it can be throttled satisfactorily at between 10 and 90 per cent of its full opening.

Since most disk valves are of the single-seat type, it should be noted that all single-seat valves have limitations which should be taken into consideration: They are impractical for throttling control if line pressures fluctuate widely, unless the controller sensitivity is very high (narrow throttling range) or the size of the valve is small. Generally, they are not recommended for throttling control above the 2-in. size, on account of the large unbalanced disk area, unless the line pressures are low. For higher pressures, the double-seated valves should be used.

Single-seat Metal-disk Valves. These, of the more common inner-valve construction, are made in both air-to-close and air-to-open types. The beveled surface of the metal disk and seat are ground to fit each other, ensuring tight closing. However, a complete shutoff cannot be expected if the valve disk and seat become eroded or corroded, or if the fluid contains

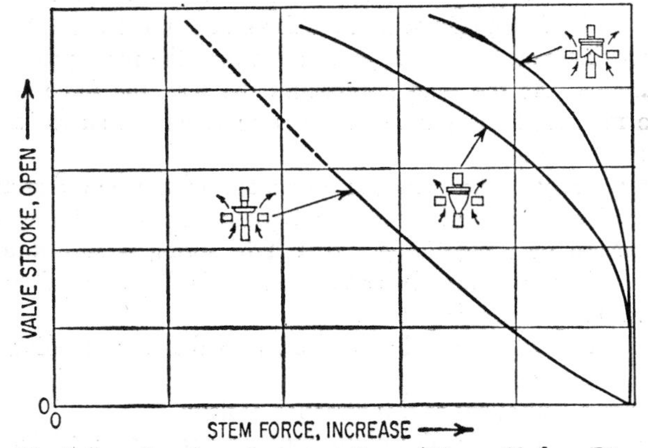

FIG. 7-3. Single-seated valve characteristics. (*Mason-Neilan Div. of Worthington Corp.*)

solid matter which may become wedged between seat and disk. For absolute shutoff service, valves with composition disks are often recommended.

Single-seat beveled metal-disk valves, in most instances, have renewable seat rings and disks.

Single-seat Composition-disk Valves. Where a leaktight shutoff is required, a composition disk is used. Its performance and general limitations are similar to those of the single-seat beveled metal-disk valves.

The seat ring has a raised surface, and the composition disk, which is held in a disk holder, seats on this surface. The seat-ring surface should never be ground.

Most composition disks are made of rubber compounds, which can be varied to meet application conditions for use with air, water, steam, oil, gas, and other fluids.

To understand the reasons for the pressure-drop limitations, a knowledge of the behavior of single-seated plugs under varying strokes and pressure drops is essential.

Figure 7-3 shows typical force-versus-valve-stroke curves under constant pressure drop for three basic plug designs: quick opening, parabolic, and V port. At the seat (zero per cent lift) the stem force is the product of the seat area and the pressure drop across the valve. As the valve opens, this force drops off more or less rapidly, dependent on the type of plug.

The exact shape of the curves would be dependent on the flow coefficient of the valve body, the actual shape of the plug, etc., but the general character would be similar to those in Fig. 7-3. The important point is that the force acting on the plug diminishes with lift. The amount of force reduction is substantial, amounting to up to three-quarters of the force at the seat position.

There is never a reversal in force. It is always in the direction of flow through the seat ring. If the flow through the body is reversed, the force curves are altered slightly, but for all practical considerations may be assumed to be identical, except that the force acts in the opposite direction.

With flow direction tending to open the valve, the force produced tends toward stability. As the plug approaches the seat, the force increases to oppose the motion of the plug. There is an effect similar to "self-regulation" in control circuits.

Inherent stability is the principal advantage of the flow direction. On air-to-close valves, excess air pressure may be used to force the valve closed with no concern over chattering. If a light spring is substituted and a positioner used, the motor becomes essentially "springless," combining simplicity with high operating force.

Single-seated valves should be installed with the "flow tending to open" in the great majority of cases.

There are two limitations to this direction of flow:

1. For valves to close on air failure, the motor spring must have sufficient initial load to close the valve against the off balance.
2. On high-pressure drop-angle valve service, it would be impossible to provide a streamlined body outlet.

With flow tending to close the valve, the force on the plug increases as the valve closes. The plug therefore tends to travel more than it would with no pressure drop. This extra travel further increases the force, and, if the slope of the force curve is steeper than that of the spring in the diaphragm motor, the valve can actually become unstable. This tendency is further aggravated by the increased pressure drop, which usually occurs as the valve closes. In larger valves, this can result in a terrific hammering, and it is necessary to set conservative operating limits to avoid trouble.

Unfortunately, this flow direction is often the desirable one for two conditions:

1. In normally closed valves where the plug force naturally assists closure
2. In angle valves where flow over the plug and out gives the natural streamlined path

The shape of a plug can have some effect on its stability. The variations in the three curves in Fig. 7-3 indicate that force behavior and valve characteristics are interrelated. Little freedom in altering the force-lift relationships is allowable if the desired flow characteristics are to be maintained. A turned plug with a linear characteristic would tend to have a force curve which is linear from maximum to minimum, and would probably remain more stable to higher differential pressures than an equal-percentage turned plug of similar design and capacity.

Attempts to balance the plug force, as is sometimes done in angle valves, only shift the force curve and can even cause the valve to fly open instead of shut. In general, they add little to over-all stability.

This problem is basic and not dependent on the type of motor operator.

The only solution is to set pressure-drop limitations on this class of application which will not cause instability in service.

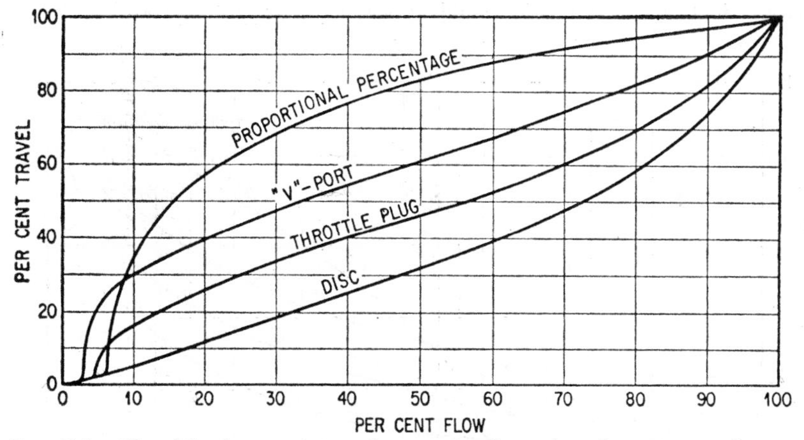

FIG. 7-4. Flow-lift characteristics of control valves plotted on rectangular coordinates. (*Taylor Instrument Co.*)

Applications demanding throttling control require the complete control system to operate as a unit. Failing this, an unnecessary burden is forced on the controller, diaphragm motor, or valve. The weakest link in the system determines the best control obtainable.

The need for "characterized" valves on low-sensitivity applications with load changes is apparent from reference to Fig. 7-4, which plots the flow-lift curves of proportional percentage, disk, V-port, and throttle-plug valves on rectangular coordinates:

First, let us assume the use of a disk valve controlling at a load condition requiring 10 units of flow. Now assume a deviation from the set point, which causes the lift to change by 1/10 unit. From the curve it is seen that the flow would change 11 units, or 110 per cent of original flow. Next assume control at a load condition requiring 40 units of flow through the valve. Now, if the same deviation occurs as in the first case, causing a 1/10 unit change in lift (an inherent feature of the controller), the result is a change of 9 units of flow, or only 22 per cent change.

This valve is not suited to applications which require the flow through the valve to be graduated to correspond to the load. The use of a disk with the characteristics outlined necessitates a change in controller sensitivity (or throttling range) to compensate for load changes, if the same character of control is to be maintained for each load condition.

It can be seen from the preceding example that the control valve should have a characteristic giving a small change in flow for a given increment in travel at low lifts (corresponding to low loads), and large changes in flow for the same increment at high lifts. Or equal increments of lift should give equal-percentage increments in flow. This so-called "percentage characteristic" is desirable for throttling control, since it allows similar control under low- and high-load conditions without changing instrument sensitivity. Such a characteristic would appear as a straight line on Fig. 7-5,

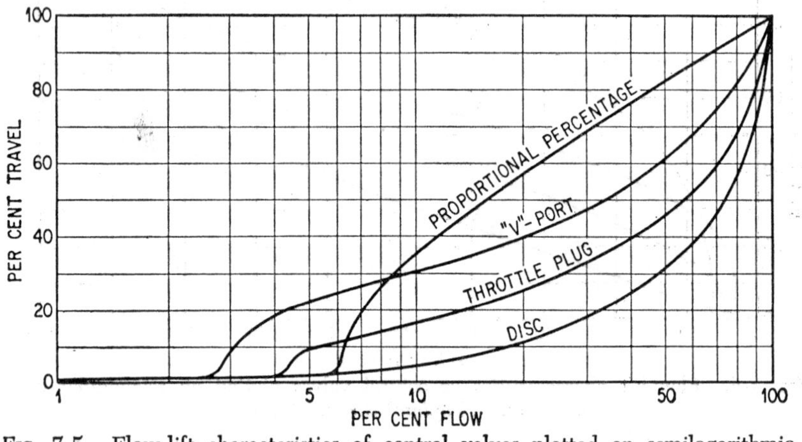

Fig. 7-5. Flow-lift characteristics of control valves plotted on semilogarithmic paper. (*Taylor Instrument Co.*)

which shows the various valve characteristics plotted on semilogarithmic paper.

One simple guide in determining when a "characterized" valve is required is found in the following rule: If the time lag, capacity, and load changes require a controller with automatic reset (automatic compensation for load changes), a characterized valve also should be used. However, the converse of this rule is not necessarily true.

Controllable Range. For any valve there is a certain minimum percentage of its total flow below which satisfactory control cannot be obtained. In throttling service, the minimum flow of disk valves should be limited to 10 per cent of their maximum flow. "Characterized" valves usually can be made to throttle down to 2 and 5 per cent of their total flow (the percentage decreasing inversely with the valve size), but they are limited by their "clearance flow."

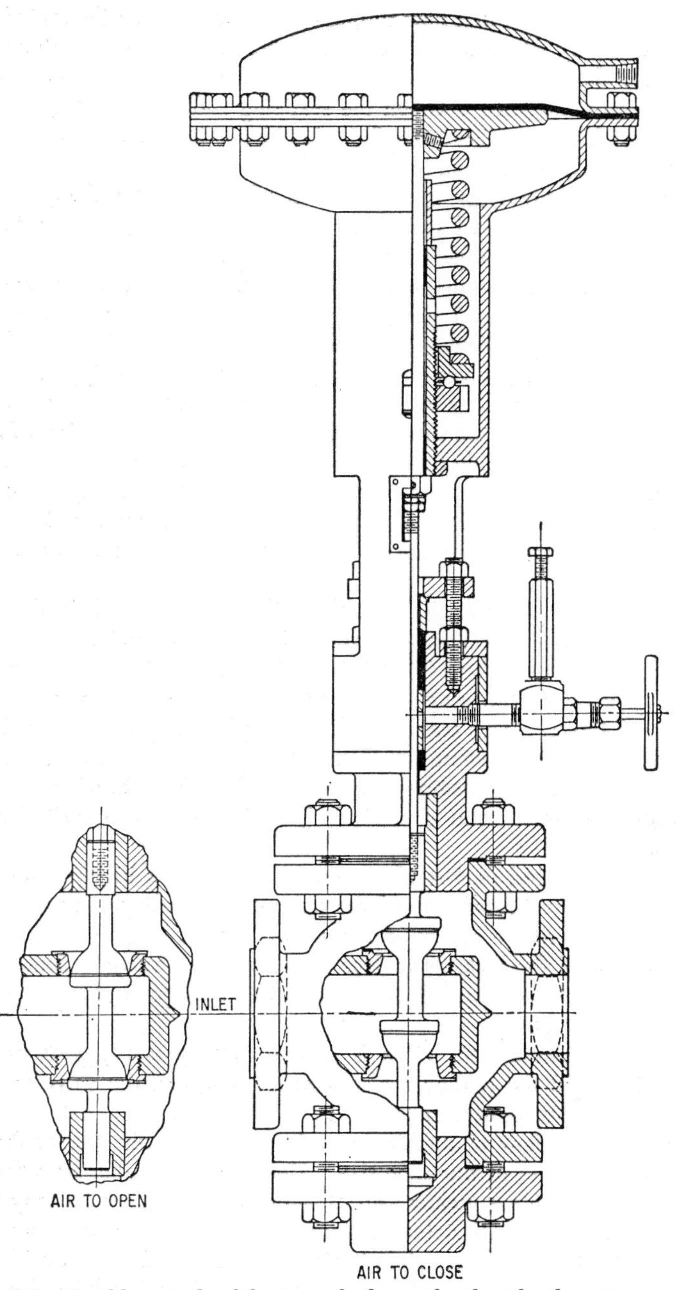

FIG. 7-6. Double-seated globe-type body with throttle-plug inner valve. Plug can be of either air-to-open or air-to-close design. (*Hammel-Dahl Div. of General Controls Co.*)

"Clearance flow" is that flow through the valve between the seat ring and the skirt or plug which occurs at low lifts before the V port or shaped plug can affect the flow-lift characteristics of the valve. This flow is due to the necessary clearance between the seat ring and skirt or plug to prevent binding or friction. Clearance flow should not be confused with leakage flow, which may result from improper valve seating.

In selecting valves for applications in which extreme changes in load may be encountered and the controlling of small flows is important, the valve's "controllable range" (maximum and minimum controllable flow) must be known. Reference to Fig. 7-5 shows that the V-port valves allow control down to about 2.7 per cent of their maximum flow and the throttle plug to about 4 per cent. These are average values of the smaller but most commonly used valves of both single- and double-seat types.

Because clearance flow increases as the valve-seat diameter, and the total flow increases as the square of the diameter, the larger-size valves can be made with increasingly better controllable ranges. Furthermore, double-seat valves, having practically twice the clearance flow of single-seat valves of the same size, cannot have such good controllable ranges as single-seated ones. Because of this, the use of double-seated valves in sizes under 1 in. should be avoided.

Double-seated Valves. Where close throttling control is demanded, double-seat valves are generally preferred in sizes above ¾ in.

On double-seated valves (Figs. 7-6 and 7-7) the upstream pressure enters the body between the seats, tending to create nearly equal upward and downward forces. This feature results in a practically balanced valve and eliminates to a great extent the effects of varying line pressures, and a reasonably constant disk position can be maintained in spite of these fluctuations. This characteristic is particularly desirable on low-sensitivity applications.

The semibalanced feature also permits operation at high-pressure drops through the valve, because the force created by the diaphragm motor does not have to overcome the full force created by the line pressure acting on the full area of the disk, as it does in a single-seat valve. Consequently, larger valves can be operated without resorting to larger and more expensive diaphragm motors.

Unless any double-seat valve is ground in very carefully at operating temperature and the seating surfaces wear uniformly, the valve cannot be relied upon to give tight shutoff.

The clearance flow of double-seat valves is necessarily greater than that of single-seat valves of the same size. Hence in the smaller sizes the clearance flow represents a greater percentage of the total flow capacity. Consequently, to obtain satisfactory performance at low lifts, it is recommended that the use of double-seat valves in sizes smaller than 1 in. be avoided.

Throttle-plug Valves. Developed for applications involving large load changes, this type of valve is so designed that each uniform increment of valve lift will result in a change in flow rate (Fig. 7-9) which is practically a constant percentage of the flow rate existing before each change in lift (Figs. 7-6 and 7-8).

The throttle plug controls rate of flow by means of variable port openings determined by the curvature of the plug proper. Its shape minimizes the tendency to spin or wire-draw, because of the absence of projections or

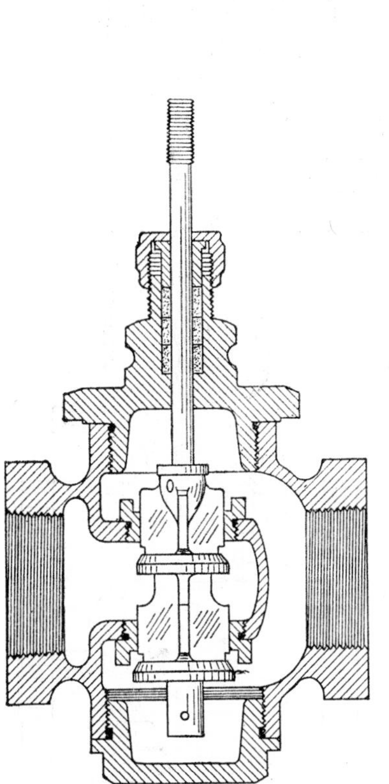

FIG. 7-7. Double-seated quick-opening inner valve with guided disks. (*Kieley & Mueller, Inc.*)

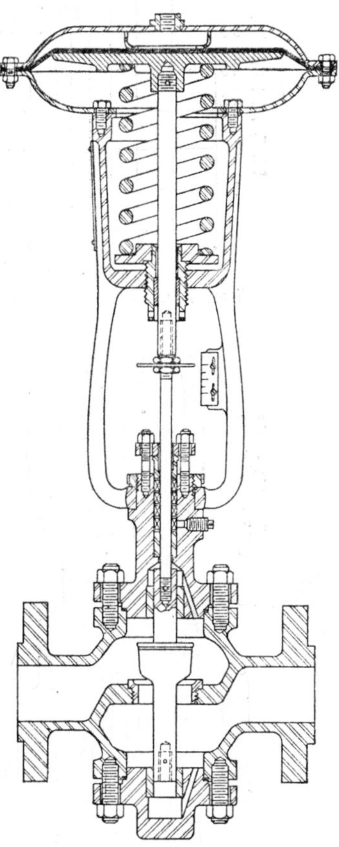

FIG. 7-8. Single-seated globe-type body with solid-plug inner valve. Characterized plug may have either "proportional percentage" or "linear" contour. (*Minneapolis-Honeywell Regulator Co.*)

wings. Because of its greater port area at low lifts, it overcomes the tendency to "coke," or to become fouled by suspended solids. It handles large pressure drops particularly well.

V Port. This inner-valve construction (Figs. 7-10 and 7-11) is adequate for the majority of applications where widely varying rates of flow are

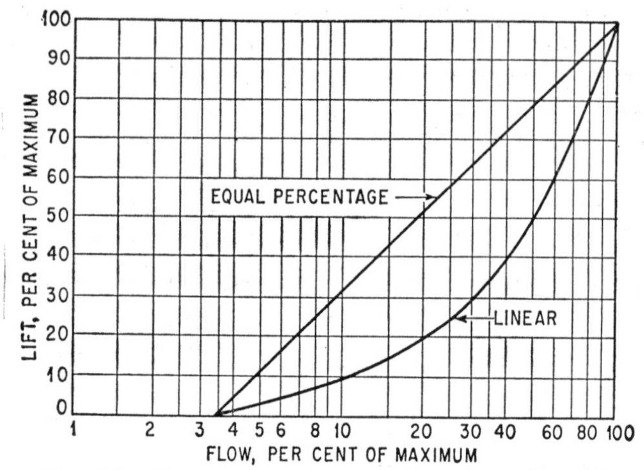

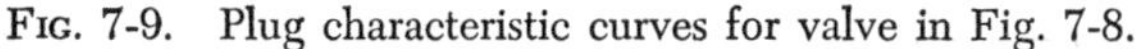

Fig. 7-9. Plug characteristic curves for valve in Fig. 7-8.

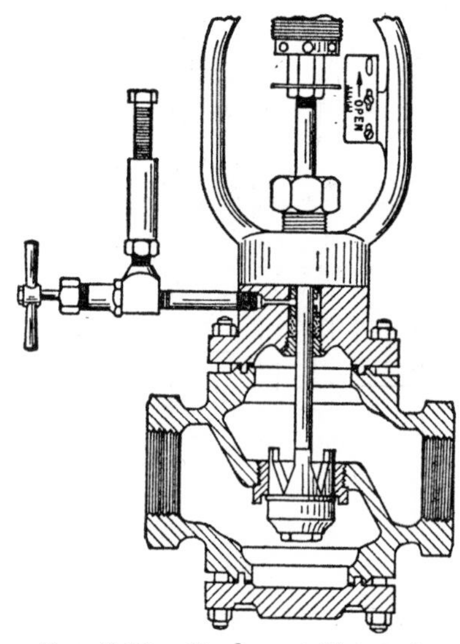

Fig. 7-10. Single-seat V-port inner valve for open-on-air-failure service. (*Fisher Governor Co.*)

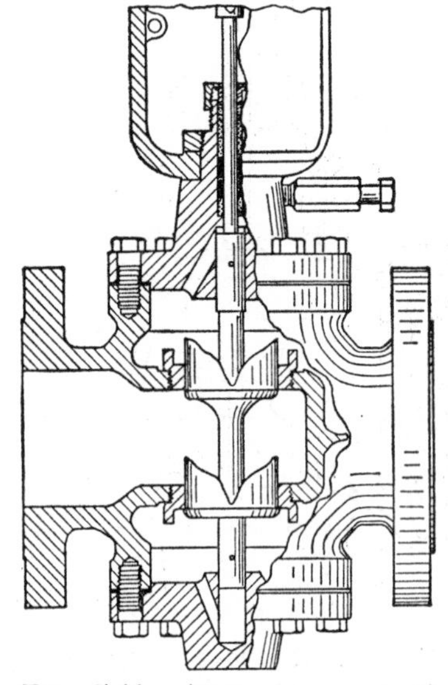

Fig. 7-11. Air-to-open percentage V-port control valve with cast-iron body. (*Mason-Neilan Div. of Worthington Corp.*)

encountered and where full throttling control over the entire flow range must be maintained. Its capacity is as high as any of the characterized valves, and it has the greatest controllable range. For specific data on flow-lift characteristics, see Figs. 7-4 and 7-5.

The improved "percentage" characteristics of the V-port over the disk construction are accomplished by the addition of a skirt having several

V-shaped ports which create a relationship of constantly increasing port openings for each unit lift—under conditions of constant-pressure drop. This combination of features makes it desirable for most difficult control applications.

Double-seat V-port valves are made in a wide range of sizes and materials. Seat rings and disks are renewable. Drain plug can be provided in the bottom if required.

Where large pressure drops are encountered, V-port valves can be top-and-bottom-guided, with stainless-steel bushings to assure proper alignment.

By the substitution of a new valve stem and disk lock nut, the standard double-seated skirt-guided V-port valve can generally be converted from the air-to-close to air-to-open type (or vice versa) in the field.

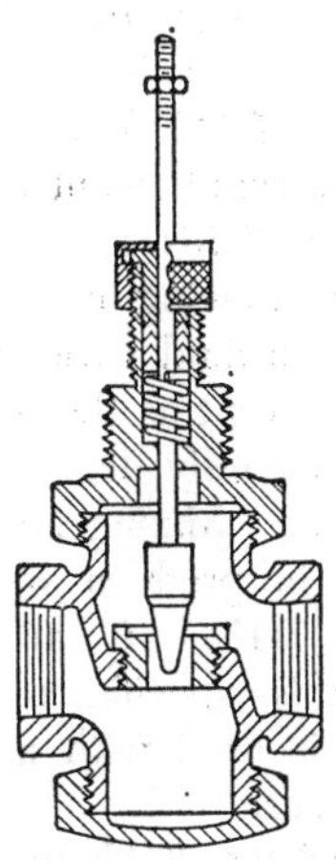

FIG. 7-12. Needle valve, air-to-close type. (*Taylor Instrument Co.*)

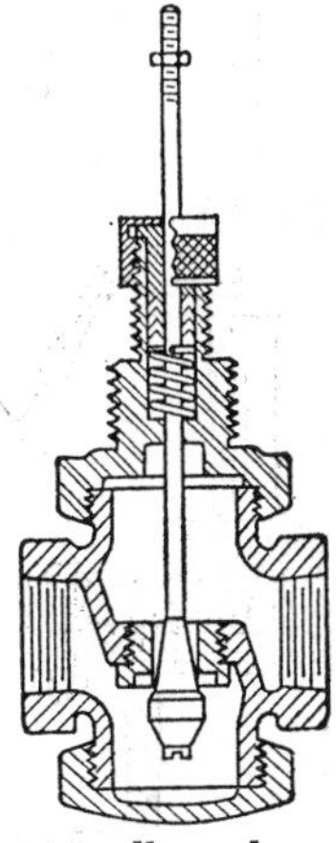

FIG. 7-13. Needle valve, air-to-open type. (*Taylor Instrument Co.*)

Single-seat Throttling. The same flow characteristics are obtainable with single-seat throttling-type valves under favorable conditions as with the double-seat type.

In sizes smaller than 1 in. they are generally preferred to the double-seat type because the clearance flow at low lifts is practically 50 per cent less. In addition, they give a tight shutoff of the controlling medium.

However, the same inherent disadvantages of single-seat valves apply to the throttling types (Figs. 7-8 and 7-10) as to the quick-opening types previously described: They are impractical for throttling control if line pressures fluctuate widely, unless the controller sensitivity is very high or the size of the valve small. They are generally not recommended above 2-in. size, because of large unbalanced disk areas, unless line pressures are low.

Small Flows. For control of very small flows, generally where the requirements are for less than ½-in. valve size, the needle-type valve (Figs. 7-12 and 7-13) is used. The conventional styles of valves are impractical in sizes below ½ in., because of the minute changes in lift necessary to obtain small changes in flow. Failure to cause proper changes in flow would result in severe "hunting" on applications requiring throttling control.

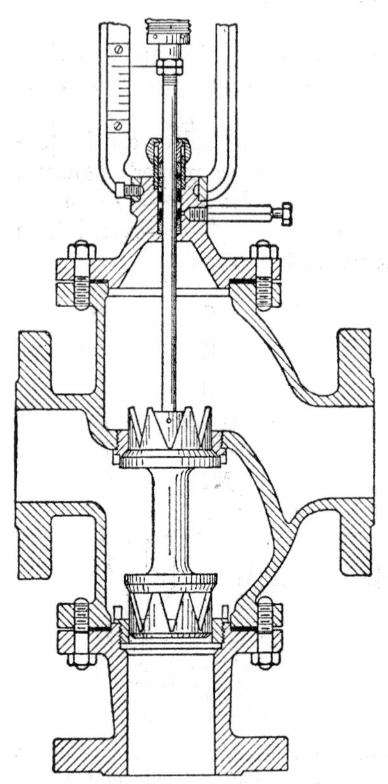

FIG. 7-14. Three-way diaphragm control valve, modified linear V-port inner valve. (*Kieley & Mueller Inc.*)

With full information on maximum and minimum flows and the range through which throttling control is necessary, practically any required flow characteristic is possible, by means of the infinite number of combinations of orifice sizes, needle tapers, and stem travels obtainable.

Three-way valves. Three-way valves (Fig. 7-14) have three primary service requirements:

1. To mix fluids so that, in the mixture flowing from the third port, one fluid varies from 0 to 100 per cent as the other varies simultaneously from 100 to 0 per cent. Mixing hot and cold water to produce warm water (Fig. 7-15*a*) is an example. For this specific service, the valve is used with a throttling controller.
2. To feed two fluids to a vessel or pipeline separately (no mixing of the two fluids in the valve body) (Fig. 7-15*b*). Sending steam to a plating tank when the solution is too cold, and cold water when the solution is too hot, is an example. For this service, the valve is used with a throttling controller.

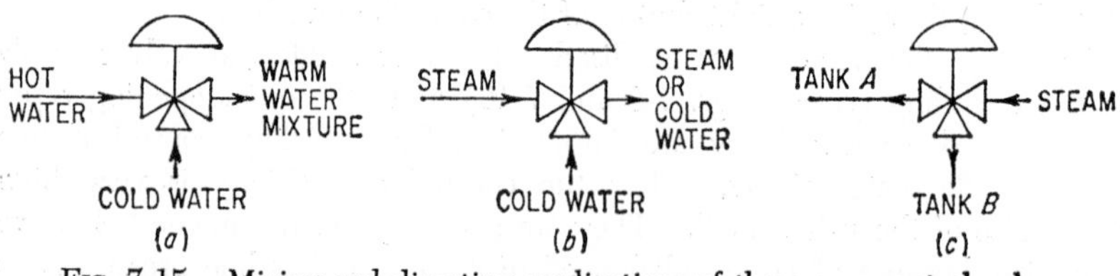

FIG. 7-15. Mixing and diverting applications of three-way control valves.

3. To divert a fluid from one vessel or pipeline to two load demands, a noncharacterized on-off disk is used (Fig. 7-15*c*). An on-off type of control instrument is required for this action.

Beveled Single-disk Type. Figure 7-16 shows this type of valve body. It is primarily for control for mixing service with the inlet at ports *A* and *B*, but where the V-port inner valve is not essential.

V-port Single-disk Type. This type of valve (Fig. 7-17) is suitable for mixing service using ports *A* and *B* as inlets and port *C* as the outlet. It is recommended where the flow characteristics of the V-port inner valve are desired.

V-port Double-disk Type. Recommended for throttling-type diversion service, this valve (Fig. 7-18) has the inlet port at *C* and outlet ports at *A* and *B*.

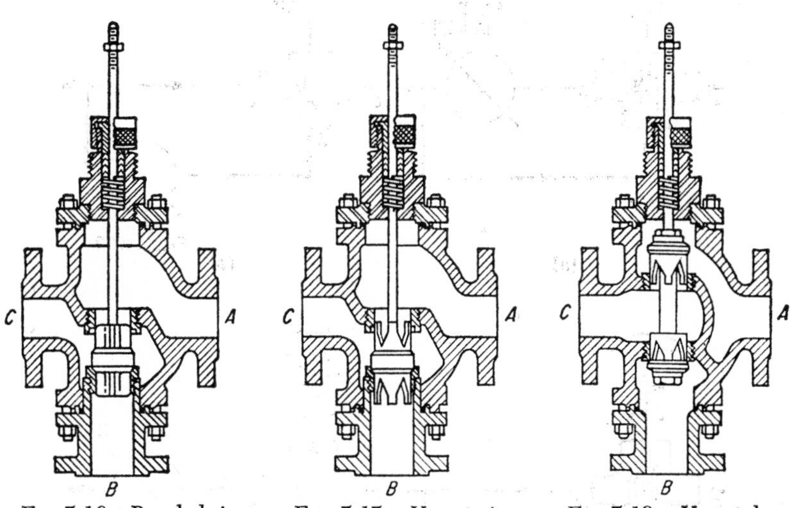

FIG. 7-16. Beveled single-disk three-way valve. (*Taylor Instrument Co.*)

FIG. 7-17. V-port single-disk three-way valve. (*Taylor Instrument Co.*)

FIG. 7-18. V-port double-disk three-way valve. (*Taylor Instrument Co.*)

One of the most common applications of three-way valves is as a bypass on a heat exchanger. The temperature of the heated fluid outlet is regulated by controlling the amount of bypass of the heating medium around the exchanger. There are several factors to be considered in selecting a three-way valve for a heat-exchanger application.

1. Location of the valve—upstream or downstream of the exchanger
2. Exchanger in series with the upper port *U* or lower port *L* to obtain desired action on air failure (Fig. 7-19)
3. Use of inside seating plug or outside seating plug to obtain flow tending to open both ports
4. Use of direct or reverse diaphragm motor to obtain maximum operating force

The first two items are a function of equipment and piping layout, whereas the last two have to do with the construction of the valve itself. In general, the three-way valve on heat-exchanger applications should be installed flow-to-open both ports, since it is basically equivalent to two single-seated valves acting in opposite directions. The important advantage of this flow direction is inherent stability. Since the maximum unbalance generally occurs across the bypass port, the air action is normally selected so that an increase in air pressure on the diaphragm will tend to close the bypass port.

Figures 7-19*a* through *d* show the four possible methods of piping up a three-way valve on a heat exchanger with flow-to-open action. *a* and *b*

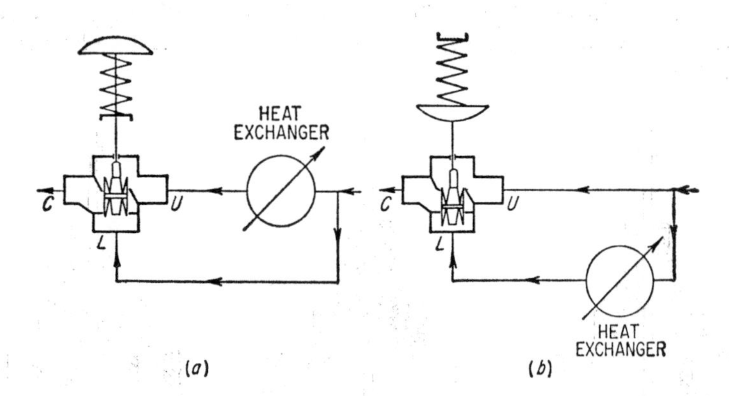

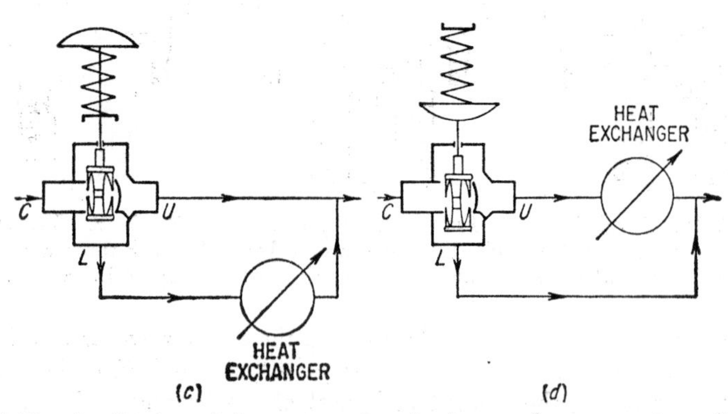

FIG. 7-19. Application of three-way valves for heat exchanger service. (*Mason-Neilan Div. of Worthington Corp.*)

show combining service. The heating medium is brought in through ports *U* and *L* and out through the common port *C*. *c* and *d* show diverting service. The fluid is brought in through the common port *C* and diverted through ports *U* or *L*. The flow direction shown in these diagrams is stable, and maximum operating force is available from the diaphragm motor to overcome off-balance. The valves are arranged so that, on air failure, the heating medium to the exchanger will be shut off.

Angle Bodies. Angle valves are generally used to facilitate piping, or where a self-draining piping system is desirable.

Since it is not possible to have bottom guiding in an angle body without restricting the flow passage, bodies are furnished with top guiding only on a plug-style inner valve, as shown in Fig. 7-20. Note that the guide bushing is longer than would normally be used in a globe body.

Since angle bodies are single-port valves, they are subject to slamming when installed with flow in at the side and out at the bottom. This should be given consideration in valve specification if pressure drop is high, or if port size is larger than 1 in. In all cases, it is preferred to have flow in at the bottom and out at the side, which results in a flow-opening installation.

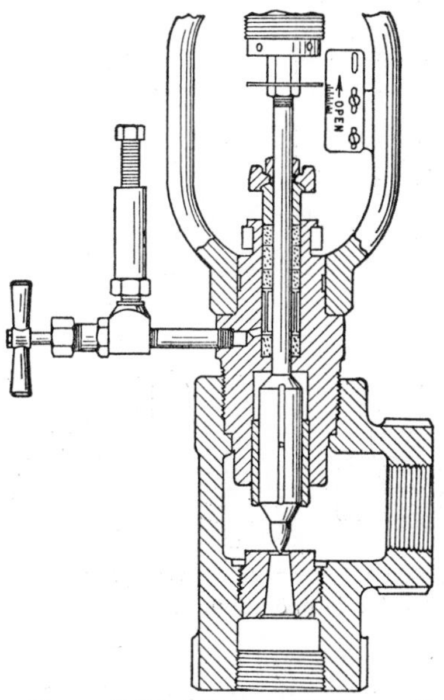

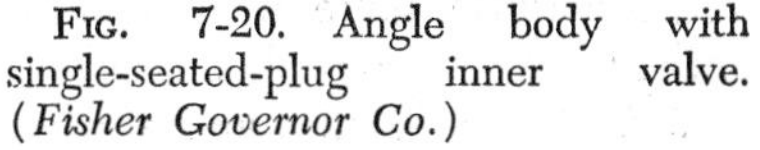

Fig. 7-20. Angle body with single-seated-plug inner valve. (*Fisher Governor Co.*)

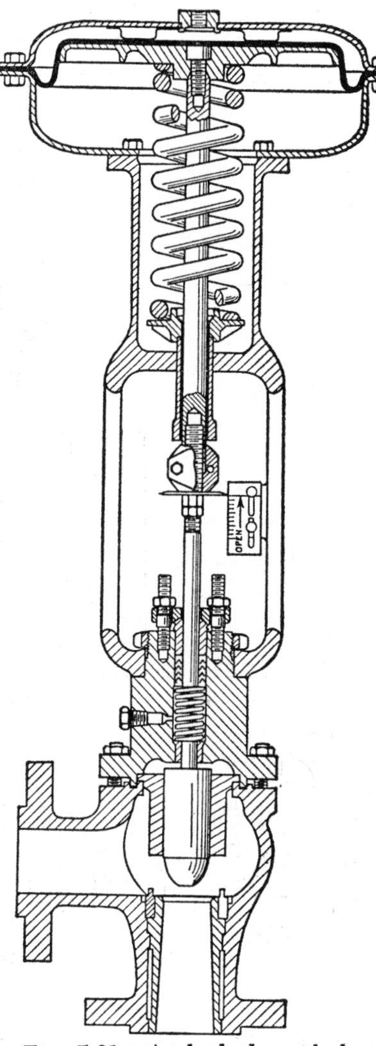

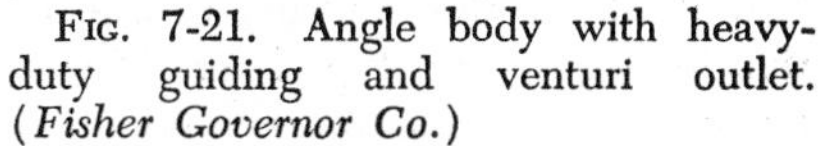

Fig. 7-21. Angle body with heavy-duty guiding and venturi outlet. (*Fisher Governor Co.*)

Angle body (Fig. 7-21) is generally used on applications handling fluids containing solids, slurries, or flashing fluids of an erosive nature. A venturi-type liner is provided in the outlet passage. This liner prevents erosion by streamlining the flow. This body is almost always installed with the

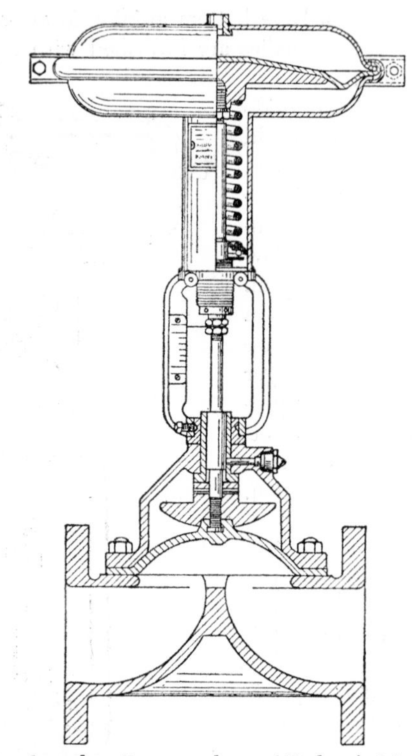

FIG. 7-22. Saunders Patent valve. (*Kieley & Mueller, Inc.*)

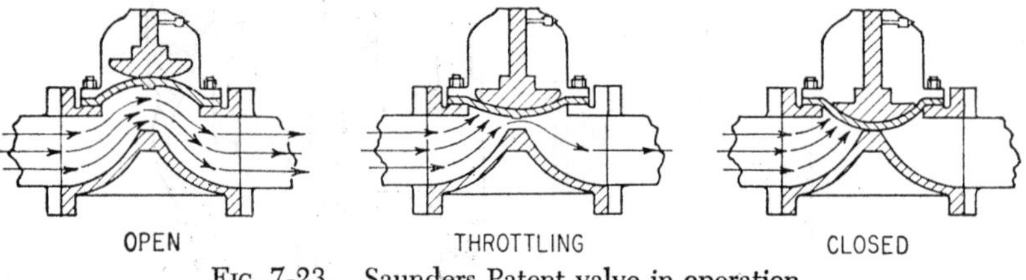

FIG. 7-23. Saunders Patent valve in operation.

flow in at the side and out at the bottom. Otherwise, there is no justification for the venturi-type liner. Consequently, the angle body should always be provided with a topwork adequate to handle the unbalance and prevent slamming.

This construction is normally furnished with a plug-type inner valve and heavy-cylinder-type guiding, as shown in Fig. 7-21.

The Saunders Patent Valve (Figs. 7-22 and 7-23). This device, equipped with a proper diaphragm motor, is especially suitable for handling corrosive fluids, slurries, and gases at moderate temperatures and pressures. It is essentially a simple pinch clamp, and closure is effected by forcing a flexible rubber or synthetic diaphragm against a weir. The diaphragm serves both as the closure member and as a packless seal to isolate the bonnet and its working parts.

One of the limiting factors in the application of this type of valve is the large stem thrust necessary to obtain tight closure. This thrust limits the pressure drop that can be handled by the valve on both air-to-close and air-to-open action.

It is difficult to shut a Saunders Patent valve tightly. First of all, there is the force required to close the valve with no pressure in the body. This force is determined by the shape and stiffness of the diaphragm, the shape

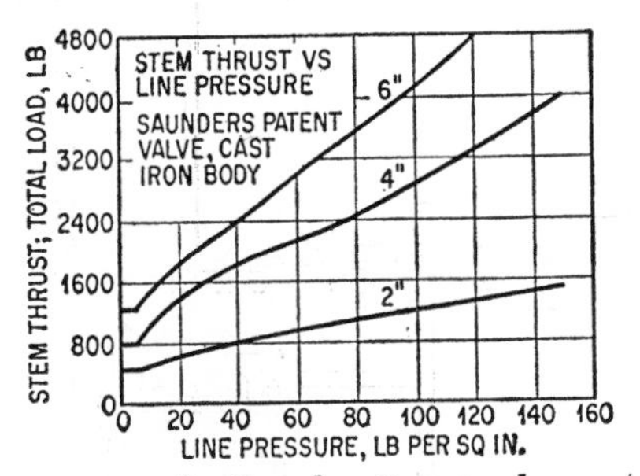

FIG. 7-24. Performance curves for Saunders Patent valves. (*Hills-McCanna Co.*)

and finish of the weir in the body, and the contour of the follower behind the diaphragm. In addition, there is the force due to line pressure acting on the underside of the diaphragm. On most applications this is the lesser of the two forces.

Figure 7-24 shows the stem-thrust-versus-line-pressure curves for 2-, 4-, and 6-in. Saunders valves. Curves of this type are used as a basis for establishing the size of diaphragm motor to be used.

Butterfly Control Valves. Butterfly valve bodies may be fitted with either manual or automatic valve operators, depending on their application. For automatic control, the butterfly-valve body may be diaphragm-, cylinder-, float-, electric-motor-, or solenoid-operated (Figs. 7-25 and 7-26).

Standard-weight bodies are light, economical valves for control of low-pressure low-velocity fluids such as furnace combustion air. Heavy-duty bodies provide control of air, gases, liquids, vapors, and semisolids. The type selected should have an allowable pressure drop equal to or greater than the actual drops that will exist across the valve.

The "butterfly" disk itself is generally held between pipe flanges by

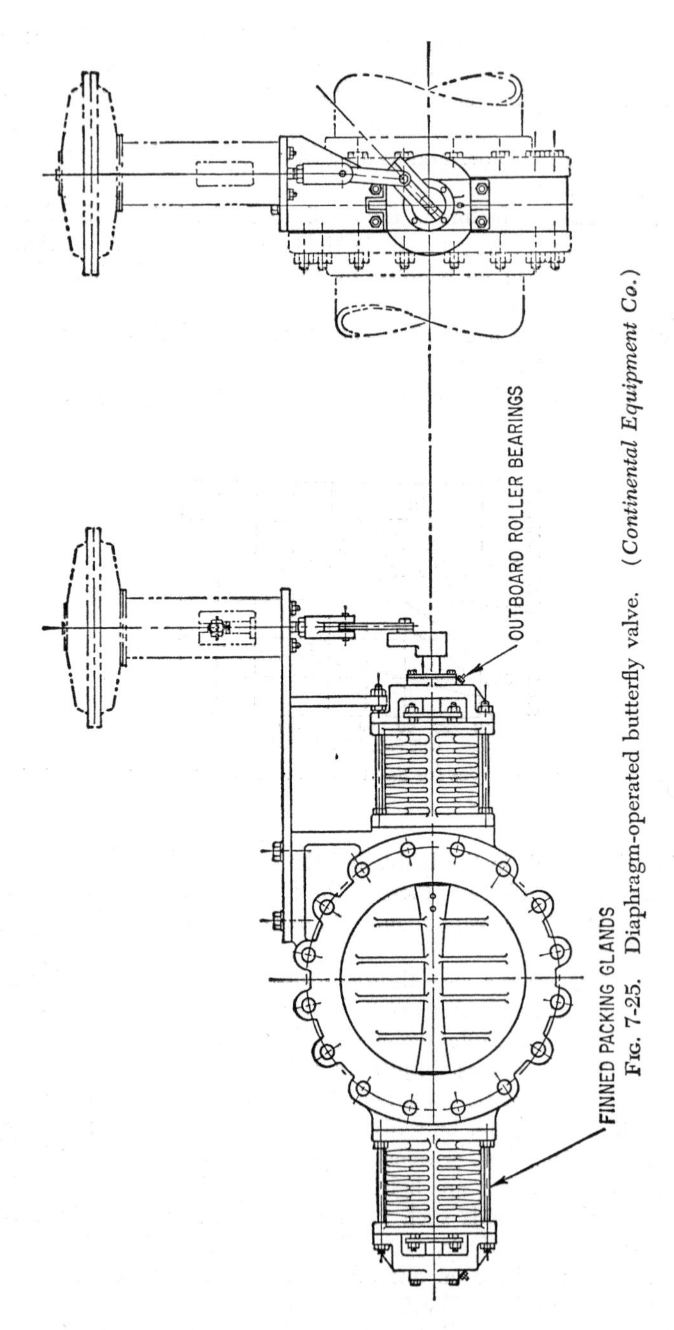

Fig. 7-25. Diaphragm-operated butterfly valve. (*Continental Equipment Co.*)

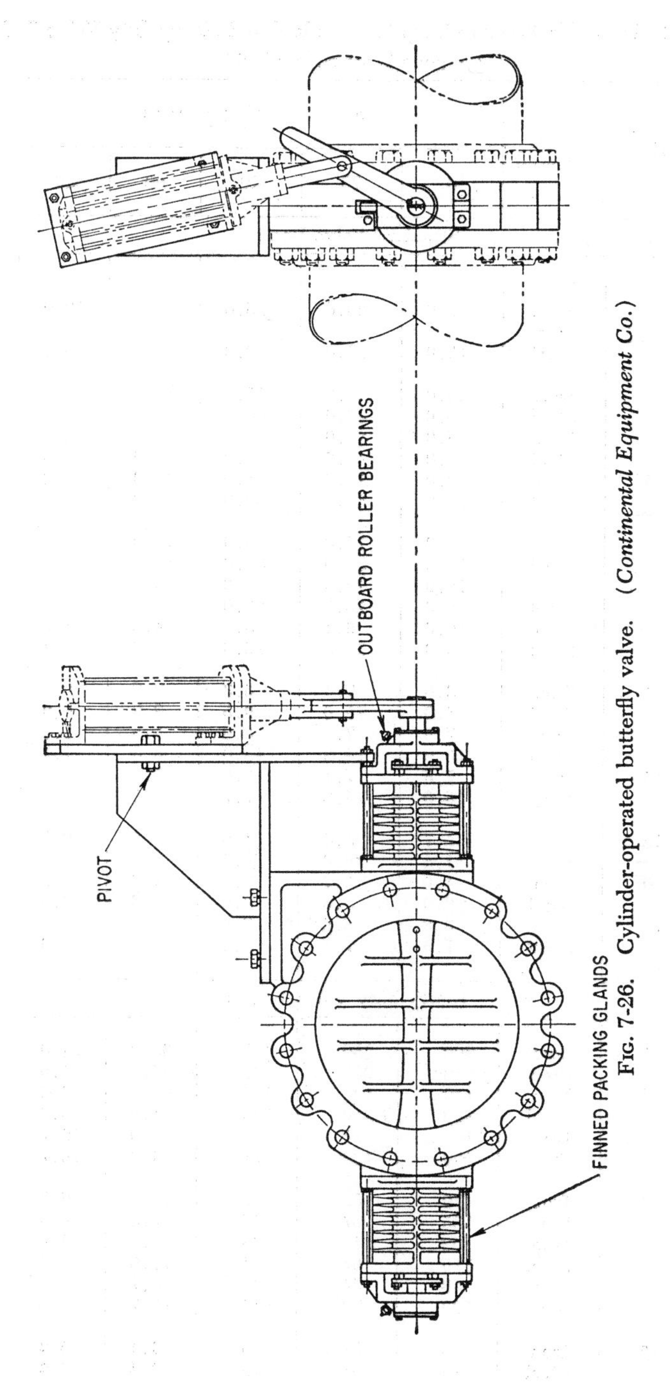

FIG. 7-26. Cylinder-operated butterfly valve. (*Continental Equipment Co.*)

Table 7-1. Allowable Pressure Drop for Standard and Heavy-duty Valve Bodies*

(In pounds per square inch)

Valve size, in.	Disk material	Angle of opening, degrees				
		0	20	40	60	80
		Standard Valve Bodies				
2	Iron	—	—	—	—	—
	Steel	25.0	20.0	15.0	10.0	10.0
2½	Iron	—	—	—	—	—
	Steel	15.0	14.0	12.0	6.0	6.0
3	Iron	25.0	20.0	15.0	10.0	5.0
	Steel	50.0	40.0	30.0	20.0	10.0
4	Iron	25.0	20.0	15.0	10.0	5.0
	Steel	50.0	40.0	30.0	20.0	5.0
5	Iron	20.0	17.0	15.0	12.0	10.0
	Steel	40.0	35.0	30.0	25.0	10.0
6	Iron	15.0	15.0	15.0	15.0	6.5
	Steel	30.0	30.0	30.0	15.0	6.5
8	Iron	12.0	12.0	10.0	8.0	3.0
	Steel	25.0	25.0	20.0	8.0	3.0
10	Iron	10.0	9.4	8.0	4.8	1.5
	Steel	20.0	18.5	16.0	4.8	1.5
12	Iron	8.0	7.4	6.3	2.8	0.9
	Steel	16.0	15.0	9.5	2.8	0.9
14	Iron	7.5	7.1	6.0	3.5	1.1
	Steel	15.0	14.2	12.0	3.5	1.1
16	Iron	7.5	7.1	6.0	2.3	0.7
	Steel	15.0	14.2	7.7	2.3	0.7
18	Iron	5.7	5.4	4.6	1.5	0.5
	Steel	11.4	10.8	5.4	1.5	0.5
20	Iron	4.0	3.7	3.2	1.8	0.5
	Steel	8.0	7.5	6.0	1.8	0.5
24	Iron	3.8	3.6	3.0	1.0	0.3
	Steel	7.6	7.1	3.5	1.0	0.3
		Heavy-duty Valve Bodies				
3	Iron	300	250	147	63.0	24.0
	Steel	340	250	147	63.0	24.0
4	Iron	140	125	124	54.0	20.0
	Steel	300	210	124	54.0	20.0
5	Iron	100	77	71	29.0	10.7
	Steel	190	130	71	29.0	10.7
6	Iron	70	52	45	17.5	6.3
	Steel	135	87	45	17.5	6.3
8	Iron	62	47	34	13.0	4.4
	Steel	110	67	34	13.0	4.4
10	Iron	58	54	40	15.7	5.5
	Steel	100	78	40	15.7	5.5
12	Iron	54	51	25	9.4	3.2
	Steel	85	51	25	9.4	3.2

Table 7-1. Allowable Pressure Drop for Standard and Heavy-duty Valve Bodies (*Continued*)

Valve size, in.	Disk material	Angle of opening, degrees				
		0	20	40	60	80
		Heavy-duty Valve Bodies (*Continued*)				
14	Iron	52	48	31	12.6	4.5
	Steel	90	65	33	12.6	4.5
16	Iron	48	44	23	8.5	3.0
	Steel	80	47	23	8.5	3.0
18	Iron	38	35	25	9.4	3.4
	Steel	75	50	25	9.4	3.4
20	Iron	34	32	20	7.3	2.5
	Steel	70	40	20	7.3	2.5
24	Iron	32	30	19	7.0	2.3
	Steel	70	40	19	7.0	2.3
30	Iron	22	21	15	5.3	1.8
	Steel	50	31	15	5.3	1.8
36	Iron	19	17	9	3.2	1.1
	Steel	40	20	9	3.2	1.1

* Courtesy of Continental Equipment Co.

through bolts. This eliminates transfer of pipe stresses to the valve body by the flanges.

Butterfly-valve disks and shafts must withstand stresses which vary with the static-pressure drop across the valve and the disk angle of opening. These parts are rated independently of the body-rating pressure, which may be materially higher than the pressure-drop ratings.

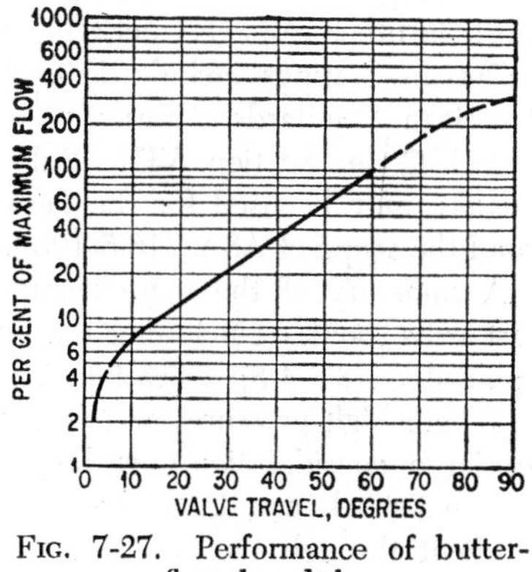

Fig. 7-27. Performance of butterfly-valve disk.

Allowable pressure drops for standard and heavy-duty designs are shown in Table 7-1. The values shown are based on carbon or stainless-steel shafts, and disks of the materials indicated. Iron ratings are for high-grade high-tensile-strength iron disks. Alloy iron or nonferrous metal disks require special ratings. Steel ratings apply to carbon or alloy steels. Ratings apply for temperatures between 0 and 500°F.

As shown in Fig. 7-27, butterfly valves have an equal-percentage characteristic for a range of approximately 12 to 1. Control valves should be sized to pass required maximum flow when open 60°. Minimum resistance is obtained using 90° opening.

CONTROL-VALVE CONSTRUCTION DETAILS

Materials.[1] The standards prepared by the American Standards Association, ASME, ASTM, and other neutral agencies provide considerable guidance regarding materials of construction for control-valve assemblies. Such sources constitute the authority for most of what follows.

The Valve Body and Bonnet. These constitute the enclosure confining the fluid and containing the valve mechanism. Mechanical strength to resist internal working pressure, rigidity to maintain the alignment of parts (when the body is subjected to pipeline strains), and resistance to corrosion are major considerations in their design.

The American Standard Steel Pipe Flanges and Flanged Fittings, ASA B16.5-1953, pertaining to cast and forged steel flanges, and cast and forged steel flanged fittings, also includes cast and forged steel flanged and butt welding end valves, covering (1) pressure ratings, (2) minimum requirements for materials, (3) minimum body-wall thickness, (4) flange facings, (5) dimensions of steel flanges, (6) dimensions of welding ends, (7) tests.

Pressure-temperature ratings are classified as class A or class B in accordance with the type of flange facing, and apply to cast and forged steel flanged-end valves with end flange, bolting, and body-wall-thickness dimensions not less than those specified in the standard for corresponding nominal size and class. Class A ratings also apply to welding-end valves having welding ends complying with details as covered in the standard, and with body-wall-thickness dimensions as specified in the standard for the corresponding nominal size and class. Valves complying with this standard must, in other respects, merit the ratings to which they apply.

Pressure ratings include 150, 300, 400, 600, 900, 1,500, and 2,500 lb American Standards. Formulas given in the ASME Boiler and Pressure Vessel Code, Section VIII, Rules for Construction of Unfired Pressure Vessels, may be used for the design of valves for operating conditions beyond the scope of ASA B16.5-1953.

A summary of the more common materials used in the construction of cast valve bodies (including the bottom closure) and bonnets with a list of applicable ASTM Specifications is given in Table 7-2.

Where high pressures and temperatures exist and valve-metal thicknesses are large, consideration must be given to thermal stresses and thermal shock. Thermal-shock conditions become serious as a result of the temperature gradient through the walls of the valve body, wherein the inside wall is heated at a rate which exceeds the thermal conductivity through the metal itself. Under these conditions the inside wall of the metal tends to expand more rapidly than the outside wall, and it is necessary to keep the stress on the inside wall below the elastic limit, to prevent deformation during the temperature change. If deformation should occur on the inside

[1] H. H. Gorrie, Vice-President, Bailey Meter Co., and W. L. Gantz, Staff Engineer, American Viscose Corp., *Practical Limitations of Current Materials and Design of Control Valves; ASME Paper 55-A-113.*

wall, when the temperatures are equalized, or under cooling conditions, cracks may occur on the inside wall, which will be aggravated with each temperature cycle until failure occurs.

Bolting. It is the function of bolting in flanged body-bonnet joints to resist the hydrostatic end forces exerted by internal working pressure under maximum operating conditions, and to maintain sufficient compression load on the gasket to prevent leakage.

Table 7-2. Common Materials Used in Cast Valve Bodies and Bonnets and ASTM Specifications

General Classification	*Applicable ASTM Specification*
Bronze	B61-52
Cast iron	A126-42, class B
Carbon steel	A95-44
	A216-53T, grade WCB
Alloy steels:	
1¼% chromium, ½% molybdenum	A217-49T, grades WC5 and WC6
2¼% chromium, 1% molybdenum	A217-49T, grade WC9
5% chromium, ½% molybdenum	A217-49T, grade C5
9% chromium, 1% molybdenum	A217-49T, grade C12
Type 304	A351-52T, grade CF8
Type 347	A351-52T, grade CF8C
Type 316	A351-52T, grade CF8M

Certain limitations are placed on the use of bronze and cast iron by the following codes:

American Standard Code for Pressure Piping, ASA B31.1-1951 and Supplement B31.1a-1953:

Bronze —500°F maximum
Cast iron—450°F maximum and for oil 300°F maximum

ASME Boiler Construction Code, Rules for Construction of Power Boilers:

Bronze —550°F maximum
Cast iron—250 psi maximum and 450°F maximum

Carbon-steel bolting is satisfactory for low temperatures, but lacks the resistance to creep necessary for high temperatures. High-strength alloy-steel studs with high physical properties at normal temperatures and good creep properties at high temperatures are used to overcome relaxation in high-temperature service.

Proper initial setup stress in the studs, to maintain residual stress in the joint under operating conditions, is essential to a tight flanged joint. High-strength surface-hardened washers may be used under the nuts to obtain correct distribution of stress.

For pressures up to 300 psi or temperatures up to 500°F, carbon-steel bolting equal to or better than grade B of ASTM Specification for Steel Machine Bolts and Nuts and Tap Bolts (A307-53T) may be used. Above these limits, alloy-steel bolting materials are required.

Packing. A valve bonnet is provided with a packing gland (Fig. 7-28) for sealing against leakage where the stem extends through the bonnet. The packing is maintained in compression by means of a gland follower and adjusting bolts and nuts.

Valve manufacturers' standard packings, in general, consist of molded rings of asbestos fiber with plastic or synthetic-rubber binder, graphite- or mica-lubricated, and various forms of Teflon. Materials with plastic or synthetic-rubber binder and Teflon are limited, generally, to fluid temperatures from 0 to 450°F, whereas graphite- or mica-lubricated asbestos materials are used, generally, for temperatures over 450 and up to 750°F.

Operating temperatures often exceed the limitations of packing materials. When the flowing-fluid temperatures exceed these specified limits, the packing material is protected from excessive temperatures by air-fin or extension bonnets (Fig. 7-29). With air-fin bonnets, asbestos packings with plastic or synthetic-rubber binder are used with fluid temperatures up to 1000°F

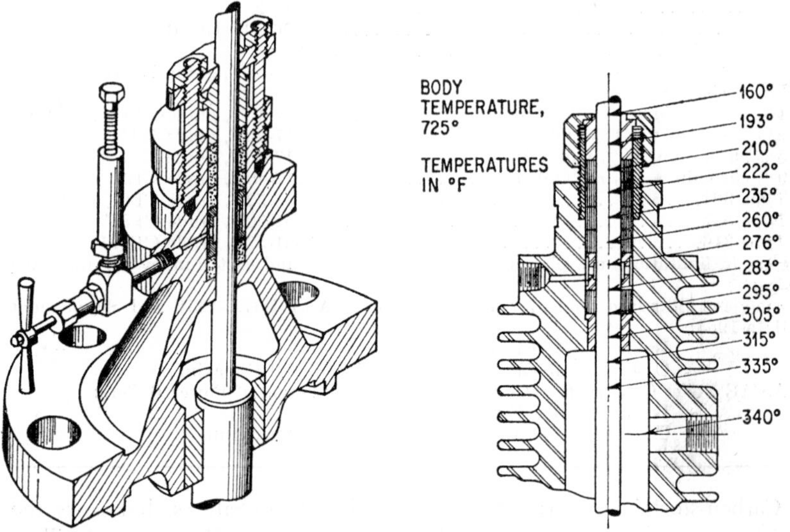

FIG. 7-28. Bolted gland-type stuffing-box head, sectional view.

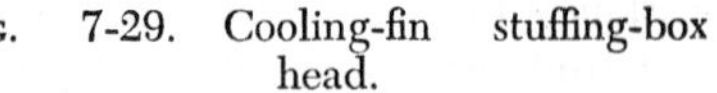

FIG. 7-29. Cooling-fin stuffing-box head.

and with Teflon up to 800°F. Extension bonnets are also used for low temperatures.

Asbestos packings are used for water, oil, and steam, and also for chemical services where they have proved satisfactory. Teflon packings may be used for services where asbestos packings are not suitable. The inertness of Teflon to chemicals is well established.

Despite the advantage of Teflon, it is lacking, in its pure form, in certain characteristics required of a packing material. Its poor resiliency and high rate of expansion make it necessary to spring-load the packing in high-temperature service.

The life of packing material is greatly affected by the stem finish, particularly Teflon. Smooth finished and polished stems are also important in maintaining tightness and low operating friction. Varying degrees of stem finishes are employed by manufacturers ranging from an average of about 4 to 12 μin.

Valve Trim. The valve plug, seat rings, and guide bushings are designated as *trim.* Resistance to corrosion and the eroding action of the flowing fluid are principal factors influencing the selection of materials for these parts.

The most commonly used valve-trim materials in control-valve manufacture are bronze, series 400 hardening-type stainless steels, series 300 austenitic stainless steels, and Monel. Other valve-trim materials are nitrided stainless steel and hard chrome-plated stainless steel.

Series 300 austenitic stainless steels are quite often faced with Stellite (Fig. 7-30), in order to increase their resistance to corrosion and/or erosion.

Valve plugs smaller than ½ in. are difficult to face with Stellite and are usually made of solid cast Stellite. Similarly, small inner-valve cages or seat rings are made of solid cast Stellite, since it is not economical and is sometimes impossible to deposit Stellite on such parts.

Large valve plugs and seat rings or cages requiring Stellite facing or seats are generally made by depositing Stellite by welding to either type 304 or preferably type 347 stabilized stainless steel.

Some manufacturers use a difference in hardness between the moving member and the fixed member of the inner valve to prevent galling. This procedure is normally followed with series 400 hardening-type stainless steel, and a difference of 50 Brinell is commonly employed.

In most cases the choice of the inner-valve material is the result of empirical data obtained from field tests under various operating conditions. For the vast majority of applications these materials are entirely satisfactory and have been used with considerable success.

There are, of course, a great many inner-valve designs, and all are affected differently by the material limitations. Consequently, there seems to be little correlation by different manufacturers between the maximum temperature and pressure-drop limitations for given materials, except that, for the most severe conditions, austenitic stainless steel and no. 6 Stellite seem to be the most desirable.

For high-pressure drops with gas, sintered-carbide materials have been used with considerable success.

Single-seated valves are normally used where control valves are required for tight shutoff conditions. Because of the limited power of most commercial diaphragm motors, single-seated valves are usually limited to low-pressure drops or to small sizes, although single-seated balanced valves are available for some services. Regardless of the size or the pressure-drop conditions, it is important that there be sufficient power in the diaphragm motor to provide enough seating pressure to prevent leakage in the closed position. If the valve does not seat tight in the closed position, leakage will occur, which will gradually cut the valve seats and damage them to the extent that they will either have to be lapped into a good seating condition or completely replaced.

Double-seated valves, mentioned earlier in this section, are used to provide a control suitable for larger sizes and/or high pressure drops with a relatively low-powered diaphragm motor.

WELD

WELD

METHOD NO. 1—ON THE SMALL SIZE VALVES (1/2"-3") IT IS NECESSARY TO SEAL WELD THE RING FROM THE OPEN ENDS OF THE BODIES. THEREFORE, THE UPPER RING WILL BE SEAL WELDED AROUND THE FULL INLETS, THE LOWER RING WILL BE SEAL WELDED FROM THE BOTTOM

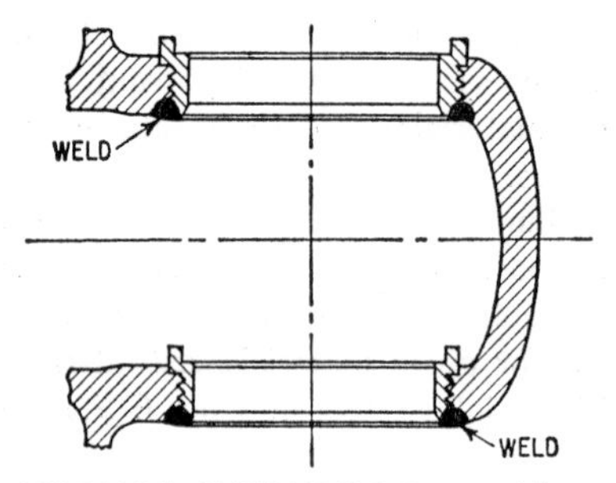

METHOD NO. 2—IN THE LARGE SIZE VALVES (4" AND UP) IT IS POSSIBLE TO SEAL WELD AT THE LOWER END OF THE RING

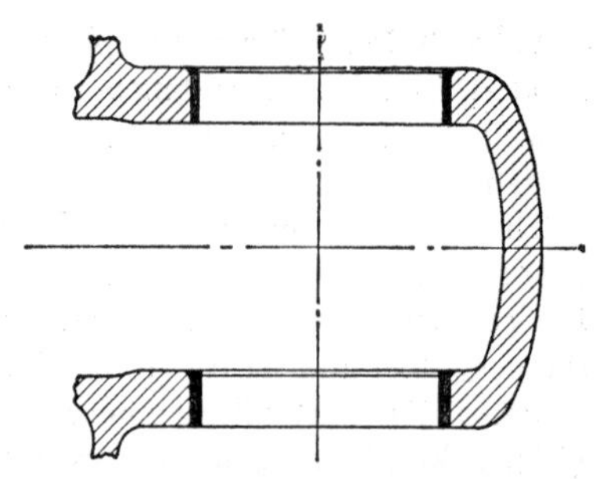

METHOD NO. 3—STELLITE LAID IN BODY IN LIEU OF RINGS. EXPERIENCE HAS SHOWN US THAT WE CANNOT SATISFACTORILY DO THIS WORK IN SIZES SMALLER THAN 1" OR IN SIZES GREATER THAN 4". THEREFORE, METHOD NO. 3 IS LIMITED WITHIN THOSE SIZES

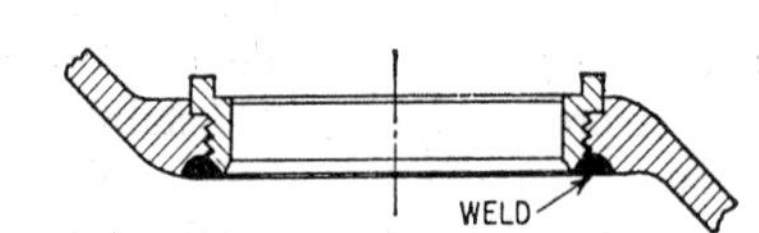

METHOD NO. 4—WHERE SEAL WELD RING SPECIFICATIONS ARE REQUIRED FOR SINGLE PORT BODIES, METHOD NO. 4 IS USED

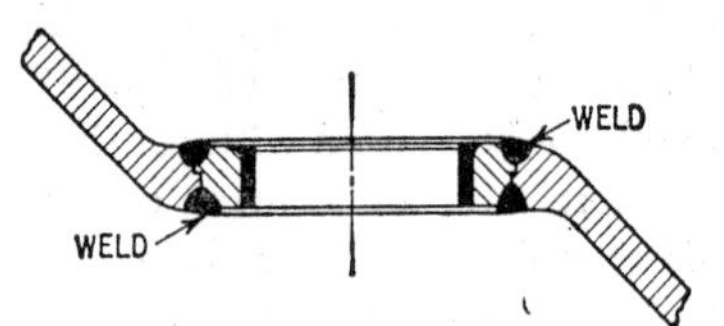

METHOD NO. 5—THIS METHOD CAN BE USED ONLY ON RESTRICTED SEAT RINGS OF SINGLE PORT BODIES

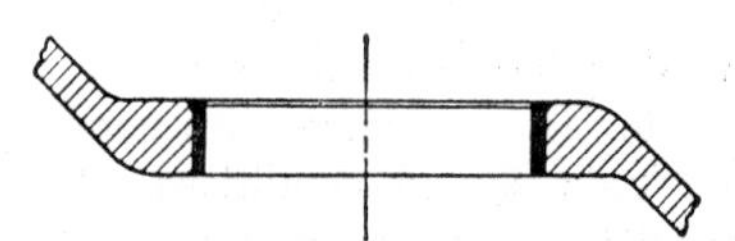

METHOD NO. 6—THIS METHOD IS USED IN PLACING STELLITE ON THE BODY IN PLACE OF SEAT RINGS. IT IS CONFINED TO SIZES 1" TO 4" FOR THE SAME REASON AS MENTIONED ON THE DOUBLE PORT CONSTRUCTION

FIG. 7-30. Methods of applying Stellite trim to valve seats.

Double-seated valves should never be used for tight shutoff conditions. The rangeability of operation of a double-seated valve will vary between 40:1 to 100:1, depending on the design of the valve, the pressure-drop conditions, and the temperature of the flowing media.

Double-seated valves cannot be tight in the closed position because of the difference in expansion between the seats of the moving valve-plug element and the body seats (which are normally an assembly fixed to the valve body), distortion of the valve body under pressure, and other factors.

Double-seated valves are made in two basic types, namely, the top-and-bottom-guided valve and the skirt-guided valve.

Top-and-bottom-guided valves (Fig. 7-6) are equipped with guides above and below the plug elements so that no guiding is necessary at the point of seating. This has the advantage of providing a plug which can be opened without restriction at the ports to an area equivalent to pipe area. This feature is important where pressure drops must be kept at a minimum under maximum flow conditions.

The skirt-guided valve (Fig. 7-14) uses the skirts (which are part of the control-valve plugs) as guides. In this type of valve, the clearances are determined by the machining tolerances of the seats and the plugs.

This diametral clearance between the moving element and the fixed element of the valve is a very important consideration. It not only affects the flow regulation near the closed position but also controls the position of the moving element under the considerable forces imposed on the valve during regulation. This is particularly important under severe pressure-drop conditions, when, if the clearances are too great, the moving element can oscillate or vibrate within the confines of the guides, resulting in the ultimate failure of the valve.

It is important for the user to appreciate that the plug element of a control valve is a precision device machined to a fine finish and to close tolerances for satisfactory control purposes. Many control valves have been ruined at the outset because the pipelines have not been cleaned properly of weld spatter, scale, dirt, sand, and other foreign materials resulting from line fabrication. This material, once lodged in the guide surfaces, causes sticking and galling, requiring repair and sometimes replacement of the plug and seats immediately after the plant has been placed into operation.

Another important feature of either the top-and-bottom-guided or the skirt-guided valve is the ratio of the guide area to the plug diameter. Where mass-flow conditions or pressure-drop conditions are severe, it is important to have a large guide area to withstand the forces which act upon the plug. Under these circumstances tremendous amounts of energy are regulated, and the guide surfaces must be such as to keep the stresses in these areas well below the elastic limit if satisfactory performance is to be obtained.

POWER OPERATORS

Pneumatic-diaphragm Motors.[2] The most common type of pneumatic power unit is the *diaphragm motor* shown in Figs. 7-31 and 7-32. The *spring type* has a flexible diaphragm with a pressure-tight chamber on one side, to which air output from the pneumatic controller is connected. On

[2] Minneapolis-Honeywell Regulator Co.

the other side of the diaphragm is placed a compression-spring assembly. A rod rigidly fastened to the metal plate under the diaphragm is directly attached to the valve stem or lever.

Air pressure applied on the top of the diaphragm creates a downward force equal to the air pressure multiplied by the area of the diaphragm. As the diaphragm moves downward, the spring further compresses until it creates an equal, opposing upward force, resulting in a balanced position of the valve stem.

For proportional control, the diaphragm motor offers a sensitive means for intermediate positioning of the final control element in accordance with each value of controlled air pressure from the instrument. For two-position control, the same principle is used; in this case, however, controlled air pressure is either at a maximum or a minimum value to position the final control element in one of two extremes of travel. Diaphragm motors for two-position control are, therefore, designed for a relatively small movement of

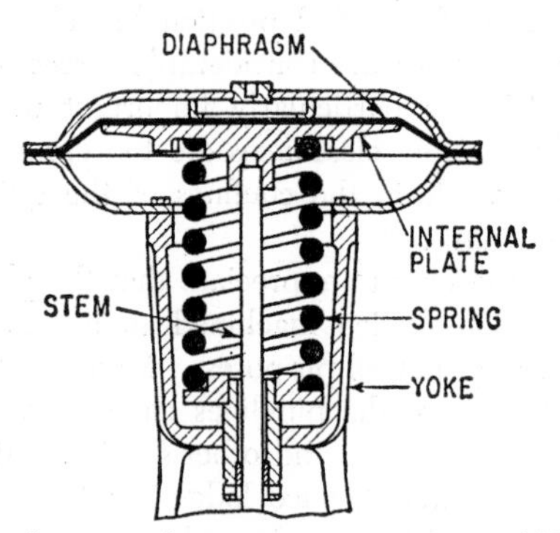

Fig. 7-31. Schematic diagram of spring-type operator. (*Minneapolis-Honeywell Regulator Co.*)

the valve stem through its full travel. This is accomplished by suitable selection of the compression spring, and mechanical travel stops in the diaphragm-head casting.

The compression springs in two-position as compared to proportional-control motors are selected to provide a markedly different relationship between the instrument-controlled air pressure and the stem travel. In two-position control, it is advantageous to reserve most of the force from the controlled air pressure for the end of the stem travel, where the valve plug or vane must be seated or unseated against the maximum force, owing to fluid pressure of the control agent. In proportional control, the diaphragm motor must provide positioning of the stem over as much of the controlled-air-pressure range as possible.

A modification of the diaphragm-motor design is the springless type (Fig. 7-33). In this motor, the underside of the diaphragm is statically loaded through a manually adjustable regulator to some suitable air pressure, from 5 to 25 psi. The top side is automatically loaded through a valve positioner to any pressure up to 30 psi required to hold a position dictated by the control instrument. Positioning is obtained by the differential air pressure on either side of the large diaphragm.

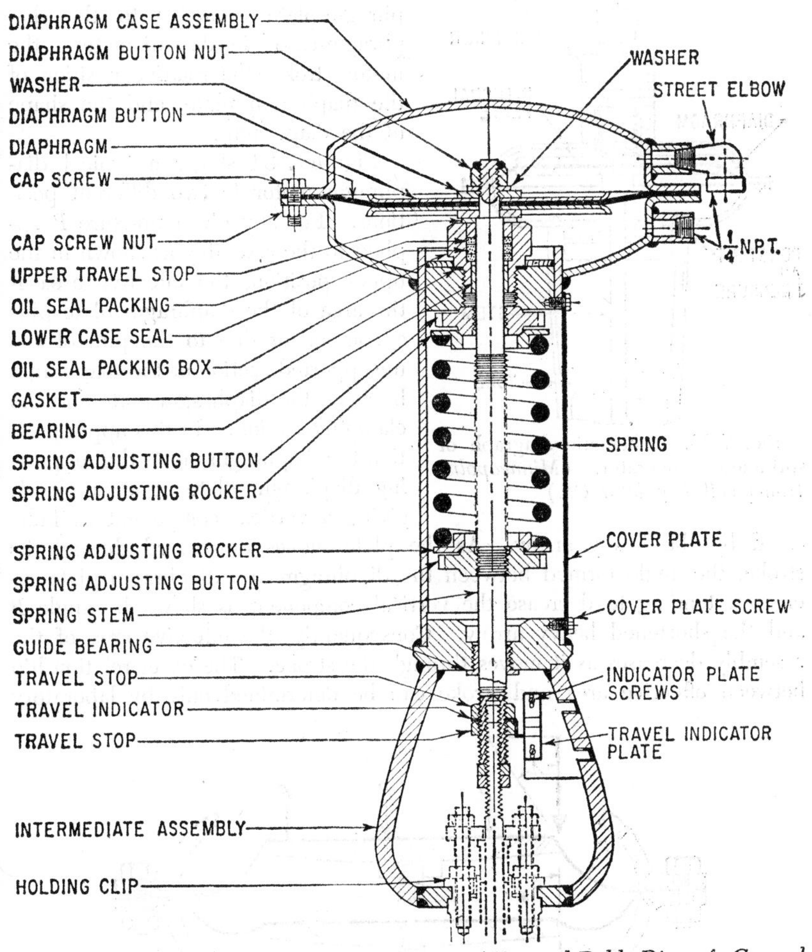

FIG. 7-32. Reversible diaphragm operator. (*Hammel-Dahl Div. of General Controls Co.*)

The initial static loading on the underside of the diaphragm is selected in accordance with the amount and direction of the load or thrust on the positioning stem. Forces of greater magnitude than are available with the spring-type diaphragm motor are consequently possible. The unit is particularly suited for sensitive positioning of heavy dampers, louvers, and large control valves.

Diaphragm Effective Area.[3] While it is common practice to refer to diaphragm motors as having certain "rated effective areas," it is important to realize that these figures are average values. The actual effective area of a conventional spring-diaphragm motor varies with stroke. The change is a function of the ratio of the diaphragm-plate diameter to the diaphragm-case inside diameter, the motor stroke, the relative position of the diaphragm plate, and the shape of the diaphragm.

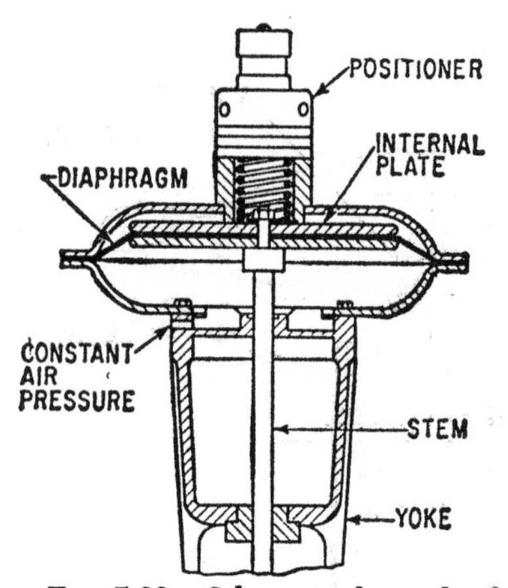

FIG. 7-33. Schematic diagraph of springless operator. (*Minneapolis-Honeywell Regulator Co.*)

Figure 7-34 shows a typical diaphragm motor in two different positions. For any given pressure P applied to the assembly as shown in the upper position, the effective area is the area of the diaphragm plate plus a component due to the pull of the unsupported section of the diaphragm between the diaphragm plate and the clamping section. In this upper position the diaphragm meets the supporting diaphragm plate at an angle A, giving a vertical component as indicated by the heavy arrow. As the plate moves downward through its stroke, the angle formed between the diaphragm and diaphragm plate increases, tending to decrease the vertical component as shown by angle B and the shortened heavy arrow. Consequently, the effective area of the assembly decreases as it moves through the stroke. The exact relationship between effective area and stroke can be determined only by laboratory

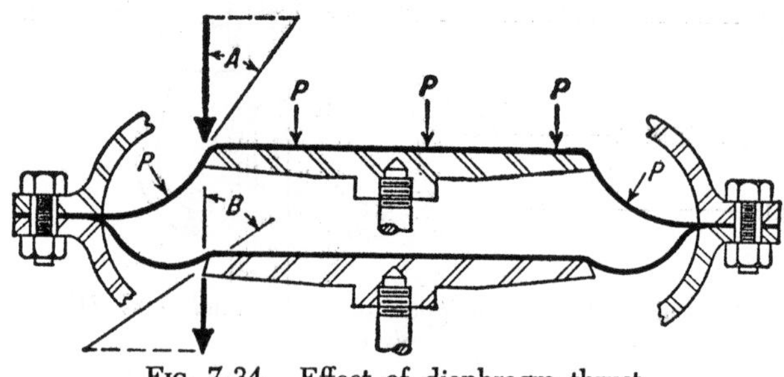

FIG. 7-34. Effect of diaphragm thrust.

tests since it is widely affected by the slackness of the diaphragm and other intangible factors such as diaphragm stretch.

The use of molded diaphragm ensures repeatability of the effective-area curve but does little to decrease the magnitude of the change. Figure 7-35

[3] Mason-Neilan Div. of Worthington Corp.

shows the effective-area-versus-stroke plot of standard diaphragm motors. These curves are average values since the curves vary with pressure.

The thrust developed by a spring-diaphragm motor at any point in the stroke is the product of the effective area at that point and the difference between the applied air pressure and nominal pressure under no-load conditions. Generally, a calculation of the thrust available at either end of the

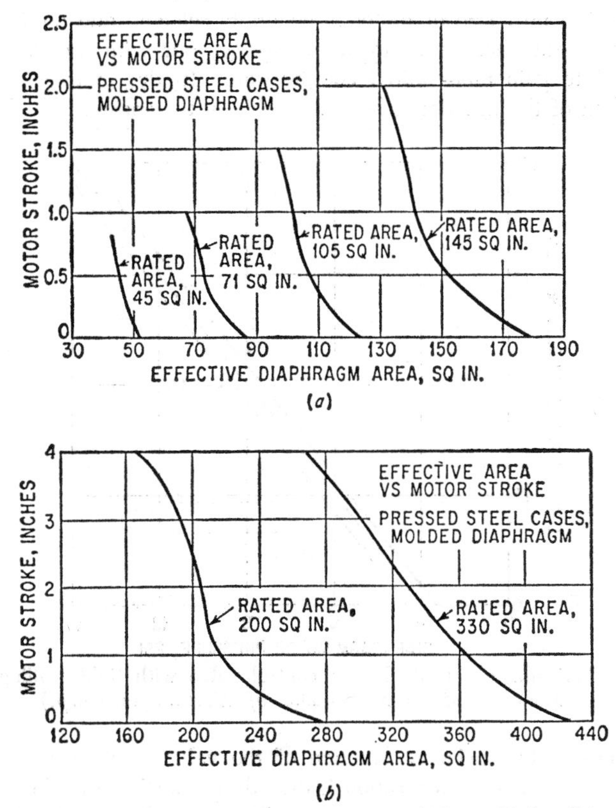

FIG. 7-35. Characteristics of diaphragm motors. (*Mason-Neilan Div. of Worthington Corp.*)

stroke is all that is required in selecting a motor for a given valve application.

If the net thrust at either end of the stroke is known, a particular motor can be selected for use with single-seated valves, either air-to-close or air-to-open, to take full advantage of this available power. The thrust at any other point in the stroke can be estimated from the curves in Fig. 7-35.

High Initial Thrust. Many single-seated valves are specified to close on air failure. This means that the initial spring compression must provide the force for closing, since the valve would normally be installed with the flow tending to open. Initial spring compression is generally expressed in terms of pounds per square inch on the diaphragm. The standard 3- to 15-psi range for motor valves indicates a spring compression equivalent to 3 psi times the initial diaphragm effective area. Similarly, the standard 6- to 30-psi range indicates a spring compression equivalent to 6 psi times the initial effective area.

If a standard motor with a 6- to 30-psi range does not develop enough net thrust, the next size of motor is generally selected and equipped with the maximum-rated-stroke 6- to 30-psi spring, but with the difference in stroke used to gain more initial compression. In this manner, advantage can be taken of the additional force available in the spring.

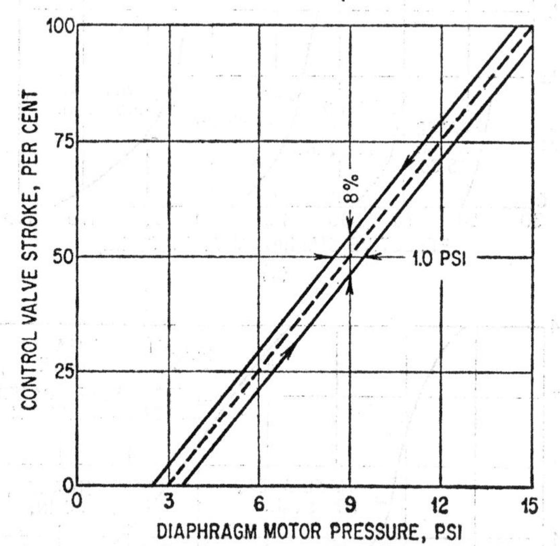

FIG. 7-36. Performance of diaphragm control valve with tight stuffing box, no positioner. (*Mason-Neilan Div. of Worthington Corp.*)

Positioners.[4] The primary function of the valve positioner is to ensure that the control-valve-stem position is directly proportional to the respective value of instrument output pressure regardless of stuffing-box friction, spring hysteresis or off-balance forces on the valve plug. When a positioner is applied to a valve, the valve is no longer operated directly by the instrument output pressure. The positioner utilizes the instrument output pressure only as a signal, and controls an independent pressure supply to apply whatever pressure is required to force the valve stem to the correct position. Thus, the positioner is a relay capable of applying maximum force to position the valve stem correctly.

In addition to their primary function, positioners provide a convenient means of changing the effective output pressure range of a controller and also of changing the valve action.

[4] Mason-Neilan Div. of Worthington Corp.

The positioner varies an independent source of constant pressure, to supply whatever pressure is required to the diaphragm of the control valve, to satisfy the relationship that valve-stem position be directly proportional to instrument output pressures. The general requirements for the air pilot are: (1) operation on small motion, (2) reasonable capacity (i.e., speed of operating control valve), (3) relatively low air consumption, and (4) simplicity. Some positioners utilize a second relay or booster pilot, but, in general, a single directly operated pilot is used in the most recent designs.

The accuracy of a positioner is best expressed as the per cent of total instrument output-pressure range required to change the positioner-pilot output a given amount. The actual accuracy of positioning depends, of

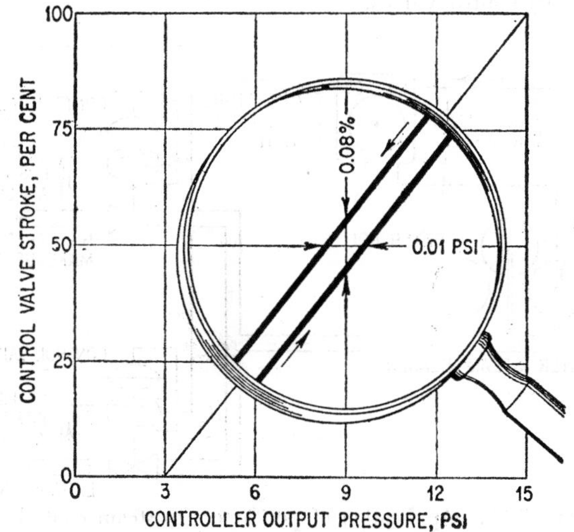

FIG. 7-37. Performance of valve in Fig. 7-36—after addition of a positioner. (*Mason-Neilan Div. of Worthington Corp.*)

course, on the forces which must be overcome in the valve, i.e., the change in positioner output pressure required.

Figure 7-36 shows a graph of valve-stem position versus diaphragm pressure for a valve having a very tight stuffing box. The graph is idealized by the assumption that the valve-stem position is directly proportional to the applied diaphragm pressure and that the "flat spot" or "dead zone" is caused entirely by a constant stuffing-box friction. Note that a 1-psi change in diaphragm pressure is required to reverse the valve-stem movement. For a standard valve (i.e., 3 to 15 psi) this would represent a flat spot of approximately 8 per cent of the total valve stroke, and on many control applications this would cause severe cycling.

Figure 7-37 is a graph of valve-stem position versus instrument output pressure for the valve just described after the addition of a positioner.

Since this positioner requires only an 0.08 per cent change in instrument pressure in order to change the pilot output 1 psi, the instrument pressure must now change only 0.01 psi (i.e., 0.08 per cent of 12 psi) to effect reversal of the valve, a reduction in flat spot to 1/100 of its previous value.

The primary function of the positioner is, as mentioned earlier, to position a control valve in the presence of appreciable stuffing-box friction, or where large unbalanced forces are present across the valve plug owing to high-pressure drop. The more difficult the application, and thus the wider the instrument proportional band required, the greater the necessity for a positioner for optimum control. Many temperature applications fall into this category.

Probably the next widest application for positioners is on "split ranging" of two (or three) control valves.

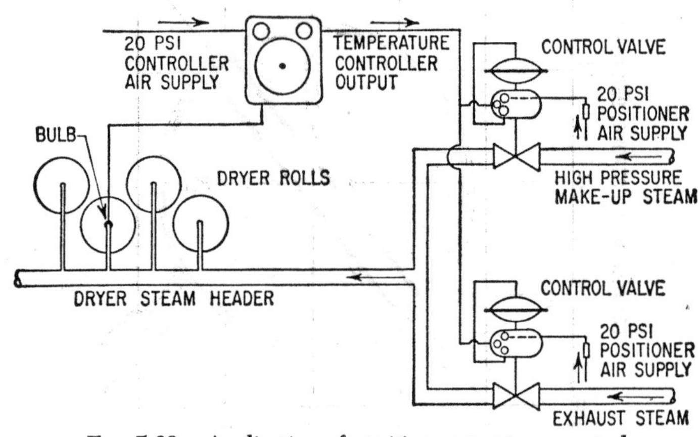

FIG. 7-38. Application of positioners to steam control.

As an example of double split-range operation, consider the paper-dryer-temperature application of Fig. 7-38. It is normally desired to maintain dryer temperature from exhaust steam (from a turbine), but necessary to admit additional high-pressure make-up steam in the event that available exhaust steam is insufficient to maintain dryer temperature. In this application, it is convenient to utilize the controller output to operate positioners on the two standard valves controlling exhaust and make-up steam inlet. The valves would be normally closed (i.e., air-to-open) and the controller would have reverse action (i.e., decreasing output pressure with increase in temperature), so that both valves would be normally closed in the event of air failure. By using double split-range positioners and adjusting the initial-positioner force-balance spring loading, it is possible to have the valve on exhaust-steam inlet operate from 3 to 9 psi and the valve on make-up-steam inlet operate from 8 to 14 psi. Thus, as temperature decreases, the controller output pressure gradually increases up to 8 psi, at which pressure the exhaust-steam inlet valve is nearly open. At this point, the make-up-steam inlet valve starts to open. Thus, dryer temperature is normally maintained by the exhaust steam, but make-up steam is cut in as the exhaust-steam valve approaches full-open position, and the controller continues to control dryer temperature by throttling the make-up-steam valve with the exhaust-steam valve in its open position.

By using a positioner having a nonstandard instrument range (say 6 to 60 psi) it is possible to operate a standard valve (3 to 15 psi) with an instrument having a nonstandard output (i.e., 6 to 60 psi).

In a similar manner, a positioner with nonstandard output can be used to operate a nonstandard control valve (say 5 to 25 psi required control-valve diaphragm pressure) with a standard (3 to 15 psi) controller.

By using a reverse-type pilot, it is possible to reverse the action of the valve. This necessity frequently arises on split-range applications to cover the requirements for valve position (i.e., open or shut) in the event of air failure.

When the control valve is a considerable distance from the controller (over approximately 125 ft), a positioner will speed up valve operation. With a positioner, the air pilot of the controller must change only the pressure in the positioner bellows instead of the pressure in the relatively large volume of the diaphragm motor.

It should be emphasized that, in all the above special applications, the positioner is still performing its primary function of positioning the valve precisely in accordance with the controller output pressure.

The force-balance positioner consists, essentially, of two opposing forces balanced one against the other to maintain a control-valve position depending upon the magnitude of the forces.

With reference to Fig. 7-39, when the positioner is at balance, the force exerted upward on the balance beam by the loading bellows is equaled by the force exerted downward by the positioning spring. The force exerted by the loading bellows depends on the control loading pressure to the bellows. The control loading pressure is that which has been established by the control system in accordance with demand for the controlled medium.

The force exerted by the positioning spring depends on the position of the valve stem (which, actually, is the position of the inner valve), and the shape of the positioning cam. The valve stem is connected to the positioner drive arm by means of the drive rod so that, for every position of the valve stem, the drive arm takes a corresponding position. As shown in Fig. 7-39, the drive arm is geared to the positioning cam. The cam moves the spring-beam assembly to give more or less tension to the positioning spring. The tension of the spring, then, depends on the position of the drive arm and, thereby, on the position of the inner valve. The positioning cam enters into determining the spring tension in that its shape may be whatever is required to give a desired characteristic of inner-valve position versus control loading pressure.

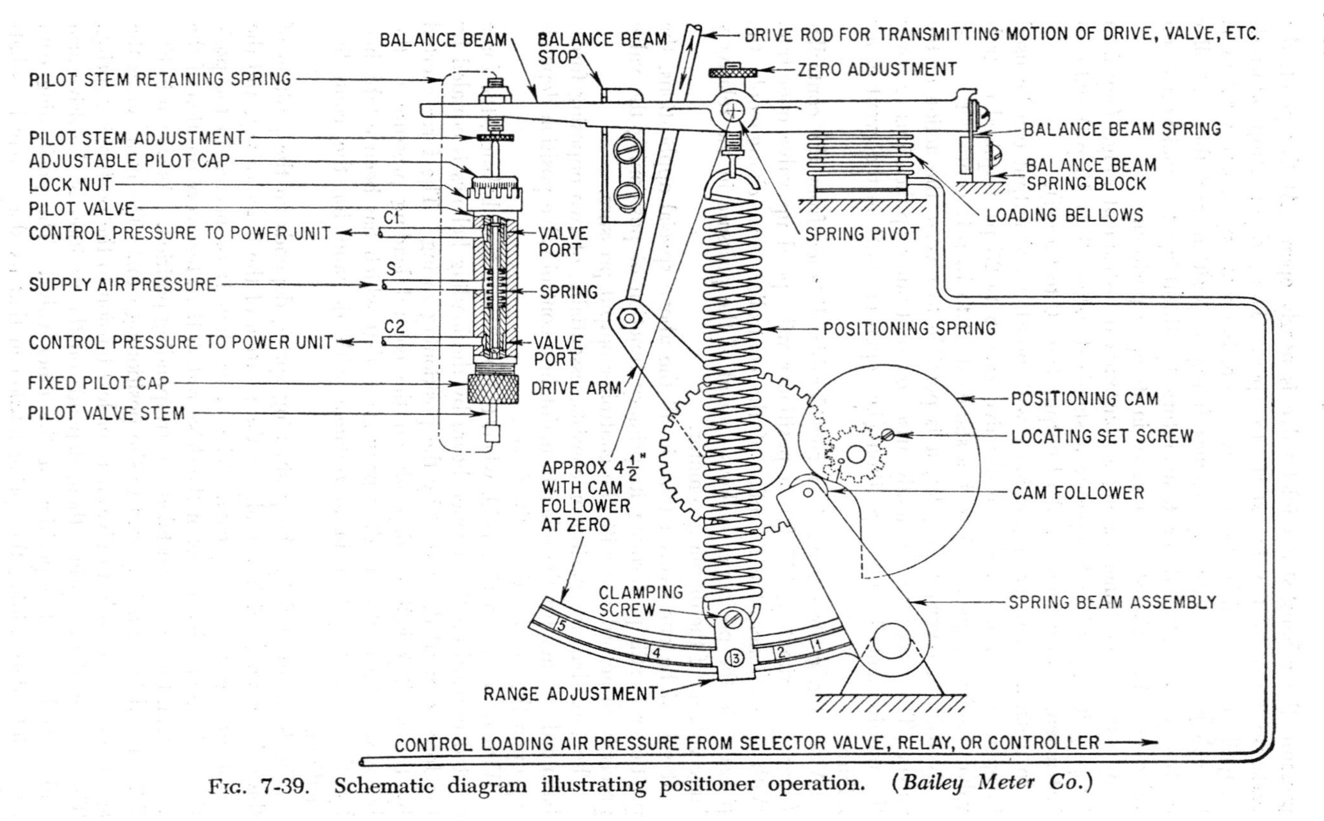

Fig. 7-39. Schematic diagram illustrating positioner operation. (*Bailey Meter Co.*)

With the forces exerted by the loading bellows and positioning spring at balance, the balance beam holds the pilot-valve stem in that position which will hold the position of the control valve constant.

When the control system indicates, by a change in control loading pressure, that the control valve should be repositioned further open or further closed, the force exerted on the balance beam by the loading bellows is increased or decreased. This will raise or lower the balance beam and pilot-valve stem to increase or decrease the control pressure to the diaphragm motor. As the control-valve stem is moved by the diaphragm motor, the positioning-spring tension is increased or decreased until the forces exerted by the bellows and spring are at balance again, the pilot-valve stem assumes a new position, and the motion of the control valve has stopped in its new position.

The difference among the applications shown in Fig. 7-40 with regard to assembly and operation of a typical force-balance positioner are as follows:

Figure 7-40*a*: Direct-action diaphragm motor, direct loading. An increase in control loading pressure to the loading bellows raises the pilot stem, increases the control pressure to the diaphragm motor, moves the control-valve stem into the valve body, lowers the positioner drive arm, and turns the positioning cam clockwise to increase the tension of the positioning spring to balance the forces on the balance beam.

Figure 7-40*b*: Reverse-action diaphragm motor, direct loading. An increase in control loading pressure to the loading bellows raises the pilot stem, increases the control loading pressure to the diaphragm motor, moves the control-valve stem out of the valve body, raises the positioner drive arm, and turns the positioning cam counterclockwise to increase the tension of the positioning spring to balance the forces on the balance beam.

Figure 7-40*c*: Direct-action diaphragm motor, reverse loading. An increase in control loading pressure to the loading bellows raises the pilot stem, decreases the control pressure to the diaphragm motor, moves the control-valve stem out of the valve body, raises the positioner drive arm, and turns the positioning cam counterclockwise to increase the tension of the positioning spring to balance the forces on the balance beam.

Figure 7-40*d*: Reverse-action diaphragm motor, reverse loading. An increase in control loading pressure to the loading bellows raises the pilot stem, decreases the control pressure to the diaphragm motor, moves the control-valve stem into the valve body, lowers the positioner drive arm, and turns the positioning cam clockwise to increase the tension of the positioning spring to balance the forces on the balance beam.

Power-cylinder Operators. For the positioning of dampers or other final control elements which require more power and a longer stroke than are available with diaphragm motors, the *power cylinder* is used. This unit utilizes air power in both directions of travel and, for proportional control, incorporates a positioner similar in function to the valve positioner to ensure that the piston will take a definite position for each intermediate value of controlled air pressure.

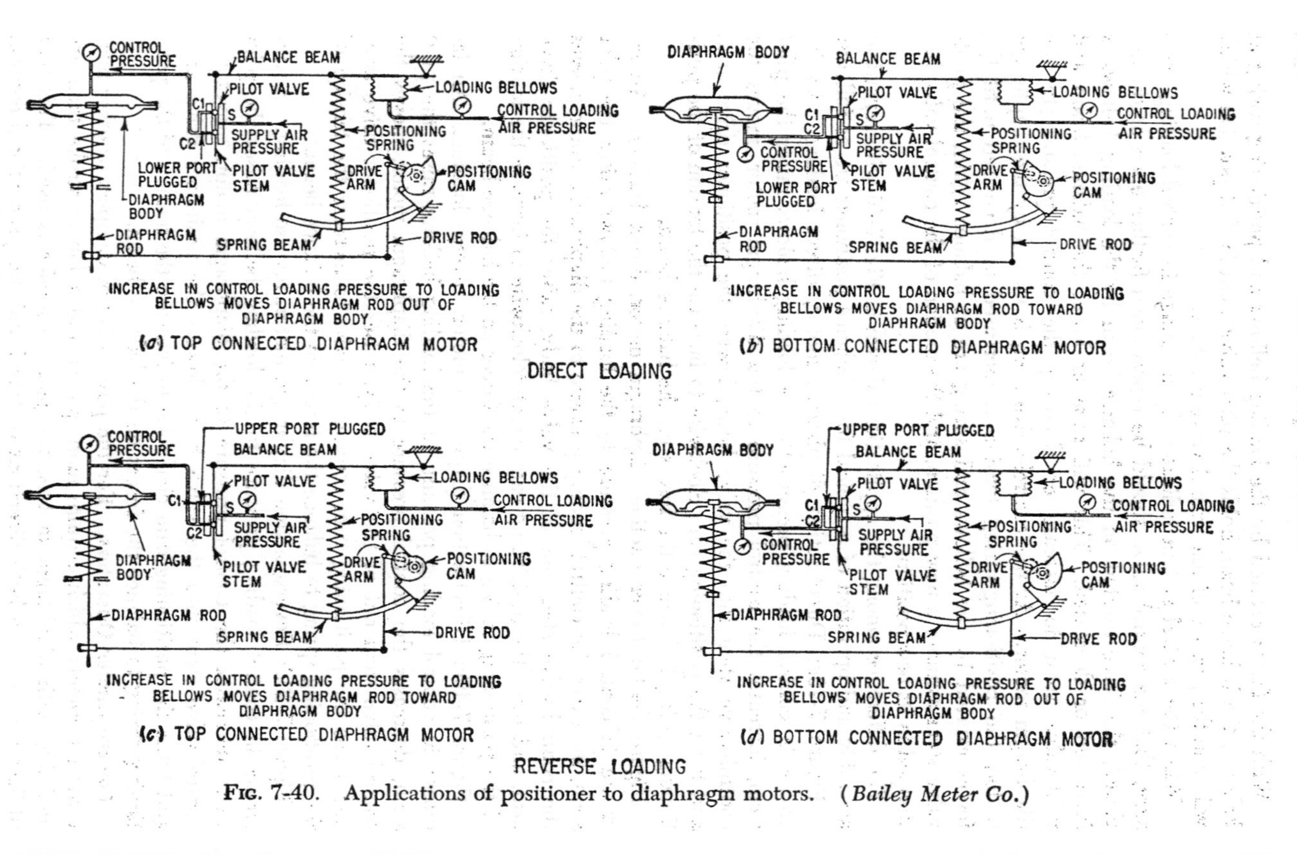

FIG. 7-40. Applications of positioner to diaphragm motors. (*Bailey Meter Co.*)

For two-position control, no positioner is necessary, but, since air pressure is required for both directions of travel, a valve is necessary to divert full air pressure from one end of the cylinder to the other and to bleed air from the opposite end. The cylinder can be operated by either an electric or a pneumatic two-position controller by the use of a suitable electric-pneumatic or pneumatic relay which provides a four-way supply and exhaust action.

Operation of a typical pneumatic design is as follows: Referring to Fig. 7-41, the power-unit portion of the control drive is a double-acting air

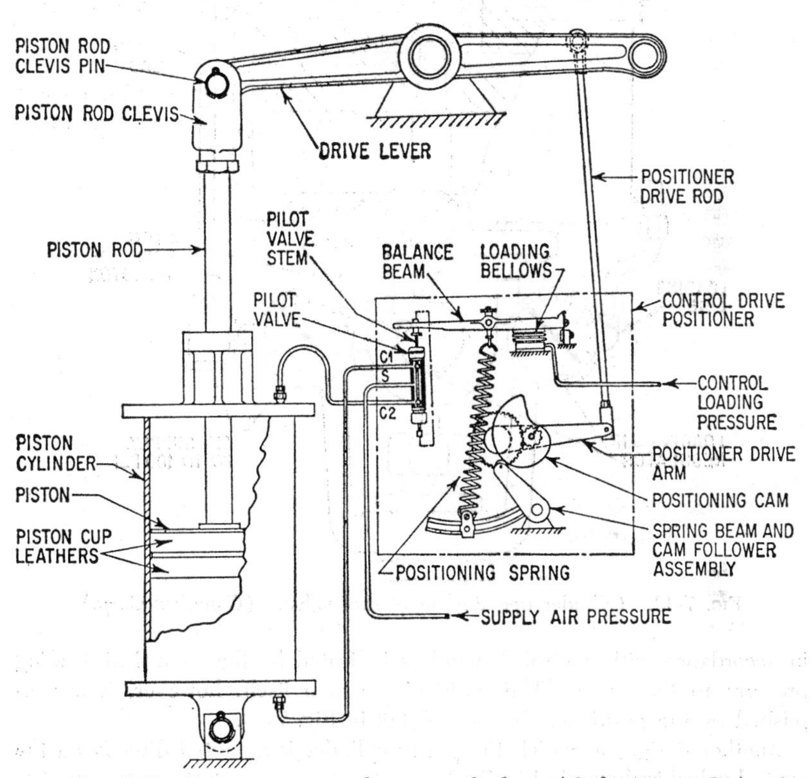

FIG. 7-41. Positioner-operated power cylinder. (*Bailey Meter Co.*)

cylinder and piston assembly. The piston is moved in the cylinder in accordance with the differential in air pressures on each side of the piston. That is, if the air pressure under the piston is greater than that on the other side, the piston will move up (the piston rod will move out of the cylinder). Opposite action of the piston will take place if the greater pressure is on top of the piston; and no motion will take place if the pressures on both sides of the piston are equal.

The control-drive piston is positioned, as required for control regulation,

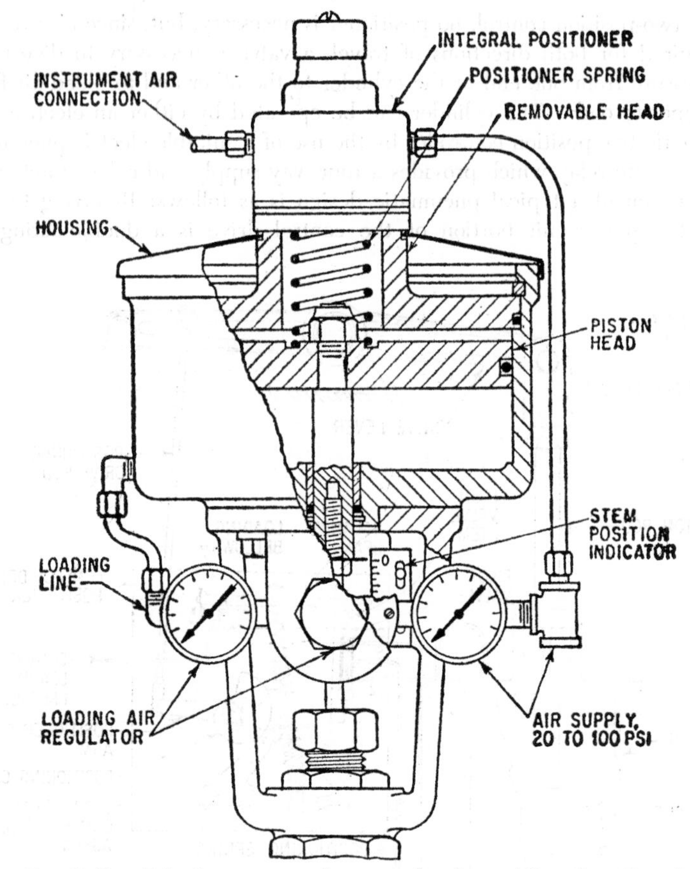

FIG. 7-42. Cylinder operator for control valve. (*Conoflow Corp.*)

in accordance with control demand as indicated by the control air-loading pressure to the drive. The positioning action itself, however, is accomplished by the positioner shown in detail in Fig. 7-39.

Another design, in which the power cylinder is mounted directly on the control valve, is shown in Fig. 7-42.

ELECTRIC ACTUATORS[5]

Electric actuators are electrically powered devices which are operated by some form of electric-control system and are used to position control elements. They are most frequently used to position control valves of

[5] Abstracted from R. E. Fishburn, Minneapolis-Honeywell Regulator Co., "Electric Actuators," in D. M. Considine (ed.), "Process Instruments and Controls Handbook," McGraw-Hill Book Company, Inc., New York, 1957.

various types. However, they also are used to position numerous other devices such as dampers, burners, rheostats, or adjustable autotransformers for the regulation of motor speeds or electric-heating loads, mechanical adjustable-speed power transmissions, adjustable-feed and proportioning pump strokes, and dry feeders. The electric actuator and the control valves are frequently supplied by the manufacturer as an integral assembly.

Electric actuators may be basically classified as (1) electric-solenoid-operated types and (2) electric-motor-operated types. Solenoid-operated actuators provide only two-position operation. Motor-operated actuators may provide two-position, multiposition, floating, or proportioning operation, depending on their construction and the control system with which they are used. Automatic-reset and rate-time responses when used are usually accomplished by the associated control system which controls the positioning of the actuator.

All electric actuators present an inductive load to the electrical circuits which control them. It is important that this characteristic be taken into consideration in evaluating the current-carrying capacities of control-system switches and relays which will carry the actuator operating currents, if normal contact life is to be obtained. Large actuators, or their control systems, are frequently provided with discharge resistors or capacitors shunted across the load contacts, to absorb the inductive kick when the circuit is opened.

Solenoid-operated actuators will move to their deenergized position on electric-power failure and may, therefore, be applied in such a manner as to provide fail-safe operation in the event of power failure or burning out of the actuator. Most motor-operated actuators, however, in the event of power failure, will remain in the position they were immediately preceding the failure. A few actuators are provided with an integral or an external spring, which drives the actuator to one extreme of its travel in the event of power failure, thus permitting the actuator to be applied in such a manner as to provide fail-safe operation. A disadvantage of this arrangement is that a considerable percentage of the actuator's force must be used to compress the spring. Many control systems which use a nonspring-return electric-motor-operated actuator achieve fail-safe operation by also using a solenoid-operated actuator to provide limit protection.

Many electric-motor-operated actuators can be supplied with a means of manual operation, such as a handwheel, which may be operated manually in the event of a power failure.

An electric-solenoid-operated actuator consists basically of an electromagnet and a movable armature, which is linked to the device to be positioned. It provides two-position operation only and is not adaptable to proportional positioning. Armature travel or stroke of most available designs is made relatively short, to permit obtaining the maximum amount of force for a given solenoid size and current flow.

Electric-solenoid-operated actuators are generally available for operation on 24, 115, or 230 volts at 50- or 60-cycle single-phase a-c and on 6, 24, or 115 volts d-c.

Figure 7-43 shows current flow and force (pull) curves for a typical a-c electric-solenoid-operated actuator. It will be noted that the total pull of the armature is the sum of the solenoid pull and the air-gap pull. Variations in design and construction of the armature and the iron-clad solenoid enclosure modify these curves considerably. The iron armature and solenoid enclosure for a-c solenoid actuators are of laminated construction. Shading coils may be incorporated in the construction to increase the pull of the armature and reduce the humming.

Characteristics of d-c solenoid actuators are similar, except that current flow is greatest after the armature has pulled in. Figure 7-44 shows the

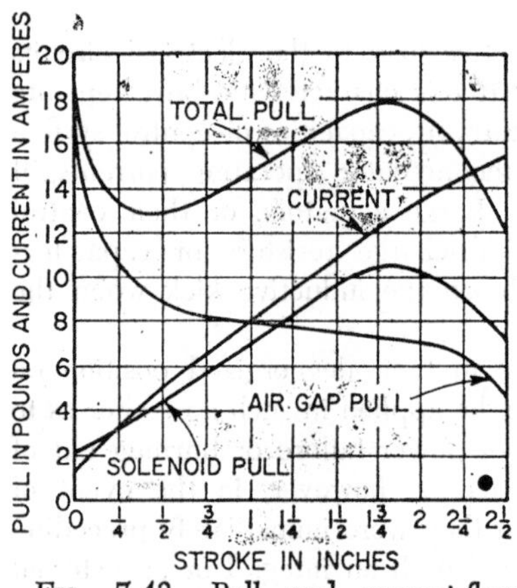

FIG. 7-43. Pull and current-flow curves for a typical a-c electric-solenoid-operated actuator.

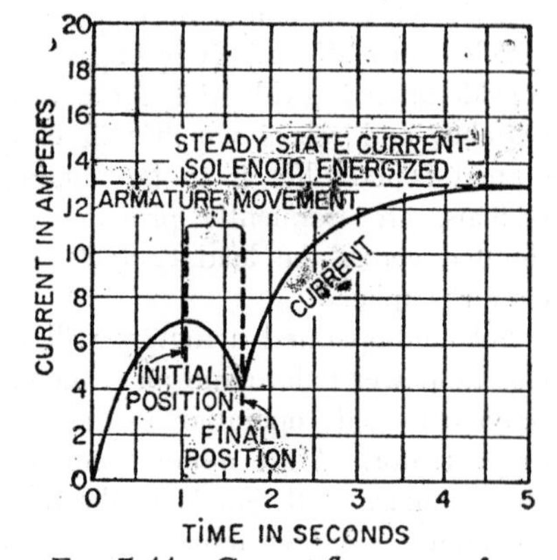

FIG. 7-44. Current-flow curve for a typical d-c electric-solenoid-operated actuator.

current flow of a typical d-c solenoid actuator. Inductance of the solenoid limits the current flow when the actuator is initially energized. As the current increases, the magnetic force becomes sufficient to start the armature to move. As the armature moves toward its final position, the counter-electromotive force resulting from its movements results in a decrease in current flow. After the armature reaches its final position, the current again increases to a steady-state value which is determined by the ohmic resistance of the actuator and the supply voltage.

In order to obtain the maximum force from a given d-c electric-solenoid actuator and still prevent excessive heating when it is energized over extended periods of time, switching mechanisms are frequently incorporated in the design to insert a resistor in series with the solenoid or to place additional windings of the solenoid in the circuit when the armature reaches its energized position, so as to reduce the continuous current flow to a safe value.

Although electric-solenoid-operated actuators are most frequently used to operate so-called magnetic or solenoid valves and are usually built integral with the valve body, they are also used to position other mechanisms and devices where relatively short travels and two-position operation only are required. For this type of service they are generally less expensive than motor-operated actuators.

Electric-solenoid-operated actuators are generally available in various sizes which will provide from a few ounces to approximately 50 lb of pull.

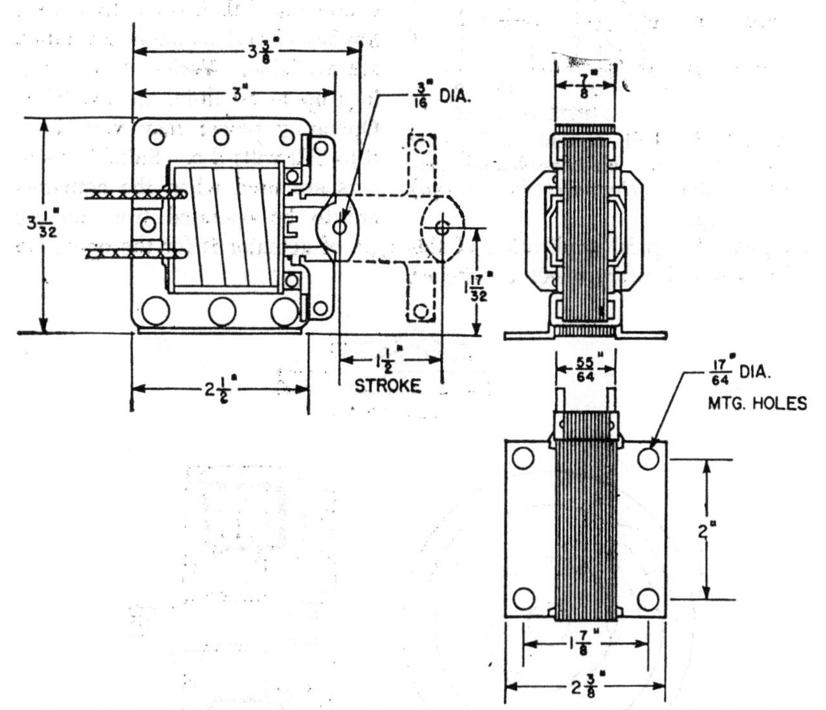

Fig. 7-45. Typical a-c electric-solenoid actuator. (*Dormeyer Sales Corp.*)

Because heating is considerably greater when solenoid-operated actuators are energized continuously than when they are operated intermittently or for only short periods of time, they are usually rated for either continuous duty or for a specified per cent duty. This rating should be taken into consideration in using electric-solenoid-operated actuators. Figure 7-45 shows a typical a-c electric-solenoid-operated actuator. Figure 7-46 shows performance curves for this actuator.

Rotary Electric-solenoid-operated Actuators. Rotary electric-solenoid-operated actuators convert straight pull to a rotary motion by utilizing the mechanical principle of the inclined plane. Figure 7-47 shows an actuator

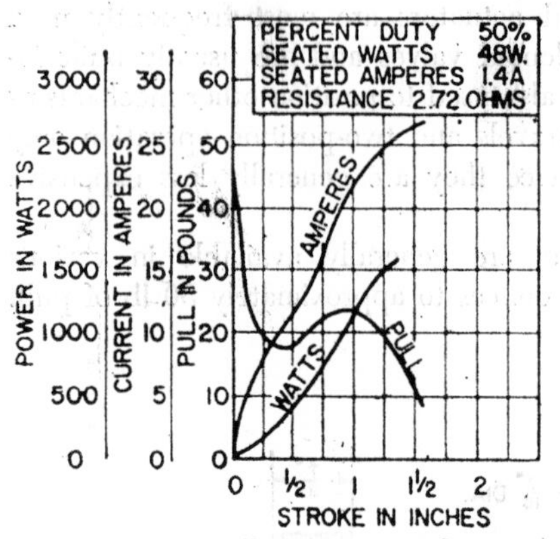

Fig. 7-46. Pull, current, and power curves of an a-c electric-solenoid-operated actuator. (*Dormeyer Sales Corp.*)

of this type. The electromagnetic pull attracts the armature. Since the armature is supported by three ball bearings that travel around and down in three inclined ball races or grooves, the armature is, therefore, forced to rotate by a cam action. Rotation continues until the balls have traveled to the deep ends of their respective races. Strokes of various angular rotation are available. Various torque ratings up to 50 lb-in. are available. Operating power may vary from 6 to 550 volts d-c. Suitable rectifiers are used when the actuators are to be operated from an a-c supply. A typical application of this type of actuator is for the operation of valves and rotary selector switches.

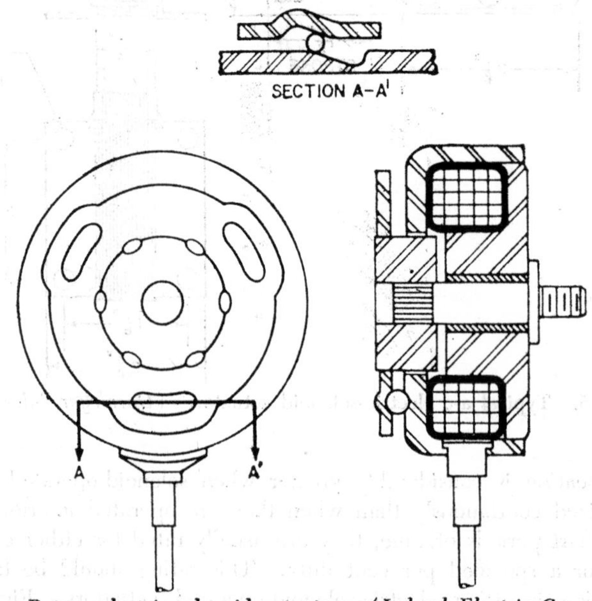

Fig. 7-47. Rotary electric-solenoid actuator. (*Leland Electric Co. of American Machine Foundry Co.*)

Electric-solenoid-operated Valves. An electric-solenoid-operated valve, frequently referred to as a magnetic valve, consists of an integral assembly of a solenoid-operated actuator and a valve body. These valves are available with practically all types of conventional valve bodies and construction and trim materials. However, body designs which use relatively short lifts or stem travels are usually used because a solenoid actuator of a given

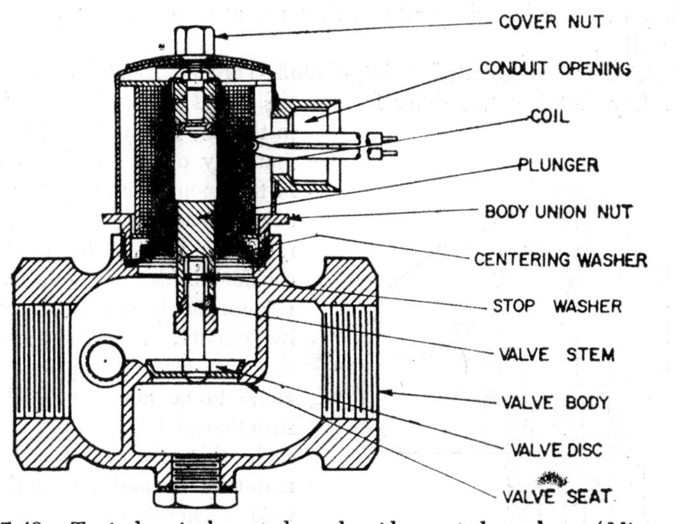

FIG. 7-48. Typical single-seated solenoid-operated valve. (*Minneapolis-Honeywell Regulator Co.*)

size can produce a greater force if designed for a short armature travel. Solenoid valves provide basically only two-position operation. They are available in both normally closed and normally open constructions. Figure 7-48 shows a typical single-seated solenoid-operated valve.

These valves are generally available for operation on 24, 115, 230, or 440 volts a-c at 25, 50, or 60 cycles as well as on 115 or 230 volts d-c. Because the operating force which is obtainable from a solenoid actuator of practical size is limited, there is a limit to the pressure drop across the valve against which proper valve operation can be obtained. Operating pressure ratings of specific valves should be observed in applying solenoid-operated valves. Figure 7-49 shows a manual-opening device which may be provided with this type of valve.

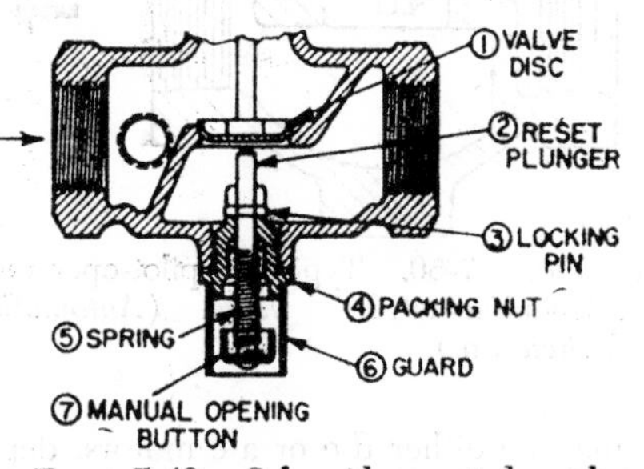
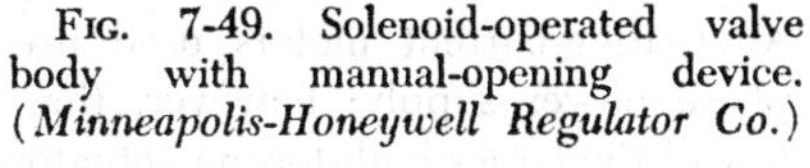

FIG. 7-49. Solenoid-operated valve body with manual-opening device. (*Minneapolis-Honeywell Regulator Co.*)

Where the required operating pressures cannot be handled by single-seated body constructions, double-seated or pilot-operated constructions are available. Figure 7-50 shows a typical pilot-operated electric-solenoid-actuated valve. Two-, three-, and four-way valve-body constructions are also available. Most of these are of small pipe sizes up to ¾ in., and are for operation on pressures up to approximately 250 psi.

Many of the solenoid-operated valves are available with explosion-proof construction.

Electric-solenoid-operated Safety-shutoff Valves. Electric-solenoid-operated safety-shutoff valves differ from conventional solenoid-operated valves in that the safety-shutoff valves are manually opened and are latched in the open position by the electric solenoid as long as the solenoid remains energized. When the solenoid is deenergized, the valve is immediately and positively closed by gravity. Safety requirements do not consider spring closure alone to be adequate. The construction of these valves is usually such that the valve cannot be manually opened unless the solenoid is energized. The usual application of safety-shutoff valves is in fuel lines to burners or in flow lines handling hazardous liquids or gases. They are generally operated by safety controls such as temperature- or pressure-limit devices or flame-failure protective equipment.

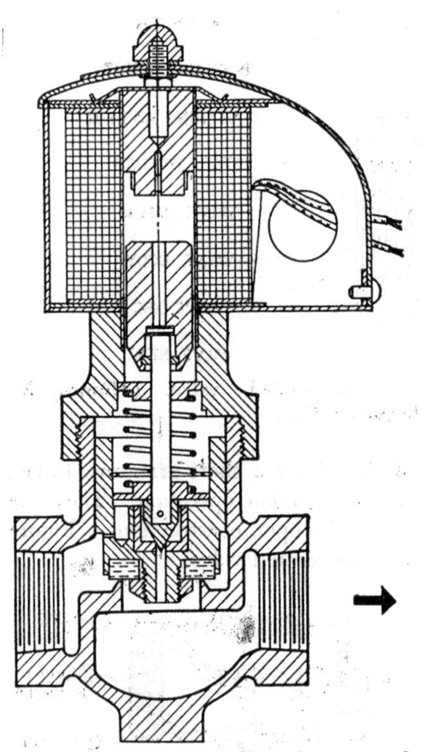

FIG. 7-50. Typical pilot-operated solenoid-actuated valve. (*Automatic Switch Co.*)

An electric-motor-operated actuator consists basically of a suitable driving motor, a reduction gearing, and an output shaft or a lever arm incorporated in a common housing. Driving motors are of various sizes, depending on the power-output requirements of the actuator. They may be either d-c or a-c motors, depending on the available power supply. Alternating-current motors used usually are for operation from a single-phase power supply; however, three-phase driving motors are used on a few of the larger high-torque actuators.

The driving motors of most actuators are of the reversible type. Single-phase a-c driving motors may be of the induction or the synchronous type. A synchronous motor produces a constant output speed up to its stall torque, whereas the output speed from an induction motor decreases as the load increases. With properly designed induction motors, the small decrease in output speed which occurs with increasing load is generally not detrimental. Synchronous motors with permanent magnetic fields have an inherent magnetic braking action when they are deenergized. Induction-motor-operated actuators are frequently provided with some form of magnetic or electric brake to prevent coasting. Induction motors are generally less costly and produce greater torques than synchronous motors of comparable size. Induction motors are the type most frequently used in electric actuators.

Types of Motor-operated Actuators. There are four general types of electric-motor-operated actuators: two-position types, multiposition types,

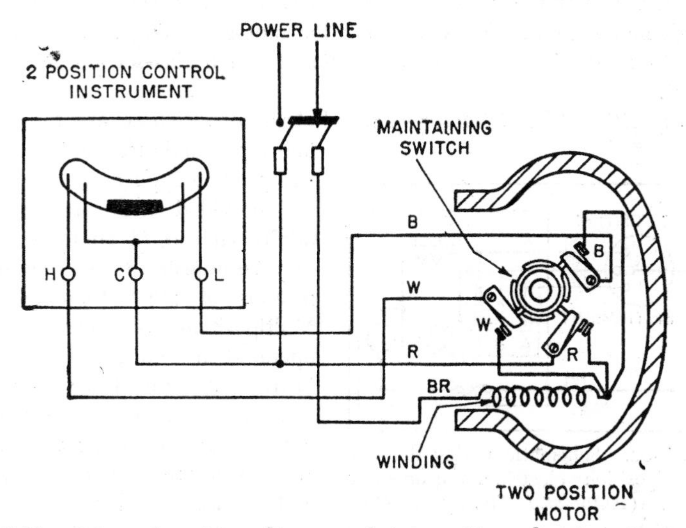

FIG. 7-51. Schematic wiring diagram of two-position electric-motor-operated actuator. (*Minneapolis-Honeywell Regulator Co.*)

reversing-floating types, and proportioning types. Selection of the proper type depends on the manner in which the final control element, operated by the actuator, is required to function, and on the control system with which it is used.

Two-position Electric-motor-operated Actuators. These may be of either the unidirectional or the reversing type. In either case they travel toward and stop only at either of the extremes of travel. This action is accomplished by holding and limit contacts which are a part of the motor circuit.

Figure 7-51 shows a schematic wiring diagram of a typical two-position electric-motor-operated actuator and its associated control instrument. The motor is the unidirectional type. The control instrument has a single-pole double-throw switching action with a dead neutral, usually adjustable in width.

When the instrument makes contact between *C* and *L*, current flows from C through L, to switch finger contact B, and from B to the motor winding. The motor being thus energized rotates in a clockwise direction. When the motor shaft has rotated a few degrees, the holding contact R is closed, after which contact B is opened, and the motor winding remains energized through contact R, even though the controller contact between C and L may be opened. The motor continues to run until it has rotated one-half revolution, at which time contact R opens and contact W closes. At this position, rotation of the motor stops, regardless of whether the instrument contact between C and L is open or closed. Subsequent closing of the instrument contact between C and L will not move the motor from this position.

When the instrument contact between C and H is closed, current flows through motor contact W, energizing the motor winding and causing the motor to rotate in a clockwise direction through the second half of its rotation.

Multiposition Electric-motor-operated Actuators. These are the reversing type, and are similar to the two-position type except that they are provided with one or more additional sets of contacts, which may be adjusted to operate at any position of the actuator's travel. Multiposition actuators must be operated by multicontact control instruments or control systems.

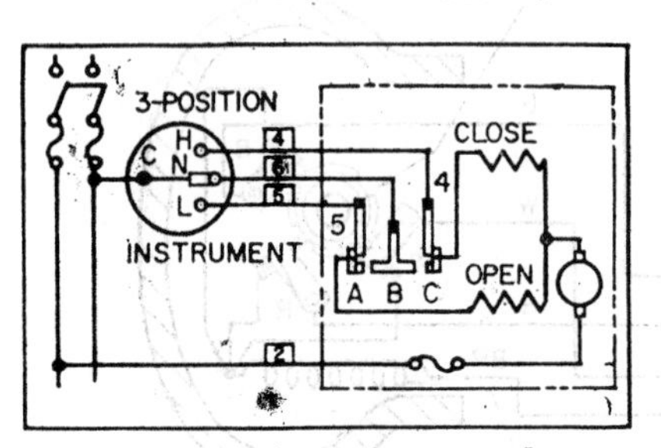

FIG. 7-52. Schematic wiring diagram of multiposition electric-motor-operated actuator. (*Automatic Timing & Controls, Inc.*)

Figure 7-52 shows a schematic wiring diagram of a multiposition electric-motor-operated actuator and its associated control instrument. The actuator is shown in a mid-position at which the contact segment B has been set to operate.

If the instrument makes contact from C to L, current flows through the closed contact between 5 and A, energizing the open winding of the driving motor. After the actuator has moved a few degrees in the open direction, contact segment B makes contact with contact segments C and 4. The actuator continues to move until it reaches the "open" extreme of its travel, at which time the contact between 5 and A is opened and the driving motor is deenergized.

If the instrument now makes contact between C and N, current flows through the contact between B and C, energizing the close winding of the driving motor. The actuator is driven toward its closed position until it reaches the mid-position at which contact B has been set. At this position the contact between B and C is opened and the driving motor is deenergized. If the instrument makes contact between C and H, the close winding is energized, and the actuator moves to the close extreme of its travel.

Floating-control Electric-motor-operated Actuators. These are reversing type and use a reversing driving motor. Limit switches are usually provided at each extreme of travel. However, the limit switches may be omitted where the actuator is required to make more than one complete revolution, if limit switches or other suitable means of protection are provided on the device being positioned by the actuator.

Figure 7-53 shows a schematic wiring diagram of a floating-control electric-motor-operated actuator and its associated control instrument. The control instrument has a single-pole double-throw switching action with a dead neutral, usually adjustable in width. This type of control usually requires a relatively slow-speed actuator. Occasionally an interrupter or percentage timer may be interposed in the common lead to the motor to obtain even slower speeds of operation.

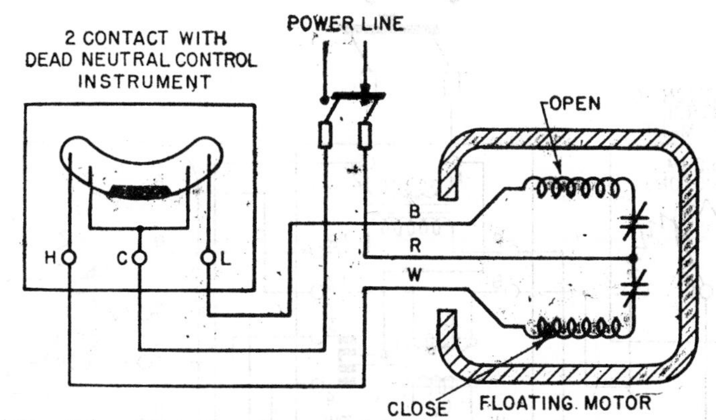

FIG. 7-53. Schematic wiring diagram of floating-control electric-motor-operated actuator. (*Minneapolis-Honeywell Regulator Co.*)

If the instrument makes contact from *C* to *L*, the open winding of the motor is energized, and the actuator will continue to move toward its open position until the result of its control action or other influences cause the contact between *C* and *L* to open, at which time the motor is deenergized and the actuator will remain at that position until contact is again made from *C* to *L* or from *C* to *H*.

If the instrument makes contact from *C* to *H*, a similar action will occur, except that the actuator will move toward the close position.

If the instrument continues to maintain contact from *C* to *L* or from *C* to *H* when the actuator reaches either of its extremes of travel, one or the other of the limit switches will be open and deenergize the motor winding.

Proportioning-control Electric-motor-operated Actuators. These are of the reversing type and use a reversing driving motor. They are provided with limit switches and a motor slidewire or control resistor. The position of the slider of the motor slidewire is related to the position of the actuator and moves from one end of the slidewire to the other while the actuator moves from one extreme of its travel to the other.

Figure 7-54 shows a schematic wiring diagram of a proportioning-control electric-motor-operated actuator and its associated control instrument and balancing relay. The control instrument is provided with a proportioning-control resistor. The position of its slider is related to the position of the pen or pointer of the instrument with respect to the position of the set-point index. When the pen or pointer position coincides with the index position, the slider is at the mid-point of the instrument-control resistor. If the pen or pointer is displaced down scale from the index, the slider will be positioned closer to the low end of the resistor by an

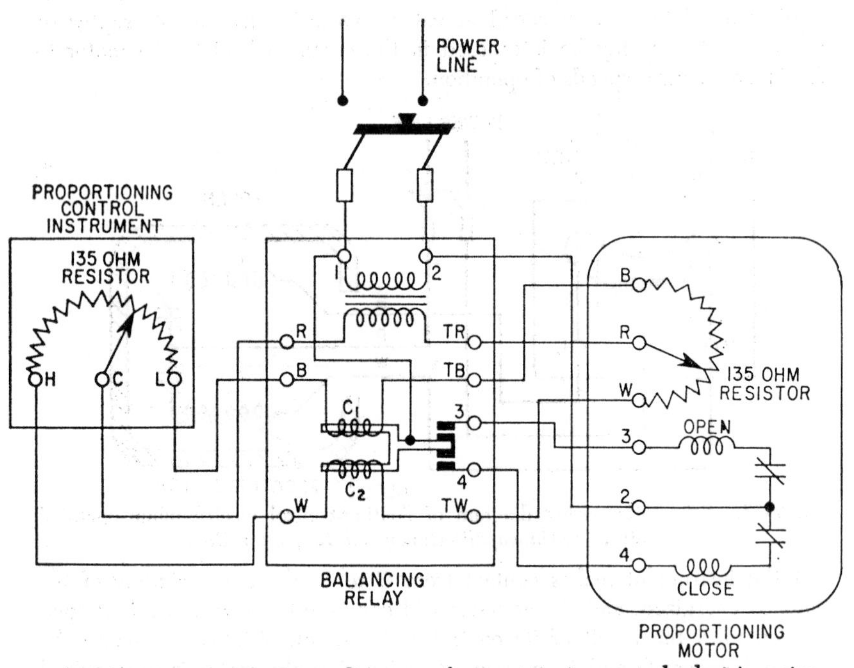

FIG. 7-54. Schematic wiring diagram of proportioning-control electric-motor-operated actuator. (*Minneapolis-Honeywell Regulator Co.*)

amount depending on the proportional-band relationship or setting of the instrument-control mechanism. The balancing relay illustrated is of the electromagnetic type. More elaborate electronic balancing relays are frequently used to provide greater sensitivity and flexibility. Referring to Fig. 7-54, it will be noted that the circuit is essentially a resistance-bridge circuit, with the instrument resistor and the actuator resistor comprising the resistance arms of the bridge, and the balancing relay being the unbalance detector. The system as shown is at a condition of balance. If, because of movement of the instrument pen and pointer or a change of index position, the slider of the instrument-control resistor is moved to a new position closer to the "high" end of the instrument-control resistor,

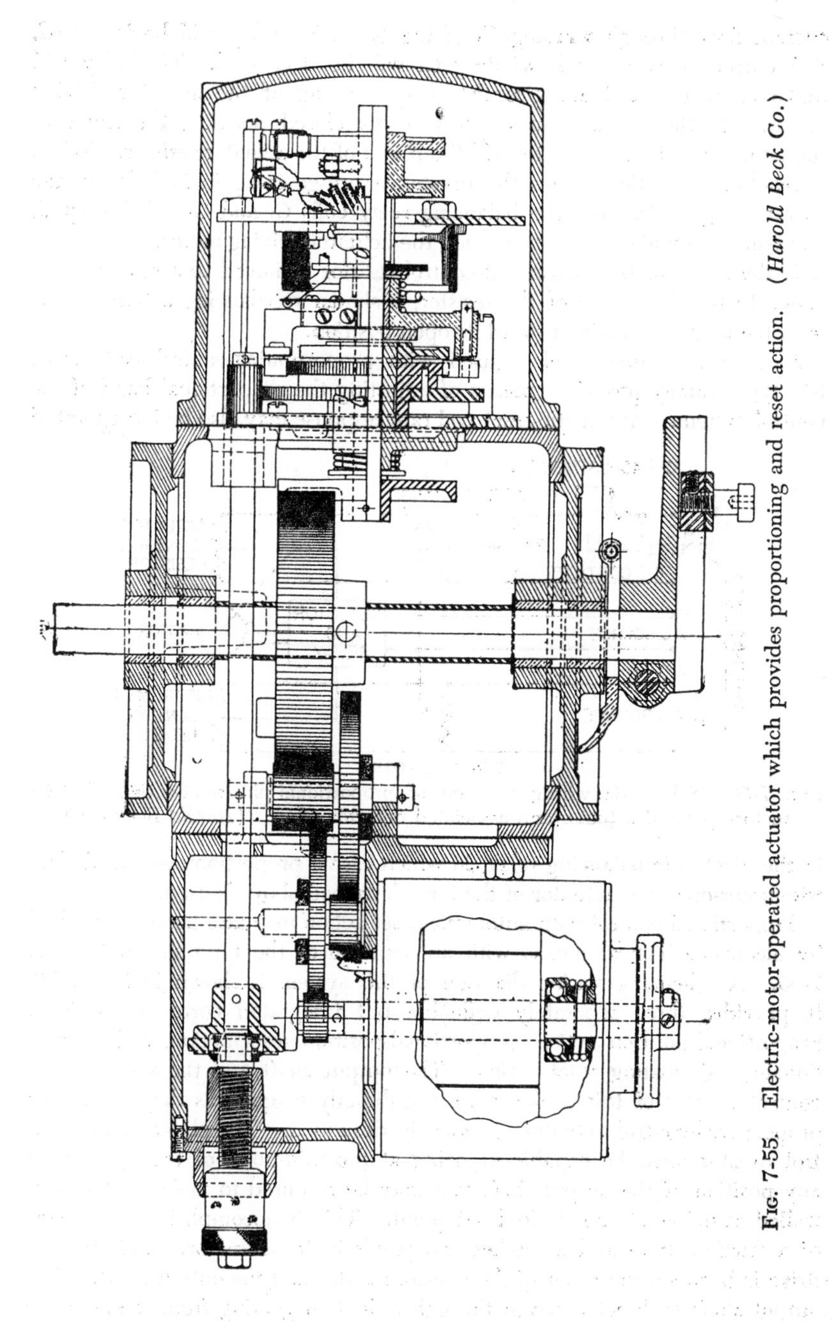

FIG. 7-55. Electric-motor-operated actuator which provides proportioning and reset action. (*Harold Beck Co.*)

current flow through winding C_2 of the balancing relay will be increased, and current flow through winding C_1 will be decreased. The relay will make contact 4 and energize the close winding of the actuator driving motor. As the actuator moves toward its closed position, the slider of the actuator resistor moves toward the *B* end of the actuator resistor. When it reaches a position where the resistance bridge is again in balance, the current flowing through the balancing relay coils C_1 and C_2 will be equal, and contact 4 will open, deenergizing the actuator driving motor.

If the slider of the instrument-control resistor is moved to a new position closer to the "low" end of the resistor, a similar rebalancing action occurs, with the actuator moving toward its open position.

Electronic balancing relays used with proportioning-control systems of this type usually provide a means of adjusting the proportional band of the control system. Automatic-reset and rate response may also be incorporated

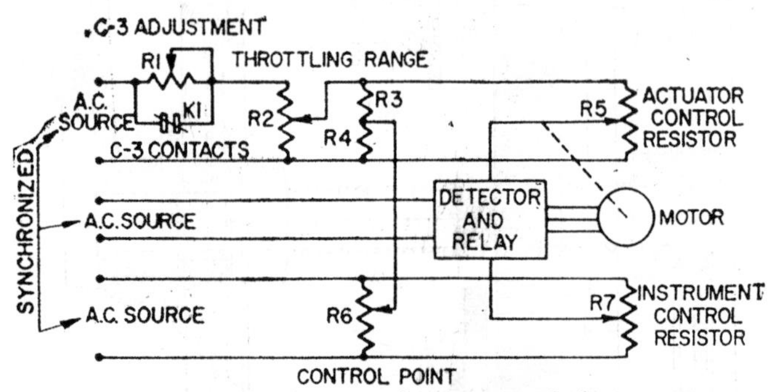

FIG. 7-56. Schematic wiring diagram of proportional-plus-reset-action electric-motor-operated actuator and associated detector relay. (*Harold Beck Co.*)

in the electronic balancing relay, in which case a proportioning-control electric-motor-operated actuator of the type described above is used.

Proportional control with automatic-reset action may also be accomplished by electromechanical means with an actuator of the type shown by Fig. 7-55. A schematic wiring diagram of this system is shown in Fig. 7-56. It provides three separately definable but integrated corrective actions: proportional-position action, proportional-position reset action, and proportional-speed floating reset action. The output shaft and the slider of the control resistor of this actuator are not directly coupled as they are in the proportional-control actuators previously described. The slider of the control resistor must be capable of being at the mid-point of the resistor for any position of the output shaft that may be required to maintain the controlled variable at the desired set point. This is accomplished by means of a friction drive and a ratchet and pawls in the actuator. This friction drive is between the slider of the control resistor and the output shaft. The output shaft is directly driven through reduction gearing from the actuator operating motor.

The friction drive and ratchet and pawls assembly functions in such a manner as to move the output shaft and the slider of the control resistor away from the mid-position of the control resistor at the same rate of speed. However, when they return toward the mid-position of the control resistor, the slider is moved at a faster rate than the output shaft, by an amount determined by the setting of the droop corrector. Thus, when the slider of the actuator-control resistor has returned to its mid-position, the actuator output shaft will have returned to a new position, somewhat displaced from its last position. This displacement is proportional to the magnitude of the initial deviation from the mid-position, and is in the direction which will tend to move the variable being controlled toward its set point. This action provides proportional-position reset.

The proportional-speed floating reset action acts as a vernier adjustment on the proportional-position reset action. It is accomplished by the action of contact C_3, shown in Fig. 7-56, which is made periodically, usually for one second of every minute. Its action effectively narrows the proportional band for the period of closure. If both the upper and lower bridges are independently balanced, as they will be if the instrument-control resistor is at its mid-position, there will be no change in signal to the electronic detector, and no movement of the actuator when contact C_3 is made. The narrowing of the proportional band changes the signal to the electronic detector and causes the actuator to move in the direction which will cause the variable being controlled to move toward its set point. When contact C_3 opens, the slider of the actuator-control resistor is moved toward its mid-position at a faster rate than that of the actuator output shaft. Thus, when the slider of the actuator-control resistor has returned to its mid-position, the actuator output shaft will have returned to a new position, somewhat displaced from its last one, and in the direction which will tend to move the variable being controlled to its set point.

Operation of this system is such as to maintain the slider of the actuator-control resistor at its mid-position, regardless of the position of the output shaft.

Ratings of Electric-motor-operated Actuators. Electric-motor-operated actuators are generally available for operation on 24, 115, or 230 volts at 50- or 60-cycle single-phase a-c. Some are also available for operation on 6 or 440 volts at 50- or 60-cycle single-phase a-c; 25-, 40-, or 400-cycle a-c; and 6, 24, or 115 volts d-c. A few high-torque electric-motor-operated actuators are supplied for operation on 230 or 440 volts at 50- or 60-cycle, two- or three-phase a-c.

Except for a few high-torque actuators which are operated by motors from ⅙ to ¾ hp, operating-power requirements vary from approximately 10 to 100 watts for most electric-motor-operated actuators.

Torque ratings of electric-motor-operated actuators are usually expressed as rated load torques. These will vary from approximately 20 to 600 lb-in., depending on the size of the actuator and its speed and angular rotation. Stall torques are usually several times the rated load torque. In evaluating torque ratings, actuator speed and angular rotation should always be considered. High-torque actuators will provide torques from approximately 100 to 1,500 lb-ft.

Speeds of electric-motor-operated actuators are usually expressed in seconds for full travel. Actuators are available in a variety of speeds from approximately 2 to 240 sec for full travel.

Angular rotation of electric-motor-operated actuators differs with different actuator designs. Rotations of 100° and 180° are, however, the most common.

The electrical resistance of the actuator-control resistor of most proportioning-control electric-motor-operated actuators, which is of considerable importance in their performance in control systems, is the accuracy with

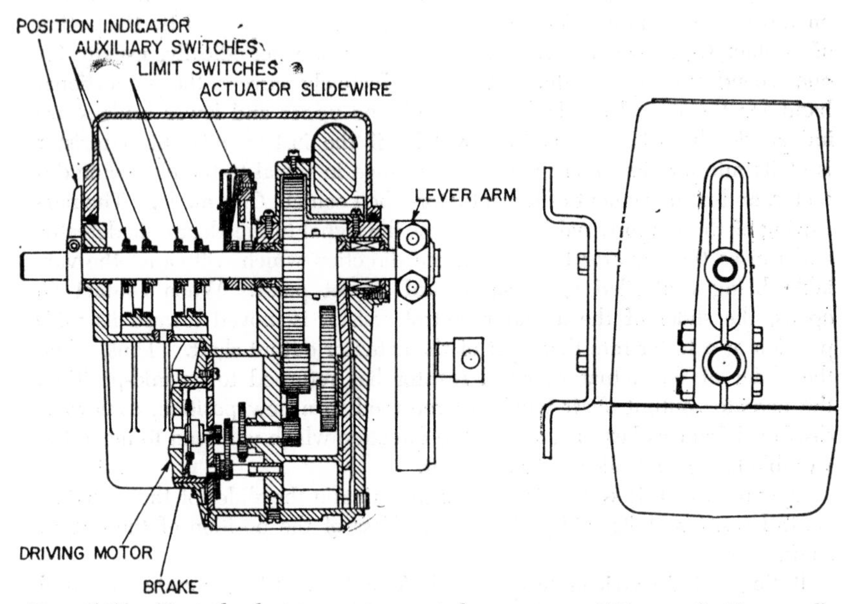

FIG. 7-57. Typical electric-motor-operated actuator. (*Minneapolis-Honeywell Regulator Co.*)

which they are capable of positioning a control element. In this connection, overtravel or coast after the motor has been deenergized is an important factor. Some form of mechanical or electrical braking is frequently used to minimize overtravel. The number of convolutions on the actuator-control resistor of a proportional-control actuator is a factor in limiting the accuracy with which proportioning-control actuators can be positioned.

Features of Electric-motor-operated Actuators. Some electric-motor-operated actuators are available in explosion-proof constructions.

Limit switches, frequently of an adjustable type, are provided in the construction of most electric-motor-operated actuators to deenergize the driving-motor winding at the limits of travel of the actuator to prevent damage to the actuator or the operated device.

Some actuators are available with adjustable auxiliary switches. These switches are used to operate signal lights or alarms, or to actuate other control devices.

Auxiliary or retransmitting slidewires are available in some actuators. They are used to operate position indicators or to position additional actuators.

Manual-operating devices, usually in the form of handwheels, are available on some actuators. They permit manual positioning or closing of the valve or other device, normally operated by the actuator, in the event of a power failure or control-system failure.

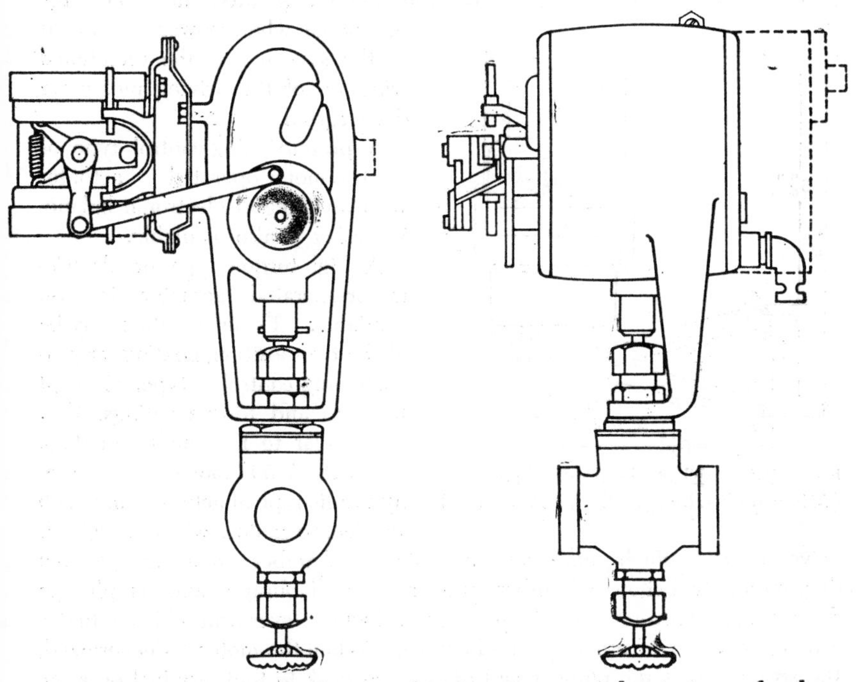

FIG. 7-58. Integral-valve yoke-type electric-motor-operated actuator and valve. (*Minneapolis-Honeywell Regulator Co.*)

Figure 7-57 illustrates a typical electric-motor-operated actuator and shows the driving motor, centrifugal brake, reduction gearing, slidewire, limit switches, and auxiliary switches.

When electric-motor-operated actuators are used to operate valves which must seat and remain seated in the closed position with a positive pressure, some form of strain-release and overtravel mechanism is built into either the actuator linkage or the valve to assure tight closure and still prevent damage to the actuator or valve body.

An integral-valve yoke type of electric-motor-operated actuator and valve is shown by Fig. 7-58. A strain-release and overtravel mechanism, in the form of a compression spring, is built into the connection between the stem and the valve disk of the slip-stem valve. The same function is accomplished for the butterfly valve by a scissors and spring mechanism in its linkage.

An electric-motor-operated actuator with an integral spring-return mechanism is also available. This actuator may be of either the floating-control or the proportioning-control type. The actuator construction incorporates an electric-solenoid-operated brake, which is engaged when the solenoid is energized and the motor directional windings are deenergized. Thus, on a failure of power, the brake releases, and the spring moves the actuator to its closed position. The balancing relay or control system used with this actuator must be arranged to energize the solenoid-operated brake to engage the brake when power is applied to the system and the directional windings of the driving motor are deenergized.

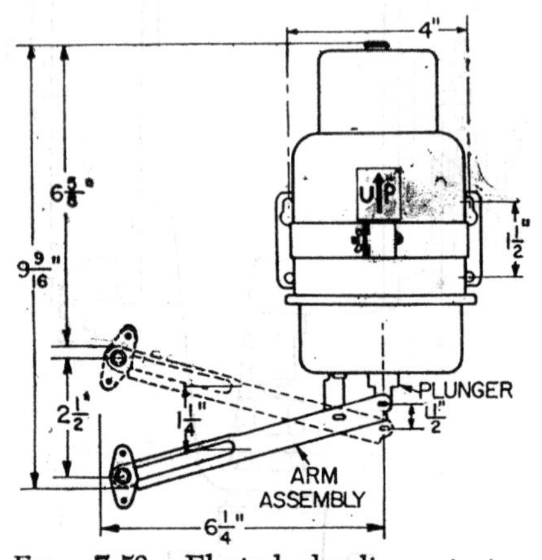

Fig. 7-59. Electrohydraulic actuator. (*Minneapolis-Honeywell Regulator Co.*)

A proportioning-control type of electric-motor-operated actuator with an integral electronic balancing relay is also available.

A high-torque type of electric-motor-operated actuator is also available. These actuators may be of the two-position, floating, or proportioning-control type. Except for size and torque ratings, they are similar to the small actuators.

Figure 7-59 shows an electrohydraulic type of actuator, in which an electric motor, when energized, drives a pump which pumps oil from a storage chamber into a spring-loaded diaphragm or bellows chamber, pushing the diaphragm and its plunger downward. Movement of the plunger positions a lever arm which actuates a damper, valve, or other operated device. When the motor is deenergized, the spring forces the plunger and diaphragm back to their original position, and the oil is returned to the storage chamber.

DIRECT-ACTING CONTROL VALVES

The direct-acting control valve or regulator requires no outside source of air or electric or hydraulic power. It utilizes the controlled medium itself, to provide required control action. Thus its main advantages are relative simplicity, low cost, and little, if any, maintenance. If properly selected, these regulators can be expected to do an efficient control job.

Direct-acting control valves are used for pressure reducing and control, temperature control, overflow regulators, pump governors, or combinations of these applications.

Pressure Regulation. Typical design (Fig. 7-60) uses an internal pilot. The piston above the valve, where dirt and sediment will not affect its operation, rides in a cylinder liner. The pilot-valve cage is screwed into the top cap as a complete unit, with valve and spring. Directly above the pilot valve is the diaphragm held down by the adjusting spring case.

The main valve is opened by high pressure acting on the large piston directly above it. The pilot valve, working in conjunction with the diaphragm, actuated by any unbalanced effect of the adjusting spring and low pressure beneath it, controls accurately the necessary pressure to the piston. Thus the pilot valve, sensitive to the secondary pressure, opens and closes the main valve in a proportionate degree, to maintain the desired constant low pressure at all times.

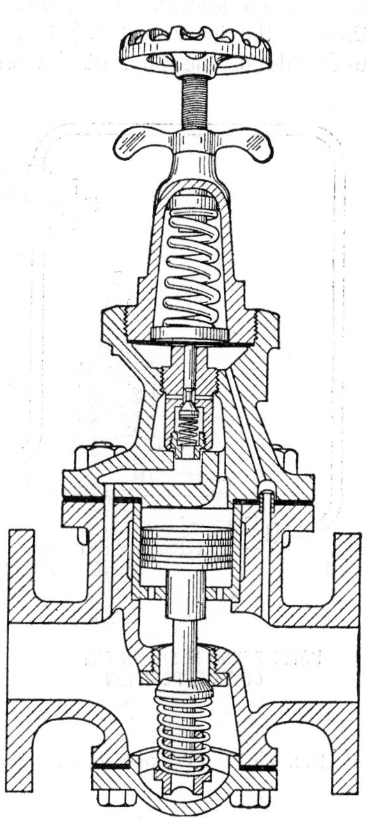

FIG. 7-60. Internal-pilot-piston pressure regulator. (*O. C. Keckley Co.*)

Sudden fluctuations in initial pressure are prevented from reaching the secondary side by the expansion-chamber effect of the piston cylinder.

Constant pressure on the secondary side will be maintained from zero or dead end to maximum flow capacity of the valve, regardless of variations in the primary or high-pressure side of the line.

To operate the regulator, all tension is removed from the adjusting spring by turning the handwheel to the left, which closes the pilot valve and the main valve of the regulator. Then the stop valve is opened wide on the high-pressure side of the line. Tension is increased slowly on the adjusting spring by turning the handwheel to the right until the desired low pressure is reached, after which the stop valve in the low-pressure side can be opened. When the regulator is in operation, both stop valves are kept fully open and lock-nut on, adjusting screw-tight.

The pressure regulator can be adjusted to any desired pressure within the range of the regulating spring and diaphragm by turning the handwheel at the top of the regulator to the right for higher secondary pressure, and to the left for lower secondary pressure.

Another design is shown in Fig. 7-61. When this unit is placed in operation, the fluid enters under the main valve (1) and passes through the port (2) into the pilot valve and through the tubing, discharging to the top of the main diaphragm (7) faster than it can bleed through the fixed orifice (9), thus causing intermediate pressure on top of the diaphragm, which in turn compresses the main spring and opens the main valve. This allows the fluid to pass through the main valve and build up pressure in the downstream system. The increasing downstream pressure is transmitted through the inner port (5) to the lower diaphragm chamber and to the underside of the pilot diaphragm (13). When the pressure under the

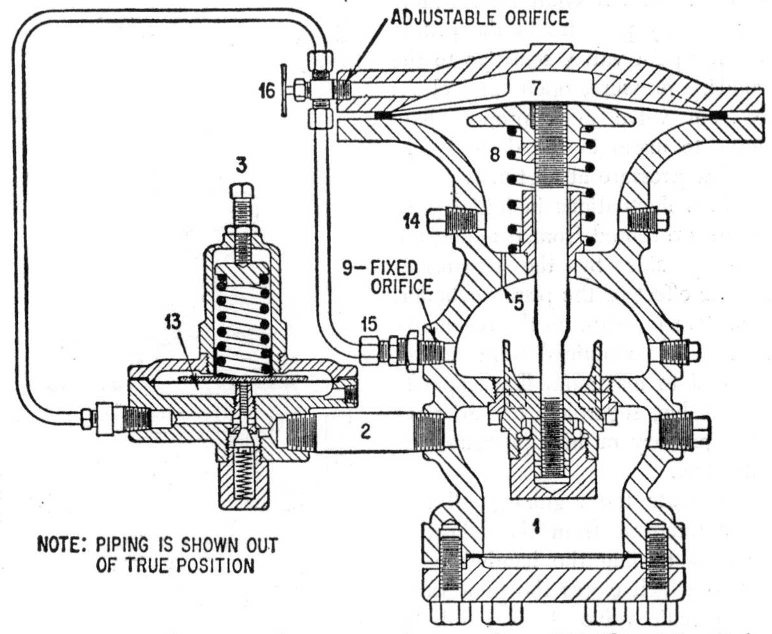

FIG. 7-61. Pilot-operated pressure-reducing valve. (*McAlear Mfg. Co.*)

pilot diaphragm reaches a point where it balances the loading of the pilot-control spring, the pilot valve begins to close, thus restricting the supply of intermediate pressure to the top of the main diaphragm (7) until the supply pressure is the same as the pressure of the flow through the orifice (9). Meanwhile the pressure in the lower diaphragm chamber (8) is building up until the differential across the main-valve diaphragm balances the loading of the main-valve spring. The main valve is passing just enough fluid to maintain a downstream pressure that will balance the loading of the pilot-control spring. When the demand changes, a slight change in the downstream pressure occurs, which will result in the throttling of both the pilot and the main valve to supply the new demand.

Reduced pressure may be increased or decreased by turning the adjusting screw (3). Tightening the screw will increase the reduced pressure. The orifice (16) may be adjusted to dampen sudden impulses and pulsations to ensure steady control.

Diaphragm regulators (Fig. 7-62) are often used to keep water or steam pressures below maximum allowable.

The maximum allowable pressure on plumbing fixtures, for example, is generally about 40 lb (Fig. 7-63). When the supply is from high-pressure mains, or from mains that may, in case of fire, be under high pressure temporarily, it is necessary to use pressure-reducing valves to protect the fixtures and ensure satisfactory operation. In industrial plants, water regulators frequently are required for processing and other specific purposes.

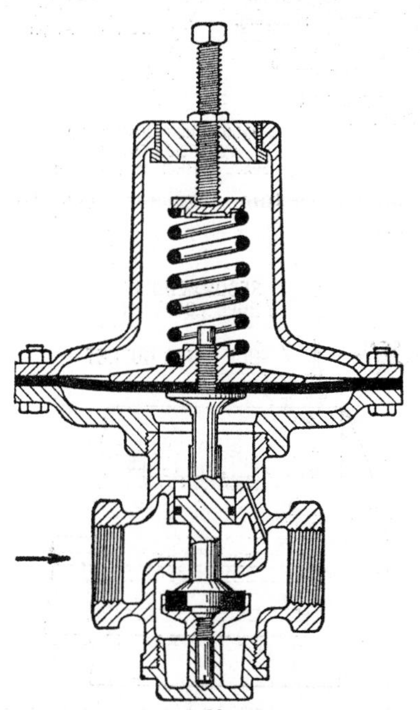

FIG. 7-62. Self-acting-diaphragm pressure regulator. (*Foster Engineering Co.*)

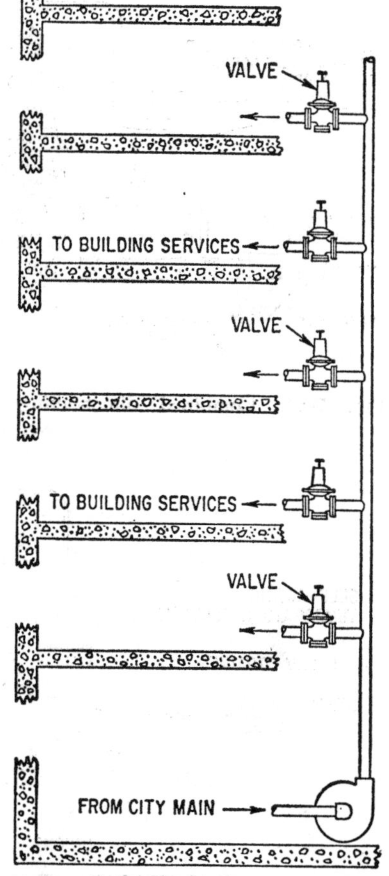

FIG. 7-63. Domestic water pressure control for tall buildings is generally provided with a diaphragm-type regulator. (*Spence Engineering Co., Inc.*)

For steam service a metal diaphragm is used, and for air a rubber diaphragm with fabric insertion.

The regulator is normally held open by the spring tension, and the steam or air enters the diaphragm chamber through the port on the delivery side of the valve, the pressure under the diaphragm forcing the diaphragm upward against the tension of the spring, causing the main valve to close,

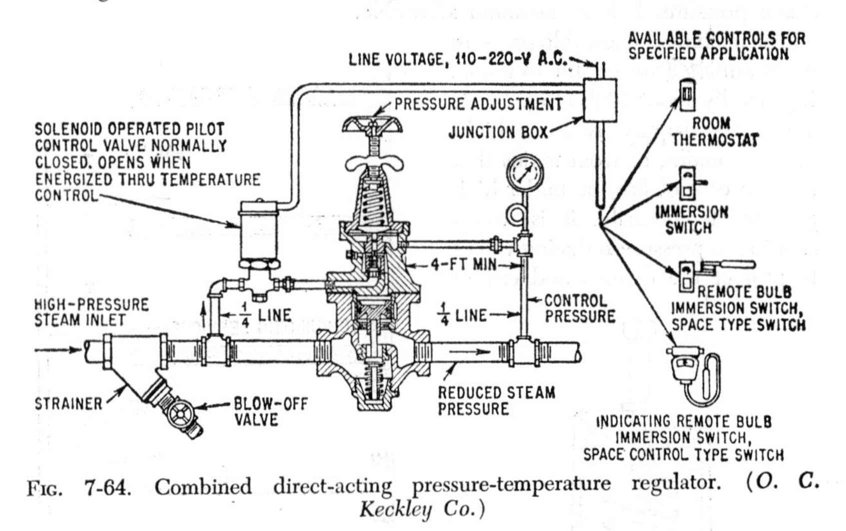

Fig. 7-64. Combined direct-acting pressure-temperature regulator. (*O. C. Keckley Co.*)

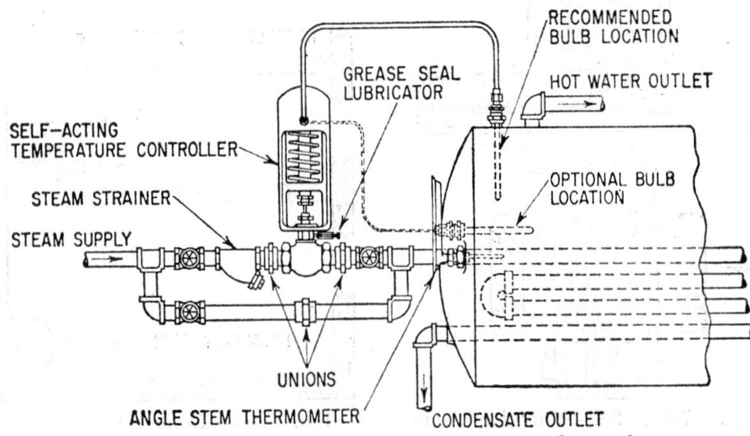

Fig. 7-65*a*. Self-acting temperature regulator installed on hot-water tank. (*Taylor Instrument Co.*)

forming a balance between the delivery pressure and the tension of the adjusting spring. Turning the adjusting screw into the top cap increases the reduced pressure.

Temperature Regulation. Figure 7-64 shows a combined direct-acting pressure-temperature regulator.

Steam is admitted to the pressure regulator through a ¼-in. solenoid

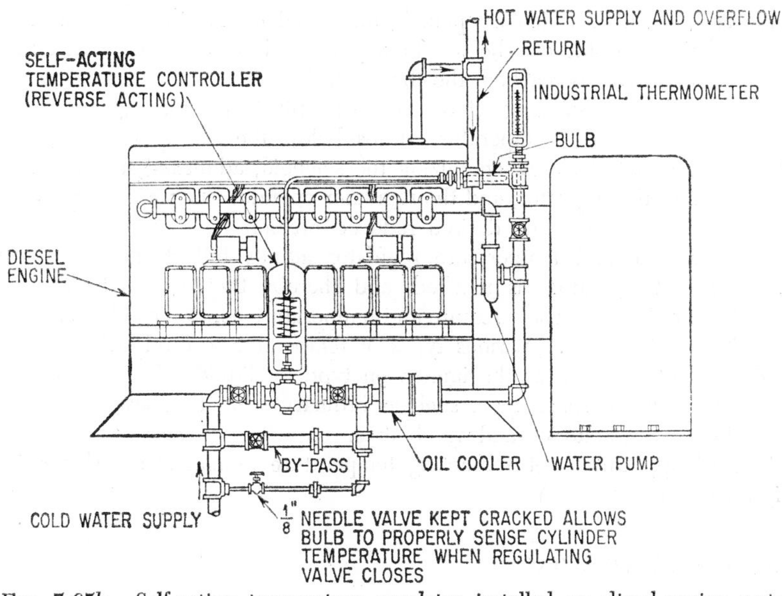

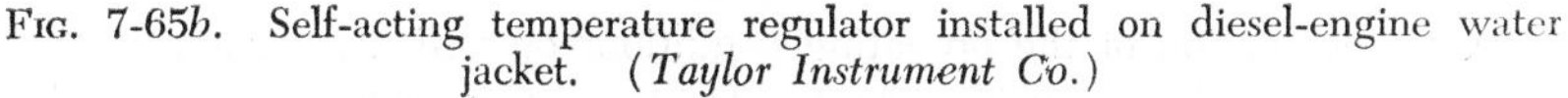

FIG. 7-65*b*. Self-acting temperature regulator installed on diesel-engine water jacket. (*Taylor Instrument Co.*)

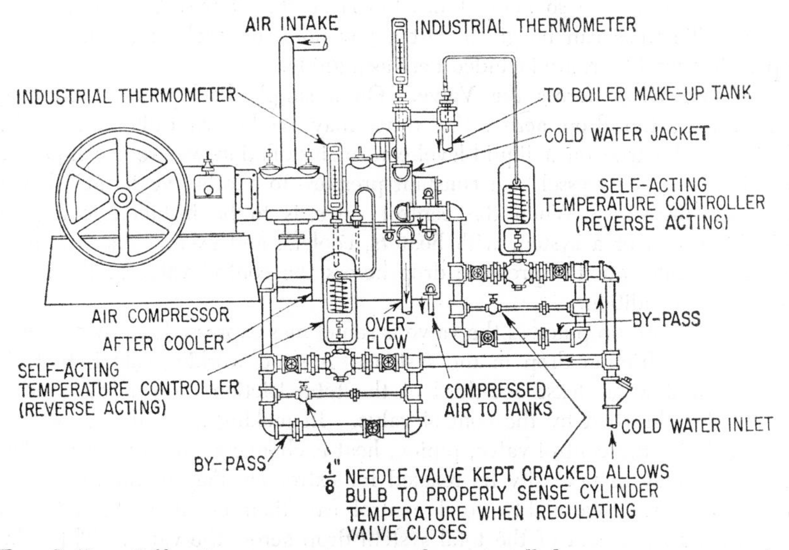

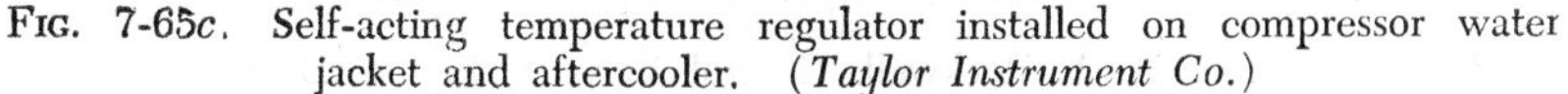

FIG. 7-65*c*. Self-acting temperature regulator installed on compressor water jacket and aftercooler. (*Taylor Instrument Co.*)

valve as illustrated, which is opened and closed by the thermostatic control. The thermostat can be set to any desired degree, and, when the proper temperature is reached, the thermostatic control opens, deenergizing the solenoid valve, shutting off the steam supply to the pilot valve of the regulator. When the temperature drops below the set temperature, the thermostatic control closes, energizing the solenoid, admitting steam to the pilot valve of the regulator, and the regulator operates as a pressure-regulating valve, maintaining any desired reduced pressure.

This pressure and temperature regulator admits full steam flow until the desired temperature is reached, and thereby the time of heating is materially reduced. The pressure-and-temperature-control combination is unlimited in its scope. Various types of temperature-control switches are readily adaptable such as the immersion type for liquids, the fin type for air ducts, and the space type, such as ordinary room thermostats for the control of offices, factories, and warehouses.

Typical applications of self-acting temperature regulators are shown in Figs. 7-65*a–c*.

CONTROL-VALVE SIZING[6]

Proper evaluation of the operating conditions on which the control-valve size will be based is very important. Most errors in sizing are due to incorrect assumptions concerning actual flowing conditions. Generally speaking, the tendency is to make the valve too large and be on the "safe" side. A combination of several of these safety factors actually can result in a valve which will be so many times too large that it may be troublesome. No specific rules can be given, but a few notes on each of the necessary operating conditions are included here as a guide.

Pressure Drop across the Valve. On a simple back-pressure-reducing application, the drop across the valve may be known quite accurately. This may be true on a liquid-level-control installation, where the liquid is passing from one vessel at a constant pressure to another vessel at a lower constant pressure. When the control valve is to be installed in a long piping system or a system with heaters, coolers, and exchangers, it is necessary to calculate the pressure drop across the control valve at the maximum flow condition.

The pressure drop across the valve is sometimes expressed as a percentage of the total friction drop in the system. A good working rule is that, at maximum flow, at least one-third of the total friction drop of the system should be absorbed by the control valve. In making an analysis, the system includes the control valve, piping, heat exchangers, and mixing nozzles. This rule may of necessity be relaxed for extremely long or high-pressure-drop systems. Reasonably good control can then be attained, with not less than 15 per cent of the total system drop across the valve. This rela-

[6] Abstracted from R. A. Rockwell, Control Valve Sizing, in D. M. Considine (ed.), "Process Instruments and Controls Handbook," McGraw-Hill Book Co., Inc., New York, 1957.

tively low-percentage drop is permissible only when the variation in flow is small.

In many instances, it is necessary to make an arbitrary choice of the pressure drop across a control valve because of the meager process data available. When the valve is in a pump-discharge line, a drop of 10 to 25 psi may be assumed to be sufficient if the data indicate that the pump-discharge line is not extremely long, or complicated by large drops through heat exchangers or other equipment. The tendency should be to use a higher figure, particularly when the pump-discharge line is rather long. If a valve figures out to be line size, the calculation for the line size should be reviewed. On more complicated systems, consideration should be given to both maximum and minimum operating conditions.

In systems involving high-pressure centrifugal pumps, the drop in the pump characteristic at maximum flow becomes a major factor, and must be taken in account in figuring the per cent of total system drop that exists across the valve.

In the interests of economy, the engineer wishes to keep the control-valve pressure drop as low as possible. On the other hand, a valve can regulate flow only by absorbing and giving up pressure drop to the system. When the drop across the control valve is reduced to a small percentage of the total system drop, the ability of the control valve significantly to increase the flow rapidly disappears.

A control valve generally figures out smaller than the line size.

Flowing Quantity. The maximum flowing quantity capable of being handled by the valve should be 25 to 60 per cent above the maximum flow required. One must make certain that no other "factors" have already been applied.

On many systems, a reduction in flow means increase in pressure drop, and the valve-port-area range may be much greater than would be suspected. If, for example, the maximum operating conditions for a valve are 100 gpm at 10 psi pressure drop, and the minimum conditions are 10 gpm at 40 psi pressure drop, the port area is the product of the ratio of maximum to minimum flow and the square root of the ratio of maximum to minimum pressure drop. In this case,

$$\frac{100}{10}\sqrt{\frac{40}{10}} = \frac{20}{1}$$

There are many systems in which the increase in pressure drop across the valve for this same change in flow is proportionally much greater than in this example.

Specific Gravity. The actual specific gravity of the liquid or gas is relatively unimportant. If the actual gravity is not known accurately, a reasonable assumption will suffice. The use of 0.9 specific gravity, for example, instead of 0.8 would cause an error of less than 5 per cent in valve capacity.

The greatest potential error in valve sizing is the incorrect interpretation of the operating data. After the required C_v of the valve has been obtained, it is then necessary to select a nominal size, and a review of the operating data should be made to guide in this selection.

Sizing Accuracy. Determination of control-valve size for a given application is not an exact science. It should be recognized that present valve-sizing procedures are based on assumptions that are not strictly valid, and on theory which in many instances is inadequate. However, without the use of simplified theory and reasonable assumptions, the practical problem of sizing becomes hopelessly complicated. Although it is based on experimental data, and corrected for body losses and efficiency of the port and passages, the experimental determination of the valve C_v does not take into account the variations in the characteristics of the fluid in question, the changes in the state of the fluid, or the wide variations of pressure drop and velocity which may exist in various installations. It is in this region that the rule of thumb, based on experience and rational assumptions, becomes necessary. Improved techniques and more exact determinations of the behavior of fluids under various conditions may improve the theoretical accuracy, but the over-all accuracy will still be limited somewhat by the difficulty of evaluating or predicting operating conditions.

With the exact maximum flow known, it is customary and desirable to provide a safety factor, and to determine the valve size on an arbitrary flow, usually well in excess of the known maximum. This excess is dependent on the operating conditions of the application, on the valve size, and on the construction of the valve under consideration.

Sizing procedures are predicated on reasonably exact knowledge of the pressure drop across the valve. In many instances, the actual pressure drop used is at best only an approximation. Therefore, care must be exercised to select the pressure drop that is consistent with operating conditions at the maximum flow. Although sizing formulas are based on the assumption that the flow varies as the square root of the pressure drop, there is some question concerning the validity of the application of the square-root law to such a complex restriction as a control valve, especially under extreme operating conditions. Except in unusual circumstances, the error resulting from the use of the square-root law is seldom of sufficient magnitude to justify complicating the sizing procedure by corrections, which, in many instances, may be extremely difficult to correlate to the case at hand.

The formulas for steam and gases include a density correction, which is a function of the outlet pressure P_2. This is an assumption which is not strictly true, as the exact correction should be some intermediate value between the inlet and outlet pressures. The use of the outlet pressure is a conservative simplification, tending to underrate the valve capacity up to approximately 15 per cent at the critical-pressure ratio. Although the use of the outlet pressure P_2 in the compressible-flow formulas may somewhat underrate the valve capacity, there is, as yet, no general agreement of test procedures, nor are there sufficient data by independent investigators to generalize on the extent of the actual deviation.

Valve-flow Coefficient C_v. Use of the valve-flow coefficient C_v has materially simplified the problem of control-valve sizing. This C_v coefficient provides a simple method applicable to a wide variety of valve constructions, valve sizes, and field services. Because of its importance, an understanding of the development, scope, and limitation of the C_v is essential.

Valve-flow coefficient C_v is the number of U.S. gallons per minute of water which will pass through a given flow restriction with a pressure drop of 1 psi.

It is a capacity index. Any flow problem for gas, steam, or liquid can be converted to the equivalent valve-flow coefficient. In other words, the flow of any fluid through a restriction with a known pressure drop may be expressed in terms of gallons per minute of water at 1.0 psi pressure drop. The flow restriction can be an orifice plate, a length of pipe, a hand valve, or, more specifically for control purposes, an automatic-control valve. Published values of C_v for control valves are at the full-open position unless otherwise noted. The C_v does not involve any specific flow restriction but is based on the fundamental law of fluid flow ($v = \sqrt{2gh}$). Most sizing formulas, slide rules, alignment charts, and tables are based on this simple law.

For liquids, flow coefficient C_v is equal to the flow in gallons per minute divided by the square root of the pressure drop in pounds per square inch, and is corrected for specific gravity by multiplying this expression by the square root of the actual flowing gravity.

For steam and gases, the use of C_v is almost equally simple. In the gas and steam equations, suitable constants permit the use of C_v and convenient units for flow, pressure, temperature, and specific gravity.

A control valve which has a flow coefficient C_v of 10 means that its effective port area (in the fully open position) is such that it passes 10 gpm of water with a 1-psi pressure drop. This is the approximate C_v capacity of a 1-in. control valve. The numerical value of C_v simply expresses the number of gallons per minute of water flow which would result from a 1-psi pressure drop. The flow capacity of a valve at maximum lift is a complex function of the port area, plug design, and body design. Valve efficiencies are dependent on the valve design, the flow pattern, and the ratio of the port area to the body area. In view of the wide variety of designs and the extreme difficulty in determining the valve efficiency, it is generally impractical to calculate the valve-flow coefficient from design data. All capacity data should be determined experimentally.

The C_v rating of a valve is always specified at full lift or in the wide-open position. Valve-flow characteristics are frequently plotted in terms of lift versus C_v. The valve capacity is the C_v at 100 per cent lift, and the use of C_v for all other lifts is a convenient way of expressing flow at a constant drop.

Control Valve C_v at Rated Lift. Control-valve capacity ratings, C_v at full-open position, are available from the valve manufacturer. These ratings are generally determined by test, and, because of the wide diversification in valve design and types of construction it is essential to use the

C_v rating of the valve under consideration in all valve-sizing calculations.

Ratings are based on the pressure drop taken in the piping close to the inlet and outlet of the valve body. No allowance is made for pressure drop in the adjacent piping.

High-efficiency valves such as the venturi type produce pressure recovery in the valve body at low drops and low velocities. The pressure recovery is negligible at high drops and high body velocities. This variation in pressure recovery indicates a higher C_v rating at a low-pressure drop than can be obtained at a high-pressure drop. The actual capacity rating for this type of valve is given for the high-pressure drop, and, therefore, the C_v rating is suitable for high-pressure-drop services and conservative for low-pressure-drop services.

For estimating purposes, the conventional 1-in. double-seated control valve has a C_v of 10. All other sizes can be approximated by multiplying 10 times the square of the nominal body size, giving the ratings shown in the table below.

Double-seated valve size, in.	C_v rating full open	Double-seated valve size, in.	C_v rating full open
¾	5.6	3	90
1	10	4	160
1¼	16	6	360
1½	23	8	640
2	40	10	1,000
2½	62	12	1,440

The ratings given in the table are representative, but, in sizing a particular make of valve, the figures published by the manufacturer should be used. Angle valves, butterfly valves, Saunders-type valves, single-seated valves, and special reduced-port valves have C_v ratings which cannot be estimated and must be obtained from the manufacturers' test data.

Procedure to Determine Valve Size and Valve Capacity. Following are the basic procedures for determining valve size and valve capacity:

To Determine Valve Size

Data given: All flowing conditions
Solve for: C_v and then select valve size for type of valve under consideration from manufacturers' tables of C_v rating versus valve size

To Determine Valve Capacity

Data given: C_v and flow conditions
Solve for: Capacity V, in U.S. gpm of liquid
Capacity Q, in cu ft/hr of gas at 14.7 psia and 60°F
Capacity W, in lb/hr of steam

Liquids. The basic equations for valve size and valve capacity for liquids are

$$C_v = V\sqrt{\frac{G}{\Delta P}} \tag{7-1}$$

$$V = C_v\sqrt{\frac{\Delta P}{G}} \tag{7-2}$$

where V = flow, gpm (U.S.)
P = pressure drop at maximum flow, psi
G = specific gravity (water 1.0)
C_v = valve-flow coefficient

When flowing temperature is above 200°F, use specific gravity and quantity at flowing condition.

When the viscosity exceeds 100 SSU (seconds Saybolt Universal) or 20 centistokes, check the viscosity correction.

Gases. The basic equations for valve size and valve capacity for gases are

$$C_v = \frac{Q\sqrt{GTa}}{1{,}360\sqrt{(\Delta P)P_2}} \tag{7-3}$$

$$Q = \frac{1{,}360C_v\sqrt{(\Delta P)P_2}}{\sqrt{GTa}} \tag{7-4}$$

where Q = quantity, cu ft/hr at 14.7 psia and 60°F
ΔP = pressure drop at maximum flow, psi ($P_1 - P_2$)
P_1 = inlet pressure at maximum flow, psia
P_2 = outlet pressure at maximum flow, psia
G = specific gravity (air = 1.0)
T_a = flowing temperature absolute (460 + °F)
C_v = valve-flow coefficient

When P_2 is less than $\frac{1}{2}P_1$, use the value of $P_1/2$ in place of $\sqrt{(\Delta P)P_2}$.

Steam. The basic equations for valve size and valve capacity for steam are

$$C_v = \frac{WK}{3\sqrt{(\Delta P)P_2}} \tag{7-5}$$

$$W = \frac{3C_v\sqrt{(\Delta P)P_2}}{K} \tag{7-6}$$

where W = lb/hr of steam
ΔP = pressure drop at maximum flow, psi
P_1 = inlet pressure at maximum flow, psia
P_2 = outlet pressure at maximum flow, psia
K = 1 + (0.0007 × °F superheat)
C_v = valve-flow coefficient

When P_2 is less than $\frac{1}{2}P_1$, use the value of $P_1/2$ in place of $\sqrt{(\Delta P)P_2}$.

The steam formula has been set up using $1/0.00225P_2$ in place of the specific volume, to eliminate the need for steam tables.

The flow of compressible fluids through a restriction reaches a saturation velocity when the differential pressure is increased to approximately 50 per cent of the inlet pressure. This critical-pressure ratio varies with the composition of the fluid. The average value of one-half the absolute inlet pressure is well within the tolerance established by the formulas.

Vapors Other than Steam. The fundamental equations (weight basis) for valve size and valve capacity for vapors other than steam are

$$C_v = \frac{W}{63.4}\sqrt{\frac{v_2}{\Delta P}} \tag{7-7}$$

$$W = 63.4C_v\sqrt{\frac{\Delta P}{v_2}} \tag{7-8}$$

where W = lb/hr of vapor
ΔP = pressure drop at maximum flow, psi ($P_1 - P_2$)
v_2 = specific volume, cu ft/lb at outlet pressure P_2
P_1 = inlet pressure at maximum flow, psia
P_2 = outlet pressure at maximum flow, psia
C_v = valve-flow coefficient

When P_2 is less than $\frac{1}{2}P_1$, use the value of $P_1/2$ in place of ΔP, and use v_2 corresponding to $P_1/2$.

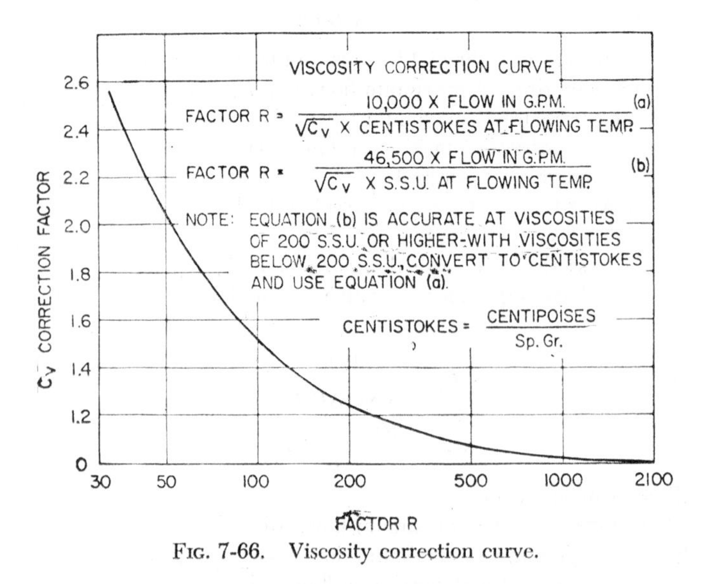

FIG. 7-66. Viscosity correction curve.

Viscosity Corrections for Liquids. The liquid-sizing formulas are idealized to the extent that no viscosity corrections have been considered. Correction for viscosity is a complex function involving the Reynolds number which is dependent on velocity, valve size, port area, and fluid viscosity. The correction has been resolved by use of factor R in a simple graph (Fig. 7-66). Factor R is an approximation of the Reynolds number under the flowing conditions.

Below 100 SSU or 20 centistokes (equals centipoises per specific gravity), the viscosity effect may be disregarded. Above these values, the correction should be found and multiplied by the C_v as determined by the formula.

To determine the viscosity correction, the following procedure should be used:

1. Solve for the C_v, assuming no viscosity effect.
2. Solve for factor R from Eq. (a) or (b) (Fig. 7-66).
3. From the curve (Fig. 7-66), read the correction factor at the intercept of factor R.
4. Multiply C_v, as determined by the formula, by the correction factor.
5. Use this corrected C_v to select the valve size from C_v versus valve-size tables.

Flashing Mixtures of Water and Steam. When water at or near saturation temperature flows through a control valve with pressure reduction, thermodynamic considerations indicate that a mixture of water and steam will exist at the outlet of the valve. In many instances, it is possible to have a condition of temperatures and pressures such that the fundamental laws of liquid flow must be disregarded, and, therefore, conventional valve-sizing methods are not applicable.

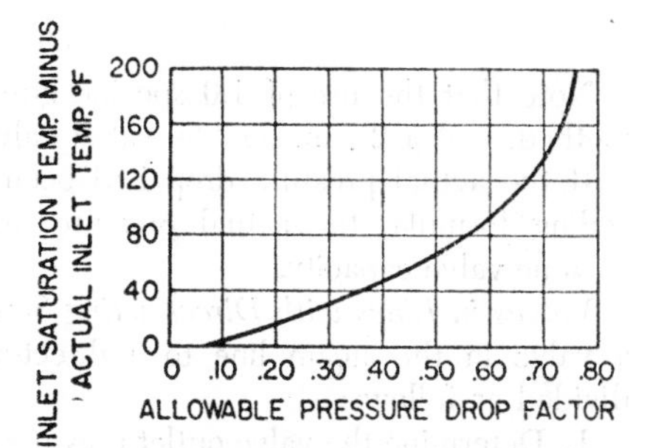

FIG. 7-67. Pressure-drop factor for water in liquid state. (*G. F. Brockett and C. F. King, Sizing Control Valves Handling Flashing Liquids, Texas A&M Symposium, Jan. 23, 1953.*)

The flow of compressible fluids (steam, gas) through a valve or an orifice reaches a maximum velocity when the differential pressure is increased to approximately 50 per cent of the inlet pressure. This point is called the critical-pressure drop, and the ratio of the outlet pressure to the inlet pressure is called critical-pressure ratio. Water entering a control valve at or somewhat below the saturated condition exhibits similar behavior. While gas and steam have a relatively constant critical-pressure ratio of 0.5, test data indicate that water up to temperatures of approximately 300°F has an apparent critical-pressure ratio that varies between 0.15 and 0.88.

This apparent critical-pressure ratio is a function of the difference between the saturated temperature of water at entering pressure and the actual temperature of the water at the valve inlet. In Fig. 7-67, test data are plotted to show this relationship for temperature differences up to 200°F with respect to the allowable pressure-drop factor (1 minus critical-pressure ratio). The allowable pressure drop is obtained by multiplying the inlet pressure by the allowable pressure-drop factor.

To determine the required control-valve capacity, use the liquid formula

$$C_v = V\sqrt{\frac{G}{\Delta P}} \tag{7-1}$$

In place of the actual pressure drop, use the allowable pressure drop as computed, or the actual drop, whichever is smaller; for example:

Data given: Flow, 400 gpm
Inlet pressure, 120 psia
Inlet temperature, 300°F
Actual pressure drop, 60 psi

The saturated temperature of water at 120 psia is 341°F. The temperature difference between saturated temperature and actual temperature is 341 − 300°F, i.e., 41°F. From Fig. 7-67, it is noted that the allowable pressure-drop factor is 0.38, and the allowable pressure drop ΔP is 0.38 × 120 psi, i.e., 46 psi.

$$C_v = 400\sqrt{\frac{1.0}{46}}$$
$$= 59$$

Note that the use of 1.0 specific gravity is within practical tolerance of method, and a 2½-in. double-seated valve will have ample capacity.

If the actual pressure drop had been less than 46 psi, then, in the preceding formula, the actual pressure drop would have been used to determine valve capacity.

Valves in Lines with Direct-acting Steam Pumps. The method for sizing a valve in the steam line to a direct-acting reciprocating pump may be divided as follows:

1. Determine the valve outlet pressure P_2
2. Determine the steam consumption.
3. Select the proper valve.

The data required for solution are (1) size of pump, (2) type of pump, (3) quantity of liquid pumped, (4) pump suction and discharge pressure, in psi, and (5) initial and exhaust steam pressure, in psi.

Control-valve Outlet Pressure P_2. The control-valve outlet pressure (steam inlet pressure to pump) is a direct function of the mean effective steam-cylinder pressure.

The pressure downstream of the control valve is determined by the expression

$$P_2 = \frac{(d_f)^2(P_{net})}{(d_s)^2} + P_{exh} + K \tag{7-9}$$

where P_2 = pressure on downstream side of valve, psi
d_f = diameter of fluid cylinder, in.
d_s = diameter of steam cylinder, in.
P_{exh} = steam-exhaust pressure, psi
P_{net} = difference between pump suction and discharge pressure, psi
K = over-all correction factor (Fig. 7-68)

Correction factor K is plotted for the two most common types of pumps as a function of the pump stroke in inches.

Steam Consumption. This quantity is dependent on the work done by the pump, the stroke of the pump, and the mean effective cylinder pressure. If the maximum pump discharge is given in gallons per minute at the flowing condition, and the net head is in pounds per square inch, then the hydraulic horsepower of the pump is given by

$$\text{Hydraulic horsepower} = \frac{\text{gpm } P_{net}}{1{,}715}$$

The net head is the algebraic difference between the pump-discharge pressure and suction pressure.

The steam consumption per hydraulic horsepower may be taken from Fig. 7-69. This figure multiplied by the calculated hydraulic horsepower gives the approximate total steam consumption of the pump.

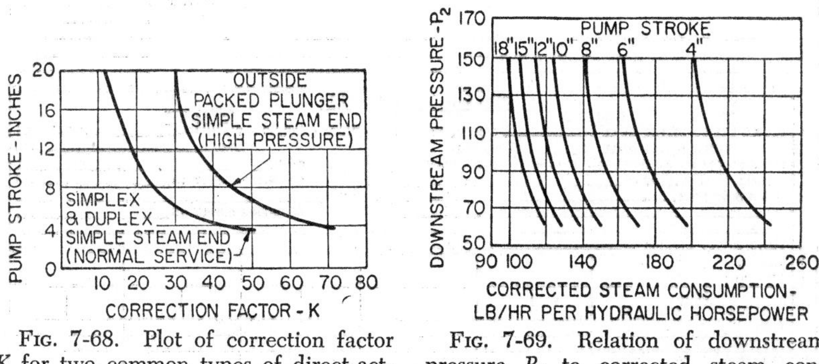

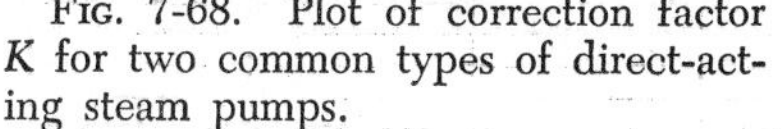

FIG. 7-68. Plot of correction factor K for two common types of direct-acting steam pumps.

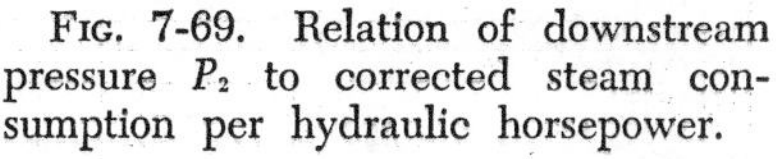

FIG. 7-69. Relation of downstream pressure P_2 to corrected steam consumption per hydraulic horsepower.

The values given in Fig. 7-69 are based on atmospheric-exhaust pressure. As the back pressure increases, the steam consumption will increase quite rapidly. Exhaust-steam pressures above 20 psi, however, are very uncommon, and an increase of about 20 per cent in steam consumption will generally compensate for the average case where the back pressure is over 5 psig.

Selection of Valve Size. With the steam consumption, inlet pressure, and outlet pressure known, the selection of the valve size may now be made by the use of the sizing formula (7-5).

Slide Rules and Nomographs. The mechanics of valve sizing can be materially simplified by the use of valve-sizing slide rules or nomographs (alignment charts).

The valve-sizing slide rules are designed to solve Eqs. (7-1) through (7-8) directly. Various types are available from control-valve manufacturers and are designed to give the answer in terms of valve size and, some types, to give both the valve size and the required C_v.

Nomographs consist of a series of flow-data scales arranged so that, by a series of straight-line intersections, the valve size or the valve coefficient C_v can be obtained. By adjusting the location of data scales and spacing of scale markings, it is possible with alignment charts to include various corrections that are generally impractical in simple sizing formulas and valve-sizing slide rules.

CONTROL-VALVE SPECIFICATIONS

Control-valve specification forms developed by the Instrument Society of America may be used as a standard, or may be modified to suit special needs (Fig. 7-70).

CONTROL VALVES

SPECIFICATION SHEET

SHEET NO. ______
TAG NO. ______
DATE ______
REVISED ______
BY ______

GENERAL SPECIFICATION

BODIES

1 DOUBLE SEATED BODIES SHALL BE ______ GUIDED, SINGLE SEATED BODIES SHALL BE ______ GUIDED

2 END CONNECTIONS AS SPECIFIED BY LINE 12. BODY JOINTS SHALL BE MFGR.STD. OR ______

3 OTHER BODY FORM ______

OPERATORS

5 PNEUM.: SPRING & DIAPHRAGM ______ .SHALL FULL STROKE ______ PSI
OTHER ______
OPERATING SUPPLY IS ______ PSI

6 OTHER ______
OPERATING SUPPLY IS ______

4 FLUID UNITS: LIQUIDS IN ______ GASES IN ______ STEAM IN ______

7	TAG NO.					
8	LINE NO.					
	BODY					
9	BODY SIZE | PORT SIZE					
10	FORM					
11	MATERIAL					
12	END CONNECTIONS					
13	BONNET					
14	LUBRICATOR | ISOLATING VALVE					
15	PACKING OR SEAL					
16	SPECIAL GUIDES					
17						
	TRIM					
18	MATERIAL					
19	NO.OF PORTS					
20	PLUG FORM					
21	PLUG & SEAT MATERIAL					
22						
	ACTION					
23	CLOSE @ | OPEN @					
24	FAILURE POSITION					
25						
	POSITIONER					
26	REQUIRED					
27	BYPASS | GAUGES					
28	FOR INPUT SIGNAL OF OUTPUT SHALL BE					
	ACCESSORIES					
29	FILTER & REGULATOR					
30	HANDWHEEL					
31						
	SERVICE CONDITIONS					
32	FLUID					
33	QTY. MIN. | QTY. MAX.					
34	QTY. NORM. @ F.T.					
35	PRESS.MAX. IN | NORM. OUT					
36	ΔP MAX. | ΔP SIZING					
37	TEMP. MAX. | NORM.					
38	SP.GR. @ 60°F | @ F.T.					
39	VISCOSITY @ F.T.					
	NOTES:					

FIG. 7-70*a*. Control-valves specification sheet. (*Instrument Society of America.*)

CONTROL VALVE INSTRUMENTS
Specification Sheet Instructions

Prefix numbers designate line number on corresponding specification sheet.

1) Write in type of plug guiding, such as top and bottom, seat ring, top, etc.

2) If special body joint required, specify requirement, such as Ring Type Joint, Small Tongue and Groove, etc.

3) If body type other than double seated or single seated flanged head required, specify, such as Saunders Patent, butterfly and give general description.

4) Specify fluid units, i.e., GPM at F.T., SCF/H, #/Hr., etc.

5) For pneumatic spring and diaphragm state instrument output range, such as 3-15 PSI, 3-27 PSI, etc. Under other specify Springless Diaphragm, Piston, etc. Advise air pressure available for positioners, boosters, etc.

6) Other operators, i.e., hydraulic, electric motor, etc. (give details). Give operating supply such as PSI, Volts, Cycles, etc.

7) Identification of Item Number.

8) State Pipe Line Number.

9) Specify nominal body size and port size in inches.

10) Specify body form; Glove, Angle, 3-way, etc.

11) Specify body material, such as bronze, cast iron, steel, etc.

12) Specify end (line) connections; Screwed NPT, 150# RF, 300# RF, 300# RTJ, etc. Flange ratings ASA unless otherwise specified.

13) Specify type of bonnet; standard, radiating fin, plain extension, etc.

14) Write "Yes" if lubricator and/or isolating valve required.

15) Specify packing required or type of seal, such as Bellows, Diaphragm, etc.

16) If special guide required, specify type material and/or hardness.

17) For additional specifications not covered by Lines 9 through 16.

18) Specify trim material. (Trim generally includes plug, stem, seat ring(s), grease ring and packing follower. Write in any deviations on Line 22). Specify manufacturer's standard, bronze, type stainless steel, etc.

19) Specify double or single port.

20) Specify plug form; Linear, Percentage, and whether V-port or Parabolic, etc.

21) Specify plug and seat material, if other than shown on Line 17.

22) For additional specifications not covered by Lines 18 through 21.

23) Specify conditions((PSI, Volts, Cycles, etc.), which valve shall be "closed" and "open".

24) Specify position valve is to take on air, hydraulic, or electric failure.

25) For additional specifications not covered by Lines 23 and 24.

26) Specify by "Yes", or name of Manufacturer, it valve positioner required.

27) Specify by "Yes" for positioner bypass. For gauges, specify quantity.

28) On top line specify input signal band and on bottom desired corresponding positioner output band.

29) Write in if individual units or combination type desired, or if to be purchased separately.

30) Specify type of handwheel if required, such as top or side mounted, etc., - consult manufacturer.

31) For additional specifications not shown on Lines 29 and 30.

32) Show flowing fluid; liquid, gas or steam.

33) Give minimum and maximum quantity of fluid valve is required to pass.

FIG. 7-70*b*. Control-valve instruments specification sheet instructions. (*Instrument Society of America.*)

CONTROL VALVE INSTRUMENTS

Specification Sheet Instructions

34) Expected normal flow at flowing temperature.

35) Maximum pressure is required for determining maximum body working pressure.

36) Show maximum ΔP and ΔP at which valve is to be sized.

37) Show maximum temperature for valve rating and normal operating temperature.

38) Show specific gravity at 60 F and at flowing temperature.

39) Show viscosity at flowing temperature.

Fig. 7-70*c*. Control-valve instruments specification sheet instructions. (*Instrument Society of America.*)

Section 8

MEASUREMENT AND CONTROL OF pH AND CONDUCTIVITY

The measurements under consideration in this section are concerned with properties of water solutions, and the measured values depend on the particular chemical nature and concentration of these solutions. For those unfamiliar with the basic concepts of chemical solutions, a review is given here. It is recommended that this be read first by those who do not have an acquaintance with such terms as *ions, neutralization,* and *buffer action,* which are commonly employed in this field.

Many industries have already benefited from pH and conductivity control. Water-treatment plants, and plants requiring cooling-tower and boiler-feedwater monitoring, find pH control invaluable. The same is true of waste and sewage treatment. Wherever water is used in industry, control of pH and conductivity can mean better product quality and more efficient operation, saving time, labor, and materials.

CHEMISTRY OF ELECTROCHEMICAL MEASUREMENTS[1]

Solution Concentration. It is well known that water easily dissolves many materials to form a clear solution. Common table salt, for example, forms a saline solution with water, and the amount of salt dissolved determines the *concentration* of the solution. The various methods of expressing solution concentration in electrochemical measurements require knowledge of only a few basic facts about solids forming solutions, as outlined in the following definitions and illustrated in Fig. 8-1.

Percentage by weight is one of the simplest methods and a fairly common means of expressing solution concentration. A 10 per cent aqueous solution of table salt, for example, contains 10 g (or lb) of the dry salt in 100 g (or lb) of solution. It could be prepared by placing 10 g of salt in a container on a scale and adding water until the total weight (less the container, of course) equaled 100 g.

For other methods of expressing solution concentration, it is necessary

[1] Minneapolis-Honeywell Regulator Co.

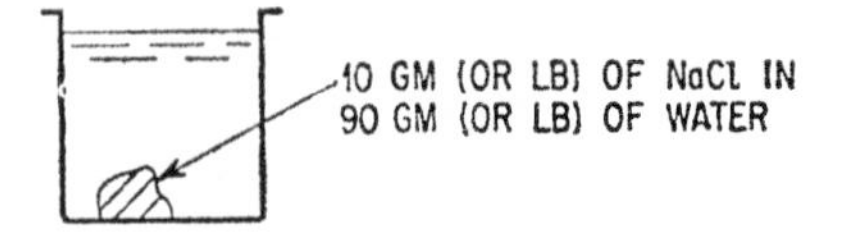

PER CENT BY WEIGHT

EXAMPLE: 10% NaCl SOLUTION

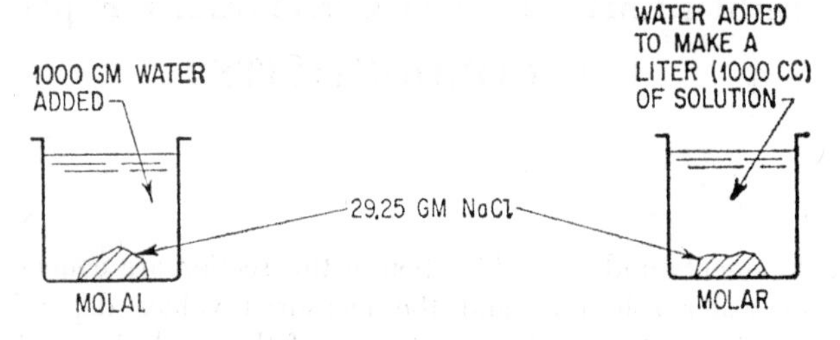

MOLAL OR MOLAR

EXAMPLE: 0.5 OR 1/2 MOLAL OR MOLAR NaCl SOLUTION.
1 GRAM–MOLECULAR WEIGHT (MOLE) = 58.5 GRAMS.
$0.5 \times 58.5 = 29.25$ GRAMS

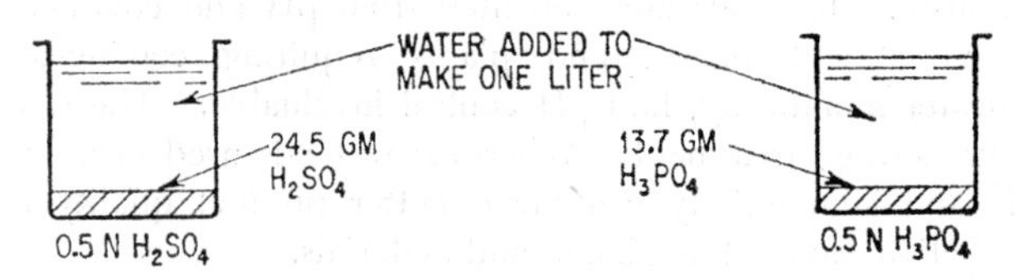

NORMAL

EXAMPLE: 0.5 N NaCl SOLUTION
(SAME AS 0.5 MOLAR FOR SALT: 1/2 GRAM–EQUIVALENT WEIGHT, 29.25 GRAMS, WITH WATER ADDED TO MAKE ONE LITER OF SOLUTION)

EXAMPLE: 0.5 N H_2SO_4 (SULFURIC ACID)
GRAM-EQUIVALENT WEIGHT = 98/2 OR 49 GRAMS (APPROX)
$0.5 \times 49 = 24.5$ GRAMS

EXAMPLE: 0.5 N H_3PO_4 (PHOSPHORIC ACID)
GRAM-EQUIVALENT WEIGHT = 82/3 OR 27.4 (APPROX)
$0.5 \times 27.4 = 13.7$ GRAMS

FIG. 8-1. Common methods of expressing solution concentration. (*Minneapolis-Honeywell Regulator Co.*)

to delve into the nature of chemicals. The chemist visualizes every grain of the salt considered above as comprising thousands of invisible particles, which he terms *molecules*. Each molecule, in turn, is made up of two smaller elements, termed *atoms*, which determine the properties of the salt. One is a sodium atom (symbol Na) and the other, a chlorine atom (symbol Cl)—the two being bonded together in a definite proportion by weight to form the sodium chloride molecule (symbol NaCl).

All the chemical elements have been assigned arbitrary *atomic weights* with oxygen as the standard having a weight of 16 units. On this basis, the atomic weight of sodium is approximately 23, and of chlorine 35.5. Thus, when sodium and chlorine are combined, the *molecular weight* of the salt is 23 plus 35.5, or about 58.5.

The weight units used can be grams, ounces, or pounds—so long as they are applied consistently. If, as is commonly the case in chemistry, the unit selected is the gram, 58.5 would be known as the *gram-molecular weight* (abbreviated *mole* or *mol*) of sodium chloride. This term forms the basis for the *molal* and *molar* methods of expressing concentration, as defined below.

Molal concentration is defined as the gram-molecular weight (or mole) of a substance dissolved in 1,000 g of solvent, whereas *molar* concentration is defined as a mole of a substance dissolved in a liter (slightly over a quart) of solution. For example, a 0.5 molal solution of sodium chloride in water comprises 58.5 times 0.5 or 29.25 g of salt in 1,000 g of water, while a 0.5 molar solution is 29.25 g of the salt with sufficient water added to form a liter of solution (Fig. 8-1). The latter is often expressed as a 0.5 *M* solution and, also, is sometimes referred to as a *formal* solution.

Another very common method of expressing the amount of a substance in solution is termed *normal concentration*, defined as the *gram-equivalent weight* of the substance dissolved in a liter of solution. The equivalent weight of a substance is that weight which will involve in a chemical reaction one atomic weight of hydrogen, or that weight of any other element or portion of a substance which, in turn, will involve in reaction one atomic weight of hydrogen.

For example, the chlorine atom of NaCl is also found in hydrochloric acid (HCl) in combination with one hydrogen atom (symbol H); therefore the gram-equivalent weight of NaCl is 58.5, the same as its gram-molecular weight. A one normal (often abbreviated 1 *N*) solution of NaCl would thus contain 58.5 g/liter of solution, and a 0.5 *N* solution 29.25 g—just as in the 0.5 *M* solution defined above.

Molar and normal solutions of the same numerical value represent different concentrations in compounds containing elements which will involve more than one atomic weight of hydrogen in reaction. For example, sulfuric acid (H_2SO_4) has two hydrogen atoms combined with one sulfate radical (SO_4), so that the gram-equivalent weight of this substance would be its molecular weight in grams (approximately 98) divided by 2, or about 49 g, and a 1 *N* solution of sulfuric acid would contain 49 g/liter.

Similarly, the gram-equivalent weight of phosphoric acid (H_3PO_4) would be its gram-molecular weight (approximately 82) divided by 3; that of calcium hydroxide $Ca(OH)_2$ its mole value divided by 2; and so on. The advantage of using solutions on the normal basis is that exactly equal volumes of two solutions which react chemically can be used.

Ion Concentrations. In electrochemical measurements, the primary concern is not the solution concentration alone, but rather the concentration of electrically charged particles termed *ions* which exist in aqueous solutions of acids, bases, and salts. The sodium and chlorine elements which remain bonded together in the salt crystals, for example, break apart or *dissociate* in solution into positively charged sodium ions (symbol Na^+) and negatively charged chlorine ions (symbol Cl^-).

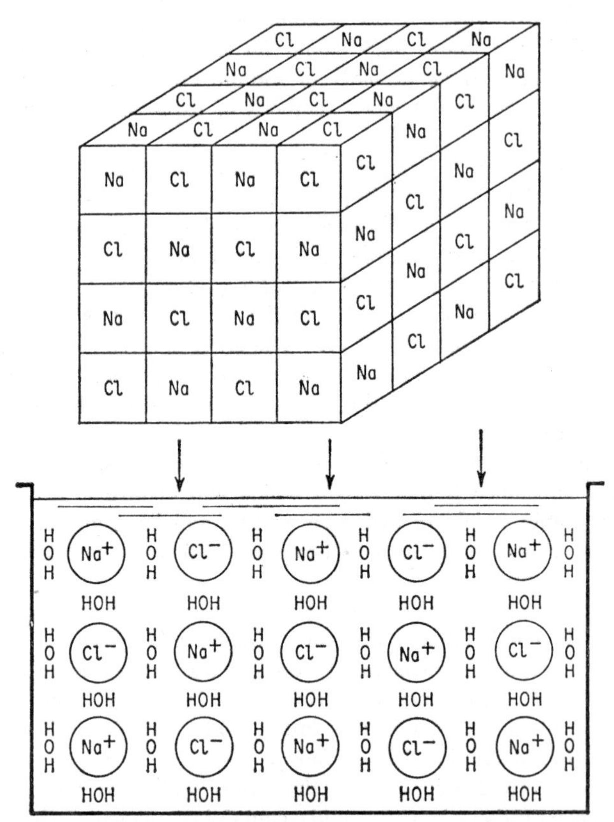

FIG. 8-2. Symbolic picture of salt (NaCl) crystal dissolving in water (HOH) to form sodium (Na^+) and chlorine (Cl^-) ions. (*Minneapolis-Honeywell Regulator Co.*)

The water may be visualized as a nonconducting barrier, which keeps apart the naturally attracted unlike electrical charges of the two ions (Fig. 8-2). This occurrence is expressed in the chemist's language as follows:

$$\underset{\text{Sodium chloride}}{NaCl} = \underset{\text{Sodium ion}}{Na^+} + \underset{\text{Chlorine ion}}{Cl^-}$$

The chemical properties of an ion are very different from the properties of the corresponding atom or molecule. For example, sodium ions are stable in the salt solution, whereas sodium atoms react violently with water, evolving hydrogen and forming a solution of sodium hydroxide. It is the *hydrogen ion* in an acid solution and the *hydroxyl ion* in a base solution which determine the properties commonly attributed to these two types of solutions.

Hydrochloric acid dissociates or *ionizes* in the following manner:

HCl	$\rightleftarrows$	H^+	$+$	Cl^-
Hydrochloric acid		Hydrogen ion		Chlorine ion

Sulfuric acid yields two hydrogen ions for each molecule by ionizing as follows:

H_2SO_4	$\rightleftarrows$	$2H^+$	$+$	$SO_4^=$
Sulfuric acid		Two hydrogen ions		Sulfate ion

A base or alkali, such as sodium hydroxide, ionizes in the following manner:

$NaOH$	$\rightleftarrows$	Na^+	$+$	OH^-
Sodium hydroxide		Sodium ion		Hydroxyl ion

In each of the expressions given above, it will be noted that the arrows between the molecule and its dissociated ions point in opposite directions. This means that the reaction is occurring continuously in both directions—that is, ions are also combining again to form the undissociated molecule. The relative amounts on each side of the arrows reach an equilibrium, with the *degree of ionization* depending on the specific substance involved and the temperature.

Because such substances in solutions will conduct an electric current if two wires are placed in them and connected to the positive and negative terminals of a battery, they are known as *electrolytes.* Solutions containing substances which ionize to a large extent are known as *strong electrolytes,* and those containing compounds which dissociate only slightly are termed *weak electrolytes.*

In a relatively dilute solution, the degree of ionization for a given compound at a given temperature is constant and is expressed by the chemist, for hydrochloric acid for example, as follows:

$$\frac{[H^+] \times [Cl^-]}{[HCl]} = K$$

The brackets around each component mean "the concentration of" (usually expressed as molar). Thus, this expression states that the product of the hydrogen- and chlorine-ion concentrations divided by the hydrochloric acid concentration in a solution is always a constant value K. If more HCl is added to the solution, a portion of it dissociates into H^+ and Cl^- ions immediately until the relative amounts are again the same.

Common-ion Effect. When a solution comprises a mixture of a weak acid (or base) in the presence of one of its salts, the effect of the ion common to both the two electrolytes has a profound effect on the resultant ion concentrations. Acetic acid, for example, is a weak electrolyte since it ionizes only slightly in the following manner:

HAc	=	H^+	+	Ac^-
Acetic acid		Hydrogen ion		Acetate ion

If one of the salts of acetic acid, such as sodium acetate (NaAc), is added to the acetic acid solution, the additional common acetate ions cause practically all the hydrogen ions to reassociate into the HAc molecule. This is seen to be true from the following equation:

$$\frac{[H^+] \times [Ac^-]}{[HAc]} = K$$

Since the product of H^+ and Ac^- ion concentrations divided by the HAc concentration must remain constant, the additional Ac^- ions cause a repression in the number of H^+ ions present in solution in the first place, and for all practical purposes the hydrogen-ion concentration in the solution is negligible.

Thus, the fact that an acid is in solution does not always mean it will have a measurable hydrogen-ion concentration. This is seen to be quite pertinent in pH measurement which involves a determination of only the hydrogen ions *actually present* in the solution.

Neutralization. It is well known that pure water itself, although made up of H^+ ions and OH^- ions in the molecule HOH or H_2O, does not exhibit the sour taste which H^+ ions produce nor the bitter taste which OH^- ions produce. The reason is that the two ions dissociate or ionize to only a very small extent, and this explains the process of *neutralization.*

When an acid and base solution is mixed, the ions combine to form a salt and water, as shown by the following expression:

BASE		ACID		SALT		WATER
Na^+OH^-	+	H^+Cl^-	→	Na^+Cl^-	+	HOH
Sodium hydroxide (dissociated)		Hydrochloric acid (dissociated)		Sodium chloride (dissociated)		(Largely undissociated)

Equivalent amounts of the acid and base thus are seen to yield an excess of neither H^+ ions nor OH^- ions, so that the resultant solution is said to be neutral. This is strictly true, however, only when strong acids and strong bases react. *Hydrolysis,* which occurs when a weak acid or base is involved, is explained below.

Hydrolysis. When a salt formed by a weak acid and a strong base, such as sodium acetate, is present in an aqueous solution, the solution is slightly alkaline because some of the H^+ ions from the water are bound in the relatively undissociated acetic acid, leaving an excess of OH ions, thus:

Na^+Ac^-	+ HOH →	HAc	+	Na^+OH^-
Sodium acetate (dissociated)	Water	Acetic acid (largely undissociated)		Sodium hydroxide (dissociated)

Similarly, ammonium chloride (NH_4Cl), the salt of a weak base and strong acid, hydrolyzes to form the relatively undissociated ammonium hydroxide (NH_4OH), leaving an excess of H^+ ions. Thus, not only acids and bases but also the salts of weak acids and bases exhibit acid and basic properties. A neutralization process, therefore, does not always produce an exactly neutral solution, when equivalent weights of the reacting compounds are present.

Buffer Action. A solution may possess a definite resistance to change of its acidity or alkalinity when either an acid or base is added to it. This resistance may be either intentional or inherent in the process involved, and is created by the presence of substances known as *buffers.*

Buffer action in a solution is caused by the mixture of a weak acid and one of its salts, or a weak base and one of its salts. Again, an explanation of this property falls back upon the theory of ionization and the fact that weak electrolytes ionize only slightly.

A common buffer solution comprises a mixture of acetic acid and its salt, sodium acetate. If an acid such as HCl is added to this solution, the hydrogen ions contained in the HCl react with the acetate ion of the dissociated sodium acetate to form undissociated acetic acid, thus:

$$H^+ + Ac^- \rightarrow HAc$$

If an alkali, such as sodium hydroxide, is added to the buffer solution, the alkali is neutralized by the acetic acid, thus:

$$OH^- + HAc \rightarrow H_2O + Ac^-$$

The tendency for such a solution to resist changes in its alkalinity or acidity is, of course, limited by the amount of the buffer materials present. With the continued addition of acid without a corresponding addition of sodium acetate, all the acetate ions would be consumed, and any excess acid would greatly increase the hydrogen-ion concentration.

Advantage of this buffer action is taken when the electrometric pH-measuring system is standardized with the aid of a buffer solution supplied with the equipment.

Definition of pH. Broadly speaking, pH is a measure of the *effective* rather than *total* acidity or alkalinity of a solution. More exactly, it is a measure of the concentration of hydrogen ions dissociated in a solution. Solution concentrations may be expressed in various ways on a weight or volume basis; the hydrogen-ion concentration might, for example, be expressed in terms of *gram equivalents* or *moles per liter.* This concentration in actual practice is of such a small order of magnitude, however, that it is more conveniently expressed in terms of pH.

In order to develop the concept of pH, consider, first of all, the nature of pure water which is usually present to form the solutions where pH is measured. Water contains hydrogen and oxygen with the formula commonly written as H_2O. Actually, it comprises a *hydrogen* atom (H) and a *hydroxyl* group (OH) bonded together in a molecule more correctly written as HOH. Furthermore, it has been determined that an *extremely small number* of hydrogen and hydroxyl groups dissociate from this molecule to form ions, as follows:

HOH	=	H^+	+	OH^+
Water molecule		Hydrogen ion		Hydroxyl ion

It has been established that the product of the concentrations of these two ions at a definite temperature (25°C or 77°F) equals a constant, extremely small value of 10^{-14}. Because there are exactly the same number of H^+ and OH^- ions present in pure water, each has a value of 10^{-7} mole/liter. Expressed as equations, these relationships are

$$[H^+] \times [OH^-] = 10^{-14} = 10^{-7} \times 10^{-7}$$

or

$$[H^+] = 10^{-7} \text{ mole/liter}$$

The hydrogen-ion concentration of water at 77°F is thus seen to be 10^{-7} mole/liter. This rather awkward expression caused the Danish biochemist Sorensen, in 1909, to coin the term pH. This unit can be considered simply as the negative exponent (or power) of the hydrogen-ion concentration (in the case of water −7) with the negative sign removed. In other words, the pH of neutral water at 77°F is 7.

pH *Values of Acid and Base Solutions.* All acids supply H^+ ions, more or less, depending on the concentration and degree of ionization. Likewise, all bases (or alkalies) supply OH^- ions. When one or the other of these ions is added to pure water, what happens to the pH?

First, assume that an acid is added, supplying sufficient H^+ ions to increase their concentration in the solution from 10^{-7} to 10^{-4} mole/liter. The product of $[H^+]$ and $[OH^-]$ present in the solution must still equal 10^{-14} by the law stated above. Thus, the $[OH^-]$ must diminish to 10^{-10} (by reassociating with some H^+ ions to form HOH), in accordance with the equation

$$\text{pH} = 4$$
$$[H^+] \times [OH^-] = 10^{-14}$$
$$10^{-4} \times 10^{-10} = 10^{-14}$$

Thus, the sum of the exponents of $[H^+]$ and $[OH^-]$ must always equal −14. If still more acid is added to bring $[H^+]$ to 10^{-2} (pH = 2), then $[OH^-]$ must become 10^{-12}. Thus it is seen that addition of acid *lowers* the pH below 7, since the negative exponent grows smaller as the $[H^+]$ increases.

In exactly the same manner, addition of OH^- ions from a base causes the hydrogen-ion concentration to decrease in order to maintain the product of $[H^+]$ and $[OH^-]$ constant. The pH values *increase* above 7 with addition of bases to water, since the negative exponent of $[H^+]$ increases to represent smaller concentrations. From this interrelation of $[H^+]$ and $[OH^-]$, it is also evident why pH is equally suited as a measure of alkalinity or acidity.

As the solution becomes more acid with H^+ ions, the pH is seen to approach zero; as it becomes more alkaline with OH^- ions, the pH is seen to approach 14. Actually, however, a pH of zero represents an H^+ ion concentration of only 10^0 or 1 mole/liter, which is not too concentrated a solution. Acids can easily be prepared to yield H^+ concentrations greater than this, but, in practical applications of pH measurement, most requirements fall above the zero limit.

Similarly, pH values above 14 are possible, but are likewise of interest more from a theoretical point of view than from a practical one. In this respect, the diagram of Fig. 8-3 will be found useful in comparing the intensity of acidity or alkalinity to an analogous scale for temperature which evaluates the intensity of heat. In this comparison, a temperature of 50°F is arbitrarily compared to the neutral pH point of 7. In contrast to the temperature scale, moreover, it will be noted that, because each pH unit represents a power of 10, a change of *one* such unit represents a *tenfold* change in the hydrogen-ion concentration.

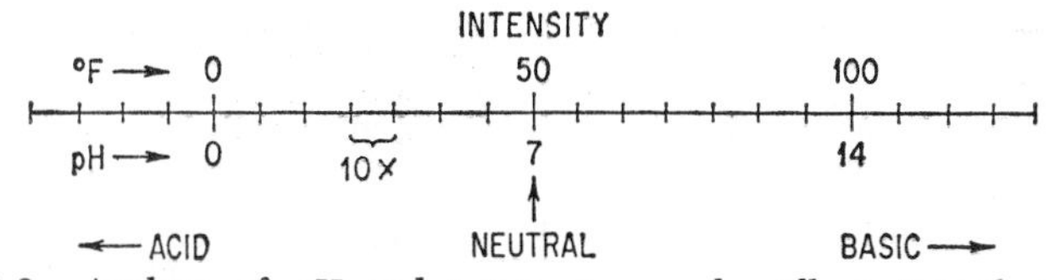

FIG. 8-3. Analogy of pH and temperature scales, illustrating that pH is a measure of intensity of acidity or alkalinity. (*Minneapolis-Honeywell Regulator Co.*)

Theory of Electrometric Measurement of pH. The electrometric system for measuring pH utilizes electrical principles, in contrast to the *colorimetric* method which has been commonly used for years. The latter employs various chemical dyes, called "indicators" (Table 8-1), which have the property of exhibiting a definite change in color at a certain pH value or narrow range of values. Thus, a measured amount of test solution to which a definite quantity of a suitable indicator is added has a color that can be compared by eye with a "standard" to determine the pH. Limitations of this method are: (1) the time delay in sampling, (2) the fact that it is applicable only to relatively clear solutions, and (3) the fact that the accuracy seldom exceeds 0.2 pH.

The basic principle of the electrometric system, illustrated in Fig. 8-4, is that an electrolytic cell, composed of two electrodes immersed in the solution being measured, develops voltage in relation to the pH of the solution. A *measuring electrode* generates its own potential in accordance with the solution pH, while the *reference electrode* has a constant potential. The liquid between these electrodes completes the electrical circuit. The cell voltage is the algebraic sum of the two electrode potentials.

Use of Amplifier and Recorder. As given in Table 8-5, at 25°C the glass-electrode potential varies from about −456 mv at 3 pH to −43 mv at 10 pH, while the calomel-electrode potential remains constant at about

+245 mv. The net emf output of the cell at 25°C is thus seen to change approximately 59 mv for each pH unit (at 100°C, this net change is 74 mv, and, at 0°C, 54 mv).

A conventional voltmeter cannot be used to measure the cell output because current must flow to actuate such an instrument, and, when an appreciable current flows, polarization effects at the electrodes cause changes in the cell voltage. Use of a *null-balance* potentiometer system makes it possible to measure the cell voltage with no appreciable current flow.

Table 8-1. Chemical pH Indicators—Range of Use*

Indicator	pH range	Indicator	pH range
p-Naphtholbenzein	0–0.8	3′,3″-Dibromothymolsulfonephthalein (bromothymol blue)	6–7.6
Picric acid	0.1–0.8	Brilliant yellow	6.6–7.9
Malachite green oxalate	0.2–1.8	Neutral red	6.8–8
Quinaldine red	1–2	Phenolsulfonephthalein (phenol red)	6.8–8.4
4-Phenylazodiphenylamine	1.2–2.9	*o*-Cresolsulfonephthalein (cresol red)	7.2–8.8
m-Cresolsulfonephthalein (meta-cresol purple)	1.2–2.8	*m*-Cresolsulfonephthalein (meta-cresol purple)	7.4–9
Thymolsulfonephthalein (thymol blue)	1.2–2.8	Ethyl bis(2,4-dinitrophenyl)-acetate	7.5–9.1
p-(*p* Anilinophenylazo)benzenesulfonic acid sodium salt (orange IV)	1.4–2.8	Thymolsulfonephthalein (thymol blue)	8–9.6
o-Cresolsulfonephthalein (cresol red)	2–3	*o*-Cresolphthalein	8.2–9.8
2,4-Dinitrophenol	2.6–4.4	Phenolphthalein	8.3–10
3′,3″,5′,5″-Tetrabromophenolsulfonephthalein (bromophenol blue)	3–4.7	Thymolphthalein	9.4–10.6
Congo red	3–5	5-(*p*-Nitrophenylazo)salicylic acid sodium salt (alizarin yellow R)	10–12
Methyl orange	3.2–4.4	*p*-(2-Hydroxy-1-naphthylazo) benzenesulfonic acid sodium salt (orange II)	10.2–11.8
3-Alizarinsulfonic acid sodium salt	3.8–5	*p*-(2,4-Dihydroxyphenylazo)-benzenesulfonic acid sodium salt	11.2–12.7
Propyl red	4.6–6.6	2,4,6-Trinitrotoluene	11.5–13
3′,3″-Dichlorophenolsulfonephthalein (chlorophenol red)	4.8–6.8	1,3,5-Trinitrobenzene	12–14
p-Nitrophenol	5–7		
5′,5″-Dibromo-*o*-cresolsulfonephthalein (bromocresol purple)	5.2–6.8		

* From D. M. Considine (ed.), "Process Instruments and Controls Handbook," McGraw-Hill Book Company, Inc., New York, 1957.

The electrical circuit of the pH cell is characterized as one having a very low current flow (10^{-12} amp) and a very high resistance (1,000 megohms or more at the thin membrane of the glass electrode). Because standard potentiometers are not designed to handle such a circuit with sufficient voltage sensitivity, a special amplifier is used to increase the current value and decrease the voltage to a range suitable for measurement by the conventional recorder. Essentially, the amplifier unit is an electronic electrometer operating on a null-balance principle, which is inherently suited to measurements of such small currents at high resistances.

As shown schematically in Fig. 8-4, the electrometric system essentially involves the continuous balancing of the cell voltage output E_{pH} by a variable current I through a fixed resistor R, the IR value constituting the opposing voltage $E_{BAL.}$. In this circuit, any voltage unbalance due to pH changes at the cell is greatly amplified to form the output current I, which is accurately linear with respect to the input voltage. This same output current flowing through the resistor serves to supply the potentiometer with its actuating voltage for pH recording, as shown.

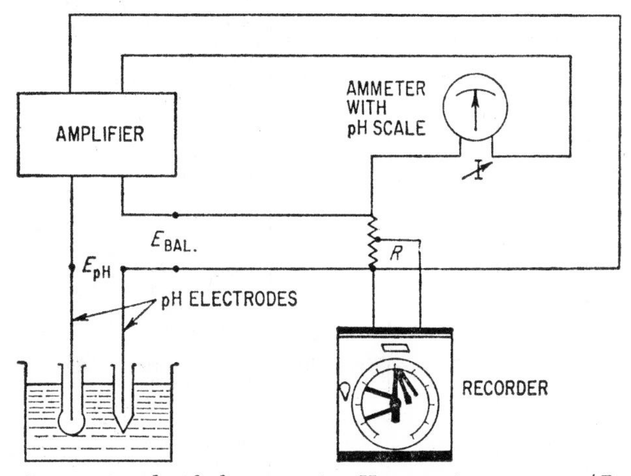

FIG. 8-4. Basic principle of electrometric pH-measuring system. (*Beckman Instruments, Inc., and Minneapolis-Honeywell Regulator Co.*)

For use with the electronic recorder, resistor R, located in the amplifier, provides an output to the recorder system of 5 mv per pH, or 0 to 35 mv for the standard pH ranges of 7 units, namely, 0 to 7, 3 to 10, and 6 to 13. The specific range is determined by a *calibrating resistor* in the amplifier and can be easily changed in service, if desired.

Actually, the amplifier provides a linear emf output corresponding to 1 pH unit below and 2 pH units above the standard amplifier range of 7 pH units, so that a standard recorder can be supplied for a range of −5 to +45 mv. This is equivalent to a range of 10 pH units, and the recorder actually is available with a chart-and-scale calibration of 2 to 12 pH units for this range.

Measuring Electrodes.[2] The basic purpose of the measuring electrode is to measure the hydrogen-ion concentration of an unknown solution. This value is expressed as pH. For all practical purposes, the measurement is

[2] Beckman Instruments, Inc.

Table 8-2. Characteristics of pH-measuring Electrodes*

Electrode type	Operating range			Limitations	Advantages
	pH	Temp., °C	Pressure, psi		
Glass (pH-sensitive)	0–13	0–100	0–100	Has high internal resistance; requires shielding, excellent insulation, and electrometer-type voltmeter. Error occurs in high conc. of alkali. Attacked by fluoride solutions	Wide pH and temperature range. Not affected by oxidizing or reducing solutions, dissolved gases, or suspended solids. Not affected by moving liquids except at high velocity
Antimony (pH-sensitive)	4–11.5	0–60	Not limited	Electrode poisoned by Bi, As, Cu, Ag, Hg, and Pb. Affected by some oxidizing and reducing solutions. Tartrates and citrates cause errors. Dissolved O_2 must be present to maintain pH-sensitive oxide coating. Active surface must be periodically scraped and reformed	Very rugged and durable for use in abrasive slurries. Has low cell resistance; shielding and special voltmeter not required
Quinhydrone (pH-sensitive)	0–8.5	0–37	Not limited	Limited pH range, "salt error," cannot be used in presence of oxidizing and reducing agents, "protein errors," quinhydrone may change pH of unbuffered solution	Simple electrode, low resistance
Hydrogen (pH-sensitive)	Not limited	Not limited	Atmospheric pressure	Cannot be used in presence of oxidizing or reducing agents. Cannot be used in presence of elements below hydrogen. Slow to reach equilibrium; large samples required	Standard of reference, no alkaline error
Calomel (reference)	Not limited	Life shortened at high temperatures	Atmospheric pressure or below except for special designs	No interferences except from contamination from high-pressure test solutions	Can be used with any pH-sensitive electrodes
Silver–silver chloride (reference)	Not limited		Atmospheric pressure or below except for special designs	Interference by contamination from high-pressure solutions	Mercury-free, may be used with any pH-sensitive electrodes

* From D. M. Considine (ed.), "Process Instruments and Controls Handbook," McGraw-Hill Book Company, Inc., New York, 1957.

accomplished by determining the potential or voltage that is developed between the measured solution and the standard solution contained within the electrode. This voltage is developed only because of the differences in the hydrogen-ion concentrations of the two solutions (Table 8-2).

The *glass electrode* (Figs. 8-5 and 8-6) is the most common means of measuring the hydrogen-ion concentration, or pH, of a solution. The glass electrode itself consists of a glass tube closed at the bottom by a membrane of special pH-sensitive glass. In contact with the inner wall of the

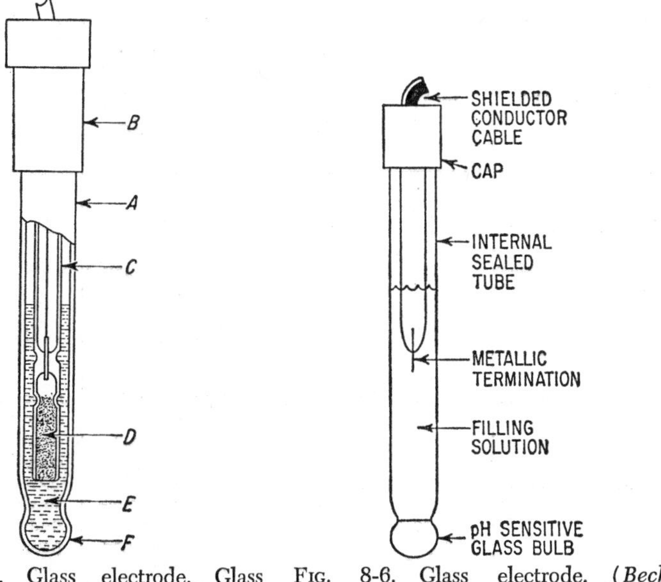

FIG. 8-5. Glass electrode. Glass tube *A*, sealed to cap *B*, protects internal element *C* containing calomel *D*. Buffer solution *E* fills sensitive bulb *F*. Voltage exists between opposite surfaces of this bulb. (*Leeds & Northrup Co.*)

FIG. 8-6. Glass electrode. (*Beckman Instruments, Inc.*)

glass membrane is a buffered chloride solution in which also is immersed a silver-coated wire. The wire penetrates an inner glass tube and is joined by the insulated and shielded conductor cable that travels from the electrode to the pH meter. It is not possible to explain definitely the exact mechanism by which a surface potential is developed on the pH-sensitive tip of the electrode. Bates[3] says that "the most plausible explanation, combines ion exchange with proton transfer. When a freshly blown bulb is first immersed in a solution, hydrogen ions from the solution exchange with

[3] R. G. Bates, "Electrometric pH Determinations," p. 181, John Wiley & Sons, Inc., New York, 1954.

alkali-metal ions from the glass membrane, finding points of high stability on the surface. The conditioned electrode behaves thereafter as a . . . site for proton transfer between the solution and the reservoir of protons in the glass surface; the potential of the surface changes as protons are acquired or lost. The glass electrode process is not one of electron exchange, that is, of Oxidation-Reduction. Hence, the glass electrode is the only hydrogen ion electrode not disturbed by oxidizing and reducing agents."

Table 8-3. Types of Reference Electrodes*

Characteristics	Ground-glass sleeve	Asbestos fiber	Palladium annulus
Flow rate (6-in. head)	5 ml/24 hr (approx.)	0.1–0.01 ml/24 hr	0.01–0.001 ml/24 hr
Electrical resistance	Less than 1,000 ohms	4,000 ohms (approx.)	6,000 ohms (approx.)
Day-to-day stability of liquid-junction potential†	±0.06 mv (±0.001 pH)	±2 mv (±0.03 pH)	±0.2 mv (±0.003 pH)
Effect of pressure	Not suited to pressure operation; flow too high	Under pressure, flow may reach 10–50 ml/day at 10 psi	Under pressure flow may reach 5 ml/day at 10 psi. Little clogging
Remarks	Very reproducible potential. Difficult to clog; can be easily flushed if clogged or contaminated by viscous material or sediments. Chemically inert	Good for general use. Viscous material or suspensions should be avoided due to tendency to clog. Chemically inert	Shows little tendency to clog in viscous material and suspensions; is easily cleaned by heating in water. Gives very reproducible potential, yet flows so slowly that contamination of sensitive media is minimized

* Courtesy of Beckman Instruments, Inc.
† In KCl or buffer solutions.

Basically, in order to complete an electrical circuit, it is necessary only to immerse a wire connected to the same amplifier that receives the glass-electrode conductor cable. However, this wire immersed in the solution may develop a variable potential of its own, depending on the type of solution in which it is in contact. To avoid this interfering variable, a somewhat more elaborate means of maintaining electrical contact with the solution is used. This consists of a reference electrode containing a tube packed with silver and silver chloride. Surrounding this inner electrode is a solution of a strong inorganic electrolyte.

Normally, this consists of a saturated solution of potassium chloride. This serves as an intermediate solution relatively constant in character between the inner electrode itself and the solution to be measured. Contact between the potassium chloride solution and the solution to be measured is accomplished by one of several types of liquid junctions.

Reference Electrodes. The reference electrode (Figs. 8-7 and 8-8) is

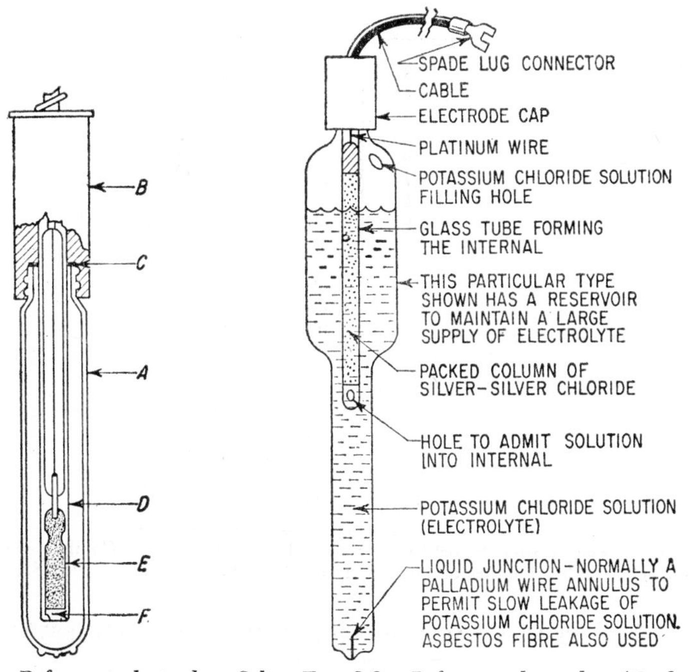

Fig. 8-7. Reference electrode. Salt bridge tube *A*, containing replaceable KCL conducting solution, screws into cap *B* against Neoprene washer *C*. Internal element *D* contains calomel *E*, which can be replaced through cotton plug *F*. Voltage exists between mercury calomel and KCL solution. (*Leeds & Northrup Co.*)

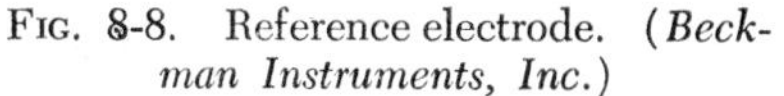

Fig. 8-8. Reference electrode. (*Beckman Instruments, Inc.*)

an important member of the electrode pair required to measure pH. This electrode permits the measurement of the potential developed at the glass electrode due to the hydrogen-ion concentration. This is its only purpose. Thus, this electrode must be stable and must not be the source of any false, variable potential.

The reference electrode normally consists of a silver–silver chloride or a mercury–mercurous chloride (calomel) internal element surrounded by an electrolyte (or salt solution) of known concentration. The silver–silver

Table 8-4. Millivolt Potential Developed at the Glass Electrode per pH Unit (or Tenfold Change in Hydrogen-ion Concentration) versus Temperature*

The values listed reveal the primary reason for the need of temperature compensation within the pH-measuring system

Temperature		Potential per pH unit, mv
°F	°C	
32	0	54.2
41	5	55.2
50	10	56.2
59	15	57.2
68	20	58.2
77	25	59.2
86	30	60.1
95	35	61.1
104	40	62.1
113	45	63.1
122	50	64.1
131	55	65.1
140	60	66.1
149	65	67.1
158	70	68.1
167	75	69.1
176	80	70.1
185	85	71.1
194	90	72.1
203	95	73.0
212	100	74.0

* Abstracted from Report to the National Research Council Committee on Constants and Conversion Factors of Physics, December, 1950, *Phys. Rev.*, vol. 82, p. 555, 1951.

Table 8-5. Millivolt Output of pH Cell versus pH at 25°C

pH	Millivolt output (approx.)		
	Glass	Colomel	Net*
3	−456	+245	−211
7	−220	+245	+ 25
10	− 43	+245	+202

* Approximately 59 mv per pH unit.

chloride type is preferred for commercial use since it can tolerate higher temperature levels. A strong solution of potassium chloride has been chosen as a suitable electrolyte within this electrode (Fig. 8-8). This solution contacts the internal metal–metallic salt and forms a conductive bridge between this half cell and the solution outside of the electrode.

To maintain electrical communication from the reference electrode into the solution to be measured, it is necessary to expose the electrolyte to the solution. This is accomplished in one of several ways.

Because of the advantages of the palladium-junction type of electrode, this type is commonly chosen for many applications. However, since palladium can be attacked by some strong chemicals, the fiber type of electrode should be selected in such applications. Specifically, the palladium junction should not be used in strong solutions of powerful oxidizing or reducing agents or metal complexing agents.

Stability of the pH Measurement. There are numerous potential problems that one can encounter in employing continuous pH measurement. Normally, these problems can be solved by understanding the conditions that must be provided for a successful pH measurement. In the majority of cases, when difficulties are experienced, they can usually be located at the site of the electrodes themselves. The following is a review of primary

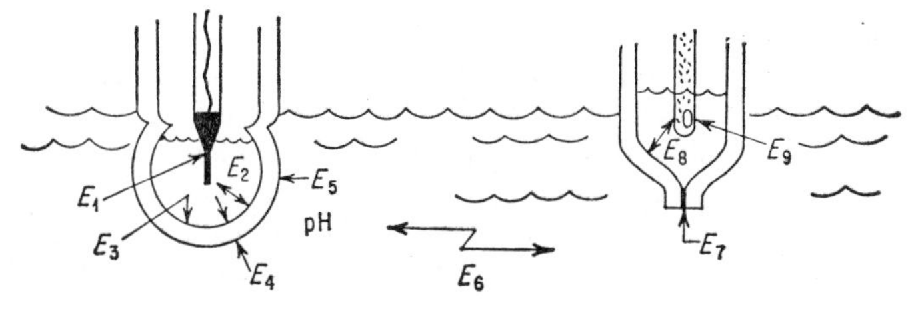

FIG. 8-9. Electrodes immersed in test solution.

reasons why such difficulties can develop. Figure 8-9 schematically portrays the glass electrode and the reference electrode immersed in a test solution. It can be seen that various potentials and resistances are present within electrodes and their surroundings. A successful pH installation will ensure that all these potentially variable factors will remain constant except the potential due only to the hydrogen-ion concentration of the test solution itself.

Beginning with the internal of the glass electrode, there exists a potential E_1 between the internal metallic electrode and the internal standard pH solution. The potential drop or resistance between the internal electrode and the inner surface of the glass membrane through the internal solution is represented by E_2. A potential E_3 exists between the internal solution and the inner surface of the glass membrane. The relatively high resistance of the pH-sensitive glass membrane itself is represented as a potential drop E_4. This high resistance, until recent years, was the chief factor in retarding widespread industrial pH development. The potential E_5 between the outer surface of the glass electrode and the solution being measured is the desired potential that varies in accordance with change in hydrogen-ion concentration, or pH, of the test solution. E_6 represents the potential drop across the test solution. E_7 represents the liquid-junction potential between the test solution and the solution within the reference electrode. The liquid junction is the opening or path of communication whereby the potassium chloride electrolyte maintains electric contact with the test solution. E_8, the potential drop within the reference electrode, is due to the resistance of the electrolyte within the electrode. E_9 represents the potential developed at the reference electrode itself in its surroundings of saturated potassium chloride solution. The following list evaluates the importance of each of these factors.

E_1: Normally, this can be considered a constant potential at constant temperature. Extremely high temperatures can dissolve the silver chloride coating from the metal contacting the internal solution.

E_2: This is a negligible factor and normally can be ignored, with the present-day types of internal filling solutions.

E_3: The stability of both the internal solution and the internal surface of the glass-electrode membrane is sufficient to assume that this value is a constant.

E_4: This will remain a constant as long as the glass membrane does not undergo serious mechanical or chemical alterations.

E_5: This is a variable, depending on the hydrogen-ion concentration of the solution to be measured.

E_6: The potential drop across the test solution normally is negligible except in dealing with very low-conductivity water or essentially nonaqueous solutions.

E_7: The liquid-junction potential initially is insignificant; yet clogging or external pressure against the junction can seriously influence this element.

E_8: The filling solution has a negligible resistance as long as it does not become contaminated.

E_9: This potential can be considered constant if the filling solution is uncontaminated.

It can be seen from the above that certain qualifications are necessary in order to assure that a constant potential or potential drop can be expected. The basic cause of most of the difficulties encountered in continuous pH measurement is due to overlooking one or more of these qualifications. Thus, some of the following situations that sometimes are encountered can now be seen to influence the pH measurement.

1. Coating or abrasion of the glass-electrode membrane (affecting E_4 and/or E_5)
2. pH measurement of high resistance or nonaqueous test solutions (affecting E_6)
3. Clogging or mechanical failure of the liquid junction at the reference electrode (influencing E_7)
4. Dilution or contamination of the reference-electrode filling solution by inflow-outflow of test solution (affecting both E_8 and, most important, E_9)

5. Failure of the calomel-type internal electrode due to high temperature (influencing E_0)

Much of pH equipment consisting of a variety of electrodes and the mechanical facilities for locating and retaining the electrodes in the solution has been designed to overcome the possible conditions enumerated above. Because of this, great emphasis must be placed on the importance of components for a specific pH application.

A third element must be included, in conjunction with the electrode pair, to automatically compensate for the temperature coefficient inherent in the pH measurement. To illustrate this factor, it can be shown that a potential of approximately 59 mv per pH unit is developed at the glass electrode in a test solution at 77°F. However, at 212°F, a potential of approximately 74 mv per pH unit is developed in the same test solution (Fig. 8-10). Since

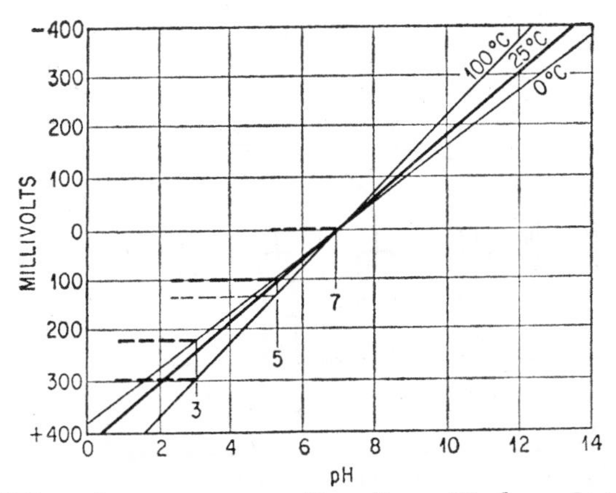

FIG. 8-10. Effect of temperature on pH reading. (*Beckman Instruments, Inc.*)

the measurement is read in pH units on a pH meter or recorder, it is necessary to compensate for this millivolt change per unit temperature change, so that the meter properly indicates the true pH, regardless of the prevailing temperature of the test solution. The resistance thermometer or thermo-compensator automatically makes this adjustment. As the temperature of the solution changes, the circuit constants are altered accordingly.

To provide a readable measurement of the potential developed at the glass electrode, a highly refined amplifier is required. The pH meter itself must be capable of overcoming the high impedance offered by the glass electrode, and must measure very accurately the potential developed at the glass electrode on a meter calibrated in pH units. Such a meter, if desired, also can provide a signal suitable for continuously recording the pH measurement.

Electrode Chambers. Application requirements usually dictate the type of electrode assembly necessary for each specific pH installation.

The three basic pH-measuring systems that satisfy most applications use sensing elements which are mounted:

1. Within a tank, trough, or exposed stream
2. Within a piped sample stream
3. Within the piped main stream

To satisfy the first category, an exposed electrode assembly is employed if the unit will not be submerged. When it is necessary to submerge the assembly (due to varying level, etc.) a submersion assembly is used.

For the second condition, it is usually possible to measure at atmospheric pressure, and, for this purpose, a variety of atmospheric-pressure flow chambers exist. A choice is available since various materials of construction are required to satisfy the maximum number of applications. When the sample

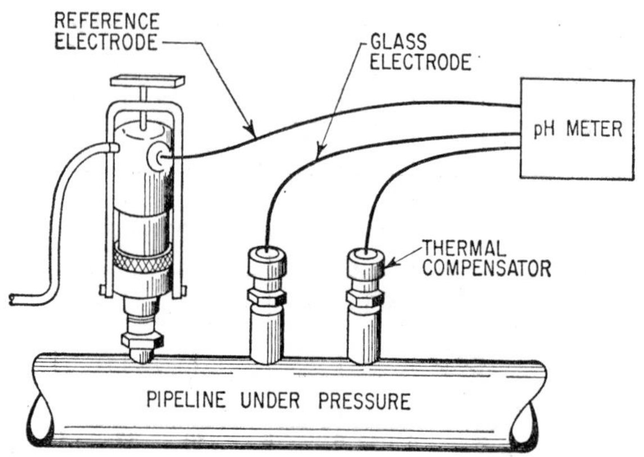

FIG. 8-11. Individual electrode glands are designed for direct installation into pipelines or through tanks. These units provide for dependable measurement of process stream pH over a wide operating range of pressure and temperature. (*Beckman Instruments, Inc.*)

stream must remain under positive pressure, a pressurized-flow chamber is used.

The third group of applications normally requires that a pressurized-flow chamber be employed since the piped stream is usually under pressure. Where the total flows are relatively small in volume, the flow assembly receives the total stream. For large pressurized flows, individual electrode gland units can be used most effectively (Fig. 8-11).

CONTINUOUS-CONDUCTIVITY MEASUREMENT

Continuing advances in boiler design with higher operating pressures and temperatures have further emphasized the necessity for generation of pure steam. In addition to this factor, the high capacity and reliability demanded of a modern steam generator has changed the status of continuous-recording and automatic-control equipment from an operating convenience to a necessity. Measurement of electrolytic conductivity as a check on dissolved solids in such systems, for example, has become almost mandatory.

In most instances, naturally occurring water is unsuitable for boiler plant use because of salts such as magnesium and calcium chlorides, sulfates, and bicarbonates, which it usually contains. It is well known that these salts have a tendency to decompose and form scale under the conditions of high pressure and temperature encountered in a steam power station. Were such water used in a modern steam generator without adequate chemical treatment, the magnesium and calcium bicarbonates would quickly decompose to form their respective carbonates, which would, in large part, deposit as hard scale on the inside tube surfaces. Additional scale would be deposited from calcium sulfate, as the concentration of salts increased during evaporation of the water. Hydrolysis of the magnesium salts would produce an acid condition in the water with a resulting corrosion of metal parts. Scale deposit and corrosion would combine to reduce the over-all efficiency of the steam generator and add to the danger of tube failure caused by overheating.

To safeguard a high-pressure steam generator and reduce the frequency of outages, it is important that high-purity water be used in the system. Improvements in boiler design to increase steam purity and reduce foaming or priming, together with the use of boiler feedwaters of exceedingly low total dissolved-solids content, have made possible efficient operation of modern power plants. Electrolytic-conductivity instrumentation contributes materially to the maintenance of this operating efficiency.

Basic Theory of Conductivity.[4] Electrolytic conductivity is basically the electrical-conducting ability of ions present in solution when an emf is impressed across two electrodes immersed in the solution. The term *electrolytic* is used because the solution contains an *electrolyte.* In contrast to pH measurements, conductivity is affected by *all* ions dissociated in a solution because they all conduct current to some extent.

Conduction of Current. Current flow in solutions is caused by the migration of ions to the electrodes bearing electrical charges opposite to those of the ions. In a salt solution, for example, the Na^+ ions migrate to the negatively charged electrode, while the Cl^- ions are attracted to the positively charged electrode. Upon reaching the respective electrodes, positively charged ions acquire electrons, while negatively charged ions release electrons and become neutralized.

This exchange of electrons completes the electrical circuit through the solution and establishes a current flow which depends on (1) the number and type of ions present in the solution (determined by the *concentration* of an ionizable chemical present and its *degree of ionization*), (2) the effective area of the electrodes, (3) the potential difference and distance between the electrodes, and (4) the temperature of the solution. Thus, for a given chemical if items 2, 3, and 4 are maintained constant, conductivity is a

[4] Minneapolis-Honeywell Regulator Co.

direct measure of the number of ions in solution, and, for many chemicals which dissociate almost completely into ions, it is a direct measure of concentration.

The standard or reference temperature at which specific conductances are computed varies with the electrolyte, but is usually about 77°F (25°C). As indicated above, however, conductivity is affected by the temperature of the measured solution, the effect being to change the measured value approximately one per cent per degree Fahrenheit. Thus, where appreciable temperature variations are anticipated, particularly where automatic control is involved, continuous temperature compensation is generally required.

Polarization of Electrodes. A prime consideration in the continuous measurement of conductivity is polarization of electrodes. This results, with a direct current, when the ions attracted to an electrode become so concentrated that their collective electrical charge almost equals that of the electrodes. If such a condition is permitted to occur, it becomes increasingly difficult for an ion to reach the electrode, because the opposing forces (including the mutual repulsion of ions of like charge) tend to establish a state of dynamic equilibrium.

For example, if Na^+ ions build up around a negatively charged electrode, there is a tendency for the like positive charge around the electrode to repulse additional Na^+ ions. Furthermore, inasmuch as the charge of the ionic field surrounding the electrode nearly equals the electrode charge, the net potential is very low and has little power to attract additional ions of opposite charge from the solution. Because of the large measuring error which would be introduced by polarization of the electrodes, alternating rather than direct current is employed in conductivity-measuring systems, usually with a frequency of 60 cycles.

Units of Measurement—the Cell Constant. The conductance of a solution, as with current in a wire, is the reciprocal of resistance and, similarly, is expressed as the ratio of current flowing to the difference in potential impressed across the electrodes. It is measured in *mhos* (reciprocal ohms), one mho being the conductance of a solution through which a potential difference of one volt will cause a current flow of one ampere. For convenience in expressions of conductance with more dilute solutions, one-millionth of a mho, or the *micromho,* is commonly used.

The standard unit of measure for electrolytic conductivity is *specific conductance.* This unit is defined as the conductance in mhos of one cubic centimeter of solution, as measured between two electrodes one centimeter square located one centimeter apart. Practical considerations in the design and application of industrial conductivity cells, however, dictate the use of electrodes which vary in size and spacing from these standards. This avoids measurement of extremely high or extremely low resistances encountered in practice, but requires a correction factor, termed the *cell constant,* to relate the actual measured value to the standard value.

Mathematically, the cell constant is the *ratio of specific conductance to measured conductance;* or, in other words, it is the constant by which the measured conductance is multiplied to give the value of the measurement in terms of specific conductance. Numerical values of the cell constant vary widely, from fractional units for poorly conducting liquids like almost pure distilled water (0.1), to higher numbers above one for highly dissociated electrolytes like sodium hydroxide (20).

Scope of Measurements. Independent of the measuring system employed, several factors affect the practical application of conductivity measurements. These include the relationship between conductivity and concentration for a given chemical, and the effect of mixtures upon such relationship. The three typical curves of Fig. 8-12 serve to illustrate the first consideration, showing the conductivity in mhos for varying concentrations of hydrochloric acid (HCl), sodium hydroxide (NaOH), and sodium chloride (NaCl).

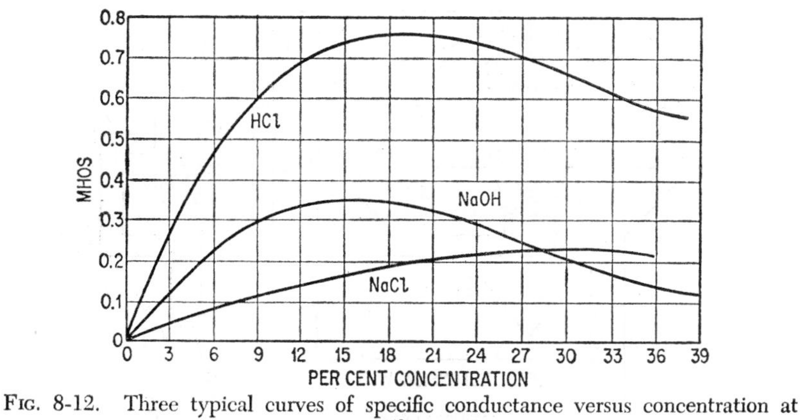

FIG. 8-12. Three typical curves of specific conductance versus concentration at 65°F.

Reference to the three curves shows that none is linear throughout the full range of concentrations. As the basis for measurement and control, the slope of the curve in the desired range of concentrations is thus an important consideration. Measurement of HCl concentration in the vicinity of 9 per cent, for example, is seen to be quite feasible, but at 18 per cent, where the curve reaches a maximum, it is obviously not suited, because decreasing conductivity values might be indicative of either decreasing or increasing concentration. Beyond this peak, it is again possible to select a suitable measuring range, say, around 30 per cent concentration of HCl.

In mixtures of electrolytes, it is usually desirable to test the solution experimentally, to determine whether or not a useful relationship between conductivity and concentration is present. If, for example, the ratio of the electrolyte concentrations remains constant, measurement of conductivity is practical. In general, however, the component of prime interest in the mixture should be present to a much greater extent than any others. Suppliers of conductivity-measuring equipment have collected considerable data on the conductivity of many commonly used electrolytes and mixtures of them; it is recommended that they be consulted with reference to the suitability of measurements.

Electrometric Measurement of Conductivity. Because measurement of electrolytic conductivity basically is measurement of resistance, a conventional type of recorder with an a-c Wheatstone-bridge circuit is adapted simply to its measurement. In addition to this instrument, all that is required in the measuring circuit is a suitable primary element, termed the *conductivity cell,* which contains the two electrodes to establish the measured resistance.

Basic Measuring Circuit. Shown schematically in Fig. 8-13 is the basic measuring circuit of the electrolytic-conductivity system. The conductivity

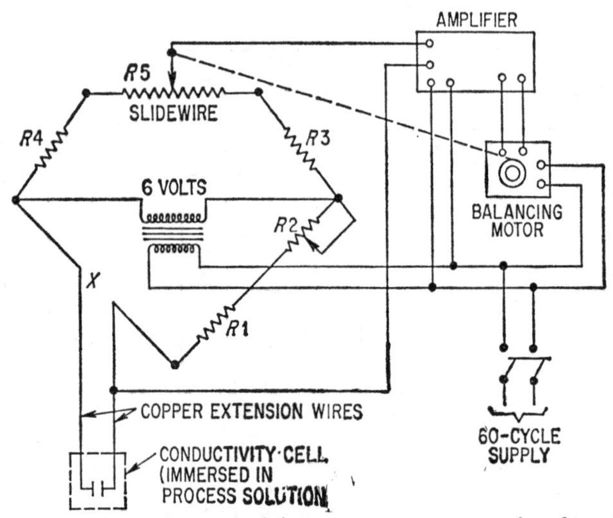

FIG. 8-13. Schematic diagram of basic measuring circuit for electrolytic conductivity. (*Minneapolis-Honeywell Regulator Co. and Industrial Instruments, Inc.*)

cell is seen to form a variable resistance *X* in one leg of the Wheatstone bridge. Resistors *R*1, *R*2, *R*3, and *R*4 in the other legs of the bridge can all be considered fixed for the moment, with *R*5 being the instrument slidewire used to rebalance the circuit. Upon a change in conductivity of the solution under measurement, resistance *X* changes, and the electronic detecting system immediately reacts to move the slider on *R*5 to a new position of balance. With this rebalancing action, the instrument pen moves a corresponding amount to record the new conductivity value.

Manual Temperature Compensation. Resistor *R*2, shown in Fig. 8-13, is installed to afford a means of manual compensation for conductivity changes which are due to variations in solution temperatures from a reference temperature. It is a variable resistor with an adjusting knob calibrated in terms of temperature, so that it can be set for the actual temperature of the solution. The adjustment alters resistance in one leg of the bridge circuit so that it compensates for the effect of temperature on the measured conductivity.

For example, if the reference temperature is 77°F, a solution temperature of 87°F would tend to give a reading about 10 per cent higher than standard. If it is raised to 87°F, the compensating adjustment removes sufficient resistance at $R2$ (Fig. 8-13) to correct for the lower value of cell resistance X (higher conductivity).

Automatic Temperature Compensation. Although such manual compensation is entirely adequate for many applications, appreciable variations in temperature usually dictate the use of continuous automatic compensation—particularly where control is required. One such method, known as the Bishop method, involves the use of two conductivity cells. The second, or "compensating," cell differs basically from the measuring cell only in that it contains a *fixed* concentration of a standard solution, and its electrodes are sealed from the process solution. It is immersed in the process solution

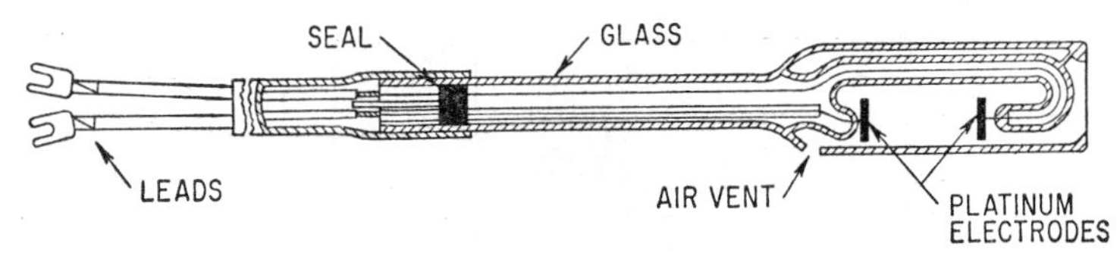

FIG. 8-14. Dip-type conductivity cell.

adjacent to the measuring cell and connected in the Wheatstone-bridge circuit in place of the manual compensating resistors $R1$ and $R2$. Subject to the same changes in conductivity (i.e., resistance) with temperature as the measuring cell, the second cell serves as a continuous and automatic compensator for temperature variations in the process solution.

A second means of automatic compensation involves the use of a temperature-sensitive resistance element connected in one arm of the bridge circuit of Fig. 8-13 in place of the manual compensating resistor. The element is immersed in the solution on or near the measuring cell, and automatically corrects for temperature variations in much the same manner as described for the previous types of compensators. This method has the advantage over the above described automatic method in that it doesn't require specially filled compensating cells, thereby making it more versatile in application.

Conductivity Cells.[5] The primary element in electrical-conductivity measurement is the conductivity cell. These cells are simple in basic struc-

[5] Abstracted from R. Rosenthal, chief chemist, Industrial Instruments, Inc., "Electrical Conductivity Measurements," in D. M. Considine (ed.), "Process Instruments and Controls Handbook," McGraw-Hill Book Company, Inc., New York, 1957.

ture, consisting typically of two metal plates or electrodes firmly spaced within an insulating chamber. Examples are shown in Figs. 8-14 and 8-15. This arrangement permits isolation of a portion of the liquid under test, and serves to make the measured resistance, in the case of an immersion cell, independent of sample volume and proximity to conductive or nonconductive walls and surfaces.

In laboratory-type cells, platinum electrodes mounted in Pyrex glass structures are commonly employed for both their excellent chemical resistance and ease of working. For work of the highest accuracy, fill-type cells, to eliminate capacitive shunting, must be used. Such cells are used in high-accuracy laboratory work.

For all practical work where errors up to ½ per cent can be tolerated, design considerations are much less critical, and dipping or immersion cells are almost universally employed. They possess the advantage of greater ease in handling, and permit measurements to be made without transferring solution. A cell of this type with cell constant close to 1.0 is shown in Fig.

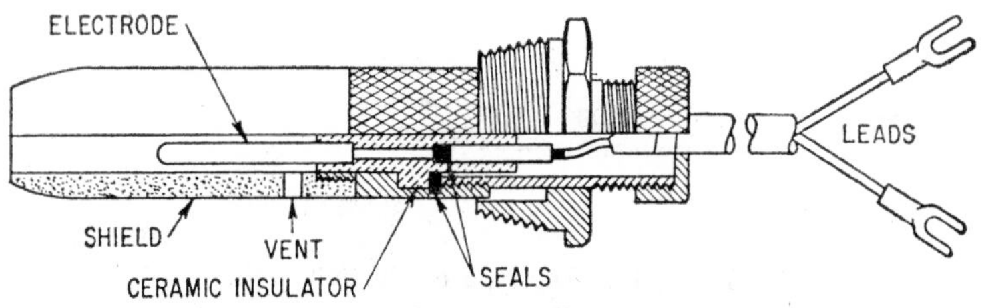

FIG. 8-15. Screw-in conductivity cell for high-pressure service.

8-14. For plant use, where rugged construction and heavier electrodes are desirable, structural parts of suitable resistant metal, plastic, or ceramic and electrodes of stainless steel or nickel are utilized.

Though much variation is found in design of both laboratory and industrial conductivity cells, this is mostly attributable to mechanical considerations rather than to any special requirements of the service in regard to the conductivity measurement itself.

Conductivity cells suitable for plant and control work can be conveniently grouped into four types.

Dip Cells. As the name indicates, these cells are designed for dipping or immersing into open vessels. Materials of construction differ widely, including glass, rubber, Neoprene, polystyrene, and Teflon as the insulating shield and body, and stainless steel, nickel, platinum, gold, and platinum-plated metals as the electrodes. Figure 8-14 shows a typical dip cell. The shield is usually perforated both to increase circulation and to provide a means of venting air to ensure its filling with the liquid under test. During both calibration and use there should be allowed a clearance of at least ¼ in. between all parts of the conductivity cell and the containing vessel. In addition, care must be exercised to ensure immersion of the uppermost air vent at least ¼ in. below the surface of the liquid. Temperature and chemical and mechanical characteristics of the materials of construction establish the service limits.

Screw-in Cells. For permanent installation in pipelines and tanks, cells provided with threaded fittings are used. Temperature and pressure limitations vary with the materials and the construction. Heavy-walled glass-bodied conductivity cells, sealed by means of compressed rubber or Neoprene gaskets in 1-in. NPT stainless-steel fittings, are suitable for continuous service up to about 50 psi at 212°F. For higher-pressure service, the construction shown in Fig. 8-15 permits operation at 500 psi and 300°F.

Insertion Cells with Removal Devices. These cells are designed to permit removal of the element without closing down or depressurizing the line in which they are installed. For inspection or repair the conductivity-cell element is drawn through a packing gland past the normally opened gate

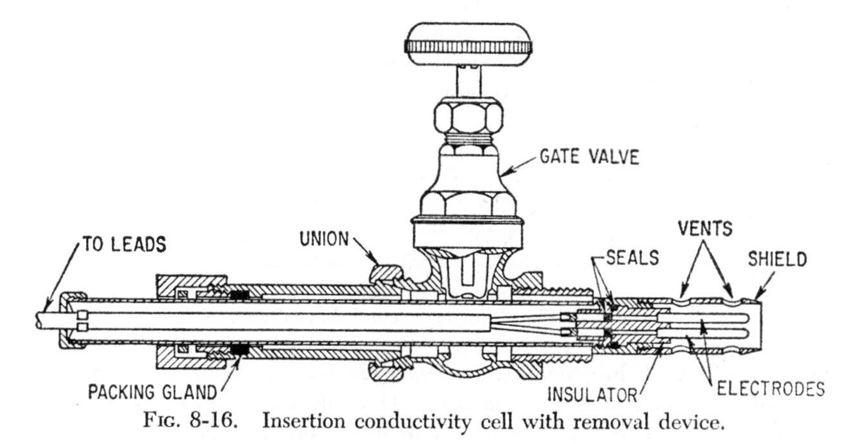

FIG. 8-16. Insertion conductivity cell with removal device.

valve (usually 1¼ in. NPT) and reaches a stop, whereupon the gate valve is closed by hand, and the element is removed. Typical construction, shown in Fig. 8-16, for service up to 200 psi at 200°F employs brass and bronze fittings, Bakelite or ceramic insulator, and gold-sheathed or nickel electrodes. For higher pressures or more corrosive service, similar cells of stainless steel with Teflon insulation are suitable up to 300 psi at 250°F.

Flow Cells. Built in sections of plastic or glass tubing with bore from several millimeters to 1 in. and more, these cells have internal electrodes, usually platinum rings, mounted close to the wall to offer little resistance to flow. Where the electrodes are in line along the axis of the tubing, it is frequently necessary to utilize three electrodes, connecting the outer two in common and to ground. This will eliminate almost completely the pickup of spurious a-c voltages resulting from current leakage from stirrers, pumps, and other equipment, which may be part of the system, and will also eliminate the effect of any shunt electrical path which might exist. In the smaller sizes, these cells are connected to the piping system with rubber or plastic tubing, and in the larger sizes by standard pipe flanges. A small glass flow cell is shown in Fig. 8-17.

Proper Location of Cell. Most industrial conductivity cells are provided with a means for permanent mounting in a pipeline or through a tank wall. The following requirements govern proper cell location:

1. Circulation must be good.
2. Representative sampling is necessary.
3. Linear velocity at the point should not be great enough to cause distortion or damage to the cell.
4. Where linear velocity is low, it is advisable to use a flow-type conductivity cell, or else to mount the conductivity cell so that the flow impinges on the open end of the unit.
5. Provision should be made near by for measuring the solution temperature. Location is preferable where temperature variations are small.
6. The cell may be mounted at any angle, provided only that it cannot become air-bound or gradually fill with solid particles.

Inspection and Replatinization of Conductivity Cells.[6] Surfaces of the electrodes of all conductivity cells are normally supplied coated with plati-

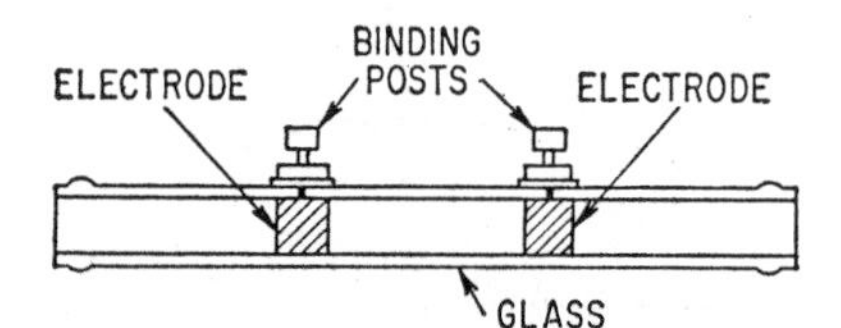

FIG. 8-17. Flow-type conductivity cell.

num black. Cells in general will operate satisfactorily only when this coating is present and in good condition. The only common exception to this is cells of low level constant, in service in very poorly conductive liquids such as good-quality distilled and demineralized water and high-purity steam and condensate. Under these conditions operation will not be affected adversely by the removal of the platinum black coating. It is recommended that conductivity cells in service be inspected at regular intervals. Conditions of service will determine the frequency of inspection that is necessary. Any unusual behavior of the Wheatstone bridge unattributable to known variations in the system being measured should be taken as a possible indication that the cell requires inspection, and cleaning or replatinization. In the absence of all indications, an initial inspection interval of one month is suggested.

Periodic inspection should include a check on the following points:

1. Are there any cracks or chips in the glass cells, or do the plastic cells show any appearance of wear or deterioration?
2. Is the platinum black deposit still present over most of the electrode surface?

[6] Industrial Instruments, Inc.

3. Is there any foreign coating on the electrodes, or any discoloration?
4. Is the shield in position and intact?
5. Are the vent holes in the shield clear and free from obstruction?
6. Is there any indication that excessive liquid velocities have caused changes in the position of the electrodes?

If replatinization is necessary, the following procedure is recommended for glass, plastic, and all other dip or insertion cells. Note that the hard-rubber shield should not be removed when replatinizing a Neoprene cell.

1. Dip the electrode chamber of the cell in 10 to 15 per cent hydrochloric acid for 2 to 5 min. If soapy material is present, wash in 10 per cent hot sodium hydroxide first.
2. Rinse thoroughly in running water.
3. Immerse the cell in a platinizing solution containing 3 g of platinic chloride and 0.02 g of lead acetate per 100 ml of distilled water.
4. Connect the cell leads to two 1½-volt dry cells in series, or to any source of 3 to 6 volts d-c. Reverse the polarity every 15 sec. Continue plating for 3 min, or until a dense black coating covers the entire electrode area.
5. Rinse the cell for one-half hour in running tap water. For cells of constant 0.1, this rinse should be followed by soaking for 15 min in several changes of distilled water.

   One hundred milliliters of the platinizing solution will be sufficient to platinize about 25 cells.

Where the electrodes have become badly fouled or coated with foreign matter, the glass cells should be immersed in warm aqua regia for a minute or two before replatinizing in order to strip completely the fouled platinum black.

Pipeline Installations. Cells for mounting in pipelines are usually supplied with tapered threaded pipe fittings. The size of the fitting is governed by the dimensions of the cell element and varies from ½ to 2 in. NPT, depending on the cell model. A proper position for mounting the cell should meet the following requirements:

1. Good circulation.
2. Representative sampling.
3. Linear velocity at the point should not be great enough to cause distortion or damage to the cell.
4. Where linear velocity is low, it is advisable to mount the cell so that the flow enters the open end of the cell.
5. Provision should be made near by for measuring electrolyte temperature.
6. The cell may be mounted at any angle provided only that it may not become air-bound.

   A typical installation is shown in Fig. 8-18.

Tank Installations. For installation of a conductivity cell in a tank, two procedures are used. The simpler way is to mount the cell at the top of the tank so that it projects down into the liquid. It must be set in deep enough so that it is immersed at least 12 in. whenever measurements are made, but it must not rest on the bottom. These restrictions are necessary to ensure that the conductivity cell is in contact with a representative sample of the liquid. If there are any moving parts in the tank, as in a bottle-washing machine, the cells can be installed very easily by using a length of 1- or 1¼-in. iron pipe to keep the cells away from the moving parts.

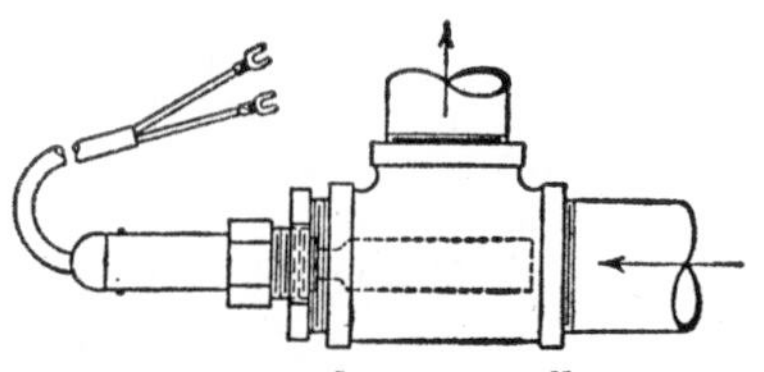

Fig. 8-18. Conductivity cell in pipe.

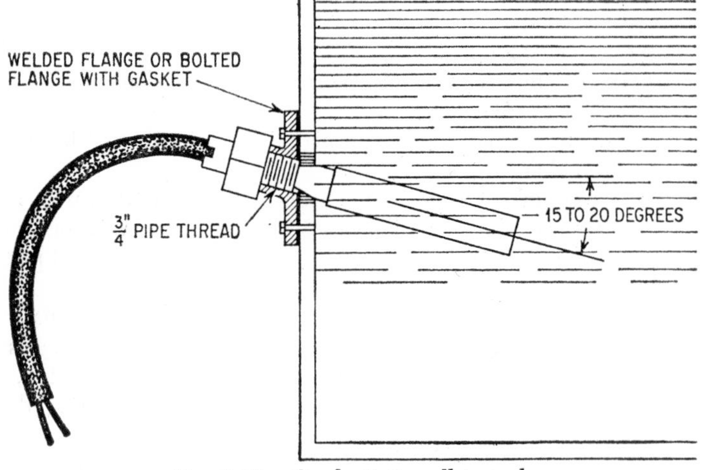

Fig. 8-19. Conductivity cell in tank.

When it is desirable to introduce the cell through the side of the tank, this can be done readily by drilling the proper size of hole and then welding or bolting a special flange fitting to the tank wall. The cell screws into this fitting, which is so designed that the free end of the cell points down at an angle of 15 to 20° from the horizontal. This avoids entrapment of air bubbles in the cell. This type of mounting requires that the tank be emptied to below the level of the cell for original installation or for replacement. Figure 8-19 illustrates this method of mounting.

Sample Coolers and Cell Holders. For certain applications, particularly in the field of boiler-water and steam testing, it is necessary to cool, or to condense and cool, samples and to reduce the pressure to atmospheric before the liquid is admitted to the conductivity cell. This procedure permits the use of standard conductivity cells, readily removed for cleaning or replacement since the cells themselves operate at low temperature and pressure. Temperature measurements also are necessary for accurate interpretation of conductance measurements. Where there is possibility of entrainment of oil as from boilers, engines, or turbines, provision should be made to prevent its reaching the electrodes of the conductivity cell.

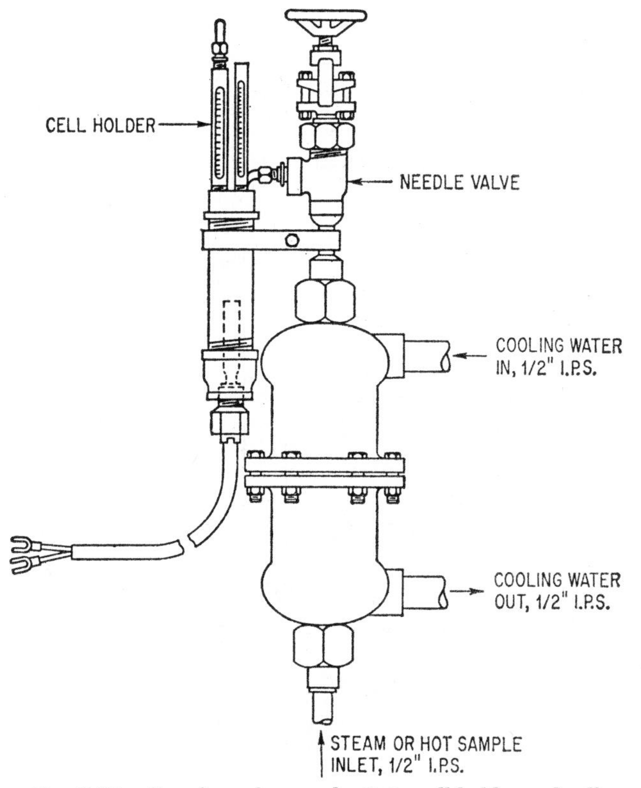

FIG. 8-20. Sample cooler, conductivity cell holder and cell.

Cell holders and sample coolers over a wide range of ratings are available. Figures 8-20 and 8-21 indicate the constructional details and the manner of connection.

The sample cooler is a water-cooled type, rated to withstand 600 lb pressure, and to cool the sample flow to 70°F at this pressure. The casing is a machined bronze casting with a cupronickel cooling coil and needle valve to regulate the sample flow. The needle valve is placed at the output end of the cooler so that the sample is condensed and cooled under pressure, while the cell holder is operated at atmospheric pressure and low temperature.

The cell holder is arranged to provide a continuous sample flow over the electrodes of either a Neoprene cell or a glass cell. The cooled sample is directed through the conductivity cell itself, and escapes to waste through the effluent tube outside the cell. The bulb of a self-contained glass thermometer is placed directly in the incoming flow stream so that the temperature reading is at all times representative of the sample being measured.

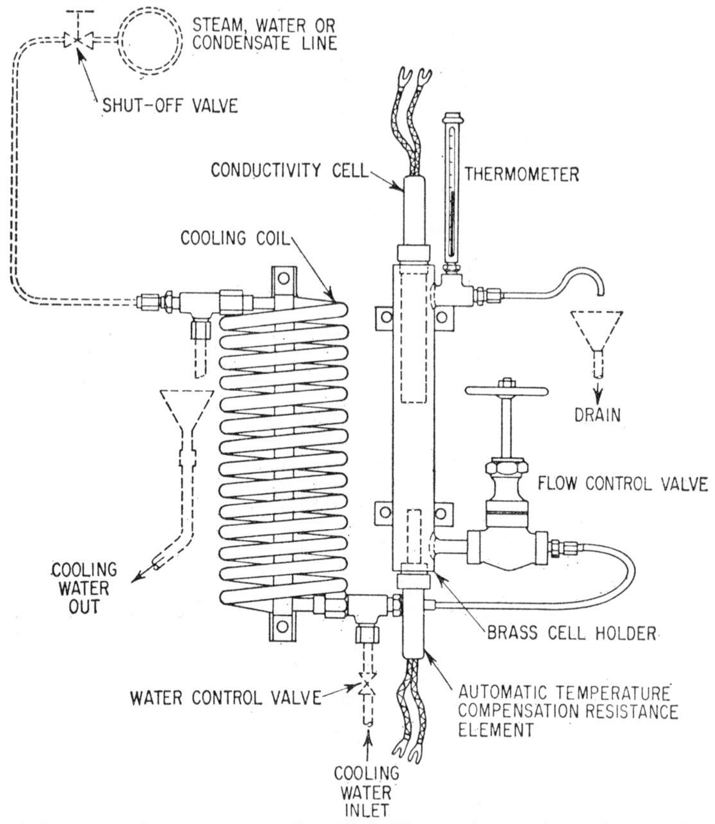

FIG. 8-21. Sampling system with cell holder and sample cooler, used where fluid temperature is 200°F or higher. (*Bailey Meter Co.*)

In many installations, small quantities of lubricating oil are carried through the system, and may interfere with conductivity measurements by coating the electrodes unless provision is made to eliminate the oil before it reaches the cell. Entrained oil collects at the top of the cell holder and becomes visible in the gage glass through which there is normally no flow. When necessary, the small outlet valve may be turned to permit entrapped oil to be blown through the gage glass to waste.

APPLICATIONS OF CONDUCTIVITY MEASUREMENT

Steam and Condensate Purity. Efficient boiler-plant operation generally requires continuous high-purity steam. Conductance measurement offers a simple and accurate means of maintaining surveillance over this important factor. The conductivity of the steam sample is a direct measure of its electrolyte content.

For periodic testing, only a Wheatstone-bridge instrument and an inexpensive dip-type conductivity cell are necessary. A sample of condensed steam is collected in a small glass beaker, the dip cell immersed, the temperature compensator adjusted, the bridge balanced, and, depending on the model, the parts per million of total electrolytic solids as NaCl or the specific conductance in micromhos is read off the scale. No calculations are necessary.

For continuous surveillance of steam purity with automatic warning, the simplest installation consists of a conductivity controller and a conductivity cell mounted permanently in a flowing condensed-steam sample. A lamp, bell, or horn, connected to the controller terminals, will automatically warn the operator of increased solids content whenever the conductance of the condensed-steam sample rises above any preset maximum within the range of the instrument. The controller, in this type of installation, may be located on a central control panel at any distance from the conductivity cell and warning device.

For permanent recording of steam purity, as well as continuous surveillance and warning of contamination, a single- or multi-pen conductivity recorder and corresponding permanently mounted conductivity cells are required. Cells are best installed in cooled sample-flow lines, while the recorder may be located anywhere in the plant.

Boiler Water. A control of the solids content of boiler water is recognized as a factor of major importance in maintaining high steam purity and clean heat-exchange surfaces. Conductance values have been found to correlate closely with the total dissolved-solids content of most boiler waters. Consequently the advantages of easy measurement and ready control of this electrical quantity have been applied to the improvement of boiler operations.

As in the measurement of condensed-steam conductance, for occasional testing of boiler-water samples, only a Wheatstone bridge and dip-type cell are needed.

Where a constant check on the boiler-water conductance is desired, a controller and a conductivity cell installed in cooled sample flow are required. With this instrument connected to a lamp or bell, a warning is given when the total dissolved solids (or the conductance) exceeds the desired maximum. Readings may be taken at any time without disrupting the signaling system. For permanent recording of boiler-water conductance, a number of recorder calibrations are available.

Raw Water and Feedwater. For efficient boiler operation, it is usually necessary to vary the feedwater treatment to compensate for changes in the composition of the raw water. A generally useful index to changes in composition is the electrical conductance of the raw water. For example, seasonal changes in the dissolved-solids content of river, lake, or spring water, or changes in salinity of tidal waters, are most easily and even automatically detected and measured by means of conductivity equipment. Where steam condensate is reused for boiler feedwater, the purity of the return is readily determined by its conductance. In this way, for example,

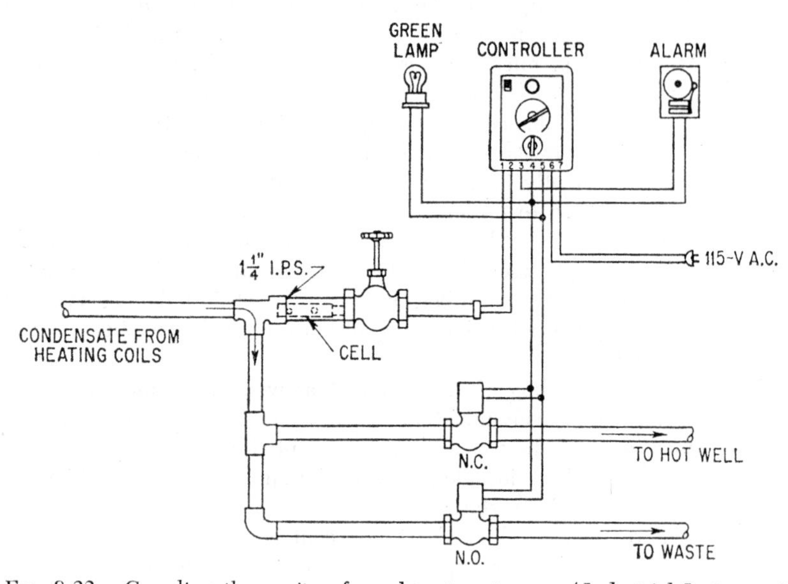

FIG. 8-22. Guarding the purity of condensate returns. (*Industrial Instruments, Inc.*)

it is possible to detect even small leaks in condenser tubes, particularly where sea water or salt water is used as the cooling agent.

Conductivity bridges and cells appropriate for these measurements, whether for manual or automatic testing or recording, are available commercially.

Contamination of Return-steam Condensate. Wherever steam used for heating tanks or kettles, internally through coils or externally through jackets, is condensed and returned to the boiler, there is always present the danger of leakage of tank contents into the steam line, and so into the boiler. The consequences of this possibility are so serious that, in some plants, hundreds of thousands of gallons of good condensate are yearly run into the sewer rather than risk the return of contaminated condensate to the boiler.

It is apparent that periodic chemical analysis of return-condensate samples is at best an inadequate safeguard. At worst, such analyses may be misleading. For example, a small break may permit no leakage while the steam pressure is up, and yet, when the steam is turned down, the bath contents may be drawn into the return line. Chemical analysis during the

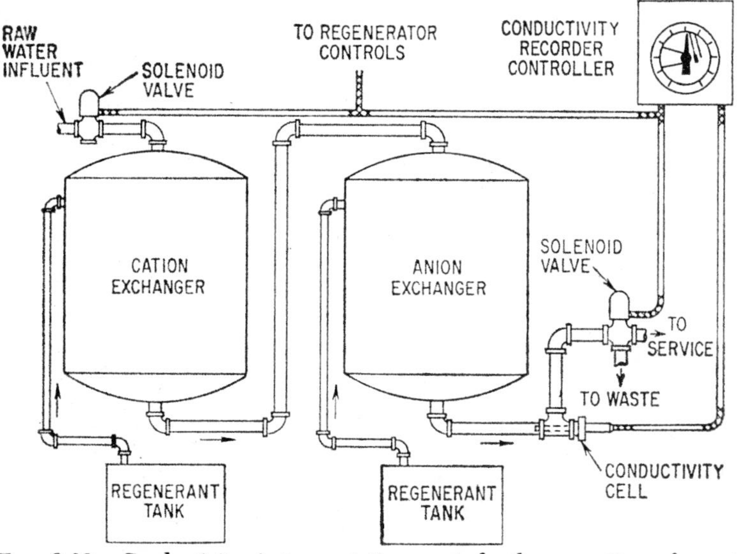

FIG. 8-23. Conductivity instrumentation controls the operation of an ion-exchange water-treating system. (*Minneapolis-Honeywell Regulator Co. and Industrial Instruments, Inc.*)

period that the steam pressure was up would give no indication at all of the break. On the other hand, a conductivity controller, with a cell mounted in the return line and wired to a solenoid or motorized cutoff valve, would stand 24-hr guard on the purity of the returns. Any contamination in excess of the present maximum would cause the controller to react instantaneously, divert the flow to waste, and sound an alarm. A schematic diagram illustrating this type of installation is shown in Fig. 8-22.

Section 9

BOILER INSTRUMENTATION AND CONTROL SYSTEMS

In steam-generating practice,[1] combustion control is assumed to include all factors required to provide complete boiler control with the exception of feedwater-level control. In fact, in many large installations, control devices are interconnected in such a manner that it becomes a simple matter to make boiler-feedwater control a part of the integrated system.

The problem of boiler control is one of coordinating five factors: steam pressure, fuel quantity, combustion-air quantity, removal of waste gases, and feedwater supply. Every steam-boiler-regulating system must recognize these factors, and must proportion and coordinate the last four factors at all times. At the same time, steam pressure must be kept approximately constant at all times.

Because of this, most commercial boiler-control systems use the tendency for steam pressure to deviate from its set point as a master-control impulse. A tendency of steam pressure to drop will cause an increase in firing rate of the boiler, whereas a tendency of steam pressure to rise above the set point will cause a decrease in firing rate.

Efficient combustion depends on how three factors, fuel, combustion air, and waste gases, are handled. Some means, direct or indirect, must be provided for proportioning fuel and combustion air, and for maintaining furnace draft at proper value. The means selected for a given installation depends on (1) money available, (2) size of installation, (3) type and number of fuels burned, (4) magnitude and frequency of load variations, (5) type and number of boiler auxiliaries, and (6) type, size, and method of controlling equipment already installed.

In order to properly control fuel, combustion air, waste gas, and feedwater level, they must be measured quantitatively as the basis for control action; or some other effect, which can be relied on as being proportional to quantitative measurement, must be used.

Check backs commonly used to control these factors are:

For Coal: (1) Speed of stoker or feeder drive shaft; (2) position of

[1] A. C. Wenzel, chief project engineer, Republic Flow Meters Co.

coal-feed lever, on either a coal feeder or its variable-speed drive; (3) position of valve on a hydraulic or pneumatic coal-feeding mechanism.

For Oil: (1) Measurement of differential pressure across an orifice located in the fuel-oil line; (2) oil pressure ahead of burners (useful with constant tip-size burners); (3) control of back pressure on return-type burners; (4) positive-displacement type of measuring device.

For Gas: (1) Measurement of differential pressure across an orifice located in the fuel-gas line; (2) gas pressure measured between regulating valve and burners.

For Air Flow: (1) Differential pressure across orifice or venturi in forced-draft duct, or drop across air side of air heater; (2) suction (static pressure) at forced-draft inlet cone can be used on variable-speed forced-draft fans, driven by turbines, magnetic or hydraulic couplings, or variable-speed motors; (3) differential pressure between wind box furnace (with oil or gas burners when oil or gas pressure is used as a check back for fuel measurement); (4) indirect air-flow measurement by using draft loss through boiler setting, gas side of air heater, or across dust collector; (5) variable furnace draft on boilers burning oil or gas and lacking forced-draft fans (This is equivalent to taking draft loss from wind box to furnace where oil or gas pressure is a check back for fuel measurement.); (6) constant furnace draft (when air is indirectly measured, as in item 5, and controlled by boiler outlet damper). The forced-draft fan is used to supply combustion air in the amount governed by the furnace-draft regulator.

For Waste Gases: (1) Measurement of draft loss across boiler passes on flue-gas side of air heater or through dust collector; (2) furnace draft held constant at all rates of steam generation by controlling boiler outlet damper.

For Feedwater: (1) Single-element control, using drum level only as controlling impulse; (2) two-element control using steam-flow impulse (measured across orifice in steam line, or pressure loss through superheater), producing a water-valve position corrected by drum-level impulse (Note: This type of regulation is satisfactory only when feedwater header pressure is constant or differential pressure across the feedwater-control valve is held constant.); (3) three-element control comprising steam-flow versus water-flow control with correction from feedwater level.

Boiler controls can be classified by function and use—from the simplest to the most complex. General classifications and their variations follow:

Positioning Control: The simplest form uses a single regulator with a boiler, and requires one fuel and stack draft control only. The regulator controls the boiler outlet damper and fuel (stoker, oil, or gas) by position only.

Modified positioning control uses two regulators with a single boiler, single fuel, and forced-draft fan. Steam-pressure regulator controls fuel and forced draft; furnace-draft regulator operates boiler outlet damper.

Control for multiple-boiler installations uses a master regulator and two positioning regulators per boiler, with one fuel and forced draft. This system provides for manually setting the fuel-air ratio, but depends on positioning of fuel and air controls with no check back to fuel or air measurements to assure constant efficiency or uniformity of results.

Sequential Control: One type, with metered air and fuel, for multiple installations uses a master regulator with fuel-air proportioning and furnace-draft regulators. This type has a tendency to smoke on increasing rating. Fuel flow must change before air flow can follow to satisfy combustion conditions.

Another sequential system, with metered air and fuel, for multiple installations uses a master regulator, boiler-damper positioner, air-fuel proportioning, and furnace-draft regulators. This arrangement uses indirect air-flow measurement provided by draft losses through boiler passages. Forced-draft air is governed by furnace draft. There is a tendency to smoke on decreasing rating, since air flow can follow. A safety factor is provided because fuel follows boiler-draft loss, and thus the control prevents overloading of the furnace with fuel.

Still another sequential system has pressure check back for fuel and air and can be arranged for fuel pressure versus wind-box pressure.

The same sequential system can also be arranged for wind-box pressure versus fuel pressure.

Premetered Control: For multiple installations, this system uses master, fuel, air-flow, and furnace-draft regulators. Fuel and air are controlled simultaneously by measuring regulators. A very satisfactory system, it is free from smoke and, with fuel and air in proper ratio, provides good boiler efficiency over a wide load range. Air-flow measurement is taken in forced-draft duct, free from all furnace disturbances.

Combustion Corrector: For single- or multiple-boiler installations, this is really a positioning system with combustion correction using steam flow as an indirect measure of instantaneous Btu heat input to the furnace. The steam-flow–air-flow relationship is the measure of combustion efficiency. The system produces simultaneous impulses to fuel and air with correction to maintain the steam-flow–air-flow relationship constant. Effective when fuels are inconsistent, this system is seriously affected by feedwater disturbances.

A variation might be to locate the corrector regulator in the fuel circuit instead of in the air-flow master circuit. Operation with such arrangements, however, has not resulted in satisfactory combustion conditions.

Uncoordinated System: In this system, there are three independent sets of controls on each boiler. The master-control system regulates only the fuel. Air flow is controlled from steam flow only. Furnace-draft control is also independent. This system works well when fuel is consistent, the number of burners is constant, and there are no serious feedwater disturbances.

Each type of control has its place. It cannot be said that any one system is universally better than any other. Before deciding which of the basic systems will give the most economical results for a given installation, it is necessary to study and evaluate several factors: size of boilers, load conditions, regulating characteristics of all auxiliaries (characteristic curves should be obtained), intelligence and attendance of operating personnel, money available, and returns expected from the investment.

Final selection usually is governed by the operating characteristics of the controlled equipment, particularly the burners and fuel-feeding mechanism, the money available, and probable investment returns.

In order to obtain highest combustion efficiency from a boiler unit and control system, the basic plant equipment selected (fans, fuel-burning equipment, etc.) must be of types that lend themselves readily to control. It is also important that good measurements for air, fuel, and steam flow be provided.

FEEDWATER REGULATION[2]

During the past decade, the use of automatic feedwater control has reached a point where a power boiler or large heating boiler without feedwater control is an oddity. This wide acceptance is partly due to the development of controls having a high degree of reliability, even under adverse conditions, and partly to the necessity of automatic control for handling the sometimes severe requirements of modern boiler plants.

Erratic flow of feedwater to many of the newer boilers has a serious effect on steam-temperature control, as well as on feedwater heaters and pumps. Unit systems having extraction heaters are badly upset if the entire water system is not kept in balance.

Many boilers are operating with steam temperatures near the maximum permissible for presently known metals, and temperature variations caused by poor feedwater control are intolerable.

Modern boiler designs have a high steam-release-to-water-storage ratio, which requires good control for safety. Controlled-circulation boilers may further increase the ratio, making good control even more essential. Steaming economizers and boilers with drums having a high steam release per unit area of water surface contribute to the demand.

Despite these severe requirements, commercial control equipment is available to meet all the demands for foolproof systems, operating under a wide variety of conditions.

Specifically, there are four types of feedwater-control systems in common use today.

1. Single-element self-operated
2. Single-element pilot-operated
3. Two-element pilot-operated
4. Three-element pilot-operated

The single-element self-operated feedwater regulator has been used for a great many years, and is still in use on small low-pressure boilers having relatively stable flow characteristics. One type of regulator falling into this category is illustrated in Fig. 9-1.

[2] C. H. Barnard, application engineer, Bailey Meter Co.

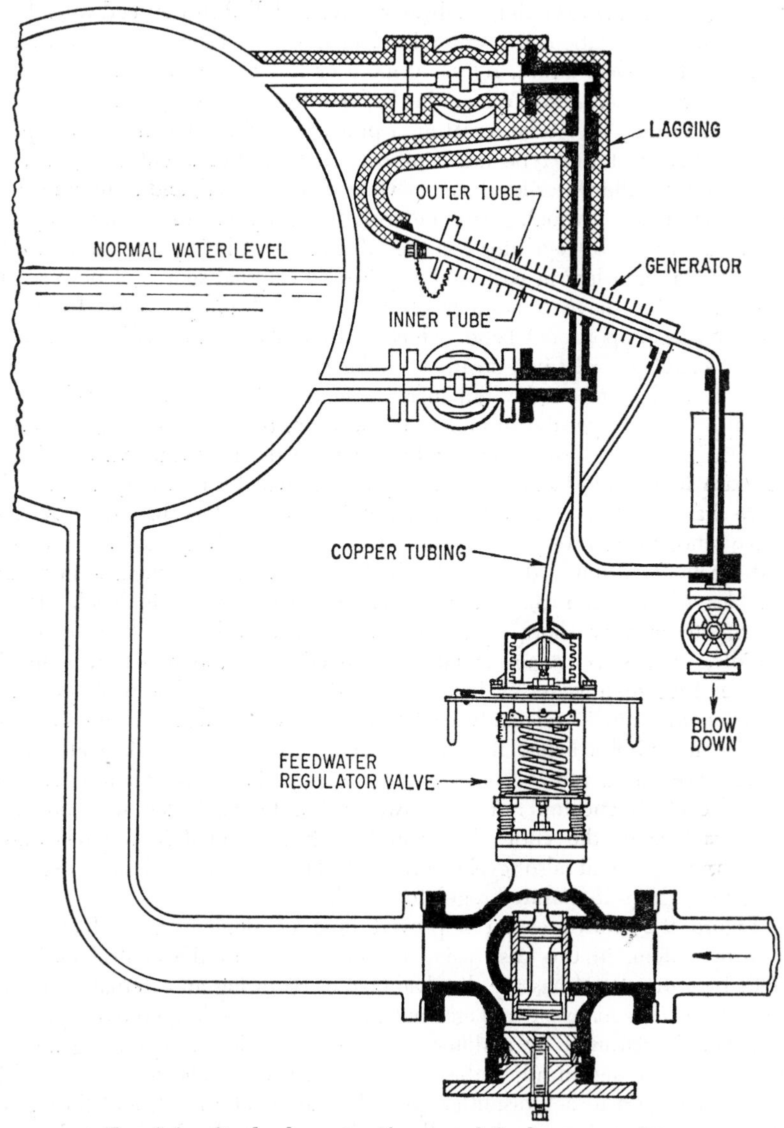

Fig. 9-1. Single-element self-operated feedwater regulator.

A generator, consisting of two concentric tubes, is installed at a suitable location on the boiler drum, the inner tube connected at points above and below the drum-water level so that the level in the inner tube corresponds with the level in the drum.

The annular space between tubes of the generator is filled with water when the generator is cold, and is connected by means of tubing to a bellows-operated spring-closing feedwater valve installed in the feedwater line.

In normal operation, the water level in the inner tube changes with the water in the boiler drum, releasing more or less heat to the outer tube as the water level varies. When the boiler-water level drops, more heat is released to the outer tube, creating a pressure, and forcing the valve open.

A regulator of this type has a proportional band of about 6 in., so that at low ratings the water level is somewhat above normal, and at high ratings somewhat below normal. This range is necessary to ensure stable operation. This type of regulator has the advantages of being extremely simple in operation and low in cost, and requires no external source of power. It does, however, have a number of distinct disadvantages aside from the rather wide proportional band, which restricts its use to small boilers, and to those having a relatively steady steaming rate.

As is well known today, an increase in steaming rate causes release of steam bubbles below the surface of the water in the boiler, raising the water level in the boiler drum, and tending to close the feedwater valve. If an increase in steaming rate has taken place, it will be necessary to provide more water, not less, in order to maintain a satisfactory balance. As the valve tends to close on an increase in rating, the relatively cool water entering the boiler drum is decreased, which causes a further increase in steam bubbles below the surface and a corresponding increase in level. However, after a very short period, the water level will tend to drop quite rapidly, and, since steam is being removed from the boiler at a much greater rate than water is being supplied to it, the valve will open, increasing the rate of relatively cool feedwater, which will collapse some of the steam bubbles below the surface, and cause a decrease in water level at an accelerated rate. Consequently, on any change in steaming rate, a severe cycle is encountered in feedwater flow to the boiler, upsetting the heat balance of the entire boiler and heaters installed in the feedwater system, which will also cycle severely. This action is characteristic of single-element self-operated regulators of all types.

Figure 9-2 illustrates an improved type of single-element feedwater-control system, in this case using compressed air as the control medium. This type of control has the disadvantage of requiring an external source of power, but has many advantages not present in the self-operated type.

Level is measured by a differential-measuring device, operating an air-pilot valve having an adjustable proportional band. The output from this pilot valve goes to a transmitter, providing automatic reset, and through a hand-automatic selector valve to the feedwater-control valve installed in the feedwater line.

Inasmuch as automatic reset is provided, a much wider proportional band can be used on the level controller, minimizing the cycling which takes place on a change in load, and the automatic reset provides for a constant level at all loads. Another advantage in this type of system is that it permits the installation of a hand-automatic selector valve at the control center, so that remote manual control is provided during startup and abnormal operating conditions. A pilot-operated valve positioner can also be installed on the feedwater-regulating valve, permitting matching the valve characteristics to the requirements of the system.

A further improvement is the two-element pilot-operated system as

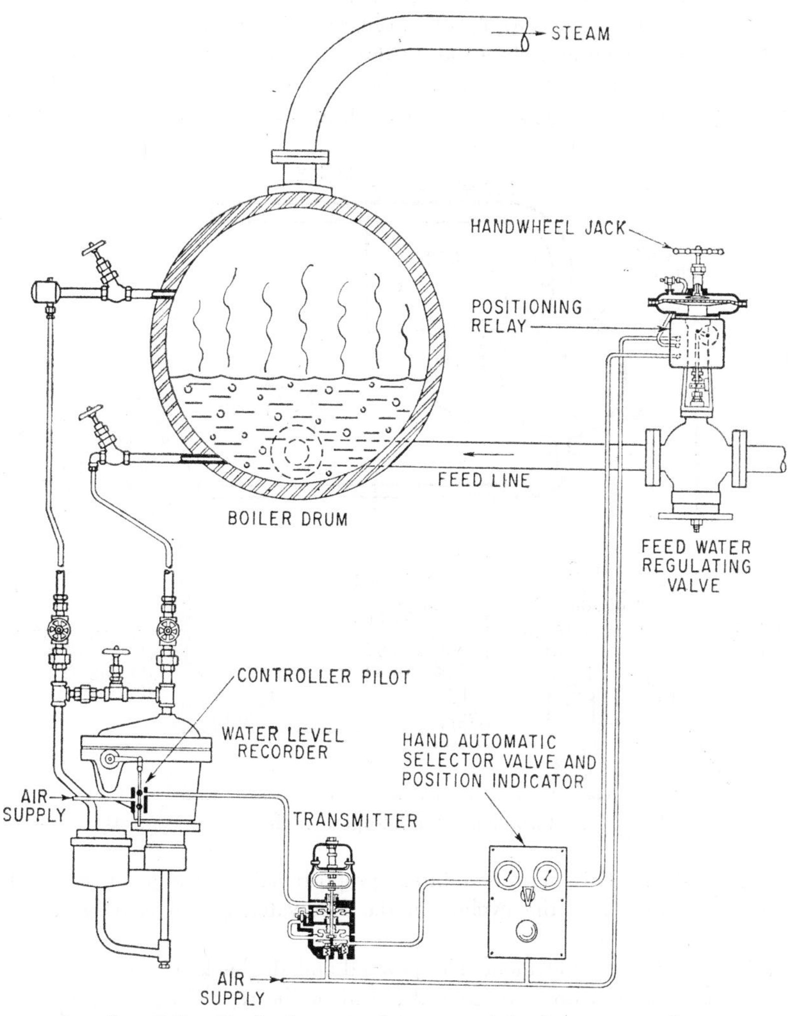

Fig. 9-2. Single-element pilot-operated feedwater control.

illustrated by Fig. 9-3. This system has all of the advantages of the single-element pilot-operated system, and, in addition, a stabilizer in the form of a steam-flow controller.

Level is measured by a differential device connected to the boiler drum, and steam flow is measured by a flow meter connected to an orifice or nozzle installed in the main steam line. Each of these devices is equipped with a pilot valve having proportional-band adjustment, the outputs going through an averaging relay. From the averaging relay, the air signal goes through a hand-automatic selector valve, and to the feedwater-regulating valve installed in the boiler-feedwater line.

With this type of control, the primary influence is from steam flow, which acts to open and close the feedwater-regulating valve as the demand for steam increases or decreases. The level controller is then used as a readjusting control only, to maintain the level within the boiler drum at the

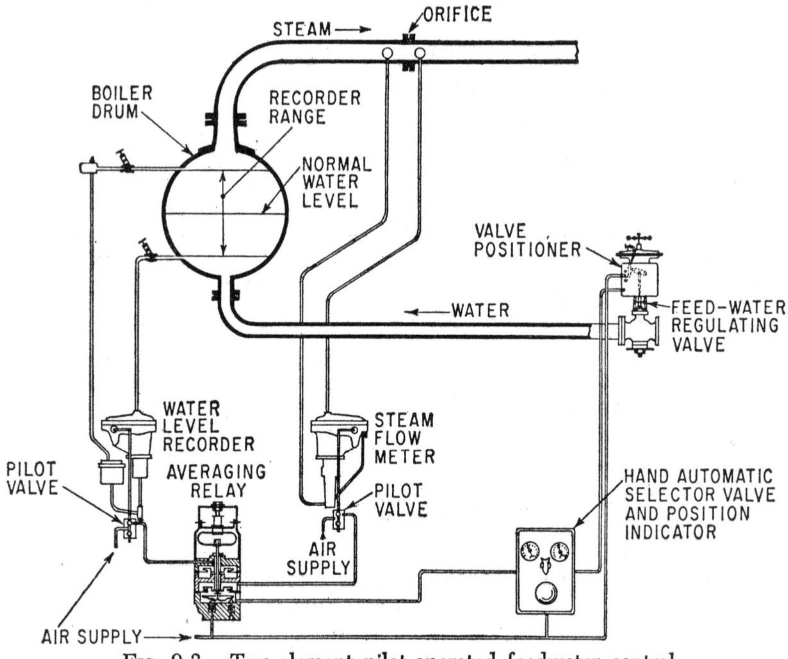

FIG. 9-3. Two-element pilot-operated feedwater control.

proper value. This permits a broad proportional-band adjustment on the level controller, and the cycling in the feedwater system is reduced considerably.

As with the single-element self-operated and single-element pilot-operated systems, feedwater-pressure control ahead of the regulating valve is essential for optimum operation. A change in pressure ahead of the regulating valve will cause a corresponding change in feedwater flow to the boiler, which reacts on the level and back through the system in order to readjust the valve. Where erratic feedwater pressure exists, the level control will be called upon to do an abnormal amount of adjusting, with a consequent upsetting of the entire feedwater system.

The system illustrated in Fig. 9-4 is a three-element pilot-operated system. To the two-element system, a water-flow meter has been added, measuring the feedwater flow to the boiler by means of an orifice or flow nozzle, and a water-flow-measuring device.

Primary control is from the ratio of steam flow and water flow, acting through a pilot valve having proportional-band adjustment, and through the transmitter and hand-automatic selector valve to the feedwater-regulating valve. In addition to moving the valve in the proper direction on a

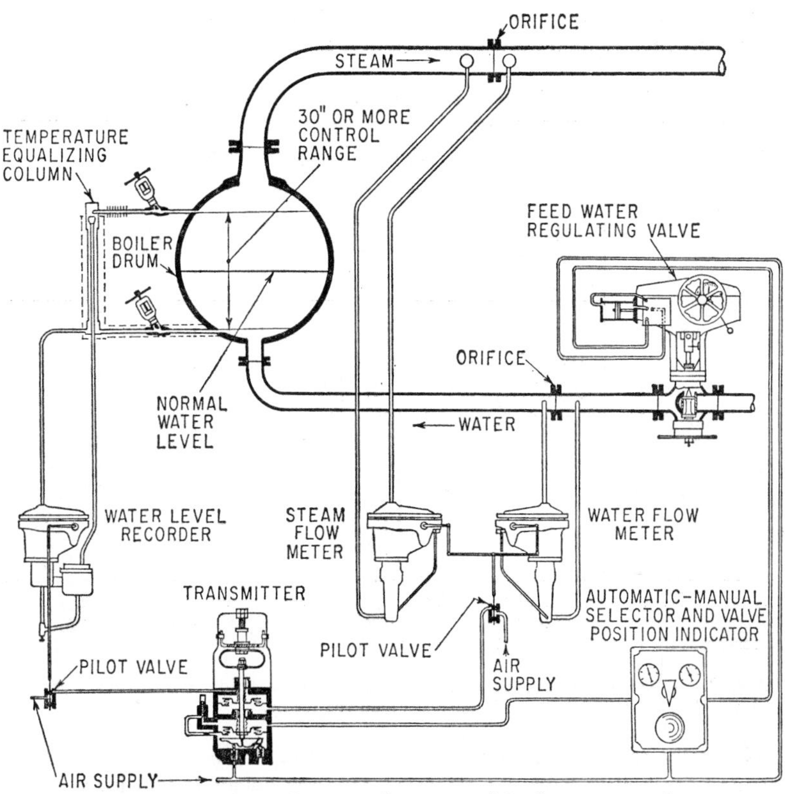

FIG. 9-4. Three-element pilot-operated feedwater control.

change in steam demand, the water-flow measurement enters the control system to position the valve so that the water-flow input is equal to the steam-flow output from the boiler. Any variations in feedwater pressure ahead of the valve, causing a change in water flow, will be immediately detected by the water-flow meter, and the valve will be repositioned to maintain the proper water flow before any effect on level is experienced.

A very broad proportional-band adjustment can be used on the level controller, minimizing the effect of swell to a point which is negligible, except upon very large changes in rating.

Basically, this is the type of control used on practically all the large utility boilers and large industrial boilers in operation today.

Selection of Proper Feedwater-control System. Many factors enter into the selection of the proper feedwater-control system. The design of boiler, concentration of solids in the boiler drum, type of load, and type of pump or feedwater-pressure control all enter the picture.

Table 9-1 is a selection chart based on broad experience in applying all types of feedwater control to all types of boilers and loads. Obviously, no chart such as this can cover all individual cases, and all factors should be thoroughly investigated before a selection is made.

Table 9-1. Boiler-feedwater-regulator Selection Chart

Boiler capacity	Type of feedwater regulator		
	Self-operated single-element	Relay-operated single- or two-element	Relay-operated three-element
1. Below 75,000 lb/hr	For steady loads (building heating or continuous processes)	For irregular loads (batch processes, hoists, rolling mills, etc.)	
2. 75,000–200,000 lb/hr	Use only in special cases	For all steady and fluctuating loads	For extreme load and water conditions and boilers with steaming economizers
3. Above 200,000 lb/hr		Use only on steady loads	For all types of loads

Note 1. Excess pressure ahead of feedwater regulator should be at least 50 psi.

Note 2. Excess pressure should be controlled by regulation of feed pump. Use excess-pressure valves only when excess pressure varies more than plus or minus 30 per cent.

Note 3. Where drum level is unsteady owing to high solids concentration or boiler feed or other causes, use next higher-class feed regulator.

Generally speaking, single-element self-operated control can be recommended for boilers having a steaming capacity below 75,000 lb/hr, where the load on the boiler is relatively steady, such as for building heating or continuous processes. Where irregular loads are encountered, as for presses, rolling mills, etc., pilot-operated single- or two-element control should be selected.

For boilers having a steaming rate between 75,000 and 200,000 lb/hr, the self-operated single-element control can be recommended only in very special cases. For all steady and reasonably fluctuating loads, pilot-operated single- or two-element control is satisfactory. For extreme load and water conditions, and boilers with steaming economizers, three-element control should be used.

Where the drum level is unsteady owing to high solid concentrations, or where boilers are not equipped with modern steam-separating devices, the next higher type of feed control should be used.

Present-day Practice. As stated previously, nearly all large boilers, both utility and industrial, are using three-element pilot-operated systems for control of feedwater. The illustrations cited employ a control valve located in the boiler-feedwater line for regulation of feedwater flow.

In many of the larger plants, heat-balance considerations have minimized the use of turbine-driven variable-speed boiler-feed pumps, and constant-speed motor-driven pumps are used. With the use of the three-element

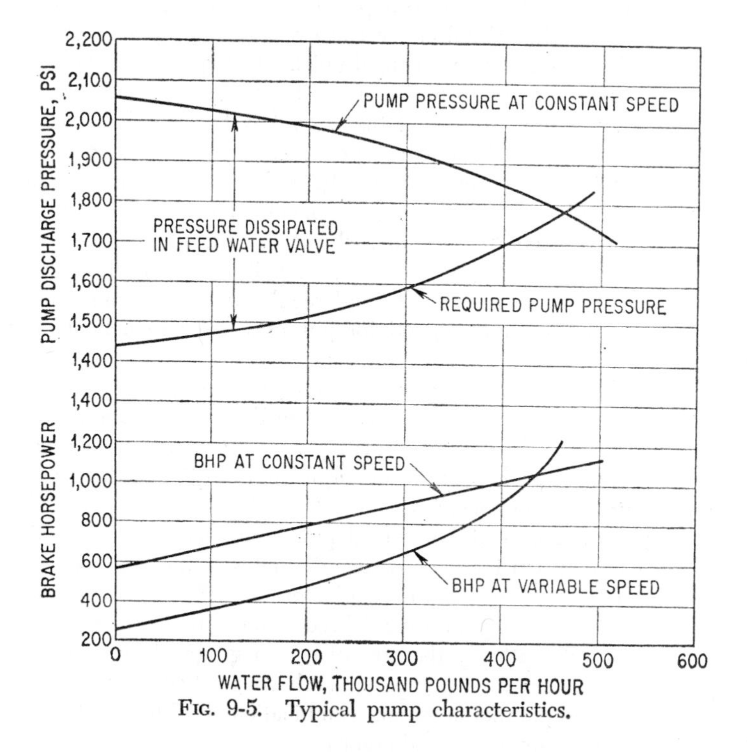

FIG. 9-5. Typical pump characteristics.

system, it is not necessary to use throttling valves at the discharges of the pumps in order to maintain a constant pressure, since, with a properly designed feedwater valve, the three-element system will function satisfactorily. However, pumping cost at loads under maximum are excessive because of the excess pressure which must be dissipated in the feedwater valve, and wear on the valve itself is materially increased. In order to overcome these objections to constant-speed pumps, variable-speed couplings, installed between a constant-speed motor and the pump, have come into fairly common usage on large boilers.

Figure 9-5 illustrates a typical variable-speed pump operation, compared to constant-speed operation. At very low loads, more than 700 lb pressure must be dissipated across the feedwater valve, and it will be noted that something in excess of 300 additional horsepower is required at the input to the pump to furnish this excess pressure, which serves no useful purpose, and creates additional wear on the pump as well as on the control valve. As the rating on the plant increases, the excess pressure to be dissipated decreases quite rapidly, and the power input to the pump approaches that required for constant-speed operation, so that, at the maximum condition, no power saving is experienced, and little if any excess pressure is dissipated across the feedwater valve.

Typical Installations.[3] These examples of feedwater-control systems include applications to horizontal and vertical reciprocating pumps as well as to turbine- and motor-driven centrifugal units. In some of these cases, the control operates the pumps to maintain constant excess pressure; in others, variable excess pressure. In some, the control system can be adjusted to give either constant or variable excess pressure.

The plants, with two exceptions, are 1,200-lb central station. Plant *H* (Fig. 9-10) and plant *I* (Fig. 9-11) are included because they illustrate the versatility of the control system and because the pumps are unusual. Also, in all the plants except one, plant *G*, the boiler-water-level controls are made up from the same type valves and regulators as the pump controls. Excess-pressure controls are applied for purposes other than boiler-water-level control.

The principal functions of an excess-pressure feed-pump control are:

1. To maintain automatically an excess of feedwater pressure over steam pressure, at the feedwater-regulating-valve inlet (either a constant amount above steam pressure or an amount varying with the rate of steam output of the boiler)
2. To operate a group of boiler-feed pumps automatically at minimum power consumption for various steaming rates of the boiler (these are large pumps, in some cases driven by prime movers of high capacity, and any power saving is a material reduction in the cost of operation)
3. To control automatically the operation of a group of large high-pressure high-capacity feed pumps, of both constant-speed and variable-speed types, in parallel in various combinations, and to maintain stable division of load among several pumps

As a result of excess-pressure pump control, the feedwater-regulating valve maintains its most effective regulating position and controls water level with a minimum of valve motion.

In all but plants *G* and *H*, the boiler-water-level-control systems are similar. A water-level regulator, actuated by water-level variations through a mercury-differential-measuring element, operates the oil-power cylinder of a turbine-type valve in the feedwater line. Hydraulic power, thus applied to the valve, positions it accurately, regardless of the pressure differential across the valve. It is not necessary to employ separate elements to cancel the effects of pressure differential on level regulation.

[3] A. C. Wenzel, chief project engineer, Republic Flow Meters Co.

It does not matter whether the pressure differential across the valves varies or not. In some cases, pump control is designed to vary this pressure differential with boiler output; accordingly, when a change in boiler-water level necessitates a change in feedwater flow, the pump control causes the pumps to change the pressure drop across the valve to give the correct flow and minimize valve travel. In other cases, however, pressure drop across the level-regulating valve is held constant, and flow is regulated by movement of the level-regulating valve. The important thing is that, when this type of water-level control is installed, either method of pump control can be successfully employed, the selection depending on other factors.

Constant-speed motor-driven pumps usually have discharge pressure controlled by a valve in the pump-discharge line actuated by a regulator. Pumps are also driven by variable-speed motors or steam turbines. The

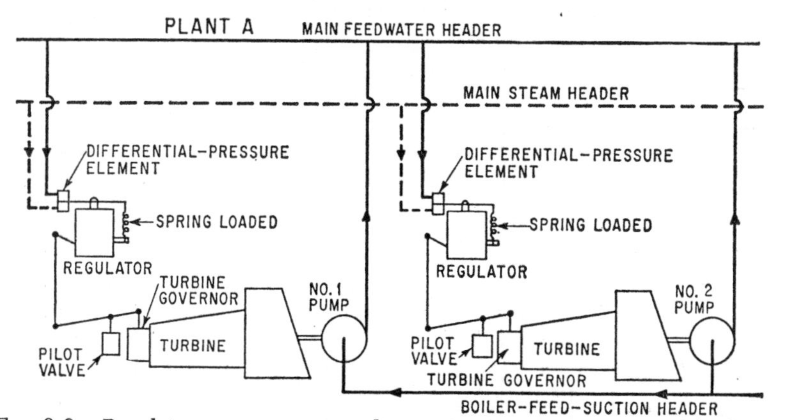

Fig. 9-6. Regulators vary pump speed to maintain constant excess of feedwater pressure over steam pressure. (*Alfred C. Wenzel, Republic Flow Meters Co.*)

former are controlled by regulators actuating electric contactors and rheostats. Such controls for variable-speed pumps have been used in some power plants, but most designers prefer not to use variable-speed motors. Turbine speed is varied by a regulator actuating the turbine governor, or a valve in the steam line to the turbine, the former being the preferred method.

Pumps can also be driven through hydraulic couplings, and speed controlled by regulators operating valves in the oil circuit of the hydraulic coupling, as at plant *K*. Variable-speed pumps driven by mechanical variable-speed drives are controlled by regulators operating the variable-speed-drive adjustment. Electric variable-speed drives, such as used in plant *H* for driving a horizontal reciprocating pump, are controlled by regulators operating contactors. An unusual control of vertical reciprocating pumps direct from water-level regulators has been developed for plant *I*.

The simplest form of excess-pressure pump control consists of a regulator maintaining constant excess pressure of feedwater over steam pressure by controlling the speed of a turbine-driven pump or a valve in the discharge of a motor-driven constant-speed pump. A typical example of this application is at plant *A* (Fig. 9-6). This is a 1,200-lb superposition consisting of one 275,000 lb/hr bent-tube boiler fired by a chain-grate stoker and supplying steam at 1,350 lb, 910°F, to a 10,000-kw turbine, which exhausts at 215 psig. Two 660-gpm turbine-driven pumps supply feedwater. On each pump turbine is mounted a regulator connected as shown schematically in Fig. 9-6. This regulator, in operating a valve in the turbine governor, varies pump speed to maintain a constant differential pressure between main-steam and feedwater headers.

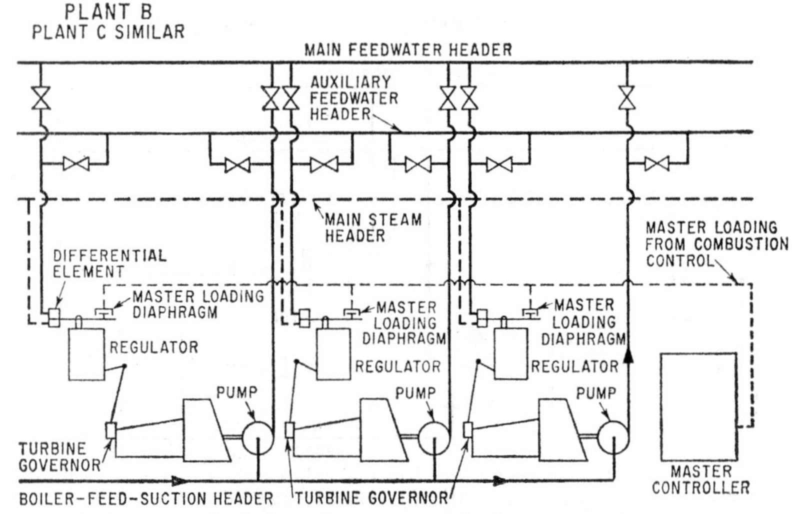

FIG. 9-7. Master-loaded regulators control feed-pump speed to increase excess feedwater pressure at increased boiler outputs. (*Alfred C. Wenzel, Republic Flow Meters Co.*)

The pump-control system shown in Fig. 9-7, plant *B*, is an amplification of that of Fig. 9-6. This system, in contrast to that of Fig. 9-6, operates the pumps automatically in parallel, to vary excess feedwater pressure in relation to boiler output, providing higher excess feedwater pressure at high steaming rates when most needed and reducing the power consumption at lower steaming rates. In this control system, the regulator arrangement, governing turbine speeds by controlling their governors, has a superimposed loading pressure. The regulators, of course, differ slightly in details of construction.

The air-loading pressure from the pump master controller, varying in a definite relation to boiler output, is applied to a regulator where it is balanced on a weigh beam against a differential-pressure element to which is supplied the steam- and feedwater-header pressures. Variations of the loading pressure unbalance the regulator, which varies turbine speed and therefore pump pressure.

Variations in loading pressure are produced by the pump master controller, which in turn is loaded from the master controller of the boiler's combustion-control system. The latter, because of its design, transmits to the pump master controller a loading pressure varying as the square of

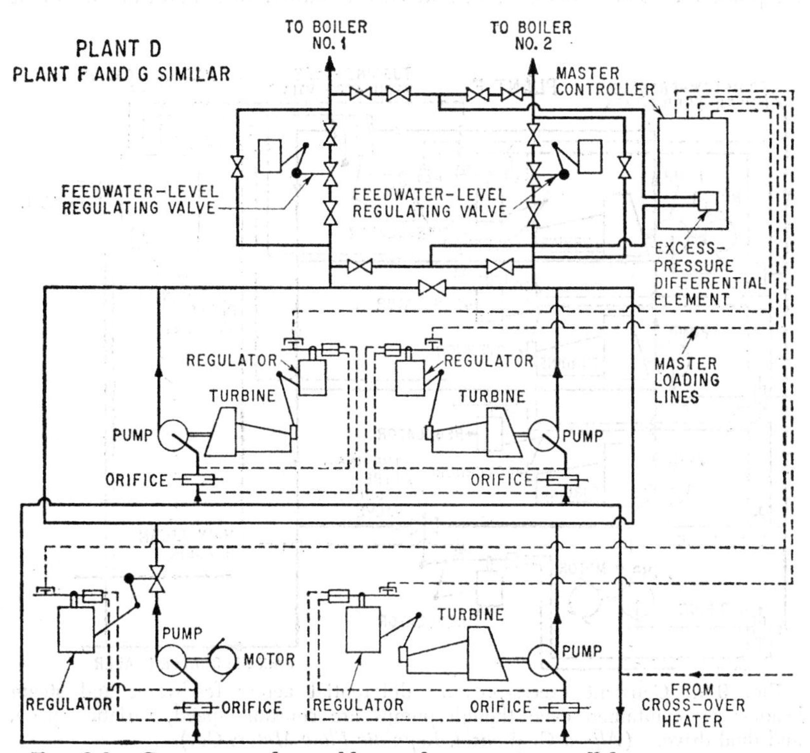

FIG. 9-8. Constant- and variable-speed pumps in parallel maintain constant pressure drop across feedwater valve. (*Alfred C. Wenzel, Republic Flow Meters Co.*)

boiler output, which is in turn transmitted by the pump master to the regulators, governing the pumps to vary excess pressure as the square of the boiler output.

Plant *C* has a system similar to that of plant *B*. In both plants, boilers operate at 1,325 lb, 900°F, serving turbines of 35,000 and 30,000 kw, respectively.

In each case three turbine-driven boiler-feed pumps supply feedwater to high-pressure boilers. At plant *B*, each pump, driven by an 1,100-hp turbine, has a maximum capacity of 425,000 lb of water per hr, a little greater than the maximum output of one boiler. Since there are two boilers, two pumps can carry full load, leaving the third for spare, and permitting any desired combination of the three pumps.

In both systems, master loading for the pump regulators is taken from a common loading line. Consequently, operating pumps divide load equally.

Excess-pressure pump controls in plants *D*, *E*, *F*, and *G* are somewhat different from those in plants *B* and *C*. Figure 9-8 shows the connections for plant *D*. The others, *E*, *F*, and *G*, are similar in general arrangement

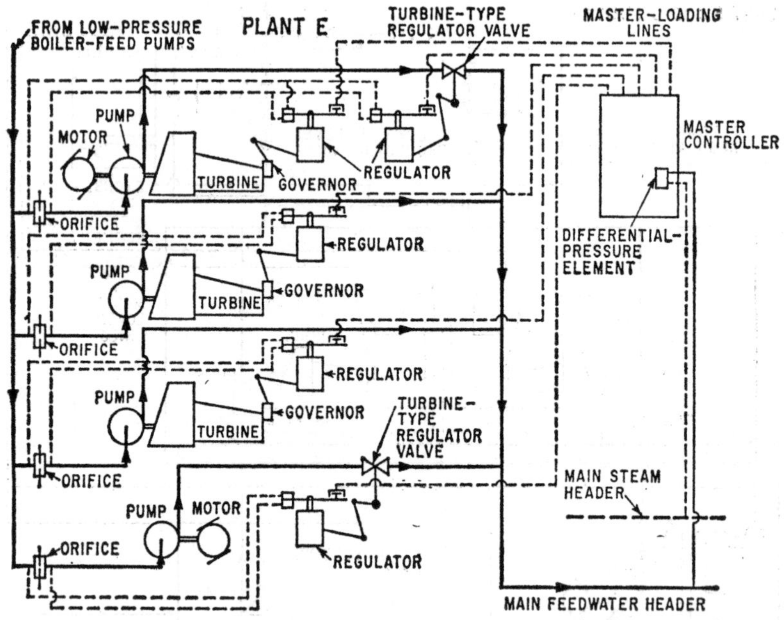

FIG. 9-9. Constant excess-pressure differential across feedwater and steam headers is maintained even though pumps are constant speed, variable speed, and dual drive. (*Alfred C. Wenzel, Republic Flow Meters Co.*)

and operate on the same principle. They are designed to maintain a relatively constant excess of feedwater pressure over steam pressure. In one case, at plant *E* (Fig. 9-9), pressure differential between feedwater header and main-steam header is kept constant. In the other three cases, plants *D*, *F*, and *G*, differential pressure across the feedwater-level-regulating valve is kept constant. In all four cases, the constant differential is transmitted to an element in a master controller, from which master-loading pressure is sent to individual regulators governing the pumps. In the regulators, this loading pressure is balanced on a weigh beam against a differential pressure taken across an orifice in the suction of each pump, this differential pressure varying with the flow.

Of course, as boiler output varies, the pressure drop through superheater, economizer, and piping also varies; hence the total feedwater pressure must change; but in plant *D*, for example, it is always in excess of steam pressure by at least the amount of pressure drop across the level-regulating valve. Using this pressure drop instead of the differential between steam and feedwater headers is preferred by some engineers, because the entire control works through a range of about 50 lb instead of through the whole range of boiler drop, say 150 to 175 lb; therefore the control regulates within closer limits. It is not affected by superheater and economizer drop. When only one or two boilers are involved, either of the above differentials (across level-regulating valve or across steam and feedwater headers) can be used.

In plants *D*, *E*, *F*, and *G*, various combinations are used: constant-speed pumps driven by electric motors, variable-speed pumps by steam turbines. This is done primarily for heat balance. Only by controlling the group with a master controller can the pumps be made to operate automatically in parallel at lowest power consumption.

At plant *E*, the pump is driven by both motor and turbine, and is controlled by two regulators, one operating a valve in the pump-discharge line, the other operating the turbine-governor mechanisms to vary speed. Both these regulators are master-loaded from the master controller, and, by varying the master loadings, through ratio adjusters, either motor or turbine drives the pump, depending on plant requirements.

In these pump controls for plants *D*, *E*, *F*, and *G*, the master controller carries manometers that show the master-loading pressure to each regulator. A ratio adjuster in each loading line can be adjusted manually to vary loading pressure, to compensate for variations in pump characteristics, and to divide pumping load in unequal proportions if desirable or necessary. Master controllers are also provided with an automatic-to-manual transfer valve and lever, so that the pumps may be controlled manually.

In any pump-control system of the types shown here, it is possible to sound a horn alarm and light signal lamps when boiler-feedwater flow goes below or above certain predetermined limits. Plant *E* uses the circuit shown in Fig. 9-12.

All the above are centrifugal pumps. Reciprocating pumps, however, are used in plant *H* (Fig. 9-10), one of the newest 1,200-lb condensing extensions. Three pulverized-coal-fired boilers total 1,300,000 lb/hr at 850°F. Feedwater will be pumped by two 5-cylinder single-acting horizontal reciprocating boiler-feed pumps, only one of which is yet installed. This pump has a capacity of 3,000 gpm at 300°F against 1,500 lb discharge pressure. Centrifugal pumps serve as standby and supplementary capacity.

The reciprocating pump is driven by a variable-speed electric drive, controlled by a potentiometer rheostat to maintain a variable excess-feedwater pressure. The drive unit consists of a 1,750-hp induction motor to furnish the base speed and a 750-hp d-c motor for variations from the base speed. The induction motor is mounted inside the d-c motor on bearings, in such a way that both stator and rotor of the induction motor rotate. The d-c motor drives, or is driven by, the induction-motor frame. Power for the d-c motor comes from a synchronous motor-generator set.

When both induction and d-c motors are operated, the latter is controlled so that the two motor speeds are added or subtracted, the resultant speed being transmitted through a 12:1 reduction gear to the pump crankshaft. Speed range of the drive is 708 to 282 rpm.

The d-c motor is controlled by adjusting the field of the motor-generator set by a motor-operated potentiometer rheostat setting. A master, loaded from the combustion control, sends a measured loading pressure, in accordance with variations in steam demand, to a regulator, which operates contactors that control a reversing pilot motor driving the potentiometer-contact arm. This pilot motor rotates a contact drive rod to cut in or out the

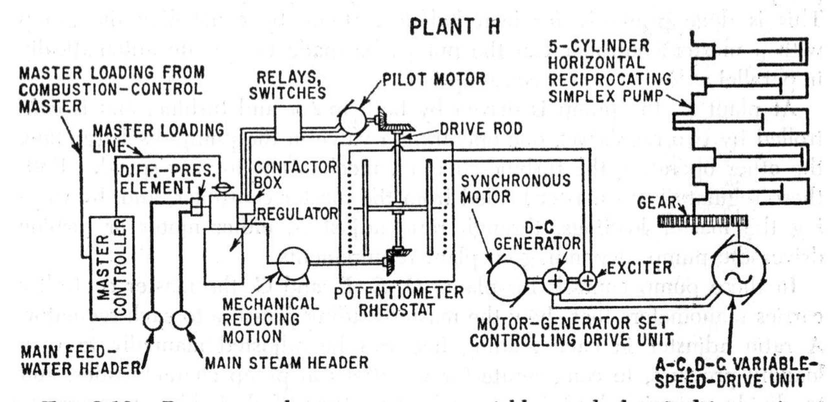

FIG. 9-10. Pump-control system governs variable-speed electric-drive reciprocating pumps. Excess pressure is varied. (*Alfred C. Wenzel, Republic Flow Meters Co.*)

necessary amount of the rheostat. Rotary motion of the potentiometer-contact-arm drive rod is transmitted back to the operating lever of the regulator through a reducing motion, as shown in Fig. 9-10, opening the contactor and stopping the pilot motor. This operation is repeated, the regulator closing one contact or the other, until the pump-discharge pressure is brought to the required value.

Another unusual control for reciprocating boiler-feed pumps has been designed for plant *I* (Fig. 9-11). A boiler-water-level regulator operates valves in the steam lines to two vertical reciprocating steam pumps, varying speed and output in direct relation to variations in drum-water level. The new pulverized-coal-fired bent-tube boiler has a capacity of 100,000 lb/hr at 500 lb, 800°F, supplying a condensing-extraction turbine-generator. Feedwater is pumped by the two pumps, capacity 150 gpm each at 12.25 double strokes per min against 500 psig. Steam cylinders operate with 500 lb steam and exhaust at 75 lb.

The water-level regulator is a standard type in which a mercury-

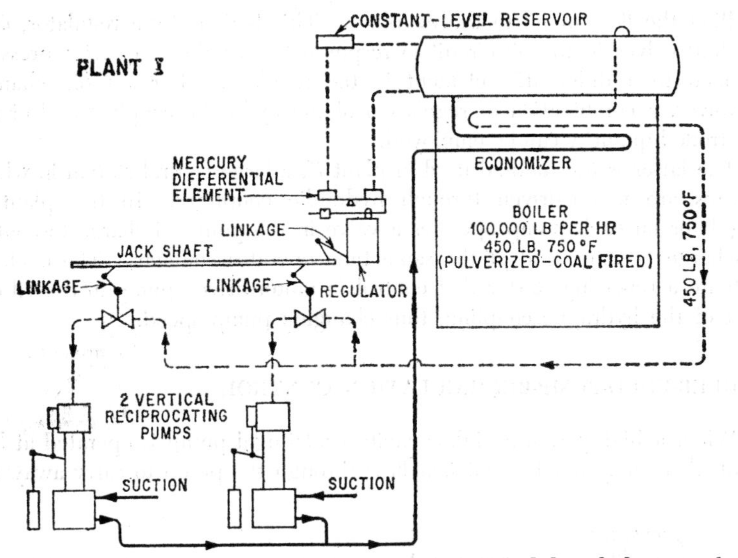

Fig. 9-11. Vertical reciprocating feed pumps are governed directly by water-level regulating system. (*Alfred C. Wenzel, Republic Flow Meters Co.*)

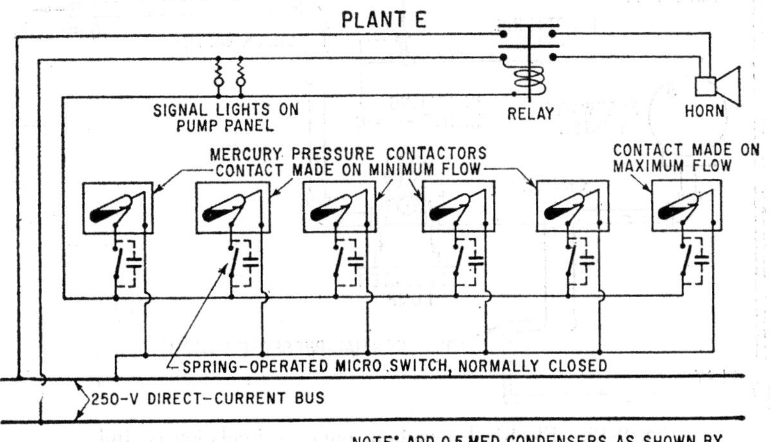

Fig. 9-12. Alarm is sounded if feedwater flow varies beyond predetermined limits. (*Alfred C. Wenzel, Republic Flow Meters Co.*)

differential-measuring element, responsive to changes in boiler-water level, operates the regulator. This turns a jack shaft, from which levers operate the pilot valves of oil-cylinder-operated turbine-type valves in the steam lines to the pumps.

In some large high-pressure installations, feed pumps are driven by motors through hydraulic couplings. Power and speed are varied by controlling the flow of oil to the coupling. This is done by a regulator, controlling valves in the oil circuit in response to variations in excess pressure applied to a differential element in the regulator. For a rapid change, however, a combination of regulation of the hydraulic coupling and of the controlled quantity can be employed.

The latter is the method used in plant *K*, a large central station in which pumps are motor-driven through hydraulic couplings. In this plant, a regulator on each pump operates a valve in the pump discharge to control discharge pressure and at the same time operates contacts. The contacts actuate a reversing starter that controls a motor-driven pump in the oil circuit of the hydraulic coupling, thus changing pump speed.

BOILER-FEED-PUMP-RECIRCULATION CONTROL

When a high-pressure high-capacity centrifugal pump is operated at low load, there may not be sufficient flow through the pump to carry away the

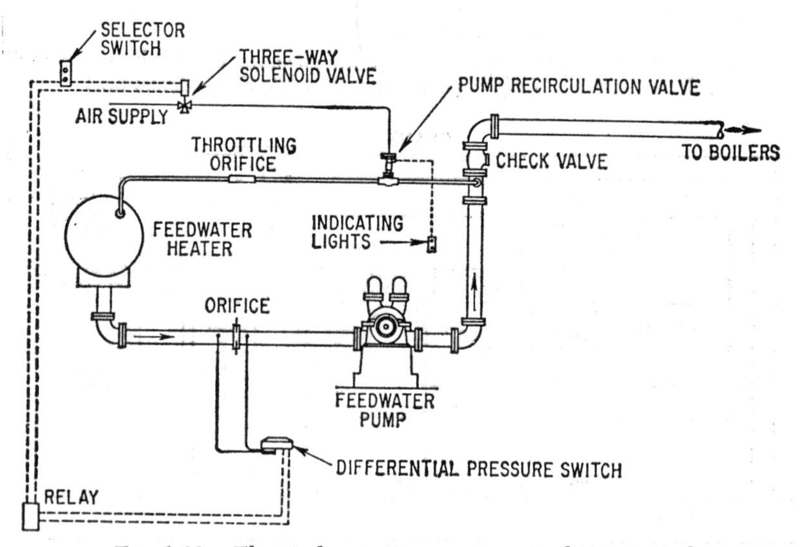

Fig. 9-13. Electrical operation, pump-recirculation control.

heat generated in the pump itself. In order to prevent an excessive temperature rise, which would lead to possible flashing or rubbing and binding of the rotating parts, it is necessary to install a recirculation line in the discharge of the pump to ensure a minimum flow through the pump.

In the operation of a centrifugal boiler-feed pump, the difference between the brake horsepower (bhp) consumed and the water horsepower (whp) developed represents the power losses within the pump itself. All this power loss is converted into heat and transferred to the feedwater passing through the pump.

For the purpose of approximating the minimum permissible flow of feedwater, it can be stated that 15°F rise is a recommended maximum value and that, to limit the temperature rise to 15°F, a boiler-feed pump should never be operated at a capacity of less than 30 gpm per 100 bhp at shutoff. This, of course, determines the minimum size and dimensions of the recirculation line for a given boiler-feed pump.

The actual amount of temperature rise for any given operating condition can be determined from the following formula:

$$°\text{F rise} = \frac{(\text{bhp} - \text{whp}) \times 2{,}545}{\text{capacity in lb/hr}}$$

This formula can also be expressed as

$$°\text{F rise} = \frac{\text{total head in ft}}{778}\left(\frac{1}{\text{efficiency}} - 1\right)$$

Electrical System, Open-and-shut Control. On startup, all electrical equipment is deenergized, and the pump-recirculation valve is open. Referring to Fig. 9-14, as flow increases, the low-flow snap-action precision switch closes at the predetermined minimum flow. On further increase in flow, the high-flow snap-action precision switch closes at a preselected point, energizing the relay, thus closing relay contacts $C1$ and $C2$. With the selector switch closed, the three-way solenoid valve is energized, allowing air pressure to close the pump-recirculation valve. On a decrease in flow, the high-flow switch opens but the relay continues to be energized through its hold in relay contact $C2$. On a further decrease in flow, the low-flow switch opens, deenergizing the relay, and opening $C1$ in the solenoid valve, which in turn shuts off the air supply to the recirculation valve and vents its diaphragm to atmosphere, causing the valve to open. With the relay energized, the valve may be opened from the control panel by opening the selector switch, thus deenergizing the solenoid valve. The selector switch will close the valve only when flow conditions permit. The built-in switch on the diaphragm top automatically operates position-indicating lights to show whether the valve is open or closed.

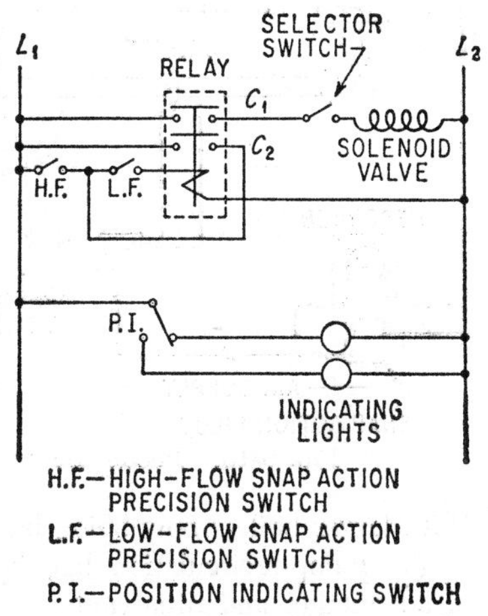

FIG. 9-14. Simplified wiring diagram of pump-recirculation control system, shown for no-flow conditions.

Pneumatic System, Open-and-shut Control. Flow through the pump is metered by an orifice or flow nozzle and pneumatic-flow transmitter (Fig. 9-15). Loading air pressures from the transmitter, which vary proportionally with the flow, are sent to a snap-acting pneumatic pilot operating a three-way air valve installed in the high-pressure air-supply line to the recirculation-valve diaphragm. When the flow drops below a desired minimum value, the pilot cuts off the air pressure to the three-way valve, shutting off the high-pressure air supply to the recirculation-valve diaphragm and venting it to atmosphere. A spring in the recirculation valve topwork and water pressure on the valve stem act to open the valve and hold it open until air-loading pressure is again applied to the diaphragm.

In the past, open-and-shut pump-recirculation-control systems have proved very satisfactory when used on average-sized feedwater systems.

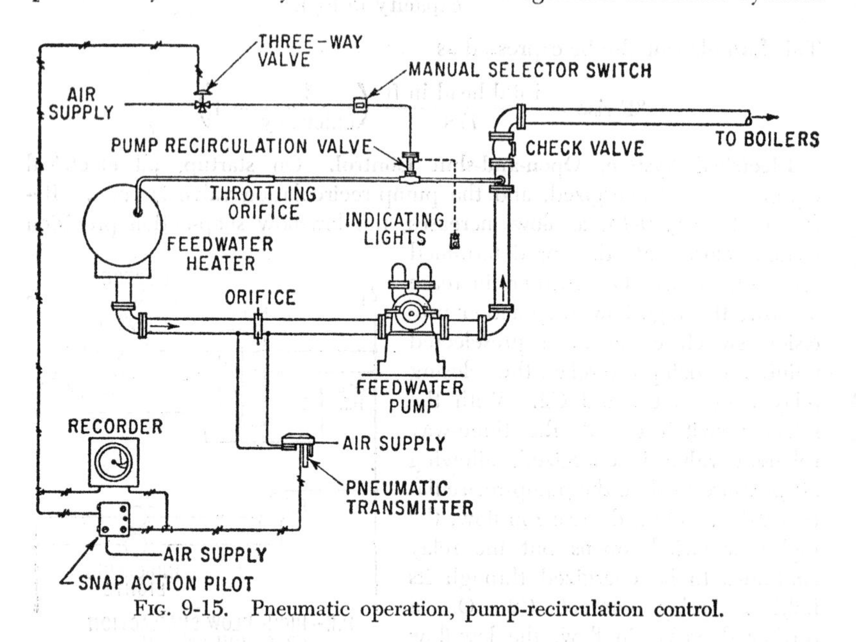

FIG. 9-15. Pneumatic operation, pump-recirculation control.

With larger boilers now being built, certain advantages of throttling control in the pump-recirculation line become important. Here is a comparison of advantages and disadvantages between the two control systems:

Open-and-shut Control. Advantages:

1. Pressure-reducing orifice reduces pressure drop across valve to negligible amount, considerably increasing valve life and lowering maintenance costs.
2. In event of air-power failure, valve "fails safe," opening and remaining open until air is restored.

Disadvantages:

1. When open, valve may pass more water than is absolutely necessary, in order to hold pump temperature within safe limits.
2. Surge of recirculation water upsets the heater or condenser and may cause plant disturbances of long duration.

Throttling Control. Advantages:

1. Exact minimum flow of water through feedwater pump is maintained with no power wasted pumping excess water.
2. Flow of recirculation water introduced into heater or condenser is gradually increased or decreased, minimizing vibrations and sudden shock.

Disadvantage:

1. High-pressure drop across valve tends to cause short service life and high maintenance costs.

COMBUSTION CONTROLS

In a steam-electric station,[4] the initial design largely fixes the maximum plant performance. Once given the steam pressure and temperature at the turbine throttle, there is little in ordinary operation that can be done by the operators to improve the performance of the steam-using and electric-generating elements of the station. The matter of attaining maximum station performance, therefore, depends largely on the efficiency of the steam-generating equipment.

The more recent steady advance in size, pressures, temperatures, and the price of fuel, along with the equally steady deterioration in quality, has made automatic combustion control even more important now than at any time in the past. In view of this, the design of a combustion-control system for a new plant is a problem worthy of much study. Some of the questions that should be considered include, in addition to the operability of the system, reliability, safety, and freedom from excessive maintenance.

Attainment of maximum reliability requires consideration not only of the conditions that exist when the plant is started up but also of the conditions to be expected years afterwards. The equipment should be designed to eliminate wear; wherever wear cannot be avoided, it should take place in a manner that is apparent to the operators long before the unit ceases to function.

Safety requires consideration of control-circuit layout, type of operating medium, and speed of operation of the equipment. Maximum freedom from maintenance requires consideration of suitability of control-system designs, ease of understanding, and simplicity of trouble shooting.

In general, the problem of automatically regulating the combustion process in a steam-boiler furnace using oils, gas, or coal involves the regulation of three related factors: (1) fuel, (2) air, and (3) products of combustion.

The rate of fuel flow is adjusted in accordance with steam-header pressure. Rising pressure indicates too high a rate of fuel supply for the load, and falling pressure indicates too low a rate of fuel supply. By converting slight variations in steam-header pressure into adjustments of the rate of fuel supply by means of the appropriate mechanism, it is possible to con-

[4] M. J. Boho, Hagan Chemicals & Controls, Inc.

tinuously proportion fuel supply to load demand, and maintain steam-header pressure reasonably constant at the desired value.

In much the same manner, control over the removal of products of combustion is attained from furnace pressure. If the pressure within the furnace chamber is rising, either too much forced draft or too little induced draft is present. If the pressure within the furnace is dropping, the boiler is being operated with too much induced or too little forced draft. A regulator responsive to furnace pressure may be utilized to convert slight variations in this pressure into adjustments of either the uptake damper or the forced-draft fan, thereby continuously maintaining the pressure within the furnace at the desired value.

The control of rate of air supply differs somewhat from the other two factors in that there is no one condition which gives an indication of the balance between existing rate of air supply at any instant and rate of air supply desired with sufficient accuracy and reliability for use in automatic control. Smoke indicators are unreliable, since there are factors other than the quantity of air supplied which affect the indication; likewise, the use of CO_2 recorders and oxygen analyzers is hampered by the difficulty of getting representative samples quickly enough to regulate the prevailing rate of flow.

The practical unavailability of a unique function for air-flow control, which corresponds to steam pressure for control of fuel flow and furnace pressure for control of removal of products of combustion, makes it necessary to control the rate of air supply in a different manner. Since the primary function of air in the combustion chamber is to support combustion of the fuel, a measurement of the rate of fuel flow should provide an indication of the rate at which air should be supplied.

In the past, measurement of fuel flow has been obtained by orifice measurement for gas and liquid fuels, stoker speed for stokers, and in several different ways depending on the type of pulverizers for pulverized coal.

In general, orifice measurements for liquid and gaseous fuels have been satisfactory. For measurement of coal-flow rate, the physical nature of the problem and its lack of easy solution have led to the development of a number of different types of measurement methods. For pulverized-coal firing, one type of measurement uses the speed of a raw-coal feeder having a constant-volume displacement per revolution. Another type often used depends on the coal-carrying capacity of air, and still another type involves the measurement of the power requirement for grinding the coal. All these fuel-measurement schemes have certain drawbacks. For example, coal which is wet may stick in the pockets of the feeder, thereby lowering the volume fed when the feeder operates at a given speed. Lump coal does not pack as does fine coal, causing a variation of coal fed per revolution. Where the carrying power of air is depended on, practical experience indicates that the fuel output is dependent on the amount of coal in the pulverizer at any given instant. Where power requirement is used as a measurement, the grindability of the coal, the condition of the pulverizer, the amount of primary air, and the moisture content of the coal all affect the indication, making it unreliable in some cases. With all of these schemes, a change in the quality of the coal may take place without an indication of the change, and requiring a manual readjustment of controls for proper results.

Classification of Systems. Boiler combustion controls can be considered, from the point of view of circuit arrangement, as falling into one of two broad classes: parallel and series systems. The series type can further be arranged in three different ways. Figure 9-16 shows these several arrangements diagrammatically.

In the parallel arrangement *C*, the initiating signal comes from steam pressure, and is transmitted in parallel to the fuel-feed and air-flow controls

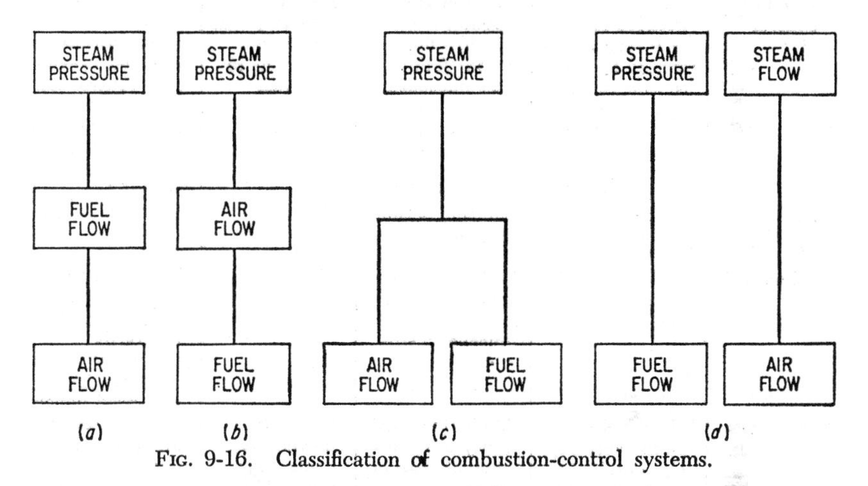

Fig. 9-16. Classification of combustion-control systems.

calling for corresponding adjustments on each. The fuel-feed and air-flow controls are each calibrated to the master signal, so that the proper amounts of fuel and air are fed to maintain steam pressure and fuel-air ratio.

In the first of the series arrangement, *A*, steam pressure initiates a signal, which controls the fuel flow required to maintain constant steam pressure. The rate of fuel flow thereby set up is measured, and a second signal, generated in accordance with fuel-flow rate, is transmitted to the air-flow control, which adjusts the rate of air flow in accordance with the fuel signal. Inasmuch as air flow follows fuel flow, it is readily apparent that any failure in fuel flow will cause a corresponding cutback in air flow. Likewise, if any limiting exists on maximum rate of fuel, this system assures that air flow will not exceed this limit.

The second of the series arrangement, *B*, is similar to the first except that the air flow and fuel feed are interchanged in the sequence. In this arrangement, pressure initiates a signal which controls the rate of air flow, and a measurement of air flow in turn initiates a signal which controls the

Table 9-2. Classification of Combustion-control Systems

Name	*A*, Series-fuel	*B*, Series-air	*C*, Parallel	*D*, Calorimeter or steam flow–air flow
Action	Temperature- or pressure-actuated master adjusts fuel rate; fuel meter adjusts air flow	Temperature- or pressure-actuated master adjusts air flow; air-flow meter adjusts fuel flow	Temperature- or pressure-actuated master adjusts fuel flow and air flow simultaneously	Pressure-actuated master adjusts fuel flow; steam flow adjusts air flow
Relative speed of control	Master adjusted for fast response because fuel-rate fluctuations caused by fluctuating pressure or temperature on master do not have correspondingly fast effect on that controlled variable	Master adjusted for slow response, because air-flow fluctuations following fast fluctuating master signal have a rapid effect on controlled variable and may cause hunting action if air-flow response is too fast	Master adjusted for slow response for same reason as in series-air	Master adjusted for fast response for same reason as in series-fuel. Steam-flow–air-flow control can be relatively rapid since steam-flow fluctuations are not so rapid as pressure variations
Used on fuels	Easily metered fuels such as oil and gas	All fuels. Oil, gas, and coal, either solid or burned in suspension	Primarily on solid fuels (grate firing)	Fuels hard to meter or fuels burned simultaneously. Commonly used on pulverized-coal-fired boilers
Advantages	When fuel may be in short supply, eliminates possibility of carrying high excess air for long period	Eliminates possibility of explosive mixture in combustion space when air fails. Eliminates need of fuel cutback for this purpose	Relatively inexpensive control system. No metering necessary	Ensures proper air-fuel ratio, even though fuel cannot be accurately metered or is of varying heat content Ensures this condition even when burning a mixture of different fuels at the same time

rate of fuel feed. With this type of system, any failure in air flow will result in automatically decreasing the rate of fuel flow to the value corresponding to the available air supply.

The last scheme, *D*, is in fact a series control arrangement similar to that shown in *A*. The steam-flow–air-flow control is direct and does not involve parallel corrector circuits. In this arrangement, steam pressure initiates a signal which controls the fuel feed required to maintain constant steam pressure. By utilizing the boiler as a calorimeter, the steam output from the boiler is the measure of the rate of heat input to the furnace. Measurement of steam flow provides the signal for regulation of air flow. This system is applicable to installations where fuel is burned in suspension, such as gas, oil, or pulverized coal. Where a reliable measure of fuel feed exists, as may be the case with gas or oil firing, arrangement *A* may be preferred. However, where either pulverized coal or a multiplicity of fuels is used, scheme *D* has certain advantages.

Table 9-2 summarizes these four basic types of combustion-control-system arrangements.

Methods of Operation. *On-off Control or Operation between Set Limits.* Some smaller domestic units, such as oil burners in homes, are satisfactorily controlled by this means. This is called on-off control. As the name implies, flow of fuel and air to the furnace is intermittent, and, when flowing, the fuel and air do not vary with respect to each other. The amount

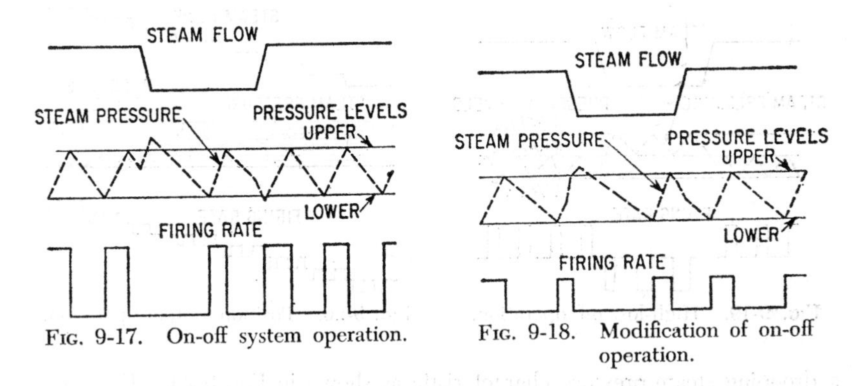

FIG. 9-17. On-off system operation.

FIG. 9-18. Modification of on-off operation.

of steam generated is controlled by the length of the on and the off periods.

During the off periods, steam supplied by the boiler is from the water-stored energy, which is released by the flashing brought about by decrease in pressure. During the on periods, the heat released by the fuel burning in the furnace generates the steam drawn from the boiler, and also replenishes the stored energy withdrawn from the boiler water during the off period. This energy storage is effected by increasing the pressure of the water and steam in the boiler.

Figure 9-17 shows how the on-off system operates between two steam-pressure levels. When the pressure drops to the low level, fuel and air are delivered to the furnace. When the pressure rises to the upper level, these flows are shut off. Because the system operates between two levels, close pressure control cannot be realized, and the steam pressure overshoots and undershoots the operating range with a load change.

A modification of the on-off system control is shown in Fig. 9-18. Here the control system operates between two pressure levels, but, instead of on-off, the fuel and the air flows are regulated between high and low points.

The high rate of flow releases more heat in the furnace than is required to meet the boiler-steam demands, whereas the low rate of flow is insufficient to meet these demands. However, it can be seen that this modification provides for smoother control than the basic on-off system, especially for boilers that have a continuous, even though not steady, demand.

A second modification of this type of control, which is a combination of the on-off and high-low, is the high-low-off shown in Fig. 9-19. If even smoother control should be desired, more modifications could be made to the control so that there would be several pressure-level operating points, each of which would result in a predetermined flow of fuel and air being delivered to the furnace.

Position Control. If a sufficient amount, or at least if very many such pressure-level settings for fuel and air flow should be employed, a so-called "positioning"-type control would be developed. The boiler is operated on a drooping steam-pressure characteristic as shown in Fig. 9-20. For every steam pressure in the boiler outlet header, there is a definite corresponding setting for the fuel flow and for the air flow. It is possible to make these settings fine enough (or to flatten out this pressure curve) so that the pressure droop is only 2 to 3 per cent of the full-load operating pressure of the boiler. Combustion-control manufacturers have developed equipment that will result in a perfectly flat curve for all boiler loads. Thus, equipment that will reset the boiler operating pressure to a fixed level after every load change is available.

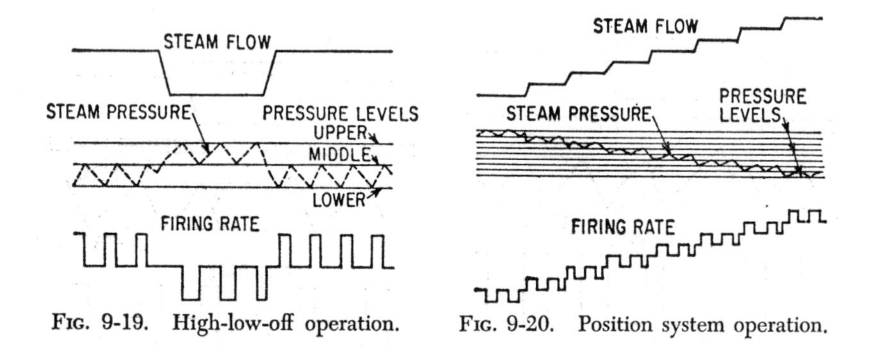

FIG. 9-19. High-low-off operation.

FIG. 9-20. Position system operation.

Metering Control. This is a further development of the positioning control. Instead of relying on the assumption that, for any given boiler load, there is one position for a boiler-auxiliary-control handle, air and fuel flows are actually measured, and final adjustments are made accordingly. The initial settings of the fuel and air flows are made by the primary element sensitive to steam pressure. These initial settings are modified or readjusted by secondary elements, which are sensitive to the actual fuel and air flows which are, of course, ultimately to be controlled. Therefore, this is a resetting type of control that is capable of "looking" at the results of its own adjustments, and making corrective readjustments to itself. This control, because of its reset characteristic, is capable of holding steam pressure and fuel-air ratios much closer to the desired settings, and is, therefore, a much better control than the positioning type.

In draft-regulated furnaces, the furnace "pressure" or draft must be maintained slightly below atmospheric. This is done by regulation of the forced- or induced-draft-fan flow, whichever is not being used to control the flow of air to the furnace, and which is usually considered a part of the combustion-control system. Whether the combustion control is positioning or metering (on-off control is seldom if ever used with a draft-regulated furnace), the furnace-draft controller is the same. It is sensitive to the furnace pressure and makes adjustments to its fan flow accordingly. This controller must, however, be very sensitive to small pressure changes and be capable of making rapid responses to these changes, for no flywheel effect is available here, such as is provided by the water-stored energy in the steam drum for fuel- and air-flow change responses. Some manufacturers make this control sensitive to boiler-outlet-steam-pressure changes as well as furnace-pressure changes, to anticipate required readjustments.

Pneumatic, Electric, and Hydraulic Mediums. The signals which are sent from the primary and secondary elements to the fuel- and air-flow regulators, and the medium of power used to operate these regulators, can be air, electricity, or a hydraulic fluid. Each has its basic differences and advantages. One difference is the speed of operation. If the transmitter of a signal (the primary or secondary element or a relay) should be considerably distant from the ultimate receiver of the signal (a relay or a regulator), there would be some lag between the time the signal is sent and the time it is received, if it is pneumatic. An electrical signal is, for all practical purposes, received at the same instant it is transmitted, no matter how great the distance of transmission.

On the other hand, a pneumatic signal has the advantage of being proportional to the change. For example, in comparing a large increase to a small increase in change in steam demand, the large increase will cause a comparatively larger drop in steam pressure, which, in turn, will cause the primary element to send out a proportionately stronger pneumatic signal. This proportionately stronger pneumatic signal will cause a pneumatic piston-operated regulator to move at a speed proportional to its magnitude. Thus, a pneumatic system has the desirable feature of being able to cause change in fuel and air flows at a speed proportional to the boiler load change. In an electric system, the primary element must close a contact before the regulators will move. Sometimes the primary elements are equipped with double contacts for two-speed operation, so that a strong primary signal will cause the regulator to run at a faster speed. However, the speed of an electrically powered regulator is established by the regulator-motor design.

A hydraulic medium is seldom used for transmitting signals, but it is used as a driving medium for regulators. Hydraulic fluid can move a regulator at a rate proportional to the strength of the signal received, the same as the pneumatic medium.

With the pneumatic and hydraulic regulators, however, a failure in the source of supply of air or hydraulic fluid may cause the regulator to move without a change in primary or secondary signal. This would result, in all probability, in unstable firing conditions, and would require almost immediate attention from a human operator to prevent damage. Sometimes, locking devices are installed on these pneumatic or hydraulic regulators which hold the regulators in whatever position they are at the time of power-medium failure. Electrically driven regulators will not change position upon power failure, but will remain exactly as they are, without the aid of locks.

Reasons for Pressure Regulation.[5] A question often raised when installation of automatic combustion control is considered is the accuracy of pressure regulation. As a premise, steam pressure cannot be held constant at all times if the basis for operation of a control system is change in steam pressure.

Several factors which determine limitation of pressure variation are:

1. Character of the plant load
2. Character of the steam generator
3. The control equipment itself

Load is important from the standpoint of pressure control. Yet in most cases it must be taken "as is" and handled in the best manner possible. Often better control may be established by using steam for other purposes, to secure a more favorable load condition. Anything that can be done in this direction is an advantage in efficiency as well as in the control of pressure. This refers especially to the reduction of peak loads.

The more constant the load, the more uniformly the steam pressure can be maintained. The load that varies widely, but gradually, over a long period of time is more easily controlled than one that fluctuates considerably at frequent intervals. The rapidity and extent of these variations are determining factors in how uniformly steam pressure can be maintained as far as the load character is concerned.

Base loads in power-generating plants, and in some types of continuous industrial processes, are constant and easily controlled. Building-heating loads, independent or municipal electric-light plants, city water-pumping loads, and some industrial plant loads increase gradually, and may then remain constant, and finally fall off slowly later on as demand decreases.

Industrial plant loads, such as loads in mines and quarries and test loads where the equipment being tested is loaded and unloaded rapidly and frequently, swing widely and require fast, accurate control.

Generally, satisfactory pressure control can be secured regardless of load character, since the tendency today is to select steam-generating, fuel-burning, and auxiliary equipment suitable to the load.

The design of the steam generator (type of boiler, setting, fuel, and fuel-

[5] The Hays Corp.

burning equipment) enters into the ability of the unit to respond to adjustments made by the control in response to steam-pressure changes.

The greater the water-storage capacity of the boiler, the more slowly it will respond to load changes owing to its greater heat-absorbing capacity. Modern boilers which approach a "flash" type unit respond very rapidly, compared to the older large-tube large-water-capacity units.

Generally speaking, boilers burning fuels in suspension will respond more quickly to changes in fuel input than will those burning solid fuels entirely on grates. However, other factors such as availability or cost of fuel will also enter into selection of firing equipment.

Solid-refractory settings, because of their greater heat-retaining capacity, are less responsive to changes in heat input than are air-cooled or thin, insulated walls, having lower heat capacity. Water-cooled walls hasten the response of the unit to changes in firing because of the rapid change in heat absorption that takes place, and because of lack of stored heat.

With the proper combination of boiler, fuel, fuel burners, and settings, even the most difficult load may be controlled. Experience indicates that load changes of 15 to 20 per cent of capacity of the unit per min can be handled without loss of efficiency or pressure control.

It is assumed, in making the above statements, that a complete metering system of control is being considered, and, therefore, best efficiency, as well as pressure control is to be secured.

It is possible to secure close pressure regulation by allowing efficiency of combustion to change with fuel feed.

It is also possible on stoker-fired boilers to control the fuel manually and regulate the air automatically from steam-pressure changes, varying the efficiency as well as the rate of combustion. In fact, the boiler dampers can be opened and closed frequently and never assume an intermediate position corresponding to the load, with the result that a uniform steam pressure will be maintained with sacrifice of efficiency.

With oil and gas, as well as pulverized fuel, it is possible to control fuel by steam pressure (and not the air at all), and to regulate pressure by varying not only the amount, but also the efficiency of combustion.

It is assumed, however, that control of combustion efficiency, as well as control of steam pressure, is desired. This being true, control of both fuel and air must be used. Pressure regulation may then be secured by regulating rate of combustion at the best possible efficiency.

The steam-pressure controller may be one of two general types—proportional-position or proportional-plus-floating (or reset). For boilers operating at fairly low pressures and having considerable water- and steam-storage capacity, a proportional-position type may be used. In this controller, movement of the power unit is directly proportional to change in pressure (between lower and upper limits). The pressure change between the limits produces 100 per cent movement and is known as the proportional band. A narrow proportional band can be used and good pressure regulation obtained on boilers with high storage capacity.

The proportional-plus-floating master controller is used on high-pressure boilers having relatively little storage capacity. This controller has proportional-position action, which produces an immediate change in the position of the power unit, with an accompanying change in pressure. In addition, it has reset action, which gradually applies an additional correction until the pressure has been returned to the set point. Use of this controller makes it possible to operate with a fairly wide proportional band to take care of sudden swings, while still retaining good pressure regulation.

Depending on the nature of the load, steam pressure should be controlled to within about plus or minus 2 per cent of set point with proportional-position action, and plus or minus 1 per cent with proportional-plus-floating action.

All combustion-control systems operate to maintain steam pressure within very close limits of a constant value, and steam-pressure changes are always used to cause the control system to function. Yet why not use changes in steam flow for this purpose, since the steam flow is the load on the plant, and is the main purpose for which it is operated? Or why not use feedwater flow since the water that is fed to the boilers is evaporated into steam, and the flow of water might, therefore, indicate load on the boiler?

Steam flow is referred to as the load, and, if flow of water and steam increase, rate of combustion must also be increased to carry the load and maintain steam pressure constant. But this is not the complete story.

It is fundamental that heat input must equal heat output, or steam pressure will vary.

Heat input is heat from the fuel liberated in the furnace.

Heat output is this same energy converted or distributed into one of the following outlets:

1. Steam demand
2. Loss in flue gas
3. Heat added to feedwater
4. Loss to ashpit
5. Loss by radiation
6. Loss by convection
7. Loss by conduction

It is apparent that steam flow is not the entire heat output of the boiler. All energy does not go into the steam, and flow of feedwater does not reflect the entire load at any one time. So neither of these forces is suited to the responsible task of initiating combustion-control movement.

Several of the heat-output factors vary with time and therefore cannot be compensated for in an attempt to use either steam flow or feedwater flow for control operation.

An analysis of the heat output indicates that any variation in quantity of heat distributed to any one of the outlets will affect steam pressure, and that variation in pressure immediately indicates an unbalance between heat input and output. For instance, all other conditions remaining constant, steam pressure drops if steam load, flue-gas temperature, feedwater flow, or any of the losses increases. If the opposite change takes place in any one, or all, of these conditions, the steam pressure rises.

These variables may or may not change in the same direction or in the same amount at the same time; therefore compensating action may or may not take place.

Instances have been experienced in which, for some reason or other, feedwater flow fluctuated so violently that the steam-pressure variations

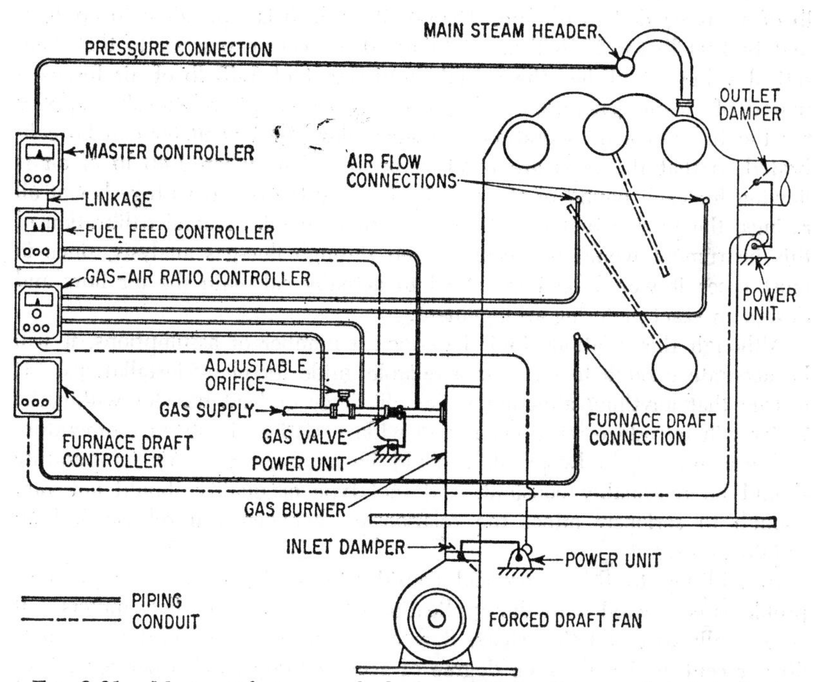

FIG. 9-21. Most satisfactory method of controlling all types of fuel feed is by steam pressure, often modified by steam flow. Master control unit normally aids in maintaining the efficiency by interaction with fuel, air-ratio, and draft control.

were many times greater than would occur with the slight variations in the load at the time.

The action of these variables makes it impossible to use either steam flow alone or feedwater flow alone as a means of operating combustion control. Since a constant pressure is desired and these variables definitely affect it, it is logical to operate combustion control from steam-pressure changes.

A measurement of steam flow, in addition to pressure, is often used to finally establish proper air flow (Fig. 9-21).

Air-flow–Steam-flow Meters. Air-flow–steam-flow meters (boiler meters) are based on the assumption that a fixed amount of air is required to produce a pound of steam at a given pressure and temperature. If steam at

600 psig pressure and 600°F temperature containing 1287 Btu/lb is generated from feedwater at approximately 300°F having 268 Btu/lb, the difference of 1019 Btu/lb between the total heats must be supplied by the boiler. Since this total heat must come from the fuel, the relationship between heat absorbed in the boiler and heat released by the fuel must be considered. Therefore, assuming an efficiency of 75 per cent, a total heat of 1358 Btu must be supplied by the fuel for each pound of steam produced. If the fuel is a typical bituminous coal, Fig. 9-56 shows that 7.7 lb of air is needed to produce 10,000 Btu, which is equivalent to approximately 1.04 lb for 1358 Btu. Adding 40 per cent excess air to this theoretical value establishes the actual requirement of 1.46 lb of air for each pound of steam generated. Then, assuming that this relationship applies for the full range of operation, the steam-flow–air-flow meter can be calibrated so that the pens are together on the chart when 1.46 lb of air is flowing for each pound of steam. By keeping the two pens together at all ratings, the proper fuel-air ratio can be maintained. Actual calibration of this instrument would be accomplished through flue-gas analysis, since in most cases it would be impractical to actually calculate the air flow and design an exact metering arrangement.

Although this relationship is based on a number of assumptions, it will be accurate enough to serve as a reliable guide for many installations. It is true that anything affecting boiler efficiency or heat transfer within the boiler will introduce some error, and wide variations in steam temperature and pressure or feedwater flow could be troublesome. Air temperature should be reasonably constant, and fuel type or quality should not vary enough to seriously affect the relationship between Btu release and air requirements.

In addition to the theoretical considerations, there are some practical problems connected with the application of air-flow–steam-flow meters. It is generally assumed that steam flow can be measured accurately down to 20 per cent, and with some degree of accuracy down to 10 per cent. Air flow must be measured with corresponding accuracy if the relationship is to mean anything. Consequently, the method of measuring air flow and the range of operation should be considered. Since air pressure will vary as the square of the flow, it is important that the differential at maximum rating be high enough to ensure accurate readings at minimum rating.

Combustion-air Flow.[6] Some factors that affect accuracy of measurement of the flow of air for combustion are: (1) variations in size and shape of the orifice used, (2) variations in temperature of the measured air (or the combustion gases), (3) change in quantity of dirt that accumulates in the orifices, and (4) change of pressure of the measured air or combustion gases.

In addition to these considerations, the conditions under which the boiler is to operate under automatic control must be studied to assure automatic operation over as wide a range and under as varied conditions as possible.

[6] O. W. Riggs, vice-president, The Hays Corp.

Satisfactory results may be measured by the CO_2 or O_2 content of the flue gases, or the relationship of air flow to steam flow.

Fans, stack, economizer, preheater, and forced-draft duct work must be analyzed as to their capacities and draft or pressure losses or differentials. This analysis will permit the best selection of location and method of measurement.

Whenever a boiler meter, which measures the relationship of air flow to steam flow, is used as a guide to efficient combustion, selection is already made of the means of measuring air flow. The air-flow record of the boiler meter should be taken from the same connections that are used by the air-flow controller for air measurement, so that both the meter and the controller receive the same measuring impulses.

In selecting the best location for an orifice or method of measuring air flow for any particular boiler on which an air-flow controller is to be used, the following methods are considered.

Draft Loss across Boiler Passes. By far the most frequently used method of measuring air flow is measuring the draft loss across boiler passes, as shown in Fig. 9-22. Most boilers should have sufficient draft loss, and the taps can be installed in the correct location in almost every design of boiler. This method measures the total volume of combustion gases, including air that has leaked into the furnace.

The boiler passes themselves remain fairly clean in most cases because of the operation of soot blowers. When the taps are located so as not to include the first pass or slag screen at the boiler, the differential remains fairly uniform. Temperature error due to varying temperatures at different locations in the boiler passes can be reduced to a minimum by compensation for temperature effect (stack action) with proper vertical location of draft taps (except on horizontally baffled boilers).

In some boilers, however, draft loss is insufficient to permit a controller to regulate the air flow closely. Although this method is the most common, in some cases it may not give the best measurement because of variables such as cleanliness of the passes and temperature of the gases. Also, this method measures air that has leaked in through the boiler setting or where it does not enter into combustion. It is therefore desirable to use a method of measurement that will be independent of these variables and consequently give more satisfactory results.

Pressure Loss across Air Side of Air Preheater. When the boiler is equipped with an air preheater, the pressure loss across the air side may be an excellent means of measuring the flow of air for combustion (Fig. 9-23).

In practically all air preheaters, draft loss on both air and gas sides is sufficient to operate the air-flow controller. However, the gas carries dirt, which tends to lodge in the tubes and change the size of the orifice, while the air side remains clean since it carries fresh air. It is usually better to use the pressure loss across the air side of the heater as a measurement of air flow.

Temperature variations in the air preheater are at a minimum since it delivers an almost constant-temperature air for all loads. The "stack" effect in a vertical air preheater is compensated for by making the connections either to it or in the controller.

In practically all cases, the preheater carries the total air including primary, overfire, and secondary air. Sometimes tempering air is also taken

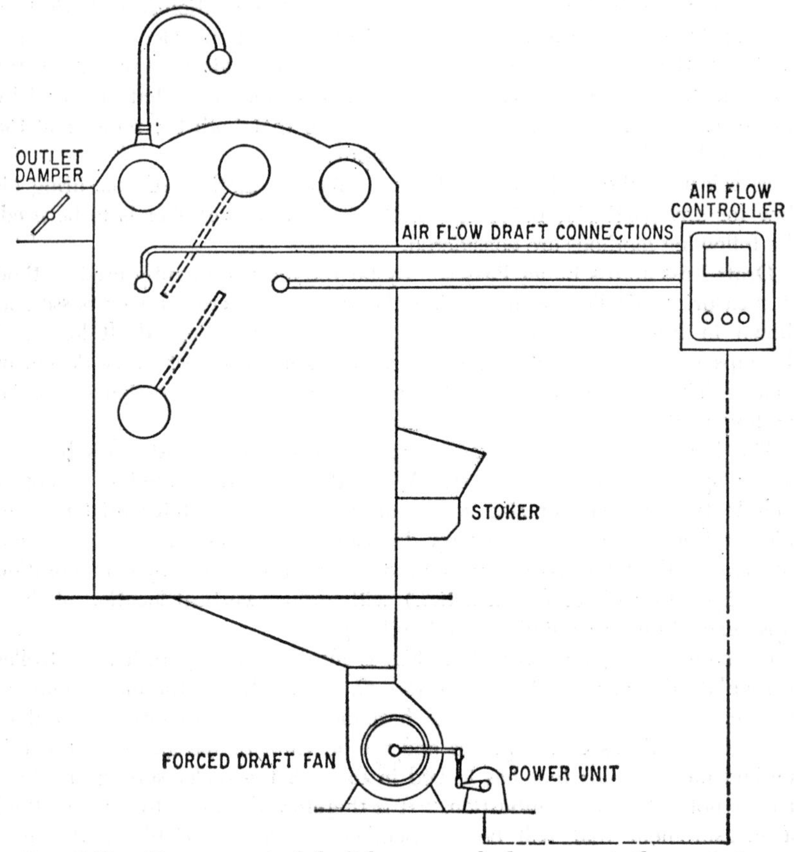

Fig. 9-22. Measurement of draft loss across boiler passes is the most common method of measuring the air flow. In some cases, loss is too low for accurate control use. (*The Hays Corp.*)

in, but in most cases this is negligible and, since it is generally proportional to the primary air, not important.

In some instances it may be desirable not to measure the overfire air. For example, when it is used for smoke elimination, it is "extra" air for burning the volatiles and reducing the smoking conditions.

Using the air preheater as an orifice places no additional load on any fan since the pressure loss that normally exists is simply measured. The air side of the preheater, therefore, is an almost ideal orifice, and is recommended wherever feasible for air-flow-measuring purposes. It is essential, however, that all air used for combustion should pass through the air heater so that it will be measured.

Loss across an Orifice in the Forced-draft Duct. Theoretically, measurement of the pressure loss across an orifice provides the best means of

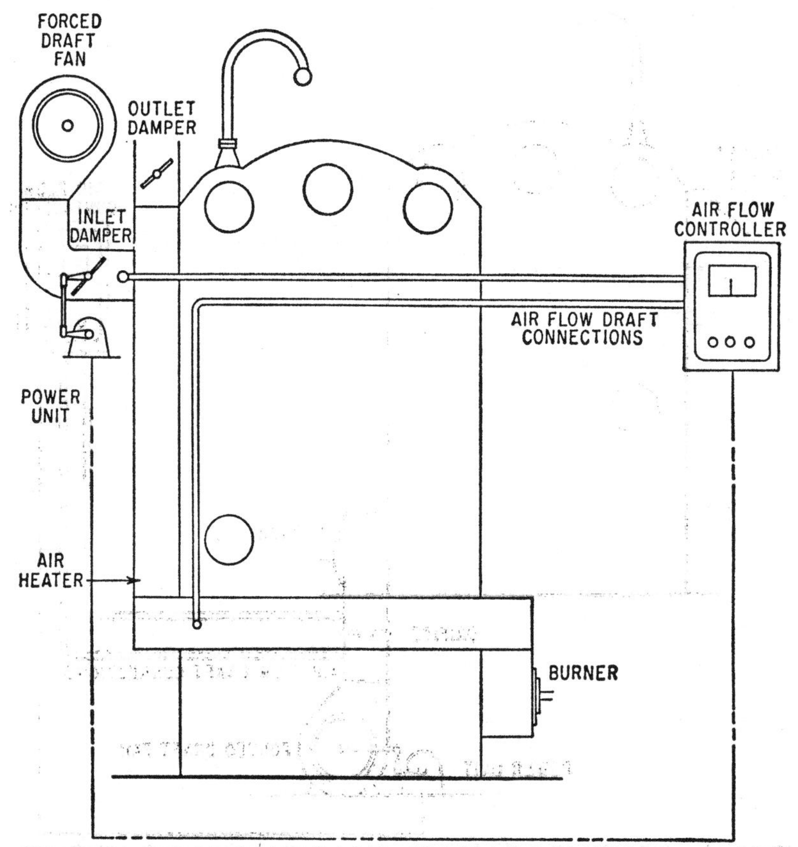

FIG. 9-23. Loss across air side of an air preheater is usually adequate, and the openings remain clean enough to operate a controller. No load is added to blowers. (*The Hays Corp.*)

measuring air flow, and should be used wherever practical conditions will permit (Fig. 9-24). This method should be employed if possible whenever the boiler has insufficient draft loss. At least 0.50 in. water draft loss should be available at maximum load. Sometimes there is an extra forced-draft damper in the duct which can be used as an orifice.

In many cases, layout of the air ducts makes the satisfactory installation of an orifice impractical. In addition, a small part of the static-pressure output of the fan is absorbed in creating the differential pressure across the orifice, slightly increasing the power consumed by the fan.

In addition to an orifice in the forced-draft duct, there are other means of creating the required pressure loss in a forced-air duct for air-flow measurement. Since forced-air ducts are most frequently rectangular in shape, a segmental type of orifice is generally used. It is most easily formed by

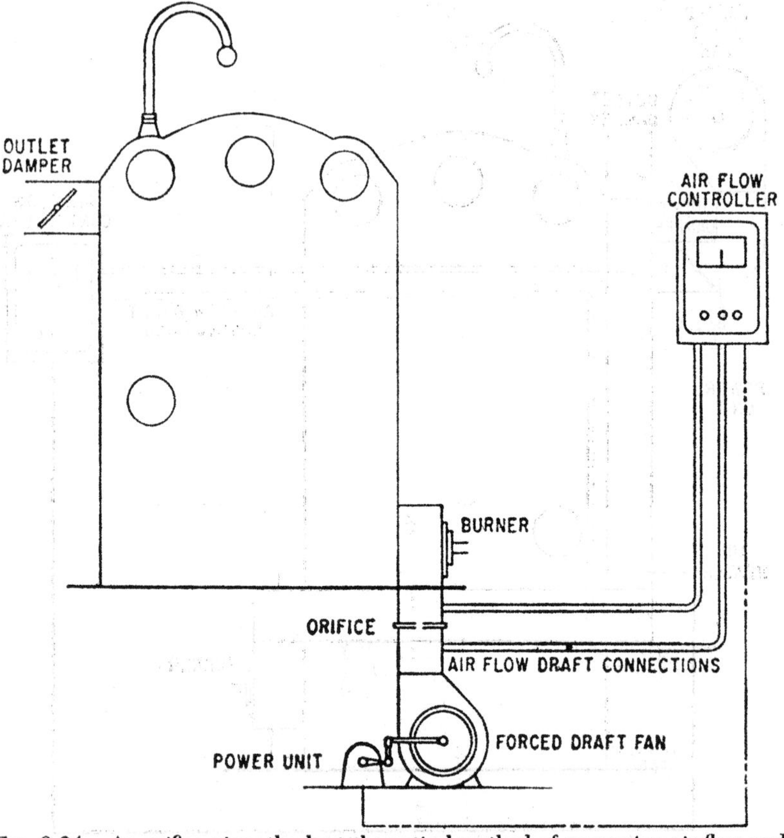

Fig. 9-24. An orifice gives the best theoretical method of measuring air flow and should be used when the boiler has insufficient draft loss. (*The Hays Corp.*)

providing a slot in three sides of the duct, into which a flat steel plate is inserted as far as is required, to create the desired loss of pressure with flow of air. The plate is then bolted or welded in place.

Another way to create loss of pressure is by means of a venturi tube or modified form thereof. When the forced-draft air ducts are round, a section is generally formed into a true venturi. The temporary pressure loss between the upstream pressure connection and the pressure connection at the vena contracta is used to measure the flow of air. In the rectangular ducts the sides are depressed on one or more sides to form a modified venturi tube. The cost of a venturi is generally higher than that of a segmental orifice. It has the advantage, however, of creating less permanent loss of air pressure for the same flow of air than is created by the orifice with the same measured differential. Therefore, the cost of creating the same usable static pressure of the air downstream from the orifice is greater than with the venturi, and this cost must be borne for the life of the equipment.

Most large steam generators have divided air ducts. It is impossible to measure the flow of air in one duct and depend on it to represent total air, by assuming that an equal flow is taking place in the other duct. Therefore, two means of measuring the air flow must be provided, one in each duct. These two air flows must be metered and totalized in order to obtain a true total air-flow measurement.

In general, location of an orifice in the breeching, economizer, the gas side of air heaters, and similar points is not considered to be good means of measuring air flow because of the variations in cleanliness and temperatures. In addition, available draft is generally not sufficient to permit the use of any additional restrictions to flow of combustion gases, as would be created by the use of an orifice. However, orifices can be used when no better means is available. Their use, however, may make more frequent checking and adjusting of the air-flow controller necessary, to compensate for the variables.

Pressure Loss across Burners Used as an Orifice. The use of burners as an orifice, by measuring the loss in air pressure between the burner wind box and the draft in the furnace, is an excellent means of measuring air flow. They may be used in preference to measuring loss across an orifice in the forced-draft duct, and are recommended for the wide-range return-line type of oil burners and oil- and/or gas-fired packaged boilers (Fig. 9-25).

Where the installation consists of several burners, pressure loss remains the same, regardless of the number of burners in operation. This is due to the fact that, when a burner is in service, the air louvers are always left wide open, and therefore constitute a fixed orifice. When the burner is out of service, the louvers are closed. The air that passes through each set of louvers is that required for use with the oil consumed by that particular burner. The pressure drop across one or several burners in operation, therefore, remains the same. Changes in tangential dampers for directing the flame have less effect than is generally presumed, and such changes can be rectified at the controller and with less disturbance to the controller than if methods measuring total air flow are used.

Temperature variations and errors due to variable conditions of cleanliness are negligible with this method of air-flow measurement. It must be used with wide-range oil burners to keep the air-fuel ratio constant with a variable number of burners in operation. This method cannot be used with more than one steam- or mechanical-atomizing burner per boiler.

When two or more such burners are used, one of the other methods of air-flow measurement must be employed.

Control of Air Flow. Little need be said concerning the means of controlling the rate of air flow for combustion. Positioning of dampers (single- or multiple-leaf, or vortex types) may be used and controlled with equally satisfactory results. The dampers may be at either the inlet or the outlet of the forced-draft fan.

Speed of the forced-draft fan can be used as a means of controlling the air flow. Control of fan speed may be by a valve in the steam line to a

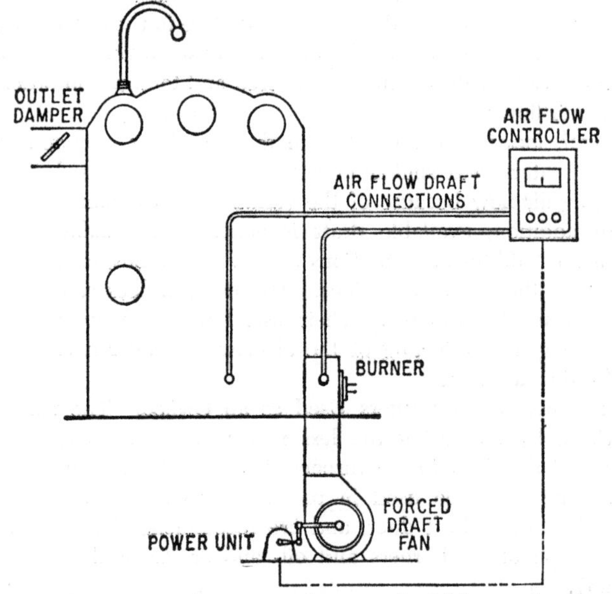

FIG. 9-25. The pressure drop across burners should be used with wide-range oil burners to keep the air-fuel ratio constant when the number of operating burners is variable. (*The Hays Corp.*)

turbine drive, or by actuating a lever-controlled governor at the turbine. The fan speed may also be regulated through a hydraulic coupling of either the scoop type or the reversing-pump type. It may also be controlled through the operation of an electric-type coupling.

When extremely wide range of control is required, or when other conditions would dictate its use, a system of cascading the operation of damper and fan speed may be employed. With such an arrangement, air flow would be controlled by regulating fan speed from maximum load conditions down to the lowest satisfactory speed; then control would be automatically transferred to the operation of the damper, which is positioned to regulate air flow for the lower load conditions. The control would be automatically transferred back to fan-speed variation with an increased load.

Control of Forced Draft. Because air for combustion comes from the forced-draft fan, it is generally most desirable to control the amount of forced draft for air volume and maintain the furnace draft constant by controlling the induced draft.

How Different Firing Methods Are Controlled. *Spreader Stoker.*[7] Although the method of burning fuel employed by spreader stokers has been used to some extent for many years, recognition of the merits of the method and application of this type of stoker have been greatest in the last few years. By using specially designed grates with improved feeders, modern spreader stokers produce decidedly better results than those secured with the earlier types.

Spreader stokers utilize methods different from those used by other stokers for (1) feeding fuel into the furnace and distributing it on the grates, (2) controlling air distribution, and (3) removing ash from the grates. These functions are performed by two separate parts of the stoker, namely, (1) coal-feeding and distributing unit, and (2) grates and air-distribution system.

Essentially, the coal-feeding and distributing unit starts with a hopper from which coal falls by gravity into the feeding mechanism. Here a fuel pusher, moving back and forth at uniform frequency, forces coal toward the furnace along a distribution plate. The length of the fuel-pusher stroke can be varied, either manually or automatically, through the linkage and lever provided, and this adjustment determines the quantity of fuel being fed into the furnace.

Coal pushed toward the furnace moves along a manually adjustable feeder or fuel-distribution plate. The position of this plate in relation to the rotor of the feeder determines the distribution of fuel lengthwise on the grates. As coal falls off the edge of the feeder plate, it drops into contact with a constantly revolving rotor. This rotor is fitted with blades which, because of their shape, throw coal into the furnace to cover the grates evenly over their entire width. Some of the fine fuel particles burn in suspension, while larger particles fall and are burned on the grates. Some stokers are arranged to feed fuel to the furnace pneumatically.

Generally, each feeder unit is provided with a combustion-chamber door directly beneath the feeder at grate level, for inspection, cleaning of fires, or hand-firing in an emergency. Each feeder also has an ashpit door below the combustion-chamber door. Several feeders may be used for one steam generator, and in this case grates and wind box are divided. Each feeder throws coal on a seperate grate section, air being supplied to that section through an individual manually operated damper.

Several types of grates such as stationary, hand-operated dumping, and power-operated dumping are available. In addition, another type, consisting of bars with holes in them formed into a chain which moves at a uniform rate, can be applied when continuous ash discharge is desired. Openings permitting air to pass through the grates are usually of the small or pinhole type, or of some special design. The large number of these

[7] The Hays Corp.

grate openings reduces air-flow resistance and provides even air distribution to the fuel. Each feeder usually receives an additional supply of air at several inches of water pressure to cool the feeder parts. This air flows into the furnace as overfire air to help support combustion.

After a quantity of coal is burned, the ash must be removed. Before this is done, the fuel feed is stopped, and the small amount of fuel on the grates at the time is burned out. Then the ash is removed by hand or by operating the dump grates. If continuous-moving-type grates are used, fuel feed need not be stopped as ashes are removed automatically.

In operation, fuel is thrown evenly into the combustion chamber, where fine particles burn in suspension while larger pieces of coal fall and are consumed on the grates. Since the grates are flat, the fuel bed is likewise flat and thin, and about 3 minutes supply is the most that is on the grates at any time. Volatiles are distilled by the heat from the fine particles burning in suspension, while fixed carbon burns on the grates through which air is passing, both processes taking place simultaneously. When all combustible in the fuel has been consumed, the ash which is left remains on the grates until it is dumped or otherwise removed.

Since a comparatively uniform burning rate, of both volatiles and fixed carbon in the fuel, takes place over the entire fuel bed, air-flow rate through the grates and fuel-feed rate are also uniform. Maximum combustion efficiency is secured only when the correct rates of fuel feed and air flow are maintained and distribution of fuel and air is complete and uniform over the entire grate area. It is necessary to have sufficient temperature for combustion to take place, time for it to be complete, and turbulence for mixing the air and gases.

To hold steam pressure constant, fuel-feed and air-flow rates must be proportional to the load. Furthermore, the ratio of fuel and air must be correct in order to maintain uniform combustion efficiency. Lack of air may cause smoke or CO in the flue gas, while too much air causes low CO_2 and loss of efficiency. The small amount of fuel in the combustion chamber at any time and the rapidity with which it burns make accurate control of both fuel and air important. As a change in the fuel supply almost immediately changes the CO_2, both air flow and fuel feed should be controlled automatically.

Uniform fuel distribution in the combustion chamber and on the grates is important, since air distribution through the grates is uniform. Adjustment for fuel distribution must be made manually, and for each quality and size of fuel the setting must be checked. If the adjustment is such that too much fuel is spread to the front or rear of the grates, uniform CO_2 will not be produced from the entire fuel bed, efficiency will be impaired, and smoke may be produced.

All three types of automatic-combustion-control systems, namely (1) off and on, (2) positioning, and (3) metering, are used with spreaders.

Off–on-type controls are employed only for very small steam generators. These systems do not produce very good efficiency, will cause smoke because of the deficiency of air while the forced draft is off, and are quite likely to allow the fire to go out if the off period is of long duration.

Positioning-control systems, especially the remote type (Fig. 9-26), are well suited to application for all sizes of steam-generating units and give satisfactory results. This system of control consists of a master steam-pressure controller, which responds to changes in steam pressure and, by means of power units, actuates the forced-draft damper to control air-flow rate, and the lever on the stoker to adjust fuel-feed rate. Steam-generating units with this type of control generally have constant-speed forced-draft fans equipped with dampers or inlet vanes, which are positioned to control the air for combustion. The boiler outlet damper may be controlled to regulate the air for combustion, but the former is the preferred method.

Furnace-draft controllers are used to maintain a low combustion-chamber draft. This reduces air infiltration through the setting to a minimum and retards any tendency for fly ash and small particles of unburned fuel to carry over into the boiler passes.

Positioning systems of automatic combustion control will hold steam pressure constant within very close limits. They are calibrated to maintain desired air-fuel ratios through proper alignment of levers and connecting linkage between the power units and the damper and fuel-feed levers which they operate. Variations in arc of angular travel of power-unit levers and damper or fuel-feed levers are used to compensate for the movement characteristics of the fuel-feed- and air-damper-control levers. In addition, the system can be provided with a convenient means for manual operation from a central point. This may be utilized for changing distribution of loads between boilers and making adjustments in fuel-feed rate to compensate for changes in fuel quality.

Metering-type combustion control (Fig. 9-27) should be used if air is not available at constant pressure or if fuel-feed rate and Btu input vary widely because of different sources of supply, moisture, or irregular sizing. If air flow only is variable, a metering type of air-flow controller can be substituted for the damper-control unit in the remote-positioning system. If both fuel and air vary, a system of control in which fuel-air ratio is readjusted from the air-flow–steam-flow relationship is recommended as most desirable.

Liquid-fuel Firing.[8] *Atomization* is the mechanically breaking up or separation of a liquid into small individual particles. When burning oil, the purpose of atomization is to allow air (oxygen) for combustion to quickly surround each particle of oil. Then rapidity of ignition and subsequent combustion determines to a large extent the maximum flame temperature and the amount of excess air required.

The burners for liquid fuels fall into three general types:

1. Air atomization
2. Mechanical atomization
3. Steam atomization

[8] Hagan Chemicals & Controls, Inc.

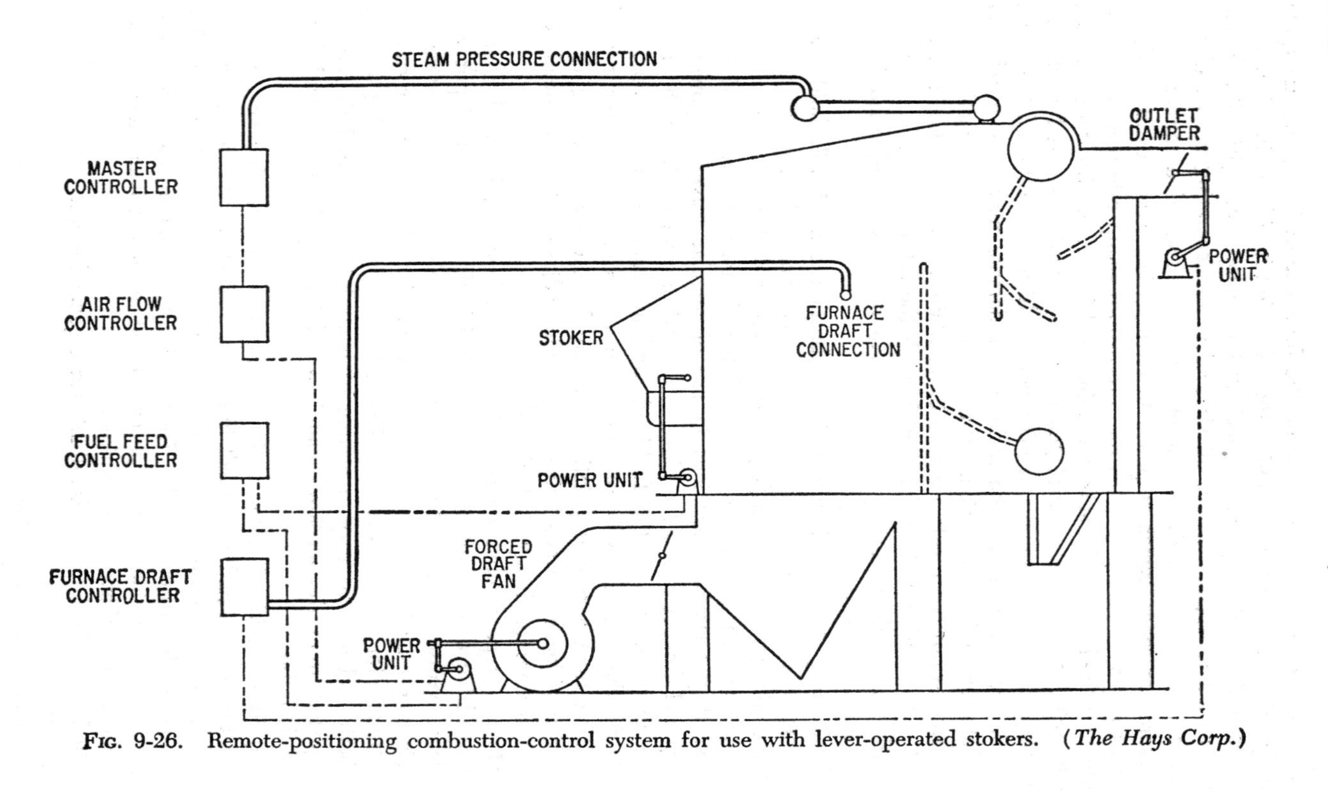

FIG. 9-26. Remote-positioning combustion-control system for use with lever-operated stokers. (*The Hays Corp.*)

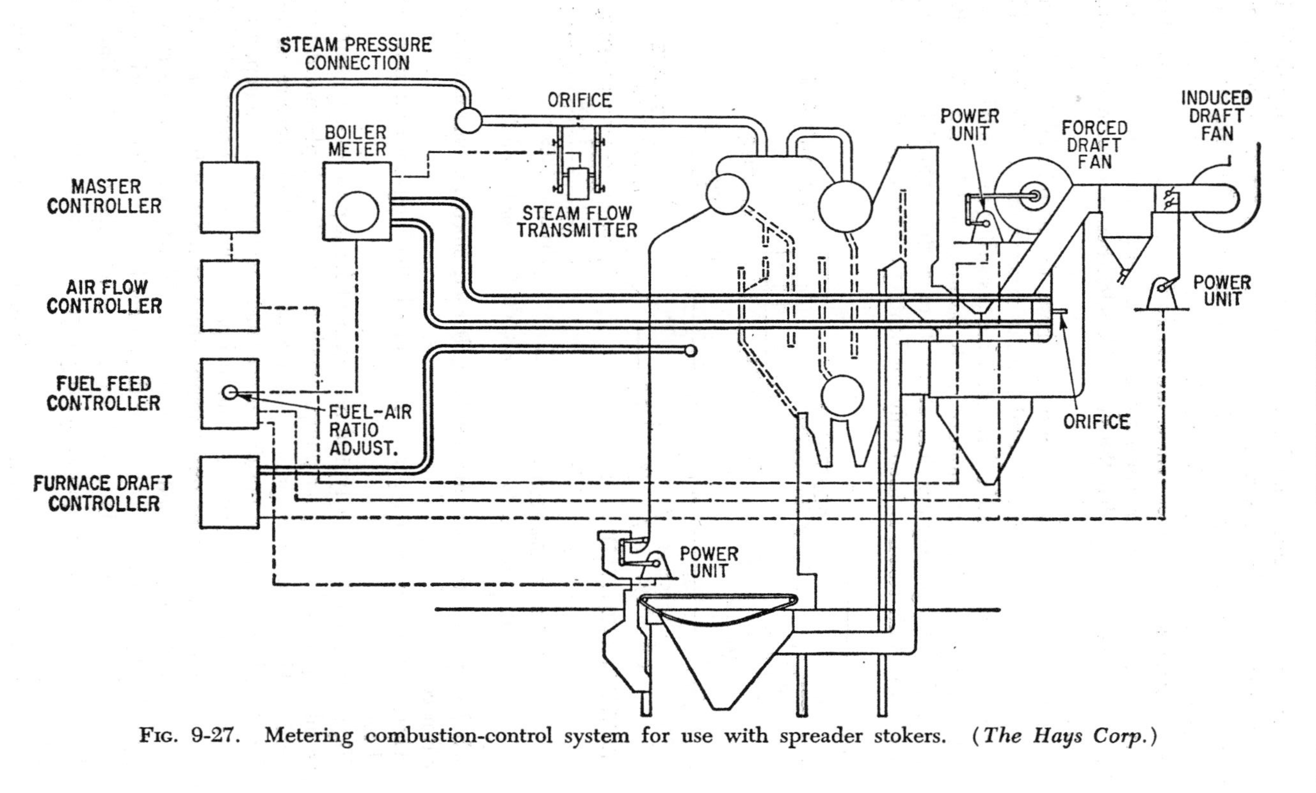

FIG. 9-27. Metering combustion-control system for use with spreader stokers. (*The Hays Corp.*)

These are three different methods of getting the fuel into finely divided form so that it presents as much surface as possible for the combustion reaction. The methods are listed in reverse order of their importance in use, the first being the least widely used.

AIR-ATOMIZING BURNERS: These are rarely found in plants of any size because of the high power consumption needed to compress the atomizing air. In general, this type has the same characteristics as a steam-atomizing burner.

MECHANICAL-ATOMIZING BURNERS: In these, the liquid fuel is led to a nozzle chamber where it is subjected to a fast rotary motion. As the liquid issues from the nozzle into the furnace, it has a swirling motion which tends to atomize it. Mechanical atomization is favored for high-capacity units because this design requires less steam than a steam-atomizing burner, since in this unit steam is used only for heating. This type also requires

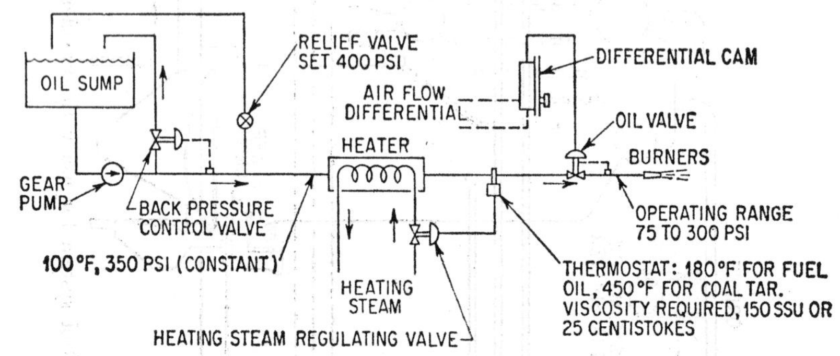

FIG. 9-28. Pressure-type burner for 5,000 lb/hr, or less oil flow.

much less power than an air-atomizing burner, since no compression of air is necessary.

There are two general types of mechanical-atomizing burners: pressure and return.

Figure 9-28 illustrates a typical arrangement for a pressure-type burner. The oil is pumped by a gear pump, which is a positive-displacement type of unit. The downstream pressure is regulated by a back-pressure-control valve, but, because of the positive-displacement feature of this pump, a relief valve is provided for safety. The fuel passes through a heater, which is supplied with steam. The outlet temperature of the fuel from the heater is regulated by a thermostat controlling a regulating valve in the steam-supply line to the heater. Ahead of the burner is an oil valve, which regulates the pressure to the burner in accordance with the input signal to the valve diaphragm. This input signal is generated by a flow signal shown in the illustration.

Since the oil flow through the burner is proportional to the square root of the pressure ahead of the burner, it is apparent that the signal from the flow-signal transmitter should be proportional to the air-flow differential in order to match air flow with fuel flow. The relationship between burner pressure and oil flow is shown in Fig. 9-29 for a multiple-burner installation.

In Fig. 9-30 is shown the relationship between total air flow through the burner registers and the differential from the wind box to the furnace. It can be observed that the same relationships hold for both fuel and air flow

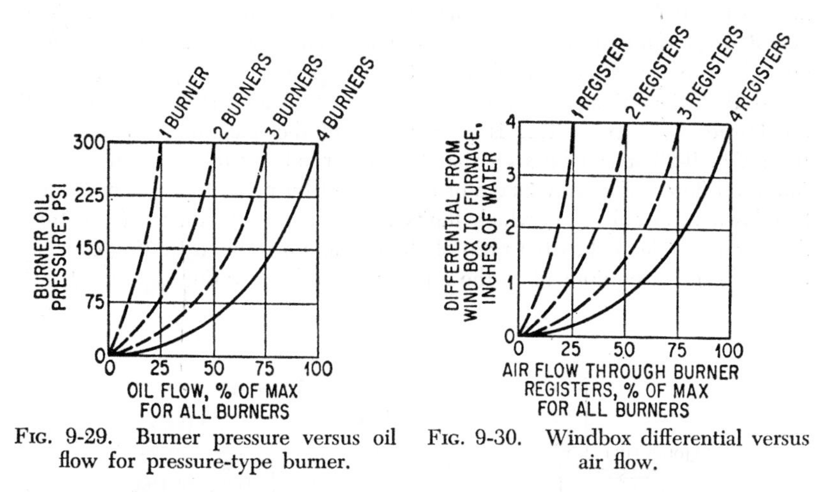

Fig. 9-29. Burner pressure versus oil flow for pressure-type burner.

Fig. 9-30. Windbox differential versus air flow.

for any number of burners, providing the oil pressure ahead of the burner is compared to the differential of the air between the wind box and the furnace.

Figure 9-31 shows the desired relationship between burner oil pressure and air differential. To obtain this relationship, it is obvious that a differential cam should be used in the variable-ratio regulator shown in Fig. 9-28.

From the study of Figs. 9-29 and 9-30, it can be seen that, if air flow were measured by taking the differential across a restriction in the forced-draft duct (not across the burner registers), or by measurement of a differential of the gas flow through the boiler setting at any point, it would be necessary to make adjustments on the variable-ratio regulator every time the number of burners were changed. The reason is that, when the number of burners is changed, the total area of oil orifices is, in effect, changed. Should the connection be located across the fixed orifice in the air duct or through the fixed orifice provided by any part of the boiler setting, then the ratio of total areas of oil orifices to air orifices would be changed every time we changed the number of burners, and it would be necessary to make an adjustment in the air-fuel relationship. When the

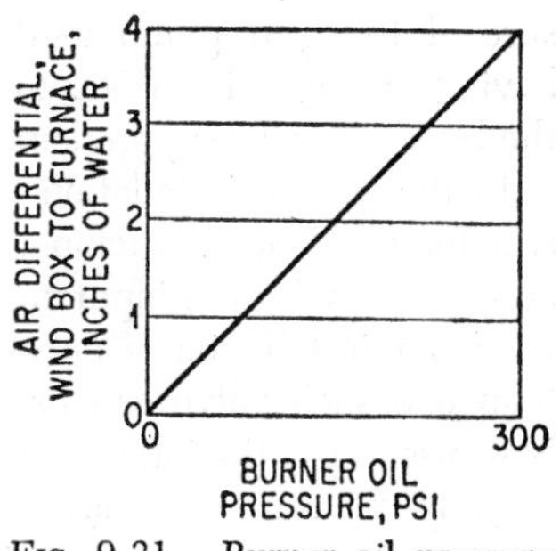

Fig. 9-31. Burner oil pressure versus air differential.

connection for air flow is made from wind box to furnace, at the time that a burner is pulled or added a corresponding change is made in the area of the air orifices. No correction in the variable-ratio regulator is then needed.

In Fig. 9-28, it is seen that the installation has a pump-discharge pressure of 350 psi and a maximum pressure ahead of the burner of 300 psi. This 50-psi drop through the line, heater, and control valve will usually give enough drop across the control valve itself for good regulation. It is further noted that the minimum oil pressure which will give stable operation at the burner is 75 psi. This is 25 per cent of the maximum pressure, but from Fig. 9-29 it can be seen that this still produces 50 per cent of maximum oil flow. A disadvantage of this type of burner is that this makes the turndown ratio only 2:1, which limits the burner load range. (The turndown ratio is defined as the ratio of the maximum stable operating capacity of any piece of equipment to the minimum stable operating capacity.)

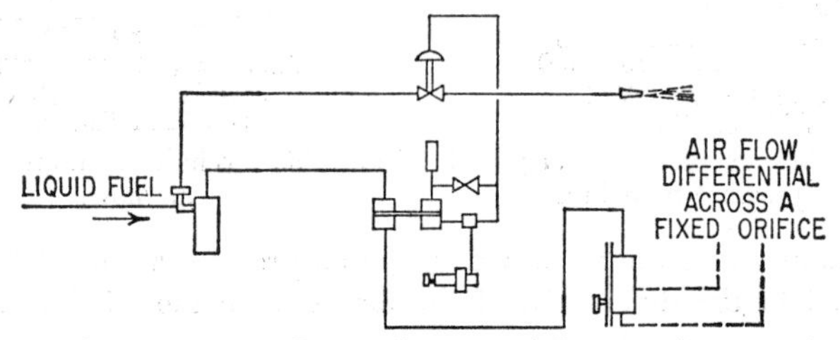

FIG. 9-32. High-capacity installation showing oil-flow regulation with positioning-type oil valve.

It should be further noted from Fig. 9-29 that, while the discharge pressure of the gear pump in this case is 350 psi, installations will be found having pressures from 300 to 1,000 psi on the pump discharge. Naturally, the burner-regulating pressures will vary accordingly.

In the typical installation of Fig. 9-28 there is shown an oil valve. This is a force-balanced machine, which actually controls the outlet pressure in accordance with the input signal. Since this valve is limited in its capacity, when an installation is encountered above the capacity requirement more than one such valve can be placed in parallel. A more common practice is to use a positioning-type valve and to put a meter in the oil-flow line to actually measure the flow of liquid fuel. This signal is then matched against a resetting totalizer in which the input signal is the signal of air flow. In this case, the air-flow differential must be taken across a constant-area air orifice, because the oil-flow orifice will now be constant, regardless of the number of the burners in service. To secure a constant-area orifice for air, the differential is taken across the air preheater, a restriction in the forced-draft duct across a portion of the setting or any other convenient part of the air-flow or gas-flow circuit which gives a usable differential. It would be common practice to use flow cams in the air-flow meter and in the oil-flow meter. The two signals matched across the resetting totalizer would both be flow signals. A portion of this typical arrangement is shown in Fig. 9-32.

Another possible arrangement of control is shown in Fig. 9-33. In this arrangement, the burner pressure must be 200 psi or less. Here the positioning-type oil valve is used, but, instead of a flow-signal transmitter to convert the burner pressure into an air-loading signal, a ratio totalizer is used with a 0.3-sq-in. bellows in the controlled-variable chamber. The reset action on this totalizer will assure that the pressure ahead of the burner is held proportional to air differential. Since oil flow is being measured by pressure ahead of the burner, it follows that air flow must be measured by pressure drop from the wind box to the furnace, so that, if the burners are changed, the same relationship will exist between air and fuel. Furthermore, since oil flow is being measured by burner pressure, which is a square-root relation to oil flow, a differential cam is used in the flow-signal transmitter on air flow.

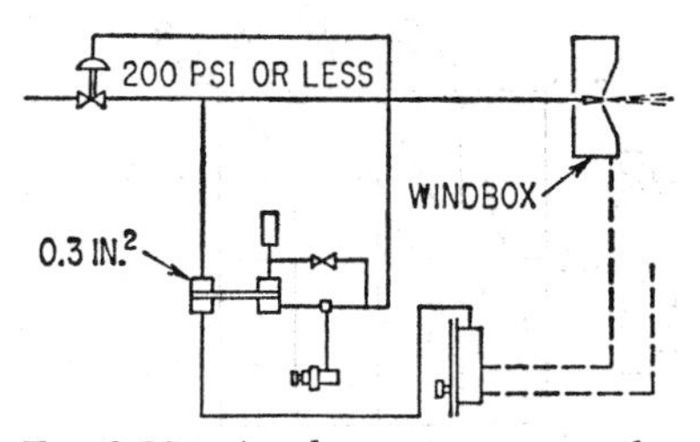

FIG. 9-33. Another arrangement of a positioning-type oil valve.

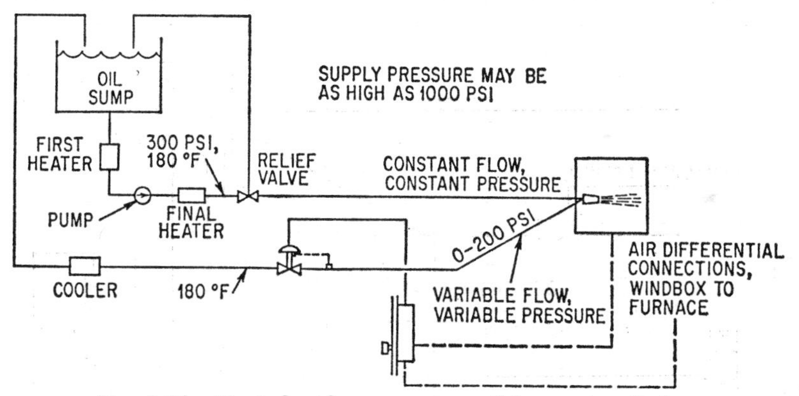

FIG. 9-34. Typical wide-range-return oil-burner installation.

For high turndown ratios, a mechanical atomizing burner of the design shown in Fig. 9-34 may be used. With this type, oil is delivered at a constant rate of flow and a constant pressure to the burner proper. The fuel passes through tangential slots into a whirling chamber at high velocity. This chamber has openings into the furnace, and a passage to an oil line, which returns the oil to the sump in the illustration. Because the oil is always delivered to the whirling chamber at constant pressure on the input, the velocity of the oil entering the furnace is practically constant; hence the atomization is good, no matter how much oil is being delivered to the furnace.

With this type of burner, the oil flow to the furnace is directly proportional to the pressure in the return line, as shown in Fig. 9-35. To regulate flow into the furnace, an oil valve is placed in the return line, as shown in Fig. 9-34. It is seen that the valve must be so placed that the upstream pressure is controlled, as this is a spill-over type of installation. (In a "spill-over" arrangement, a control valve regulates its upstream pressure by discharging or "spilling" whatever amount of fluid is necessary. The downstream pressure will be variable.)

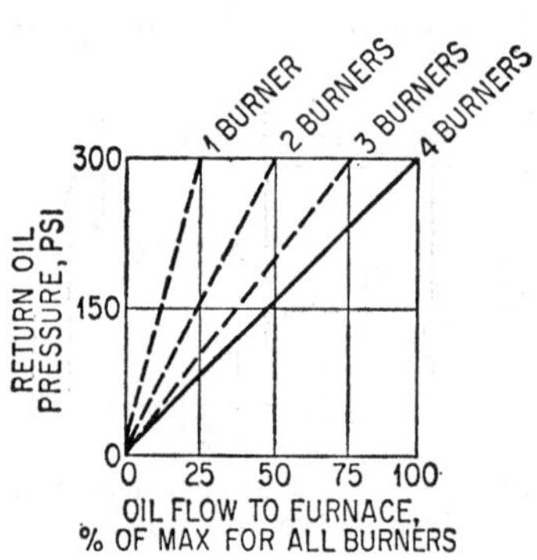

Fig. 9-35. Relation of return-oil pressure to oil flow to furnace.

The oil or tar is heated to its final burner temperature after passing the pump, so there will be no vapor lock in the suction line. At low ratings with a large portion of the heated fuel being returned to the sump, this would tend to raise the sump temperature too high. To forestall this, a cooler is placed in the return line.

In Fig. 9-34, it is assumed that this is a series-air system in which air is

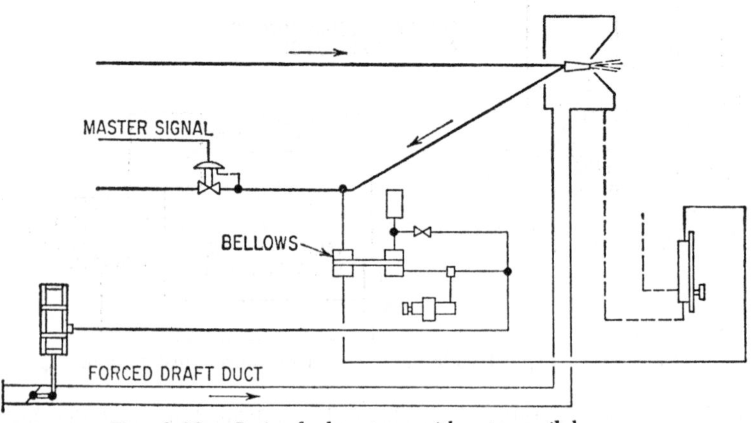

Fig. 9-36. Series-fuel system with return-oil burner.

metered to control fuel. The air-differential connections are taken across the wind box. Since air flow is proportional to the square root of this differential, a cam is provided in the variable-ratio regulator, which will extract the square root and send out a loading signal to the oil-control valve, which is directly proportional to air flow. Therefore, a flow cam is used.

Since return-oil pressure is directly proportional to oil flow to the furnace, as shown in Fig. 9-35, this properly proportions air and fuel.

Figure 9-36 illustrates a portion of a series-fuel system. This is less commonly used than the series-air arrangement. With this system, the master signal is used to regulate return-oil pressure. Since the master signal from either a temperature master or a pressure master is directly proportional to fuel requirement, this will properly regulate fuel.

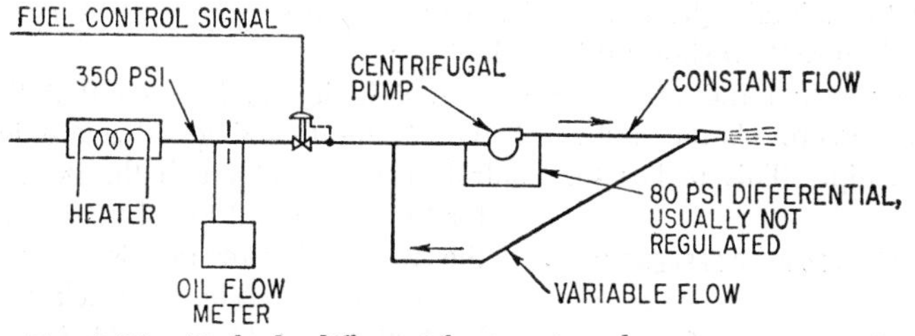

FIG. 9-37. Peabody differential-return-type burner arrangement.

By using the proper size of bellows in one chamber of a resetting totalizer, a set point for air flow is established, because oil pressure is also a measure of fuel flow. A variable-ratio regulator is then used to establish the controlled variable or air-flow signal. The output signal from the resetting totalizer is used to control air flow. The variable-ratio regulator must have a flow cam, since it is measuring differential, whereas the set-point signal is in direct relation to the flow.

When it is desired to have an actual record of oil flow, the control and metering arrangement with this type of burner becomes expensive. Accurate metering requires a meter in the return-oil line as well as in the fuel oil-supply line. The difference of the two signals can be obtained by a totalizer to measure the amount of oil flowing into the furnace.

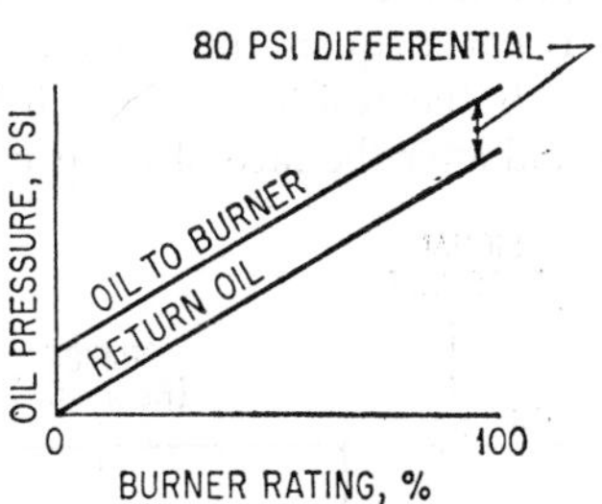

FIG. 9-38. Peabody differential-return-type burner characteristics.

The Peabody constant-differential-return-burner arrangement is shown in Fig. 9-37. With this arrangement, a centrifugal pump supplies a constant amount of oil to the burner. The unused oil is recirculated without returning to the supply sump. The make-up oil is controlled by an oil valve. One meter in the make-up oil line then measures the amount of oil actually being used by the burner. Figure 9-38 shows the characteristics of oil-pressure and burner rating or flow into the furnace for this type of burner. This flow is linear with either return-oil pressure or with burner pressure, since their differential is constant. The Babcock & Wilcox Company have a similar arrangement, except that they use a differential-regulating valve instead of a centrifugal

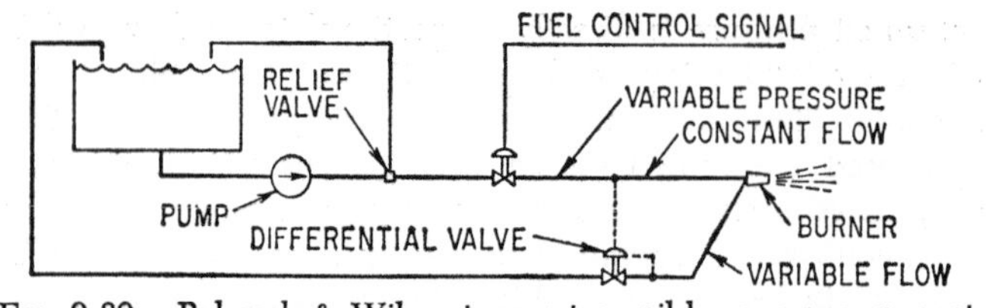

Fig. 9-39. Babcock & Wilcox-type return-oil-burner arrangement.

pump to maintain the differential between the supply and return oil pressure. Figure 9-39 shows this method.

STEAM-ATOMIZING BURNERS: In Fig. 9-40 it will be seen that the flow-versus-pressure characteristics of the steam-atomizing valve are in linear relationship. This makes the control arrangement much the same as that for the return-oil type of burner in which the return-oil pressure is controlled to regulate flow. The Babcock & Wilcox Y-jet atomizer (Fig. 9-41) is a typical arrangement of this sort of burner.

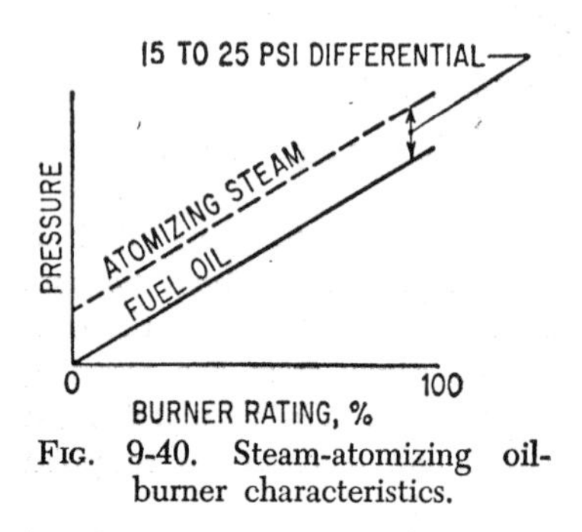

Fig. 9-40. Steam-atomizing oil-burner characteristics.

One of the main differences in this steam-jet atomizer is that an orifice, called a "resistor," is placed ahead of the burner, so that it requires more fuel-pressure change for a given flow change at high ratings. This tends to make a finer manual adjustment possible.

The addition of this resistor makes the peculiar characteristics of pressure versus flow shown in Fig. 9-42. This is not a square-root curve.

If the resistor is used, it is necessary to use a linear cam on a machine and alter the face of the cam to fit the characteristic of this burner.

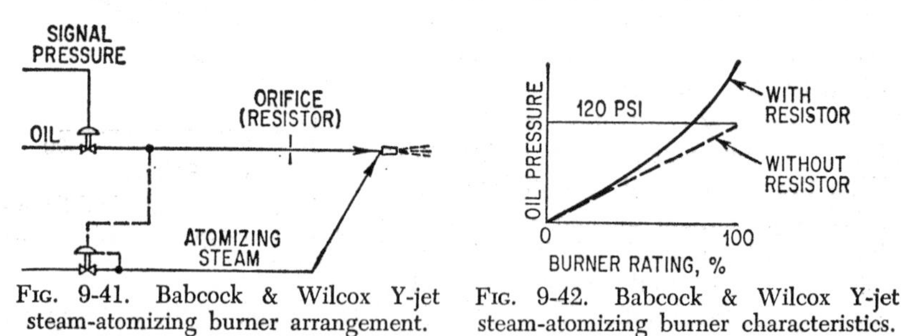

Fig. 9-41. Babcock & Wilcox Y-jet steam-atomizing burner arrangement.

Fig. 9-42. Babcock & Wilcox Y-jet steam-atomizing burner characteristics.

Table 9-3 summarizes the methods of liquid-fuel atomization.

Air Registers. In order that combustion may take place, all burners have registers which serve the purpose of admitting and directing air so it will come in contact with the oil. The design or construction of the register has no effect on the method of control used since the vanes, spreaders, or louvers are arranged for manual operation and should be allowed to remain so. This applies to both natural-draft and forced-draft air-flow arrangements.

Natural-draft Operation. When air for combustion is supplied by natural draft, the preferred method for controlling air flow is to operate the boiler outlet damper. This assures the lowest possible furnace draft for all ratings, and therefore reduces air infiltration to the minimum. Furnace draft must increase with rating to provide the necessary additional air. Rotary-cup burners nearly always operate on natural draft and have dampers directly beneath the burners to control the secondary-air volume. With this arrangement, constant furnace draft can be maintained while load variations are compensated for by operating the secondary damper.

Table 9-3. Summary of Atomization Methods

Method of atomization	Method of controlling oil to furnace	Control characteristic	Advantages	Disadvantages
Air	Supply valve	Linear	Low oil-supply pressure. High turndown ratio	High power requirement
Mechanical	Supply valve	Square root	Low steam and power consumption	Low turndown ratio
	Return valve	Linear	As above	As above
Steam	Supply valve	Linear*	Low oil-supply pressure. High turn-down ratio	High steam consumption and high make-up

* Except for Babcock & Wilcox Y-Jet which is between square root and linear.

Forced-draft Operation. Where forced-draft operation is employed, the air for combustion is supplied under pressure. The flow of air is controlled by positioning the inlet vanes or outlet damper at the forced-draft fan or by regulating the speed of the forced-draft fan.

If the boiler and its furnace are designed for operation with a draft in the combustion chamber, a furnace-draft controller must be used to maintain a constant low furnace draft. This is accomplished by independently positioning the boiler outlet damper, dampers, or vanes at the induced-draft fan or by controlling the fan speed.

Boilers which are operated with a pressure in the furnace and gas passages do not require the use of a furnace-pressure controller.

Off-On Control. Most off–on-type control systems are actuated by pressure switches or thermostats. Safety devices, such as flame-failure and programming controls, low-water cutout, stack-temperature switches, low-oil-pressure and low-oil-temperature and atomizing-steam-pressure switches, also are incorporated in these systems. Many times, off-on controls have no provision for regulating the combustion-chamber draft or operating the boiler outlet damper, although the boiler and burner are designed for operating with a constant low draft in the combustion chamber for best results. This lack can be compensated for by using a furnace-draft controller with a sequencing feature, which maintains a constant furnace draft both while the burner is off and when it is in operation. This does, however, by means of the sequencing feature, open the boiler damper widely before the starting period, and hold it open until the burner is in successful operation. Any such device must be safe in operation, however, since all installations of this type are governed by regulations that do not permit the use of an outlet damper unless it is positioned automatically by an approved controller.

Off-on control systems, when on, must of necessity generate more heat than is required to meet the demand, and less than is required when off. This inherently results in a lower efficiency of operation, wider fluctuation of steam pressure, and other undesirable conditions, most of which are overcome by use of proportioning-type controls such as the positioning and metering methods.

Positioning Control. Positioning control is used only for continuous operation of the smaller-size oil-fired steam generators. When burners are of the natural-draft type, only a steam-pressure controller operating from steam pressure is required to position the oil-control valve and boiler outlet damper.

When forced draft is used, a draft must be maintained in the combustion chamber, and a furnace-draft controller is required in addition to the steam-pressure controller. Usually this type of system is arranged so that the steam-pressure controller operates the oil-control valve and the forced-draft-fan damper, while the furnace-draft controller operates the boiler outlet damper. Although the positioning method of control will do reasonably well as far as steam pressure is concerned, manual adjustments are frequently required in order to maintain the fuel-air ratio within desirable limits. Manual adjustments are also required to correct for changes in sprayer-plate size, number of burners in operation, changes in position of the air louvers, difference in valve and damper characteristics, etc. It is therefore advisable to include in the system a means of conveniently making the necessary fuel-air ratio adjustments, for the operator may not make them if they are not easily accomplished. A combustion guide such as an oxygen or CO_2 analysis of the flue gas is therefore often used for the operator's guidance.

Metering Control. The metering method of control is used mostly with steam- or pressure-atomizing-type oil burners. This is due to the fact that these burners are used almost exclusively on larger steam generators. For economic reasons also, this method is used with these larger boilers as it obtains the best results.

Generally the system consists of a steam-pressure controller, which responds to changes in steam pressure as indicating a change in load on the steam generator. This controller adjusts a fuel-feed controller, which establishes the required oil pressure at the burners. A ratio controller, which measures the flow of oil and the flow of combustion air, maintains the desired fuel-air ratio throughout the range of operation.

Steam-atomizing burners require the use of a differential valve to control the pressure of atomizing steam in proper relationship to the oil pressure at the burners. Burners of this type, almost without exception, require forced-draft air. The inlet vanes or dampers at the fan or the speed of the fan are controlled to regulate the air flow. When forced draft is used, a draft must be maintained in the combustion chamber, and a furnace-draft controller is required. It maintains a constant draft by operating the damper at the boiler outlet or induced-draft fan, or by regulating the fan speed.

Fully Automatic Control. With the advent of the package boiler, fully automatic combustion control has come into common use on this type of steam generator. Off-on control for light loads is combined with proportioning control for heavy loads. The proportioning control is either of the positioning type, generally used on the smaller-size boilers, or of the metering type used on the larger-size boilers. The proportioning type of control regulates the rate of combustion, according to the load on the boiler, from maximum capacity of the boiler down to the lowest rate at which good and efficient combustion can be maintained. For all loads below that which can be carried by this low-fire condition, the combustion is cut on and off at this low fire.

In order to operate so completely automatic, it is essential that the system of control have incorporated in it all the flame-failure and programming equipment and safety devices required for complete protection against any hazardous conditions that might arise. It is obvious that, since so much depends on the automatic equipment in a system of this sort, high-quality reliable elements should be used for every part of the system.

Although safety devices are considered an inherent part of a fully automatic-control system, they are also highly desirable as a part of the continuous-operating, positioning, or metering systems of control. A load-limiting controller that limits the rate of combustion (and therefore, the load which the boiler can carry) to prevent an unsafe furnace draft is a worthwhile protection against too low, or loss of, induced draft. A fuel-air ratio-limiting controller, which prevents too high a rate of fuel flow for the flow of combustion air, protects against insufficient air for combustion, as would result from failure of the forced-draft fan. Low oil pressure and similar safety devices also can be added to the system for complete protection against unexpected operating irregularities.

Figures 9-43 and 9-44 show typical combustion-control-system arrangements for return-flow and steam-atomizing fuel-oil burners.

Gas Firing.[9] Natural gas is practically free from noncombustible gas or solid residue. It is very uniform in heat value and burns rapidly, leaving

[9] The Hays Corp.

no refuse, which eliminates the disposal problem associated with solid-fuel firing. Since natural gas is readily adaptable to constant-pressure control, it is being used in all sizes of steam-generating plants and for numerous industrial applications where solid fuels cannot be used successfully.

Various by-product gases, such as coke-oven gas, refinery gas, and blast-furnace gas, as well as manufactured gas, principal product of gas producers, also are employed for heating and steam generation. Although these gases usually vary in character more than natural gas, they are easy to use and to control, either individually or in combination with other fuels.

Types of Burners. Both low-pressure (approximately ⅛ to 5 psig) and high-pressure (up to 25 psig) burner designs are available for use with

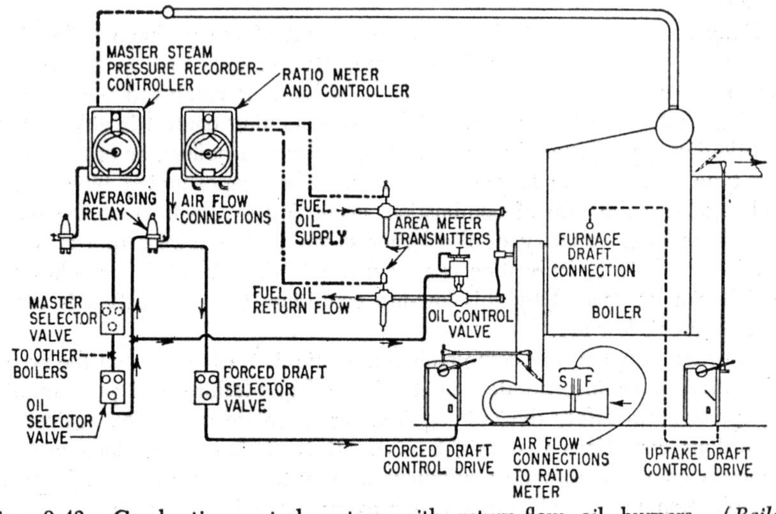

Fig. 9-43. Combustion-control system with return-flow oil burners. (*Bailey Meter Co.*)

different types of gas and for a wide variety of applications. Most steam-generating units utilize the latter design and operate at pressures of 2 to 25 psig. These high-pressure burners may be classified in four main types, namely: (1) gas ring, (2) center diffusion tube, (3) turbine, and (4) tangential.

Gas burners can be obtained for either natural- or forced-draft operation. In both types, any adjustments, such as position of gas nozzles or rings and burner louvers for the control of air distribution, are designed for manual operation. Some natural-draft types do include louvers or dampers at the burners for regulation of the combustion-air volume, however, and these may be controlled automatically.

Gas flow to all the burners under each steam generator is controlled by means of a suitable valve in the main gas line. This valve must be correctly sized and should have proper characteristics for the control system with which it is to be used, in order that accurate flow regulation can be obtained over a wide range of operation. Preferably, gas flow should be under the jurisdiction of the master steam-pressure controller, either directly or through some intermediate medium such as a gas-pressure controller, flow controller or valve-positioning controller. This applies to both natural- and forced-draft types of burners.

When the air for combustion is obtained by natural draft, the volume may be controlled by holding the furnace draft constant and adjusting the burner louvers. The preferred method, however, is to increase the furnace

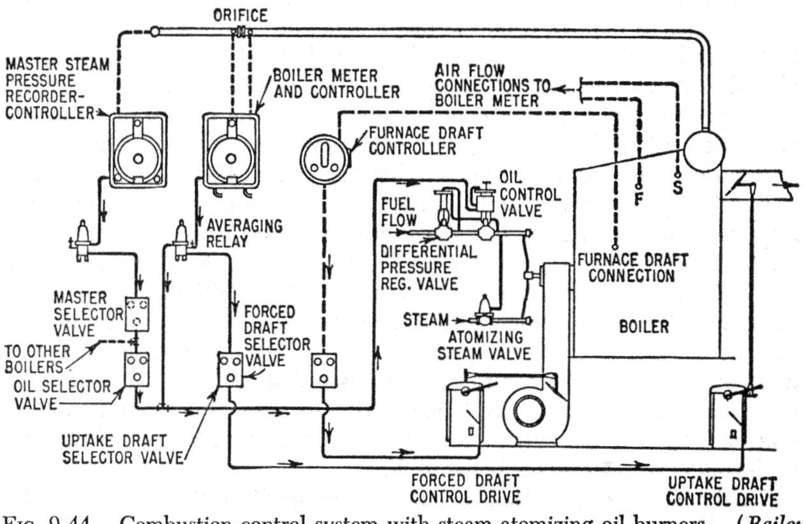

FIG. 9-44. Combustion-control system with steam-atomizing oil burners. (*Bailey Meter Co.*)

draft by opening the boiler outlet damper when greater quantities of gas must be burned to satisfy the load demand. Some burner designs use the flow of gas to induce a flow of air proportional to the flow of gas without further control.

Most systems control the air for combustion from calibrated measurements of the air flow. This arrangement is more desirable than depending on direct positioning of dampers or louvers to give the desired results, since variable draft conditions and damper characteristics make it difficult to establish a definite relationship between damper position and air flow. With forced-draft-type burners, constant furnace draft should be maintained by a separate controller operating independently of the controls for steam pressure, gas, and air flow.

Off–on-type Control. Often gas is burned intermittently, with off–on-type controls for regulation. However, this arrangement is used only on small steam generators because it is limited in its ability to control steam pressure and efficiency. In operation, the pressure switch opens, and closes a valve in the gas line when the steam pressure reaches predetermined set limits. In some instances, louvers are operated simultaneously to control most of the air used for combustion, but means should be provided for control of the outlet damper so that minimum furnace draft will be maintained both while the burner is in operation and during the off period. With controlled furnace draft, air supply can be held at the required value for best efficiency during the on period without causing excessive air leakage during the off period.

Diaphragm-valve Control. One common method of gas control is the use of a diaphragm valve. Steam pressure, either directly or through a pilot control, is impressed on the diaphragm valve, which controls gas flow to the boiler in accordance with the load demand. This method is employed only on small continuously fired boilers because it does not hold steam pressure within close limits or provide for the control of air flow in proportion to gas flow. Thus, high efficiency is not obtained unless a fuel-air-ratio controller is added to complete the combustion-control system.

Positioning Control. Combustion-control systems of the positioning type may take various forms, but the basic principles remain the same. A master steam-pressure controller responds to changes in steam pressure in the main header and actuates a power unit that positions the gas-control valve and air-control medium simultaneously. The air-flow-control medium may be the boiler outlet damper, the burner louvers, or both. If forced-draft burners are to be controlled, a furnace-draft controller is required to maintain constant furnace draft, in addition to the master controller. This unit may operate the outlet damper or inlet vanes on the forced-draft fan or the outlet damper on the boiler itself.

Positioning-control systems maintain very uniform steam pressure, but the combustion efficiency achieved depends on the correct mechanical matching of valve and damper movements and characteristics. Even when the setting is fairly accurate, efficiency is affected by variable gas-pressure and draft conditions. Then manual changes in the linkage are required to obtain optimum performance.

Metering Control. Metering controls are preferable to the other types previously mentioned, and are used frequently on large steam generators because they provide more uniformly high efficiency over the entire range of operation. This method of operation features control of burner-gas pressure, metering of gas and air, base-loading control, adjustments for load distribution and fuel-air ratio, and remote manual control.

In application, these systems utilize one master steam-pressure controller with either proportional or proportional-plus-floating action. This unit measures pressure in the main-steam header and, through fuel-feed controllers, establishes the steam output of each boiler connected to the system.

Individual adjustable-pressure fuel-feed controllers position the gas valve to each boiler through a power unit. The master controller establishes the boiler rating by setting the fuel-feed controllers for the gas pressure to be maintained at the burners while the fuel-feed controllers measure the gas pressure and balance it against master loading. Control of gas pressure is not dependent on valve position alone. The correct minimum gas pressure is maintained in a multiple-burner system, regardless of variations in gas-supply pressure or of the number of burners in service. This affords protection against loss of ignition. Furthermore, an adjustment is provided to increase or decrease boiler output with respect to master loading.

In the metering system, fuel-air ratio controllers proportion air flow to gas flow by controlling the forced-draft dampers. These controllers have two differential elements. One measures the pressure differential across an adjustable orifice in the gas line, while the other measures the pressure differential across an orifice in the air duct, or draft loss through the boiler where practical. In both cases the differential pressures vary as the square of the flow, and can be balanced directly in the controller. Any unbalance, resulting from a change in gas flow or air flow, causes the controller to adjust the forced-draft damper.

Since fuel-air ratio controllers automatically proportion air flow to gas flow, the proper fuel-air ratio can be maintained with the gas valve being controlled either manually or automatically. If a valve for a fan-turbine drive, a lever on a hydraulic coupling, or a rheostat on a magnetic coupling is to be controlled instead of the damper, a proportional-plus-floating controller is substituted for the proportional-speed floating controller. A combination of fan-speed and damper control also can be obtained.

In addition to the controllers mentioned, each boiler is equipped with a separate furnace-draft controller that maintains constant minimum furnace draft. This unit operates independently of the master, fuel-feed, and ratio controllers, but its action for damper or fan-speed control is similar to that of the ratio controller. Each furnace-draft controller includes a calibrated draft set-point adjustment.

Additional controllers can be furnished for limiting the fuel and air input to the available draft, for limiting fuel to the available air supply, and for shutting off the fuel if the limiting action is ineffective after a definite time interval. Manual-to-automatic transfer switches can be interlocked so as to necessitate a definite sequence of transferring power units from manual to automatic control. Additional interlocks for starting up are also available.

Typical Gas-fired Systems.[10] *Parallel Control System, Natural-draft Boiler.* The natural-draft type of boiler may be equipped with a simple parallel-operated combustion-control system. If more than one boiler is to be automatically controlled, a common steam-pressure controller ensures balancing of the plant load between boilers.

As shown in Fig. 9-45, the steam-pressure controller establishes an output loading signal in accordance with variations in the plant steam-header pressure. This signal is transmitted directly to the boiler relay station for each of the controlled boilers. Where automatic operation is limited to a

[10] Hagan Chemicals & Controls, Inc.

single boiler, the relay station provides the boiler operator with a means of regulating the boiler combustion rate manually. If two or more boilers are automatically controlled, the relay station permits proportioning the load between the boilers to obtain balanced operation. The signal, repeated or modified by the boiler relay station, is transmitted directly to the gas-valve positioner, and to the air-fuel ratio relay station. The gas-valve positioner is a receiving regulator mechanically linked to the control-valve operating

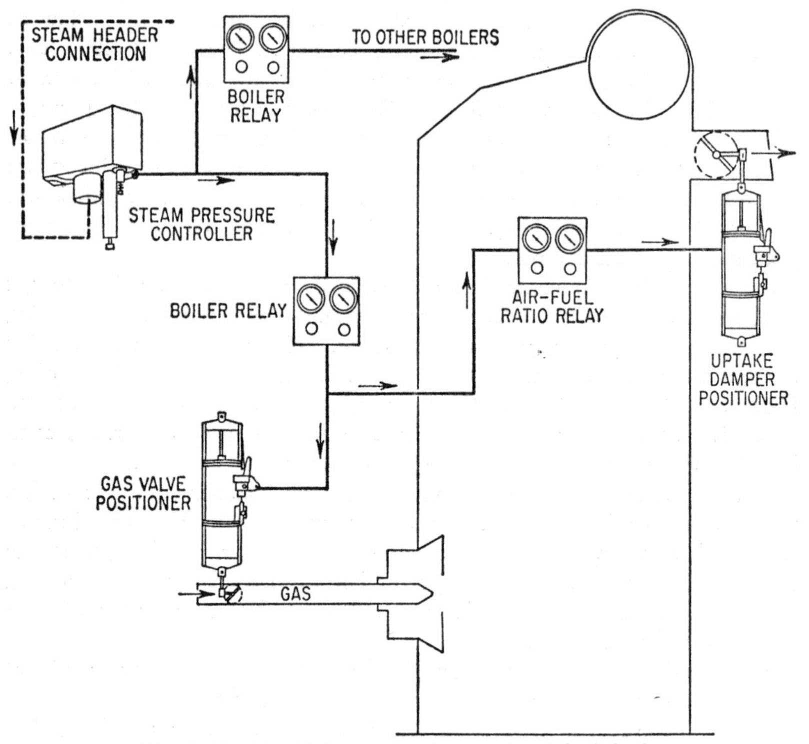

Fig. 9-45. Parallel control system, natural-draft boiler.

lever. This regulator positions the valve to a definite opening for every pneumatic signal received, thus regulating the flow of gas to the burners.

The ratio relay station, set to automatic, provides the boiler operator with a manual means of adjusting the desired air-fuel ratio. This ratio then remains constant through all variations of boiler load. When set to manual, the relay station provides a remote means of positioning the uptake damper manually, thus establishing a fixed flow of combustion air through the boiler.

The ratioed pneumatic signal established by the ratio relay station is transmitted directly to the uptake-damper positioner, a receiving regulator, which positions the uptake damper to a definite opening for every signal received. By characterizing the compensator bars or cams of the uptake damper and gas-control-valve-receiving regulators, proper paralleling of the air flow to the gas flow can be obtained.

Metering equipment for a boiler of this type includes a steam-flow–air-flow boiler meter, as well as a multipointer draft gage for indicating gas pressure, furnace pressure, and uptake draft.

Direct-positioning System, Forced-draft Boiler. The basic combustion-control system described for a natural-draft boiler can be adapted as shown

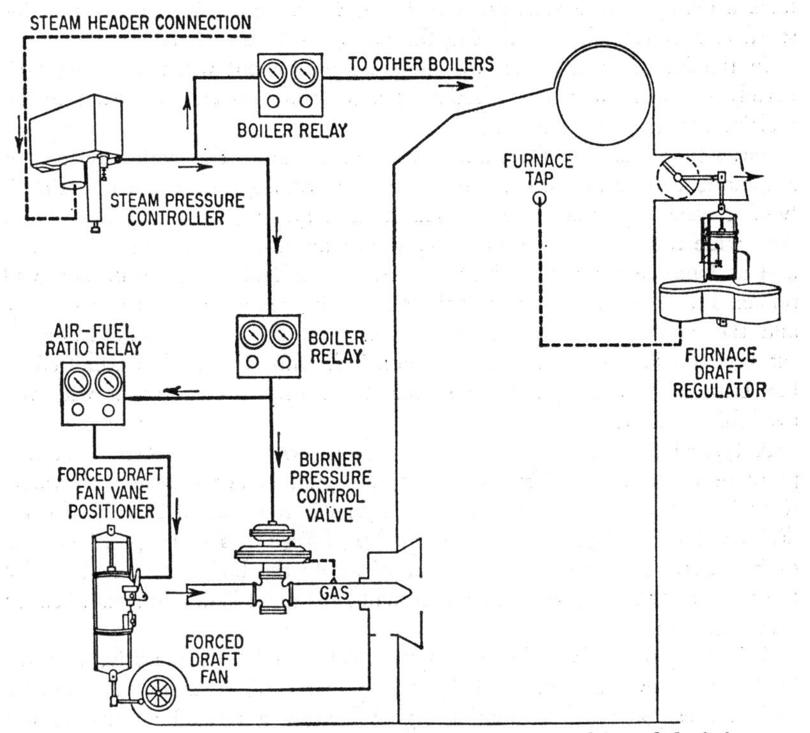

Fig. 9-46. Direct-positioning system with constant-speed forced-draft fan.

in Fig. 9-46 to suit the operation of a forced-draft type of boiler. The operation of the steam-pressure controller, boiler relay, gas-valve positioner, and air-fuel ratio relay remains the same.

In a forced-draft boiler installation, the regulating signal established by the ratio relay station is transmitted directly to the forced-draft-fan positioner. This positioner, a receiving regulator, is mechanically linked to the forced-draft-fan-damper operating lever, which positions the inlet vanes or the outlet dampers to a definite opening for every signal received, thus regulating the flow of combustion air to the boiler.

With the addition of forced draft, it is necessary to add a furnace-draft regulator to maintain the furnace draft at a desired setting. The furnace-draft regulator shown in Fig. 9-46 is of the balanced-float type and is mechanically linked to the uptake-damper operating lever. In an automatic-control system, this regulator repositions the uptake damper so as to maintain the substantially constant furnace draft necessary for best performance.

As an alternative, Fig. 9-46 shows a balanced-diaphragm type of gas-control valve in place of the power-cylinder-positioned valve shown in the preceding drawing. In the alternate installation, the control valve maintains a definite burner-gas pressure for each pneumatic signal received, instead of mechanically positioning the gas valve to the signal.

In the forced-draft boiler, the metering equipment will remain approximately the same with the addition of a wind-box-pressure indicator to the multipointer draft gage used.

Series-type System, Constant-speed Draft Fan. The series system of automatic combustion control has certain advantages over the direct-positioning systems just described. With a series system, it is possible to obtain very wide turndown; burner and registers may be cut in and out of service, and as long as a minimum burner pressure is maintained, the burners will remain lit. A series system has the added advantage, from both the safety and the combustion efficiency standpoint, that the flow of combustion air controls the flow of fuel to the burners. In this way, fuel is admitted to the burners only in quantities sufficient for proper combustion with the available air flow.

A typical installation of this system is shown in Fig. 9-47. A steam-pressure controller establishes an output signal in accordance with variations in steam-header pressure. This signal is transmitted through the individual boiler relays to the corresponding forced-draft-vane positioner for each boiler. By controlling the flow of air to the boiler, as called for by the master controller, the boiler then maintains plant load and steam-header pressure.

A fuel regulator (variable-ratio type), connected to be sensitive to combustion-air flow as measured by the pressure drop across the burner registers, establishes a regulating signal bearing a ratioed relation to the air flow measured. This signal is transmitted through a manual-control station to the loading head of the burner-pressure-control valve, where it is used to regulate the fuel pressure to the burners. By using the handwheel mounted on the fuel regulator, the boiler operator may change the ratio of the output signal generated to the actual air flow measured, thus changing the fuel-air ratio setting. If at any time it becomes necessary to regulate the burner pressure manually, the manual-control station is set to interrupt the automatic signal generated by the fuel regulator, and substitute for it a manually preset signal, providing the operator with a remote manual means of setting up a burner-gas pressure.

A pressure-reducing valve, set at some predetermined value, is tied into the signal line between the fuel regulator and the burner-pressure-control valve. The reducing valve is used as a minimum pressure regulator to maintain a minimum signal to the gas valve at low boiler loads. In this way, satisfactory burner operation is assured at all loads.

The burner-pressure-control valve is usually of the double diaphragm type, acting to balance burner-gas pressure against the input-signal pressure receiver. A valve of this type is suitable for installations where the burner-gas pressure is higher than 4 psi at maximum load.

The furnace-draft control may be a balanced-float regulator, as described for the installation shown in Fig. 9-46. When it is desired to provide for

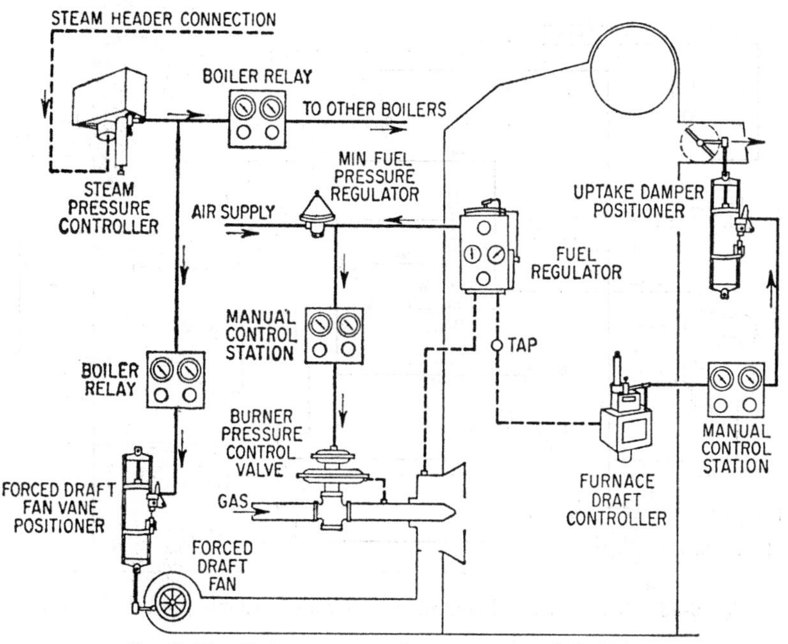

FIG. 9-47. Series system with constant-speed draft fan.

remote manual positioning of the uptake damper, the draft-control installation may consist of a diaphragm regulator, acting through a manual-control station, and the uptake-damper positioner, for maintaining furnace draft at the established set point. The manual-control station is similar to the unit used on the fuel-control system, and the uptake-damper positioner is a receiving regulator similar to the forced-draft-vane positioner previously described.

Series Metered System, Constant-speed Draft Fans. In boilers equipped with a metered type of control system, the air flow through the forced-draft fan is controlled in accordance with steam-header pressure, similar to the series system previously described. The induced-draft fan operates in a like manner, to maintain the desired furnace draft.

The fuel-control system is changed, as shown in Fig. 9-48, to balance a fuel-flow signal against an air-flow signal. Any difference between the two flow signals establishes an output-regulating signal to control the flow of gas to the burners, and thus restore the balance of flows.

The air-flow transmitter is connected to measure the pressure drop across the burner registers as a function of actual combustion air flow. The gas-flow unit in turn measures the pressure differential created by the flow of gas through an orifice in the supply line. Both units then transmit output signals proportional to the flows measured.

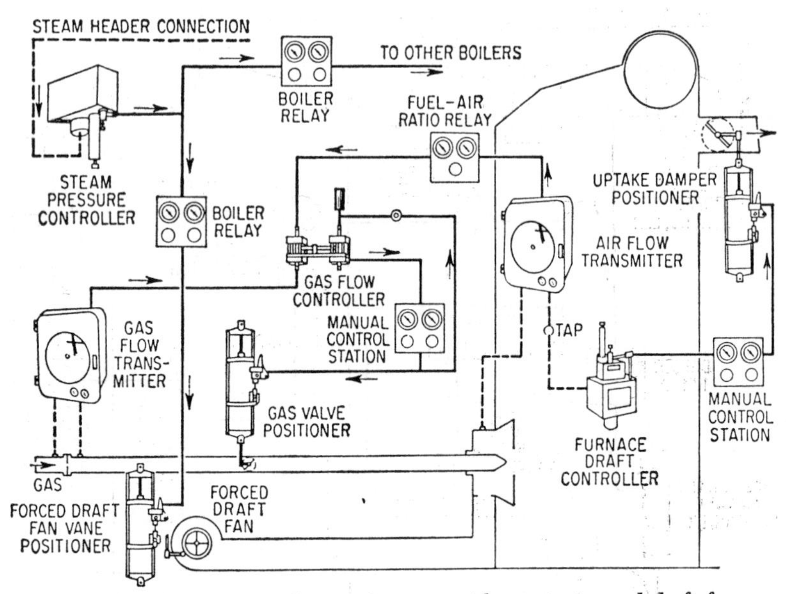

Fig. 9-48. Series metered control system with constant-speed draft fans.

The pneumatic signal from the air-flow transmitter is relayed through a ratio relay station to one chamber of the gas-flow controller (ratio totalizer), where it is used as the demand for gas flow. The use of the ratio relay station permits the boiler operator to change the fuel-air ratio setting by changing the point at which the air-flow and gas-flow signals are in balance.

The signal from the gas-flow transmitter is transmitted to an opposing chamber of the flow controller, where it is used as the indication of actual gas flow.

The ratio totalizer, in this installation, is used as a flow controller, and generates a signal which remains constant at some existing value, as long as the forces on the two input chambers are in balance. Any unbalance between the input signals causes the signal generated by the totalizer to change in the proper direction, and to the extent necessary to restore the balance by regulating the flow of gas to the burners. A feedback connection, through a needle valve and volume tank to the remaining chamber of the totalizer, provides reset action for the control system.

The regulating signal from the gas-flow controller is transmitted through a manual-control station to a gas-valve positioner, or to the loading head of the gas-control valve.

Pulverized-fuel Firing.[11] All pulverizers must perform certain similar functions. These duties are to (1) feed fuel into the mill in proportion to the demand for pulverized fuel, (2) separate foreign matter from the fuel, (3) pulverize the fuel, (4) separate fuel which is sufficiently fine from that which is too coarse, and return the coarse portion to the mill for further grinding, and (5) remove pulverized fuel from the mill, and send it to the burners. The mechanical elements required to accomplish these various functions may or may not be combined in one unit. Equipment arrangement will be determined by the pulverizer capacity, the physical layout of the plant in which the equipment is to be installed, and the design and construction of the pulverizer utilized.

Two common methods are used to control the quantity of pulverized fuel flowing from the mill to the burners. One method is to regulate raw-fuel-feed rate into the mill according to the demand for heat. The other method is to control the rate of primary-air flow through the mill since this air carries the pulverized fuel. Hence, its flow determines pulverized-fuel-flow rate.

Various methods are used for feeding fuel into the mill or grinding chamber. One method utilizes a rotating feeder table, with a plow or knife to scrape fuel off the rotating table, from which it falls by gravity into the mill below. Both plow position and table speed affect the quantity of fuel flowing to the mill. This type of feeder controls the quantity of pulverized fuel leaving the mill by controlling the quantity of fuel entering the mill. The knife is set by hand, and feeder-table speed is varied to proportion the fuel to the heat required. When d-c variable-speed motors or adjustable-speed a-c motors of suitable design are used for this service (standard slip-ring motors are not suitable), the rate of feeder-table speed or fuel feed is determined by positioning the motor-control rheostat. Lever-operated variable-speed mechanical transmissions with constant-speed drives also can be used, feeder-table speed being set by the lever position.

Another type of fuel-feed device consists of a spider, which forms a series of pockets revolving around a shaft, each pocket discharging a given fuel volume into the mill. In addition to the drives mentioned, this type of unit can be operated by a constant-speed drive through a lever-operated speed changer with free-wheeling clutch or pawl and ratchet on the feeder shaft. The position of the lever determines the rate of fuel feed to the mill, and therefore the rate of pulverized fuel leaving the mill.

A pulverizer discharge damper and sometimes a recirculating damper are provided as part of the mill, and may be automatically controlled simultaneously with the fuel feed. This arrangement assures proper primary-air

[11] The Hays Corp.

flow to the burner to obtain optimum conditions of mixture richness for best burner ignition, and decreases effective time lag in the mill.

Some installations use constant-speed feeder tables that are started and stopped automatically by a controller furnished as part of the mill. This controller is generally one of two types. One is actuated by fuel level in the mill and operates to maintain the level within close limits. The other measures the differential between primary-air-pressure drop due to flow of air and pressure drop across the grinding chamber in the mill. When fuel flow into the mill is controlled in this manner, automatic combustion controls have no direct jurisdiction over the fuel fed into the mill, but rather govern pulverized-fuel flow from the mill to the burners. This is accomplished by controlling the rate of primary-air flow through the mill, and thus the rate of fuel flow, in accordance with the demand for heat.

Air-temperature Control. Generally primary air to the mill is preheated, and for best results the primary-air or coal-air-mixture temperature leaving the mill must be within established limits. This temperature can be controlled either automatically or manually by proportioning the quantities of preheated and tempering air. Too high a temperature may result in pulverizer difficulties, and too low a temperature does not produce sufficient drying during pulverization.

Mill-pressure Control. Some types of mills are designed to operate under a pressure which may vary and need not be controlled. Other mills operate best when pressure in the mill is maintained at a slightly negative value. When primary air is under natural draft or atmospheric pressure, no control of mill pressure is required. However, in most cases primary air is taken from the forced-draft duct, and pressure in this duct varies approximately as the square of the boiler load. Therefore, a primary-air-pressure controller or mill-suction controller is used to maintain air pressure or suction to the mill at a constant value. Usually this is accomplished by regulating a primary-air-duct damper.

Primary-air-flow Control. Primary-air-flow control is important with all types of mills. Although it is not a critical adjustment on that type of mill in which fuel-flow rate is determined principally by fuel feed into the mill, it does favorably influence ignition characteristics and range of operation. A damper at either the inlet or outlet of the mill exhauster is operated simultaneously with the fuel-feed control for regulation of primary-air flow, with either the same or a parallel controller.

For mills where primary-air-flow rate determines pulverized-fuel flow to the burners, control of primary-air flow is very important. It must be accurately proportioned to both load on the boiler and to the secondary-air flow in order to maintain the correct air-fuel ratio and maximum efficiency. Usually the mill exhauster or blower operates at constant speed, and dampers are modulated for primary-air control.

Methods of Pulverization. All types of pulverizers use one or more methods for reducing coal from the variable sizes in which it may be received to a powder of extreme fineness. The three most common methods for grinding or pulverizing fuel are: (1) ring-roll method, where the lump of fuel is crushed or ground between metal rollers or balls under spring pressure and a metallic grinding ring or race, (2) impact method, in which lump fuel is crushed by being struck as with a hammer or falling metal balls, and (3) attrition method, where lump fuel is reduced from a granular form to a powder by attrition or rubbing of the granular particles against one another or against rapidly moving specially designed parts of the pulverizer.

The various methods used have in themselves very little effect on the automatic-combustion-control system, since the mills all run at a constant speed and are not connected to the automatic combustion controls. The more important factors are the methods used to control pulverized-fuel flow to the burners.

Methods of Separation. Usually, the process of separating out fuel that has not been sufficiently well pulverized for satisfactory combustion is accomplished in one of two ways, namely, (1) mechanical separation, where coarse fuel is thrown out of the stream of pulverized fuel by the revolving element of the mill as it leaves the last stage of grinding, and (2) classifier separation, consisting of passages through which the stream of pulverized fuel must pass and be deflected, velocity reduction, and other such methods that will reject those particles too large for satisfactory combustion and return them for further reduction. Automatic combustion controls have no direct control over this function, which is manually set and allowed to remain fixed until tests for fineness show that adjustments are needed.

Primary-air Fans. A fan normally removes the pulverized fuel from the mill. It may be on the primary side and force air through the mill under pressure, or on the outlet side and pull coal-air mixtures through the mill. In both cases, coal-air mixtures are conveyed through piping to the burners.

Burners. Although several different types of pulverized-fuel burners are available, the burner design does not affect the method of automatic combustion control employed. In fact, all adjustments provided at the burner are designed primarily for manual operation since they affect fuel-air distribution and flame character, and not the ratio of air to fuel. In addition to the points of control mentioned in connection with the pulverizers, it is also necessary to control air for combustion and maintain constant furnace draft with a constant-suction controller that positions the induced-draft-fan damper.

Method of Control. The exact method of automatic control selected will depend on the type of pulverizer, number of pulverizers, and characteristics of the auxiliaries. Figure 9-49 illustrates a typical boiler-efficiency-compensated control system designed to control two ring-roll-type pulverizers and constant-speed fans. This system features electrical transmission, adjustments for load distribution and for fuel-air ratio, boiler base-loading control, and remote manual control. One pulverizer, or two pulverizers at equal or unequal ratings, can be controlled automatically. In addition, automatic readjustment of fuel-air ratio by the boiler meter corrects for

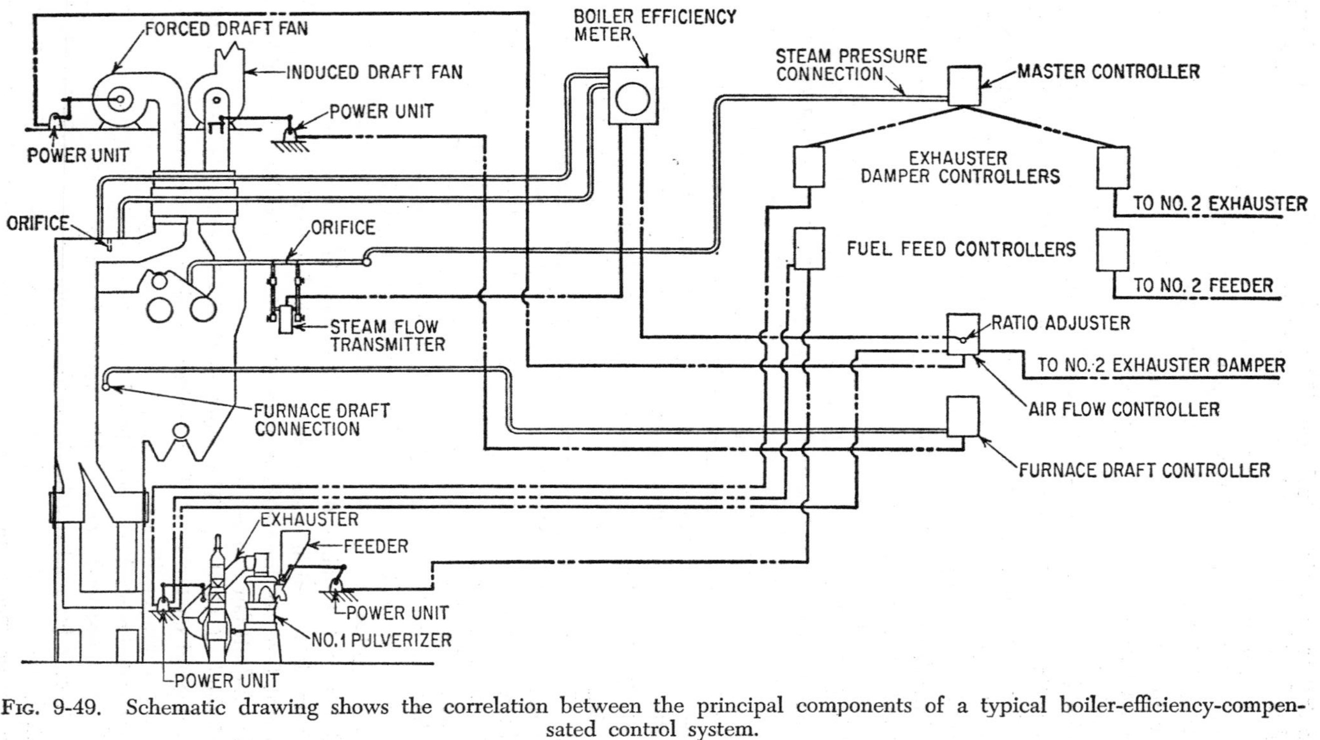

FIG. 9-49. Schematic drawing shows the correlation between the principal components of a typical boiler-efficiency-compensated control system.

irregularities in fuel supply or quality. The master controller measures steam pressure. It loads the primary-air and fuel-feed controllers which establish the load carried by the boiler.

Each primary-air controller and its power unit operate an exhauster fan damper to regulate pulverized-coal flow to the burner. The transmitting potentiometer for the controller is mounted on the master controller, while the receiving potentiometer is adjusted by the exhauster-fan-damper power unit. A manually adjusted biasing potentiometer in the controller changes the ratio of coal flow to master-controller loading, thereby changing load distribution in a multiple-boiler system, or division of load between pulverizers.

Each fuel-feed controller adjusts a lever on the pulverizer feeder to proportion coal input to primary air. The transmitting potentiometer is adjusted by the exhauster-damper power unit, and the receiving potentiometer is operated by the fuel-feed power unit. A biasing potentiometer in the controller permits changing the ratio of coal feed to primary-air flow.

A totalizing air-flow controller with its power unit controls the forced-draft-fan damper to proportion air flow to total fuel flow. Two transmitting potentiometers are a part of the controller. One potentiometer is adjusted by each primary-air-damper power unit, and it establishes a voltage corresponding to damper position and coal flow. The sum of these voltages is balanced against the voltage on a potentiometer operated by the forced-draft damper power unit. The forced-draft damper position or air flow is proportional to total fuel flow with either or both pulverizers under automatic control. When a pulverizer is shut off, a resistance is cut out of the totalizing circuit to indicate zero rather than minimum coal flow on that particular pulverizer. The biasing potentiometer in the controller serves as either a manual or an automatic fuel-air ratio adjustment for the system.

With the boiler meter, proper combustion conditions exist when air-flow and steam-flow pens are recording the same value on the chart. If the pens separate, indicating a change in combustion efficiency, the controller acts on the fuel-air ratio adjustment in the totalizing controller until the two pens have assumed like positions and the controller set point has been reached.

In addition to the basic controllers shown, additional equipment can be provided to (1) limit fuel and air input to the available draft, (2) limit fuel input to the available air, and (3) shut off fuel if the limiting action is ineffective after a definite time interval. Manual-to-automatic transfer switches can be interlocked to necessitate a definite sequence of transferring power units from manual to automatic control. Additional interlocks for starting up are also available.

Multiple-fuel Firing.[12] By far the largest number of multiple-fuel-fired installations are equipped for burning only two fuels.

The selection of the fuels is usually governed by the geographical area in which the plant is located. For instance, in the Middle West, many plants are equipped to burn coal and gas, or coal and oil. In Texas and

[12] A. C. Wenzel, chief project engineer, Republic Flow Meters Co.

along the Gulf Coast, the choice is gas and oil, either one at a time, or in any combination, utilizing boiler-control equipment to maintain the ratios of fuels and the proper proportions of fuels and air for combustion.

In some large industrial plants, waste fuels of limited availability present special problems. Local plant conditions usually determine the sequence or combination in which the fuels are burned, with careful consideration given to maximum economy, safety, and continuity of operation at all times.

The Fuels Which Are to Be Burned. The selection of fuels is usually governed by local conditions of availability and economy. Fuels are selected which are readily available at reasonable prices. Since the continuity of supply is important, fuels should be selected so that the curtailment or stoppage of delivery of one of the fuels, for any reason whatsoever, does not affect the continued availability of the other fuel. When the selection of the fuels has been made, the sequence of burning the fuels must be established.

Sequence in Which Fuels Are to Be Burned. Since the elaborateness of the control system is largely governed by the sequence in which the fuels are to be burned, it is important that the sequence of burning be clearly outlined before design of the control system is begun. For a boiler designed to burn either gas or oil separately, but not in combination, a relatively simple control system can be provided, which will control the firing of one fuel at a time. However, should plant conditions require the burning of both fuels simultaneously and in any combination over wide load ranges, it will then be necessary to provide a relatively elaborate control system together with complex safety devices.

In many installations, the sequence and combination of fuels are governed by seasonal availability. In the Middle West, natural gas is available to many steam-generating plants. However, it is available in large quantities for power generation only during the summer months, when the domestic demand for natural gas is at a minimum. Thus, in installations in this area, the controls are usually designed to burn natural gas in combination with coal or oil, in any sequence or combination, in order to take full advantage of the availability of the low-cost natural gas. Frequently insufficient gas is available to carry the entire load on a plant, and, under these conditions, the natural gas is burned to the extent of its availability, and the load swings are carried by varying the rate of combustion of the second fuel.

In any case, when the selection of the fuels has been made, and the sequence of burning has been established, the designer of the fuel-burning equipment must establish a third factor: the arrangement of burning equipment, air ducts, and boiler auxiliaries.

Arrangement of Fuel-burning Equipment, Air Ducts, and Boiler Auxiliaries. The designer of the fuel-burning equipment must select the proper burner equipment for the fuels. He must also select this burning equipment to provide the desired load range.

While frequently the available space governs the arrangement of air ducts, the designer of the fuel-burning equipment and duct work must provide a means for accurately measuring the air for combustion. It is far more satisfactory to take a direct measurement of air flow before the air enters the burners or combustion chamber, or even before it enters the air heater, than to take an inferential measurement of air flow by measuring the draft loss through some portion of the boiler setting.

The air for combustion should be measured at either the inlet or the discharge of the forced-draft fan, where the air is at relatively constant temperature. This results in a far more accurate and consistent measurement than one taken inferentially, by measuring the products of combustion by taking the draft loss either through a portion of the boiler setting or across the gas side of the air heater. These latter measurements are subject to errors created by temperature difference, by fouling of the gas passages, and by the action of soot-blower elements, which periodically upset the measured pressures. The efficiency of combustion depends just as much on an accurate combustion-air-flow measurement as it does on an accurate fuel measurement.

Having established the fuels to be burned, the sequence in which they are to be burned, and the arrangement of the fuel-burning equipment, duct work, and auxiliaries, it then becomes the problem of the combustion-control designer to arrange a control system which will provide the desired features, with a minimum amount of equipment and at a competitive price.

Boiler-control Systems Involving Multiple-fuel Firing. In the descriptions and the diagrams which follow, no attempt has been made to incorporate all the interlocks and the safety devices which are desirable, and sometimes essential, for the completely automatic operation of the boiler-control equipment.

The arrangement of controls illustrated in Fig. 9-50 is typical of a great many installations of smaller size, where gas or oil, but not both simultaneously, is burned. The control is simple and effective, provided the gas and oil burners have the correct characteristics for this type of arrangement.

It is assumed that both gas and oil burners will utilize the same wind box and the same air-control louvers.

The control consists of a steam-pressure controller of the proportional-pressure-with-reset type. This steam-pressure controller produces a loading impulse, either pneumatic or electric, which is transmitted to two separate subpanels. One of these subpanels is provided for the control of gas fuel, and the other for the control of oil fuel. Each subpanel is equipped with a transfer valve or switch and a hand-loading device, together with synchronizing gages and some means of ratio control so that the change-over from one fuel to the other can be accomplished smoothly. The steam-pressure-controller output is transmitted either to the gas-control valve or the oil-control valve. In the system we have illustrated, these valves are controlled purely by position, and have no flow check back. This assumes that the gas pressure is constant and the oil pressure is constant ahead of the control valves.

An air-flow regulator is provided. This air-flow regulator is equipped with a force-balance type of measuring element, with a gas-pressure diaphragm. Adjustable weigh beams are provided so that the gas-pressure–air-pressure ratio may be set for best combustion conditions. Since this is a combination measuring element involving the measurement of gas pressure and oil pressure, it is evident that the air-flow regulator cannot function properly unless only one fuel at a time is being burned. This air-flow regulator is of the proportional-speed full-floating type. In burning gas, the sequence of operation is as follows: The steam-pressure controller

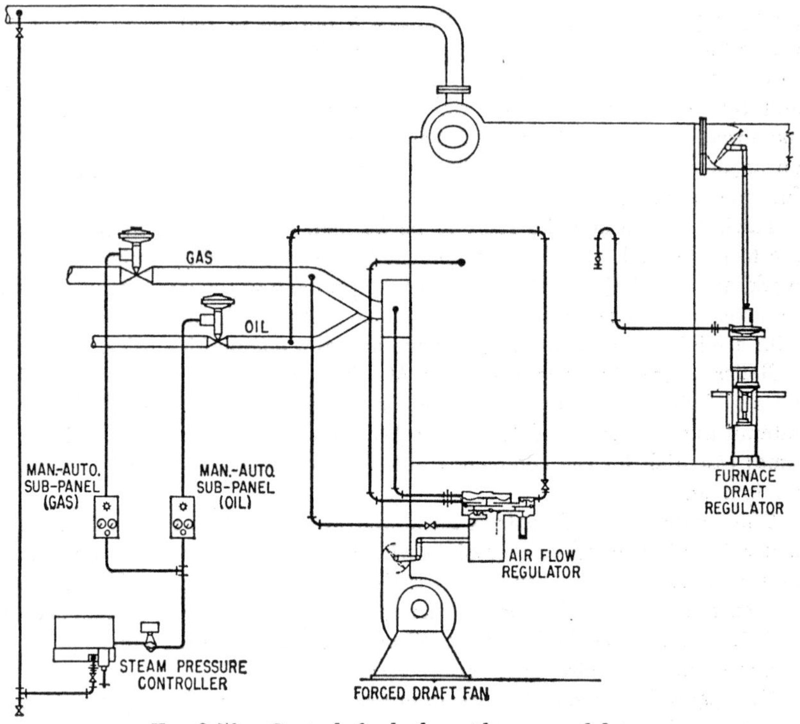

FIG. 9-50. Controls for boiler with gas or oil firing.

sends out a loading impulse through the gas subpanel, positioning the gas valve, and maintaining a supply of fuel to the boiler to keep the steam pressure constant. The gas pressure ahead of the gas burners and downstream of the gas-control valve is assumed to be a measure of gas flow. This gas pressure is conducted to the gas-pressure diaphragm of the air-flow regulator, where the force produced by it is balanced on a lever system against the force produced by the pressure differential between the wind box and the furnace of the boiler as measured across the burner louvers. Thus, in effect, gas pressure is balanced against wind-box pressure. Any change in gas pressure will affect the air-flow regulator, causing it to move the forced-draft damper in such a way as to provide a wind-box pressure, which balances the gas pressure. This arrangement will work satisfactorily, regardless of the number of burners in operation, provided that the operators close the louvers on those burners which are not in operation.

Similarly, when oil is being burned, the oil valve is positioned from the impulse received from the steam-pressure controller. The oil pressure then represents a measure of oil flow, and the force created by the oil pressure acting on the oil-pressure diaphragm of the air-flow regulator is balanced against the force produced by the differential measurement across the louvers between the wind box and the furnace.

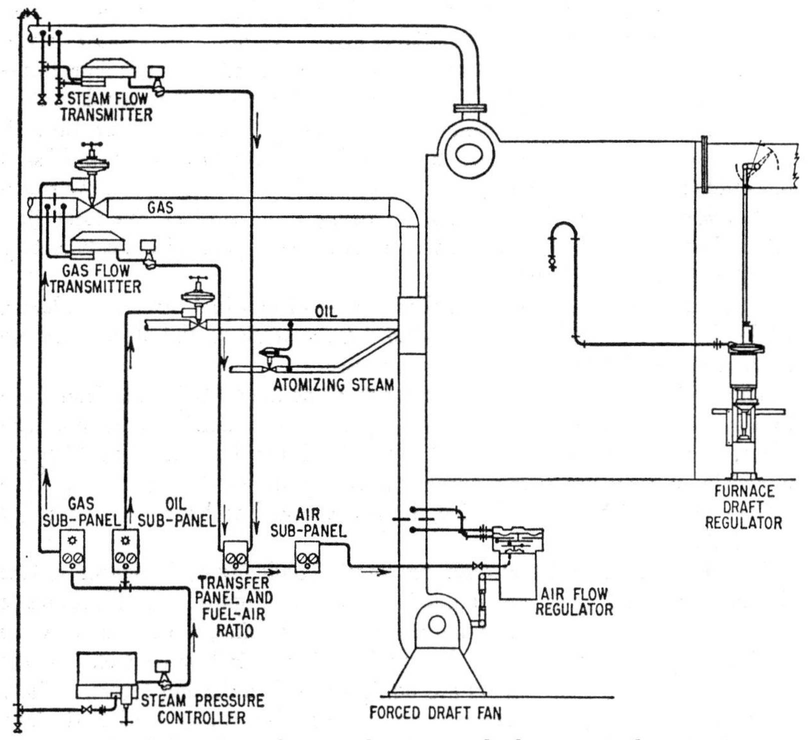

Fig. 9-51. Control-system firing gas and oil in any combination.

Therefore, this system has a sequential control, wherein the steam-pressure controller operates the fuel valve, and the change in fuel pressure acts as master loading for the air-flow regulator. The air-flow regulator then creates a wind-box pressure proportional to the fuel pressure, thus providing proper fuel-air ratio for good combustion efficiency.

The furnace-draft regulator is a separate device operating the boiler outlet damper to maintain constant furnace draft at all times.

Gas and Oil in Any Combination. Figure 9-51 illustrates a simple arrangement frequently utilized in small utility or large industrial plants.

This arrangement lends itself to complete flexibility, where fuels can be burned singularly or in any combination. The system requires some manual manipulation in changing from gas to oil, or from gas alone to a combination of fuels, but, when once set up, will operate satisfactorily. When burning gas alone, the steam-pressure controller transmits its loading impulse to the diaphragm-operated gas valve, and positions it in proportion to the requirements of the steam-pressure controller to maintain constant steam-header pressure. When gas alone is being burned, the gas-flow transmitter is sent to the air-flow regulator. This is accomplished by means of a transfer valve. Thus the air-flow regulator will receive an impulse from the gas-flow transmitter, and will balance this impulse on a force-balance system against the differential across the combustion-air-flow orifice. In effect, this is a sequential system with the output of a gas-flow transmitter establishing the master-loading pressure for the air-flow regulator.

When oil is burned, or when a combination of gas and oil are burned, the transfer-panel transfer valve is set so that the air-flow regulator follows the output of a steam-flow transmitter rather than that of a gas-flow transmitter. In this way the steam flow, being a measure of the total Btu, establishes the combustion-air flow and consequently provides a satisfactory impulse for multifuel firing.

This arrangement also permits the burning of one fuel at constant loads, while the second fuel can follow the requirements of the steam-pressure controller to maintain a constant steam pressure.

This system requires manual readjustment of the fuel-air ratio from time to time under these latter conditions, since true totalizing does not take place, and the air requirements of gas and oil may be somewhat different, especially if they are being burned in widely different proportions. The system is practical and effective, however, and has been used in many installations.

Two Pulverizers and Gas with Automatic Totalizing and Low-rating Gas Safety Combined with Maximum-fuel Safety. Figure 9-52 illustrates a control system for two coal pulverizers and gas firing, where all fuels are fired through a common wind box with a single air-flow control. In the system illustrated, coal feeders are equipped with adjusting levers. These may be variable-speed transmissions or variable-stroke feeder devices, wherein the position of the control lever is an approximate measure of the coal fed to the pulverizer. To control these feeders, positioners are equipped with cams. These cams can be cut in the field so that the loading pressures received by the positioners are relatively close measurements of the fuel fed to each pulverizer.

The gas flow is measured by means of a transmitter having a linear output, and the output of the transmitter is balanced against a master-loading impulse received from the steam-pressure controller. The gas-flow regulator is equipped with minimum gas-pressure control, and a separate minimum gas-pressure-stabilizing device, so that it is not possible to lower the gas pressure below the point of stable ignition for the gas burner.

The steam-pressure controller produces an impulse which is transmitted simultaneously through the boiler manual-automatic subpanel and through the maximum-fuel safety device to three manual-automatic subpanels, one for each pulverizer and one for the gas firing. Each subpanel is equipped with transfer device, synchronizing gages or meters, and hand loaders, as well as ratio knobs for proportioning the loadings of the pulverizers and also for proportioning the quantity of gas to be burned. The impulses which are transmitted to the coal-feeder positioners are also transmitted

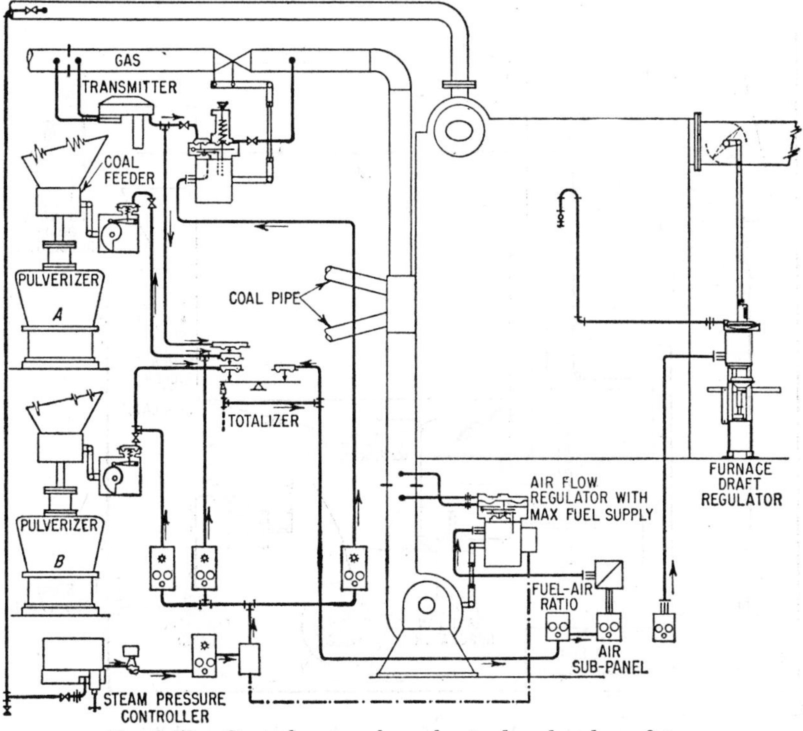

FIG. 9-52. Control system for pulverized coal and gas firing.

to two diaphragms of a three-element totalizer. The third element of the totalizer receives loading pressure from the output of the gas-flow transmitter, and, consequently, the three impulses are totalized to produce a loading pressure proportional to the total fuel. This loading pressure is transmitted through the fuel-air ratio subpanel to the combustion-air-flow regulator through a squaring device in order to provide the proper mathematical relationship.

While the totalizer is illustrated with three diaphragms, it will be noted that the gas-flow diaphragm is larger than those for each of the pulverizers. Thus, by providing the correct sizes of diaphragm and by adjusting the individual lever arms of each diaphragm, it is possible to provide an output pressure from the totalizer which truly represents the proper quantity of combustion air for burning the fuels in any combination.

With the control system illustrated, it is possible to burn either of the fuels at constant rating, with the other fuel responding to the requirements of the steam-pressure controller to maintain constant steam pressure, or the fuels may be burned singly or in any preset combination. The combustion-air-flow regulator is equipped with a range-end safety device, which will

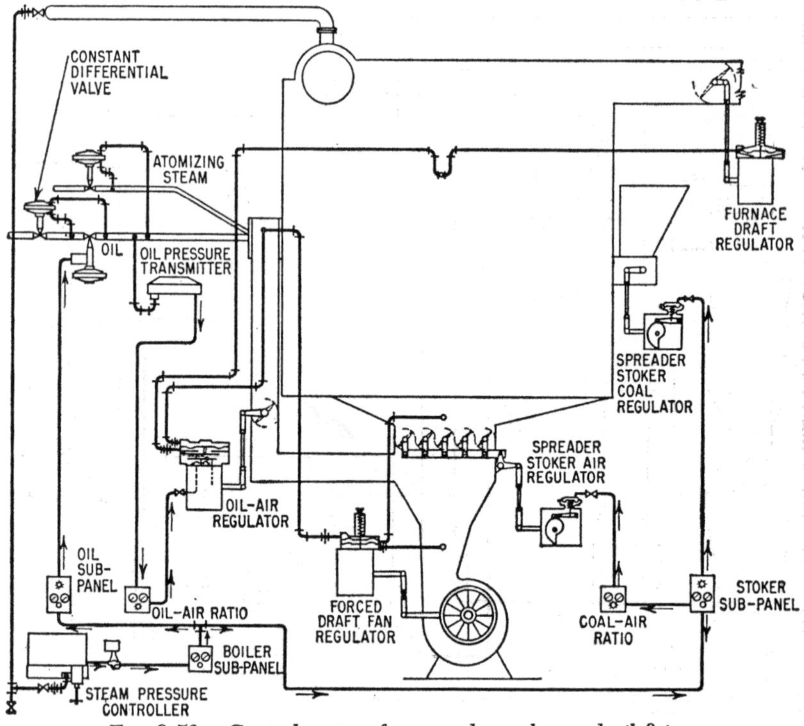

FIG. 9-53. Control system for spreader stoker and oil firing.

cut back on the fuel supply in the event that the air-flow regulator reaches its maximum opening. Thus, with this range-end safety, it is impossible to overload the furnace with fuel.

Spreader Stoker and Oil Firing. Figure 9-53 illustrates a combination of fuels with which it is necessary to have a separate air-supply duct for the stoker and another air duct for the oil burners. Both the stoker and oil-burner ducts are supplied with combustion air by means of a common forced-draft fan. The arrangement illustrated has been used with considerable success on several installations. The steam-pressure controller sends out an impulse through a boiler subpanel to an oil subpanel and a stoker subpanel. The supply of coal to the spreader stoker is controlled by means of a positioner equipped with a cam, which can be cut in the field in order to provide a loading pressure proportional to the supply of coal. Simultaneously, air-loading pressure is transmitted through the coal-air ratio subpanel to another positioner also equipped with a cam. Thus, for the control of the stoker, two positioners simultaneously operate the coal-feed mechanism and the stoker dampers for the combustion of the coal.

The oil subpanel transmits master-loading pressure to a diaphragm-operated valve equipped with a valve positioner. In order for the position of this valve to represent some measure of oil flow, a constant-differential oil valve maintains constant pressure across the master-controlled oil valve.

In this system the oil pressure, as measured ahead of the oil burners, is used as an oil-flow measurement, and is measured on an oil-pressure transmitter. The output of this transmitter passes through an oil-air ratio subpanel, and the impulse is transmitted to a force-balance combustion-air regulator, where the output of the oil-pressure transmitter is balanced against the wind-box pressure. In order that the pressure drop across the oil-burner air louvers be a relatively true measurement of air flow, a furnace-draft connection is also provided to this air-flow diaphragm, so that the air-flow measurement is not affected by changes in furnace pressure. The oil-air regulator operates a separate damper in the air duct to the oil burners, and, consequently, any changes in oil pressure ahead of the burners produce proportional wind-box-pressure changes, thus proportioning the combustion air for burning the oil to the flow of oil, as described for Fig. 9-50.

Therefore, stoker control is achieved by means of positioners and an oil control with suitable oil-air measurement. The furnace-draft regulator is again an independent unit operating the boiler outlet damper or induced-draft-fan controls to maintain constant furnace draft.

The forced-draft-fan regulator is equipped with a double diaphragm element with spring loading. This element is so designed that it will maintain a constant-differential pressure, across which every damper requires the higher pressure. Under normal conditions the stoker would require the higher pressure, and, consequently, when both fuels are being burned, a constant differential is always maintained across the spreader-stoker damper. If the stoker is shut down, the forced-draft-fan regulator would then maintain the constant differential across the oil-air-regulator damper without requiring any hand adjustment or manipulation. The regulator is completely automatic, always maintaining the constant draft drop across which every unit requires the higher pressure.

This arrangement permits the burning of either fuel singly or in any combination, with either fuel maintained at constant rating while the other fuel follows the requirements of the steam-pressure controller to maintain constant steam pressure.

Two Gases Burned to the Limit of Availability of One Gas, the Second Gas Being a Make-up Fuel. Control Is Equipped with Steam-flow-load Control and Steam-pressure Correction. Figure 9-54 illustrates an arrange-

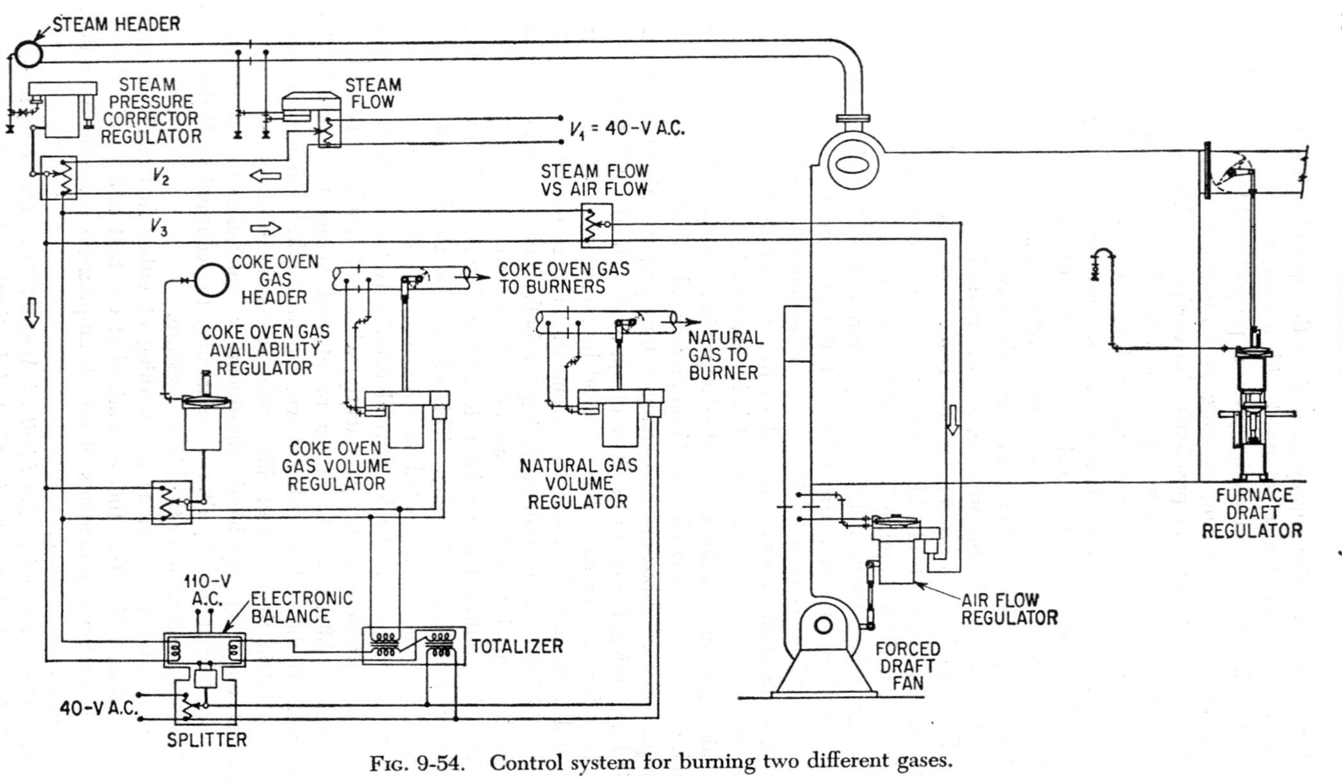

FIG. 9-54. Control system for burning two different gases.

ment for a single-boiler single-turbine installation, where coke-oven gas is to be burned to the limit of its availability, with natural gas as the make-up fuel. Load swings are wide and rapid, since the major portion of load is a rolling-mill load. In order to provide the necessary speed of response and accuracy of control, the steam-flow-load-control principle is used in which the output of a steam-flow transmitter provides the master impulse for actuating the control system. The impulse is transmitted through a steam-pressure corrector unit with a ratio device to increase or decrease the flow impulse a small amount in order to maintain constant steam pressure. The output of the steam-flow transmitter, after leaving the steam-pressure corrector, is transmitted through a fuel-air-ratio-adjusting subpanel to the combustion-air-flow regulator. Here the impulse is balanced against the differential pressure as measured across the combustion-air-flow orifice, thus providing steam-flow–air-flow control. The impulse leaving the steam-pressure-corrector regulator is also transmitted to a "splitter" device, actuated by the coke-oven-gas availability regulator.

The availability regulator is actuated from coke-oven-gas pressure. When ample coke-oven-gas pressure is available, the availability regulator holds its biasing control at one end of the stroke, so that the total impulse from the pressure-corrector regulator is transmitted directly to the coke-oven-gas volume regulator. Thus, when ample coke-oven-gas pressure is available, only coke-oven gas is burned.

In the installation in question, the splitter device is an electronic balance in which the impulse from the steam-flow transmitter is balanced against an impulse proportional to the sum of the Btu in the coke-oven gas and the Btu in the natural gas supplied to the furnace as fuel. When only coke-oven gas is being burned, the total Btu in the coke-oven gas balances the total Btu requirement as indicated by the steam-flow transmitter, and the splitter does not call for any impulse from the natural-gas control. However, when there is a deficiency of coke-oven gas, as indicated by a tendency of the coke-oven-gas pressure to fall, the coke-oven-gas availability regulator cuts back on the coke-oven gas being burned in order to maintain constant coke-oven-gas-header pressure. The electronic balancer immediately feels this falling off of the impulse because of the cutting back of the coke-oven-gas volume, the splitter comes into operation and sends an impulse to the natural-gas regulator in an amount to make up the difference in Btu not provided by the coke-oven-gas volume control. Thus, the splitter continues to send out an increasing impulse until the electronic balancer is satisfied and until the natural-gas regulator is putting in sufficient Btu to make up the total balance.

Although it is not shown in the diagram, the system is equipped with switching means and subpanels so that either gas can be burned at constant volume while the other gas takes the load swings as required by the steam-flow transmitter and pressure corrector.

In actual practice, the system is also equipped with minimum gas-pressure safety, so that it is not possible to lose ignition at low loads from unstable load conditions. This system illustrates the flexibility which can be provided by means of electronic computing devices.

FLUE-GAS ANALYSIS

Combustion Principles. Basically, good combustion involves the proper combination of fuel and air, temperature, turbulence, and time. In most fuels, the principal elements will be carbon and hydrogen, although some will contain small amounts of sulfur, while others, such as blast-furnace and producer gas, will include relatively high percentage of CO.

In the burning of any fuel, the main objective is to utilize all variable combustible without losing valuable Btu up the stack or in the ashpit. Enough oxygen must be supplied to consume all the fuel. Any excess oxygen and nitrogen that appears in the combustion gases, however, must be heated, and, since heat contained in gases leaving the boiler is lost, air supply should be held to a practical minimum. Furthermore, increasing the volume of gases also increases the velocity with which they pass through the furnace and boiler. As heat transfer depends partially on the length of time that gases are in contact with heat-absorbing surfaces, increased velocity reduces heat transfer and causes higher gas temperatures.

There is considerable difference between practical and theoretical fuel-air ratios, and actual requirements will depend on the fuel, the furnace shape and size, the ash-softening temperature, the method of ash disposal, and other factors of boiler and fuel-burning equipment design. In some cases, where coal and gas are burned in the same furnace, it may be necessary to consider superheat temperature in arriving at the proper fuel-air ratio. Some excess air will always be required, and, after the practical ratio for best over-all performance has been determined, taking into consideration combustion efficiency, maintenance, and steam production, efforts can be directed toward maintaining proper conditions.

Carbon is one of the two main elements in fuels. When pure carbon is burned, one molecule of carbon combines with one molecule of oxygen to produce one molecule of CO_2. Since air consists of 20.9 per cent oxygen and 79.1 per cent nitrogen by volume, each molecule of oxygen is accompanied by 3.78 molecules of nitrogen, which carries through the boiler without any change. Therefore, the combustion gases contain one molecule of CO_2 and 3.78 molecules of nitrogen, which means that CO_2 constitutes 20.9 per cent of the total volume. This corresponds to the percentage of oxygen in air and is correct since all oxygen in the air has combined with carbon. Adding carbon to oxygen does not change the volume of the gas but does change its weight. A molecule of CO_2 has the same volume as a molecule of oxygen but is heavier.

Carbon is at one end of the fuel scale with various forms of coke and anthracite coal approaching this limit although all contain some hydrogen. At the other end of the range normally encountered in industry is methane or CH_4, one of the main constituents of natural gas. Burning methane involves combining the carbon with oxygen to form CO_2, and the hydrogen with oxygen to form water. One molecule of methane, two molecules of oxygen, and 7.56 molecules of nitrogen produce one molecule of CO_2, two molecules of water, and 7.56 molecules of nitrogen. In order to reduce gas measurement to some standard base, it is common practice to consider all flue-gas analyses on a dry basis, and all water vapor is removed or canceled out. Consequently, when burning methane, CO_2 will be 11.68 per cent of the total volume of combustion gases with water removed. The equations representing these two extreme theoretical conditions are shown in Fig. 9-55. Most common fuels should fall within these two limits, and

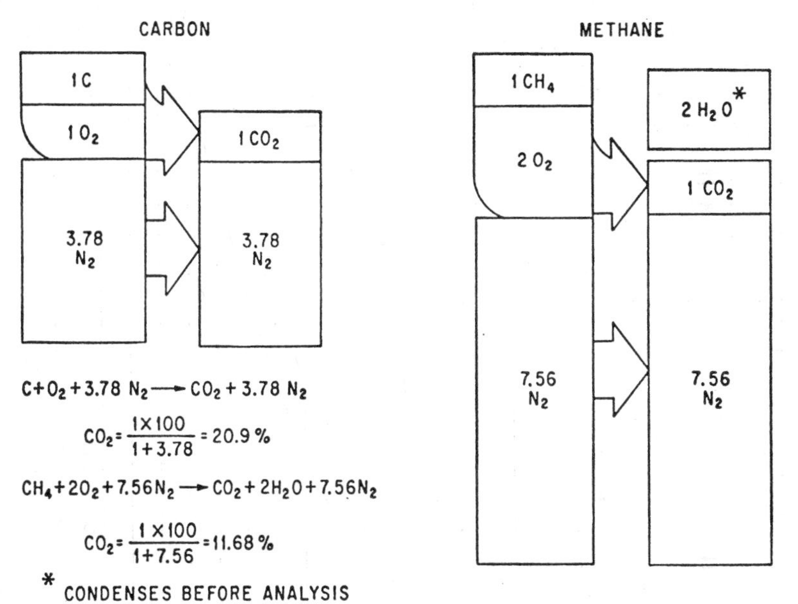

Fig. 9-55. Combustion equations for carbon and methane.

the percentage of CO_2 between 11.68 and 20.9 will depend on the hydrogen-carbon ratio.

The term *hydrogen-carbon ratio* is used to indicate the relationship between the hydrogen and carbon content of the fuel. It is the ratio of hydrogen by weight to carbon by weight. Consequently, the hydrogen-carbon ratio of pure carbon is zero and that of methane is 0.333 (Fig. 9-56).

In Fig. 9-57 the values shown on the left are the purely theoretical calculations previously mentioned, and do not take into account the fact that excess oxygen or excess air will be required to ensure complete combustion. If 40 per cent excess oxygen or air was used during the burning of carbon, the oxygen and nitrogen percentages would be increased by 40 per cent. Since only the same amount of oxygen is utilized for burning the fuel, 0.4 molecule of oxygen and approximately 5.28 molecules of nitrogen would

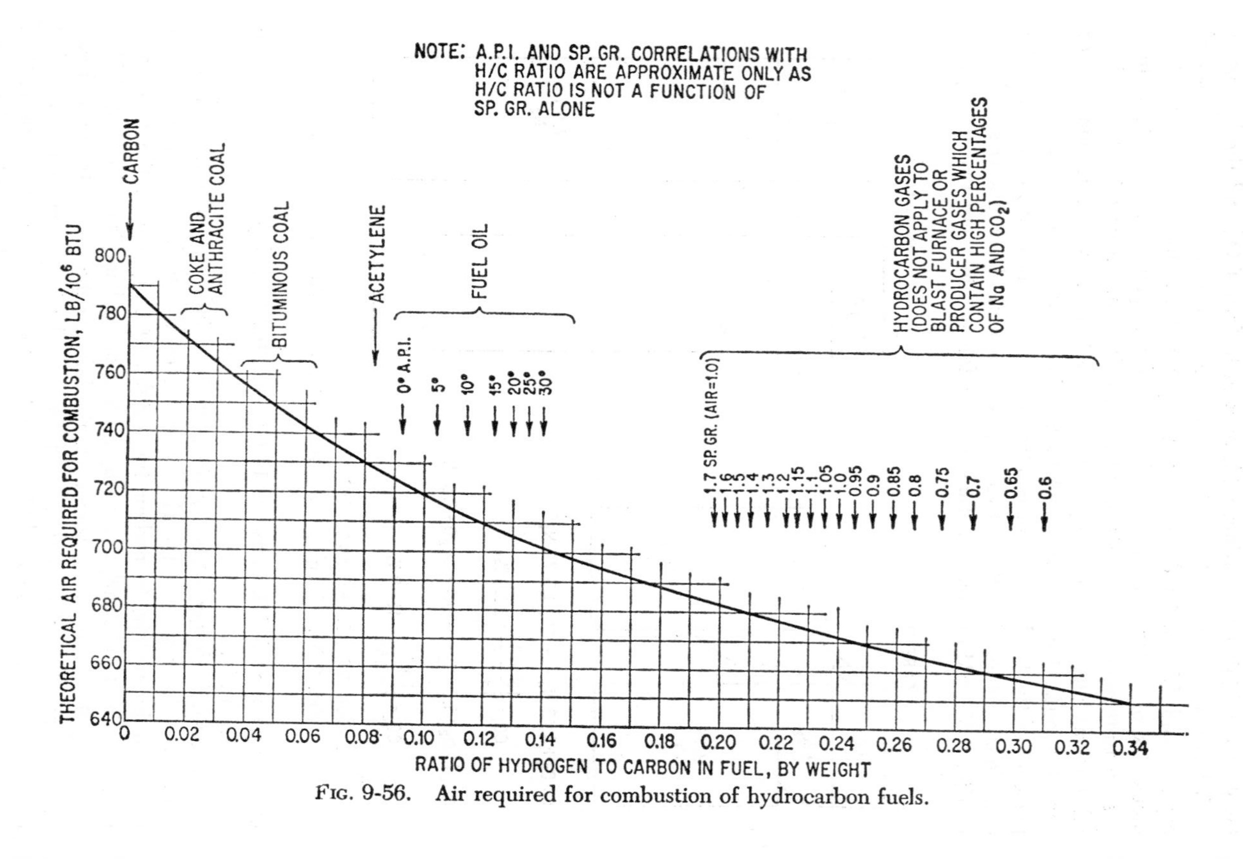

FIG. 9-56. Air required for combustion of hydrocarbon fuels.

appear in the combustion gases. Under these conditions, the CO_2 content would be reduced to 14.9 per cent, but a new element, excess oxygen as approximately 6.0 per cent O_2, would also appear in the gases.

If there is a deficiency of oxygen, all the carbon does not oxidize to CO_2, and some CO occurs in the exit gases. Presence of any CO represents a real loss, since 1 lb of carbon burning to CO yields only 4,350 Btu, whereas 1 lb of carbon burning to CO_2 yields 14,550 Btu. Thus the second atom of oxygen produces a much greater heat release than the first, and insufficient oxygen causes a substantial heat loss.

Figure 9-58 shows the relationship among oxygen, excess air, CO_2, and hydrogen-carbon ratio for various fuels. The short diagonal line in the

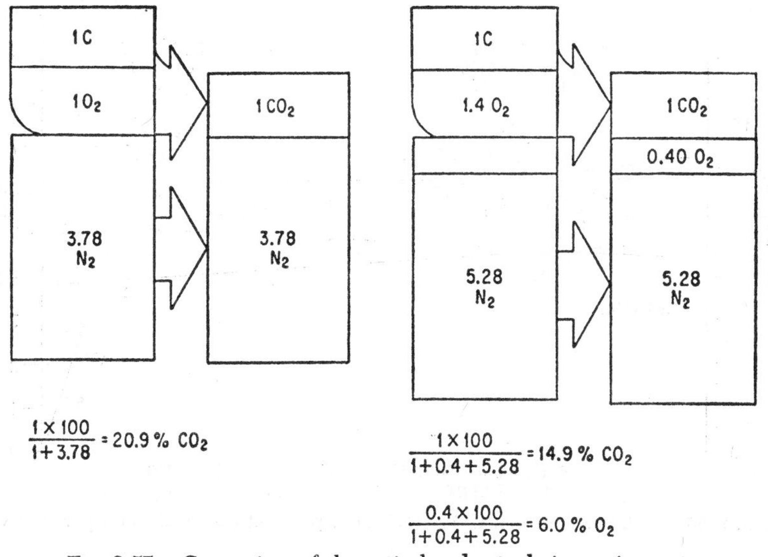

$$\frac{1 \times 100}{1 + 3.78} = 20.9\%\ CO_2$$

$$\frac{1 \times 100}{1 + 0.4 + 5.28} = 14.9\%\ CO_2$$

$$\frac{0.4 \times 100}{1 + 0.4 + 5.28} = 6.0\%\ O_2$$

FIG. 9-57. Comparison of theoretical and actual air requirements.

left center, calibrated from zero to 0.33, is marked in terms of hydrogen-carbon ratio. As previously mentioned, carbon has a zero ratio, and methane a ratio of 0.333, with most of the other conventional fuels falling between these two points (Fig. 9-56). Some bituminous coals will fall at about 0.07 on the scale, while some fuel oils will have an approximate 0.11 hydrogen-carbon ratio. If hydrogen and carbon are essentially the only two combustibles present and the hydrogen-carbon ratio of the fuel is known, the percentage of CO_2 and O_2 can be determined for any percentage of excess air. For a fuel having a hydrogen-carbon ratio of approximately 0.07 with 40 per cent excess air, drawing a line through the proper points on the respective scales gives 12.7 per cent CO_2 and 6.2 per cent O_2. It would, of course, also be possible to determine the approximate hydrogen-carbon ratio of a fuel and the percentage of excess air if the percentage of CO_2 and O_2 were known and no combustibles were present.

Although CO_2 percentage will vary widely, O_2 percentage will remain very nearly the same for any given amount of excess air, regardless of the variation in hydrogen-carbon ratio. For example, with 30 per cent excess air, the oxygen content is approximately 5 for bituminous coal and 5.25 for methane, while CO_2 content varies from 8.8 to 13.7. This relationship is significant when several fuels of different hydrogen-carbon ratio are burned.

Orsat Analysis. Just how much excess air to admit with a given set of conditions can be determined only by testing the flue gas for CO_2 percentage, the object being to secure the highest amount of CO_2 by using the

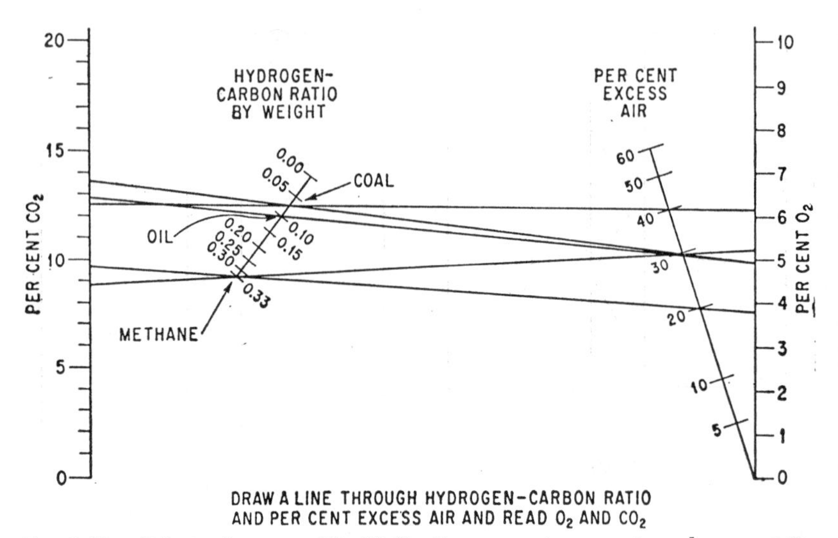

Fig. 9-58. Relation between CO_2, H:C ratio, per cent excess air, and per cent O_2.

smallest amount of excess air without producing CO. With coal burning on a grate, the best results are obtained by admitting around 40 per cent excess air, which produces approximately 15 per cent CO_2. With oil or gas, considerably less than 40 per cent excess air is necessary to ensure proper mixture and efficient combustion, with CO_2 approximately 16 per cent for oil and 12 per cent for gas.

The most efficient percentage of CO_2 will vary with (1) the hydrogen-carbon ratio; (2) the physical condition of the fuel, whether solid, liquid, or gaseous; (3) the relation of volume of the combustion chamber to the grate area; (4) the method of firing; (5) the available draft; etc. But for every fuel and condition there is a percentage of CO_2 that will give maximum combustion efficiency, for the conditions prevailing.

The most reliable method of measuring the percentage of CO_2 in the flue gases is by using a gas analyzer or Orsat (Fig. 9-59). A sample of flue gas is drawn into the analyzer by means of a hand bulb. It is trapped and accurately measured. Then the sample is passed through a chemical (potassium hydroxide) that has an affinity for carbon dioxide. After the CO_2 has been absorbed by the chemical, the sample of flue gas is again measured and its volume compared to that of the original sample. The

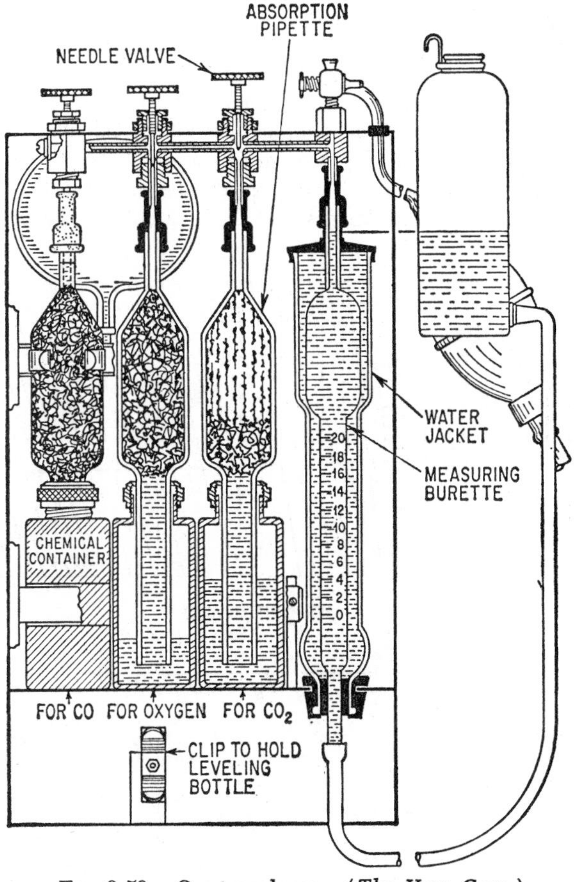

FIG. 9-59. Orsat analyzer. (*The Hays Corp.*)

difference in volume before and after absorption gives an exact measurement of the percentage of carbon dioxide in the gas sample. If further analysis of the remaining gases (carbon monoxide and oxygen) is desired, the residue sample is exposed successively to the carbon monoxide absorbent (cuprous ammonium chloride) and the oxygen absorbent (pyrogallic acid) and the shrinkage in volume noted after each absorption. The results of the analyses are read from a scale etched on the side of the measuring burette.

Calculation of Total Air from Orsat Readings. The per cent **total air** used for combustion can be calculated as follows from each Orsat analysis:

$$A_p = \frac{N_2}{N_2 - 3.78O_2} \quad \text{or} \quad \frac{N_2}{N_2 - 3.78(O_2 - \frac{1}{2}CO)}$$

where A_p = per cent total air (theoretical plus excess)
N_2 = per cent nitrogen in gases = 100 − (% CO_2 + % O_2 + % CO)
O_2 = per cent oxygen in gases
CO_2 = per cent carbon dioxide in gases
CO = per cent carbon monoxide in gases

When a number of different samples are taken, calculate the total air from each sample, and take the average of these values.

The total air for a coal-fired boiler can also be determined directly from Fig. 9-60 by applying Orsat analysis of each sample. The ordinate lines

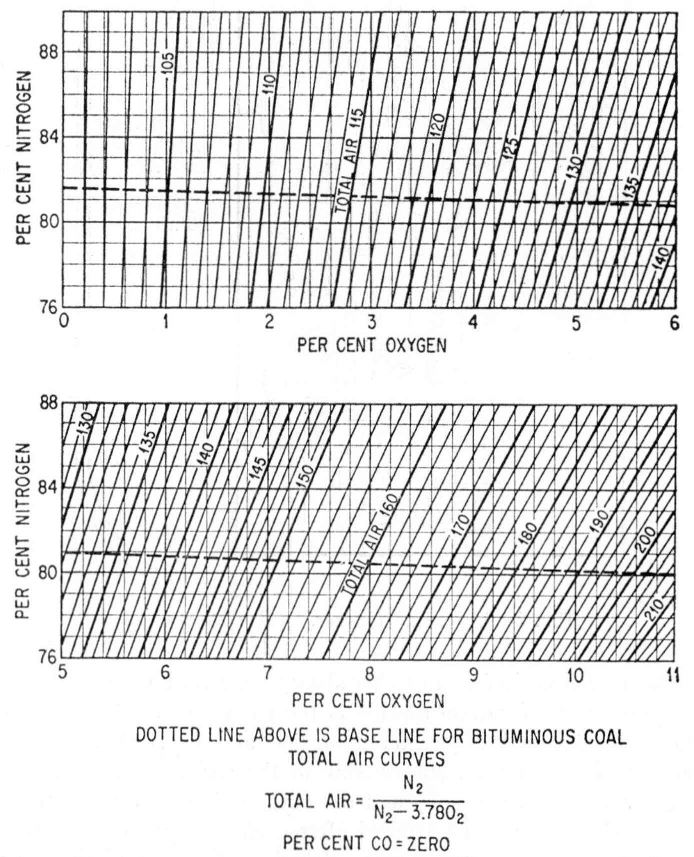

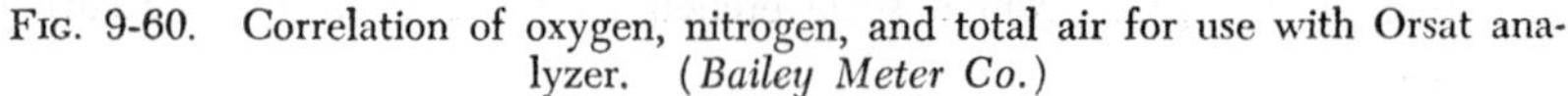

PER CENT CO = ZERO

FIG. 9-60. Correlation of oxygen, nitrogen, and total air for use with Orsat analyzer. (*Bailey Meter Co.*)

running right and left represent nitrogen ($100 - CO_2 - O_2$), the diagonal lines running from top to bottom are the per cent total air, and the abscissa lines are percentages of oxygen. For example, take a sample of flue gas showing 13.2 per cent CO_2, 4 per cent oxygen, and no CO. The intersection of the 82.8 ordinate (interpolated between 82 and 83) and the abscissa of 4 gives approximately 122.4 per cent total air, as read on the vertical diagonal lines.

A third method for determination of total air is by the use of Fig. 9-61. For this, only the percentage of oxygen need be used. Using the example

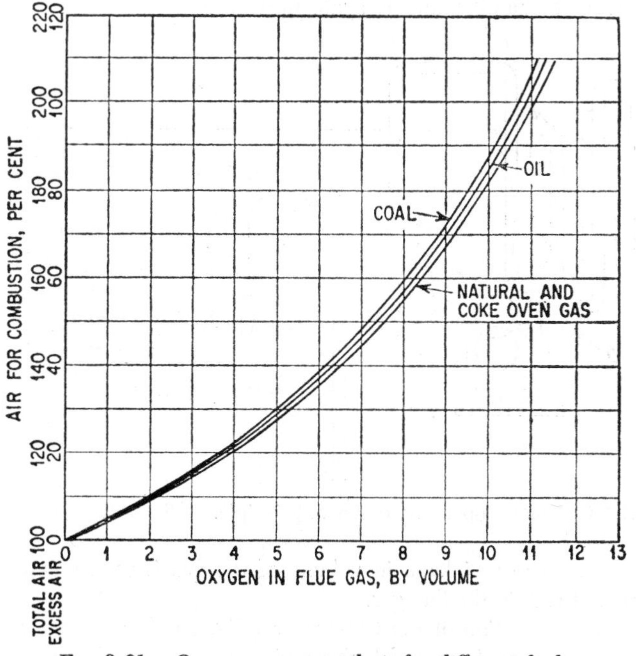

FIG. 9-61. Oxygen versus total air for different fuels.

given in the paragraph above (4 per cent oxygen), read the point where the abscissa of 4 crosses the *A* (coal) curve. This gives a reading of approximately 122.4 per cent total air on the ordinate lines.

CO_2 Recorders. Most modern boiler rooms are equipped with a permanently mounted, continuously operating type of CO_2 measuring device known as a combustion meter or CO_2 recorder. Instead of showing the result of a single spot analysis such as given by the hand Orsat, it reveals combustion conditions as they exist throughout the entire 24 hr, as a complete analysis is made each 2 min throughout the day and night, and without any attention from the operator.

Furthermore, a permanent record is made which reveals to those interested just when efficiency in firing is being attained by each shift of operators each day. Besides the CO_2 record, a record of furnace-draft and flue-gas temperature may also be obtained.

There are three common methods of measuring carbon dioxide, the CO_2 recorder being the least expensive and most commonly used instrument of this type. One of these methods employs chemical absorption similar to the hand Orsat, another compares thermal conductivity of the gases with that of air, and a third measures specific gravity.

Typical Chemical-absorption Meter.[13] An operating cycle consists of extracting a definite volume of gas and passing it through a solution of caustic potash, where the CO_2 is absorbed. The remaining gas passes into a displacement chamber and is again measured, at the same pressure and

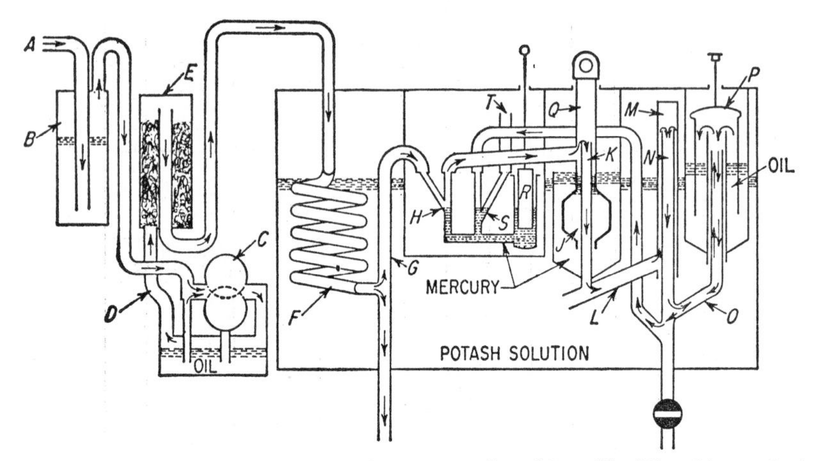

FIG. 9-62. CO_2 meter operating on Orsat principle. (*Republic Flow Meters Co.*)

temperature. The shrinkage in volume of the gas is then recorded as the percentage of CO_2 in the flue gas.

During the operation of the CO_2 meter (Fig. 9-62), a continuous sample of flue gas is drawn from the boiler and discharged to the atmosphere, at the rate of about 4 cu ft/hr. Referring to Fig. 9-62, the gas from the sampling tube in the boiler enters *A* and flows through indicator bottle *B*, which contains water, so that any dust that may have been carried through the sampling tube is washed out. The indicator bottle also provides a convenient means of checking the flow of gas to the meter. After it leaves indicator bottle *B*, the gas is drawn through a motor-driven pump *C*, which also draws a small amount of oil through it, to act as a seal for the pump. From pump *C*, the gas passes through the pump-oil chamber, which acts as a separator, where the entrained oil is liberated, and out through tube *D*. Tube *D* enters into bottle *E*, which is filled with rock-wool, through which the gas passes and which removes the finer oil particles, allowing clean gas to pass through the analyzing section of the

[13] Republic Flow Meters Co.

instrument. The gas then passes through temperature-control coil *F* and vertical tube *G*, one end of which is open to atmosphere. Coil *F* and vertical tube *G* are immersed in the potash solution, so as to bring the temperature of the gas to that of the solution. This is done to eliminate any possible variations in the volume of gas, due to temperature or saturation changes between the initial and the final measurement of gas.

The flow of gas is continuous, up to this point, with the gas in tube G kept slightly above atmospheric pressure.

At intervals of 100 sec, a volume of gas is drawn from tube *G* through intake valve *H* to be analyzed for per cent CO_2. This cycle of operation is accomplished by the rise and fall of unit *Q*, which moves measuring chamber *J* in and out of a body of mercury.

As unit *Q* starts on its upward stroke, the level of mercury in the measuring chamber *J* begins to fall and creates suction, which induces a flow of gas up tube *G* through intake valve *H* and into the measuring chamber *J*. When unit *Q* starts on its downward stroke, plunger *R*, actuated by cam on the motor-transmission assembly, moves to its bottom position, causing mercury to rise in intake valve *H*, which traps a definite amount of gas in the measuring chamber. As the pressure continues to increase, owing to the rise of the mercury, it forces the gas out of measuring chamber *J*, down through the center tube *K*, and through the potash solution in baffle tube *L*. As the gas travels up the flat inclined ceiling of the baffle tube, the CO_2 in the gas is absorbed by the potash solution.

The remaining gas passes out through separator tube *M*, down into collector tube *N*, and up through tube *O*, where it raises positioning bell float *P*, which is sealed in oil.

The wiper arm on the transmitter or indicator and the pen arms on the mechanical-recorder reading instrument are positioned by the height of bell float *P* and read directly in per cent of CO_2. After the arm has been positioned by the rise of bell float *P*, it is held in place by a brake (not shown) operated by a cam connected to the shaft, which raises and lowers unit *Q*. The arm is held at this position until the completion of the succeeding cycle, at which time it is released momentarily by the brake to permit any change in the height of bell float *P* to reposition the arm, where it is again held by the brake. In this manner, a continuous reading of percentages of CO_2 is obtained.

A turbulence in the caustic potash solution is caused by the flow of gas through baffle tube *L*. This prevents stagnation and ensures maximum effectiveness.

The cycle of operation is completed when plunger *R* rises to its top position, allowing mercury to fall in relief valve S and opening the end of relief tube *T* to the atmosphere, which releases the trapped gas and permits the bell float to come to rest.

Figure 9-63 shows the meter installed.

Typical Thermal-conductivity Meter.[14] Thermal-conductivity-type CO_2

[14] Bailey Meter Co., Hays Corp., Leeds & Northrup Co.

recorders operate on the principle that every gas has a somewhat different rate of heat transfer. Figure 9-64 shows the effect of passing several different gas samples through a thermal-conductivity type of instrument that produces a deflection of 100, for a 100 per cent CO_2 sample. The vertical

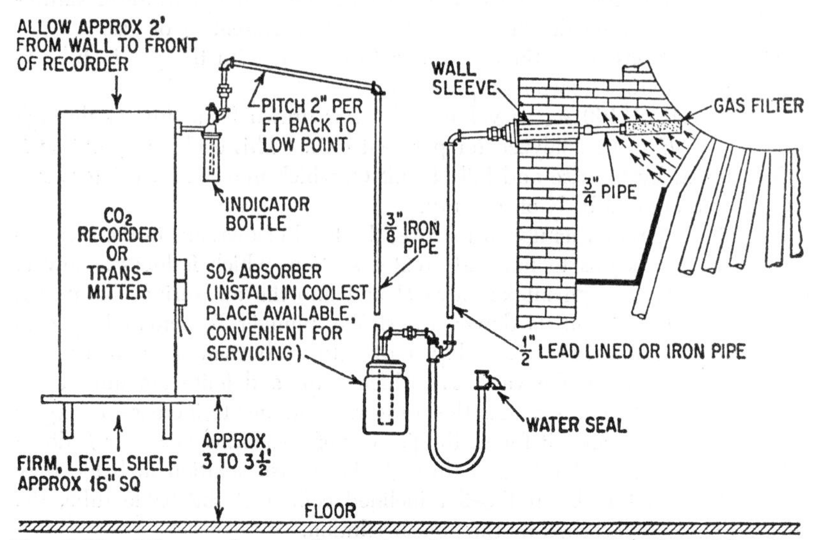

FIG. 9-63. CO_2 recorder installation based on Orsat principle. (*Republic Flow Meters Co.*)

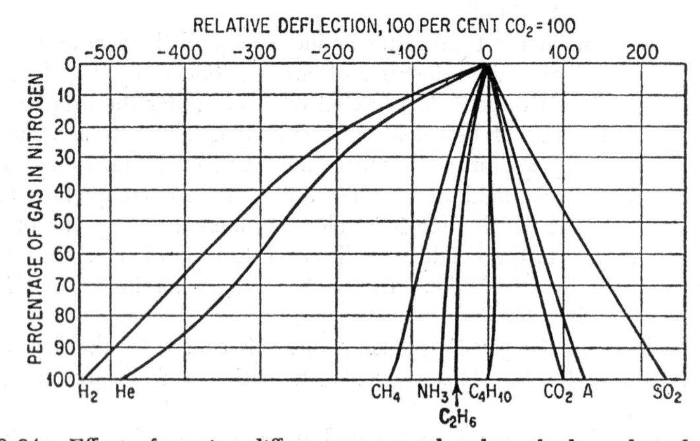

FIG. 9-64. Effect of passing different gas samples through thermal-conductivity instrument.

scale shows the relative deflection obtained with various gas samples, while the horizontal scale gives the percentage of each gas contained in nitrogen. Nitrogen was used as the base or zero reference line, but air would be practically the same. Generally speaking, in combustion gases CO_2, nitrogen, and some oxygen would be the principal gases present. Small quantities of other gases such as argon and even small percentages of SO_2 would have little effect on the reading, since they produce a deflection in the same direction as CO_2 but of slightly different magnitude. Sizable quantities would introduce an error, but the amounts normally encountered could be calibrated out since there will be a definite relationship between their percentages throughout the range of the instrument.

Hydrogen will produce a substantial error in the reading because it has a very pronounced deflection in the opposite direction from that of CO_2. In some cases this may prove to be advantageous, as hydrogen may begin to show up when a deficiency of air exists, and even a small amount will reduce the CO_2 reading. A sharp drop in the record when operating at high CO_2 values would indicate the presence of hydrogen.

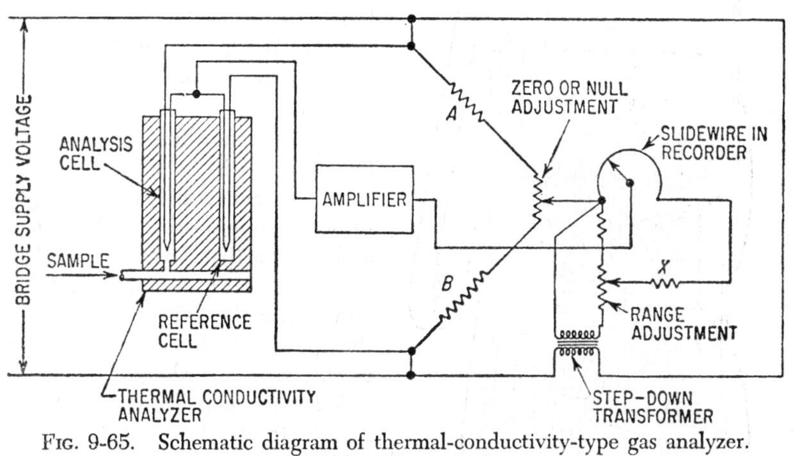

FIG. 9-65. Schematic diagram of thermal-conductivity-type gas analyzer. (*Bailey Meter Co.*)

Figure 9-65 shows the cell-block assembly used in one thermal-conductivity CO_2 recorder. There are two identical cells, each containing a heated resistor forming part of a resistance-bridge circuit. The measuring cell receives gas samples while the isolated comparison cell contains air, both being saturated to provide an analysis on a standard basis. The heated resistors are small in diameter and have very little mass, and so there is very little heat inertia present and temperature changes are rapid. As the gases conduct heat away from the headed resistor in the left-hand cell, the temperature decreases, the amount of change being dependent on the thermal conductivity of the gas in the cell. Meanwhile air occupies the right-hand cell at all times, and keeps the resistor on this side at a fixed temperature and, therefore, a fixed resistance. Since these two heated resistors form legs of a resistance bridge, the difference between the two elements can be determined, thereby showing the thermal conductivity of the gas in the cell.

Typical Specific-gravity Meter.[15] The simple operating principle of a specific-gravity CO_2 analyzer is illustrated in Fig. 9-66. The instrument chassis consists of two hollow cylindrical chambers. Each chamber contains a motor-driven impeller and an impulse wheel, similar to the arrangement used in modern automotive "fluid drives." The lower impeller draws in a continuous sample of the gas to be tested, and spins it at high speed against the vanes of the companion impulse wheel, creating a torque proportional to the gas density.

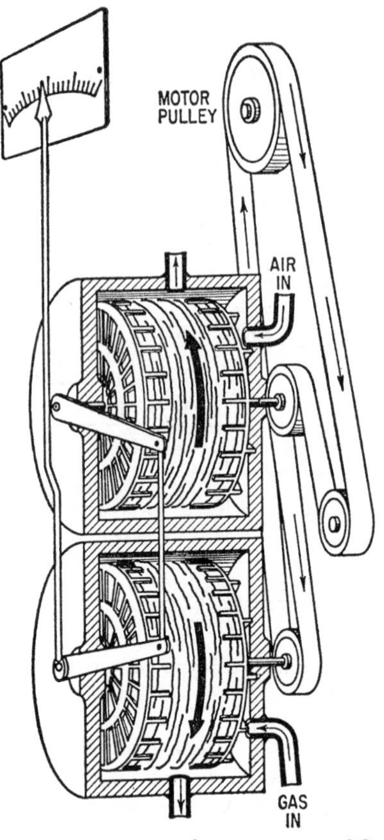

FIG. 9-66. Specific-gravity-type CO_2 gas analyzer. (*Ranarex Instrument Div., Permutit Co.*)

Similarly, the upper impeller draws in a continuous sample of ambient air, and spins it at the same speed but in opposite direction to the gas, creating on its companion impulse wheel a torque proportional to the air density.

The difference between the opposing torques is a measure of the specific gravity, and is transmitted through impulse-wheel pivot shafts and a sensitive lever and linkage arrangement to the pointer, which moves over the indicating scale, and the pen, which moves over the clock-driven recording chart.

Since the impeller draws the gas sample to the instrument, separate aspirators or pumps are not required. The use of ambient air in the reference chamber automatically compensates for variations in ambient temperature and barometric pressure.

The gas sample and reference air are usually discharged directly to the surroundings. When hazardous gas is tested, however, it must be discharged through a vent pipe leading to a safe area at atmospheric pressure.

To prevent errors in measurement and avoid damage to the instrument, the gas sample may require conditioning.

The gas should enter the instrument at substantially atmospheric pressure (within ½ in. water column) or at a flow rate of 10 to 25 cu ft/hr. If the gas pressure is greater than ½ in. water column, it should be reduced with pressure-reducing equipment such as a hand-operated globe or needle valve

[15] Ranarex Instrument Div., Permutit Co.

used with a manometer or rotometer—or a differential-pressure regulator with rotometer and, with higher pressures, a pressure-reducing valve.

If the gas suction is greater than ½ in. water column, the exhaust gas and air must be piped to a downstream point in the system at the same suction condition. The impeller will then draw gas in the normal manner. Maximum permissible gas suction is 12 in.

The gas should enter the instrument at substantially ambient temperature. Usually the sample piping is 15 ft or more in length, and cools or heats the gas to ambient temperature so that auxiliary cooling or heating devices are not required.

The gas and air should have the same degree of humidity or dryness as they enter the measuring chambers. It is generally more convenient to attain equalization by saturating both the gas and the air. For these installations, the instrument is furnished with a built-in double-compartment humidifier.

When the gas is initially very dry or has very low or very high density or is soluble in water or is corrosive when saturated, or when the user prefers to measure on the dry basis, the humidifier is omitted and a built-in reference-air dryer is supplied.

Suspended particles and some types of corrosive constituents are removed by suitable filters supplied with the instrument. When certain corrosive gases are to be analyzed, they must be measured dry, and the instrument is built of materials which are resistant to the gas.

OXYGEN ANALYSIS[16]

Refineries, steel mills, and other industries have for years burned a combination of fuels consisting largely of by-products from various processes. In recent years, the trend toward burning several fuels in a stream or power plant instead of a single fuel has spread to all industries. Special rates on gas at specified times, lack of coal at other times, and the desirability of burning liquid or gaseous fuels under certain load or conditions have all influenced this change.

Figure 9-67 illustrates the relationship among O_2, CO_2, and excess air for various fuels. These charts show that, for a given percentage of O_2, there is very little difference in excess-air percentage, even though the fuel is changed from coal to natural gas. When burning gas, the percentage of excess air would be less than it would be for coal, but limits can easily be established for maintaining the oxygen reading at the desired point for the particular fuel being burned. Actually, a chart calibrated in per cent of excess air can be prepared, so that the instrument will provide direct readings for any type of fuel.

Oxygen Instruments. The most important and widespread use of these analyzers is for the determination of oxygen in products of combustion.

[16] Abstracted from W. H. Pugsley, The Hays Corp., in D. M. Considine (ed.), "Process Instruments and Controls Handbook," McGraw-Hill Book Company, Inc., 1957.

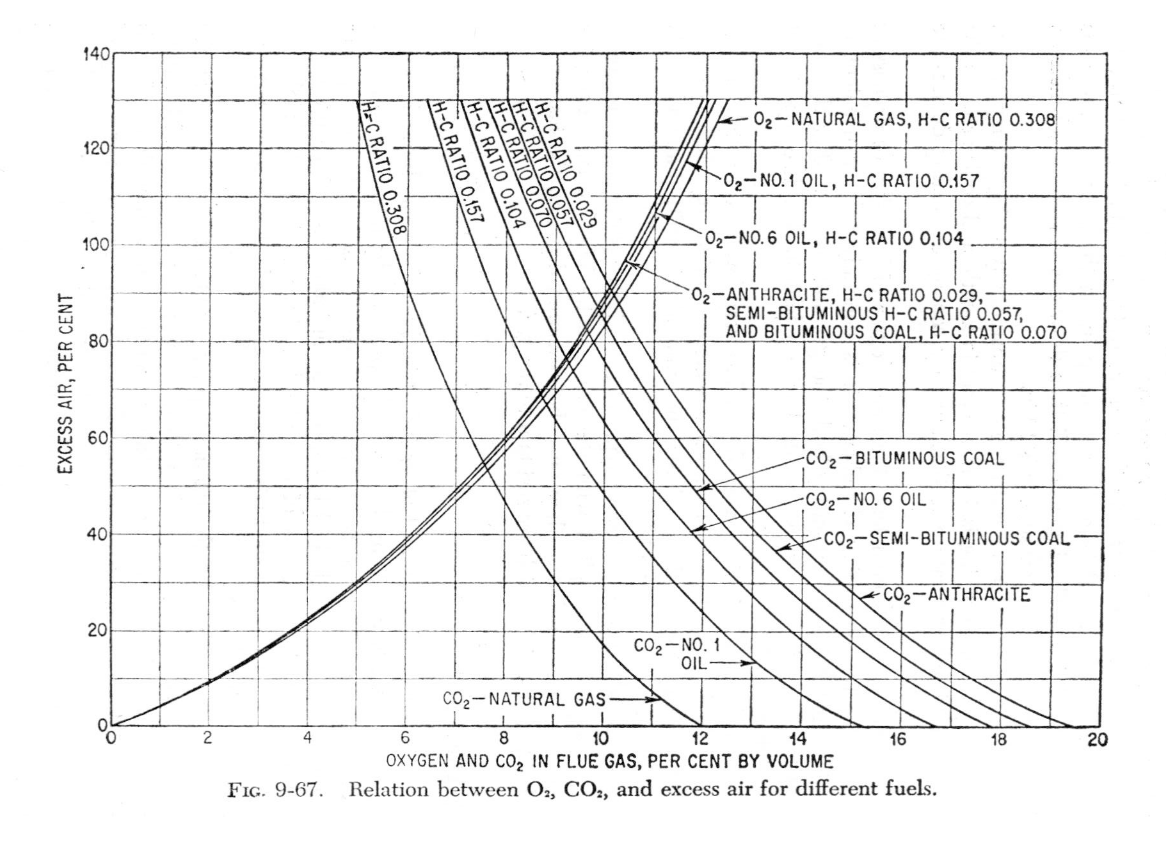

FIG. 9-67. Relation between O_2, CO_2, and excess air for different fuels.

Knowledge of the O_2 content in combustion gases, together with the CO_2 analysis, is essential for economical use of all commercial fuels in present-day heating plants and power-station applications.

A summary of the most important methods of oxygen analysis now employed in continuously recording instruments follows.

1. *Paramagnetic.* Oxygen has very large paramagnetic susceptibility compared with other common gases and vapors, causing it to displace substances of lower paramagnetic susceptibility from a magnetic field. This effect permits oxygen measurement (*a*) directly by determination of the displacement of a test body from a magnetic field by the oxygen, and (*b*) indirectly by measuring the cooling effect on a heated wire of the oxygen flow in a magnetic field. This oxygen flow arises because heated oxygen loses its paramagnetic susceptibility and is displaced by cooler oxygen.

2. *Catalytic combustion.* The heat produced by rapid oxidation, or combustion, is a useful measure of oxygen concentration under circumstances where (*a*) a known combustant, such as hydrogen, can be added in suitable excess, (*b*) the reaction can be catalytically controlled to go to completion, and (*c*) the temperature rise can be accurately measured.

3. *Electrochemical.* The electrochemical activity of oxygen permits its measurement by the magnitude of the depolarizing effect of oxygen on a special galvanic cell.

4. *Inferential thermal conductivity.* The unique combination of the very high thermal conductivity of hydrogen and the way it can be catalytically reacted with oxygen make possible inferential analyses by (*a*) adding hydrogen to sample and measuring thermal conductivity before and after a continuous combustion which removes the oxygen, or (*b*) using hydrogen to scrub oxygen from a liquid and measuring the resultant hydrogen dilution by decrease in thermal conductivity.

These methods, unless specifically designated, are primarily suited to gas analysis. Many of them can be extended to determine dissolved oxygen in liquids, by use of an auxiliary gas stream to scrub the liquid and carry the oxygen to the analyzer.

Paramagnetic Oxygen Analyzers. These instruments depend for operation on the fact that oxygen is unique among gases in being strongly *paramagnetic* (attracted into a magnetic field), whereas other common gases, with few exceptions, are slightly *diamagnetic* (repelled out of a magnetic field). In one type of instrument, paramagnetic susceptibility is measured directly by determining the change of the magnetic force acting on a test body suspended in a nonuniform magnetic field when the test body is surrounded by the sample gas. This principle is shown in Fig. 9-68. If the sample gas surrounding the test body is more paramagnetic than the test body, the gas will tend to displace the test body away from the region of maximum flux density. Conversely, if the sample gas is less paramagnetic than the test body, the test body will be drawn toward the region of maximum flux densities.

Method of Measurement. The test body used in these analyzers is shown in Fig. 9-69. It consists of a light, dumbbell-shaped element made of two small hollow glass spheres that are supported on a quartz torsion fiber. A small mirror on the test body permits optical indication of the angular position of the assembly. The test body is free to rotate in the nonuniform permanent-magnetic field and is subject to a magnetic force, which is proportional to the difference between the volume magnetic susceptibilities of the test body and the gas which the test body displaces.

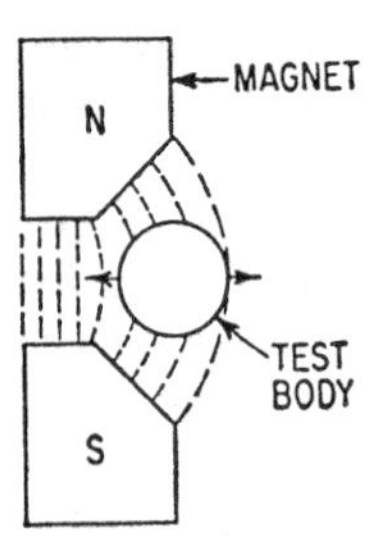

FIG. 9-68. Paramagnetic susceptibility of oxygen is measured directly by determining change of magnetic force acting on test body suspended in nonuniform magnetic field when test body is surrounded by sample gas.

These analyzers are available as either deflection or null types.

DEFLECTION INSTRUMENTS: These instruments are small, portable, and quite accurate.

The general operation of the deflection-type instruments is shown schematically in Fig. 9-69. The test body rotates to an equilibrium position at which the magnetic rotational force is balanced by the torsional restoring force of the quartz fiber. A beam of light, reflected from the mirror, is deflected in accordance with the rotation of the test body, and indicates oxygen concentration by the position of a spot of light on a calibrated translucent scale. The instrument is supplied with a minimum span of 5 per cent (40 mm partial pressure) between 0 and 80 per cent O_2, and a minimum span of 15 per cent (120 mm) above 80 per cent. Wider ranges are available up to 0 to 100 per cent (0 to 800, or 0 to 760 mm).

NULL INSTRUMENTS: Null-type instruments are used where greater sensitivity is required and also for continuous-recording applications. The operation of a recording null-type instrument is shown schematically in Fig. 9-70. In this instrument, the test body is restored to its null position by the application of an electrostatic force equal and opposite to the magnetic force produced by the gas sample. The glass test body in this instrument is coated to make it conductive, and is connected to a variable source of electrical potential. Near the test body are two electrodes maintained at fixed electrical potentials, and arranged in such a manner as to constitute, with the test body, a heterostatic electrometer. If no potential existed between the test body and the electrodes, the test body would be free to rotate to an equilibrium position where the magnetic rotational force was balanced by the torsional restoring force of the quartz fiber.

In this instrument, however, the electrical potential of the test body is automatically and dynamically varied to produce an electrostatic force equal and opposite to the magnetic force. The potential required to hold the test body in null position is a direct measure of the oxygen concentration of the gases surrounding the test body.

A beam of light reflected from the mirror attached to the test body is deflected in accordance with any rotation of the test body, owing to the presence of a magnetic sample gas. The light beam is divided between two phototubes. Any motion of the test body due to changes in oxygen concentration results in an unbalance in the light on the two phototubes. This unbalance produces an electrical signal, which is amplified and fed back to the test body to control its potential. This may be done by a

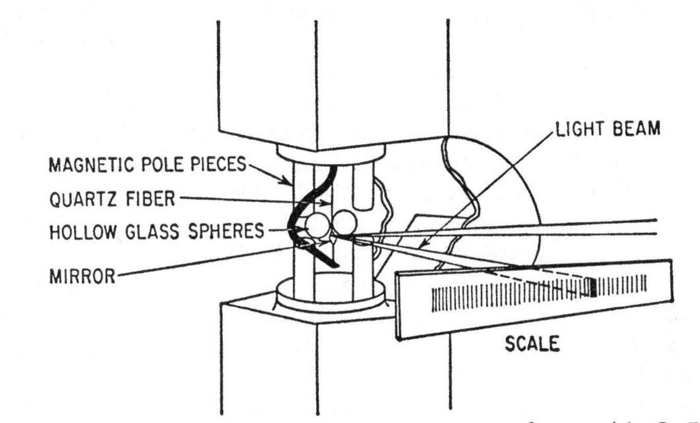

FIG. 9-69. Test body used in paramagnetic oxygen analyzer. (*A. O. Beckman, Inc.*)

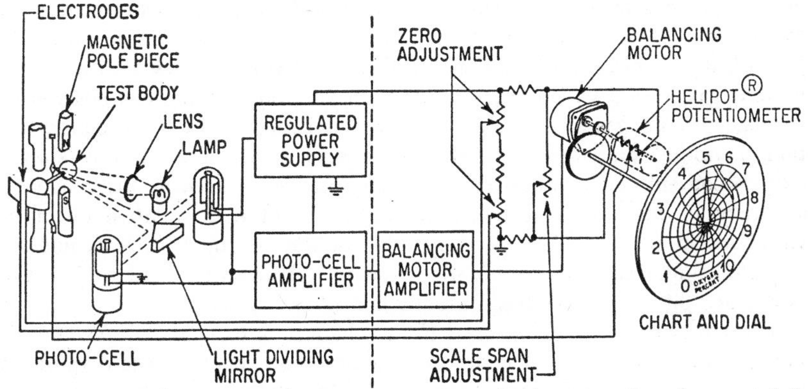

FIG. 9-70. Null-type recording oxygen meter. Analyzer portion shown at left; recorder at right. (*A. O. Beckman, Inc.*)

direct-feedback amplifier or by a motor-driven potentiometer such as that shown in Fig. 9-70. In this latter method, the electrical signal is amplified by the phototube amplifier in the analyzer and the balancing-motor amplifier in the recorder and is used to drive a balancing motor forward and backward, depending on the direction of rotation of the test body. The motor is coupled to the shaft of a potentiometer, which supplies a variable electrostatic potential to the test body. By this mechanism, the potential required to hold the test body in null position is automatically and continuously supplied. Any change in magnetic force due to a change in oxygen partial pressure is always balanced by an electrostatic force, which, in turn, is proportional to the d-c voltage delivered by the potentiometer. Thus, changes in oxygen partial pressure are accurately translated into proportional voltage changes, which can easily be recorded on commercially available recording potentiometers. This makes it possible to couple the oxygen analyzer to all types of controllers.

High accuracy is obtained by careful thermostatic control. The sample gas passes through the instrument at a very slow rate and is warmed to the same temperature as the instrument. Thus, variations of magnetic susceptibility with temperature are minimized. The instruments also employ internal-shock mounting to protect the test-body suspension, phototubes, exciter lamp, and gas passages from shock or vibration.

On installations where extreme accuracy is required or where other considerations make advisable periodic calibration checks of the instrument, such checks can readily be made either manually or by means of automatic standardization.

Automatic self-standardization frequently is desired for instruments with narrow scale span, i.e., with high sensitivity. In the very low ranges of oxygen concentration, where the cumulative magnetic effect of the other components of the gas is significant in relation to the magnetic susceptibility of the oxygen, it often is convenient to standardize the instrument with oxygen-free gas obtained by removing oxygen from the sample gas. By periodically zeroing the instrument with reference to the oxygen-free background gas, errors which might arise from variations in the composition of the background gas are avoided. In the high ranges of oxygen concentration, periodic standardization by reference to a gas of accurately known oxygen content, such as dry air or cylinder oxygen of known purity, assures accuracy of measurement by compensating for the cumulative effect of variations in any factors which might affect the accuracy, such as uncompensated fluctuations in barometric pressure. Automatic standardization is performed on a repetitive basis, usually hourly.

The initial response of the analyzer to changes in oxygen concentration is immediate. The rate of response depends on the porosity of the diaphragm between the sample duct and the analysis cell. The instrument is supplied with a diaphragm which will give a 95 per cent response to any change in oxygen concentration within 1 minute.

The analyzer and recorder are in separate metal cases, and these two units may be mounted separately or may be mounted together on a vertical panel or in a totally enclosed cubicle as desired. Weatherproof shelter is required.

Only a very small sample flow is required; as little as 50 cc/min is adequate. The gas sample does not flow directly over the test body but diffuses into the analysis cell through a porous diffusion disk. For continuous sampling, readings are independent of flow rate over a wide range, usually from 50 to 250 cc/min. No chemical treatment of the sample is required, nor is it necessary to saturate the sample or to dry it. The instrument measures the gas on an "as-received" basis. The temperature of the sample gas entering the instrument should not exceed approximately 100°F. Because the sample is warmed in passing through the instrument, condensation should not occur. Usual filtering precautions are required external to the analyzer to be sure that the sample is free from dust or oil droplets. For hot furnace gases, a primary ceramic filter is recommended. If highly corrosive gases are present which will attack Neoprene, aluminum, stainless steel, or glass, they should be removed, or a special analysis cell should be obtained to withstand the corrosive action.

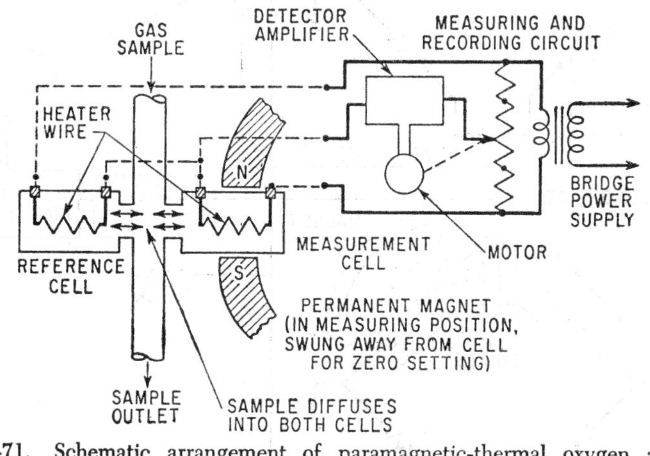

FIG. 9-71. Schematic arrangement of paramagnetic-thermal oxygen analyzer. (*The Hays Corp.*)

When desired, sampling equipment can be supplied to meet the conditions outlined. Maintenance consists primarily of keeping the sampling system in operating condition, cleaning filters if used, and periodically checking the analyzer with standardized gases or an air check with which most instruments are provided. Frequency of maintenance is determined primarily by the application.

Paramagnetic-thermal Oxygen Analyzers. The paramagnetic susceptibility of oxygen varies inversely as the square of the oxygen temperature, decreasing rapidly as the temperature is increased. Accordingly, a thermal-convection system, somewhat similar to a thermal-conductivity cell but with circuit changes and the addition of a magnetic field, can be employed for oxygen analysis. This method largely compensates for the effects of thermal conductivity and provides an output proportional primarily to the volume paramagnetic susceptibility of the sample gas, which, in most practical cases, is directly proportional to the oxygen concentration. However, the background thermal conductivity requires that an individual calibration be used for each application if maximum accuracy is desired.

Two commercially available oxygen analyzers employ variations of this basic principle, as shown schematically in Figs. 9-71 and 9-72, respectively.

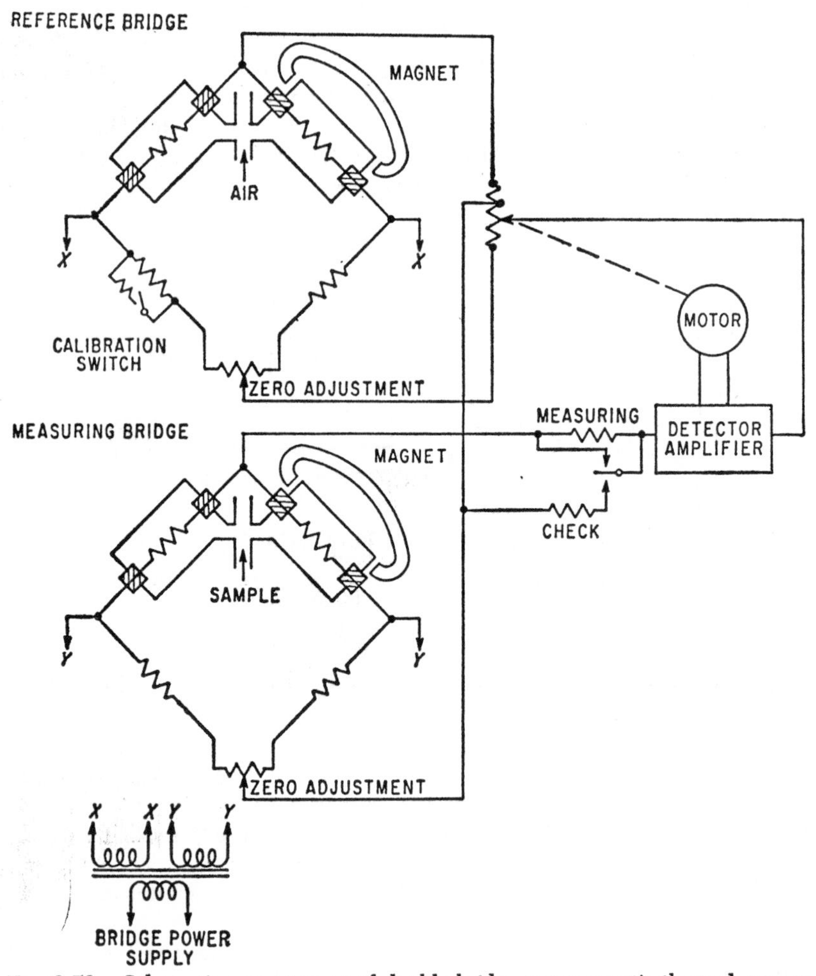

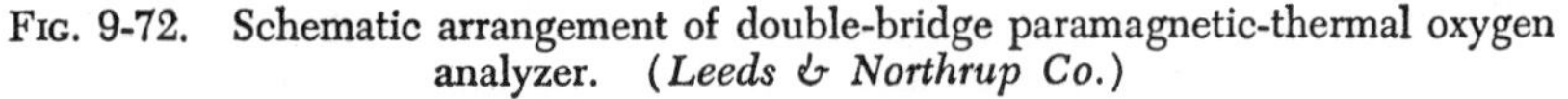

Fig. 9-72. Schematic arrangement of double-bridge paramagnetic-thermal oxygen analyzer. (*Leeds & Northrup Co.*)

The sample gas is permitted to diffuse into two similar cells, each of which contains a heater wire or element. The wires will be cooled equally by the conduction and natural-convection effects of the gas and by radiation, so that, if the wires have the same temperature coefficient of resistance, are equally heated, and are symmetrical with respect to heat losses, there will be no measured difference in their resistance. However, if a permanent magnet is swung into position so that one of the heater elements is located in a region of high magnetic flux, then oxygen, which is paramagnetic, will tend to concentrate in that region, displacing the other gases. As soon as the oxygen is heated by the wire or element, it loses much of its paramagnetic susceptibility and is displaced by cooler oxygen, causing a thermal-magnetic-convection current that cools the heater wire in the measurement cell more than that in the reference cell.

The heater wires are made of platinum or of some alloy having a high-temperature coefficient of resistance, so that a simple electrically null-balanced Wheatstone-bridge circuit can be employed to measure the difference in their temperatures. This difference is proportional to the effective magnetic susceptibility of the gas and, when oxygen is the only paramagnetic gas present, the oxygen content.

Of course, this measurement is not absolute. In the first place, the calibration depends on the physical size and shape of the cells and the heater wires, the cell-block and heater-wire temperatures, the absolute sample pressure, and the background thermal conductivity. In both the commercial instruments of this type, the analyzer case is thermostated and the heater current is carefully regulated. Sample pressure compensation is effected by one instrument by using the ratio type of measurement shown in Fig. 9-72. The analyzer system is composed of two thermomagnetic bridges, each bridge having two cells operating as previously described. The cells in one bridge are filled with air, and the electrical output of this bridge provides the voltage across the recorder slidewire. The gas sample to be measured flows through the second bridge, and its output is balanced against a portion of the slidewire voltage by means of an amplifier and balancing motor. The oxygen measurement obtained, therefore, is basically a ratio of the oxygen concentration in the sample gas to the constant oxygen concentration of the atmosphere. Since both sets of cells operate at the same pressure, the ratio will remain constant, in spite of variations of this pressure. Another instrument makes use of an auxiliary sealed bellows whose internal pressure is made to balance the sample-system pressure by heating the confined gas. A thermistor in the recorder-slidewire circuit converts the gas temperature in the auxiliary chamber into the necessary barometric-pressure compensation.

The effects of background thermal conductivity may be minimized by calibration with a sample of the nominal composition flowing through the instrument. If large variations in hydrogen content occur, however, the effects may still be appreciable.

Zero setting is easily accomplished by swinging the magnet away from the measuring cell. After the instrument has been adjusted to read zero with the magnet swung out and with a sample flowing through the cell, an up-scale point is checked to complete the calibration. On instruments whose full-scale span includes the concentration of oxygen in the atmosphere (20.7 per cent), air is admitted and the instrument is adjusted to read 20.7 per cent O_2 with the magnet swung into operating position. General practice also permits the use of air to check full scale on narrower-range instruments by switching in a calibrating resistor to make the full-scale reading correspond to 20.7 per cent O_2 so that the up-scale adjustment can be completed.

These instruments are considerably more sensitive and reproducible than the usual calibrating technique. Accordingly, although the maximum sensitivity and reproducibility correspond to about ± 0.01 per cent O_2, the error in reading oxygen content may be as great as about ± 0.25 per cent O_2 for ranges up to 15 per cent O_2 full scale and 2 per cent O_2 for wider spans.

Response time is limited by diffusion and the relatively low sample-flow rate needed to assure sample heating. In general, initial response to a change in oxygen concentration occurs in about 5 sec with a normal flow rate of approximately 0.5 cfm; 90 per cent of final response is obtained in substantially less than 1 minute.

Both the commercial instruments consist of two parts, the analyzer proper and the recorder. The analyzer is housed in a case of the same dimensions as the recorder. This case is insulated and thermostated and should not be mounted in an exposed location where very wide swings in ambient temperatures are to be expected.

Catalytic-combustion Method of Oxygen Analysis. A catalytic-combustion instrument for analysis of oxygen in a gaseous mixture is based on the addition of hydrogen to a continuous gas sample in the presence of a heated noble-metal catalytic filament, which causes a combustion process to take place. The heated filament is one leg of an a-c Wheatstone-bridge circuit which becomes unbalanced through variation in resistance of the filament due to the heat of the combustion process. The unbalance measures the volumetric percentage of oxygen present in a gas sample. The instrument is shown schematically in Fig. 9-73.

A continuous sample of the measured gas is delivered to the analyzer by the sampling system. The pressure of the measured gas entering the analyzer block is controlled by two gas-pressure-regulating valves located at the rear of the analyzer case. The valves are arranged in series and form loosely fitting pistons for the cylinders in which they are housed. At the top of each cylinder, a rectangular slot forms an exhaust port to atmosphere.

In operation, the valves float on the gas stream, rising and falling to vary the volume of gas exhausted to atmosphere so that a constant pressure is maintained to the analyzer block. The pressure in each valve chamber is determined by the weight of the valve.

Only about 1 per cent of the gas delivered to the analyzer from the sampling system actually undergoes analysis; the remainder is exhausted to atmosphere. This relatively high rate of flow from the sampling system is maintained to keep the response lag of the instrument at a minimum.

At the constant pressure established by the regulating valves, the gas sample passes through the sample orifice, which maintains a constant rate of flow. At the discharge side of the orifice the measured gas is combined with hydrogen, and the mixture of measured gas and hydrogen enters the analyzer cell at a constant temperature of approximately 160°F maintained by the heater provided in the analyzer block.

The hydrogen is obtained from a storage cylinder, as shown schematically in Fig. 9-73. A two-stage pressure regulator mounted on the cylinder reduces the hydrogen pressure to 25 psig, which is maintained to the inlet

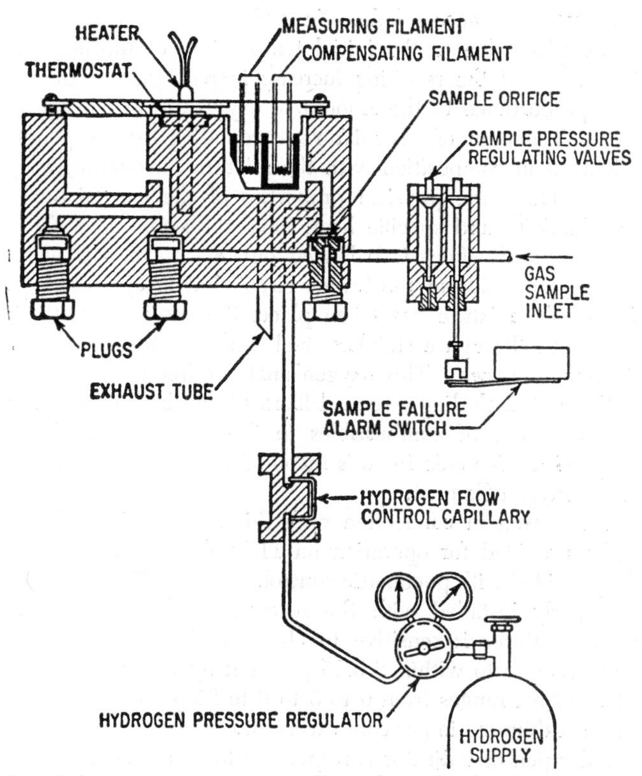

Fig. 9-73. Schematic arrangement of catalytic combustion oxygen analyzer. (*Bailey Meter Co.*)

of the flow-control capillary. The capillary provides for a constant flow of hydrogen to the analyzer block. Hydrogen consumption is 5 to 7 cu in./min. A 180-cu-ft cylinder will last about 6 weeks.

A filament assembly shown in Fig. 9-73 consists of two identical noble-metal catalyst filaments mounted on a common base. The filaments are surrounded by a shield and are separated from each other by a metallic wall. One filament chamber is covered at one end only by a screen, so that the filament is completely exposed to the gas mixture entering the analyzing cell. This filament is the measuring filament. The other filament chamber is closed on all sides, except for an access hole which has only a small percentage of the area of the screened section of the measuring-filament chamber. The access hole allows a small amount of the gas mixture to enter the chamber. The filament in this chamber is the compensating filament.

The mixture of measured gas and hydrogen enters the analyzing cell and passes through the screen into the measuring-filament chamber. A relatively small amount of the mixture enters the compensating-filament chamber through the access hole. In each chamber, combustion of the mixture takes place due to the initial temperature of the filaments. The temperature rise and the resulting increase in resistance of the measuring filament are proportional to the amount of oxygen present in the measured gas. The small amount of gas mixture surrounding the compensating filament is identical in composition with the mixture surrounding the measuring filament. Therefore, physical properties of the gas mixture, such as thermal conductivity and specific heat, have the same effect on both filaments, and any such effect is balanced out, since the two filaments are located on the same side of the bridge circuit but in opposite legs. After combustion of the mixture has taken place, the gases are exhausted from the analyzing cell through a stainless-steel discharge tube.

Combination Analyzer. This oxygen analyzer frequently is furnished as a combination unit including a second filament and bridge circuit for simultaneous measurement of combustibles in the gas sample. As shown in Fig. 9-73, provision is made for this second filament in the heated block to the left of the oxygen filament.

The recorder used in connection with this analyzer is of the electronic type; can be provided for operation on 115 volts, either 50 or 60 cycles; and can be provided with pneumatic control.

When properly installed with the recommended sampling system and recorder, the analyzer is sensitive to changes of less than 0.05 per cent oxygen and accurate to within ±0.25 per cent by volume of oxygen in the gas sample. Chart ranges from 0 to 5 to 0 to 25 per cent are available, as are also charts calibrated in per cent excess air.

The recorder pen or indicator is responsive to 63 per cent of total oxygen change in less than 15 seconds.

The analyzer should be mounted as close as possible to the sample point (each 25 ft of sampling line creates a response lag of approximately 1 sec), in a location providing convenient access to the instruments for inspection and maintenance. The ambient temperature of the analyzer should not exceed 110°F. The analyzer and recorder may be separated by distances up to 500 ft, using no. 14 or larger copper-wire size.

The recommended method of calibrating the oxygen analyzer and recorder is to apply gas samples of known oxygen content to the analyzer, and then make necessary adjustments to the recorder to obtain correct readings. The prepared gas is applied to the analyzer by disconnecting the sampling line and connecting the prepared gas sample to the gas inlet of the filter and heater assembly. If a prepared gas cylinder is not available, the calibration of the analyzer may be checked against an Orsat analysis.

Routine maintenance consists primarily of checking the operation of the sample-regulating valve and maintaining the sampling system by cleaning and renewing filter felts and checking connections for leaks.

Hydrogen-difference Method of Oxygen Determination. An oxygen analyzer employing the hydrogen-difference method of measurement is shown in Fig. 9-74.

In this device, a small continuous sample of gas is drawn through the sample system into the analyzer by means of a water-operated aspirator. The gas is cooled, water-saturated, and then mixed with a constant flow

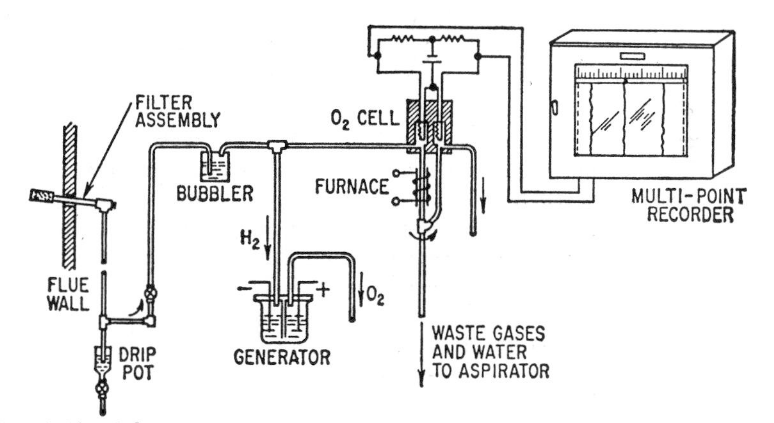

FIG. 9-74. Schematic arrangement of hydrogen-difference method of oxygen determination. (*Cambridge Instrument Co.*)

of hydrogen. An electrolytic generator produces the hydrogen. The mixture of sample gas and hydrogen passes through one side of a thermal-conductivity cell in which the thermal conductivity of the gas mixture is measured. The sample then passes through a low-temperature furnace in which all the oxygen in the gas sample is caused to combine with as much of the hydrogen as is required for forming water. The water is condensed and expelled. The remaining gas sample then passes through the other side of the thermal-conductivity cell.

Heated resistors in each of the cell cavities are cooled, and their resistance is determined by the cooling effect of the gas sample. The heated resistor in the first cell is cooled more, and, therefore, its resistance is decreased more than that of the resistor in the latter cell. The difference in change of resistance is due to the loss of hydrogen and oxygen combined to form water. This difference in change of resistances causes an unbalance of the Wheatstone-bridge circuit of the analyzer and recorder.

This unbalance is directly proportional to the concentration of oxygen in the gas mixture, and is recorded continually by the electric-type recorder calibrated to read directly in per cent of oxygen. Contacts can be provided where required to actuate an alarm at any given concentration of oxygen. Calibration at any point on the scale can be performed by the use of bottled standardized gases of known oxygen concentration. Zero calibration can be effected by merely turning off the analyzer furnace and allowing it to cool, thus exposing both cells to the same gas sample.

The O_2 analyzer is frequently combined with additional cells to form an analyzer and recorder for not only oxygen but also CO_2 and CO.

The normal full-scale oxygen range is 0 to 5 per cent, but other ranges of 0 to 1, 0 to 2, and 95 to 100 per cent O_2 are available.

The maximum sensitivity is a few hundredths per cent oxygen, and the accuracy is ± 0.15 per cent oxygen on the 0 to 5 per cent O_2 range.

A gas-sample flow of 200 cc/min is required at not over 10 in. of water suction. Cooling-water flow of 2,000 cc/min below room temperature is required for the operation of the analyzer. The instrument is not critical to wide changes in temperature, pressure, and humidity. Power supply is voltage-regulated for fluctuations between 105 to 129 volts, 60-cycle single phase. The maintenance required in addition to the normal recorder upkeep involves biweekly addition of distilled water to the gas generator and an occasional check of bridge and generator currents. The latter operation takes only a few minutes time and is accomplished by the use of switches located on the back of the analyzer.

The sampling system, depending on the type used, requires maintenance to the extent of cleaning filters and sample line and keeping the system tight to prevent leakage.

COMBUSTION SAFEGUARDS

Furnace-gas Explosions.[17] No furnace nor any kind of fuel is immune to explosion hazard. As long as the human element is involved, we will have the possibility of explosions in the small furnaces of heating boilers, as well as in the large furnaces of power boilers.

The majority of package-type boilers are either oil- or gas-fired, while those of the large water-tube type are fired with pulverized coal, oil, and/or gas. The need for employing extreme care and precaution while burning these fuels should be emphasized.

Statistics show that oil and gas are about equal in hazard and approximately four times as hazardous as pulverized coal. Although gas is a very explosive fuel, it maintains its ignition quite easily, while oil is less explosive in character, but ignition does not occur so quickly as with gas. Pulverized coal is the least explosive and the most difficult to ignite.

A large number of furnace explosions occur in starting up or lighting off. The next prevalent condition causing explosions is the stopping or decreas-

[17] Mutual Boiler & Machinery Insurance Co.

ing of the fuel feed, which causes loss of ignition, immediately followed by a resumption of the fuel feed without following the proper lighting-off cycle.

Another condition which exists is the operation of furnaces with an inadequate supply of air, thus permitting unburned fuel to accumulate in the furnace or flues. This unburned fuel can ignite when the proper air ratio is restored, or if an excessive amount of air is suddenly supplied. The latter condition sometimes occurs in hand- or stoker-fired furnaces.

Types of Operation. The methods of firing can be divided into three groups: (1) those of the fully automatic unattended type which are usually low-pressure-heating boilers, (2) in the industrial field, those of the fully automatic attended type and (3) those which are operated manually.

The low-pressure-heating boiler using gas as a heating medium usually has a low-pressure burner with a gas pilot for ignition. Heat from the pilot actuates a bimetal expansion element, which closes the electrical circuit between the pressure control and the electrically operated main gas valve. In the event of a pilot failure, the electrical circuit is opened, and the main gas valve remains closed, thus preventing gas from accumulating in the furnace.

When oil is used as the medium of heat transfer in the low-pressure boiler, an electrical ignitor is usually used. Occasionally an electrical ignitor lights a gas pilot, which in turn ignites the oil; however, this type is in the minority.

For ignition-failure protection, a stackswitch is installed, either in the flue between the boiler and the chimney, or in the last pass of the boiler. The stackswitch is really a time relay connected in series with the pressure control.

When the burner starts and ignition takes place, the temperature will increase around the expansion element of the stackswitch, causing the relay to lock in position. If ignition fails, there will be no temperature increase, and the relay will open the circuit after about 90 sec. After this cycle is repeated three times, the relay remains open, and it has to be reset manually.

The burning of gaseous fuels such as oil, gas, and pulverized coal in larger boilers is comparable to that in the low-pressure boilers, with the exception of the protective devices. The majority of the boilers in this category are not left unattended. The problem becomes one of educating the operator to follow a safe procedure.

Suggested Precautionary Measures:

1. The furnace should always be thoroughly purged of all fuel before lighting the burners.
2. Ample torch flame should be applied to each burner before the fuel is turned on.
3. The air supply in the burners should be kept low enough to prevent blowing out the flame.
4. A sufficiently rich mixture should be supplied so that ignition will take place immediately.

5. The pressure in the furnace should be slightly on the negative side at all times (unless furnace is pressurized type).
6. The correct ratio of air to fuel should be maintained throughout the operating period.
7. After stable ignition has been obtained, the lighting torches should always be shut off.
8. The boiler should be equipped with the necessary instruments or protective devices to allow sufficient speed of action to prevent an explosion in case of a power failure or other unexpected circumstances.

Interlocks. Interlocks may be used for purging a boiler before lighting off, or for shutting down equipment in case of fan failure. These applications are shown diagrammatically in Figs. 9-75 and 9-76, respectively.

The operation of a purge interlock, shown in Fig. 9-75, is as follows:

The interlock is actuated by a differential pressure proportional to air flow through the boiler, usually the differential across the boiler tubes.

When the boiler is out of service, there is no differential applied to the interlock measuring element (diaphragm), and contact *M*2 is open. Contact *R*3, which is closed, may be auxiliary fingers on a fan-motor breaker (on whichever, or both, of the induced-draft or forced-draft fans that will be used for the purging), or may be a relay actuated by contact *M*1 of a fan interlock. With contact *M*2 open and the top contact of *R*3 closed, the time-delay relay *Ri* is deenergized and contact *R*1 is open, thus opening the fuel-starting circuit to prevent the supply of fuel. Also, the red signal light, if furnished, is energized.

Before fuel may be supplied to the furnace, the fan motors must be started, to open the top contact of *R*3 and close the bottom contact, and the air flow through the furnace must be increased to, say, 60 per cent of capacity. This high rate of air-flow differential closes contact *M*2 to complete the circuit from *L*1 through *M*2 and the *R*1 coil to *L*2. Also, the amber signal light, if furnished, is energized.

The time-delay relay *R*1 allows the high rate of flow through the furnace for several minutes to purge the boiler of any explosive mixture before contact *R*1 is made to energize the fuel-starting circuit to permit fuel supply. When contact *R*2 is made, it sustains itself through one of its own contacts and the lower contact of *R*3. The air flow can then be reduced for lighting off. Upon a boiler shutdown, the lower contact of *R*3 will open to deenergize the relay *R*2 and bring about the same conditions as mentioned above when the boiler is out of service.

The operation of a fan interlock, shown in Fig. 9-76, is as follows:

Illustrated here is an arrangement of interlock contactors including a purge interlock, induced-draft-fan interlock, and forced-draft-fan interlock for use with turbine-driven auxiliaries. The boiler with which this system is used is equipped to burn gas, fuel oil, or pulverized coal, and has single forced- and induced-draft fans. This is only one of many possible arrangements.

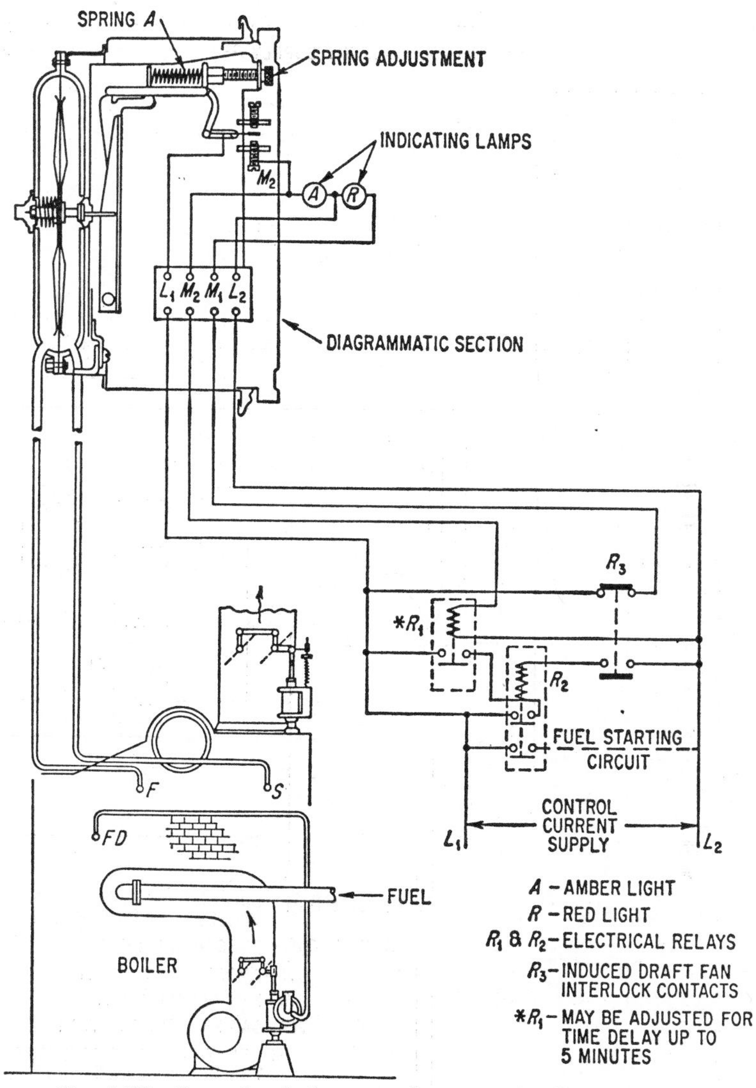

FIG. 9-75. Purge interlock wiring diagram. (*Bailey Meter Co.*)

The operation of the purge interlock is the same as outlined for Fig. 9-75.

Reduction of the differential measured by the induced-draft-fan interlock below a predetermined value (due to shutdown or fan failure) energizes relay *R*3 to interrupt the solenoid circuit of the gas- and oil-shutoff valves, the pulverizer-turbine-shutoff valve, and the forced-draft-fan-turbine-shutoff valve.

Reduction of the differential measured by the forced-draft-fan interlock

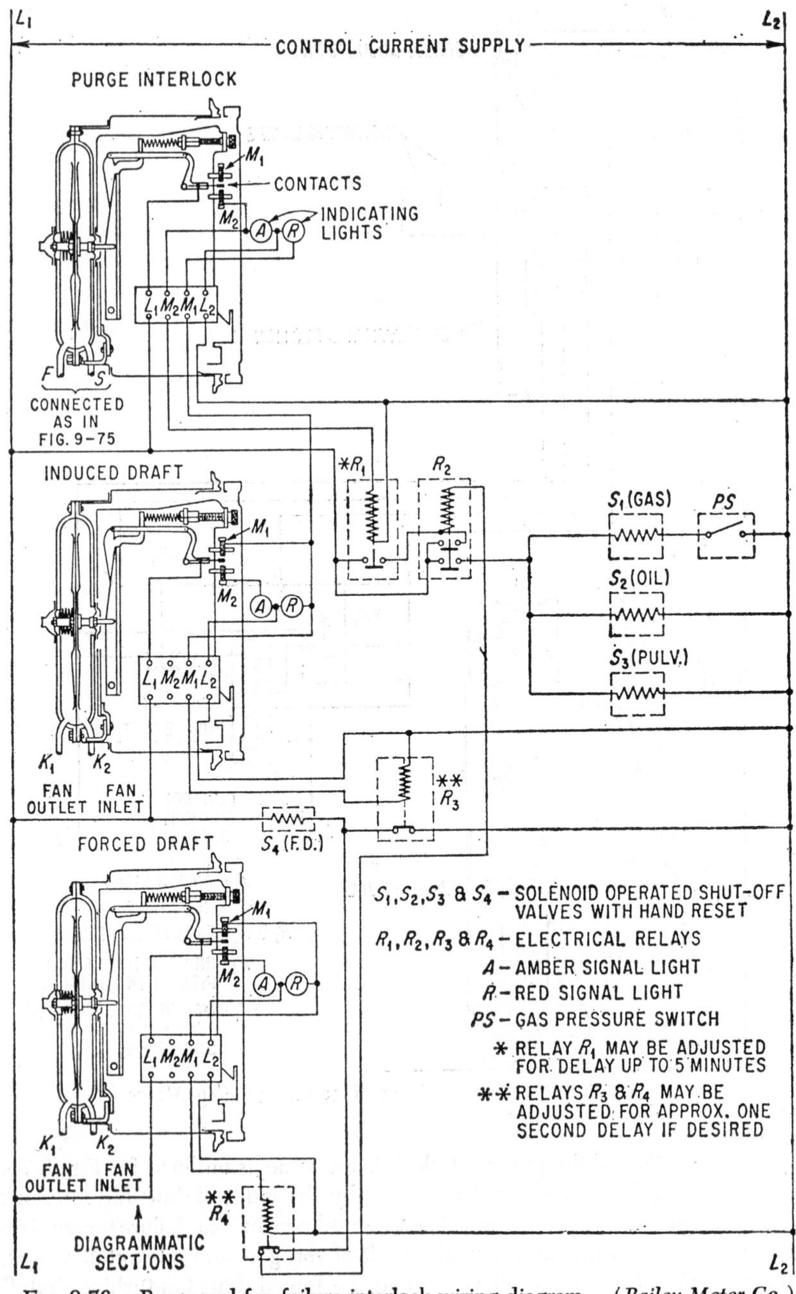

FIG. 9-76. Purge and fan-failure interlock wiring diagram. (*Bailey Meter Co.*)

below a predetermined value energizes relay *R4* to interrupt the fuel-shutoff equipment circuit.

The measuring element of a fan interlock should be connected to measure the differential across the fan rotor. The two pressure connections should be made to include the drop across the rotor and not across dampers which may be in the vicinity of the fan on the inlet or outlet side. This procedure ensures much faster and more positive operation of the contact device than the single-connection measuring forced-draft duct pressure, induced-draft duct pressure, or furnace draft. When two forced- or two induced-draft fans are used in parallel, a separate contactor is required for each fan.

Shutoff valves may be diaphragm-actuated valves with three-way solenoid-actuated pilots in the air-supply line to the diaphragms or valves which incorporate the solenoid trip elements in the operating mechanism. In any case, these should be arranged to require manual resetting; and the fuel-shutoff valves must be of single-seat design to ensure tight seating.

Requirements of Combustion-safeguard Systems.[18] For maximum safety a combustion-safeguarding system must perform the following basic functions repeatedly without error.

1. It must respond instantaneously to flame conditions so as to prevent an accumulation of unburned fuel unless a satisfactory and safe means of ignition is present.
2. When in operation, it must be able to distinguish between unsafe firing conditions and normal variations in combustion control. Nuisance shutdowns, aside from the loss of production time, have detrimental psychological effects on the operator, in that repeated unwarranted shutdowns are apt to cause him to devise ways and means of bypassing the combustion-safeguarding system.
3. It must be self-checking; i.e., the unit must automatically check the entire combustion-control system on each and every operation, and guard against all possible unsafe failures, which might otherwise permit the burner to start up and operate under unsafe conditions.
4. The sequence of operation of the combustion-safeguarding instruments and system must be such as to prevent the possibility of an unsafe start, should the operator fail to follow recommended starting procedure.
5. It must be versatile enough to provide a suitable operating sequence, as may be required for the burner and the auxiliaries of the control system.

Of these basic requirements, the first, calling for an instantaneous and positive means of detecting the absence or presence of a flame, presents the major problem in the design of a combustion-guarding system.

The principal types now used for industrial burners are the flame rod, the photoelectric cell, and the lead-sulfide cell. All operate practically instantaneously. Other types of safeguards, actuated by radiant heat or thermal

[18] Wheelco Instruments Div., Barber Colman Co., Inc.

or thermocouple action, require more time for operation. Usually, they are not considered responsive enough for commercial-size main burners.

Flame Conductivity. A flame can conduct an electric current, simply because, when fuels are burned, a disassociation of the components of the fuel takes place and leaves free, within the flame boundary, a concentration of ions, carrying electric charges. These ions are produced by splitting the neutral particles which make up the gas into positive and negative parts—the splitting-up process constantly taking place during the burning of the fuel. Since an electric current is a flow of charged particles through a substance, a flame which is made up of a concentration of charged particles may conduct an electric current. These positive and negative parts have a strong tendency to recombine, because of their mutual attraction. It is, therefore, necessary that a constant process of splitting be consummated in order to ensure the constant presence of these charged particles. This process of splitting takes place constantly during combustion, and, should combustion cease (loss of ignition or flame failure), the positive and negative particles instantly combine, because of their mutual attraction, and the conductive path disappears.

When flame temperatures are of such a value as to make it practical to insert a metallic electrode, and when the combustion chamber is relatively clear of soot, and the atmospheres are not detrimental to alloys, etc., as is generally true of gas-burning and some light-oil-burning equipment, it is possible to use the flame as a component part of an electronic circuit.

Flames are relatively high in resistance (ranging from 2 to 80 megohms). If industrial apparatus is to be controlled from such a flame, it is necessary, therefore, that a means of amplifying the minute current passing through the flame, as well as a means of distinguishing flame resistance and resistance inherent in other media, be provided. Such amplifiers may be designed to recognize either *the ohmic resistance* or *the rectified impedance* in the input circuit. (*Note:* A rectified impedance is one that passes a greater current in one direction than in the other, given the same magnitude of applied voltage.)

Flame Luminosity. Another well-defined and easily sensed characteristic of flames is luminosity. This is particularly important where it is not feasible to insert an electrode directly into high-temperature flames such as are generally encountered when burning oil. In such cases a photocell, which is mainly sensitive to the blue portion of the flame spectrum (blue predominates in a spectrum of oil flame) and extremely insensitive to extraneous light emitted from incandescent brickwork, etc., is used as a sensing element to introduce proper resistance values to the input circuit of the electronic circuit.

Typical Combustion Safeguard System Designs. *Rectified Impedance.* The system (Figs. 9-77 and 9-78) operates on the flame-rectification principle. As mentioned previously, this principle is based on the fact that either a flame or a photocell sighted at a flame is capable not only of conducting an electric current, but also of rectifying an alternating current. The system utilizes this principle by applying alternating current to either a flame electrode inserted in the flame or to a photocell sighted at the flame. The resultant rectified current, which can be produced only when flame is present, is in turn detected by the relay.

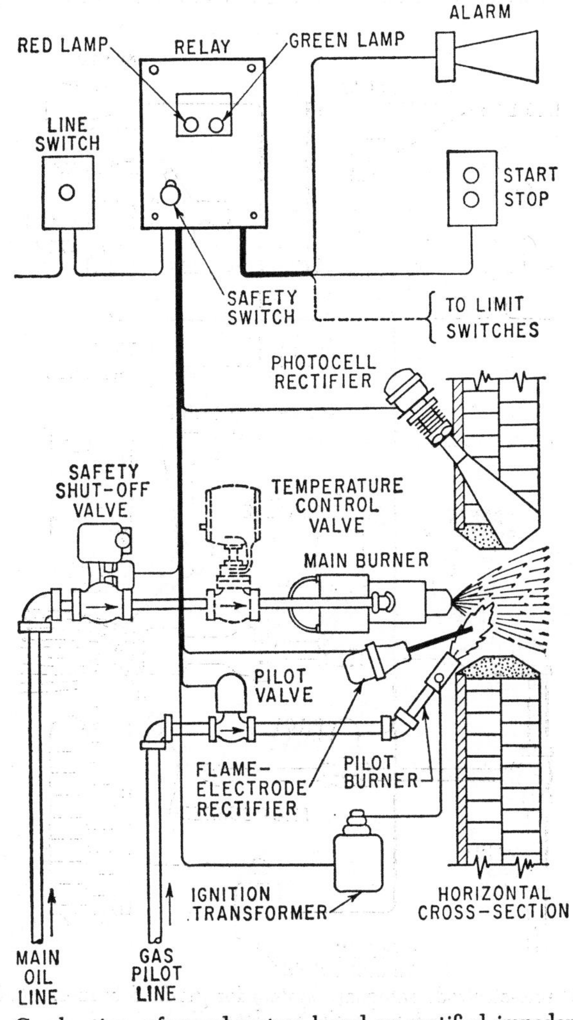

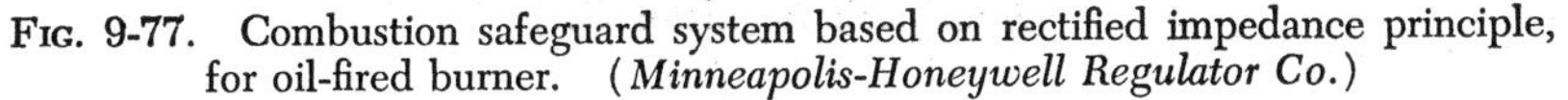

FIG. 9-77. Combustion safeguard system based on rectified impedance principle, for oil-fired burner. (*Minneapolis-Honeywell Regulator Co.*)

The actual flame-detecting units consist of flame-electrode and photocell rectifier assemblies, such as the types illustrated, and the protecting relay. The flame-electrode type is generally used for nonluminous flames, such as gas flames, whereas the photocell type is generally used for luminous flames, such as oil flames.

FLAME-ELECTRODE RECTIFIER: The electrode is inserted in the flame, and the flame itself forms a part of the electronic circuit. As the flame is a free-ion gas, because of the chemical process of combustion, it is a fairly good conductor. However, advantage is taken also of the fact that the flame will act as a rectifier if interposed between two electrodes of unequal areas;

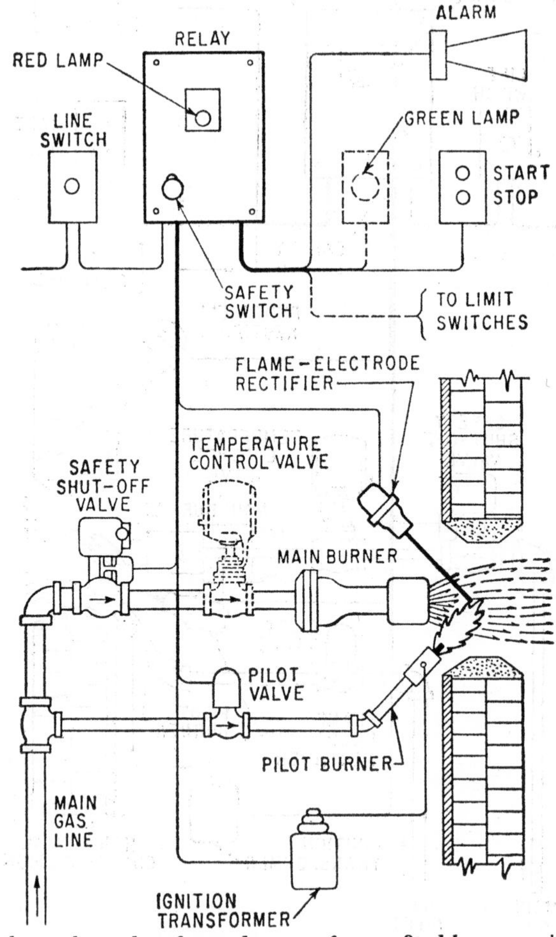

FIG. 9-78. Flame-electrode safeguard system for gas-fired burner. (*Minneapolis-Honeywell Regulator Co.*)

the burner itself, and a ground electrode, when used, will generally have an area of contact with the flame many times that of the flame electrode. Rectified current can be produced from the applied alternating current only when a flame is present.

PHOTOCELL RECTIFIER: The photocell is sighted at the flame. Advantage is taken of the fact that the photocell is a natural rectifier: When light from the flame strikes it, the photocell produces rectified current from the applied alternating current.

RELAY: The relay recognizes the presence of flame only if it receives a rectified current from either the flame-electrode or the photocell unit. When flame is present, this rectified current is fed into one grid of a double-triode electronic tube. The triodes are connected so that the output current will energize the control relays. On flame failure, no rectified current can be produced, the tube output current will drop to zero, and the control relays will be deenergized.

The system, shown in Fig. 9-78 for a gas-fired burner, provides (1) automatic cutoff on flame failure, (2) automatic safe-start component check, (3) electric ignition, pushbutton start, (4) gas pilot and gas main-flame supervision (constant pilot), (5) automatic purge timing.

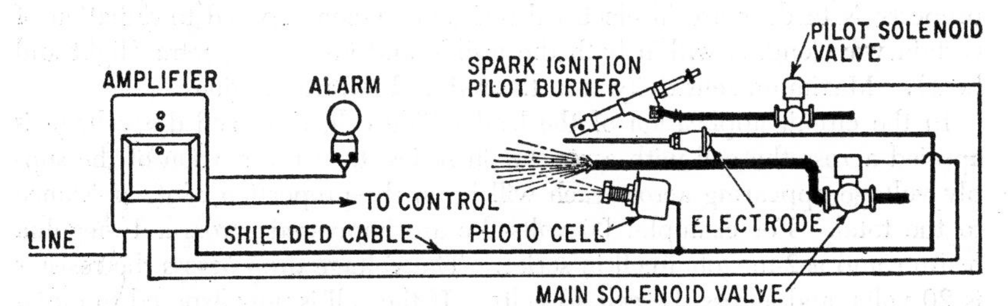

FIG. 9-79. Flame-electrode and photocell system based on ohmic resistance principle. (*Wheelco Instruments Div., Barber Colman Co.*)

A system similar to that described above, but providing gas-pilot-flame supervision on start and oil main-flame supervision on run, is shown in Fig. 9-77 for an oil-fired burner. With this system, the flame-electrode rectifier is used to supervise the gas pilot flame (constant pilot) during startup only, while the photocell rectifier is used to supervise the oil main flame. If the oil-fired burner is ignited by an intermittent pilot, the flame-electrode rectifier is not required.

Ohmic Resistance. The system (Fig. 9-79) uses a sensitive two-stage amplifier operating on alternating current, to sense only the ohmic-resistance flame characteristics and produce the required instantaneous switching action (signaled by the presence or absence of the flame). This circuit provides flexibility in choice of burner and pilot design or arrangement, and protection against nuisance shutdowns caused by varying combustion conditions.

On gas-burning equipment the system uses the flame itself as a component part of the input circuit. If a high-temperature rod is inserted directly in the flame, the flame is in contact with the flame electrode and the circuit is completed; thus the sequence control actuates to permit the flow of fuel through the valve. The resistance of the flame remains within definite limits and is distinguished from the amount inherent in other media, such as ceramics, metals, moisture, or the atmosphere.

On oil-fired equipment, equivalent resistance values are introduced into the input circuit by means of the photocell, whose resistance varies in accordance with the luminosity and spectral characteristic of the flame.

On combination fuel burners, such as oil with gas piloting, the flame electrode is used in addition to the photocell.

In the event of flame failure the input circuit is opened, and the sequence control acts to stop fuel flow to the burner, thereby preventing the accumulation of fuel and fumes in the combustion chamber—the source of explosions. In the case of excess moisture or grounding of the flame electrode, the resistance is below the defined limit of the flame, and the control will also act to prevent fuel flow.

The Lead-sulfide Cell.[19] The lead-sulfide cell is a semiconductor. The principal identifying compound of the cell is lead-sulfide (PbS), and its property is to decrease in electrical resistance when exposed to radiation of certain wavelengths within both the visible and invisible spectra (light and heat). Maximum sensitivity is in the red and infrared regions.

In the circuit application of the lead-sulfide cell, if a fixed d-c voltage is applied across the cell with a resistor in series, then the portion of the supply voltage appearing across each will be in the proportion of its resistance to the total. For example: 60 volts d-c are impressed across a 1-megohm resistor and a 2-megohm cell in series. The voltage drop across the resistor is 20 volts, and across the cell 40 volts. If the cell is now exposed to radiation, of a magnitude to cause its resistance to drop to 1 megohm, the voltage division is now 30 volts across the resistor and 30 volts across the cell. The change in voltage across the cell in this example is the difference between 40 and 30, or 10 volts, and a 10-volt signal is produced. This signal is fed to an electronic amplifier which in turn operates a relay, and thus produces an electrical switching action as a result of the cell "seeing" radiation.

It has been determined empirically that the radiation intensity of any portion of any flame is constantly fluctuating by a small percentage at frequencies ranging from one to several hundred times per second. The cell resistance thus may change by a large percentage at a very slow rate, owing to temperature change in furnace or firebox, and in addition may vary by a relatively small percentage at a relatively rapid rate, owing to the inherent fluctuating characteristics of the flame. If this composite signal is fed to an amplifier which accepts frequencies in the neighborhood of 10 cps, and which virtually rejects or filters out all other frequencies, the cell and its associated amplifier thus become responsive to flame and will discriminate against incandescent light. This, in basic principle, is how the lead-sulfide cell is used for detection of flames in combustion equipment.

The scanner's line of sight must be such that it passes through only the flame to which it is assigned (Fig. 9-80) in the zone where the flame should be located, and then hits into a target background such as a water wall,

[19] Combustion Control Div., Electronics Corp. of America.

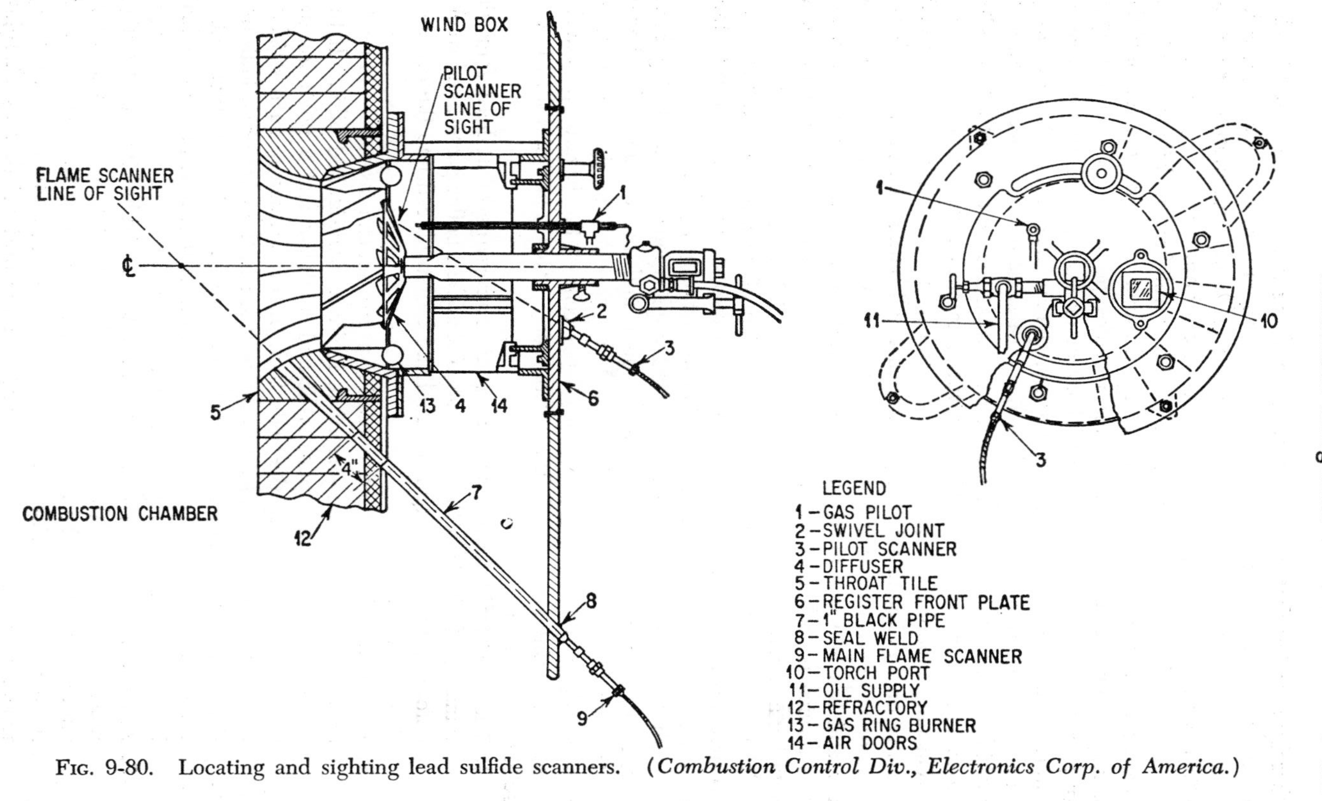

FIG. 9-80. Locating and sighting lead sulfide scanners. (*Combustion Control Div., Electronics Corp. of America.*)

thereby assuring that no adjacent or stray flame is in the line of sight. By sighting through pipes installed angularly through a wall or wind box, individual flame monitoring is achieved, and the flame is monitored for a specific desired zone of fire or distance off its burner tip.

On all makes of front-, roof-, side-, or rear-wall-fired multiburner boilers, these scanning pipes are run angularly through the wind box to provide divergent lines of sight. On circular register-type burners, the line of sight pierces the furnace wall in the vicinity of or through the throat tile—on intertube or cross-tube types of burners, the furnace end of the scanning pipes is terminated inside the wind box, just short of the furnace wall and looking through that secondary-air port offering the best line of sight. Cyclone burners are handled as single-burner combustion chambers.

On tangentially fired boilers, burners can be monitored on an individual basis or on an elevational basis, inasmuch as all burners on the same elevation, one in each corner of the same furnace, have a common source of supply and fire into one another. The scanner application for accomplishing this service is such that the water-wall tubes must be offset to allow for a sighting aperture.

The actuation resulting from the detection of an adverse condition can be used to sound an alarm, energize a light on the annunciator board, shut off fuel, or shut off other functions either in part or in whole.

With the above-described field-proved applications on monitoring both torch or semiautomatic spark-ignited pilot and main flame, the boiler instrumentation and controls industry has proceeded to develop a practical interlock system to circumvent 85 per cent of the hazards previously described during startup and firing of burners. This is accomplished in the following manner:

1. The electrical-interlock system requires prepurge of at least four air transfers of all air in the combustion chamber, boiler passes, and stack, before any pilot fuel or main fuel can be introduced into the combustion chamber.
2. At the end of purge time, operating personnel can insert a hand torch, or, through pushbutton operation, ignite a spark-ignited pilot. If the torch or spark-ignited pilot is lit and established at a point for the satisfactory light-off of the main fuel, then indication of such satisfactory condition will be given to the operator in the form of an indicator light going out. The operator can then energize the main-fuel-shutoff valve (oil or gas) or coal pulverizer and its feeder for the individual burners he is attempting to light off.
3. After this equipment is energized for the individual burner, ignition must take place within 10 sec or the interlock system will automatically shut off the main fuel and require purge time as described in step 1 above. This forestalls the operating personnel from making repeated attempts to light off or "beat the button," thereby charging the combustion chamber with unburnt fuel. This has proved the most serious cause of injurious and costly explosions.

4. After light-off of the main burner flame, the control system is so devised as to monitor only the main flame of the individual burner, since it is logical that we should look at only the main flame to see that it is maintained. In other words, we are no longer interested in the pilot, as a proved pilot is not assurance that the main flame is established or is maintained. If there is any question in anyone's mind of this fact, let him move the fuel-air ratio on the main burner off 10 per cent lean or rich with a good pilot established at the proper ignition point, and it will be readily seen that there will be no ignition of the main burner flame.

As for the practicality of this system, it does not limit the operating personnel from turning on or turning off whatever burner they choose to operate.

SMOKE-DENSITY RECORDERS

Excess smoke may be avoided, often with improvement in boiler efficiency, by the use of equipment which samples smoke automatically at the stack or duct and transmits a record to the firing aisle so that operators can watch smoke conditions closely.

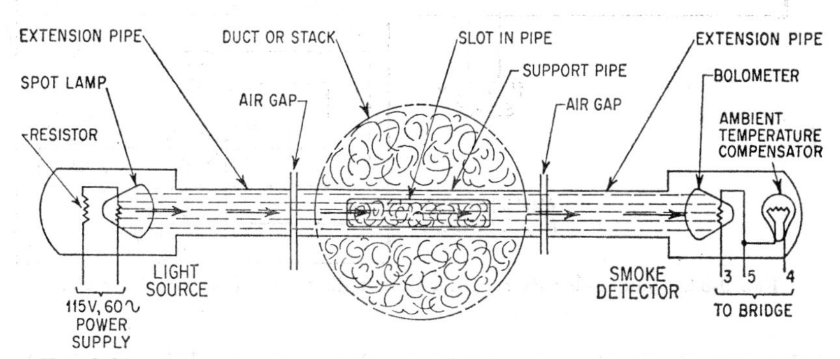

FIG. 9-81. Diagrammatic arrangement of Bolometer smoke-density detector. (*Bailey Meter Co.*)

Bolometer Type. The smoke-density recorder shown in Figs. 9-81 and 9-82 is an instrument utilizing a Bolometer for light detection. The specially designed Bolometer is essentially a total radiation device having an internal resistance which varies with the intensity of radiation. A light-beam source is directed through a variable field of light absorption and onto the Bolometer. The Bolometer measures, by means of the measuring circuit, the amount or degree of light absorption present in the field at any given time. The field, in this case, is the atmosphere within a duct or stack, and the degree of light absorption (or light cut off from the Bolometer) varies with the density of smoke present in the duct or stack.

The instrument is so calibrated that the recording pen is at minimum or zero chart when the maximum amount of light falls on the Bolometer detector. At this condition, no smoke is present, and the stack or duct is known as "clear." When the entire light beam has been cut off from the Bolometer, the smoke present being at maximum or 100 per cent density, the recording pen is at maximum or 100 per cent chart.

The smoke-density recorder circuit is an a-c potentiometer circuit which operates on the null-balance principle, in which a voltage is developed by an unbalanced Wheatstone bridge and is measured by balancing against a known adjustable voltage. In the a-c potentiometer circuit as applied to smoke-density measurement, a voltage is developed at the output of the

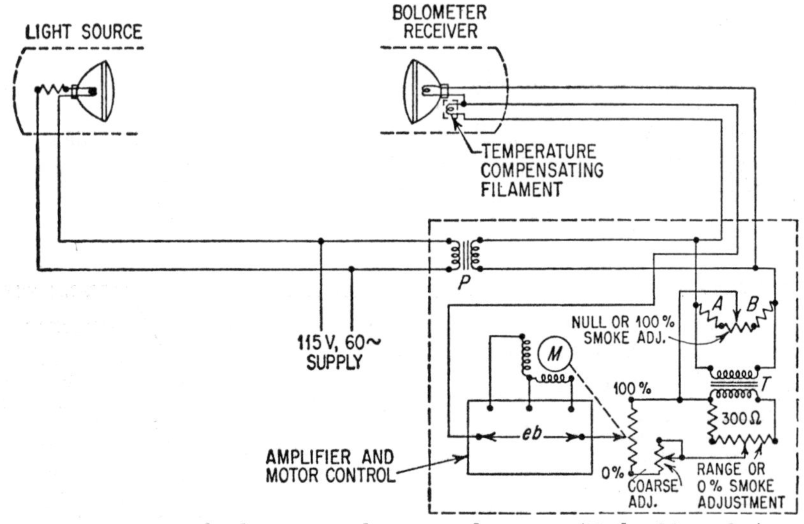

FIG. 9-82. Smoke-density-recorder circuit diagram. (*Bailey Meter Co.*)

Wheatstone bridge because of bridge unbalance, the unbalance being inversely proportional to the density of smoke present in the duct or stack. A change in smoke density produces an unbalanced voltage from the Wheatstone-bridge circuit. The magnitude of the unbalance is inversely proportional to the density of the smoke. This unbalance is applied to the electronic amplifier which amplifies the voltage and operates the slidewire motor to balance the potentiometer circuit.

The amplifier circuit consists of an electronic amplifier and, if necessary, a step-up transformer.

The motor-control circuit consists of a double triode tube, a supply transformer, resistors *A* and *B*, and the slidewire motor.

The instrument is made up of three components: (1) the recorder, (2) the Bolometer smoke detector, and (3) the light source.

The Bolometer element, in a Wheatstone-bridge circuit, is sensitive to light intensity. When rays of light (of any wavelength) are directed onto the element, the heat of such light increases the temperature of the element. The increase in temperature varies with the amount of light on the element. An increase in element temperature causes an increase in element resistance, this increase also being in proportion to the amount of light. As shown in Fig. 9-81, a beam of light from the light source is directed across a duct or stack and onto the detector. The beam is focused by the parabolic reflector of the Bolometer onto the Bolometer element, so that the element receives the greatest possible concentration of light (and, thereby, heat).

When there is no interference between the source and the detector, the maximum light is detected. When there is smoke in the duct or stack, light is cut off from the detector, the amount cut off varying with the density of the smoke.

At the condition of no light detected (maximum or 100 per cent smoke density), the Bolometer element is of such resistance that the Wheatstone bridge is at balance, and there is no voltage output. The recorder pen reads at 100 per cent chart, depicting no light or 100 per cent smoke density. At the condition of maximum light detection (no smoke in the duct or stack), the Bolometer element is at maximum temperature and maximum resistance. This resistance causes a maximum bridge-unbalance voltage output. The recorder reads at zero chart, detecting 100 per cent light or zero smoke. For intermediate degrees of smoke density, proportional degrees of light detected, element temperature, element resistance, bridge-unbalance output voltage, and recorder-pen reading prevail.

A compensating element is included in the smoke-detector assembly so that both legs of the bridge are maintained at the same ambient temperature. Thus, ambient temperature changes have no effect on the bridge, all bridge unbalance being due to light detection.

Thermopile Type. To obtain a typical sample, this design draws two streams of stack gas through separate intakes and merges them in its cylindrical measuring chamber, from which the sample flows continuously back to the stack (Fig. 9-83). Radiation from a voltage-controlled lamp, mounted at one end of the chamber, passes constantly to a thermopile (series of thermocouples) mounted at the other end, and falls on it in inverse proportion to the density of the smoke through which it passes.

The electromotive force generated by the thermopile is measured continuously by a potentiometer, a standard indicating and recording pyrometer, calibrated to 0 to 100 per cent smoke density.

The recorder can be equipped to operate signals, usually a set of lamps, each of a different color to represent progressive densities—0 to 20, 20 to 40, 40 to 60, 60 to 80, 80 to 100 per cent. As long as smoke density is below 20 per cent, one lamp burns; between 20 and 40 per cent, two lamps burn; until, at densities above 80 per cent, all five are lighted.

By drawing smoke from two points, instead of from just one, the effect of stratification is minimized. All samplers are of one size, and are supplied with intake tubes of specified length, so as to "reach far enough" into any stack to get a sample representative for operating purposes.

To keep the lamp and the thermopile lenses clean, the two entering streams of stack gas flow toward a central return tube, and away from the

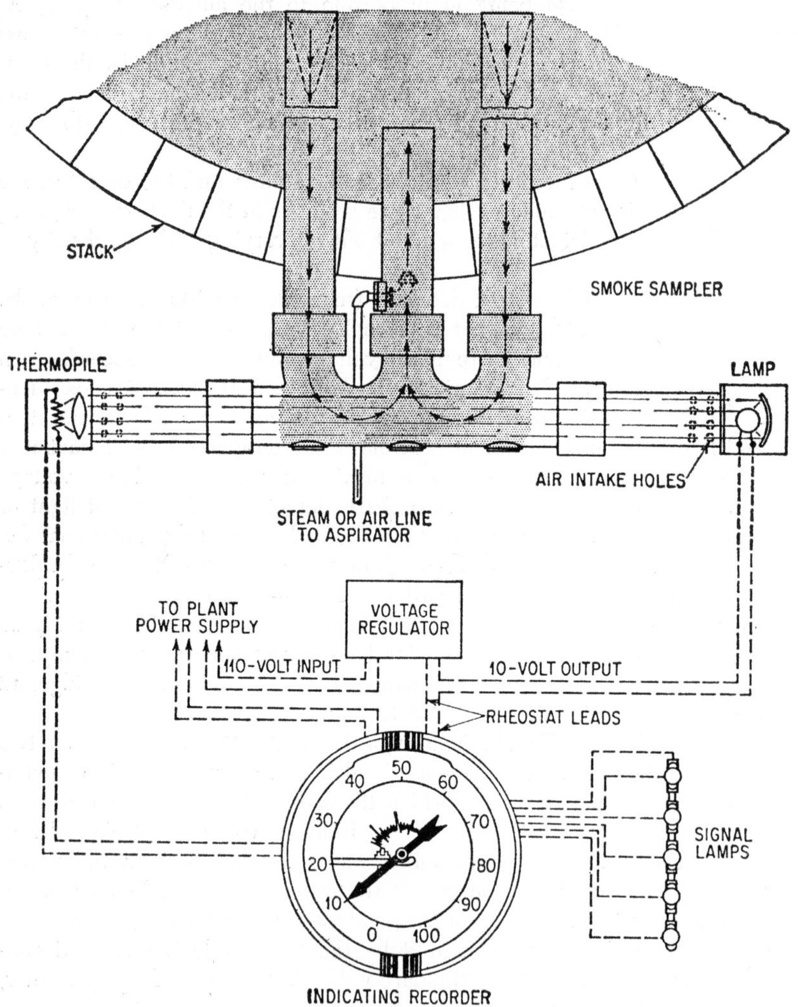

Fig. 9-83. Smoke-density-recorder utilizing thermopile detector. (*Leeds & Northrup Co.*)

ends of the chamber. An additional cleaning action is provided by tiny streams of air, which enter through small holes just in front of the lamp and thermopile lenses.

The potentiometer provides an electrical balance in which emf from the thermopile is balanced against an adjustable standard emf, and is measured

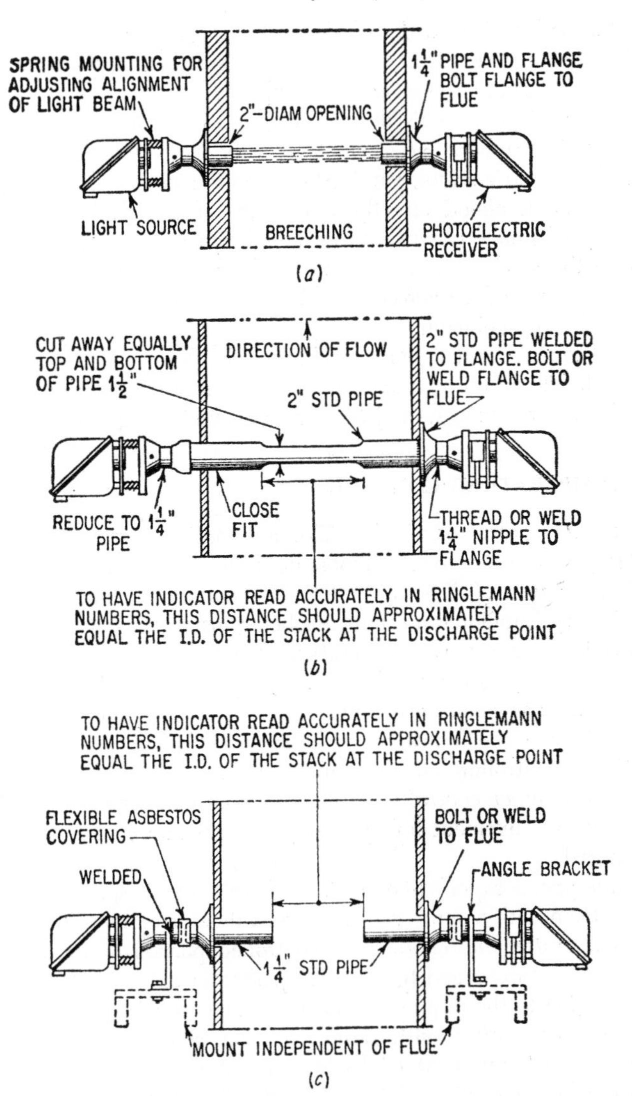

FIG. 9-84. Smoke-density-recorder utilizing photoelectric detector. Three typical mountings are shown: (*a*) brick breeching, (*b*) sheet-metal breeching, (*c*) extremely hot breeching. (*Combustion Control Div., Electronics Corp. of America.*)

in terms of this working standard. At the time of measurement, the forces are in balance. No current flows. Thus, measurement of thermopile emf is independent of other factors, such as variations in resistance caused by temperature changes along the wires connecting sampler to recorder.

Because smoke density is not measured by galvanometer deflection, but by the standard emf needed to balance thermopile emf, the galvanometer is uncalibrated. It stands at zero when the system is in balance. Deflection shows when standard emf must be adjusted.

Photoelectric Type. This design (Fig. 9-84) consists of a light source, a photoelectric scanner, and a smoke-density indicator. The light source and the photoelectric scanner are installed on the stack or breeching, so that the light beam is aimed directly at the lens of the photoelectric scanner. The density of smoke obscuring the light beam is measured and indicated directly on the smoke-density indicator in both per cent smoke and Ringelmann number, where 20 per cent smoke equals 1 Ringelmann; 100 per cent smoke equals 5 Ringelmann.

Red and green alarm lights tell when smoke density is above or below a preset level. The indicator can also be used to actuate recorders, audible alarms, damper regulators, air jets, or smoke-eliminating devices.

STEAM-TEMPERATURE CONTROL[20]

Temperature-control Methods. The automatic control of boiler outlet steam temperature was not common until the middle '30s. At that time steam temperatures of 875 and 900°F were considered tops, whereas today 1000°F is very common, and units are being installed for 1050 and 1100°F. These maximum operating temperatures are limited by metallurgy, and it has been necessary to control the steam temperatures within close limits for maximum safety as well as economy.

In recent years the following six basic methods have been employed for automatically controlling steam temperatures leaving the boiler.

1. *Bypass damper control* with single bypass damper or series and shunt damper arrangement for bypassing flue gas around the superheater as required
2. *Spray-type desuperheater control* where water is sprayed directly into the steam with a spray-water-control valve for temperature regulation
3. *Attemperator control* where a controlled portion of the steam passes through a submerged tubular desuperheater and a control valve in the steam line to the desuperheater or attemperator is employed
4. *Condenser control* with desuperheating condenser-tube bundles located in the superheater inlet header and water-control valve or valves to regulate a portion of the feedwater flow through the condenser as required
5. *Tilting-burner control* where the tilt angle of the burners is adjusted to change the furnace heat absorption and resultant steam temperature
6. *Flue-gas-recirculation control* where a portion of the flue gas is recirculated into the furnace by means of an auxiliary fan with damper control to change the mass flow through the superheater and the heat absorption in the furnace, as required to maintain steam temperature

[20] Abstracted from E. D. Scutt, Leeds & Northrup Co., "Modern Trends in Superheat and Reheat Controls."

On many boilers it is common practice to employ two of these methods for controlling superheat temperature, with sequence control transferring to the second method when the first has reached the limit of its control range.

The reheat cycle with its added economies has gained favor over the past five to ten years in central-station power plants, and it is now necessary to control two steam temperatures on these boilers. Normally, at least two of the basic methods of control are employed, and on many recent installations three of the methods are incorporated in the boiler design for controlling superheat and reheat temperatures.

Though the combustion-control equipment in the past has been considered the major automatic control on a boiler, the superheat- and reheat-temperature controls combined on some installations practically equal the combustion-control investment. The control problems encountered in regulating these steam temperatures are in many instances more difficult than those handled by the combustion control. This is primarily due to the time lags involved, and the fact that there are many variables within the boiler which can change steam temperature.

Table 9-4. Steam-generator Outlet-temperature Response Tests

	Control element	Average response time, minutes	
		First response	Full response
1. Boiler A	Spray valve	0.9	5.4
2. Boiler B	Spray valve	0.9	4.5
3. Boiler C	Spray valve	0.8	4.0
4. Boiler D	Shunt damper	2.1	8.0
5. Boiler D	Tilting burners	1.5	5.7
6. Boiler E	Tilting burners	1.4	6.0
7. Boiler E	Bypass damper	1.5	8.7
8. Boiler F	Tilting burners	1.3	6.2
9. Boiler F	Bypass damper	1.5	4.5
10. Boiler G	Spray valve	1.7	4.7
11. Boiler H	Spray valve	2.5	5.0–5.5
12. Boiler I	Condenser valve	4.1	7.5
13. Boiler J	Condenser valve	3.4	8.0
14. Boiler K	Condenser valve	2.3	7.0

It should also be noted in this respect that a change of 10° at 950°F corresponds to roughly 6 Btu/lb of steam, which is approximately one-half of one per cent of the total heat being absorbed by the water as it is being converted into steam within the boiler. While the combustion control is regulating the total Btu input to the boiler, some of which may be stored, the steam-temperature control is required to actually regulate the Btu input per pound of steam within very close tolerances.

Table 9-4 shows the average time lags which have been obtained on a number of different installations.

In order to provide good regulation, it is necessary with these large time lags to

1. Use a temperature detector with a minimum time lag.
2. Use controllers which provide good proportioning and reset action with the wide proportional-band settings required, and ones which can introduce rate action, if necessary.
3. Consider the use of anticipating devices.
4. On reheat installations apply the automatic control in such a manner as to minimize interaction between the superheat and reheat controls.

Figure 9-85 shows the construction of a high-speed thermocouple and well assembly for superheat and reheat control. This unit provides metal-to-metal contact between the thermocouple and well by means of the silver plug construction shown. Such an arrangement reduces the detection lag of the primary element to a minimum, since the mass of the metal which must change temperature is small, and the heat conduction between the well and the thermocouple is through metal.

An actual response curve for this thermocouple assembly, including well, is also shown on Fig. 9-85. It was obtained by transferring the assembly from an ice bath to a stirred-water bath at 136°F, thus giving a temperature change of 104°. The initial response of the thermocouple occurs in 3½ sec, with 50 per cent of the change in 15 sec and 90 per cent in 48 sec.

Though anticipating devices can be employed to assist the basic temperature-control action, these can only be calibrated approximately, owing to the operating variables within the boiler, which are continuously changing to affect the calibration. Because of this condition and the long time lags involved, it is important that the basic temperature-control equipment incorporate the following control actions as accurately as possible.

1. *Proportional action,* which positions the controlled device in proportion to the temperature deviation
2. *Reset action,* which moves the controlled device at a rate proportional to the temperature deviation from the control point
3. *Rate action,* which positions the controlled device in proportion to the rate of change in temperature

In order to obtain stable steam-temperature regulation, it has been necessary to employ extremely wide proportional-band settings because of the time lags involved. On most superheat-control installations, the proportional-band setting is over 100°F, and may be several hundred degrees. The control, however, is expected to hold temperature within ±5 to 10°F. Thus the basic temperature control must provide good proportional-plus-reset action, as well as rate action where it can be employed. It is not possible to employ rate action on certain installations, since it normally makes the control very active and may upset drum level and/or combustion conditions. On some installations, it is necessary to make some sacrifice in closeness of steam-temperature regulation, in order to obtain more stable boiler operation.

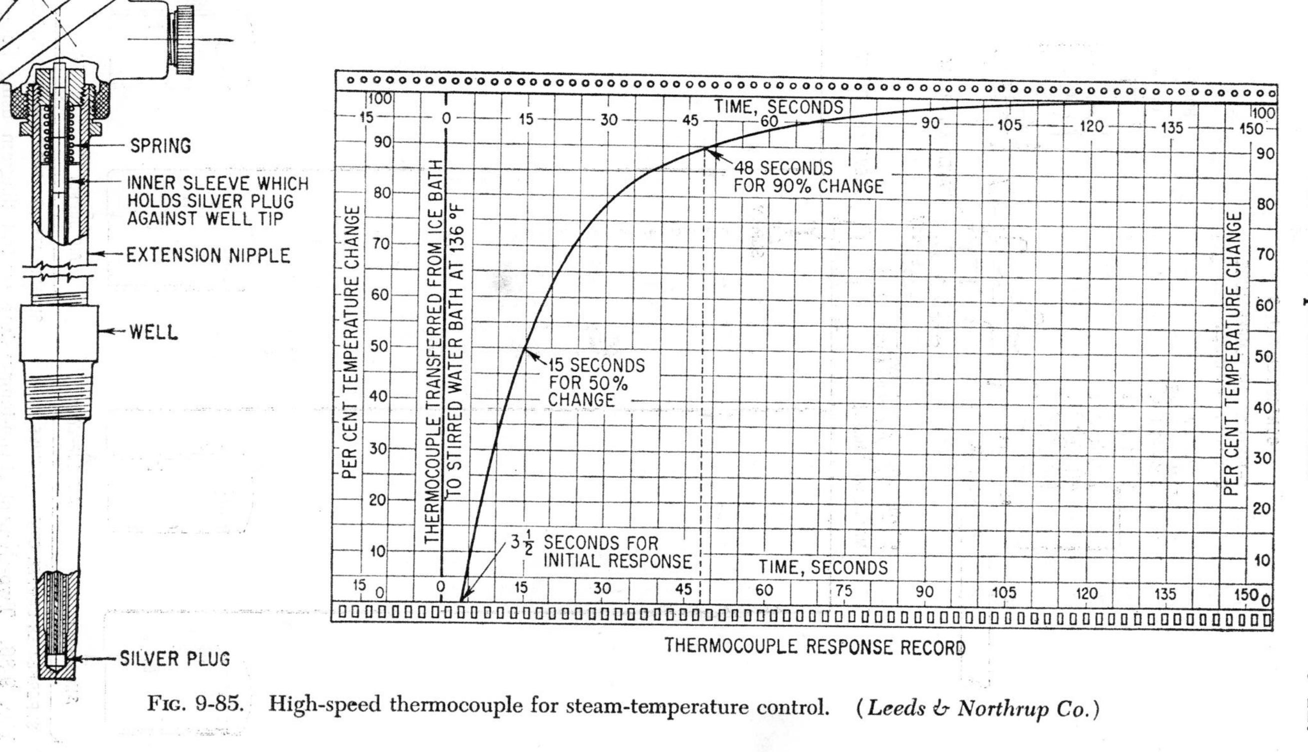

Fig. 9-85. High-speed thermocouple for steam-temperature control. (*Leeds & Northrup Co.*)

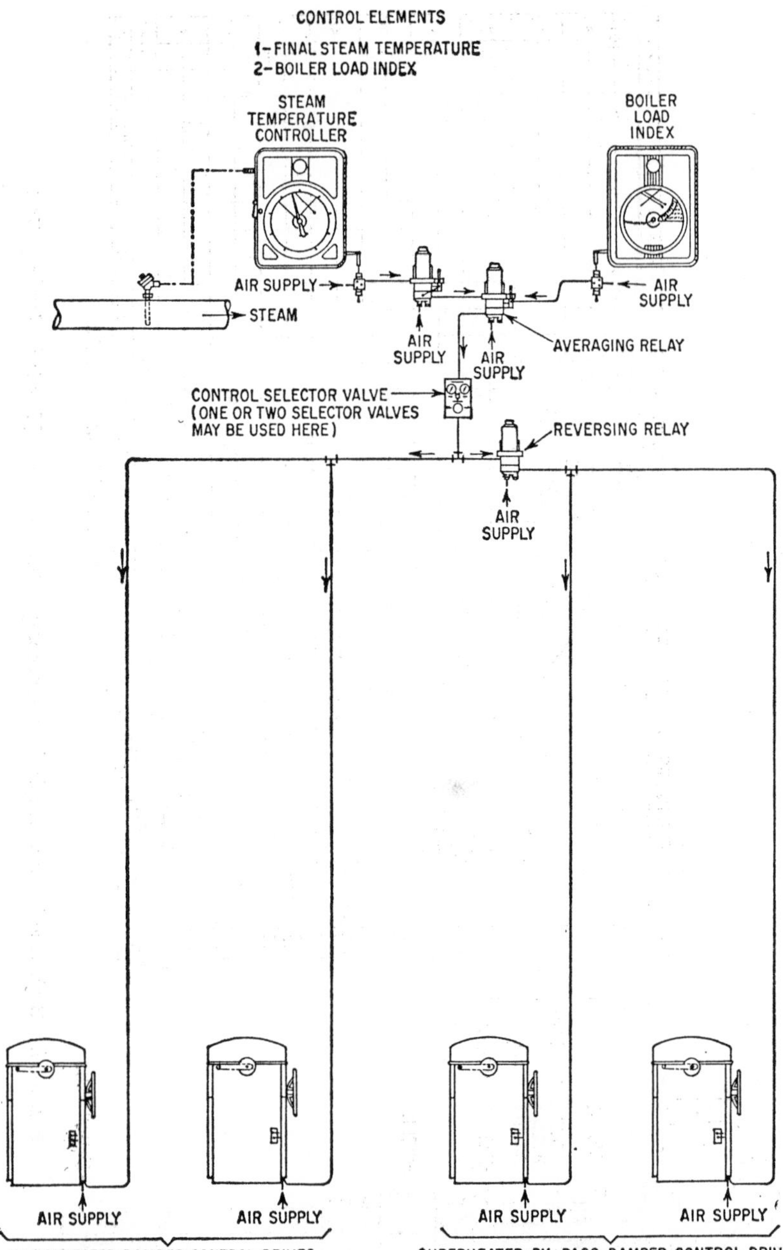

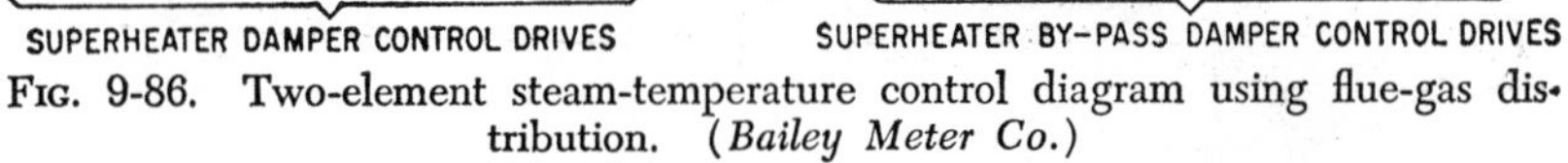

Fig. 9-86. Two-element steam-temperature control diagram using flue-gas distribution. (*Bailey Meter Co.*)

As mentioned earlier, it is common practice on many installations to have at least two means of controlling steam temperature (Fig. 9-86). This requires sequence control of two actuators from one controller, and the lag characteristics may be entirely different for these two methods of control. Suitable adjustment must, therefore, be provided in the control equipment, to accommodate these different control characteristics as the controller is transferred from one actuator to the other. On some installations three sets of adjustments have been provided to accommodate a single controller to three different operating conditions.

Anticipating Devices for Steam-temperature Controls. There are three basic requirements for an anticipating device.

1. It should always move the control in the proper direction over the control range.
2. It should have a minimum time lag.
3. It should be dependable, require low maintenance, and not be subject to failure.

Boiler air flow is often used for this anticipating action. Except for boilers equipped with large radiant superheaters, superheat- and reheat-steam temperatures normally increase with increased combustion rate. With correct combustion, boiler air flow is a direct indication of the combustion rate and can, therefore, be utilized to provide a response which assists the control in the proper direction. Air flow can also be measured with a minimum of time delay.

On boilers equipped with spray-type desuperheaters for regulating steam temperature, the temperature of the steam leaving the desuperheater changes immediately with changes in spray-water flow. These spray desuperheaters are normally located at the intermediate header, and cascade control (Fig. 9-87) has been employed where the steam-temperature lags have necessitated it. With this arrangement, controller *A* has its control point set by controller *B*, and full proportioning, reset, and rate actions are available in each of the two controls. Some control systems which employ desuperheater outlet steam temperature as an auxiliary response in the control only introduce proportioning action from this measurement, rather than all three basic control actions. Though not shown in the sketch, air-flow anticipation is also introduced in the cascade combination.

With this control scheme, any upset in steam temperature at the intermediate superheater header, due to changing furnace conditions, soot blowing, etc., is compensated for by controller *A*, which can employ a comparatively narrow proportional band because of the small time lag involved. Controller *B* is required only to compensate for the effect of these changes on the finishing stage of the superheater.

On reheat units, interaction between the two steam-temperature controls must also be given careful consideration. For example, changing burner tilts will change both the reheat and superheat temperatures, but spraying water in the desuperheater entering the finishing stage of the superheater affects superheat temperature only. Where sprays are used for both

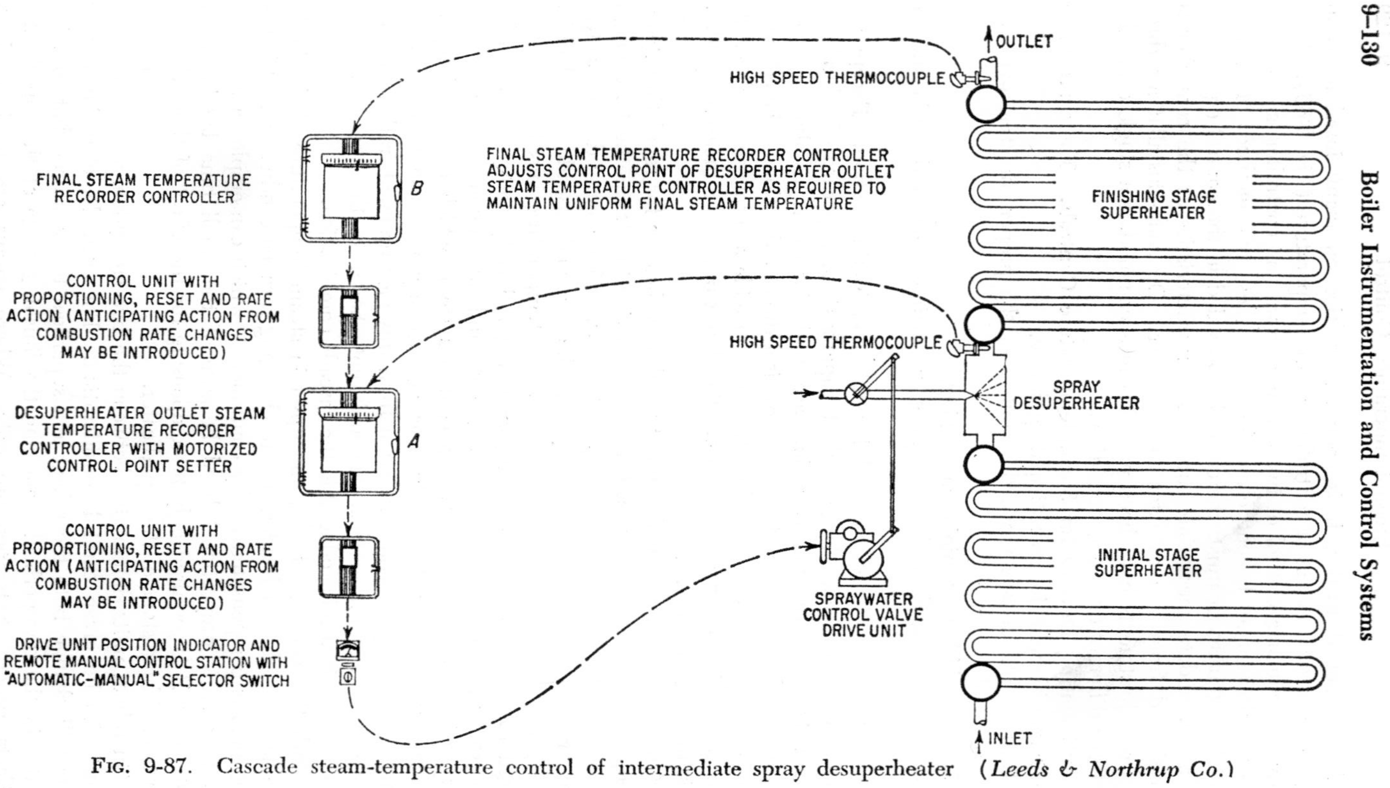

FIG. 9-87. Cascade steam-temperature control of intermediate spray desuperheater (*Leeds & Northrup Co.*)

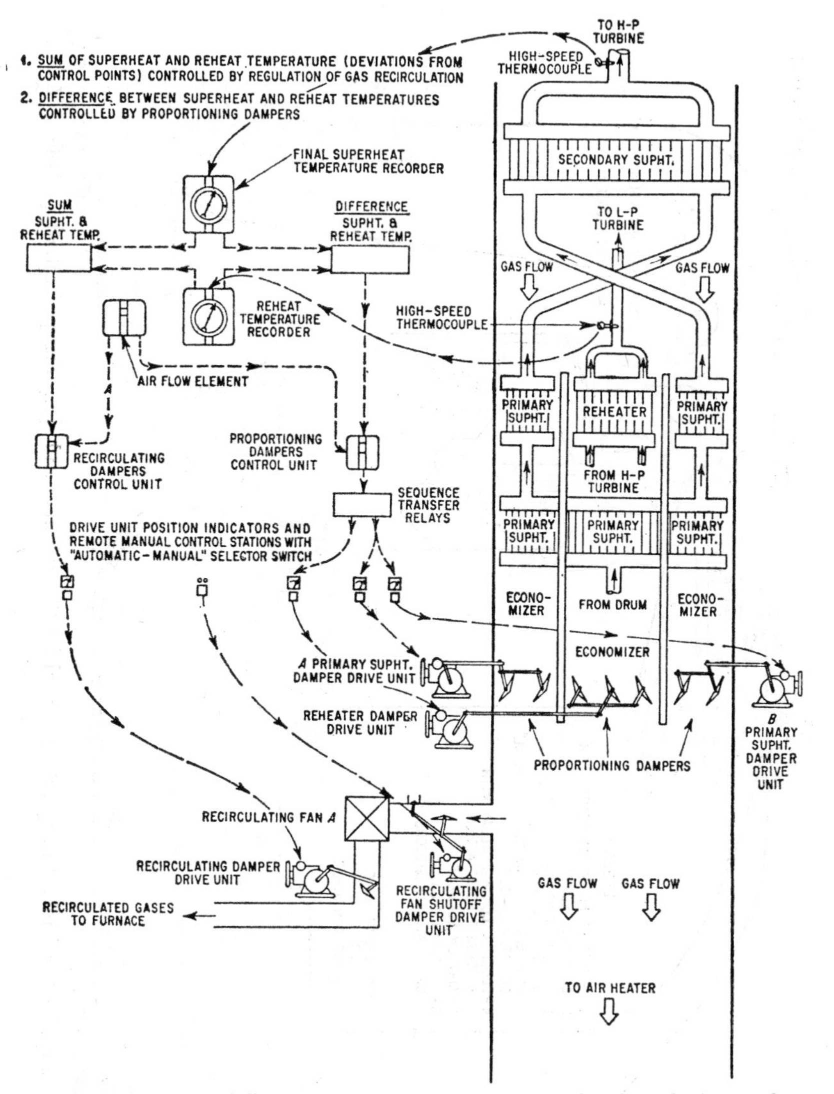

FIG. 9-88. Sum-and-difference steam-temperature control. (*Leeds & Northrup Co.*)

superheat- and reheat-temperature regulation, there is no appreciable interaction between the two controls.

On some recent reheat-boiler designs, the means available for regulating each of the two steam temperatures affects the other temperature. For example, the amount of hot gas recirculated will affect both superheat and reheat temperatures, while any movement of the proportioning dampers

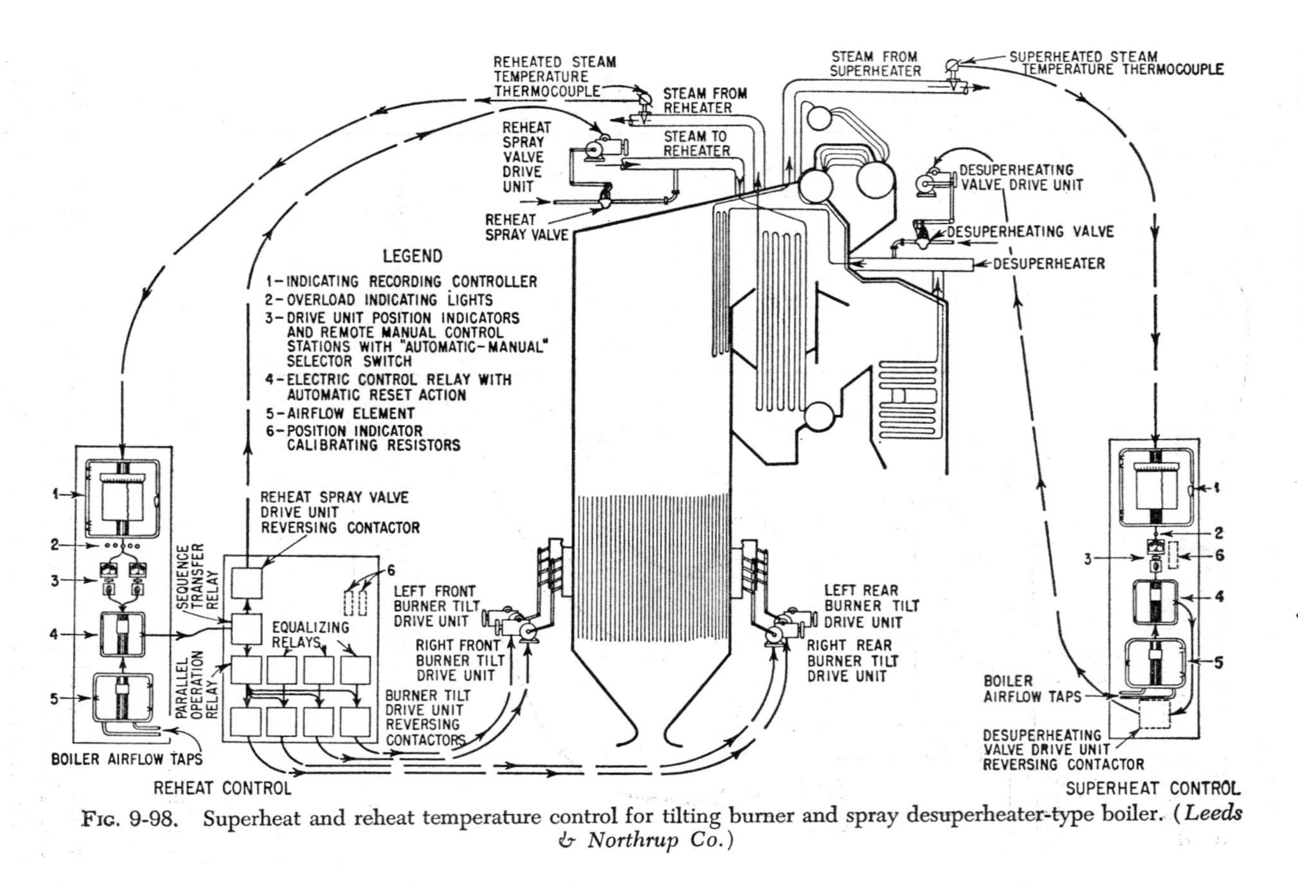

Fig. 9-98. Superheat and reheat temperature control for tilting burner and spray desuperheater-type boiler. (*Leeds & Northrup Co.*)

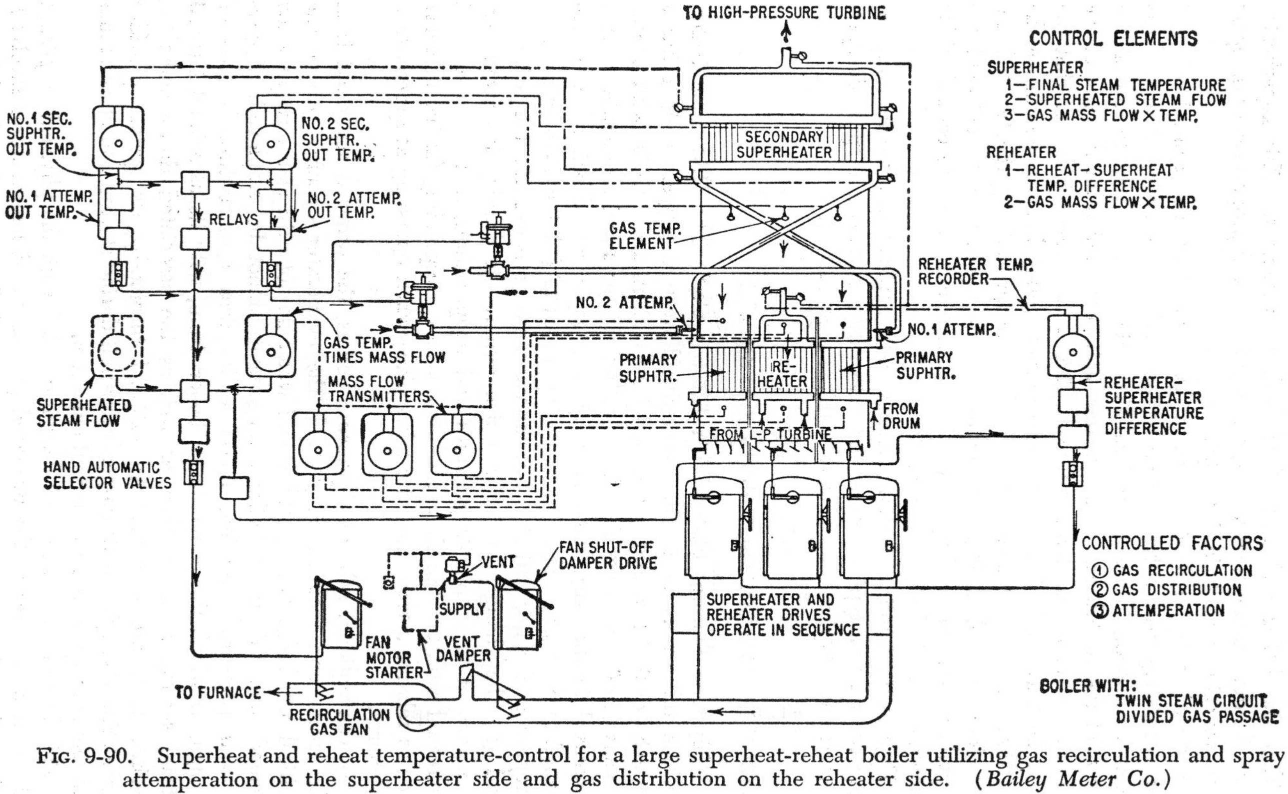

FIG. 9-90. Superheat and reheat temperature-control for a large superheat-reheat boiler utilizing gas recirculation and spray attemperation on the superheater side and gas distribution on the reheater side. (*Bailey Meter Co.*)

within the boiler will also affect both temperatures. This superheater and reheater arrangement with gas recirculation and proportioning dampers is shown schematically in Fig. 9-88.

For such a boiler arrangement, a "sum and difference control scheme" for regulating the hot-gas recirculation and proportioning dampers has been developed. Since gas recirculation determines the total heat available for both the superheater and reheater, it is controlled from the *sum* of the superheat and reheat temperatures. This measurement indicates that the total heat available is correct. The proportioning dampers determine the distribution of the available heat to the reheater and superheater and are, therefore, controlled from the equivalent of the *difference* between the superheat and reheat temperatures. This sum and difference control is also shown schematically on Fig. 9-88.

Figures 9-89 and 9-90 show the schematic application of steam-temperature control to a typical reheat unit with tilting burners and spray desuperheaters and a typical reheat boiler with gas recirculation.

LARGE BOILER-CONTROL INSTALLATIONS

Eddystone Station, Philadelphia Electric Co.[21] The Eddystone boiler-turbine units operate in the supercritical-pressure range. Unit 1 has a capacity of 350 Mw and is designed for 5,000 psi and 1200°F with double reheats of 1050 to 1050°F. As operating pressures and temperatures have increased in the past, the relative storage capacities of boilers have continued to decrease. This has been due in part to (1) the inherent physical properties of water and steam at the increased pressures and temperatures, and (2) the changes in boiler design required to provide improved performance at these higher pressures and temperatures, and the associated increased ratings. For these same reasons a corresponding further decrease in this relative storage capacity takes place when operating at supercritical pressures.

In conventional subcritical-pressure boilers, the water is separated from the steam and recirculated through the evaporator circuits. Above the critical pressure, however, steam and water do not exist as a mixture when heated, and this basic boiler arrangement cannot be employed. Supercritical pressure operation, therefore, requires a once-through type of design, in which the conventional boiler drum, with its associated internal steam separator and washer, as well as the associated downcomers for recirculation, are omitted. An external turbine-bypass system provides circulation through the steam-generator unit during startup and at minimum turbine outputs or under abnormal operation.

This basic change in boiler design, coupled with the physical properties of steam and water at the increased pressures and temperatures involved, affects the relative storage capacity of the Eddystone units. In this con-

[21] E. D. Soutt, Leeds & Northrup Co.

nection, calculations indicate a "residence time" of approximately 2 min for unit 1. This is the time required for water to pass through the unit and be converted to superheated steam when operating at full load. This compares with an average figure of 6 to 7 min for unit 2 at Cromby Station, which is a controlled-circulation twin-furnace boiler operating at 1875 psi and 1000°F. Residence time is only one of a number of factors which affect the relative storage capacity of a boiler. The over-all operation of the combustion-control system, however, is directly concerned with this relative storage and the response characteristics of the boiler and its auxiliaries. In the case of Eddystone, the control must be designed to accommodate the anticipated reduction in storage capacity.

The turbine-bypass system, as required with the once-through boiler system, also imposes unusual requirements on the combustion-control system. With this bypass arrangement it is desirable to coordinate closely the operation of boiler and turbine in order that steam will not be bypassed during normal operation. The combustion control, therefore, integrates the boiler and turbine operation as a unit, and, under normal conditions, keeps the generator output within the capabilities of the auxiliary equipment in service.

In physical appearance, the Eddystone units resemble the conventional twin-furnace boilers installed in a number of stations throughout the United States where unit capacities of 200 Mw and above are involved. It has been normal practice with these units to locate the final superheater in one of the furnaces and the reheater in the other. Because of the double reheat at Eddystone, each furnace is provided with its respective reheater, and burner-tilt control will regulate these reheat temperatures. The initial or primary steam temperature at each furnace is controlled by spray-type desuperheaters.

The twin-furnace arrangement, when compared with single-furnace operation, imposes several new requirements on the combustion-control system. Foremost among these is the need for automatically proportioning the fuel and air distribution between furnaces to provide (1) the correct heat input per furnace, and (2) the optimum combustion conditions in each furnace. Oxygen analysis for automatic control is used in the basic design of the combustion-control system for Eddystone.

It has been general practice with pulverized-coal-fired boilers to employ steam flow as a measurement in the combustion-control system. As capacities have increased, twin steam leads have been employed in many recent installations, and this has required totalizing equipment. In this connection, at Eddystone there are eight steam leads leaving the boiler, and these are reduced to four inlets entering the turbine. This design, though required by the increased capacity and operating conditions, makes it impractical to measure total steam flow from the unit.

This steam-flow measurement in present control systems has normally been employed as an indirect measure of the heat input to the boiler unit, and air flow has been proportioned to it. However, electrical-generator output provides a similar indirect measurement of boiler input and can be readily measured. Totalizing, when required, can also be more easily accomplished, since no square root is involved in the basic measurement. Generator output is therefore an important part of the combustion-control system, and the conventional steam-flow measurement has been eliminated.

The unusual plant and equipment conditions which affect the basic design of the combustion-control system for Eddystone can be summarized as follows:

1. Reduced storage capacities and related response characteristics associated with supercritical-pressure operation
2. Turbine-bypass system required with once-through boiler operation, which, in combination with the associated reduced storage capacities involved, makes coordinated operation of the boiler and turbine as a unit highly desirable
3. Twin-furnace design with associated needs for automatic control from oxygen analysis
4. Multiple steam leads with related use of generator output in the combustion-control system, since total steam flow cannot be measured practically

In principle, the combustion-control system regulates the turbine governor and the input fuel and air supplies to the boiler furnaces to give the desired electrical-generator output, while maintaining a uniform steam pressure and optimum combustion efficiency. The feedwater-input and primary desuperheater-water supplies are adjusted as required to give the desired temperatures throughout the primary steam circuits, to match the firing conditions established by the combustion control. The reheat temperatures in turn are maintained by adjusting the burner tilts in the respective furnaces.

As shown in Fig. 9-91, it has been normal practice in the past to regulate the turbine governor to give the desired electrical-generator output, and to maintain steam pressure by adjusting the input fuel and air supplies to the boiler. With this method of operation, the boiler in effect follows the turbine demand. This approach has proved satisfactory, especially where the boiler units have relatively large inherent storage capacities.

A more direct method of operation, as shown at the bottom of Fig. 9-91, consists in regulating the input of fuel and air to the boiler to give the desired electrical-generator output, with steam pressure maintained by adjusting the turbine governor. With this direct-energy-balance method, the turbine keeps in step with the boiler output, and boiler input is directly controlled to give the desired generator output. This system also inherently keeps the boiler and turbine "in step," which is most desirable with a once-through type of boiler and turbine unit.

Fundamentally, however, the combustion-control system should provide maximum flexibility to accommodate as many operating conditions as possible. While the direct-energy-balance method of control appears most desirable, actual field experience with it is limited at present. Therefore, both basic methods of operation are included, and the operator may select either method by means of a transfer switch.

As a further means of integrating over-all operation of the boiler and turbine as a unit, the combustion-control system employs a boiler-turbine governor component. It basically provides means of (1) changing boiler-turbine output in an orderly manner, and (2) keeping the output within

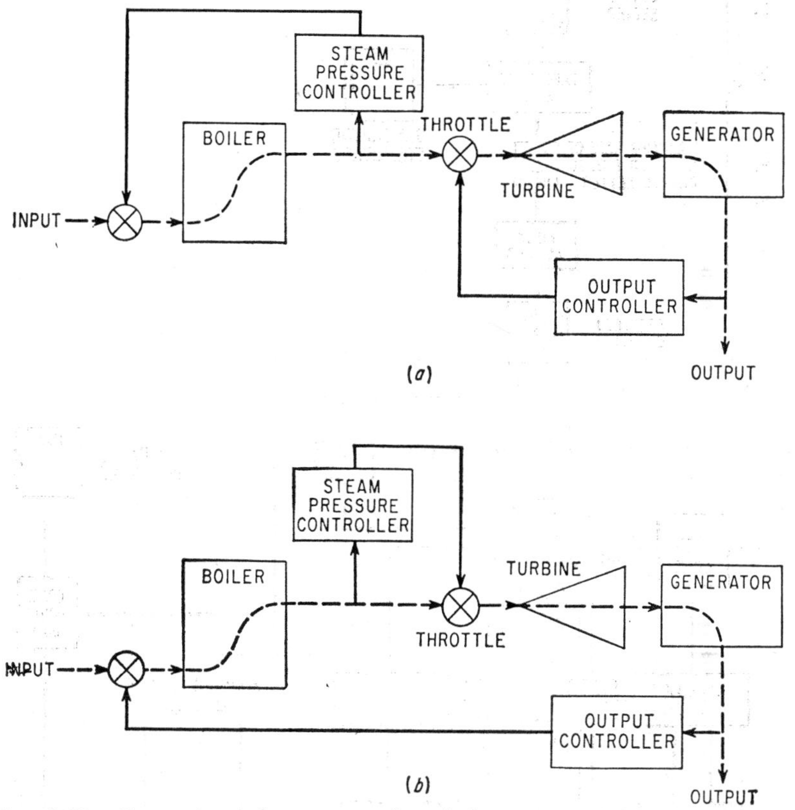

Fig. 9-91. Comparison of conventional and direct-energy-balance methods of boiler control. (*Leeds & Northrup Co.*)

the capabilities of the boiler and turbine equipment in service. This governor component is sensitive to generator frequency, which adapts the combustion control to the inherent regulation of the normal turbine governor.

The boiler-turbine governor is employed with either of the basic methods of control. It is shown schematically at the top of each of the two control schemes (Figs. 9-92 and 9-93) which cover the direct-energy-balance and conventional methods of operation, respectively.

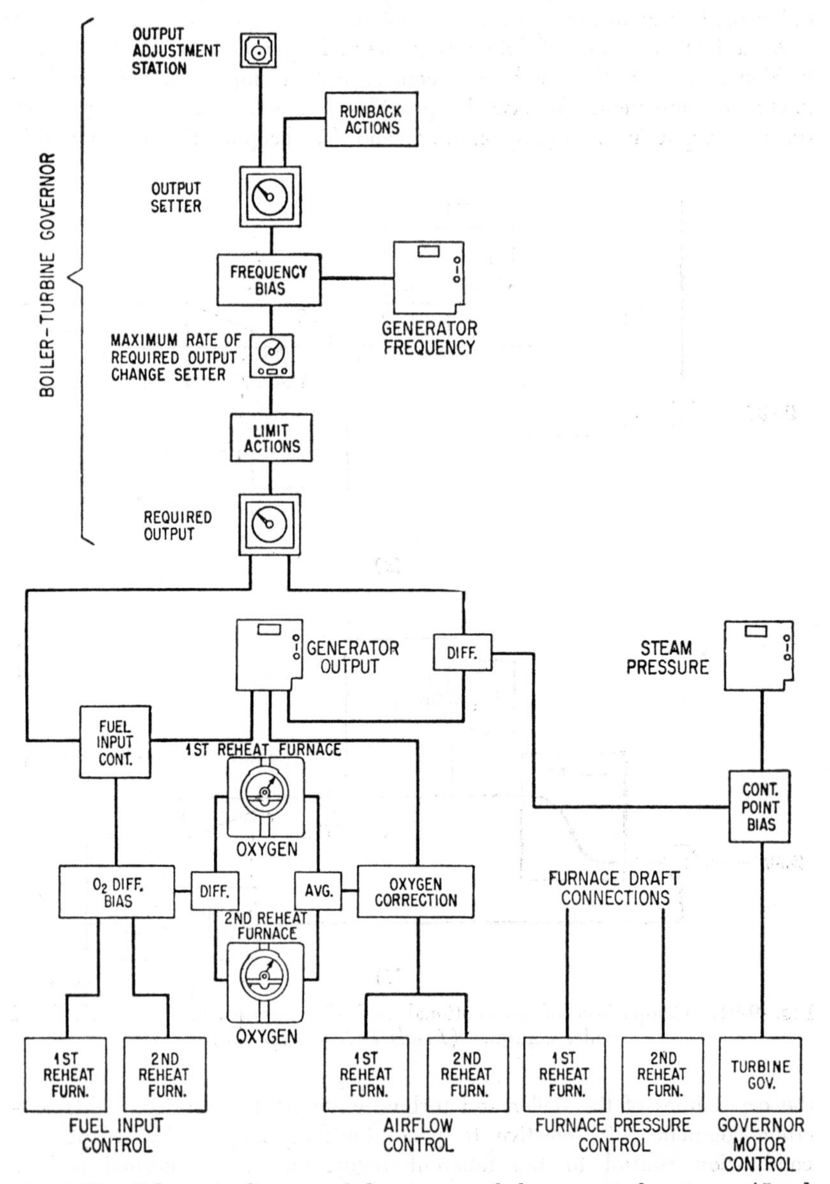

Fig. 9-92. Schematic diagram of direct-energy-balance control system. (*Leeds & Northrup Co.*)

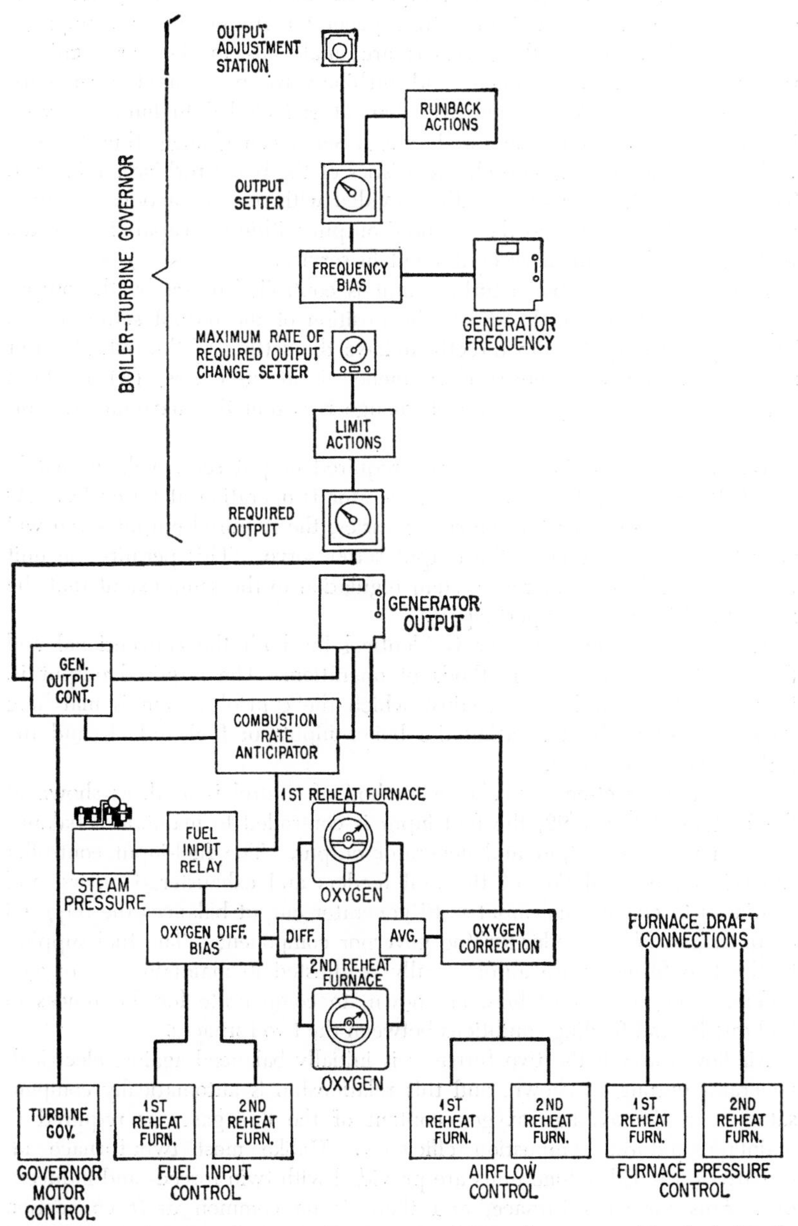

FIG. 9-93. Schematic diagram of conventional control system. (*Leeds & Northrup Co.*)

Referring to the top portion of Fig. 9-92, the boiler-turbine governor is composed basically of output setter, generator frequency, and required output service. These three servos are analogous to the synchronizing motor, speed-sensitive element, and turbine-valve position of a standard turbine governor. A rate-sensitive element is included to limit the maximum rate at which the required output servo can change, thus keeping within the inherent response characteristics of the boiler-turbine equipment. The boiler-turbine governor is also provided with runback actions and limit actions, as shown, to keep the required output within the capabilities of the boiler and turbine auxiliary equipment in service.

The output of the boiler-turbine unit is controlled or set by the output adjustment station, which adjusts the position of the output setter servos. The required output servo directly follows the setting of the output setter servo, providing (1) generator frequency is at 60 cycles, (2) no limit actions occur, and (3) the rate of change is within the maximum rate-of-change setting.

When no limit actions occur, the required output servo will ultimately match the output setter when the generator is operating at 60 cycles. As generator frequency or turbine speed varies, the required output servo will vary for a fixed setting of the output setter servo. This permits the unit to participate in the inherent system regulation to the same extent that the normal turbine governor participates.

The boiler-turbine governor is identical for both the conventional and the direct-energy-balance methods of operation. The required output is, in effect, the control point against which the control system is matching generator output by regulating the boiler inputs of fuel and air and the turbine governor valves.

When the direct-energy-balance method of control is used, as shown at the bottom of Fig. 9-92, the fuel input is controlled to maintain a balance between required output and generator output. The fuel-input controller supervises the regulation of the mill feeders and exhauster dampers, and continues to change fuel input until generator output balances the required output servo of the boiler-turbine governor component. The fuel supplies to the two furnaces are automatically readjusted to maintain zero oxygen difference or any preset bias, as shown, to compensate for differences in fuel quality and feeding conditions between the two furnaces.

Air flow through the two furnaces is initially balanced against electrical-generator output, as shown, and this relationship is automatically compensated from the average oxygen content of the flue gases as required to maintain uniform combustion efficiency. Unlike most twin-furnace installations, the Eddystone units are provided with twin forced- and induced-draft fans for each furnace, and there is no common draft connection between the two furnaces. Therefore, both the fuel and draft equipment for each furnace are independent of that for the other furnace.

The two furnace air-flow controllers regulate their respective fan equipments, to provide the proper air flow to each furnace, and individual furnace-pressure controllers maintain uniform furnace drafts.

The standard turbine-governor motor receives its regulation from a steam-pressure controller, the control point of which is automatically biased from the difference between required output and generator output servo positions, as shown. This bias occurs during load changes, and thus provides maximum response of the boiler and turbine equipment,

The basic conventional method of operation is shown schematically in Fig. 9-93. The boiler-turbine governor component functions the same as described previously for the direct-energy-balance method. With the conventional method, the standard turbine-governor motor is controlled to maintain a balance between generator output and required output servo positions.

Fuel input receives its initial regulation from changes in electrical-generator output, plus the regulating impulses applied to the standard turbine-governor motor, as introduced through the combustion-rate anticipator unit shown. This initial control action is supplemented as required from the steam-pressure master controller, to maintain a uniform steam pressure. The fuel supplies to the two furnaces are automatically biased from oxygen difference, as previously described.

Air flow through the two furnaces, as well as furnace draft, is controlled the same as previously outlined for the direct-energy-balance method.

The unusual features of the Eddystone combustion-control system can be summarized as follows:

1. Two basic methods of operation are provided: the direct-energy balance and the conventional methods.
2. A boiler-turbine governor component integrates operation of the boiler and turbines as a unit and keeps the output within the capabilities of the auxiliary equipment in service.
3. Automatic control from oxygen is provided to accomplish:
   (*a*) Bias of fuel inputs between furnaces
   (*b*) Final adjustment of air flow to maintain optimum combustion within each furnace
4. The oxygen analyzers employ average flue-gas sampling rather than a single probe measurement.
5. The steam-flow measurement normally found in combustion-control systems has been replaced by electrical-generator output.

In conclusion, the combustion-control system for Eddystone basically treats the boiler and turbine as a unit, and the final combustion conditions are automatically controlled from oxygen.

Hawaiian Electric Co., No. 8 Station.[22] The steam-generating equipment for Honolulu Number Eight is a Babcock & Wilcox 485,000-lb/hr single-drum radiant-type unit, having pressure and temperature ratings of 1,500 psi and 950°F, respectively.

[22] Bailey Meter Co.

The arrangement of the control room of Number Eight is as follows:

1. Mounted on a vertical panel are recording instruments which show not only the instantaneous value of each variable but also its trend. A total of 45 continuous records is provided.

   An operator going on duty has a complete picture of the plant's operation for at least the past 8 hr. The circular-chart records also provide a means of maintaining a day-by-day history of plant operation, and of analyzing related factors at any time.
2. Miniaturized indicating gages are mounted on the control desk. Related factors such as steam flow, water flow, drum level, drum pressure, and feedwater pressure are indicated by a series of adjacent gages. The normal operating value of each is in the same relative position on its scale. Thus the operator need only scan the indicators to observe abnormal operations in any part of the plant.
3. When action is required, the operator has remote control of all vital factors. Each of the selector stations, mounted on the control desk, provides a selection switch for changing the control from automatic to manual, and a control knob for altering the pneumatic signals to the final control elements.

Figure 9-94 shows the automatic combustion-control system.

The purpose of the automatic combustion control is to supply fuel oil and combustion air in accordance with load requirements and in the proper ratio for maintaining optimum firing conditions.

The system operates in the following manner:

1. *Master control.* A master steam-pressure transmitter controls the fuel-oil flow, and the forced-draft fan and induced draft fans in parallel and in the approximate desired relationship to each other. Oil flow is regulated by a recirculating valve, and each fan by both inlet vanes and outlet dampers.
2. *Fuel-oil-limit control.* A fuel-limiting signal modifies the master-control impulse so as to retard the fuel-oil flow in event the fuel–combustion-air ratio reaches an established limit. This phase of the control system is a safety feature which minimizes the possibility of firing with an extreme deficiency of combustion air at a time of rapid load change or in the event of a sudden loss in fan capacity.
3. *Oil-flow–Air-flow ratio.* The master-control impulse is readjusted to its final control of the forced-draft fan from a signal established as a function of the ratio of the metered fuel and combustion-air flows.
4. *Furnace draft.* So as to maintain the desired balance between the induced- and forced-draft-fan capacities, a furnace-draft controller initiates a readjusting control impulse which modifies the master signal in its control of the induced-draft fan.

In addition to the master steam-pressure signal and modifying signals mentioned, a pneumatic signal from an oil-pressure transmitter controls burner retraction to maintain the desired oil-burner position.

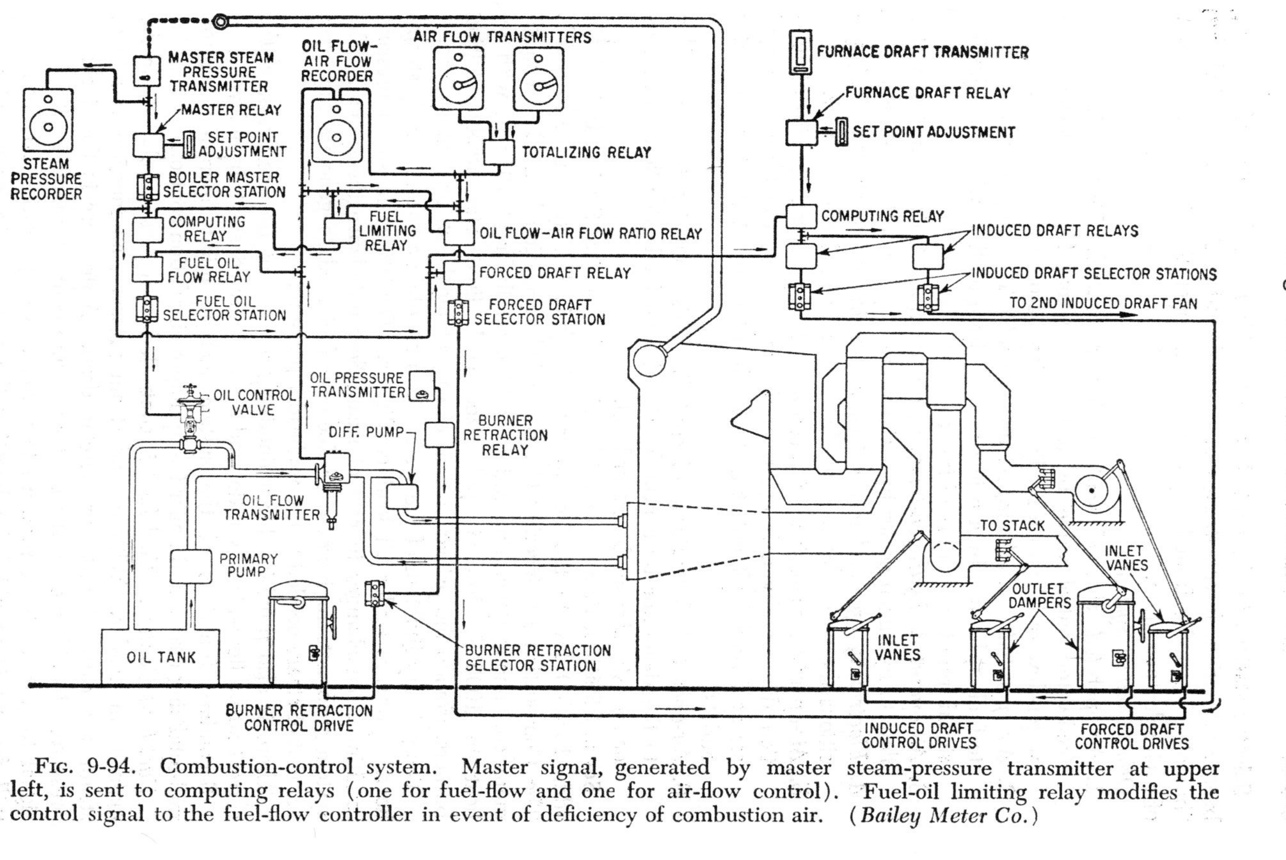

FIG. 9-94. Combustion-control system. Master signal, generated by master steam-pressure transmitter at upper left, is sent to computing relays (one for fuel-flow and one for air-flow control). Fuel-oil limiting relay modifies the control signal to the fuel-flow controller in event of deficiency of combustion air. (*Bailey Meter Co.*)

Selector stations in the control room provide manual control of oil flow, burner retraction, forced draft, and induced draft. A selection switch for "bumpless" transfer from hand to automatic operation is an integral part of each selector station.

Other Control Systems

1. *Air-operated three-element feedwater control.* This positions the feedwater-control valve by a primary signal from the steam-flow–water-flow ratio with a readjusting signal from the drum level.
2. *Air-operated fuel-oil-temperature control.* This regulates steam flow to fuel-oil heaters by a primary signal from oil flow with a readjusting signal from temperature of fuel oil.
3. *Air-operated hydrogen-temperature control.* This regulates water flow to hydrogen cooler by a signal from hydrogen temperature.
4. *Electrically operated feedwater-pump-recirculation control.* This positions the recirculation valve by a signal from feedwater flow.
5. *Air-operated air-recirculation control.* This positions the air-recirculation dampers by a signal from the average of the temperatures of air at forced-draft outlet and of exhaust gases at the two air-heater outlets.
6. *Air-operated deaerator-level control.* This regulates the water flow to the deaerator by a primary signal from feedwater flow and a secondary signal from deaerator level.

Seward Station, Pennsylvania Electric Co.[23] Seward Station, Unit No. 5, adds 138,000 kw to the Pennsylvania Electric Co. system (Fig. 9-95). The Combustion Engineering Co. controlled-circulation boiler has a maximum continuous generation rating of 900,000 lb/hr at 1055°F, with reheat to 1005°F. Fuel is pulverized coal, fired tangentially. Tilting burners provide the primary means of controlling the steam temperature.

The primary function of the system is to maintain a constant main-steam-header pressure under all load conditions. Deviations in steam pressure resulting from changes in rate of steam flow unbalance the system. The system then acts to increase or decrease fuel feed to generate the exact amount of steam needed to restore header pressure.

There is also provision for maintaining a constant furnace pressure, the correct steam-flow–air-flow ratio, and a practically constant drum level.

Secondary functions of the system include mill-temperature control, mill-totalizing circuit and boiler-feed-pump-recirculation control.

Fuel-feed Control. Fuel feed is controlled from pressure changes in the main-steam header. With normal steam pressure, an increase in rate of steam flow causes throttle pressure to drop. A decrease in flow increases pressure. In either case, the master pressure controller modifies its output signal to restore steam pressure to normal. Compensation for lag in the mill feeders is provided by boosting the master-controller output signal with an impulse from the steam-flow transmitter.

The stepped-up-air-loading signal from the master pressure controller

[23] Copes-Vulcan Div., Blaw-Knox Co.

is sent to the mill-feed controller. The output signal from this controller is piped to four auto-manual biasing stations, one for each mill. Each station relays the demand signal—plus or minus a manually adjustable bias—to the drive unit of its related fuel feeder.

The fuel-feed drive units are similar in type to all drive units in the control system. The design combines a positioner, a four-way valve, a power piston, and a feedback cam into one integrated, compact unit. Manual operation of each drive unit is provided by lever or handwheel, depending on torque requirements.

Air loadings applied to each mill-feed drive unit are totalized by computing relays and fed back to the mill-feed controller. The controller action is continuous until the totalized feedback balances the master demand resulting from throttle-pressure deviation.

Working in conjunction with each mill feeder is a precharacterized exhauster drive unit, which operates in parallel with changes in feeder speeds. Separate bias adjustment permits additional exhauster control, should this prove desirable.

Air-flow Control. For efficient combustion under all load conditions, an air-flow control is established by balancing total air flow and steam flow, the latter being used as a measure of total fuel input.

Output from the steam-flow transmitter, acting as an air-flow-demand signal, is balanced against the transmitted air signal, which is proportional to the primary-air-flow metering sections. Any unbalance of these two signals results in the forced-draft-fan dampers being moved to reestablish the balance.

The circuit is so calibrated that a 1:1 ratio between steam flow and air flow normally provides the correct rate of air flow for most efficient combustion. This ratio may be varied by a panel-mounted relay between the steam-flow transmitter and the air-flow controller. Any change in input signal to this relay causes a proportional change in output.

Compensation for inherent lag in forced-draft fans and dampers during periods of load changes is provided by a combining relay that adds a booster impulse to the air-flow-demand signal.

Furnace-pressure Control. Working in conjunction with the air-flow control, but independent of it, the furnace-pressure control positions the inlet damper of the induced-draft fan to hold furnace pressure at the correct negative value. A manually adjustable set point is balanced with an air loading directly proportional to furnace draft. Any unbalance between these two signals changes the controller output to restore the balance by acting on drive units that increase or decrease the opening of the induced-draft fan dampers.

Fuel-cutback Circuit. In case of inadequate air for complete combustion, a master overriding relay automatically cuts back fuel feed, with a resultant loss of main-steam-header pressure. This relay is between the master controller and the mill-feed controller. When mill circuits are operated manually, the cutback controller is bypassed.

Mill-temperature Control. Coal-air temperature in the mill is maintained constant, regardless of load changes, by controlling the mixture of hot air through the hot-air damper and tempering air through the weight-loaded tempering-air damper. For any fixed exhauster-damper position,

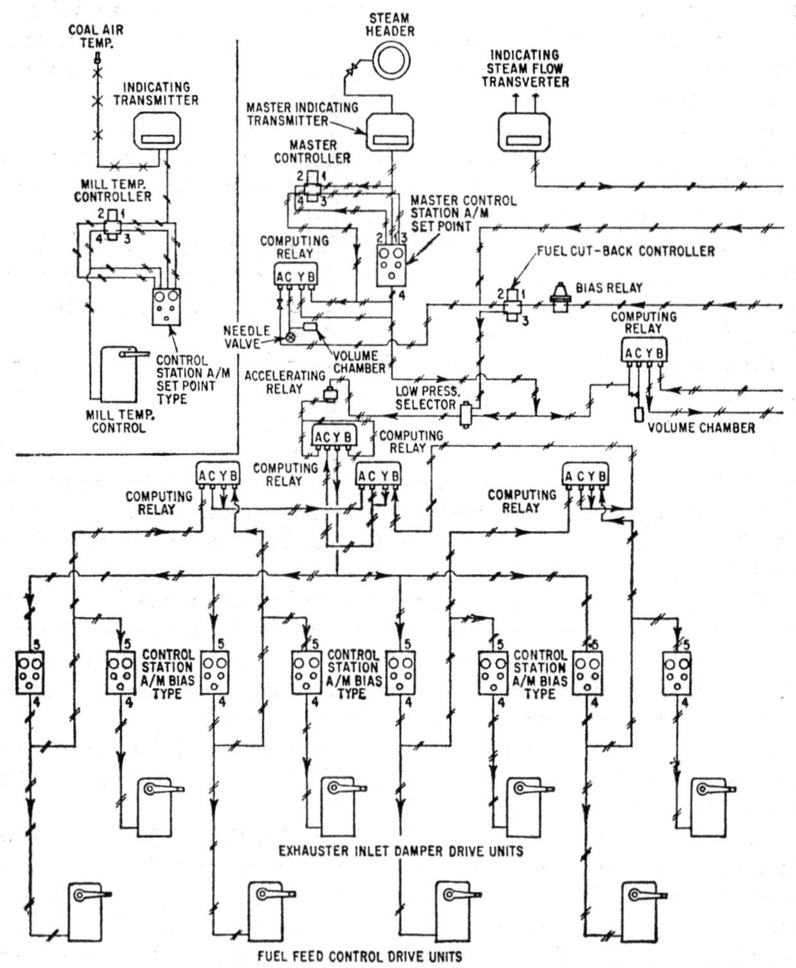

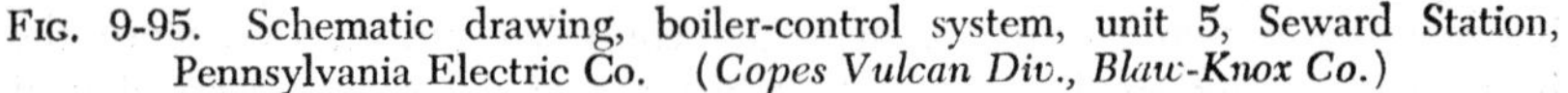

FIG. 9-95. Schematic drawing, boiler-control system, unit 5, Seward Station, Pennsylvania Electric Co. (*Copes Vulcan Div., Blaw-Knox Co.*)

the flow of tempering air increases as the flow of hot air is decreased, and decreases as hot-air flow is increased.

Feedwater Control. With the feedwater control, feed to the boiler is modulated by three influences—steam flow, feedwater flow, and drum-water level. Feed input closely matches steam output, regardless of load

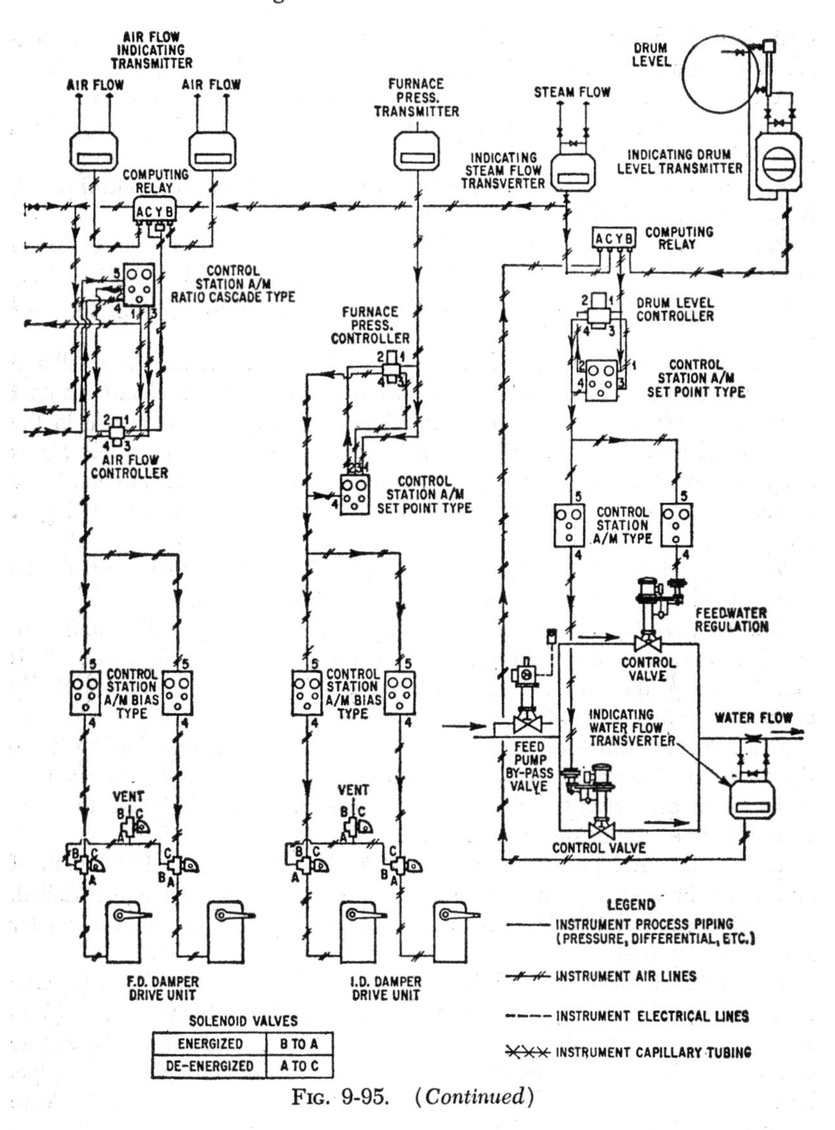

FIG. 9-95. (*Continued*)

conditions. Drum-water level is held practically constant at all times, regardless of load or feed pressure, and while blowing down or blowing soot.

The two flow influences are obtained from the primary elements of the flow meters. Each acts on a transmitter, which extracts the square root and sends a linear air impulse, proportional to flow, to the balancing relay. This relay also receives water-level influence as a straight-line air impulse. The air impulse resulting from balancing the three control influences is sent through the controller to the valve positioner, and to the auto-manual-selector station where there is continuous indication of controller output.

A relay-operated feed-control valve is installed in each of the two main-boiler-feed lines. During startup periods, high pressure drops have been handled satisfactorily.

A motor-operated valve, actuated by a switch at the boiler-control panel, is installed in the feedwater bypass line.

Boiler-feed-pump-recirculation Control. Each of the three motor-driven boiler-feed pumps is rated at 1,120 gpm against a head of 6,700 ft. Normal practice is to divide the load between two pumps, holding the third on stand-by.

Each pump has a separate low-flow-control circuit. When the discharge from any pump falls below a predetermined low limit, an electric-contact controller—responsive to the differential produced by a flow nozzle in the pump-discharge line—energizes a solenoid valve, through which air is supplied to the diaphragm-operated bypass valve. This automatically opens the bypass valve to assure sufficient flow to prevent overheating of the pump.

When the load rises above the low limit, the controller deenergizes the solenoid valve, and the bypass valve is automatically closed.

Boiler Operating Panel. All necessary processing information and control functions are indicated or recorded at the boiler operating panel. This permits firing the furnace or feeding water to the boiler automatically or by remote manual control.

Jones & Laughlin Steel Corp., Pittsburgh, Pa.[24] This new facility is designed to be integrated with the existing 160,000-lb coke-breeze-fired oil boilerhouse comprising eight 20,000-lb boilers. It is designed to more than double the older steam capacity.

Three Riley RX boilers with a total nominal capacity of 180,000 lb of steam per hr and a peak capacity of more than 225,000 lb are installed. The principal fuel is coke-oven gas, with provision for use of residual tar as an auxiliary fuel.

The residual tar is burned just like oil, with steam atomization. One problem was liquefying the tar; at ordinary ambient temperatures (0 to 100°F) the tar is a solid. The residual tar must be kept at a temperature of 250°F if it is to flow readily in sufficient quantity through a 3-in. pipe. The solution to the problem was an all-electric system in which the pipe walls themselves serve as the heating element. Peabody combination oil-gas burners handle the two fuels and the separate supply of natural gas for the pilot lights.

Maximum provision was made for failure, replacement, or maintenance of components without interfering with operation of the automatic-control system. Any two of the three burners in each boiler can maintain the maximum load. All three, however, are normally in use together. If the flame fails for any reason in any one burner, it is cut off automatically, and

[24] Hagan Chemicals & Controls, Inc.

the other two increase their use of fuel to produce the combustion rate required. Similarly, if two or three of the boilers are in use, one can be shut down automatically by reason of any one of a number of possible failures, but the remaining one or two are still in fully automatic operation.

All controls and instrumentation shown in Figs. 9-96 and 9-97 are pneumatic, except the flame-scanning system. Each of the nine burners is scanned by an independent electronic cell, set to cut off all fuel to that burner (including pilot gas) in the event of either flame failure or change in the character of the flame, which would indicate improper combustion. Electrical interlocks (not shown on the diagram) operate in conjunction with the scanning mechanism to shut off the fuel to all three burners in a

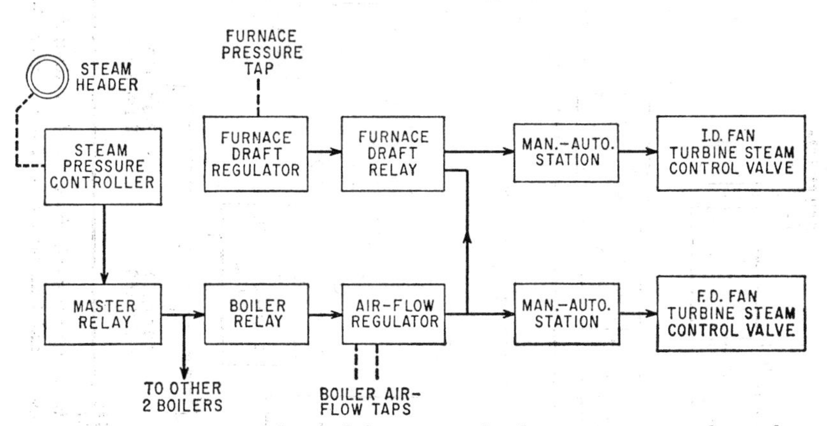

FIG. 9-96. Basic control signal from steam-header pressure controls combustion air by controlling induced-draft and forced-draft fans. Air flow controls fuel flow as shown in Fig. 9-97. (*Hagan Chemicals & Controls, Inc.*)

given boiler in the event of failure of feedwater, forced draft, or induced draft.

Basic information for the operation of the combustion-control system comes from the steam header which is fed by both the old and the new boilerhouses. Any variation in pressure from a predetermined control point generates a demand signal, which increases or decreases the flow of air provided by the forced- and induced-draft fans, both turbine-operated. The volume of combustion air is sensed by boiler air-flow taps, which generate a pneumatic signal that influences all the combustion controls, as shown in Fig. 9-97. This delivers fuel to the burners as required to convert the air to CO_2. This system differs from conventional practice (in which air is supplied proportionally to the fuel supply) in that the fuel is proportioned to the air.

In the following description of the fuel control, note that the tar demand exists only as a function of insufficiency of coke-oven-gas supply necessary to satisfy the steam-header demands. Design provides for 300 psig steam

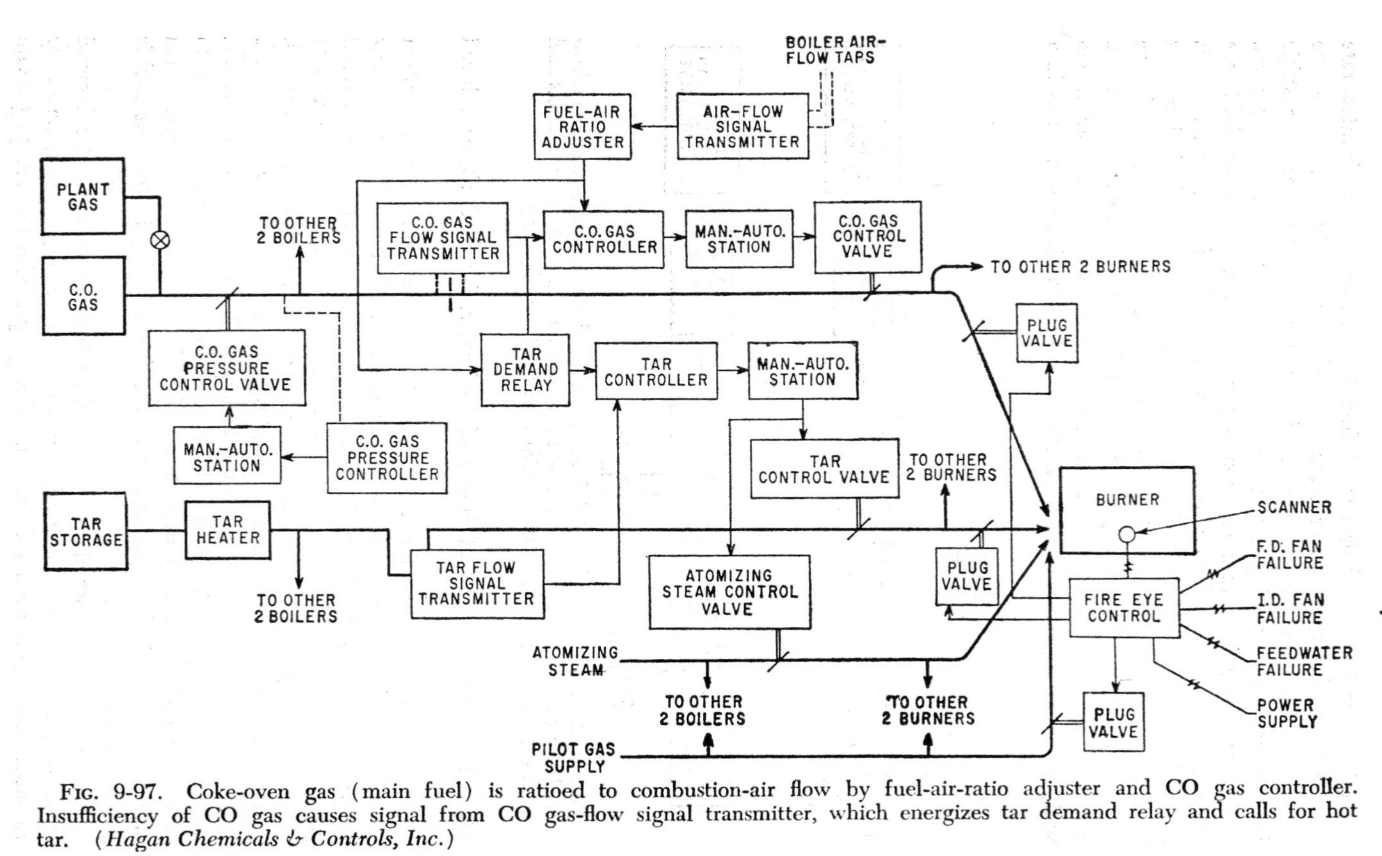

FIG. 9-97. Coke-oven gas (main fuel) is ratioed to combustion-air flow by fuel-air-ratio adjuster and CO gas controller. Insufficiency of CO gas causes signal from CO gas-flow signal transmitter, which energizes tar demand relay and calls for hot tar. (*Hagan Chemicals & Controls, Inc.*)

at 600°FTT from the new facility. Note also that plant gas is another auxiliary fuel supply. In an emergency, it could be burned with only minor adjustments.

The steam-pressure controller generates a pneumatic signal of 60 to 0 psi. This signal, which is constant as long as the steam-header pressure is at the control point, is directed to a relay which sends an equivalent signal to each of the three boiler relays.

Master relay and boiler relays provide for either automatic or manual operation. Thus the entire plant may be manually operated remotely from the master panel; or any one of the three boilers may be operated manually from its own panel. When set to manual, the automatic incoming signal is interrupted at the transfer valve, and a signal generated by the operation of the manual knob is substituted.

The air-flow regulator generates a pneumatic signal (30 to 0 psi), which remains constant as long as the air-flow-demand signal (from the boiler relay) is balanced by the indication of actual air flow measured by the pressure drop across the boiler. Any unbalance results in a signal transmitted to the furnace-draft auxiliary relay, where it acts to start the furnace-draft-regulating system in the proper direction for variations in air flow, as will be described. The same signal, however, also is transmitted through the forced-draft-turbine manual-control station and a solenoid valve (not shown) to the forced-draft-fan turbine-steam-control valve. The manual-control station is another means of operating remotely by manual control if desirable. Normally, it is set to automatic. Failure of induced draft will energize a solenoid in the safety-check system, interrupting the loading signal and substituting for it a constant 30-psi air signal. This positions the forced-draft-control valve at the minimum opening, slowing down the forced-draft turbine.

The forced-draft-fan turbine-steam-control valve is equipped with a valve positioner. A signal of 0 psi positions the valve to maximum opening. A forced-draft safety-check switch (not shown) closes contacts upon failure of the forced-draft fan, and thus actuates both the proper annunciator light on the boiler panel and the three scanning controls on the burners of that boiler, shutting off all fuel.

The furnace-draft regulator receives its information from a furnace-pressure tap. Variation from the control point causes a signal that returns furnace draft to the control point. The pneumatic signal from the furnace-draft regulator goes directly to the furnace-draft relay. This is a four-chambered totalizer, which also receives a signal from the air-flow regulator. This latter signal causes an immediate and rapid variation in combustion-air flow, in advance of the regular controlling signal from the furnace-pressure tap. This bias signal is gradually removed as air flow stabilizes.

The induced-draft-fan manual-control station, and the induced-draft-fan turbine-steam-control valve operate in the same manner as their counterparts in the forced-draft-control system.

An induced-draft safety-check switch closes on failure of induced draft, and actuates both the induced-draft-fan-failure annunciator light, on the boiler panel and the manual-reset solenoid valve. Thus, failure of induced draft cuts down forced draft and indirectly operates the fuel cutoffs. When the induced draft is returned to normal, the forced draft may be reset to automatic, merely by resetting the solenoid valve.

Figure 9-97 shows the fuel-control system. Basic information comes from pressure taps, which indicate boiler air flow (located alongside the taps for the air-flow regulator previously described). The air-flow signal transmitter generates a pneumatic signal (0 to 60 psi) proportional to the flow of combustion air as measured by the pressure drop across the boiler. This signal is transmitted directly to a fuel-air recorder (mounted on the boiler-control panel) where it indicates actual air flow; and to the fuel-air-ratio adjuster, where it is used as the fuel demand signal.

The fuel-air-ratio adjuster is a ratio relay with a panel-mounted adjusting knob, which moves the fulcrum and thus provides for changing the ratio according to requirements. The output is a pneumatic signal, which bears a ratioed relation to the air flow. This control makes possible maximum combustion efficiency and holds emitted smoke below the maximum permitted by the Pittsburgh smoke-control law. The ratioed fuel-demand signal is transmitted directly both to the coke-oven-gas controller and to the tar-demand relay.

The coke-oven-gas-flow signal transmitter is similar to the air-flow signal transmitter described previously. It generates a pneumatic signal (0 to 60 psi) proportional to the coke-oven-gas flow to the three burners of a particular boiler. This signal is transmitted directly to a fuel-air recorder on the boiler panel, to the coke-oven-gas controller, and to the tar-demand relay.

The coke-oven-gas controller is a Hagan ratio totalizer. It generates a signal (0 to 60 psi) which remains constant as long as the gas-demand signal is balanced by the indication of actual gas flow. Any unbalance changes the flow to the burners.

The coke-oven-gas manual-control station receives its signal from the controller and normally transmits it directly to the coke-oven-gas-control-valve operator. In emergencies, where remote manual operation is desired, it permits the substitution of a manually controlled signal in the manner described previously for the air-flow-control system. The valve operator is linked mechanically to the coke-oven-gas-valve operating lever. It is mounted on a floor stand and equipped with manual-positioning gear. The coke-oven-gas-control valve itself is an 8-in. V-port butterfly with a definite position corresponding to every pneumatic signal received by its operator. It controls the gas to all three burners, each of which has its own safety-check cutoff plug valve. Thus, when a single burner is cut off, the same volume of gas and air continues to be burned in the remaining two burners, without manual intervention.

Coke-oven-gas-pressure control, necessary for proper functioning of the equipment already described, is provided by an independent automatic-control system, shown in Fig. 9-97. The recorder on the master panel measures gas flow and gas pressure. This system has the usual remote-switching valve for remote manual operation and maximum control response.

Residual tar produced by the by-product plant is stored in a tank maintained at a temperature above 250°F.

In this system (Fig. 9-98) the pipe walls represent the impedance element. Heating current is 440 volts, 60 cycle, single phase, reduced to a potential of 20 to 24 volts by three 15-kva transformers, each of which serves a 300-ft section of the pipe. Cellular glass covered by Neoprene surrounds the pipe, serving as a water-resistant electric and heat insulator.

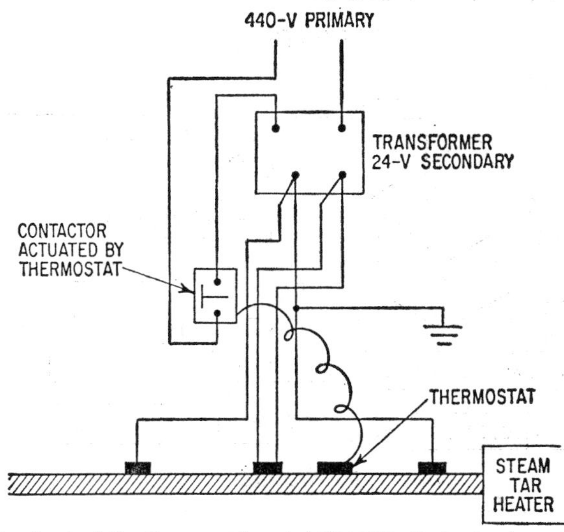

Fig. 9-98. Last of the three outdoor heating circuits heats about 307 ft of 3-in. hot-tar line; 15-kva transformer causes heating current in line itself. (*Hagan Chemicals & Controls, Inc.*)

Electrical connections of the hot-tar transmission system are shown in Fig. 9-98. Thermostats, in contact with the pipe walls, control the current used in each section, according to requirements. When plant operating conditions permit, no current at all is consumed.

The hot fluid tar is controlled like oil; steam is used for atomization in Peabody burners. The tar-demand relay generates a pneumatic signal (0 to 60 psi) proportional to fuel demand and coke-oven-gas flow. When there is insufficient coke-oven gas, a positive signal is transmitted to the tar controller, which provides enough hot tar, plus atomizing steam, to maintain the boiler load. The tar controller receives its feedback information from a tar-flow signal transmitter.

Safeguards. Electric-power failure, by itself, does not stop steam generation. A drop in the voltage of the plant power supply of a magnitude sufficient to affect the flame scanner releases a spring-loaded solenoid.

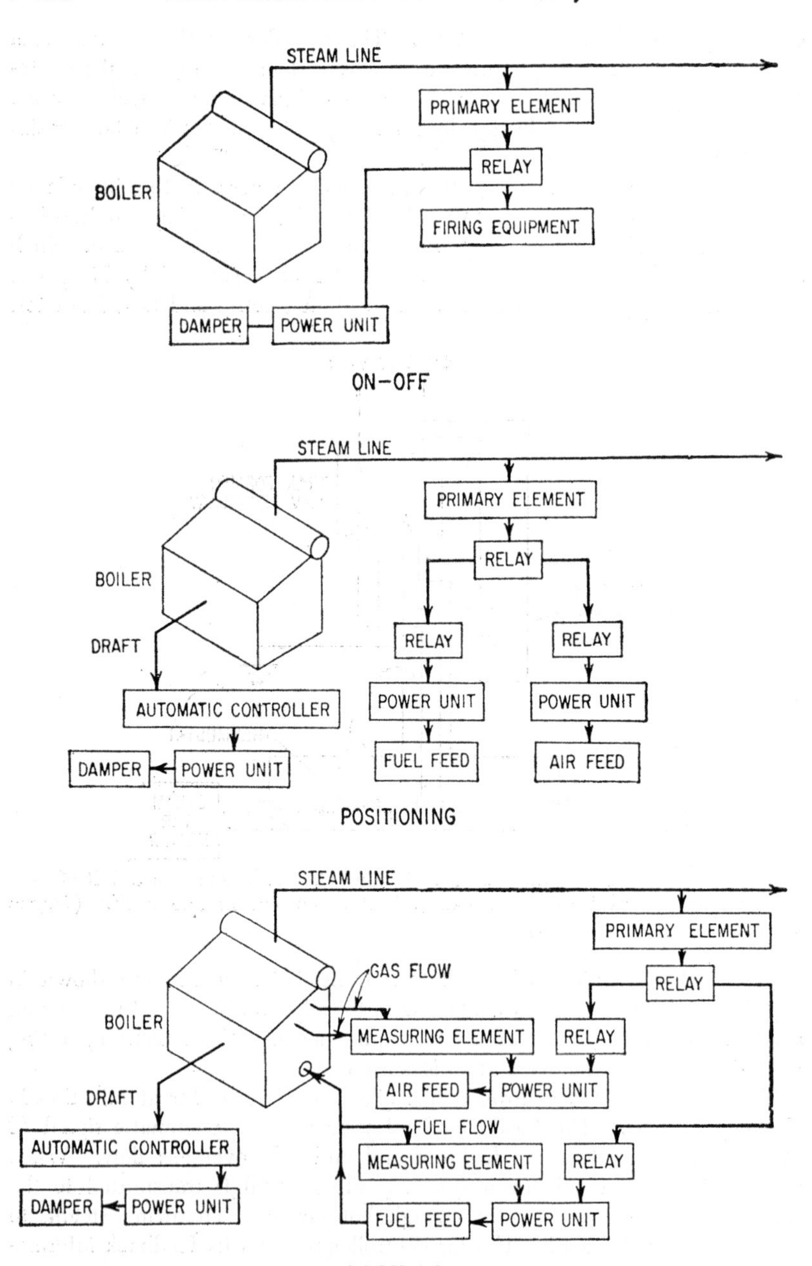

Fig. 9-99. Basic packaged boiler-control systems.

This in turn cuts off operating air to the three operators of the shutoff plug valves installed in the fuel lines to each burner. These plug valves remain in the full-open position, and the fuel supply continues to be regulated by the automatic-control system.

Failure of almost any component is evidenced by an annunciator light on the respective control panel. Flame failure or power failure causes an audible alarm (horn), which cannot be silenced until all power-failure solenoids have been reset manually (one for each burner suffering the failure) and complete monitor control has been restored. Both the annunciator lights and the horn, with other key electric interlocks, are powered by direct current, independent of the regular plant power supply.

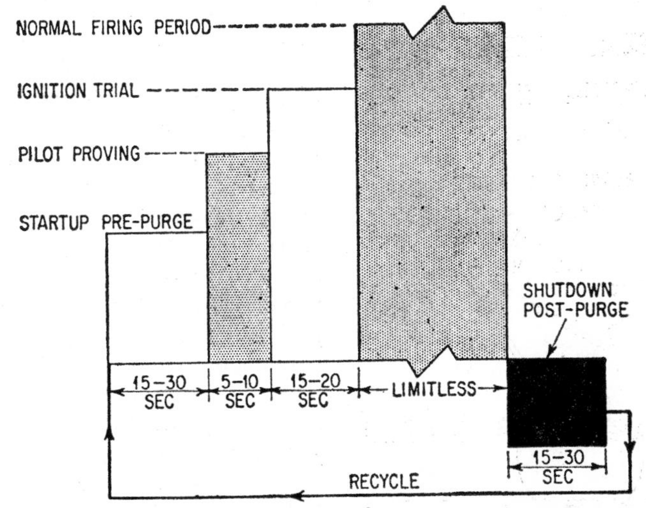

Fig. 9-100. Typical packaged boiler programming sequence.

If any boiler has all three of its burners off for any reason, an automatic-interlocked purge cycle must be accomplished before pilots may be relighted and automatic or manual control reestablished. This is the only manual operation required from the boiler operating platform when the equipment is in service.

Packaged Boiler Controls.[25] Control systems for packaged boilers generally include ignition proving, combustion control, flame-failure equipment, safety shutoff valves, and limit switches, plus startup and shutdown programming. The control senses and makes the boiler respond to changes in steam demand to the point of shutting the unit down when the steam needs drop to zero.

Three basic combustion-control schemes are: (1) on-off, (2) positioning, (3) metering (Fig. 9-99).

[25] R. C. Bellas, associate editor, *Power* (a McGraw-Hill publication).

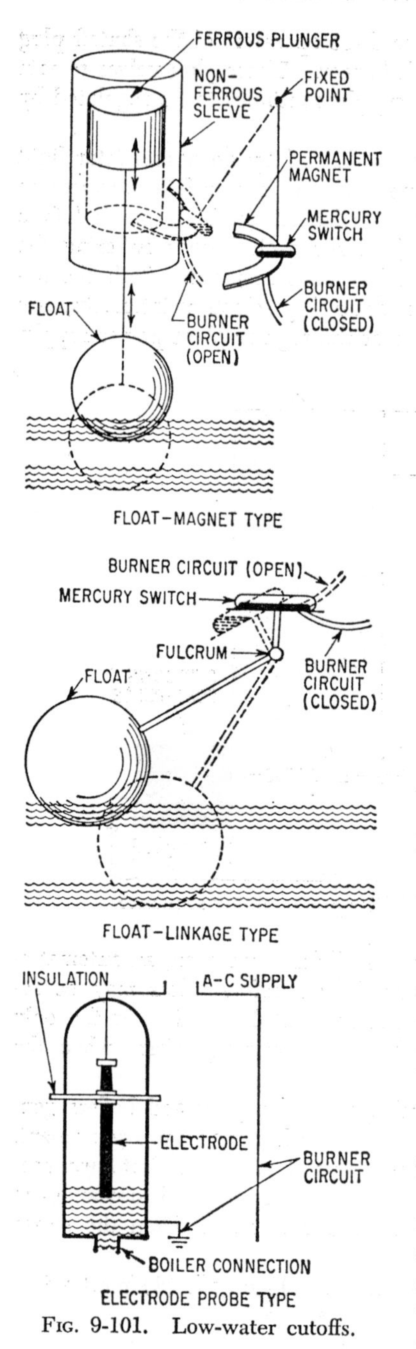

FIG. 9-101. Low-water cutoffs.

The on-off system works between two steam-pressure levels to start and stop both air supply and fuel feed.

The positioning-system response to steam pressure assumes that a given damper position will provide sufficient air for a given fuel flow to hold a constant fuel-air ratio throughout load range. Yet, at any position, air flow may be influenced by voltage variations at the blower motor, draft loss through the boiler caused by soot or slag, and other variables. Thus, though the system may hold steam pressure constant, still combustion efficiency may not stay constant.

Metering-type control goes a step beyond positioning by actually measuring fuel and air flows and draft loss. These values are balanced against signals for more or less steam pressure. The result is compensation for variables which tend to influence the positioning system's operation.

Generally, both positioning and metering systems include an on-off feature, extending low-load range beyond that of modulating.

SAFETY CONTROLS: Automatic controls, while following load demand and maintaining steam pressure, also serve a safety function.

Programming-sequence controls safeguard the boiler throughout the operating cycle (Fig. 9-100). On start-up, purge operation forces air through the furnace, and gas passes to clear out any combustible gas pockets.

Ignition is next. Depending on the fuel, the ignition system may be electric spark, light-oil–electric (a spark ignites a thin oil which in turn touches off the main fuel), or gas-electric (a spark ignites a gas pilot). A flame scanner operates during ignition. The control system will not let the main gas or fuel-oil valve open unless the scanner indicates that ignition is normal (ignition proving). After light-off, the scanner continues to supervise the burner. If the flame is lost, the scanner shuts the unit down. After shutdown, the blower again purges the furnace, and gas passes before the ignition cycle recurs.

The low-water cutoff, separate from the programming-sequence control, shuts down the boiler immediately if the water drops to a dangerously low level (Fig. 9-101).

The float-magnet type uses a ferrous plunger on one end of a float rod. The plunger slides within a nonferrous sleeve. A permanent magnet, with mercury switch affixed, is supported by a pivot adjacent to the nonferrous sleeve.

Under normal water conditions, the ferrous plunger is above and out of reach of the magnetic field. Since in this position the mercury switch lies in a horizontal plane, the burner circuit is kept closed. But, if the boiler-water level drops, the float drops, bringing the ferrous plunger within the magnetic field. Then the magnet swings through a small arc toward the plunger; the mercury switch tilts, opening the burner circuit.

The float-linkage type uses a float connected through a linkage to a plate supporting a mercury switch. The plate is horizontal in the normal water-level position, and so the switch holds the burner circuit closed. If water level drops, the float drops, tilting the plate so that the switch opens the circuit.

The submerged electrode type uses the water proper to complete the burner circuit. If the water level drops below the electrode tip, current flow is interrupted, and the burner shuts down. On fire-tube boilers, the low-water cutoff generally includes an intermediate switch that controls the feed pump.

Section 10

STEAM-PRESSURE REDUCING AND DESUPERHEATING[1]

When a steam-generating plant is extended by the addition of high-pressure boilers, flexibility in operation requires a means of reducing the high-pressure steam to the pressure and temperature for which the low-pressure system was designed. Thus the steam generated at high pressure may be supplied to the low-pressure system for prime movers or process, replacing or supplementing steam exhausted to them by high-pressure prime movers or supplied by low-pressure boilers. Pressure-reducing and desuperheating stations are the means by which this operating control may be effected.

Such control must satisfy more exacting requirements than in former years. In the modern plant, fewer main boiler and turbine units are installed for a given capacity than in older plants. The units, especially in industrial plants, must operate continuously at high efficiency over a comparatively wide range. Close control of related and interdependent pressures and temperatures thus becomes of vital importance. The necessity for coping with high and higher steam pressures, temperatures, and capacities, and with greater differentials between systems, has necessitated many important improvements in the pressure-reducing and desuperheating equipment itself, as well as in the methods of controlling it automatically. The availability of the pressure-reducing and desuperheating equipment must be high, to match that of the steam- and power-generating units whose operations they coordinate; hence the equipment must be carefully and ruggedly designed and built.

Fundamental Equipment. The fundamental equipment, shown diagrammatically in Figs. 10-1 through 10-4, consists of pressure-reducing and regulating valves, desuperheaters, and the regulators and temperature controllers that control them.

Pressure-reducing and Regulating Valves. The regulator that operates the valve is controlled to maintain the required pressure in the pipeline downstream from the valve and desuperheater. In some cases, the valve

[1] Republic Flow Meters Co.

regulator is controlled to maintain the high pressure constant, as explained below. This is known as spillover control.

Two types of desuperheaters are used in most of the applications. These are the venturi or nozzle type shown in Figs. 10-1 and 10-2 and the steam-atomizing type shown in Fig. 10-3.

A third type, the mechanical-atomizing desuperheater (Fig. 10-4), has also been developed, primarily for use in turbine-exhaust lines, or where

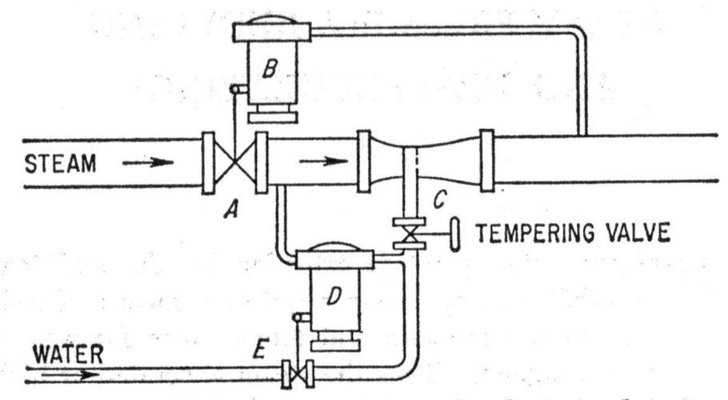

Fig. 10-1. Diagram of system consisting of reducing valve *A* controlled by regulator *B*; venturi-type desuperheater *C* with ratio control composed of regulator *D* and valve *E*.

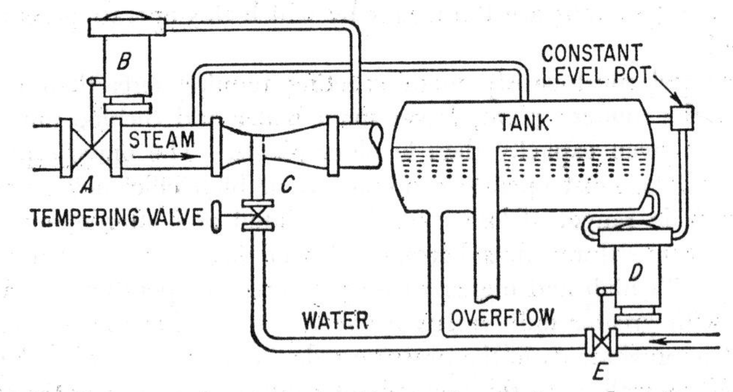

Fig. 10-2. Diagram of system consisting of reducing valve *A* controlled by regulator *B*; venturi-type desuperheater *C* with ratio-tank control composed of water tank, regulator *D*, and valve *E*.

atomizing steam is not available or where only a small amount of desuperheating water is needed. All three types desuperheat the steam by the evaporation or direct-contact method, in which water is injected into the high-temperature steam, and is evaporated by absorbing heat from it, thus reducing the temperature of the mixture.

Venturi Desuperheater. The venturi desuperheater is a venturi tube, supported by and forming part of the pipeline, and containing a water-injection nozzle located so that the desuperheating water is injected into the steam at the point of minimum area, to ensure complete evaporation of the desuperheating water.

The principal characteristic of the venturi desuperheater is that the desuperheating water supply is automatically controlled by direct mechanical forces derived from the steam flow through the venturi. This control proportions water flow to steam flow, to hold the low temperature within the allowable limits. The control is effected by utilizing in two different ways the steam-pressure drop created at the venturi throat by the steam flow.

In the first method, this venturi steam-pressure drop is used to control a regulator, which operates a valve in the desuperheater water line so as

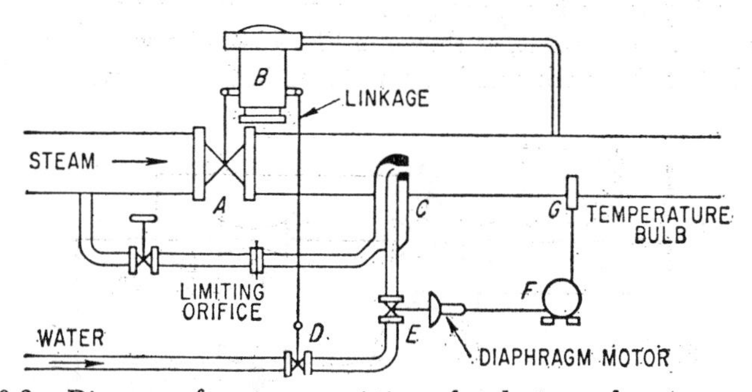

FIG. 10-3. Diagram of system consisting of reducing valve *A* controlled by regulator *B*; steam-atomizing desuperheater *C* controlled by temperature controller *F* and control valve *E*.

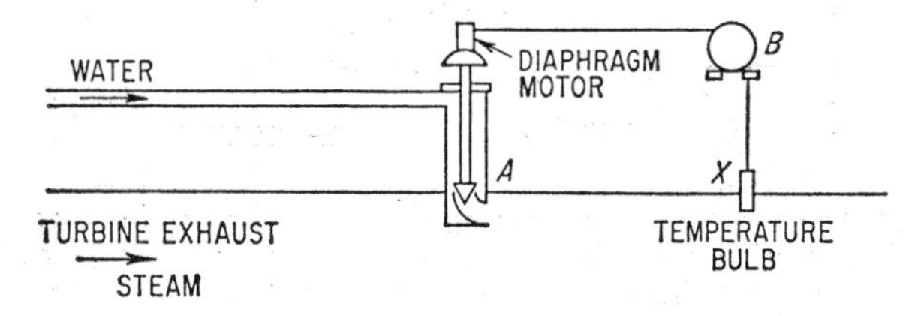

FIG. 10-4. Diagram of system consisting of mechanical atomizing desuperheater *A* controlled by temperature controller *B* operating diaphragm motor, which actuates water valve in desuperheater.

to proportion the water flow correctly to the steam flow. This is known as the ratio control, shown in Fig. 10-1.

The second method is to impress the venturi steam-pressure drop on a water tank in such a way that, as the venturi pressure drop varies with the steam flow, water is caused to flow to the desuperheater nozzle in proportion to the steam flow. This is known as the ratio-tank control, shown in Fig. 10-2.

In both these control methods, if the high-side steam and water conditions are substantially constant, no further temperature control is needed. But, if the high-side conditions vary widely, an auxiliary temperature-control device may be desirable.

Steam-atomizing Desuperheater. In the steam-atomizing desuperheater, shown in Fig. 10-3, desuperheating water is atomized by the impact of a small quantity of high-pressure steam taken from the reducing valve, or, if no such valve is installed, from some convenient high-pressure header. This steam is introduced at the nozzle in sufficient quantity and at constant velocity to produce complete atomization at all flows. This unit, therefore, has a wide range of operation. The supply of desuperheating water is controlled automatically by the final temperature.

Mechanical-atomizing Desuperheater. For a number of applications, as when high-pressure steam is not available or a comparatively small flow of desuperheating water is required, a mechanical-atomizing type of desuperheater is used. It consists, as shown in Fig. 10-4, of a water-injection valve and nozzle in a desuperheater body, installed in the pipe carrying the steam

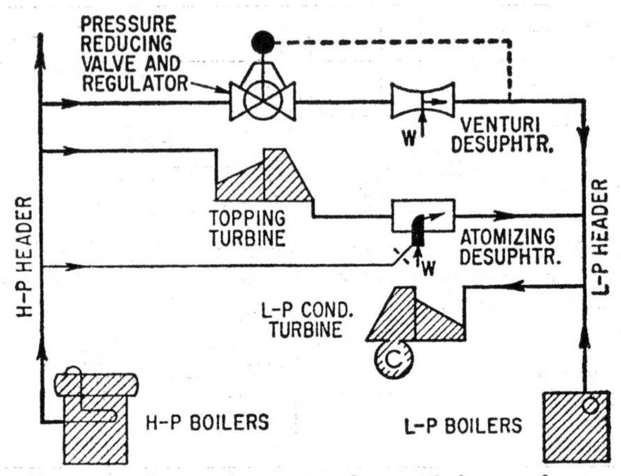

Fig. 10-5. Diagram of typical topping plant with bypass for topping turbine, showing reducing valve controlled to maintain low pressure constant and venturi desuperheater for controlling steam temperature. A turbine exhaust desuperheater is also installed.

whose temperature is to be controlled, the water-injection valve being operated by a diaphragm motor controlled from downstream temperature.

Turbine Bypasses. One of the most common modern applications of pressure-reducing and desuperheating equipment is to form a turbine bypass. A typical bypass for a straight-back-pressure topping turbine is shown in Fig. 10-5. Such systems have been installed in recent steam stations and industrial plants. The bypass steam either replaces or supplements the high-pressure steam passed through the topping turbine to the low-pressure system.

In the first case, the bypass is closed when the topping turbine is operating normally, but goes into operation immediately in case the turbine trips out, passing the high-pressure steam to the low-pressure system at the required low pressure and temperature. This confines the loss of electrical load to that being carried by the topping turbine, and permits all the rest of the plant—high-pressure boilers and low-pressure turbines—to continue operating as before. This is important in a central station, where loss of generating capacity at certain times may be a serious matter.

In the industrial plant, while the loss of topping-turbine capacity may be serious, it is usually more important to maintain steam pressure and temperature to processes using exhaust steam; otherwise, the manufacturing operations may be seriously disturbed. Such a disturbance is prevented by the operation of the turbine bypass, which passes steam from the high-pressure boilers, in place of the turbine-exhaust steam, to maintain the required process pressure and flows.

In some cases, however, especially in industrial plants, it is desired that the turbine bypass should operate to supplement the steam exhausted by the topping turbine. A common method of doing this is to operate the topping turbine on speed control and hold the back pressure constant by

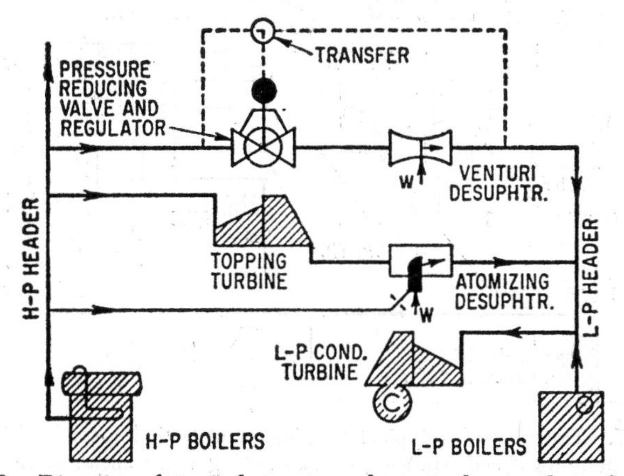

Fig. 10-6. Diagram of typical topping plant similar to that of Fig. 10-5, showing how reducing valve may be transferred either to spillover control, to maintain the high pressure constant, or to the usual low-pressure control.

means of the pressure-reducing and desuperheating station, to hold constant pressure in the low-pressure process headers. Then, if the turbine at a given electrical load does not supply the required amount of steam for processes, enough steam to make up the deficiency will flow from the high-pressure boilers through the bypass.

Spillover Control. Topping-turbine bypasses are also controlled under some conditions by spillover control. In this method, the flow through the bypass is controlled from the pressure in the high-pressure header instead of the pressure in the low-pressure header. In some plants, the bypass control is equipped with a transfer valve so that the control may be instantly changed from spillover to back-pressure control. Figure 10-6 shows diagrammatically how this is done.

Under spillover control, the high-pressure boilers are operated at constant rating, and the constant supply of high-pressure steam will be divided between topping turbine and bypass, so that the turbine will consume the steam it needs, the excess flowing through the bypass to the low-pressure system.

The most common need for the spillover control arises when, for some reason, it is desired to operate the high-pressure boilers at constant rating. A common instance of this is in starting up the topping-plant turbine. After the high-pressure topping-plant boilers are lighted off, the bypass is placed on spillover control and the reducing valve warmed up by steam from the low-pressure system. When the boilers finally reach full line pressure, and go on the line, the spillover control, will automatically open the bypass, and steam will flow to the low-pressure system, keeping the high-pressure

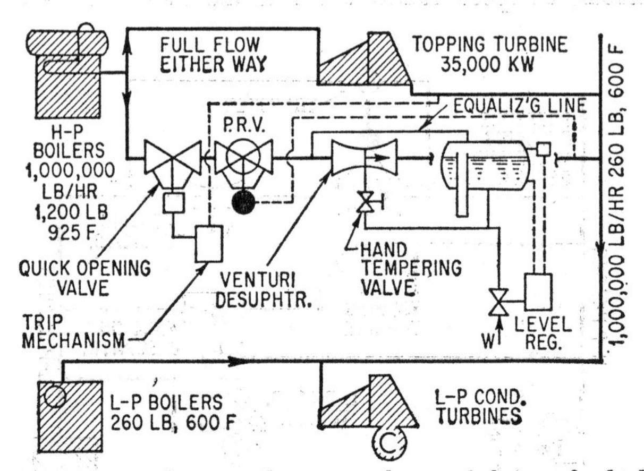

FIG. 10-7. Diagram of a typical topping plant with bypass for high capacity and high operating speed, showing quick-opening valve ahead of reducing valve and ratio-tank control of venturi desuperheater.

system at constant high pressure. In starting up the high-pressure topping turbine, it is desirable to keep the high-pressure boilers on manual control during the warming up of the turbine, permitting the bypass to remain on spillover control, so that whatever steam is not consumed by the heating of the turbine will flow to the low-pressure headers. Once the turbine is started and being brought on the line, it is usually desirable to put the topping boilers on automatic control and the bypass on low-pressure control.

High-speed High-capacity Systems. Topping-turbine bypasses in some steam stations have been designed to meet the demand for unusually high speed of operation as well as for high capacity. These requirements have been met in several plants by systems of the type shown in Fig. 10-7.

One of these installations is designed to pass over 1,000,000 lb of steam per hr; pressure is reduced from 1,200 to 260 psi, and temperature from 900 to 600°F. It was designed to open wide from fully closed position in ½ sec if the turbine trips.

it was thought best under such conditions to use two valves, one to open quickly, under the control of a rugged tripping mechanism, in case the topping turbine trips out, and the other to be always maintained, by a back-pressure control, in position to pass the amount of steam flowing through the turbine.

Thus the regulating valve would have to move only a slight amount, if at all, in case the steam flow of the topping turbine were suddenly transferred to it by the quick-opening valve. The venturi type of desuperheater with ratio-tank control (Fig. 10-2) was used to ensure instant response of the desuperheater, and to provide a method of supplying desuperheater water from storage for several minutes in an emergency.

Operating tests on one of these installations have indicated that the actual speed of operation required of these bypasses is much less than that for which they were designed. In one test, with the topping turbine operating at about one-half load, almost a minute was required after the topping turbine tripped before the pressure drop in the low-pressure system caused the quick-opening valve-tripping mechanism to open the quick-opening valve. This was a result of the cushioning effect of the piping system and the low-pressure boilers. As soon as the bypass took charge, however, the low-side pressure and temperature were maintained within the required limits. In this installation the high-pressure boilers are under automatic combustion control, and the high pressure and temperature were maintained so accurately that the desuperheater ratio-tank control held the low temperature constant without the need of a temperature controller.

Observation of this and other similar installations in operation over long periods has led to several conclusions. One is that the quick-opening valve is optional, its use depending on the cushioning effects of low-pressure piping, the number of low-pressure boilers in operation, and other factors.

A second conclusion from operating experience is that it might greatly improve operation to have the quick-opening valve controlled from the pressure in the high-pressure steam header, the regulating valve being controlled as before by the pressure in the low-pressure header. In most present installations, the tripping mechanism of the quick-opening valve is controlled from the turbine exhaust, from the low-pressure header downstream from the desuperheater, electrically from the topping-turbine-generator circuit or electrically from the turbine-throttle trip mechanism. When the quick-opening valve is controlled by either of the first two methods, if the topping turbine trips out, an appreciable time must elapse before the pressure at the turbine exhaust, or in the low-pressure header, falls to the value required to operate the quick-opening valve-tripping mechanism. During this interval, however, the pressure in the high-pressure header builds up very rapidly, and usually causes the high-pressure safety valves to open, an operation that is always undesirable.

The rise of pressure in the high-pressure header, therefore, just ahead of the turbine would form a good control medium for the quick-opening valve. The tripping mechanism of this valve could be set to open it just before the high pressure rose to the value required to open the high-pressure safety valves. This would put the turbine bypass into operation very quickly and stabilize all plant conditions more effectively.

Bypass Systems in Parallel. Another method of designing topping-turbine bypasses for high pressure, high speed, and high capacity is shown in Fig. 10-8. In some cases, two bypasses in parallel are installed; in others, three parallel systems are installed. An interesting feature of these installations is the control of the parallel systems by a master controller.

This master controller, as shown in Fig. 10-9, is fundamentally the same as the master controller of a combustion-control system. It responds to

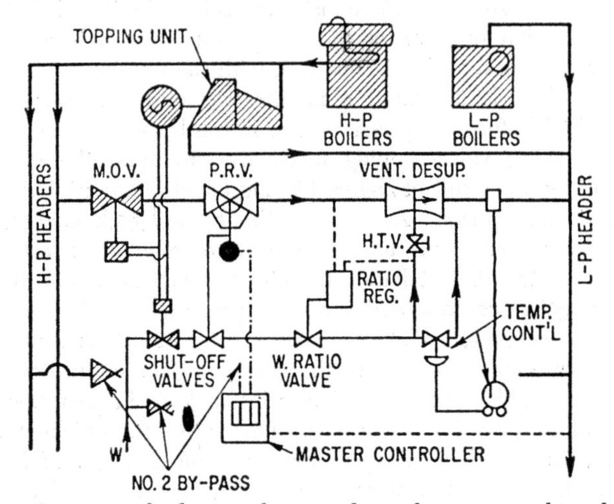

FIG. 10-8. Diagram of plant with two identical topping-turbine bypasses in parallel controlled by master controller. Only one bypass is shown in detail.

variations in the pressure in the low-pressure header and produces a master-loading air pressure, varying inversely with the variations in low pressure. This master-loading pressure is dispatched to the regulators controlling the respective reducing valves of the parallel systems. In accordance with the master-loading pressure, the regulators operate the valves so that the flow through the various systems will be at the correct value to maintain the pressure in the low-pressure header. The total flow of steam may be divided among the individual systems in any desired proportions, by adjustment of the master-loading pressures to the individual reducing-valve regulators, ratio adjusters being provided for this purpose on the master-controller panel. Once these ratio adjusters are set for a given load division among the systems, the master controller maintains the load proportions with stability throughout the normal flow range of the bypasses.

The desuperheaters following the reducing valves operate as previously described.

Very important is the provision for manual as well as automatic control of these parallel systems. One of these installations, consisting of three systems in parallel, is designed for transfer from automatic to manual operation at a central control panel. If the high-pressure turbine trips out during manual operation, automatic throw-over devices on the control panel will instantly transfer the bypasses back to full automatic control by the master controller. This system is also equipped with safety controls to prevent drawing too much steam from the high-pressure boilers and to prevent overloading the low-pressure system.

Refinements in Control. Figure 10-8 also shows a number of refinements, which have been applied to the fundamental controls shown in Figs.

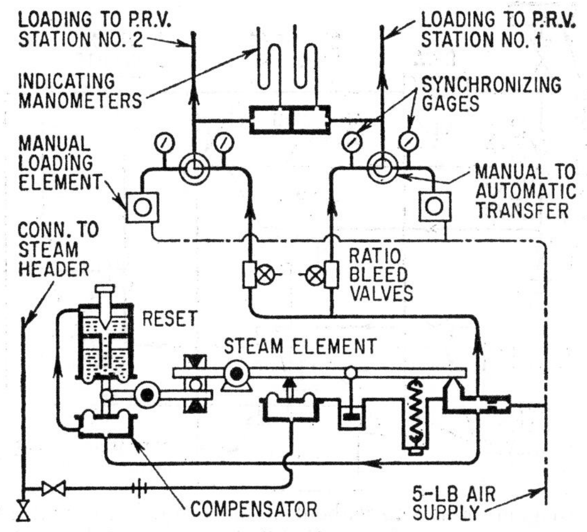

Fig. 10-9. Diagram of master controller of type shown in Fig. 10-8 to proportion system flows and hold desired low pressure.

10-1 to 10-4. As a safety precaution a water-shutoff valve in the desuperheater water line is closed by the reducing valve when that valve closes. This water-shutoff valve is sometimes built into the reducing-valve assembly. If desired, however, it may be a separate valve, lever-operated from the reducing valve as shown dotted, or it may be operated by a separate regulator. In some installations, a motor-operated quick-opening valve ahead of the reducing valve is controlled electrically from the topping-turbine-generator circuit and is also electrically connected to operate simultaneously a motor-operated water-shutoff valve in the desuperheating water line. Again, in cases where parallel systems are controlled by a master controller, to prevent the low-side temperature from going above a predetermined limit, a high-temperature safety device may be provided to operate leak-off valves in the master-loading lines leading to the reducing-valve regulators, thus limiting the steam flow to reduce the high temperature.

Parallel Systems without Master Controllers. Parallel pressure-reducing and desuperheating systems may be installed without master controllers, and the individual systems may have equal or different capacities. Without the master controller, the load will be divided according to the capacities of the respective systems. In some such cases, the regulators controlling the reducing valves are set so that the systems will go into operation in sequence.

Figure 10-10 shows a typical installation of this kind installed in an industrial plant. It consists of two reducing valves, one with a capacity of 15,000 lb of steam per hr. the other with a capacity of 45,000 lb of steam

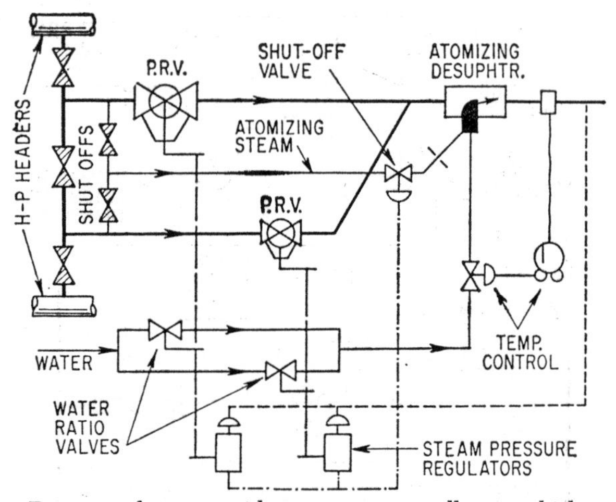

FIG. 10-10. Diagram of system without master controller, in which two reducing valves operate in sequence, discharging to single desuperheater.

per hr, operating in parallel and discharging steam to a single steam-atomizing desuperheater with a capacity of 60,000 lb of steam per hr, equal to that of the two valves together. The pressure is reduced from 410 to 120 psi, and the temperature from 625 to 380°F. The smaller valve operates first, up to the limit of its capacity; then the larger valve goes into operation automatically, and the two operate in parallel up to their maximum total capacity. The steam-atomizing desuperheater provides accurate steam-temperature control over the entire range of flow.

Turbine-exhaust Desuperheaters. As noted, desuperheaters are often used in the exhaust lines of topping turbines and other back-pressure turbines, as shown in Figs. 10-5, 10-6, and 10-12, to prevent the temperature in these lines from rising above the values that can be sustained by the equipment utilizing the exhaust steam. An increase in exhaust temperature occurs at low load in a straight back-pressure turbine.

The steam-atomizing desuperheater has proved itself well adapted for control of turbine-exhaust line temperatures, because of its wide range of operation. Occasionally, to secure the desired capacity, or to secure extremely wide range of operation (40 or 50 to 1), two or more of these desuperheaters are installed to operate in sequence.

In a large steam station, for example, the temperature of the exhaust steam from a topping turbine may be limited to 650°F by two steam-atomizing desuperheaters. One of these is located in each branch of the exhaust header leading from the turbine. They are controlled by manual operation of a valve in the water-supply line. Each desuperheater is designed to desuperheat to 650°F a maximum of 300,000 lb of steam per hr at 250

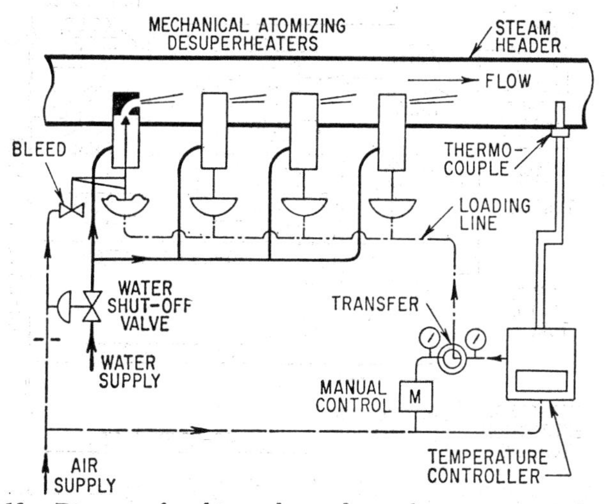

FIG. 10-11. Diagram of turbine exhaust desuperheating system, showing four mechanical atomizing desuperheaters controlled in sequence, with automatic water shutoff and provision for manual control.

psi pressure. Desuperheating water at 300 lb and 240°F is used. The range of operation required of each unit is from the maximum of 300,000 lb of steam per hr down to 150,000 lb/hr.

In another station is a turbine-exhaust desuperheater installation consisting of four mechanical-atomizing desuperheaters, of the type shown in Fig. 10-4, installed in the turbine-exhaust header as in Fig. 10-11 and operated in sequence by a single temperature controller to hold the exhaust-steam temperature at or below its maximum allowable value.

Tie Lines. Some stations have installed new generating units operating at higher pressure and temperature than the old plants but constituting additions, not topping plants. In such cases, there is usually a tie line to enable steam from the new high-pressure boilers to supplement or replace that from the low-pressure boilers, and in such a tie line a pressure-reducing and desuperheating station is necessary.

An unusual tie line installed contains two pressure-reducing and desuperheating systems in parallel. One system, of 960,000 lb of steam per hr capacity, consists of a 24-in. pressure-reducing valve, one of the largest ever built, and a steam-atomizing desuperheater, the system reducing pressure from 250 to 200 lb and temperature from 620 to 500°F. The second system, with a capacity of 50,000 lb/hr, is installed in parallel with the first and consists of a 5-in. valve and a steam-atomizing desuperheater. The two stations are controlled by a master controller; hence stable load division is maintained between them when they are operating in parallel, but either system may operate alone to meet the conditions.

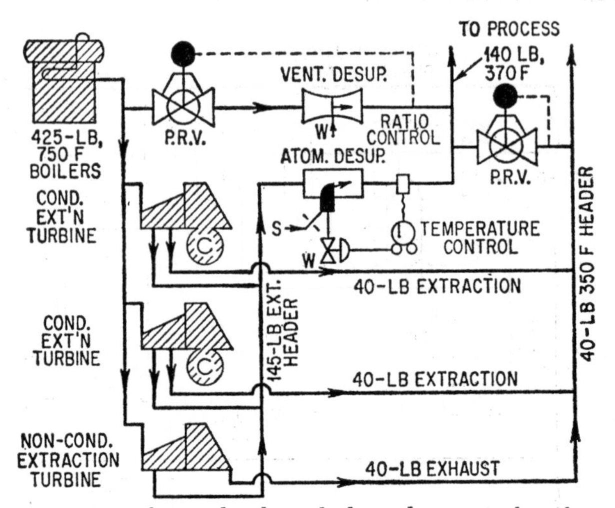

Fig. 10-12. Diagram of typical industrial plant, showing turbine bypass, valves, and desuperheaters controlling relations between steam systems.

Industrial Plants. In industrial plants, as pointed out above, the problem of connecting various pressure systems through pressure-reducing valves and desuperheaters is common. Steam must often be supplied to supplement either turbine exhaust or extraction steam. In some industrial power plants, no low-pressure boilers whatever are installed. Instead, all steam is generated at high pressure, 400 lb and sometimes 600 lb, and the required process flows and pressures are obtained by extraction or exhaust from steam turbines or through pressure-reducing and desuperheating stations. Condensing extraction turbines are frequently used for this service. Figure 10-12 shows diagrammatically a typical modern industrial plant in which pressure-reducing and desuperheating stations form vital links in the steam-supply system.

Section 11

STEAM-TURBINE INSTRUMENTATION AND CONTROL

TURBINE GOVERNORS[1]

In fundamental characteristics, the various types of pressure regulators are directly comparable to turbine-speed control. Turbine-speed-control devices are basically "floating" or "proportional action" in operation. These basic systems can be modified, to provide additional stability, to give isochronous (zero-regulation) operation, or otherwise to adapt the control to the specific demands of the application.

In any case, the three essential elements of automatic control will always be present: the sensitive element (or speed governor), the transmission (and sometimes the amplification) of the control impulse, and finally the "servo" element which resets the valve to return the controlled function to the set point.

In the field of turbine-speed control, turbine speed is of course the variable which initiates the action of the control system. The speed-sensitive element in all turbine-speed-control devices may be one of two basic types:

1. The mechanical governor
2. The hydraulic governor

The Flyball Governor. This type of governor depends on the centrifugal forces exerted by some type of rotating element whose speed is proportional to turbine speed. This centrifugal effect may be used directly *as force* or may be used as *motion* to actuate the control system.

A flyball governor of the type illustrated in Fig. 11-1 is used for speed control of mechanical-drive turbines where extreme sensitivity is not required and the inherent friction of the system is not a serious consideration.

This governor assembly is mounted on the turbine shaft. The outward motion of the weights is opposed by the adjustable governor spring and is converted into motion of the valve stem which, through linkages not shown, closes the turbine steam valve. The governor spring, housing, and weights

[1] Abstracted from "Fundamentals of Turbine Speed Control," published by Elliott Company.

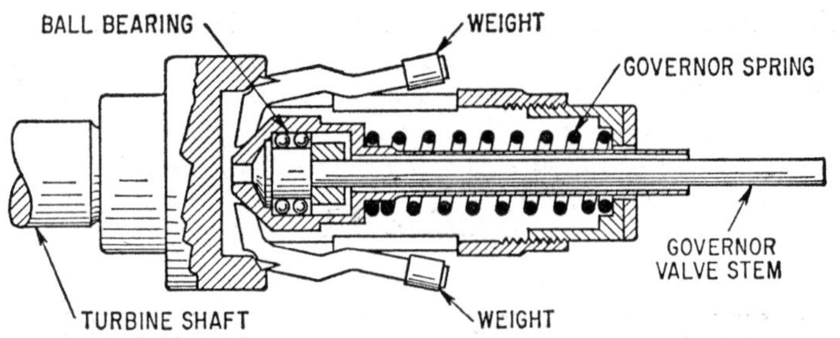

FIG. 11-1. Direct-action (mechanical-shaft) governor. (*Elliott Co.*)

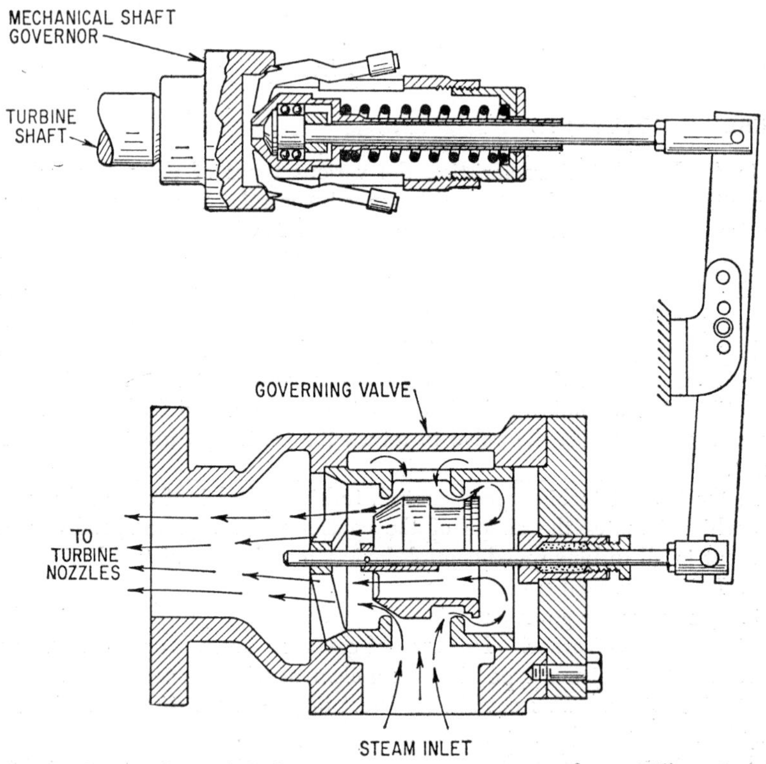

FIG. 11-2. Mechanical-shaft governor actuating steam valve. (*Elliott Co.*)

rotate with the shaft, and the motion of the weights is transmitted through the ball bearing to the stationary spindle.

The governor weights are sufficiently heavy to provide the force required to move the governing valve within a certain specified regulation.

Mechanical-speed-governor Control. The simplest type of automatic turbine-speed control is the speed governor shown in Fig. 11-1 applied as a direct-acting governor (Fig. 11-2). The horizontal, flyball-type, speed-sensitive element is mounted on the end of the turbine shaft and connected to the valve by a lever. In this application the sensitive element acts directly on the valve to restore the variable to the set point. Outward movement of the governor weights is translated into horizontal movement of the spindle. The relation between the governor weights and the scale of the governor spring determines the operating speed and the regulation of this type of control.

The simple direct-acting speed governor is suitable for many mechanical-drive steam-turbine applications. However, it is not highly sensitive and does not permit a high degree of speed control. The governor is a NEMA

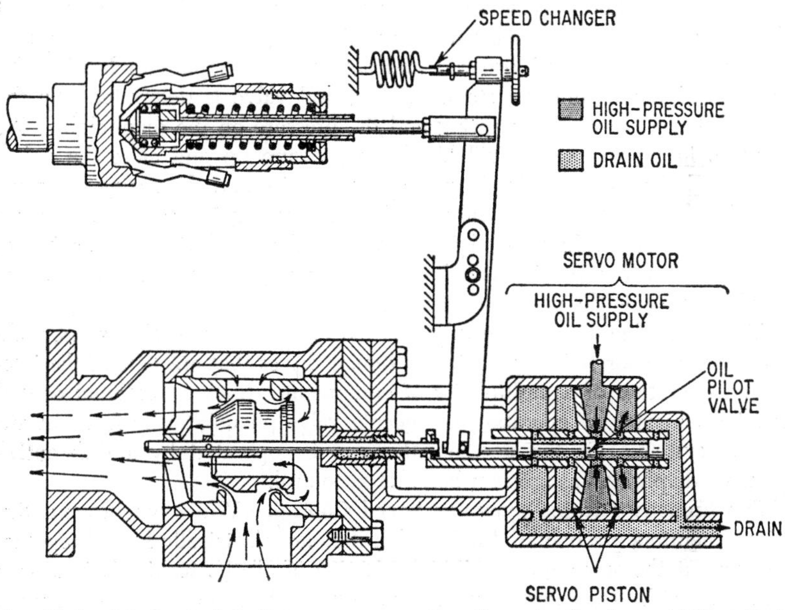

Fig. 11-3. Mechanical-shaft governor operating through oil relay. (*Elliott Co.*)

class A type, and has wide application for driving fans, pumps, and compressors in boiler rooms, refineries, and similar installations.

Oil-relay-speed-governor Control. This type of speed governor (Fig. 11-3) makes use of the same speed-sensitive element as the direct-acting mechanical governor.

However, a *servomotor* has been introduced here to move the governing valve. The only effort required of the speed governor is to move the pilot valve in the servomotor which, in turn, produces the power to move the governing valve. Thus, since the speed governor is not required to move the governing valve, the sensitivity of the entire speed-governing system is greatly increased.

This governor system gives closer speed control than that in Fig. 11-2, and corresponds to a NEMA class B. It is frequently used on small generator drives, such as auxiliary lighting sets.

The servomotor illustrated is the center-pilot type having the pilot in the center of the piston. When the pilot moves, it opens an oil port, admitting oil which forces the piston to move with the pilot. This piston motion shuts the oil off when the piston has caught up with the pilot. Note that, while the pilot motion to the left puts high-pressure oil on the right side of the piston, it also connects the left side of the piston to drain. This gives a pressure differential over the piston, positioning the valve more accurately than if the piston were on a spring return.

Variable-speed-governor Control. The speed-sensitive element of this control (Fig. 11-4) is a shaft-mounted oil pump. When discharged through an orifice (a needle valve), the discharge pressure from the pump is proportional to speed squared and acts on the diaphragm head. The position of the diaphragm stem, linkage, and valve is determined by the force balance between the spring and the pressure on the diaphragm. An increase in turbine speed will build the pressure up and close the valve. The regulation is determined by the pressure rise and the spring compression for full stroke of the valve. If a stiff spring is used, the pressure change for full stroke will be increased and the regulation will be broad. The speed of the turbine can be varied by varying the pressure on the diaphragm, using the needle valve as an adjustable orifice.

This type of governor is readily adapted to external-control devices, making it suitable for boiler-fan drive or similar drives that are part of the integrated operation of a multiplicity of equipment. Such an external control can be achieved by operating the needle valve pneumatically from an air-control system. Or the needle valve can be operated by a differential-pressure diaphragm as used for boiler-feed control.

Synchronous Turbine-generator Speed Control. When synchronous turbine-generators are operated in parallel, they are electrically locked together and run at the same speed. The regulation of all the turbine drives must be substantially the same in order for the generators to accept load changes in proportion to their ratings. The situation is similar to that of a nest of springs supporting a floating load. If the load changes, each individual spring must reach the state of equilibrium and carry its proper share of the new load. To obtain this condition of proportional sharing of the electric load, it is customary to specify that all turbines in this service have a steady-state speed-regulation value of 4 per cent as standard.

On a multivalve turbine, the steam forces on the valves are of considerable magnitude, and it takes a large servo to handle the valves. This large servo takes its position signal from the control pressure set by the sensitive type V governor. The speed-control system pictured in Fig. 11-5 shows how the control pressure acts on the bellows in the pilot servo. The oil piston in this device positions the pilot valve that controls the oil flow to the main piston for actuating the valves.

The connection from the rotating pilot to the large oil-control pilot for the main servo piston is, in effect, a solid column of incompressible oil. Therefore, the signal from the rotating pilot is transmitted instantaneously, the same as it would be through a solid rod or link.

This design conforms to requirements of "Recommended Specifications

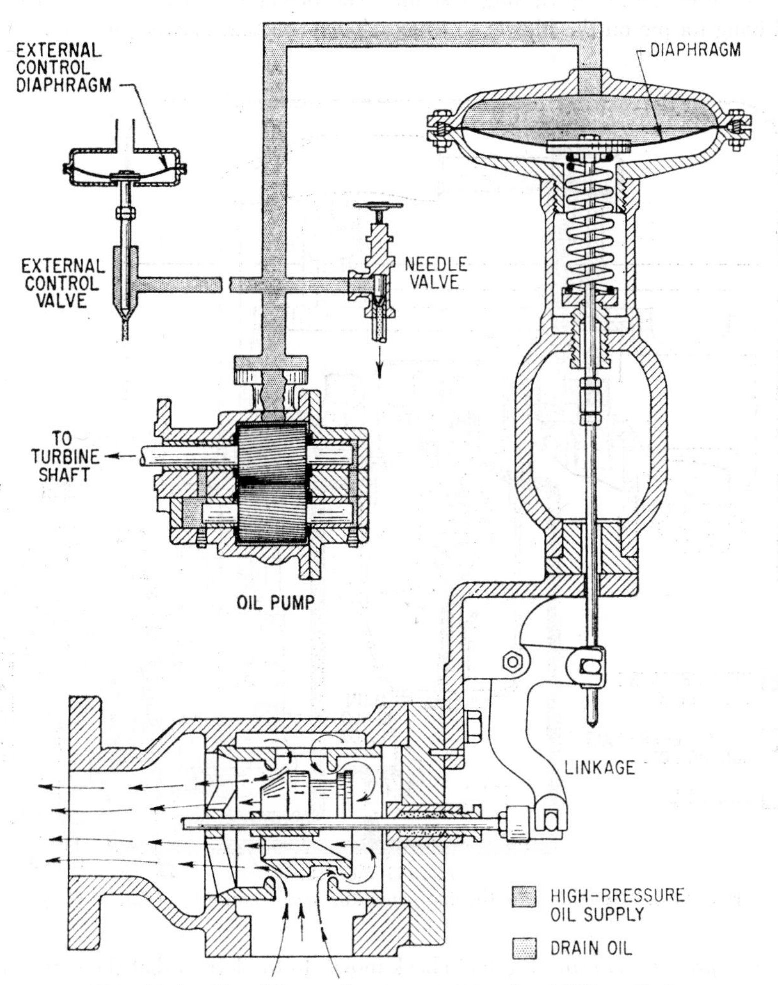

Fig. 11-4. Variable-speed-governor control. (*Elliott Co.*)

for Speed Governing of Steam Turbines Intended to Drive Electric Generators Rated 500 kw and Up"—AIEE No. 600.

Governing-valve Positioning by External Control. This so-called "vertical oil-relay governor" with horizontal valve is used for more refined control than is available with the mechanical shaft governor in Fig. 11-3 or the orifice governor in Fig. 11-4.

The external-control device pictured is used when it is required that an external impulse directly control turbine steam flow and driving torque. In the control illustrated in Fig. 11-6, the discharge pressure of a centrifugal blower is picked up by a pressure-sensitive element, and the impulse is transmitted to the governing system. The object would be to control the driving torque on the blower so as to achieve constant blower pressure.

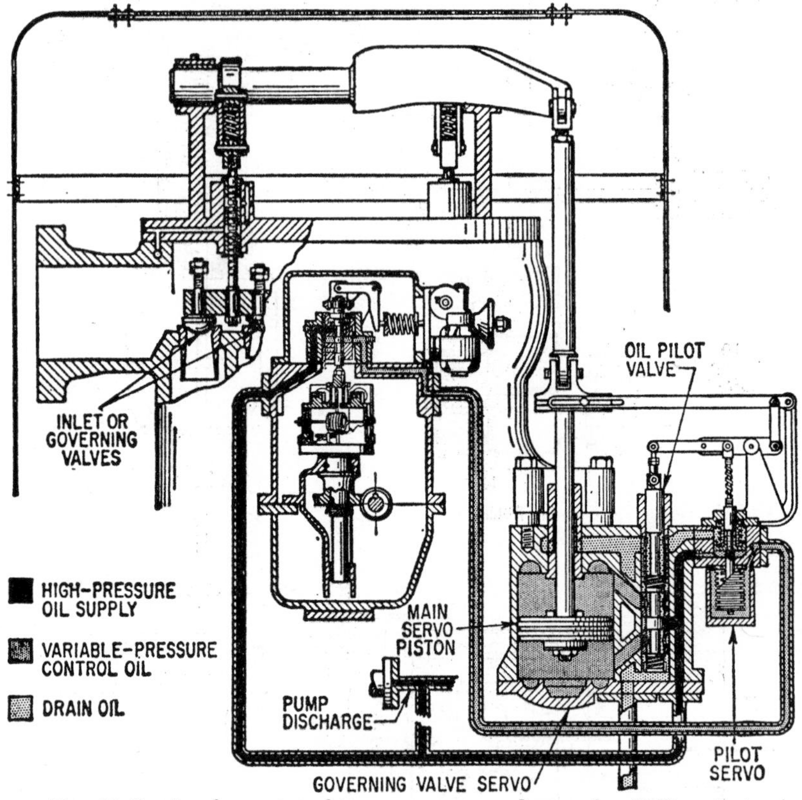

FIG. 11-5. Synchronous turbine-generator speed control. (*Elliott Co.*)

For *pressure control,* the ball check moves to the left so that the external-pressure pickup device establishes the control pressure in the governor servo. This pressure pickup makes use of hydraulic-orifice impulse transmission. High-pressure supply is throttled through an orifice, and the pressure downstream from the orifice is controlled by a relief valve. This relief valve is loaded by the sensitive element, and thus pressure under the relief valve, and in the circuit beyond the orifice, becomes a function of the actions of the sensitive element. Use of differential areas results in amplification.

In the system shown, a standard 3-to-15-psig air signal from the blower-discharge control can actuate the bellows and load the relief valve. The resultant control pressure under the relief valve—if it is *higher* than the governor-control pressure—positions the governor servo and the inlet valve without speed-governor action. This happens because the speed governor is not effective, so long as its control pressure is *lower* than the control pressure from the pressure control. Thus, blower output controls the turbine

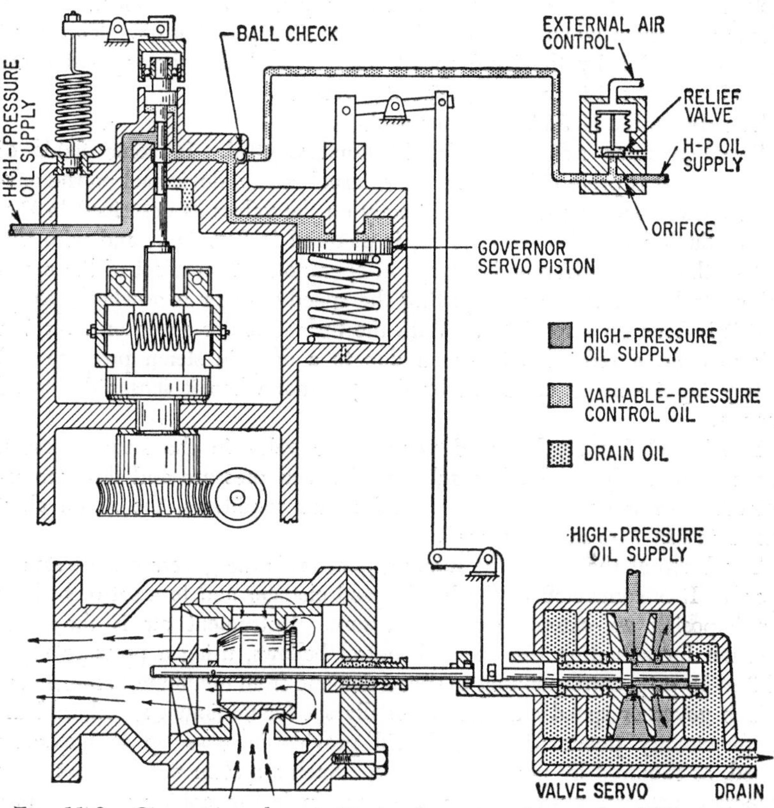

FIG. 11-6. Governing valve positioning by external control. (*Elliott Co.*)

driving torque directly and without delay. As soon as blower discharge pressure changes and actuates the air control, the governing valves are moved directly, the change in steam flow changes the driving torque, and the blower-discharge pressure is restored. During this time, the governor has been set for a certain maximum speed, below which the external control has unlimited effect on range of turbine power and speed. The external control could shut the governing valves, resulting in zero output. If the speed is increased, reaching the speed setting of the main governor, the governor pilot will open and set a control pressure that will override the pressure from the pickup. The ball check moves to the right, as shown in Fig. 11-6, and the speed governor takes control.

When the governor has taken over and the unit is under *speed control,* the control pressure from the governor moves the governor servo, and the servo piston positions the center pilot in the valve servo. The piston that moves the valve is double-acting and gives accurate positioning of the valve. With the unit under governor control in this manner, the external-pressure pickup is not effective, and the unit is subject to normal speed control. The same applies when the external-pressure pickup is shut off. The speed range obtainable by the hand speed changer, and also the regulation are the values inherent in the normal speed governor.

This design is a NEMA class B control.

Speed-governor Adjustment by External Control. In some fields of application it is preferred to have the external-control device act on the governor speed changer rather than directly on the valves as described in Fig. 11-6. Such a speed-control system is pictured in Fig. 11-7, and with this system the speed governor is always in operation. The effect of the external control is the same as adjusting the speed-changer handwheel, except that it is done by a regulator. With the turbine under governor control in this manner, a reduction in speed due to a drop in steam pressure and energy input will be corrected by the governor when it senses the speed change.

In Fig. 11-7 a single-seated venturi-type valve in the steam chest is shown, commonly used when substantial steam pressure and flow are involved. For more moderate flow and steam conditions, the steam chest shown in Fig. 11-6 might be used.

The beam at the top of the rotating pilot loads the governor and sets the speed. It is connected to the hand speed changer and to a set of bellows for imposing an external-control pressure. Instrument air is the usual medium for actuating the external-control bellows, using a pressure range of 3 to 15 psig, corresponding to speed demand from minimum to maximum.

The governor speed changer is initially set for minimum speed, and the supplementary pneumatic control will provide increase up to maximum speed, as required by process demands. Such a system may be part of an elaborate process control, where a change in instrument air pressure produces a change in turbine speed by changing the speed setting of the governor.

Compared with the external-pressure control in Fig. 11-6, this system has the advantage of the governor picking up a speed change when the energy input changes because of variations in steam pressure or vacuum, and the governor will adjust the governing valve accordingly. On the other hand, such a change in steam pressure has to be substantial before it is reflected in noticeable speed change. And, for quick, precise response to process control, Fig. 11-7 puts the signal through the governor, while Fig. 11-6 sets the valve and energy input directly, and so Fig. 11-6 is bound to respond

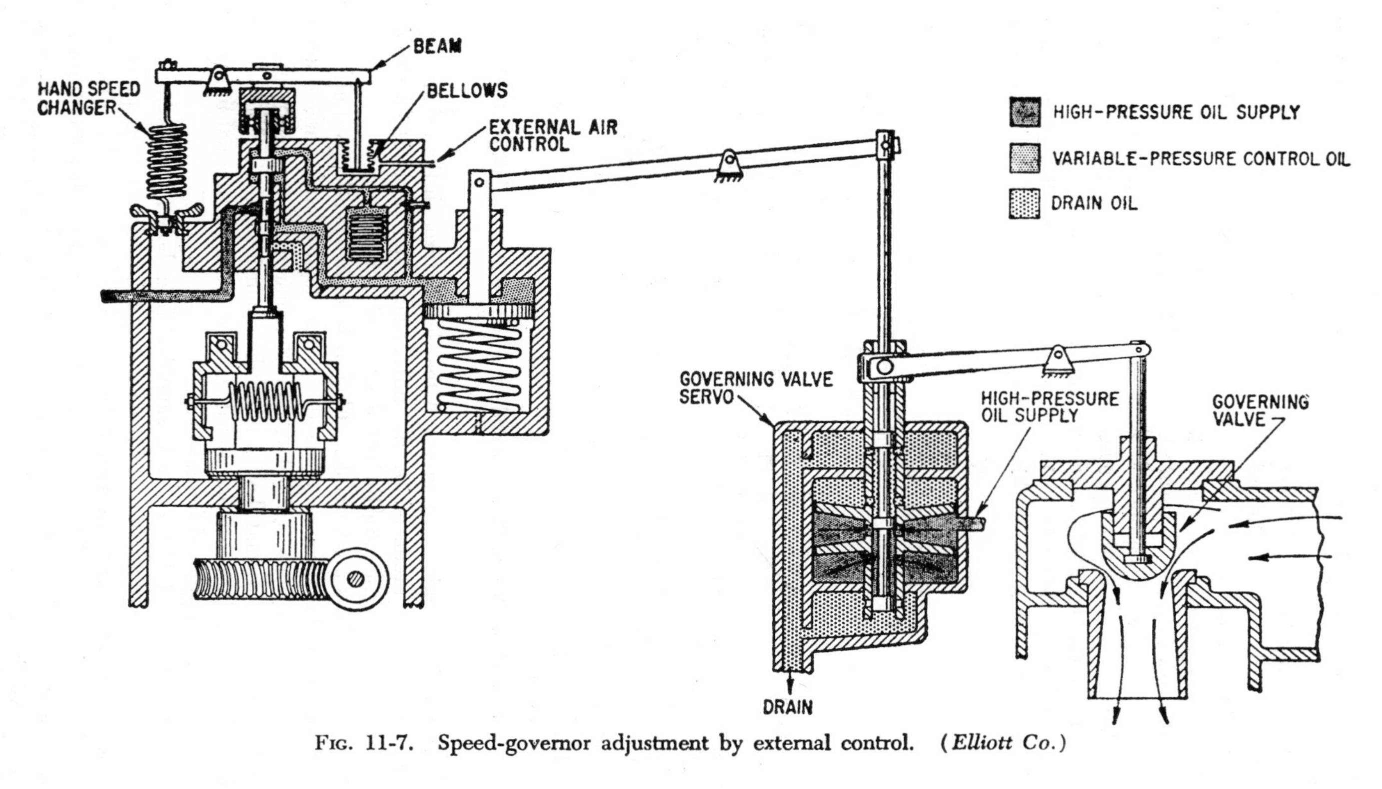

FIG. 11-7. Speed-governor adjustment by external control. (*Elliott Co.*)

faster to the external control. But the time difference is less than one second.

The control in Fig. 11-7 may also have the isochronous feature added, as shown. This is the same principle and control as in Fig. 11-6. It will maintain accurate speed for changes in both steam conditions and load. It gives the machine automatic reset.

Automatic-extraction-pressure Control. Steam at a constant pressure for process use can be supplied from a steam turbine by adding a pressure-regulating system to the speed-control system. The combined speed and pressure control is called extraction control.

Essentials of Automatic-extraction-pressure Control. The steam flow to the inlet of an extraction turbine is the total of stream *B* that leaves through

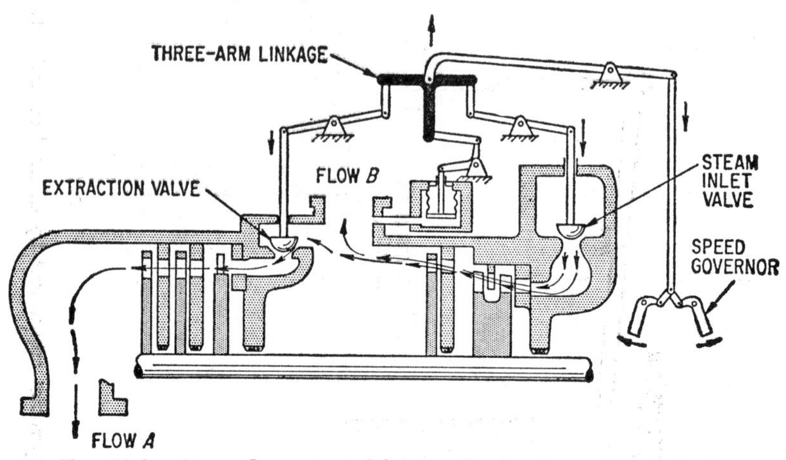

Fig. 11-8. Steam flow actuated by speed governor. (*Elliott Co.*)

the extraction opening and stream *A* that leaves through the exhaust opening. The power produced by the turbine-generator depends on the total torque exerted on the rotor by both flow *A* and flow *B*.

Both these flows are variable. Flow *B* is an *independent* variable, since it varies with the process demand. If the total power demand on the turbine-generator were 8,000 kw, flow *B* required for process might generate only 2,000 kw, and flow *A* must then contribute the additional 6,000 kw. Flow *A* is the *controlled* variable, since it can be adjusted to maintain correct power output and frequency.

Figures 11-8 and 11-9 show in simplified form how an automatic-extraction-control system responds to the varying steam demands of power and process. If the speed governor calls for reduced steam flow (Fig. 11-8), the linkage is designed to reduce the steam flow through both steam inlet valve and extraction valve. This is accomplished by lifting the three-arm linkage vertically so that it effects the same flow change through inlet valve and extraction valve. Thus, while flow *B* to the extraction opening remains unchanged, flow *A* to exhaust is reduced.

However, an extraction control must also handle changes in the process demand, flow *B*, without at the same time affecting the kw generated (Fig. 11-9). The control system must, for example, simultaneously produce a decrease in flow *B* and an increase in flow *A*, thus maintaining the total torque on the rotor constant for constant kilowatt output.

This is accomplished by the bellows, which is responsive to the extraction pressure. A reduction in extraction demand, and consequent increase in pressure, will compress the bellows. The upward motion of the bellows will rotate the T-shaped beam as shown, closing the inlet valve while opening the extraction valve. Flow *B* is thus decreased, and flow *A* increased. By properly proportioning the two horizontal linkage arms on the T beam,

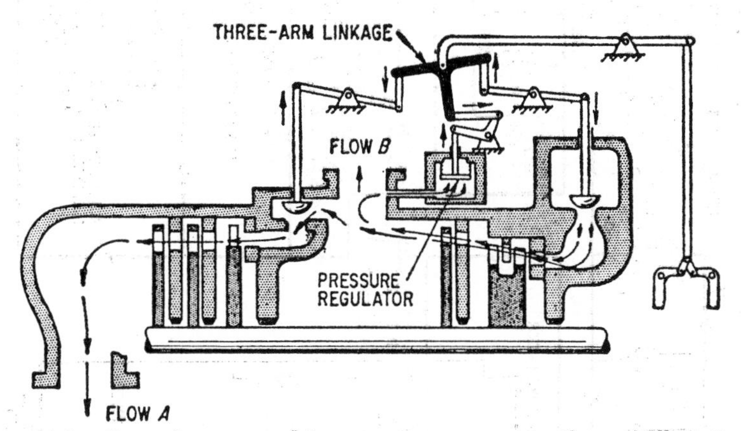

Fig. 11-9. Steam flow actuated by extraction-pressure regulator. (*Elliott Co.*)

the total steam torque can be kept constant, maintaining proper speed and kilowatt output despite fluctuation in process steam demand.

Development of Large and Superpressure Turbine Governors.[1] Power industry demands for larger-rating turbine-generator units without a proportional increase in weight require that the applied control system have excellent response characteristics. Also, superpressure turbine designs necessitate special control consideration in regard to the power requirements of the valve-operating mechanisms.

Figure 11-10 illustrates, in schematic form, a 300-psig lubrication and control system which has been developed for application on large turbine-generator units. For comparison, a similar diagram (Fig. 11-11) shows a 150-psig system which has been applied to central-station turbines.

In comparing the 300- and 150-psig systems, two basic differences are apparent: (1) the arrangement of the high-pressure auxiliary oil pump and

[1] E. G. Noyes, Jr., supervisory engineer, Westinghouse Electric Corp.

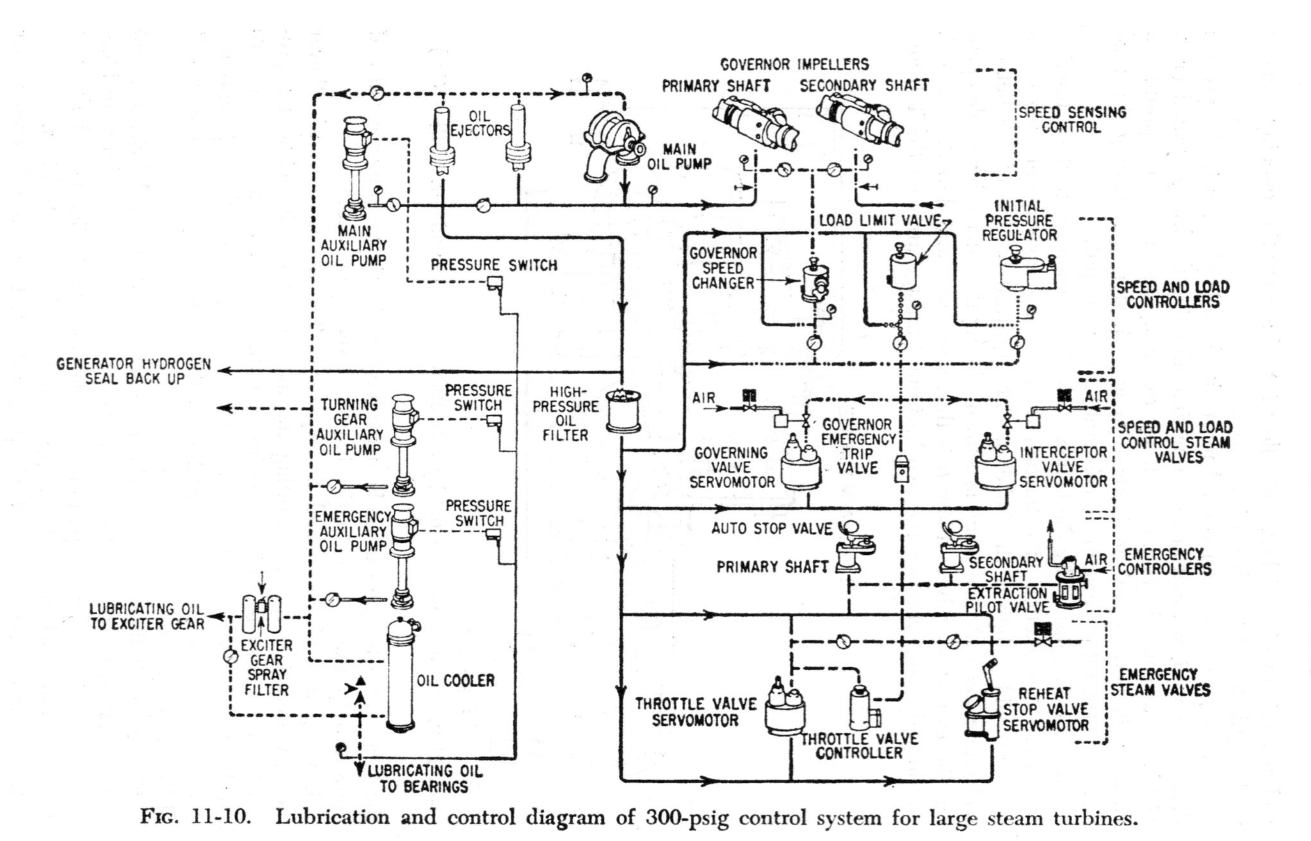

FIG. 11-10. Lubrication and control diagram of 300-psig control system for large steam turbines.

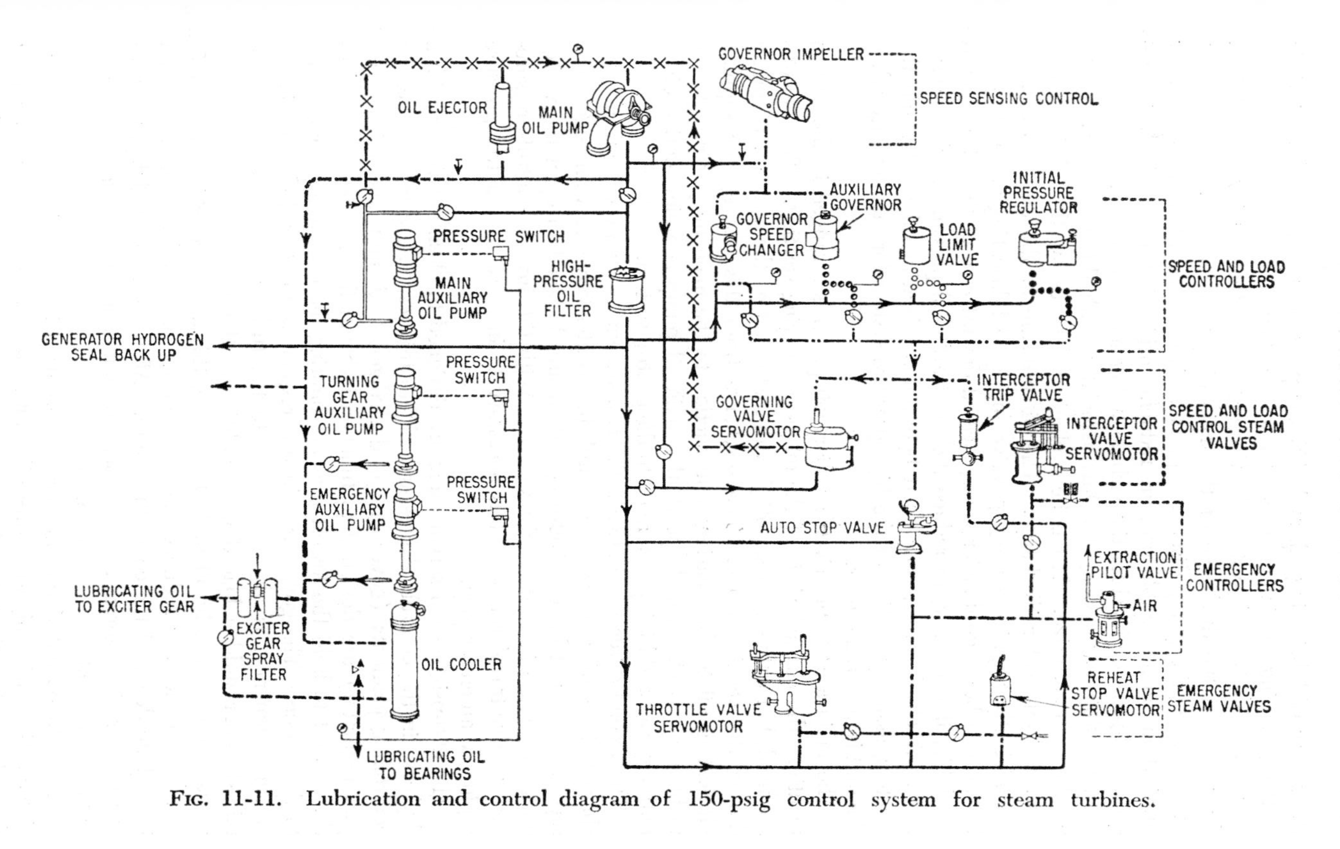

Fig. 11-11. Lubrication and control diagram of 150-psig control system for steam turbines.

the source of lubricating oil, and, (2) the pressure-responsive characteristic of the governing and interceptor-valve servomotors.

In the 150-psig oil system, lubricating oil for the bearings is obtained directly from the discharge of the main oil pump located on the turbine shaft. Since the pump-discharge pressure is in excess of that required for lubrication, an adjustable pressure-reducing orifice is used to obtain the required oil-pressure level for the turbine and generator bearings.

In the 300-psig system, the loss associated with this pressure reduction would be prohibitive. It has been eliminated by taking the lubricating oil from the oil-ejector discharge. The oil pressure at the ejector discharge is sufficiently high to assure a positive supply of lubricating oil to the bearings. The bearing-oil relief valve, located downstream of the oil coolers, is available, if necessary, to reduce the lubricating-oil pressure to the desired level. A check valve, positioned between the bearing-oil header and the ejectors, assures isolation of the high-pressure system for maintenance when the unit is on turning gear.

The main a-c motor-driven auxiliary oil pump is arranged in the system to utilize the characteristic of the oil ejector during starting. Two ejectors are used, but only one is supplied with motive oil by the auxiliary oil pump during the starting cycle. The discharge from this one ejector supplies lubricating oil to the bearings and suction to the main oil pump. As the turbine approaches rated speed, the main oil pump supplies motive oil to both ejectors and high-pressure oil to the control system.

The second difference between the two systems is the pressure-responsive characteristic of the governing and interceptor-valve servomotors. The servomotors used on the 150-psig system are designed to open the valves with a decrease in control oil pressure. With this inverse control-pressure-versus-valve-travel relation, it is necessary to increase the control pressure rapidly in an emergency.

The 300-psig-system servomotors are designed to close the valves with a decrease in the control oil pressure. With this response characteristic, the controlling elements (governor speed changer, load limit, and initial pressure regulator) are designed to function individually as relief valves, with the one set to hold the lowest pressure limiting the control-header pressure to that value. The governor, which is responsive to normal changes in the system frequency, is also responsive to the turbine-shaft acceleration obtained under sudden load-dump conditions. If the rate of speed increase exceeds approximately 50 to 100 rpm/sec, this acceleration-responsive feature instantaneously reduces the governor set pressure to zero, causing an immediate decay of the control pressure. After a brief time delay, the control pressure is restored, and the speed is maintained at a value corresponding to the normal steady-state speed regulation.

It should be noted that all servomotors used in the 300-psig system are single-acting, with high-pressure oil used as the motive means for opening the valves. For the governing and interceptor-valve servomotors, the control oil acts on a pilot relay, which in turn positions the main relay controlling the motive oil to the bottom of the operating piston. A similar arrangement is used for the stop-throttle and reheat stop-valve servomotors. However, since these are emergency valves, the autostop oil acts on the pilot relay to control the valve position. Springs are used on all valves to ensure positive closing action.

Loss of Electrical Load and Turbine-generator Overspeed. In the 150-psig oil system, a separate auxiliary governor furnishes the preemergency

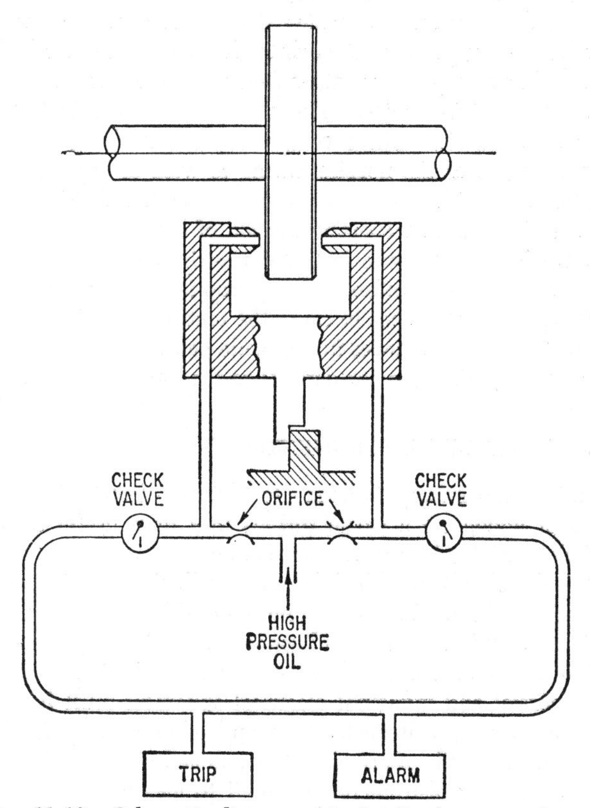

FIG. 11-12. Schematic diagram of hydraulic-thrust trip device.

protection for the turbine-generator unit in the event of loss of electrical load. Under normal-rated-speed operating conditions, the auxiliary governor will not assume control of the unit since its control pressure is less than that established by the main governor. However, with an increase in speed, the auxiliary governor pressure increases rapidly, causing the governor and interceptor valves to close.

Protective Controls. The autostop valve, which controls the position of the emergency steam valves, is a convenient single control that can be actuated by other devices to protect the turbine-generator unit. Standard features are provided on large turbines to protect the unit for conditions of overspeed, low bearing-oil pressure, low vacuum, and thrust-bearing failure. A solenoid is also provided which will permit remote tripping of the throttle and reheat stop valves.

The thrust-bearing-failure protection is a feature which has been added within the last few years and is a good example of the adaptability of the hydraulic-control system. As seen diagrammatically in Fig. 11-12, the device consists essentially of two nozzles, each placed a specified distance from either side of the disk on the turbine shaft at or near the thrust bearing. These nozzles are supplied with oil through orifices from the high-pressure oil system, and have a pressure take-off on each side through a check valve to an alarm and a trip device. The alarm device is electrical, while the trip device is hydraulic. These components are so designed that, as the disk approaches one of the nozzles, owing to wear or wiping of the thrust shoes, the nozzle-to-disk distance tends to become the limiting restriction in the oil line, and takes an increasing portion of the available pressure drop, thereby increasing the pressure in the alarm-trip line. When this pressure reaches the alarm or trip setting, the appropriate action occurs. The required axial spindle movement to cause the unit to trip is insignificant when compared to the axial clearances in the blade path. Therefore, in the event of a thrust-bearing failure, major damage to the unit is avoided.

TURBINE-SUPERVISORY INSTRUMENTS[3]

Steam-turbine-supervisory instruments indicate electrically the mechanical motions or actions of a steam turbine in a form useful to the turbine operator in determining his future course of operation.

Since a turbine is a trouble-free machine with several years between scheduled overhauls, the instrument must be able to go a long time with little trouble and without major adjustments. This long operating-life requirement means that the ordinary economic standards for the design of similar commercial and industrial electrical apparatus must be exceeded in the design of these instruments. Like any other commercial instrument, the devices must be easily adjusted and maintained.

Turbine-supervisory instruments are used to measure, indicate, and permanently record the following mechanical conditions of a turbine:

1. Shaft eccentricity
2. Bearing or shaft vibration
3. Shell expansion
4. Speed
5. Camshaft position
6. Differential expansion

Eccentricity Recorder. The eccentricity recorder indicates and records the amount of eccentricity present at a given position on the turbine shaft from turning-gear speed to synchronous speed. This instrument makes it

[3] General Electric Co.

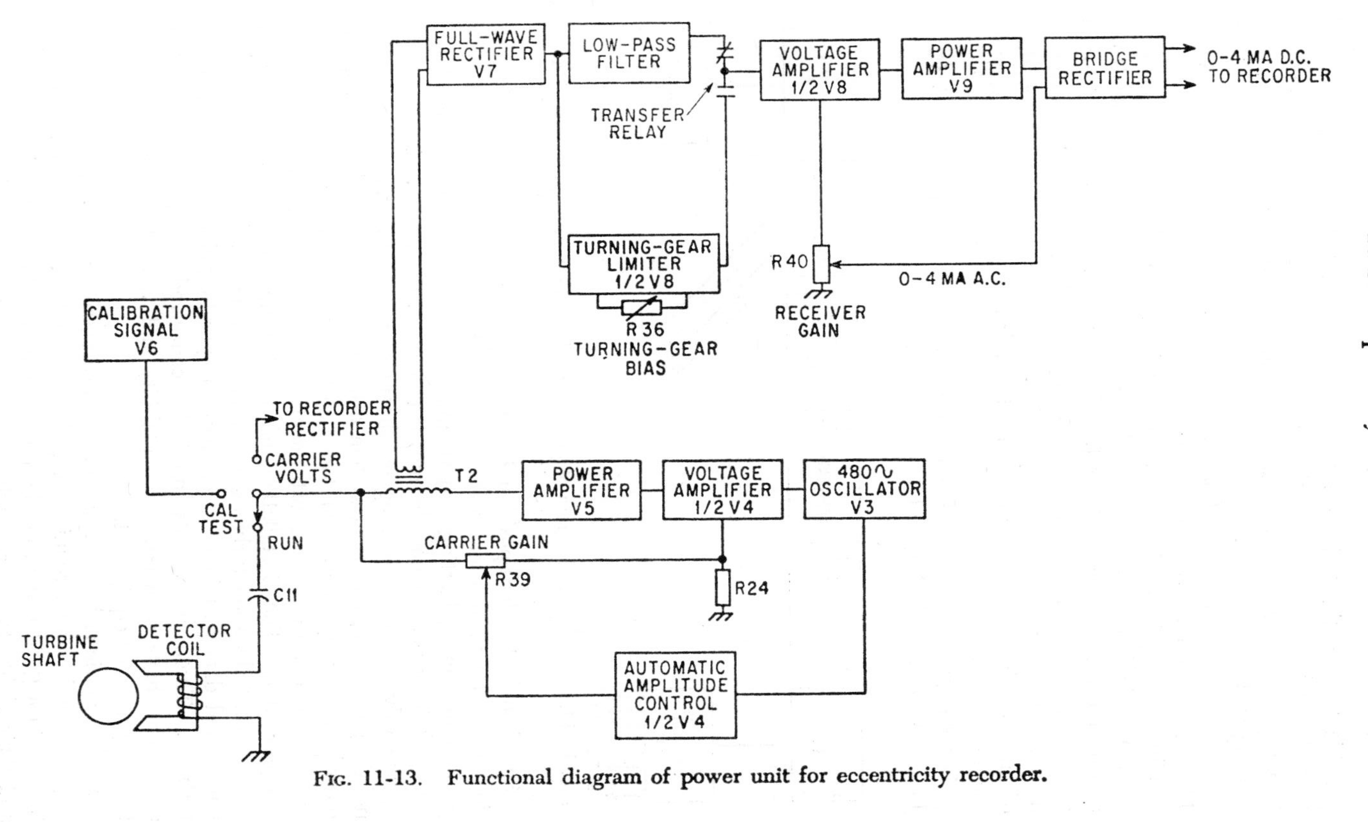

Fig. 11-13. Functional diagram of power unit for eccentricity recorder.

possible to start the turbine, after a shutdown period, in a much shorter time than when eccentricity values were inadequately known. The condition of the shaft is known at all times, and the operator can govern his starting cycle accordingly.

The equipment consists of a recording instrument, a power unit (Fig. 11-13), an alarm initiator, and a detector.

Vibration-amplitude Recorder. The vibration-amplitude recorder provides a periodic record in mils of the vibration of the bearings or shaft occurring on a turbine-generator set. The equipment assists an operator in foreseeing harmful trends of unbalance by providing records which can be used for checking a machine's operation. It also assures proper operation of large machines starting from a cold standstill by recording vibration as the machine comes up to speed.

The equipment consists of a recording instrument, a power unit (Fig. 11-14), an alarm initiator, and detector units.

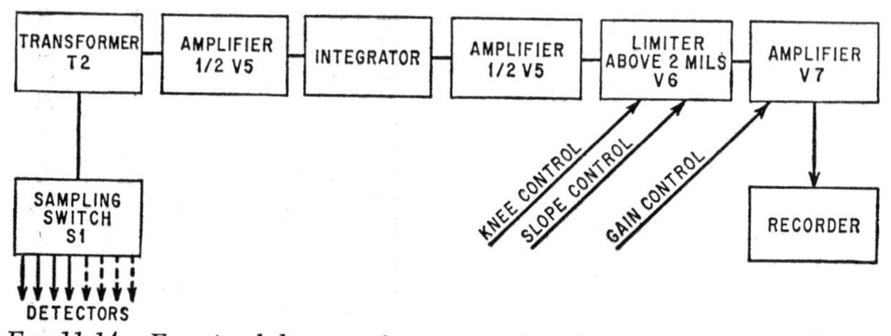

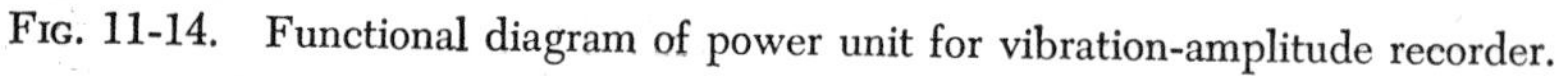
FIG. 11-14. Functional diagram of power unit for vibration-amplitude recorder.

Speed and Camshaft-position Recorder. The speed and camshaft-position recorder indicates and records the speed of the turbine in revolutions per minute during starting and stopping periods, and the rotation of the camshaft in per cent while the unit is connected to the line. The opening of the valves and the loading on the machine are functions of the cam position; therefore, effects of system disturbances undiscernible on a speed record are clearly shown on the camshaft-position record. The equipment consists of a recording instrument, a power unit, and detectors.

Application. The installation of turbine-supervisory instruments on existing central-station turbines has brought to light considerable new and unexpected data pertaining to mechanical operating characteristics of large turbines. These instruments, originally conceived for remotely operated equipment, have shown their adaptability and value to attended station-turbine-generator equipment.

In the past, in order to ensure safe and successful performance in starting, stopping, and normal running of turbine-generators, it has been necessary to rely, to a great extent, on the power of observation of turbine operators and their ability to describe what took place. Starting cycles and normal running instructions were prepared by the manufacturers, recommending the procedure to follow for individual units. The recommendations were based on observations of hundreds of machines in service, and, in general, have provided safe but probably conservative technique for individual machines.

Based on experience gained with individual units, operators have modified their practice. Unusual conditions, both inside and outside of the turbine, influenced the manner in which the turbine should be operated. To obtain the greatest efficiency from any given station, it is necessary that

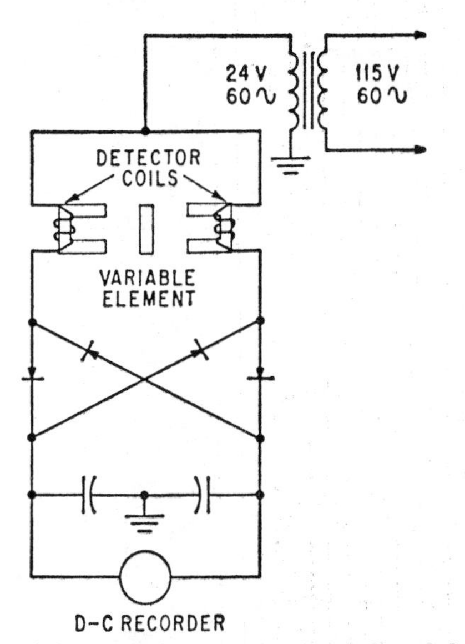

Fig. 11-15. Functional diagram of power unit for shell and differential-expansion recorder.

the individual turbine-generator units be capable of being started in the minimum length of time, and operated without danger of impairing their efficiency or mechanical performance.

Turbine-supervisory instruments provide the operator with an indication and also a permanent record of mechanical performance throughout the starting period and subsequent running time. These permanent records permit all interested individuals to review the past operation of the turbine, and to plan the future operating method to be employed, instead of having to rely on the opinion of one operator. When unusual conditions are shown, the turbine operator will be able to make changes in his operating procedure to prevent unnecessary wear, which might otherwise occur.

Record Analysis. Since the value of turbine-supervisory instruments depends on ability to analyze the records obtained, the following case is

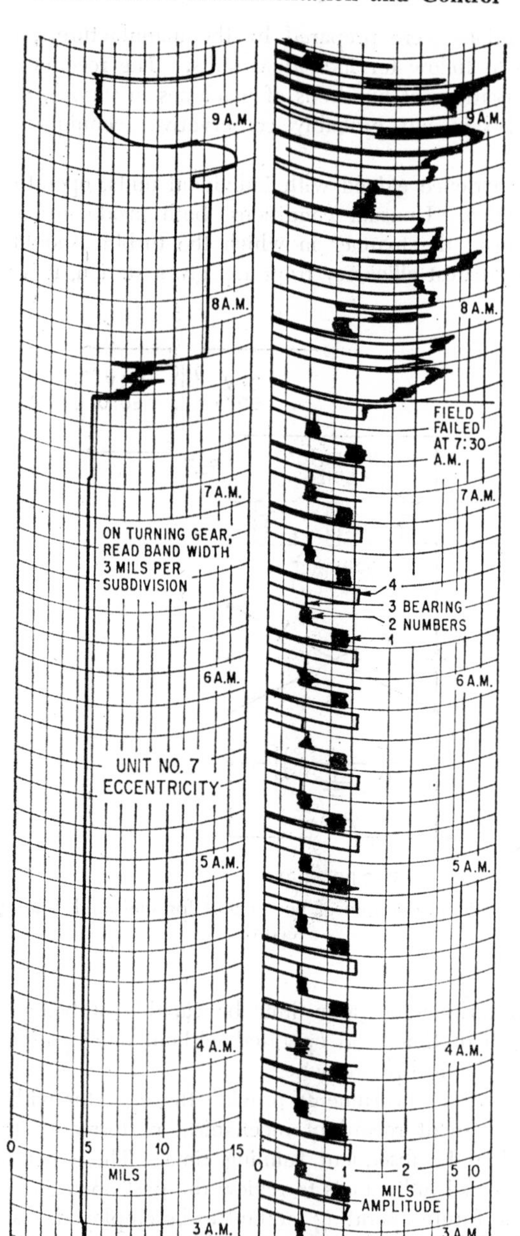

FIG. 11-16. Eccentricity and vibration records taken from a 60,000-kw 3,600-rpm noncondensing turbine-generator during field failure.

described in which the vibration and eccentricity instrument warned of impending trouble in a 60,000-kw 3,600-rpm noncondensing machine in which the generator rotating-element field coils failed to ground. The record from 3 to 7:28 A.M. indicated that the machine was normal while carrying full load (Fig. 11-16). During this period the shaft eccentricity was steady at a value of 4.75 mils, and the bearing vibration was constant with the number 4, or generator end bearing, at 1.2 mils. At 7:30 A.M. the shaft eccentricity increased very rapidly to 7 mils and oscillated. At the same time, the bearing vibration increased to about 2 mils at the number one bearing. The shaft eccentricity showed rapidly varying values until 7:42 A.M., at which time the eccentricity leveled off at 12 mils. The bearing vibration, in the meantime, was steadily increasing. At 8:36 A.M. the shaft eccentricity dipped rapidly, and then went to 14 mils. At the same time, the bearing vibration increased still further. Between 8:30 and 8:45 A.M. the unit vibrated so violently that it was taken off the line. The unusual records were caused by a field coil on the generator rotor failing to ground. This failure developed a high local heat concentration which resulted in a shaft bow . . . that became steadily worse with time. Inspection of the unit after the shutdown disclosed that all the bearings on both the generator and the turbine were wiped, and considerable damage was found on the oil deflectors.

The mechanical damage could have been minimized if the unit had been taken off the line at 7:28 A.M. when both the eccentricity and vibration recorder indicated unusual operating conditions instead of its being allowed to continue.

Figure 11-17 shows the eccentricity, vibration, and speed–camshaft-position records of a 100,000-kw 3,600-rpm tandem-compound machine during a bucket failure. In this particular installation, two eccentricity-detector coils were installed 90° apart around the circumference of the shaft at the front end of the machine, instead of the usual one. This was a special installation, which was made for the express purpose of determining the approximate direction of the major axis of the shaft-vibration ellipse. The two coils are connected to the measuring circuit through a time switch, one after the other. As can be seen from the eccentricity record, the time that one detector is being recorded is shorter than the time for the other. Consequently, it is easier to identify them,

The spikes or sudden surges of the pen partway up the scale are due to the switching of the measuring circuit from one detector to the other, and can be disregarded.

The load was increased from 65 per cent at 12:05 P.M. to about 90 per cent by 12:20 P.M. This load was held constant for about 2 hr. The records indicate that during this period conditions were normal.

At approximately 2:56 P.M., both the shaft eccentricity and bearing vibration increased sharply. Eccentricity increased from approximately 1 to 5.5 mils; vibration from 0.3 to 1 mil. The operators immediately recognized that an unusual condition existed, and made arrangements for the system to pick up the approximate 100,000-kw load this unit was carrying so that the unit could be shut down.

At approximately 3:08 P.M., unloading was started, and, at 3:58 P.M., the unit was taken off the line. During the unloading period the bearing vibration continued to increase. When the unit was shut down, inspection

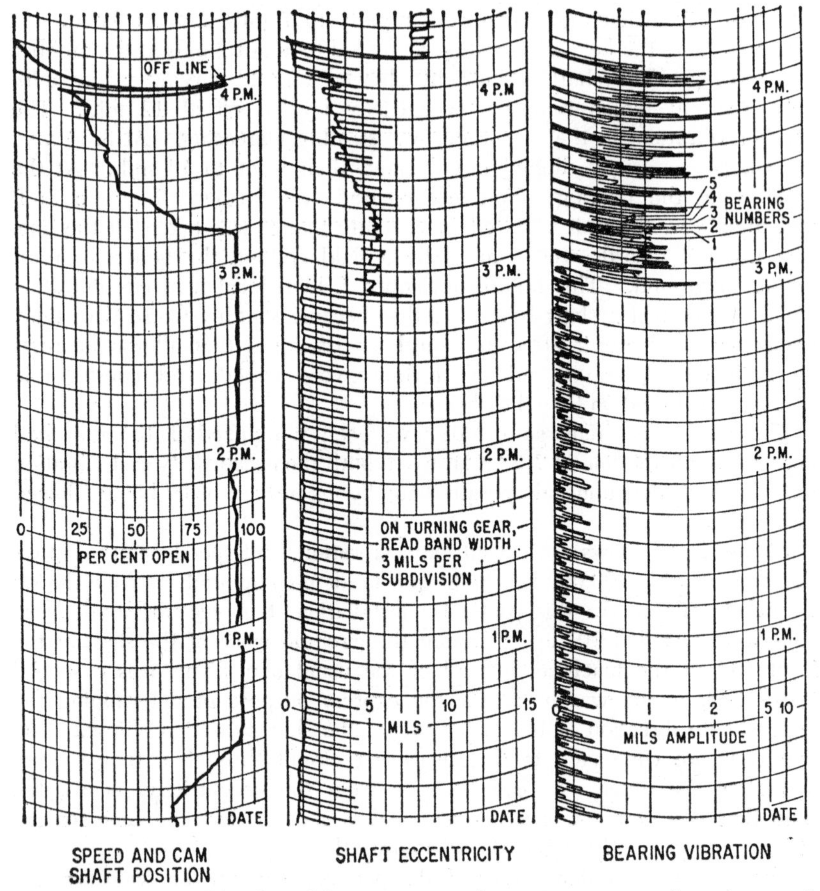

FIG. 11-17. Records taken from a 100,000-kw 3,600-rpm tandem-compound turbine-generator in which buckets failed.

revealed that a bucket failure had occurred with a minimum amount of damage. It is believed that the damage was lessened by the prompt action of the operators in shutting the machine down after observing the abnormal records. Longer operation of the machine might have resulted in further damage.

Figure 11-18 shows the eccentricity, vibration, speed and camshaft-posi-

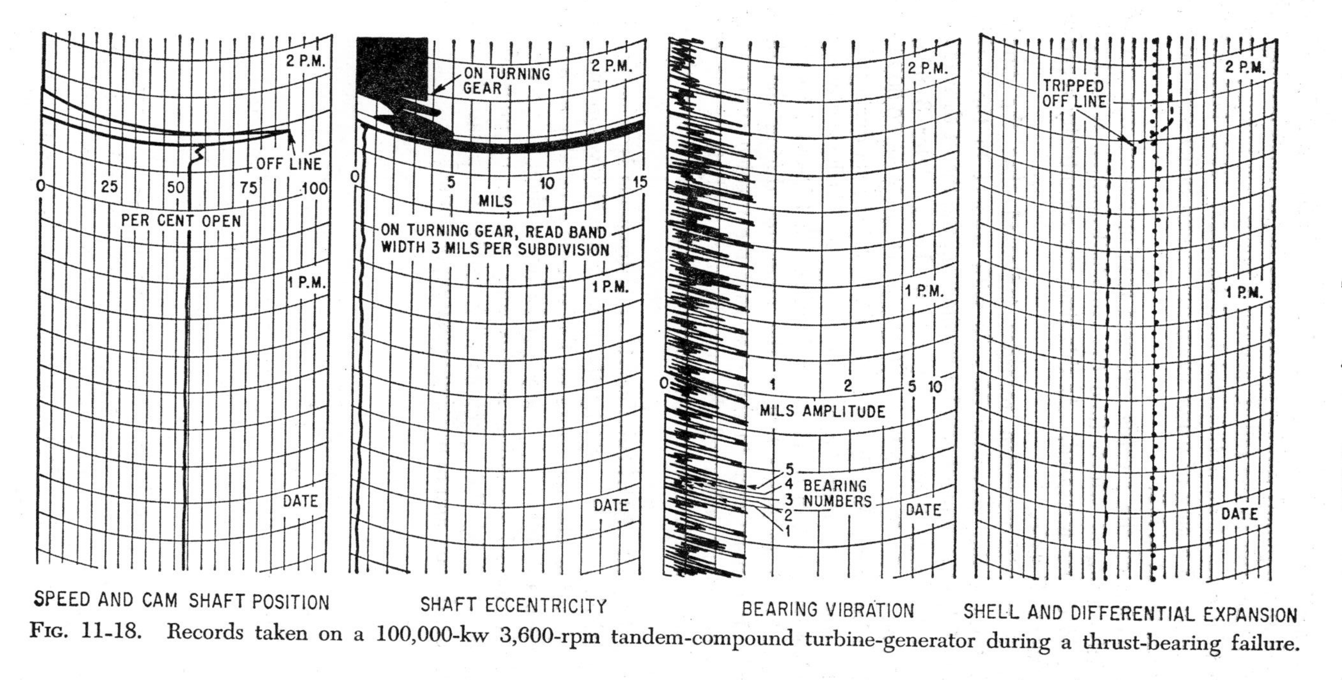

FIG. 11-18. Records taken on a 100,000-kw 3,600-rpm tandem-compound turbine-generator during a thrust-bearing failure.

tion, and shell and differential-expansion records from a 100,000-kw 3,600-rpm tandem-compound turbine generator during a thrust-bearing failure.

At the time the thrust failed, a second boiler was being placed on the line to take care of an anticipated load increase. It is believed that the turbine received a slug of water when the second boiler was cut in, which caused the thrust to fail.

Examination of the records indicates that operation was normal up to about 1:32 P.M., at which time the camshaft-position recorder indicated that the load was increased. At this time the eccentricity recorder immediately went off scale.

The differential-expansion recorder indicated that the turbine rotor moved toward the generator 50 mils with respect to the turbine shell. This was followed immediately by alarms from the thrust thermoalarm relay and high-discharge thrust-oil-temperature alarm. After 3 min, the unit was tripped off the line while carrying approximately an 80,000-kw load.

It is to be noted that the vibration did not increase during this interval, since the rotor did not go downstream enough to rub the stationary elements.

Further examination of the records after the trip-out indicates that the rotor continued to move toward the generator during the deceleration interval. It can be seen in this case that the turbine supervisory instruments immediately indicated the trouble, and analysis of the eccentricity and differential-expansion records showed that there was a thrust-bearing failure.

A complete set of turbine-supervisory instruments is valuable, since only a study of all records made by the recorders will give an over-all picture of the operating characteristics of the turbine-generator at any particular time. Since the records for any particular machine are peculiar to that unit, both normal operating records and records of unusual conditions should be used by the operators as a guide in operating, in detecting unusual conditions, and in training operating personnel.

Section 12

DIESEL-ENGINE GOVERNING[1]

Diesel-engine governors are speed-sensitive devices that automatically control or limit engine speed by regulating the amount of fuel fed to the engine. The governor adjusts the fuel-supply rate to keep the engine running at steady speed, regardless of load. This type is called a speed governor.

Governor Principles. Any change in an engine's load immediately causes an increase or decrease in engine speed. Up to an engine's capacity, power developed depends on the amount of fuel burned in the cylinders. The more fuel oil or gas injected per stroke, the more power the engine develops.

If the power developed by the engine exceeds the load, this excess accelerates engine parts, causing speed to increase. When the load becomes greater than power developed, the engine slows down.

An engine running with a fixed throttle (constant fuel-supply rate) speeds up when the load falls off and slows down when the load increases. Under some conditions the engine may "run away" if too much load is taken off, or stall if it is heavily overloaded.

Fuel Supply. To keep engine speed constant, just enough fuel must be supplied to the cylinders so that the power developed equals the load at desired speed. The function of a speed governor is to perform this job rapidly, accurately, and automatically. It first notes change in engine speed, and then adjusts fuel admission rate to suit.

Flywheels. The amount that an engine accelerates when power output exceeds load depends on the inertia of the flywheel and other rotating parts, because surplus energy is converted to kinetic energy in these parts. When load exceeds engine output, the rotating parts give up kinetic energy. In both cases the flywheel reduces the amount of speed change.

Flywheel Function. Flywheels serve two purposes: (1) They supplement the governor action by preventing too much speed change during the time the governor is changing the fuel-flow rate and the cylinders start to develop needed power output; and (2) they smooth out momentary speed

[1] Abstracted from "Diesel Engine Governors," by E. J. Kates, Consulting Engineer. Published by *Power Magazine,* McGraw-Hill Publishing Company, Inc., March–August, 1953.

changes that occur as the pistons go through their cycles of compression, expansion, and exhaust. Cyclic speed variations occur because the pistons tend to speed up the engine during firing stroke and slow it down during compression.

The constant-speed governor described is intended to maintain the engine at a single speed from no load to full load. Other types of governors are:

The variable-speed governor maintains any selected engine speed from idle to top speed. Speed-limiting governors control the engine at its maximum or minimum speed. One that holds an engine at its maximum safe speed is an overspeed governor. A speed-limiting governor does not control speed when it is within the designed limit or limits.

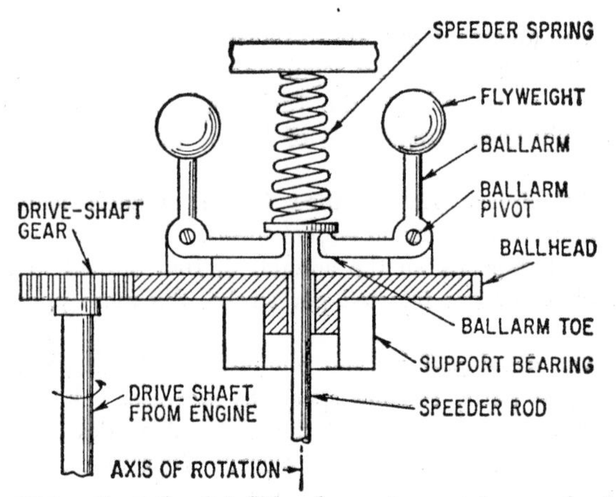

FIG. 12-1. Centrifugal ball-head speed-measuring mechanism.

An overspeed trip shuts down the engine if it overspeeds. It is a safety device only.

Load-limiting governors limit the load applied to an engine at any given speed. Their purpose is to prevent overloading the engine at whatever speed it is running. Load-control governors control the amount of load applied to the engine. Their purpose is to adjust the load to the safe power capacity of the engine at the speed it is running.

Pressure-regulating governors used on engines driving pumps maintain a constant inlet or outlet pressure on the pump. The torque-converter governor controls the speed of the output shaft of a torque converter attached to an engine when the engine runs at less than its set speed.

Speed Governors. In governing engine speed, the first step is to measure it. All governors, from the simplest to the most elaborate, include an accurate speed-measuring device.

The second step in governing an engine is to transfer the indication of the speed-measuring device (when a speed change occurs) into a movement of the governor's terminal or output shaft. This shaft connects to the control rod of the fuel-injection system, regulating the amount of fuel injected into the engine cylinders.

Speed Measuring. In almost all diesel- and gas-engine governors, speed is measured by a centrifugal ball head (Fig. 12-1). Flyweights mounted on opposite sides of a shaft are whirled by the engine through gears. These weights produce a centrifugal force opposed and balanced by the speeder ring.

The balanced position of ball heads at normal speed is shown in Fig. 12-2. Note that the ball arms are vertical. If the engine speed increases, the centrifugal force of the flyweights also increases, and they move out from their axis of rotation.

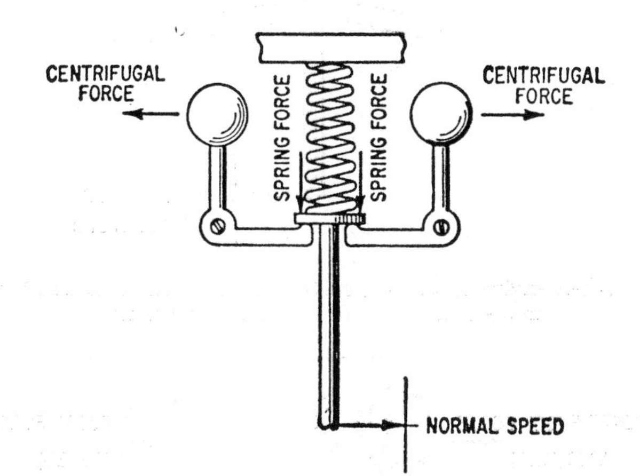

FIG. 12-2. Balanced position of ball heads when engine is running at normal speed. Spring force equals the centrifugal force.

The outward movement of weights lifts the ball-arm toes, increasing the opposing spring force. Since the spring is stiff compared to the centrifugal force, a balance is reached at a point where the spring force equals the centrifugal force at some new position of the flyweights a little further out (Fig. 12-3).

Reverse action occurs if the speed falls. The flyweight centrifugal force decreases, and the speeder spring pushes the flyweights inward until a new balance point is reached (Fig. 12-4). Thus, for any given speed, the flyweights assume a definite position at a certain distance from the axis of rotation.

Throttle Operation. The second step of governor action is to move the engine's fuel-control mechanism. With a jerk-pump fuel-injection system, the fuel-control mechanism changes pump delivery. In a common-rail type, the control mechanism changes the fuel flow to the injectors.

Speed Adjustment. In most governors the engine operator can adjust the spring force that resists the centrifugal force of the flyweights. Figure 12-5 shows a typical speed adjuster.

The upper end of the speeder spring fits into a speeder plug, that may be adjusted up or down by a knob or lever on the outside of the governor. Governors on engines in plants generating electric power often have the speed-adjusting screws operated by small reversible motors controlled from the plant switchboards.

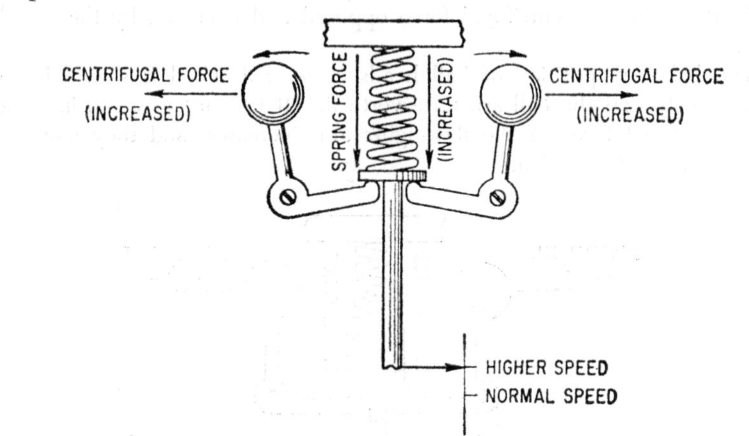

FIG. 12-3. When engine speeds up, centrifugal force increases and the flyweights move out. Toes rise until forces balance.

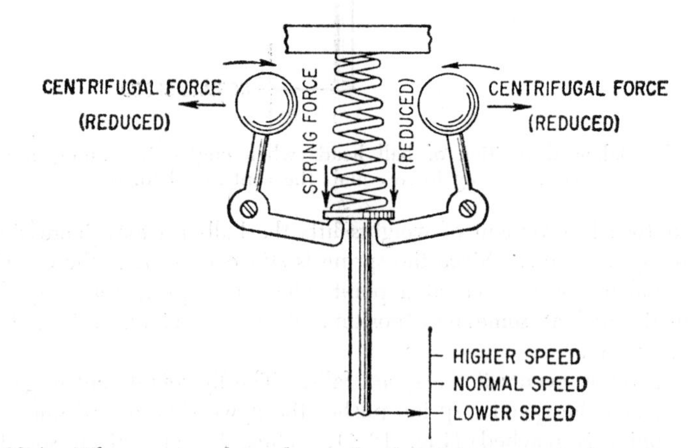

FIG. 12-4. When engine slows down, the flyweights move in, lowering the ball-arm toes. Spring force falls until balance exists.

A change in the speeder-spring force changes the engine's control speed, by destroying the balance between ball head centrifugal force and spring force. To reach a new balance, the engine speed must change.

Suppose the spring force is reduced by turning the speed-adjuster screw (Fig. 12-5) to raise the speeder plug and reduce spring compression. For the reduced spring force to balance the flyweight centrifugal force in the same vertical position of the flyweights as before, less centrifugal force is needed. This means that less engine speed is needed. So reducing spring force causes the engine to run at lower speed when carrying the same load as before.

Increasing spring force raises the engine speed for the same load because more centrifugal force is needed to balance the spring.

Many governors use an independent spring for adjusting speed. Any spring that resists flyweight force affects the engine speed at which the governor forces balance. The independent spring, usually outside the governor, is generally "softer" than the speeder spring to permit fine adjustment of speed.

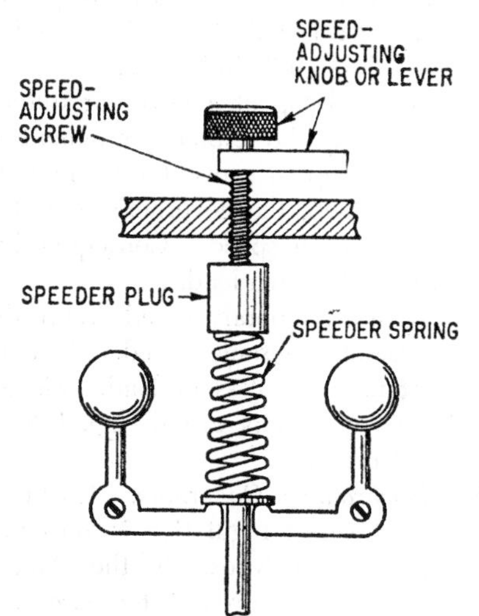

FIG. 12-5. Adjustment for speeder-spring force is made with a knob or a small motor.

Speed Regulation. Speed regulation is the change in an engine's steady speed when the load is changed from the full-rated value to zero, or vice versa, without governor adjustment.

Change in steady speed is expressed as a per cent of full-load speed. Speed regulation per cent $= 100 \times$ (no-load rpm $-$ full-load rpm) $\div$ rated full-load rpm. Speed regulation refers to a change in steady speed. This means that the governor and the engine are given enough time to reach a stable position and speed, respectively, for the given load. For partial changes in load, steady-speed change is proportional to the speed regulation.

If an engine runs at 600 rpm at full load, and at 624 rpm steady speed when all load is removed, the speed increase is 24 rpm between full and no load. Speed regulation is $(100)(24) \div 600 = 4$ per cent.

If the speed regulation is known, speed change can be computed for any partial load change. With a load change of one-quarter rated load, speed change is one-quarter of that from full to no load. For this engine it is $\frac{1}{4}(24) = 6$ rpm. The table below summarizes change of speed with load, for this engine.

Load	Speed, rpm	Speed change, %
Full 4/4	600	0
3/4	606	1
1/2	612	2
1/4	618	3
0	624	4

Speed regulation determines how two or more engines driving the same load will share any load change. An engine's speed regulation is directly related to the governor speed droop.

Speed Droop. Speed droop is the change in rotating speed of the governor that causes its output shaft (fuel-control rod) to move from full-open-throttle position to full-closed position, or vice versa. Since every speed governor is a device responding to speed changes, its droop shows how big a speed change is needed to cause the governor's output shaft to travel through its full range.

Speed droop of a governor differs from engine speed regulation in that droop may be either permanent or temporary. Governors must have speed droop to prevent false motions or overcorrection. With permanent speed droop, the governor's output shaft comes to rest in a different position for each speed. Consequently, the engine's final or steady speed is different for each load.

With a temporary speed droop, the governor's output shaft always comes to rest at the same speed. As a result, the engine's final speed remains constant, regardless of load. Since this is often advantageous, many hydraulic governors are designed to use temporary rather than permanent speed droop.

Isochronous Governors. Isochronous governors regulate engine speed so that speed regulation is zero per cent. This means that the engine's steady speed is exactly the same at any load from zero to full load. Governor speed droop is temporary.

Momentary speed changes, also termed instantaneous speed changes, are temporary changes occurring immediately after a sudden load change. These always exceed the final speed change corresponding to the speed droop. Momentary speed changes are expressed as per cent increase or decrease in speed, referred to engine speed at the instant of load change.

Promptness is the speed of governor action, which depends on its power. The greater the power, the shorter the time required to overcome the resistance.

Work capacity denotes the governor's power as shown by the amount of work it can do at its output shaft. Work capacity (in inch-pounds) equals the average force (in pounds) exerted by the governor at its terminal lever multiplied by the distance (in inches) through which the terminal lever moves in its full travel. A governor whose terminal lever exerts a 4-lb force over a distance of 3 in. has a work capacity of 12 in.-lb.

Corrective Action. Every engine governor depends on a change in speed for its corrective action, so-called because it corrects engine speed. The amount of speed change needed to produce corrective action determines the amount of resulting throttle movement. The more sensitive a governor, the less the corrective action needed after a change of engine load. Also, rapidity of control movement, which depends on promptness of governor action, influences the amount of corrective action needed when speed changes.

Both sensitivity and promptness, are, therefore, important factors in performance. Some hydraulic governors combine sensitivity with high work capacity to such a degree that they respond to a change in speed of less than 1/100 of 1 per cent. They can shift the fuel-control mechanism from full-load to no-load position, or vice versa, in less than one-quarter second.

Engine-speed deviation is any change in speed from normal. Deviation for a given load change depends on both engine and governor characteristics. Conditions determining it are, briefly, (1) time needed to correct fuel-injection rate to correspond with new load, (2) inertia of flywheel and other moving parts of the engine and its connected load, and (3) time

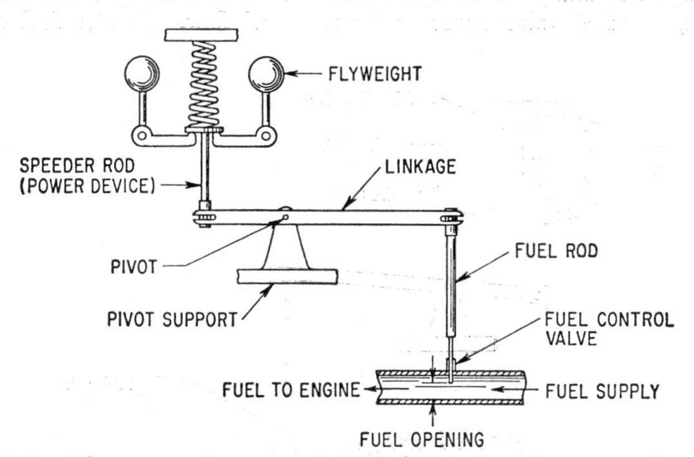

FIG. 12-6. Schematic arrangement of the fuel-control mechanism for diesel and dual-fuel engines.

needed for the engine's power output to respond to the change in fuel-injection rate.

Item 3 depends on the number of engine cylinders, the engine speed, and the type of fuel-injection system. For example, a two-cylinder 200-rpm engine takes longer to respond than an eight-cylinder 1,200-rpm unit.

Ideally, if it were possible (1) to detect instantly the amount of load change, (2) to correct the fuel-injection rate instantly and accurately, and (3) to obtain immediate power response from the engine cylinders, then engine speed would never change, no matter how rapid or great the load changes were. Actually, some delay always occurs between the moment when the load changes and the moment when the engine responds to the corrected fuel-injection rate. Flywheel inertia, by absorbing or supplying energy, reduces the speed change during the delay period, but, the longer the delay, the greater will be the speed change.

Mechanical Governors. As shown in Figs. 12-3 and 12-4, when the governor ball head speeds up, the flyweight centrifugal force becomes greater than the opposing spring force. The opposite occurs when speed falls. Mechanical governors utilize these forces directly to operate the fuel-control mechanism (Fig. 12-6).

This mechanism for diesel and dual-fuel engines consists of a power device (speeder rod) to transmit power directly from the flyweights, a linkage to connect the power device and the fuel-control valve. Figure 12-6 shows a simple gate valve instead of the elaborate design actually used in engines.

Load Increase. When load is applied to the engine, its speed decreases (Fig. 12-7). Ball-arm speed falls with engine speed, reducing the centrifugal force of the flyweights, allowing the speeder spring to force the flyweights in and push the speeder rod down. As the speeder rod moves down, the fuel valve opens to increase fuel flow. Greater fuel supply

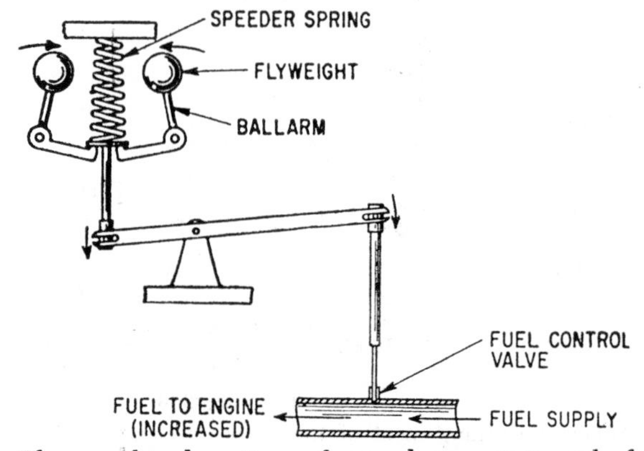

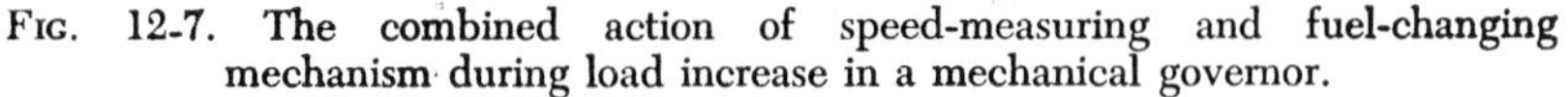

Fig. 12-7. The combined action of speed-measuring and fuel-changing mechanism during load increase in a mechanical governor.

increases the engine output to carry the larger load. The engine speed rises but does not reach the original speed. If it did, the throttle would not be open wide enough to supply the increased load. Wider throttle opening can be obtained only by the flyweights being in toward their axis of rotation because of slower speed.

Load Decrease. On dropping engine load, speed increases (Fig. 12-8). Higher ball-arm speed increases the centrifugal force acting against the speeder spring to raise the speeder rod. This closes the fuel valve to reduce the fuel supply and lower the engine output to meet decreased load. The engine speed falls, but not to the former value, because it is not possible, mechanically, to maintain reduced throttle with the flyweights back in their original position. The smaller fuel opening needed for reduced load requires that the ball arms be out from the vertical because of higher speed than before.

In this governor action the final speed of the engine is less after load is picked up and higher after load is dropped. This permanent speed droop is inherent in all mechanical governors because the power of the flyweights moves the throttle directly by mechanical means. For many services, a reasonable amount of permanent speed droop is quite satisfactory.

Mechanical governors are inexpensive and simple, and are satisfactory where it is not necessary to maintain exactly the same speed, regardless of engine load.

Sensitivity, however, is poor because the speed-measuring device must also furnish the force to move the engine fuel control. Unless the governor is extremely large, its power is relatively small. Speed droop is unavoidable, and the governor cannot provide truly constant speed where it is needed.

Figure 12-9 shows a typical mechanical governor and fuel-pump assembly. The sleeve surrounding the speeder spring acts in the same way as the

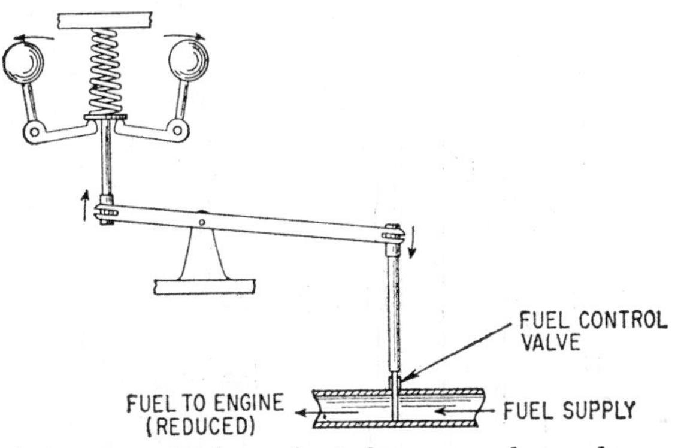

FIG. 12-8. Action in a simple mechanical governor during decrease in engine load. Governor reduces fuel supply to engine.

speeder rod in previous diagrams. The speed adjuster uses an independent set of outside springs. A flexible drive at the lower end of the governor drive shaft prevents momentary engine-speed fluctuations from disturbing the governor action.

Figure 12-10 shows a mechanical governor used on small high-speed engines. It employs a single outside spring, which acts both as speeder spring and a speed-adjuster spring. Antifriction bearings improve the governor's sensitivity.

Fuel-injection pumps of small engines often include a built-in mechanical governor for compactness. Depending on the class of service, such governors are the speed-limiting, variable-speed, or constant-speed types.

Hydraulic Governors. These have the following advantages: (1) They are more sensitive; (2) they have greater power to move the engine's fuel-control mechanism; and (3) they can be made isochronous (having identical speed for all loads).

In these governors the power that moves the engine throttle does not come from the speed-measuring device. Instead, it comes from a hydraulic-power piston, or servomotor. This is a piston acted on by a fluid under pressure, generally oil pressurized by a pump. By using an appro-

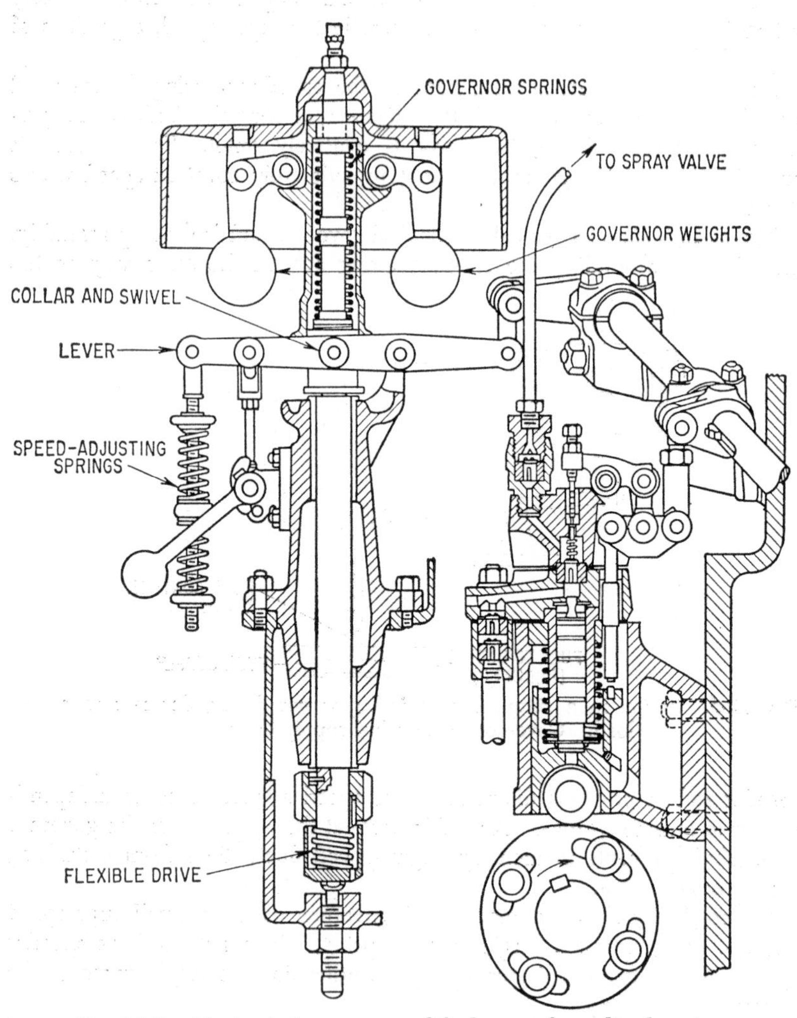

Fig. 12-9. Mechanical governor and fuel pump for a diesel engine.

priate piston size and oil pressure, the power output at the governor's output shaft can be made sufficient to operate promptly the fuel-changing mechanism of the largest engines.

The speed-measuring device, through its speeder rod, is attached to a small cylindrical valve, called the pilot valve. This valve slides up and down in a bushing containing ports, which control oil flow to and from the servomotor. The force needed to slide the pilot valve is extremely small, and a small ball head is able to control a large amount of power at the servomotor.

The elementary hydraulic governor (Fig. 12-11) has the land (raised portion) of the pilot valve equal in width to the port. When the governor operates at control speed, the land closes the port, shutting off oil flow.

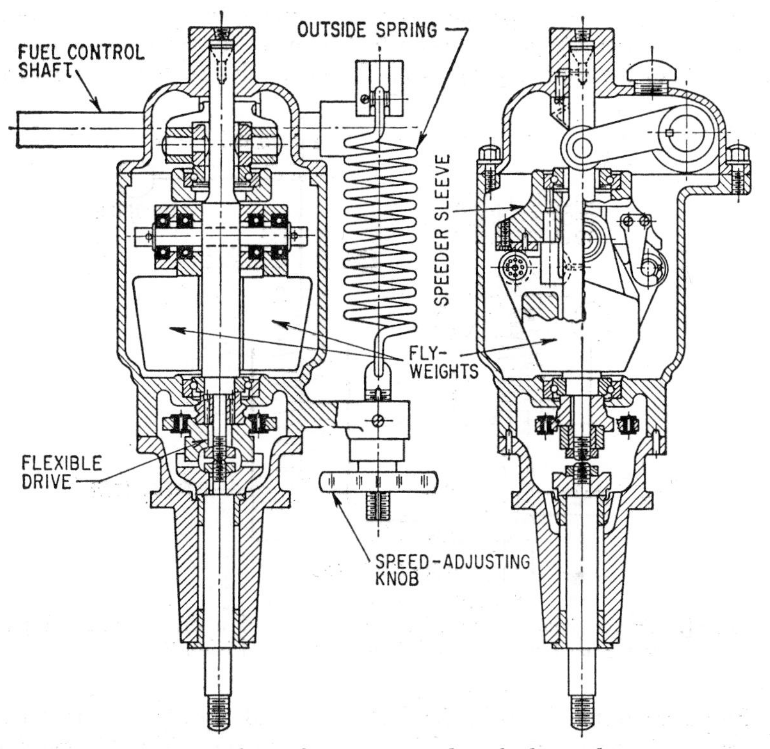

FIG. 12-10. Mechanical governor used on high-speed engines.

If the governor speed falls, because of increase of engine load, the flyweights move in and the pilot valve moves down. This opens the port to the power piston, and connects it to a supply of oil under pressure. The oil acts on the power piston, forcing it up to increase the fuel flow.

If the governor speed rises, because of decrease in engine load, the flyweights move out and the pilot valve moves up. This opens the port from the power piston to a drain leading to the sump. The spring above the power piston forces the piston down to decrease fuel supply to the engine.

The port stays closed only at one speed, and the throttle can be in any position from full load to no load at this speed. Ideally, therefore, the engine should run at exactly the same speed at any load, and the governor would be called isochronous.

Unfortunately, a simple hydraulic governor like this has a serious defect, preventing its practical use. It is inherently unstable. That is, it hunts continually, making unnecessary corrective actions.

Hunting is caused by the unavoidable time lag between the moment the governor acts and the moment the engine responds. The engine cannot instantly come back to the speed called for by the governor.

If engine speed is below the governor-control speed, the pilot valve is moving the power piston to increase fuel supply. By the time the engine speed rises to the control setting with the pilot valve centered and motion

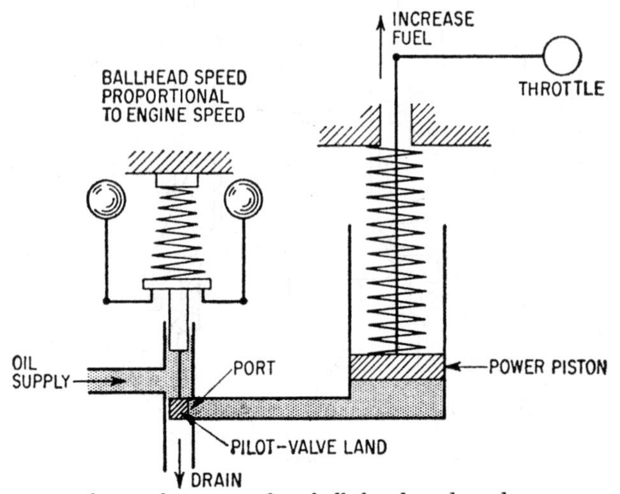

Fig. 12-11. Simple combination of a ball head and a direct-connected pilot valve has but one equilibrium position and is unstable.

of the power piston stopped, the fuel supply has already been increased too much. Engine speed, therefore, continues to rise. This overspeeding opens the pilot valve the other way to decrease fuel supply. But, by the time engine speed falls to the right value, the fuel control has again traveled too far. The engine then underspeeds, and the whole cycle repeats. Some means of stabilizing the governor must be added to provide satisfactory operation.

To obtain stability, most hydraulic governors employ speed droop. It gives stability because the engine throttle can take only one position for any one speed. When a load change causes a speed change, the resulting governor action ceases at the point that gives the amount of fuel needed for the new load. In this way, speed droop prevents unnecessary governor movement and overcorrection of fuel supply (hunting).

Note, however, that, to prevent hunting, the speed droop must be enough to take care of the unavoidable delay while the engine is responding to governor action. If the droop is insufficient, some hunting still occurs while the engine is returning to its steady speed after the first momentary speed change.

Hydraulic governors may be built with either permanent or temporary speed droop, depending on their application. Governors with permanent speed droop are not isochronous.

Permanent speed droop can be obtained in several ways. One of these (Fig. 12-12) is to connect a lever between the power piston and the speeder spring, so that the speed setting is decreased as the fuel supply is increased.

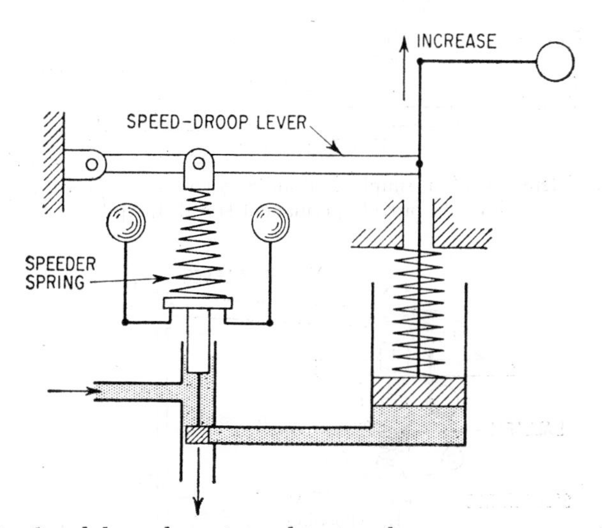

Fig. 12-12. Speed-droop lever is used to introduce permanent speed droop to stabilize the action of a hydraulic governor.

Reducing speeder-spring force lowers engine speed; increasing spring force raises engine speed. The speed-droop lever (Fig. 12-12) performs this function in this governor.

In Fig. 12-13 a simple pilot-valve plunger, attached to the end of the speeder rod, slides in a bushing having drilled control ports. Oil lines from these ports connect to both sides of a power piston on the fuel-control rod. Valve ports are just closed when the ball arms are in a vertical position with the engine running at desired speed and load.

When load is applied to the engine, its speed decreases. As the speed falls, the ball arms move in (Fig. 12-14), lowering the pilot-valve plunger. This opens the ports, allowing oil under pressure to flow through the lower port to the underside of the power piston, to force it up and increase fuel supply. Oil on top of the piston flows out of the upper port and escapes from the top of the bushing to a sump.

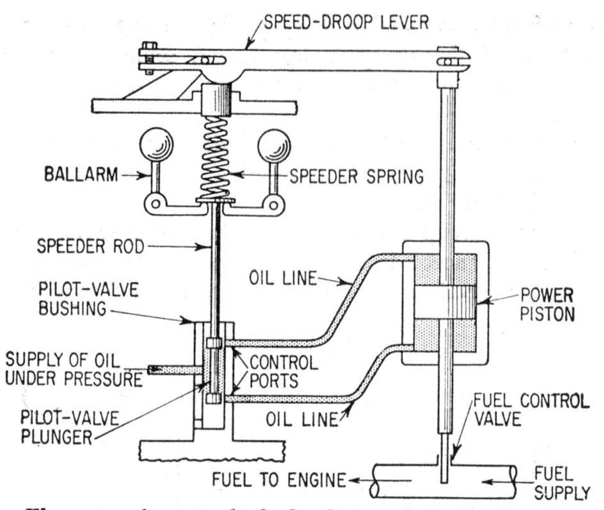

FIG. 12-13. Elements of a simple hydraulic governor fitted with speed-droop lever to provide permanent speed droop.

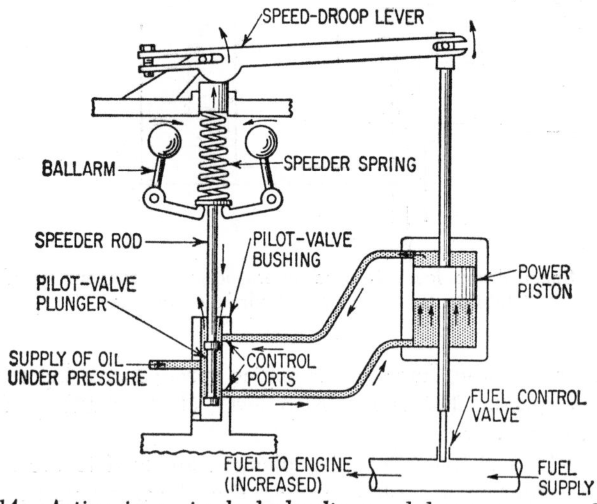

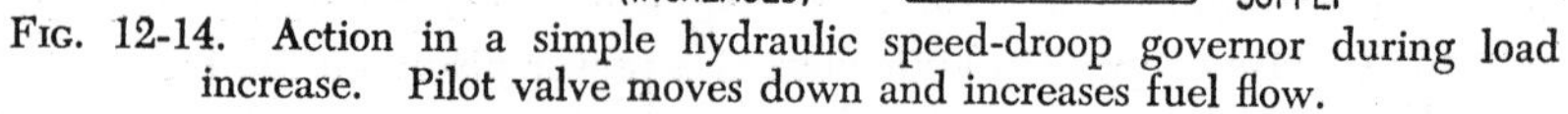
FIG. 12-14. Action in a simple hydraulic speed-droop governor during load increase. Pilot valve moves down and increases fuel flow.

As the power piston moves up, it pushes the speed-droop lever up, reducing speeder-spring force. Reduced force allows the ball arms to move out, raising the pilot-valve plunger, and slowing the further upward movement of the power piston. When the ball arms reach the vertical position, the control ports are closed, and the power piston stops moving upward.

Since the speeder-spring force is reduced as fuel or load increases, a balance is reached with less flyweight force, that is, with lower engine speed. The reduction of engine steady speed caused by the load increase is the speed droop. Note that the speed-droop lever prevents overcorrection of fuel (hunting) by stopping power-piston corrective movement before the engine returns to its previous running speed.

As load drops, engine speed rises. The governor ball arms move out, raising the pilot-valve plunger (Fig. 12-15). Raising the plunger opens the control ports so that oil under pressure can flow through the port to the upper side of the power piston. Oil pressure forces the piston down to reduce fuel supply. Oil under the piston escapes through the lower port to the sump.

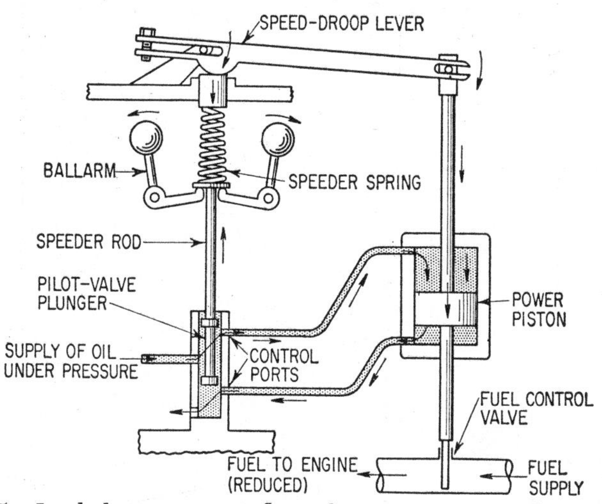

FIG. 12-15. Load decrease causes flyweights to move out and raise the pilot valve to push the power piston down and reduce fuel.

As the power piston moves down, it pulls the speed-droop lever down, increasing the speeder-spring force. The larger force pushes the ball arms in, lowering the pilot-valve plunger, and slowing the further downward motion of the power piston. When the ball arms reach the vertical position, the control ports are closed, and the power piston has stopped moving.

Since the speeder-spring force is increased as load is reduced, it requires more weight force—that is, a higher engine speed—to balance the spring force. Increase of the engine steady speed caused by the load decrease is the speed droop. It prevents hunting by stopping the corrective movement of the power piston before the engine returns to its previous speed.

Advantages of this type hydraulic governor are that it is (1) relatively inexpensive, (2) accurate and sensitive, giving good speed control, (3) simple for a hydraulic governor, having few parts, and (4) more powerful than a mechanical governor of similar dimensions.

Disadvantages are that (1) it is not isochronous, and (2) speed-droop adjustment is inconvenient because it must be made inside the governor.

Figure 12-16 shows the internal construction of a hydraulic governor employing permanent speed droop to obtain stability. In addition, Fig.

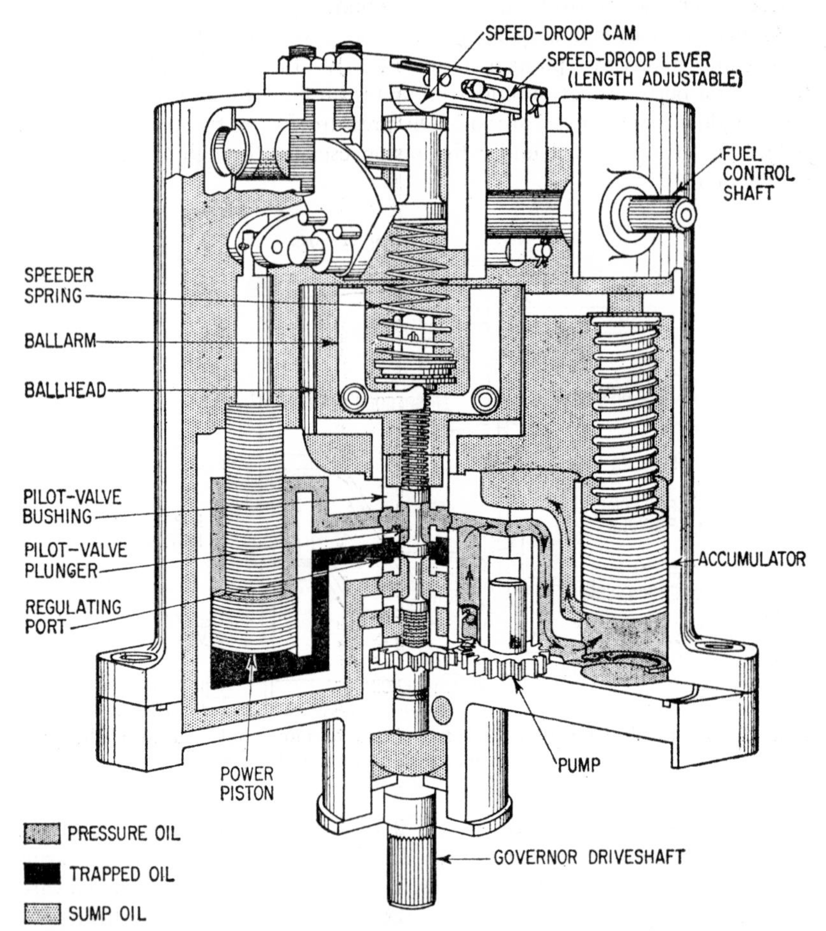

Fig. 12-16. Cross section through a hydraulic governor having permanent speed droop. The oil pump maintains a pressure of 120 psi for the operation of the servomotor.

12-16 shows the pump that supplies oil at 120 psi and the accumulator or reservoir that keeps the oil at constant pressure.

The ball head is rotated by the governor drive shaft through the pilot-valve bushing. The amount of speed droop may be changed by adjusting the length of the two-part speed-droop lever. Speed adjustment is provided by a knob (not shown), which raises or lowers the upper seat of the speeder spring.

This hydraulic governor cannot maintain the engine at exactly constant speed, regardless of load, because it employs permanent speed droop to prevent hunting. It is not isochronous.

An isochronous hydraulic governor maintains exactly constant speed without hunting. It does this by employing temporary speed droop, to give stability while fuel flow is being corrected, and then gradually removing the droop, as the engine reacts to fuel correction and returns to its original speed. Use of temporary speed droop to prevent overcorrection of fuel is called compensation.

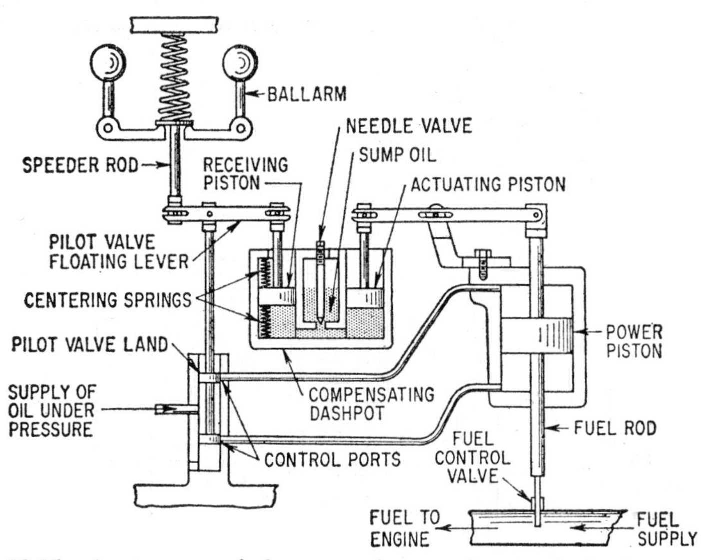

FIG. 12-17. Arrangement of the parts of an isochronous hydraulic governor. Note the actuating, receiving, and power pistons.

Compensation requires two actions: (1) a droop applicator as fuel flow is changed, (2) a droop remover as the engine responds to the fuel change and returns to its original speed.

These actions can be accomplished by a combination of two hydraulic pistons and a needle valve. Droop action is applied by the two pistons, which are connected by an oil passage. Droop is removed by allowing oil to escape from the connecting passage through the needle valve. The oil goes to a sump after leaving the passage. Centering springs return the droop-applying piston to its original position.

The parts needed in a governor to apply and remove temporary speed droop (compensation) are shown in Fig. 12-17. These are: (1) a transmitting or actuating piston to transfer motion of the fuel-changing mechanism to (2) a spring-loaded responding or receiving piston, which acts on parts of the governor to cause speed droop, and (3) an adjustable needle valve in the connecting oil passage of the two pistons, which allows oil to leak off to the governor sump.

With the engine running at normal speed under steady load, the ball arms are vertical, and the pilot-valve floating lever is horizontal (Fig. 12-17). Control ports in the pilot-valve bushing are covered by lands on the pilot-valve plunger. The receiving compensating piston is in the normal position. The power piston and the fuel rod are stationary. The position shown corresponds to about one-half fuel.

When load is applied to the engine, its speed decreases (Fig. 12-18). The ball arms move in, lowering the pilot-valve plunger to admit oil pressure under the power piston. Oil pressure moves the power piston up

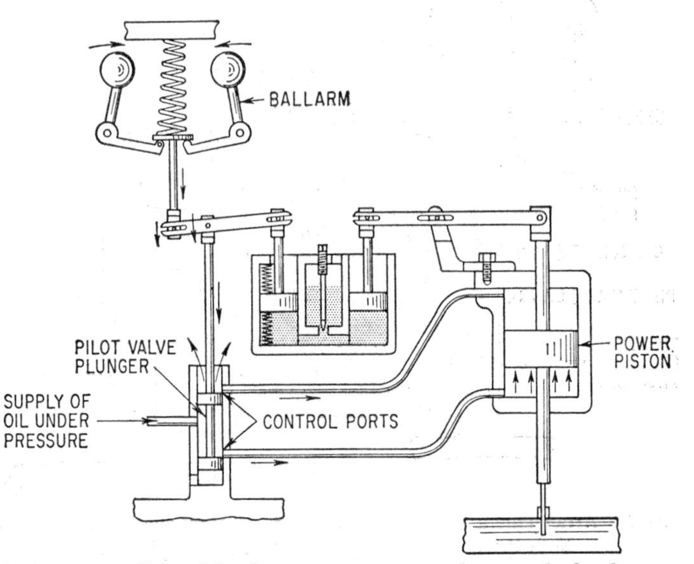

Fig. 12-18. First effect of load increase on an isochronous hydraulic governor. Ball arms move in and lower pilot-valve plunger.

(Fig. 12-19). This pushes the actuating piston down, which displaces oil to the receiving piston, forcing this piston up to compress the upper spring and lift the pilot-valve plunger. The plunger lands close the control ports to stop further movement of the power piston.

This action is rapid, and the small opening of the needle valve prevents appreciable leakage. Therefore, oil displaced by the actuating piston causes a corresponding movement of the receiving piston.

As the power piston moves up, it increases the fuel supply to bring the engine speed back to normal. As the engine responds to the fuel change, its speed gradually returns to the original speed, and the ball arms gradually resume their vertical position. At the same time, the upper spring starts pushing the receiving piston down to its normal position, and the floating lever tilts about the pilot-valve pivot pin.

The rate at which the receiving piston moves down is determined by the needle-valve opening, which is adjustable. If the opening is correct, the rate of return of the receiving piston exactly matches the rate of return of the ball arms to the vertical.

At the completion of the cycle, the engine will be running at its original speed, the ball arms will be vertical, the floating lever horizontal, the control ports closed, the receiving piston back in its original position, and the power piston up in a new position, supplying increased fuel for the increased load (Fig. 12-20).

When load drops off, the governor goes through exactly the same steps as for load increase, but all movements are in the opposite direction. The

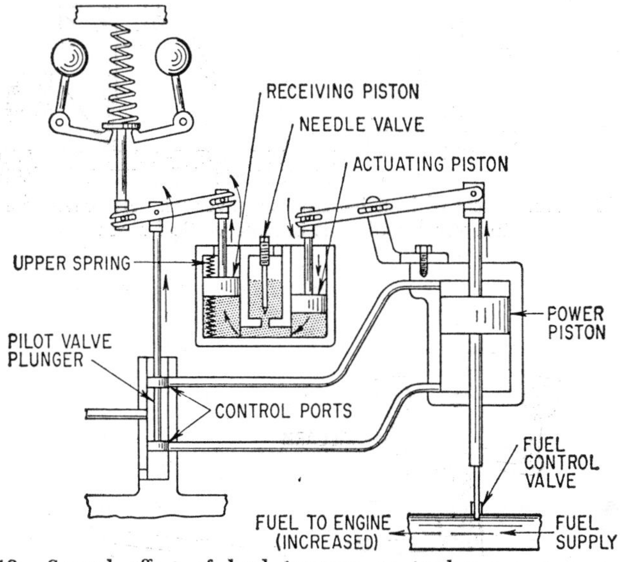

Fig. 12-19. Second effect of load increase on isochronous governor. Power piston moves up and recenters pilot-valve plunger.

sequence can be followed in Figs. 12-21, 12-22, 12-23. In Fig. 12-21, the ball arms have moved out and lifted the pilot-valve plunger. The power piston moves down (Fig. 12-22), reducing the fuel supply and lifting the actuating piston. This causes the receiving piston to move down, lower the pilot-valve plunger, and close the control ports.

In Fig. 12-23, oil leakage through the needle valve has permitted the receiving piston to return gradually to its original position, in the same time that the engine took to return to its original speed and the ball arms to return to the vertical.

Note that, in this governor, the temporary speed droop (compensation) is applied by changing the effective length of the connection between the speeder rod and pilot valve. Changing the connection length causes speed droop for the same reason that changing the compression of the speeder

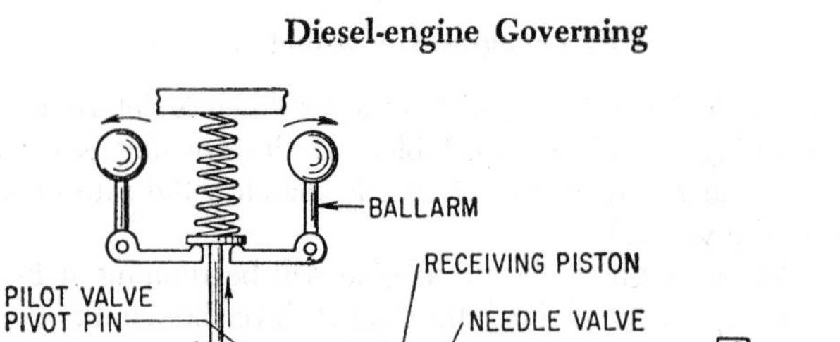
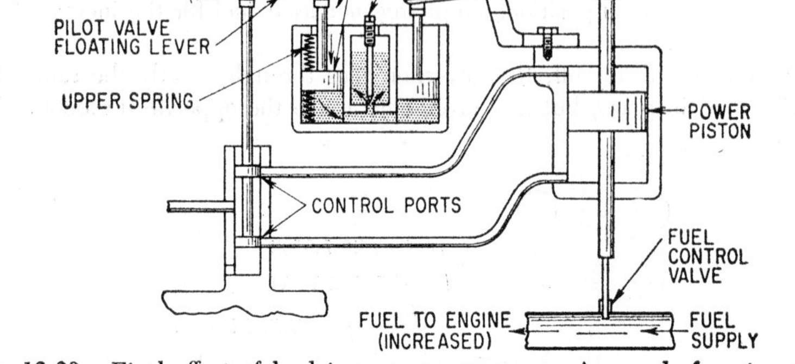

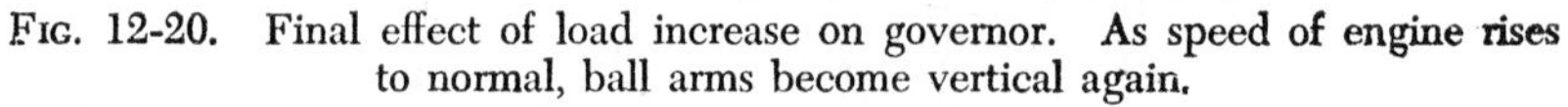

Fig. 12-20. Final effect of load increase on governor. As speed of engine rises to normal, ball arms become vertical again.

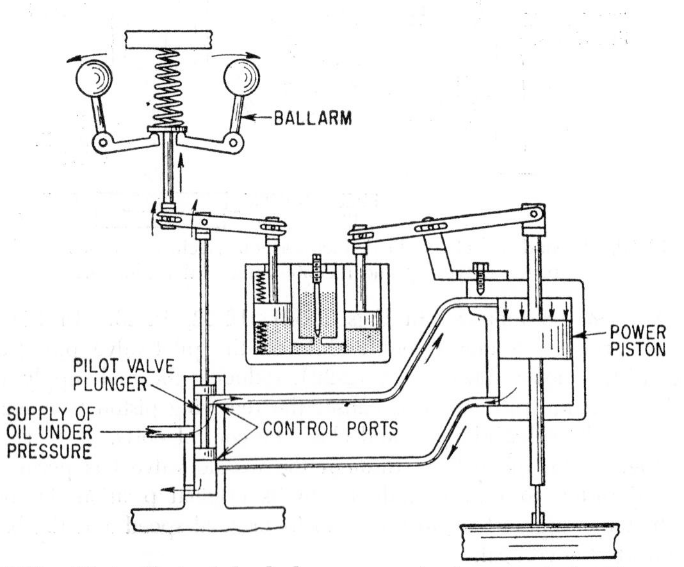

Fig. 12-21. First effect of load decrease on isochronous hydraulic governor. Ball arms move out and lift pilot-valve plunger.

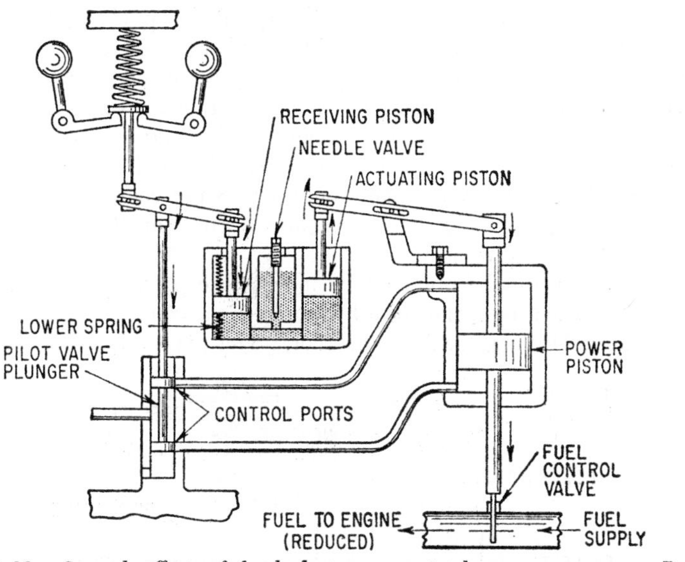

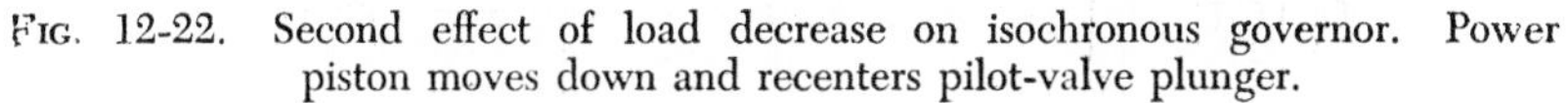
FIG. 12-22. Second effect of load decrease on isochronous governor. Power piston moves down and recenters pilot-valve plunger.

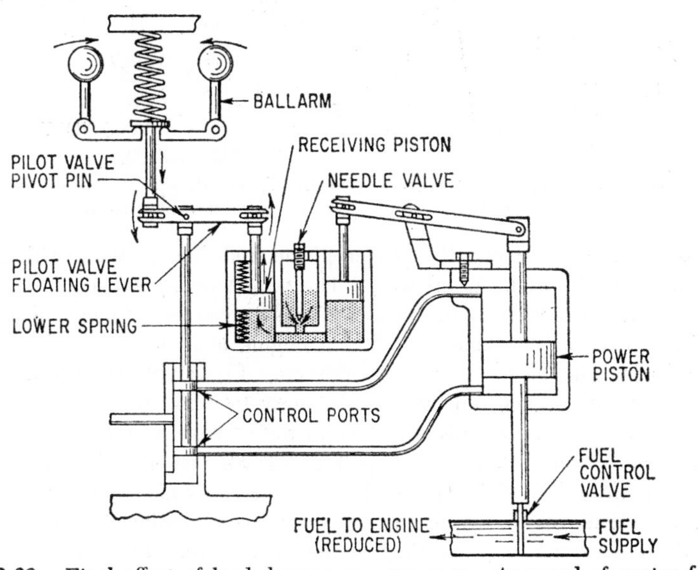

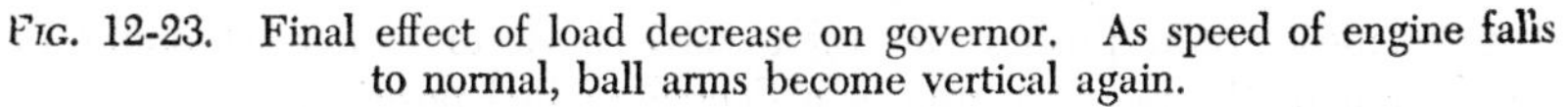
FIG. 12-23. Final effect of load decrease on governor. As speed of engine falls to normal, ball arms become vertical again.

spring does. Each method of causing speed droop changes the engine speed at which the pilot valve seals the control ports.

Only isochronous governors are true constant-speed governors. All mechanical governors, and some hydraulic governors, have permanent speed droop and are "nonisochronous" because engine steady speed is slightly different at different loads. But the speed change from full load to no load (speed regulation) is usually only 3 to 8 per cent. Since such governors hold engine steady speed substantially constant, they too are classed as constant-speed governors.

Stability can also be obtained by making the power piston act directly on the pilot valve by means of a hydraulic connection.

Figure 12-24 shows such a governor. The pilot valve controls flow of

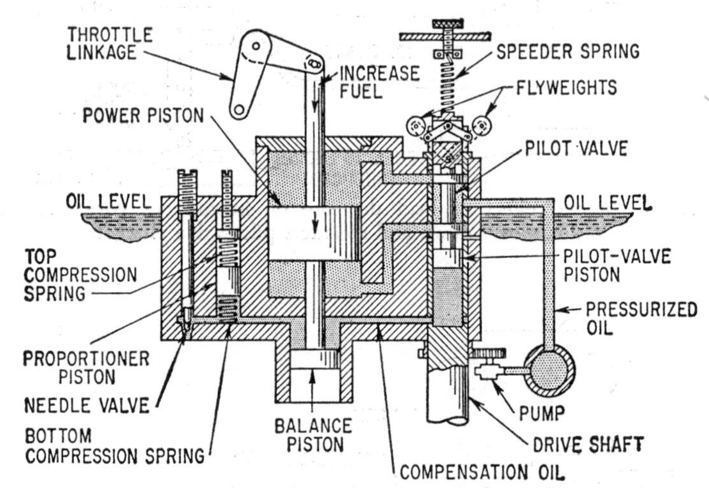

Fig. 12-24. Isochronous governor controls speed droop by an oil connection between the power piston and pilot-valve piston.

pressurized oil to move the power piston and change the position of the engine throttle in the usual way. The pilot valve, however, is recentered by an oil connection to the power piston instead of by mechanical levers.

The governor acts like this: When engine load decreases, the speed of the engine and of the governor increases. Flyweights move out and pull the pilot valve down, uncovering the ports to admit pressurized oil to the underside of the power piston. The piston moves up to decrease fuel flow.

As the power piston moves up, the balance piston displaces compensation oil upward to the place of least resistance, which is the proportioner piston. It moves up against its top spring, compressing it to a force equal to the downward force existing on the pilot valve. Further movement of the balance piston pushes the pilot valve back to center, stopping the movement of the power piston at a decreased fuel position.

The proportioner piston, being still above center, is pushed back to center by its top spring at a gradual rate, determined by the rate at which oil bleeds through the needle valve. The opening of this valve is adjusted so the pilot valve is kept centered by pressure of the compensation oil until the engine responds to the decreased fuel rate and returns to its original speed. When the compensation period ends, the engine and the governor are running at the original speed, and both pilot valve and proportioner piston are recentered. The power piston, however, is in a new position, supplying the amount of oil needed for the prevailing load.

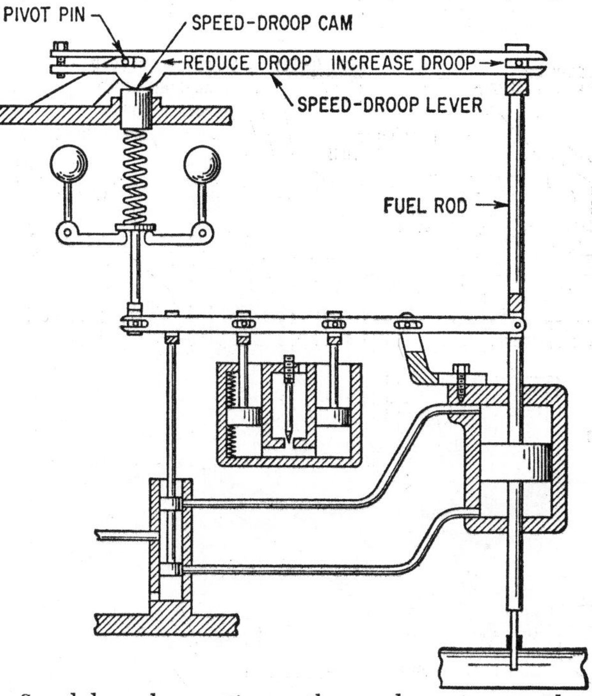

FIG. 12-25. Speed-droop lever acting on the speeder spring provides permanent speed droop, which can be adjusted by the operator.

Governor Applications. The basic governors described are often modified or fitted with auxiliary devices to meet special requirements.

Many electric-power plants contain several units generating alternating current. Each unit must take a share of the load while a constant frequency is held in the power system. This requires the use of isochronous governors fitted with a mechanism to provide adjustable permanent speed droop. Units like this employ both temporary and permanent speed-droop mechanisms. Temporary speed droop is always effective, to prevent hunting. The permanent speed-droop mechanism can be adjusted, as required, from no droop (isochronous) to about 5 per cent droop in steady speed.

Figure 12-25 shows how adjustable permanent speed droop is added to the isochronous governor described previously. The speed-droop lever introduces permanent droop by changing the force of the speeder spring.

To adjust droop, the speed-droop lever is shifted endwise, to change its leverage. With the lever shifted to the right, a given movement of fuel rod causes a greater movement of the speed-droop cam, and a greater change in spring force. Shifting the lever to the left causes the cam to have less effect on the spring; speed droop is reduced. With the lever shifted all the way to the left, the cam nose comes directly under the pivot pin, and

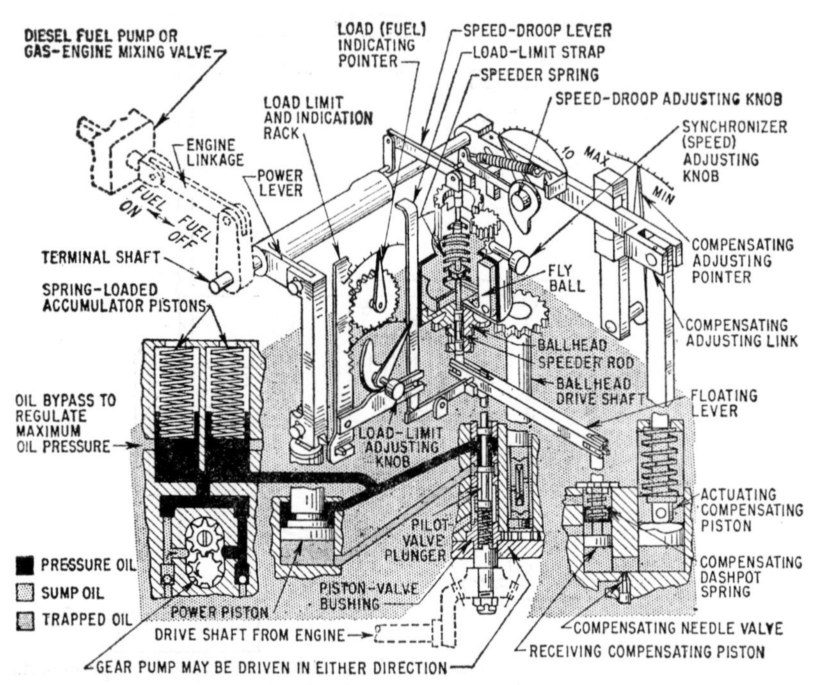

Fig. 12-26. Schematic diagram of the isochronous hydraulic governor in Fig. 12-27, showing the components used in its construction. Knobs on the exterior permit easy adjustment.

the cam has no effect on the speeder spring. This eliminates permanent speed droop; the governor becomes isochronous.

Figure 12-27 shows a cross section of an isochronous hydraulic governor with adjustable permanent speed droop; Fig. 12-26 is a schematic diagram of this unit. The speed-droop lever connected to the governor terminal shaft varies the compression of the speeder spring when the terminal shaft rotates in response to a load change. The effective length of the speed-droop lever is adjustable from an external knob, permitting permanent speed droop to be changed while the engine runs.

The floating lever changes the pilot-valve position relative to the speeder rod, introducing temporary speed droop (compensation), which makes the governor isochronous if the permanent speed droop is adjusted to zero. The compensating pistons and needle valve gradually remove the temporary speed droop.

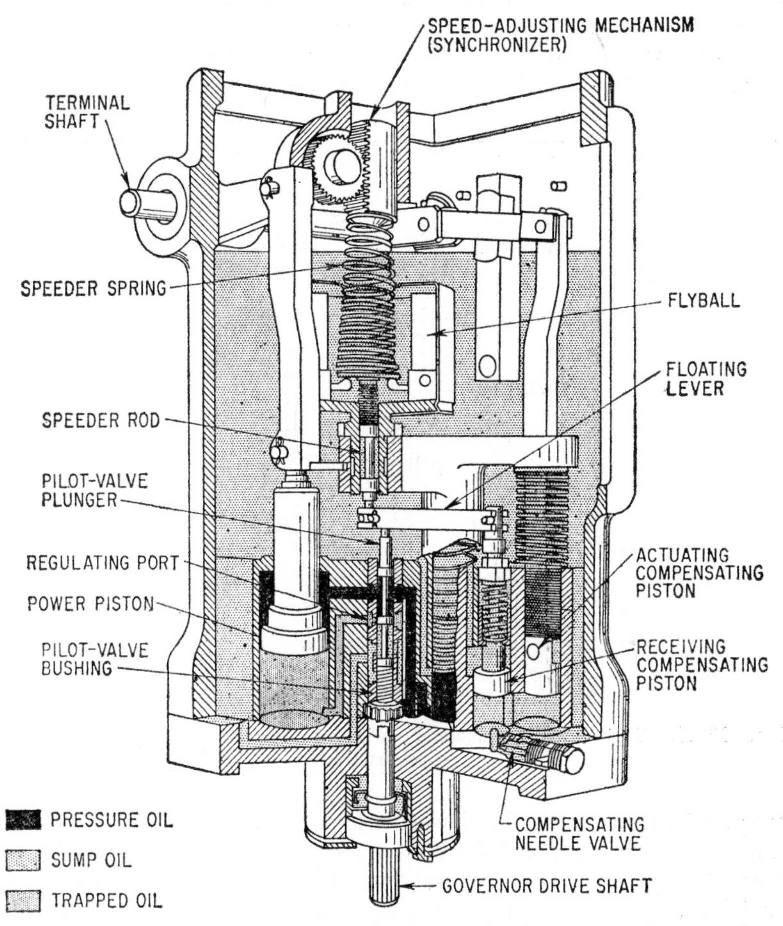

FIG. 12-27. Cross section of an isochronous hydraulic governor having adjustable speed droop. Note the speed-adjusting mechanism (synchronizer) and the compact design.

Variable-speed Governor. This is a governor that controls engine speed at several values between minimum and rated speed. It resembles a constant-speed governor in its ability to hold the engine at steady speed; it differs in that an operator can adjust the governor to maintain any one of several desired steady speeds.

Speed adjustment is obtained in most variable-speed governors by changing the spring force that resists the flyweight centrifugal force. Described earlier was an adjustment of spring force used to make minor changes in the control speed of a constant-speed governor. Speed adjusters for units of this type work through a range of only a few per cent change in speed, but the same principle is used in many variable-speed governors to permit control speeds as low as one-sixth normal.

To work well through so great a range, a special trumpet-shaped or conical speeder spring is needed, because an ordinary cylindrical spring does not give such good control at reduced speed as at full speed. At reduced speed, a cylindrical spring, which is stiff enough (large force per inch of

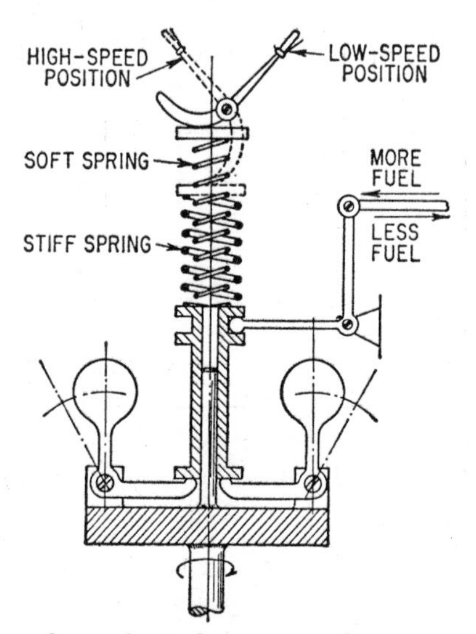

Fig. 12-28. Two-speed mechanical governor has two springs that act in combination.

compression) to match the flyweight force at high speed, is too stiff to match the greatly reduced flyweight force at reduced engine speed. The trumpet-shaped spring is so wound that, at light loads (low speeds), all the turns are active and the spring is quite soft. At high loads (high speeds), the larger turns touch each other and go out of action, leaving the smaller turns to provide a stiffer spring.

Two-speed Mechanical Governor. This controls engine speed at idling and at high speed with two different springs. One is stiff for high speeds; the other is soft to give greater sensitivity and less speed droop at low speed. The springs may be arranged to act either singly or in combination (Fig. 12-28).

In this unit, the control lever is put in the position shown if low-speed operation is desired. Then only the soft inner spring acts. For high-speed operation, the lever is pushed to the left, putting both springs in action.

Overspeed Governor. This is a safety device that protects the engine from damage which might be caused by overspeeding from any cause. When the engine has a regular speed governor, the overspeed governor functions only if the regular unit fails.

If overspeed control merely slows the engine down, but allows it to continue running at a safe speed, it is an overspeed governor. If it brings the engine to a full stop, it is an overspeed trip.

Overspeed governors and trips use the principle of resisting the centrifugal force of a flyweight by means of a spring. The spring is preloaded to a force that overbalances the centrifugal force of the flyweight until engine speed rises above the desired maximum value. When this occurs, the centrifugal force overcomes the spring force and puts into action the controls that slow down or stop the engine.

Overspeed governors generally act on fuel supply to the engine cylinders, like regular units. So do many overspeed trips. It is possible, however, for an engine to overspeed because fuel enters the cylinders independently of the fuel-injection system. For example, oil or gas might accidentally enter the engine's air intake. Some overspeed trips guard against such occurrences, and shut the engine down by holding the exhaust valves open or by closing off the air supply.

Load-limiting Governor. Too much fuel can seriously damage an engine. A governor, whether isochronous or not, responds only to speed changes. So, if an engine slows down because load is excessive, the governor increases the fuel supply. To prevent excessive fuel supply under these conditions, most engines have a fixed collar or stop on the fuel-control shaft to limit the maximum amount of fuel supply, even though the governor calls for more. This is called a *fuel-limiting stop.*

Fuel-limiting stops have their drawbacks, however. If an extremely heavy load is put on an engine and causes fuel supply to reach the limit, engine speed drops below control speed. The engine then runs at a throttle-setting fixed-fuel stop, as though it had no governor. Also, when running at reduced speed due to overload, the engine cannot develop as much power as at full speed, because power is proportional to speed. Thus, just when maximum power is demanded, developed power falls off.

Load-limiting governors prevent a loss of engine speed due to heavy load by limiting the load to the engine's safe capacity. They cannot be used for all jobs, because it is not always feasible to limit the load. An important application, however, is on diesel-electric locomotives. Here the load-limit device is an attachment to the governor. When the governor terminal shaft reaches the maximum safe fuel position, this device operates an electric control that reduces generator output. The engine runs at full speed, developing its full safe power.

Pressure-regulating Governor. This governor is for engines driving pumps whose discharge pressure must be kept constant, as in pipeline service. The ordinary variable-speed governor can be converted to this use by the addition of a device to vary speed setting according to pumping pressure.

Pumping pressure acts on a piston or diaphragm, balanced by a spring and connected to the governor's speed adjuster. If pressure falls below the desired value, diaphragm movement moves the speed adjuster to increase speed setting. (Some designs have a servomotor for more power and sensitivity.) The engine runs faster at the increased speed setting, and pump pressure returns to normal. The opposite occurs when pump pressure rises.

Governor Use. Constant-speed governors are for engines that must maintain steady speed while supplying a varying load. All mechanical governors have speed droop, and cannot maintain exactly the same engine speed at different loads. But speed regulation with this type is fairly close, 3 to 8 per cent. This is close enough for many services, such as driving machine-shop line shafts or generators supplying loads that permit some variation in voltage and frequency.

Governors for engines driving generators always have accurate speed adjusters to (1) adjust engine speed to conform with generator rated speed, (2) synchronize one a-c generator with another, (3) regulate electrical output of engines that are operating in parallel.

When engines are coupled through a-c generators, the speed adjuster is really a load adjuster, because it can adjust the load on the engine without substantially changing its speed. The reason for this is that electrical forces of a-c generators compel all coupled engines to keep in step and change speed in unison.

When the speed adjuster of one engine is turned toward higher speed, fuel supply increases. If the engine were running alone, this would increase its speed. But, since the engine is electrically interlocked with the others, its speed increases only slightly. The engine tries to push ahead, its increased torque advancing the generator's phase angle, increasing the electrical output.

Hydraulic governors are more popular for engines driving generators because frequency control is more accurate, voltage regulation is closer, and there is better load sharing. The last factor depends on the speed droop of the governors.

Speed droop determines how a common load divides among two or more engines when the load changes. A common load, whether connected mechanically or electrically, compels the engines to run at the same relative speeds.

Suppose two engines having rated full-load speeds of 800 and 1,200 rpm are running at full load, driving paralleled 60-cps a-c generators. If the load falls and engine speeds increase to raise the frequency 5 per cent to 63 cps, both engines speed up 5 per cent. The first engine will run at 840 rpm, and the second at 1,260 rpm.

If both engines have the same speed regulation because both governors have the same speed droop, say 10 per cent from full to no load, both set their fuel controls for half load because they are running at 5 per cent overspeed. Each engine continues to carry its proper share of the common load.

But, if governors have different speed droops, say 5 per cent for one engine and 10 per cent for the other, the load will not be equally shared. For instance, if a load decrease speeds up both engines 5 per cent as before, the first engine's governor cuts off all its fuel so that no power is delivered, while the second engine will develop half power. Governors of engines connected together must have equal speed droops if each is to automatically take its correct share of a varying common load.

Common mechanical governors have fixed and fairly large speed droops. So, if two or more engines have different makes of mechanical governors, their speed droops will probably differ considerably. If the engines are coupled to a common load, each will not take its proper share when the load varies.

Nonisochronous hydraulic governors have inherent permanent speed droop, which is generally adjustable. Such governors, when fitted to coupled engines, may be adjusted for equal droop to give correct load sharing. This generally requires a speed droop of 2 or 3 per cent. Since speed droop is always present in such governors, they will not keep constant frequency in a-c generating plants. Consequently, electric clocks will not keep accurate time unless a watch engineer or an automatic device frequently corrects all engine speed adjusters.

Isochronous hydraulic governors without speed droop work well on engines running singly. But they are unsuitable for coupled engines because the absence of speed droop prevents proper load sharing.

So, isochronous hydraulic governors with adjustable speed-droop attachment find wide use in multiple-unit a-c generating plants. Because their temporary speed droop gives them stability, the permanent speed droop can be made quite small and still give good load sharing.

Also, such governors permit parallel operation at constant frequency. The governor of one engine is set for zero speed droop (isochronous), and all other governors are given equal speed droop of sufficient amount to ensure correct load sharing. Load changes are automatically taken by the engine with the isochronous governor, which runs at constant speed. This compels other engines to maintain their same speeds, and so the electrical frequency of the plant stays constant. In this arrangement, engines with speed droop deliver constant output. Therefore, the engine with the isochronous governor must have enough capacity to take all load changes.

Variable-speed governors are used where the operator must be able to change the engine speed while it is running. Output of pumps and compressors is varied efficiently by varying the speed. Governors of this type not only make speed changes easy; they also keep the engine under governor control at the speed the operator selects.

Section 13

CONTROL OF HEATING, AIR-CONDITIONING, AND REFRIGERATION SYSTEMS[1]

CONTROL ADAPTED TO EQUIPMENT

The simplest arrangement for conditioning a commercial or industrial building, from the standpoint of control, is the direct-radiation heating system, with steam or hot water supplied to radiators or convectors throughout the building.

Where the same *central-fan* equipment is used for heating, cooling, and ventilation, the control system consists of a combination of suitable features for control of central-fan heating, for central-fan cooling, and for summer-winter change-over.

In split systems, summer cooling and winter ventilation are provided by the same central-fan units, but the heating load is carried by direct radiation. The control features discussed in this chapter, in suitable combination, again make up the control system.

Where the heating load is carried by direct radiation supplied from a central source of steam or hot water, and cooling is provided by independent fan-cooling systems for each zone, the heating control may follow the principles illustrated in this chapter and the zone-cooling units may be controlled by features selected to suit the particular installation.

Zones Limited by Piping. A building may be divided into any number of zones, provided only that the arrangement of piping permits setting up zones so divided that temperature conditions and requirements are approximately the same for all parts of a given zone. If piping or duct layout does not permit zoning strictly according to the ideal requirements, the parts of the building subject to the most extreme conditions should be separated from the other parts. In new construction, provision can be made for zoning as required. The factors determining these requirements, and therefore governing the arrangement of zones, are *exposure, occupancy,* and *construction.*

Exposure. Of these three factors, zoning for *exposure,* which includes

[1] J. E. Haines, Minneapolis-Honeywell Regulator Co., "Automatic Control of Heating and Air Conditioning," McGraw-Hill Book Co., Inc., New York, 1953.

direction of prevailing winds, wind velocity, solar radiation, and outside-air temperature, is usually most important.

In a typical example in the North Central states, the north space of a building, which gets little sun, requires the most heat in winter.

Similarly, although east and west fronts get the sun about the same percentage of the day, the west space is more exposed to prevailing northwest winds than the east and therefore requires more heating, although less than the north space.

The south space gets the most sun and is sheltered from the wind. This should then be a separate zone.

If necessary, the north and west spaces, most exposed to wind, can be combined into one zone, and likewise the south and east spaces.

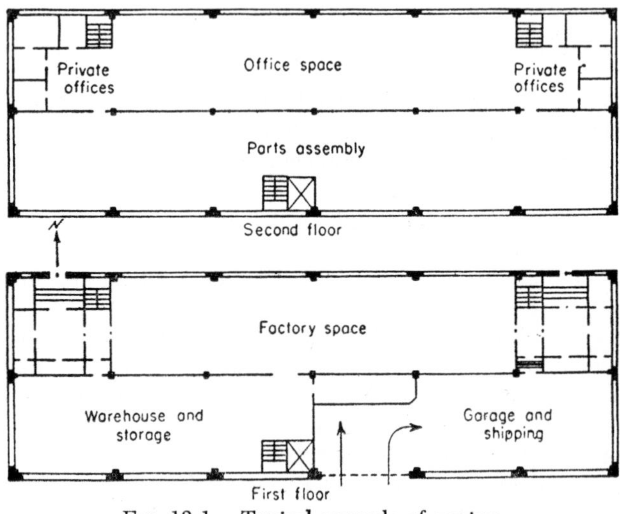

Fig. 13-1. Typical example of zoning.

In larger buildings, even more subdivision of space may be necessary. For example, an internal area, little affected by sun or wind, may require treatment as a separate zone.

Occupancy. Occupancy affects the heating requirements of different parts of a building in two ways. If part of a building is used for offices, and other parts for factory operations, higher temperatures are necessary in the offices, for the comfort of desk workers. In addition, machines and lights may reduce the heat that must be supplied from the heating plant to maintain desired temperatures in the factory.

In the second place, if some parts are occupied when others are not, it is of course uneconomical to heat both on the same time schedule.

Construction. Zoning with reference to construction—that is, vertical zoning—is one phase that has not been thoroughly analyzed. From the standpoint of economy of operation it may often be highly desirable.

When vertical zoning is applied, the distribution system again becomes the limiting factor. For example, in a building with a large central stair well or other opening throughout its height, air which is heated at any point in the building tends to collect in the top stories and escape there. Not only does heat supplied below move up, but also less cold air tends to enter above, and more below. If these effects were constant, correction by proper radiator sizing would be possible, but they vary so much with outside temperature and wind that dividing the building into zones arranged one above the other would be the only satisfactory solution except for individual-room control.

An example provides analysis of a typical job requiring zone control. Control equipment is shown later.

Figure 13-1 represents a two-story industrial building measuring 100 by 300 ft. The building is exposed on all sides, and the long frontages face north and south. The different sections of the building are used as follows:

First floor: North half, heavy factory work
Southwest quarter, storage and warehouse
Southeast quarter, garage and shipping platform
Second floor: North half, general and private offices
South half, light assembly operations

Steam heating is used. All of the building, except the offices, has direct radiation. The office space has year-round, central-fan air conditioning, and each private office has separate heating-and-cooling coils.

The following points are worth noticing in this example:

1. In general, an industrial building such as this should be zoned for occupancy rather than exposure, but in this example the two factors coincide.
2. The three sections of the first floor have different heating requirements. The storage space requires only freeze-up protection instead of heating for human comfort. The garage and shipping room require more heat than the storeroom because of frequent opening of the garage doors, but less than the factory where more workers are concentrated. Heavy factory work helps to keep workers warm, but temperatures must still be kept within a reasonable comfort range, higher than that required in a warehouse or garage.
3. The light-assembly area requires somewhat higher temperatures for comfort than the heavy-factory area, but also has lower heat loss through exposure.
4. The central-fan system for the office area must be controlled separately from the direct radiation in other areas.
5. The individual heating-and-cooling coils for the private offices must also be independently controlled.

Zone-control Systems. These systems for commercial heating fall into two general classes: those using inside thermostats exclusively to measure the heating requirements, and those depending entirely or chiefly on measurement of outside conditions for determining the necessary heating rate.

Control of Heat at Source. Since it would generally be impractical to provide a separate heating plant for each zone in a commercial building of any size, it follows that the temperature-control system regulates the *distribution* of heat, usually by means of valves. A continuous supply of steam or hot water is maintained by suitable controls in the boiler room. From the standpoint of zone control, these boiler-room controls may generally be taken for granted.

MODULATING OR TWO-POSITION CONTROL

Some of the systems illustrated here are based on two-position operation of the control valves; others are best suited to modulating control. Generally speaking, modulating control, if it can be properly applied, is more satisfactory than two-position control.

Advantages of Two-position Control. Two-position control, with the valve opening wide on demand for heat and then closing tight, has the advantage that periodically saturating and draining the radiation usually presents no problem of even distribution. Modulating control, on the other hand, requires special consideration of distribution, since with a zone valve only partly open the tendency is for the radiators farthest from the valve to get less than their share of steam.

Advantages of Modulation. Two-position control has two disadvantages, however. Because of lag between changes in thermostat demand and changes in actual delivery of heat, the tendency to temperature overshoot and undershoot is increased, especially if the control system characteristically produces rather long, infrequent valve cycles. In addition, sudden changes in steam demand, which may occur if most of the zone valves can operate simultaneously, may cause boiler pressure suddenly to drop or to rise to a dangerous level. It is sometimes necessary to provide pressure relief by means of a reverse-acting pressure control which can open one or more zone valves on an abnormal rise in boiler pressure. With modulating control, wide temperature swings can almost always be prevented, and the typically gradual changes in demand on the boiler avoid the tendency to sudden pressure changes.

Cost Factors. Considerations of cost in the selection of the control system apply in the same way as in the selection of any other equipment. The standard appropriate to the age, type, and use of the building; the saving in operating cost offered; the added return that may be realized through maximum occupancy of rental property—all these factors must be weighed.

Multiroom Buildings. Buildings divided into many small units of space, such as apartments, office buildings, and hotels, are subject to one general rule. Since a satisfactory location for a key thermostat can hardly ever be found, the choice is usually between outdoor-control systems and individual-room control.

Day-Night Control. In any zone-control system, clock-type thermostats or master clocks may be used to provide automatic lowering of the temperature-control point in each zone or in zones with the same schedule of occupancy. The master clock may be a clock thermostat in one zone serving as the master night control for all zones, or a remote program switch used in conjunction with day-night thermostats in the zones to shift to the night-temperature setting at a predetermined time.

Number of Zones. The zone-control systems represented in the diagrams in this chapter show one or two zones, but additional zones may of course be used as needed.

INSIDE-THERMOSTAT CONTROL

Steam-heating Systems. Some buildings are so arranged that an average temperature for a given heating zone exists at a certain point. In this case the method shown in Fig. 13-2 may be used.

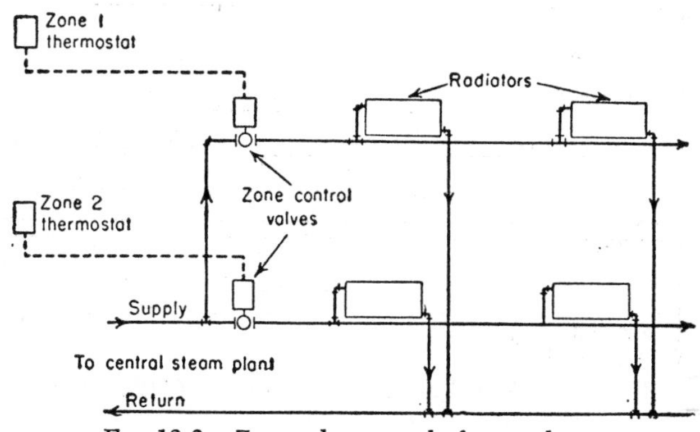

FIG. 13-2. Zone-valve control of steam heat.

Single-zone Thermostat. Here a room thermostat in a key location in each zone controls the motorized valve in the steam line serving the radiation in the zone. This system usually makes use of two-position steam valves, although modulating thermostats and valves are frequently employed successfully where adequate provision for distribution is made by proper sizing, use of orifices, and so forth. The controls may be pneumatic or electric, as the job requires.

Averaging Thermostats. Because of the difficulty of finding an accurately representative thermostat location in a large zone, an averaging thermostat system is sometimes used. The several thermostats in this system are connected together so as to control the zone valve according to the average of the several temperatures measured by the thermostats.

Individual Radiator Valves. In arranging a zone for a complete zone-control system, individual radiator valves under command of a single zone thermostat are often used, as shown in Fig. 13-3. This method frequently avoids complicated piping arrangements which would otherwise be necessary. It is therefore commonly used in zoning apartment buildings. Since no steam-distribution problem is involved, it is desirable to use modulating equipment wherever possible.

Hot-water Heating. In zoning a hot-water heating system, the same considerations of thermostat location and general arrangement should be given as in zoning a steam heating system. The schematic diagram shown in Fig. 13-2 would apply to a gravity hot-water installation. Here, of course, the motorized valves selected would be of a type designed for water service instead of steam.

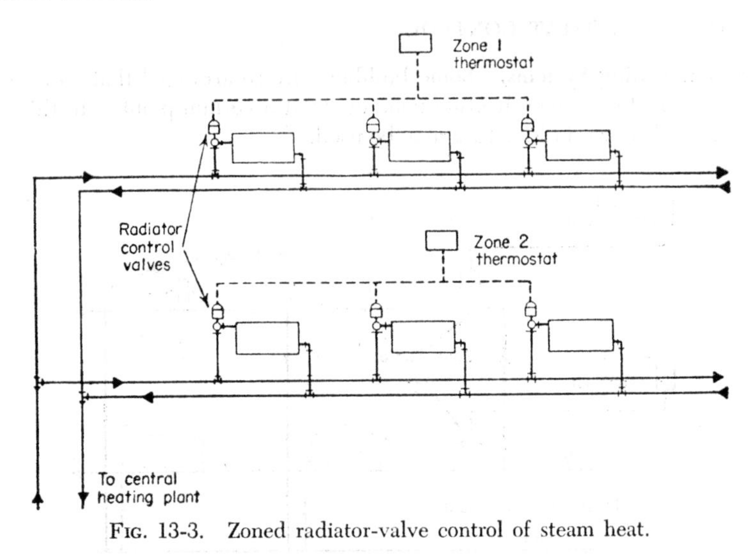

FIG. 13-3. Zoned radiator-valve control of steam heat.

Intermittent-flow Control. Where the heating system is forced hot water, a zone-control system of the intermittent-flow control type, such as that shown in Fig. 13-2, should further provide for the control of the circulating pump. Figure 13-4 illustrates how this is accomplished. Here each motorized zone-control valve is provided with an auxiliary switch. The auxiliary switches are connected in parallel to the circulating pump. A call for heat from any zone thermostat causes the respective zone valve to open and start the circulator. Thus, when any number of zone valves are open, the circulator runs. When all zones are up to temperature and all valves are closed, the circulator shuts down.

Continuous Flow. Figure 13-5 illustrates a system of zone control applied to a forced-hot-water heating system which provides modulated heat and a continuous flow of water. A modulating zone thermostat in each zone controls its respective motorized *three-way mixing valve* installed so as to proportion the amounts of heated and bypassed water flowing to the zone. As the temperature in any zone rises to approach the desired level for which the thermostat is set, the three-way mixing valve starts to close off its hot-water inlet and admits proportionately more bypassed water into the supply line to the zone. When the thermostat in any zone is fully satisfied, all water flowing to that zone is bypassing the heat source. Varying the water temperature in this manner provides very close regulation of

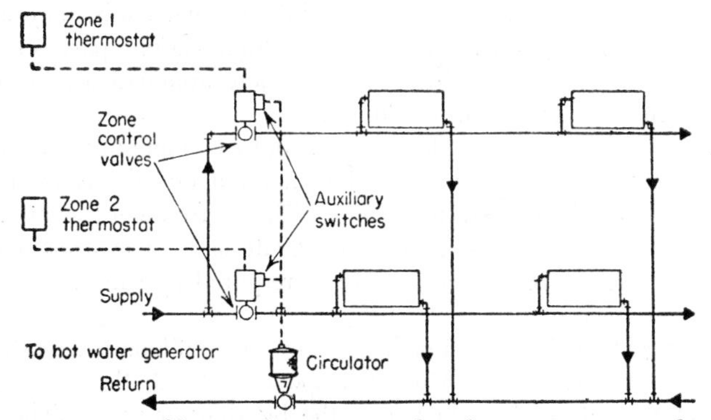

Fig. 13-4. Forced-hot-water zone control with intermittent circulation.

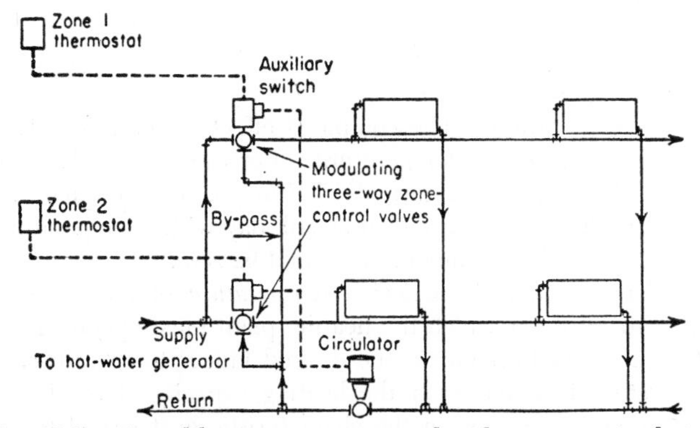

Fig. 13-5. Forced-hot-water zone control with constant circulation.

the temperature in the zones. Usually, it is not desirable to allow the circulator to run when no zone requires heat. Each zone three-way mixing valve may be provided with an auxiliary switch, as shown, to shut down the circulator when all zones are satisfied.

Indirect Systems. The mixing-valve method is often used on indirect systems where, instead of direct radiation, each zone is provided with a central hot-water coil. Each coil heats the air, which is passed to its respective zone by means of a duct system and fan. Here the amounts of water flowing through and around the indirect-heating coil are governed by the modulating zone thermostat and three-way mixing valve as required to maintain the desired zone temperature.

OUTDOOR-THERMOSTAT CONTROL

In a large building or zone, it is often impossible to find an accurately representative location for an inside thermostat. Local conditions, such as variations in occupancy, use of lights, and machine loads, may affect the thermostat so much, and so inconsistently, that various parts of the zone become underheated or overheated. For these applications, various types of outdoor-thermostat systems are in use.

Often a simple thermostat designed for outdoor mounting, or one having a remote sensing element that can be mounted outside, is used so as to measure outdoor-air temperature alone. This is frequently combined with one or more indoor thermostats which operate as limit controls or compensators, to adjust the basic heating rate as required by sun and wind effects or internal heat gain.

Measurement of Heating Load. Ideally, an outdoor thermostat should measure the actual heating load by responding to all the weather factors that affect the rate of heat loss from the building or zone. These factors are:

1. Outdoor-air temperature
2. Wind velocity
3. Wind direction
4. Solar radiation

Resultant Temperature. The combined effect of these four factors may be expressed as a *resultant temperature.* For example, a cloudy day with a 10-mile wind and a temperature of 15°F may cause the same rate of heat loss from a certain building as a cloudy, still day with a temperature of zero. In each case the resultant temperature would be zero.

Design Temperature. The *design temperature* of any building is the resultant temperature at which the heating plant must operate 100 per cent of the time in order to maintain the desired indoor temperature. The design temperature is a matter of the heating capacity of the heating plant and distribution system, and of the characteristic heat loss of the structure. Each zone in a building may have an entirely different design temperature, depending on its construction, orientation, occupancy, and radiation capacity.

Fade-out Temperature and Load Range. The resultant temperature at which *internal heat gain* is sufficient to maintain the required indoor temperature without operation of the heating plant is the fade-out point or no-load condition. For any resultant temperature between *design* and *fade-out,* the heating plant must operate part of the time, and the rate of heat input required increases as the outdoor resultant temperature falls further down in the load range. For example, if the fade-out point is 70°F resultant and the design temperature is −10°F, at a resultant temperature of 30°F the building would require heat 50 per cent of the time. And any two buildings, regardless of differences in size and construction, if they have the same fade-out and design temperatures, would under any given resultant temperature conditions require the same heating rate.

Heat-loss Thermostat. On this principle is based the operation of a *heat-loss* outdoor thermostat. A two-position thermostat is enclosed in a weather-resistant housing, along with an electric-heating element to which current is supplied during on periods of the thermostat. Adjusting the current input to the heater is equivalent to adjusting the capacity of a heating plant, and in this manner the *design temperature* of this miniature building is regulated. With the correct adjustment, the outside thermostat requires heat the same percentage of the time as the building or zone whose heat supply it controls.

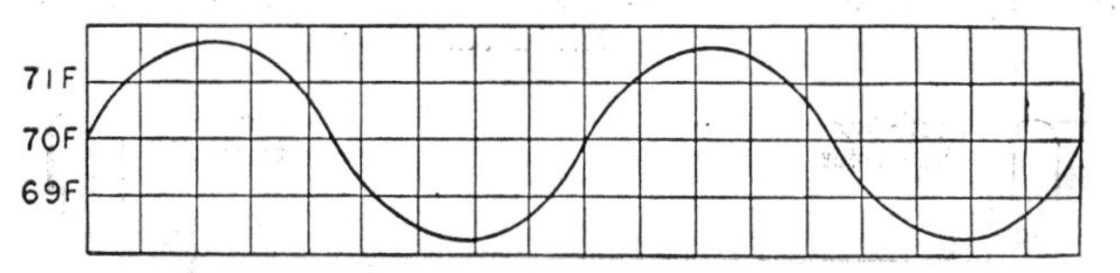

FIG. 13-6. Characteristic temperature curve for a steam-heated building with inside thermostat.

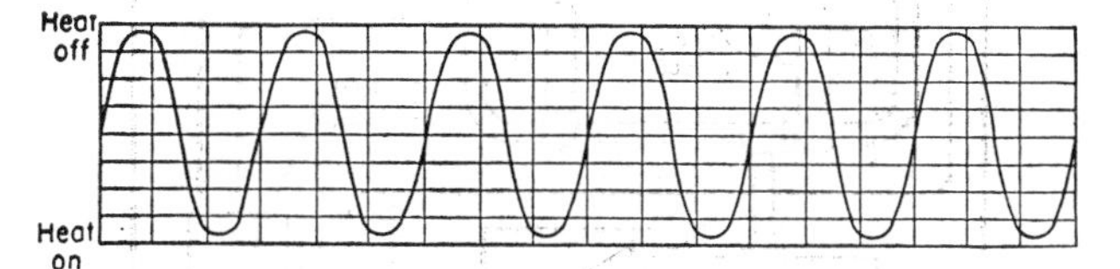

FIG. 13-7. Characteristic temperature curve for a heat-loss outdoor thermostat alone.

Effects of Slow Cycling. Figure 13-6 illustrates the results of attempting to control a large building or zone from an inside thermostat. Because of *thermal lag* or slow response, excess heat stored in the risers and radiation causes a further rise in temperature after the thermostat shuts off. Similarly, delay in starting the delivery of heat on a drop in space temperature permits *undershoot.* The diagram illustrates the resulting wide swings in temperature over relatively long periods of time.

Effect of Faster Cycling. The outdoor thermostat gains and loses heat much more rapidly, as illustrated in Fig. 13-7. When this rapid cycling rate is applied to the heat delivery in the building controlled by a heat-loss thermostat, it results in frequent shots of heat, with only slight variations in temperature of the building or zone, as illustrated in Fig. 13-8. With shorter on and off cycles, heat can be supplied for the same percentage of the time, in terms of minutes per hour rather than hours per day. In effect, heat is furnished to the zone at an almost constant rate.

Zone System. Figure 13-9 pictorially illustrates a heat-loss thermostat system for controlling the heat supply to two zones. More zones may, of course, be added. For each zone a heat-loss thermostat is required, operating through a zone-control panel to open and close a zone steam valve.

Figure 13-10 is a schematic circuit diagram for one zone in a system of the type described. Study of this diagram reveals how the heat-loss thermostat actuates a relay in the control panel, and is in turn supplied with heat whenever the relay pulls in. The relay simultaneously actuates

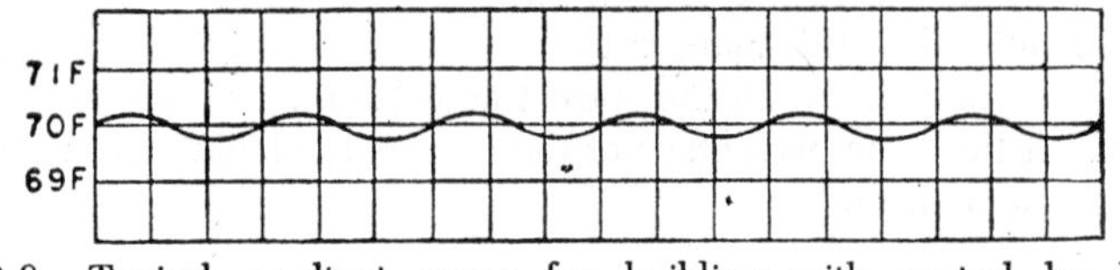

FIG. 13-8. Typical resultant curve for building with control by heat-loss thermostat.

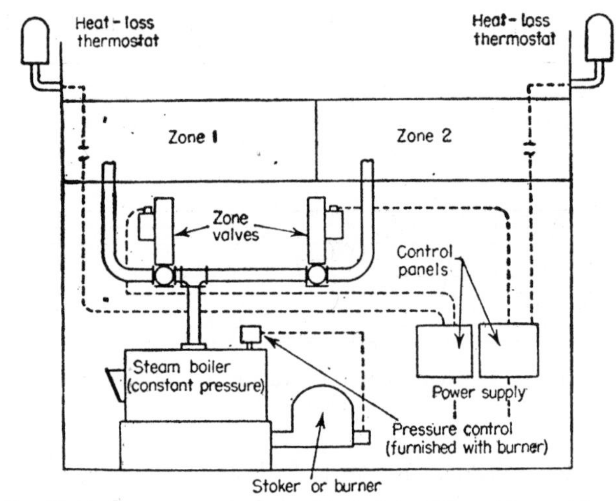

FIG. 13-9. Typical zone-control system (steam) with outdoor heat-loss thermostat.

a two-position motorized steam valve. A double-pole switch permits manual opening of the valve for rapid warm-up, or manual shutdown of the system, as well as automatic operation.

This type of control system is often used for regulating the heating system of an entire building where treatment of the building as a single zone is satisfactory. In these applications, the relay controls the operation of the burner directly, and a remote-bulb thermostat, reverse-acting, connected in series with the heater circuit, delays the delivery of heat to the outdoor thermostat until steam begins moving through the mains.

Outdoor Control of Water Temperature. Figure 13-11 is a sketch of another type of outdoor control system, particularly adapted to forced-hot-water heating. Figure 13-12 is a schematic circuit diagram of the system.

The outdoor thermostat, of the modulating type, positions the reset motor in the control panel. The motor adjusts the scale setting of a remote-bulb thermostat, also mounted in the panel, which in turn controls the position of a modulating three-way valve so as to mix hot boiler water and cooler return water from a bypass, in the correct proportions for the required heating rate. Since the outdoor thermostat measures only the basic load condition represented by outdoor-air temperature, an indoor thermostat is generally

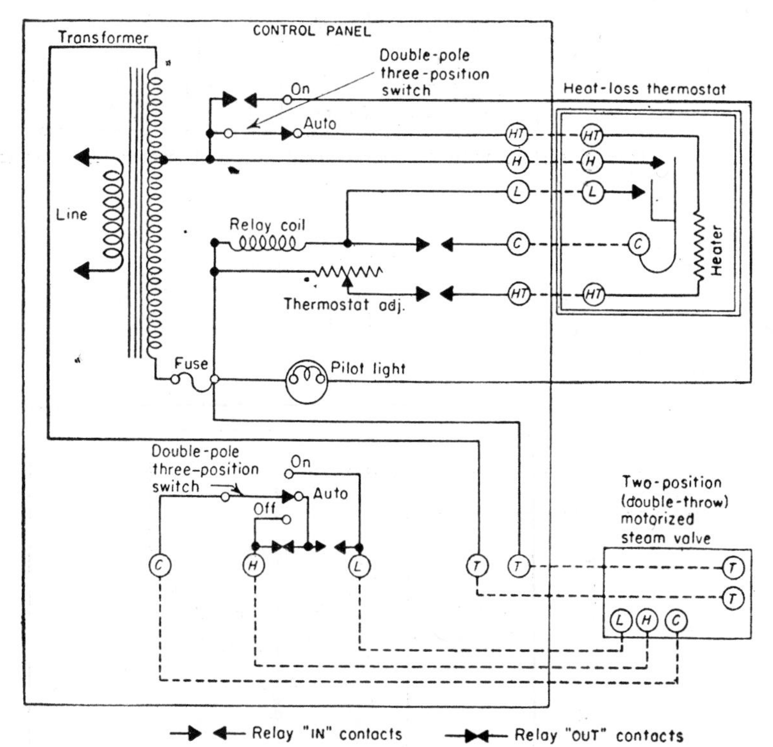

FIG. 13-10. Simplified schematic circuit of system shown in Fig. 9-9 (one zone represented).

used as indicated, to adjust the final position of the mixing valve according to variations in heating requirement resulting from other weather factors and from internal heat gains. Manual means of control and of adjustment of the control point are included.

An auxiliary switch on the reset motor (or on the valve motor) stops the circulator when no heat is needed.

An outdoor-reset system is often used effectively for *basic-rate* control, maintaining boiler-water temperature slightly higher than necessary for the maximum heating demand *at the existing outdoor temperature.* This enhances the performance of indoor zone or individual-room controls.

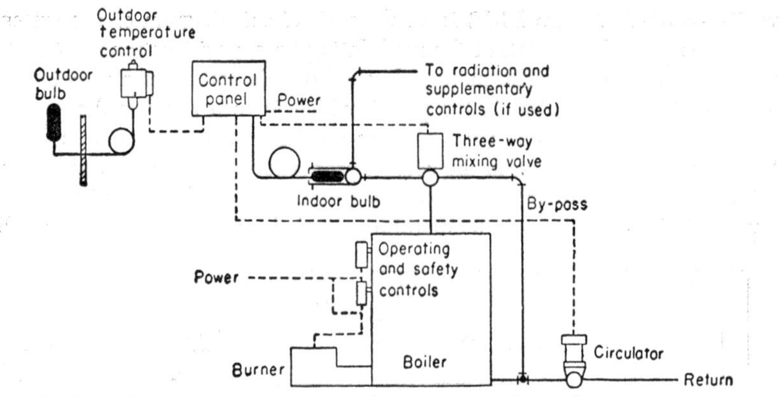

FIG. 13-11. Typical system with outdoor control of supply-water temperature.

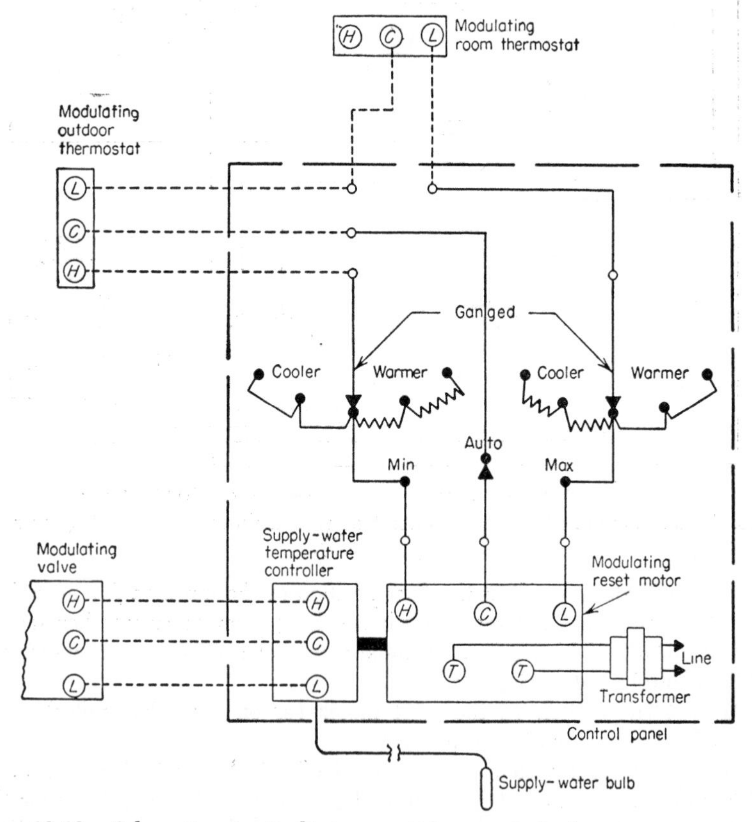

FIG. 13-12. Schematic circuit diagram—outdoor control of water temperature, with indoor compensating control.

INDIVIDUAL-ROOM CONTROL

The principle of zone control carried to the ultimate is individual-room control. If each room is provided with its own thermostat and one or more actuators (such as radiator valves), the temperature can be maintained exactly according to the needs and desires of the occupants. In multiple dwellings, it is becoming more and more common to provide a thermostat in each apartment. And in large rooms, such as offices and stores, several thermostats are frequently provided, each controlling the heat supply to a section of the space, because of variations in exposure and occupancy.

CONTROL OF UNIT HEATERS AND UNIT VENTILATORS

Unit Design of Equipment. *Unit heaters* and *unit ventilators* are alike in one respect, namely, the principle of equipment design which leads to the provision of individual conditioning *units,* as fully self-contained as possible, for separate divisions of space in a building. The differences in operation and application lead to widely different requirements in the design of the control systems. For this very reason they furnish a striking illustration, on a small scale, of the variety of control sequences that the automatic-control designer is able to provide.

Functions of Unit Heaters. Unit heaters are designed primarily for applications where the major requirement is heating, with a maximum of flexibility and capacity combined. They normally handle only recirculated air, and during warm weather are expected only to provide air movement for minor cooling effect.

Functions of Unit Ventilators. Unit ventilators, on the other hand, are designed for applications requiring a range of performance, during the heating season, from maximum heating to maximum cooling by means of outdoor air. The typical application, in fact, illustrates in extreme degree the need for individual room control of temperature. In a school building, for example, one room, sheltered from wind, exposed to sun, and full of active children, may require a high rate of ventilative cooling; and, at the same moment, another room, shaded and exposed to wind, and only partly filled with children, may require a moderately high degree of heating, in addition to a certain minimum rate of tempered ventilation as specified by health laws. Yet this relationship may change in a few minutes, with a change in wind direction and cloud cover and a shift of occupancy from room to room. Only the provision of temperature controls for each individual room, together with conditioning equipment specifically adapted to individual-room control, can begin to meet the problem of comfort created by such conditions—to say nothing of the problem of operating economy.

Unit ventilators, then, are designed to handle large volumes of outdoor and recirculated air, in varying proportions, and to provide at least enough heat for tempering the outdoor air to room temperature when necessary.

Auxiliary radiation is often used in conjunction with the unit ventilators, in *split systems*, to provide maximum heating capacity when needed; but the unit ventilators may often be required to handle the full heating load alone.

Construction of Unit Heater. A unit heater consists of a heating surface, such as a steam coil, and an electric fan forcing rapid circulation of air over the heating surface. The unit is usually suspended overhead in the conditioned space, partly as a means of saving floor space.

Construction of Unit Ventilator. A unit ventilator, on the other hand, although it includes a heating surface, normally a steam coil, also includes a centrifugal blower for maximum efficiency in air handling, and a system of dampers operating in synchronism for metering the proportions of outdoor and recirculated air. In addition, it employs a rudimentary system of ducts, if only an intake duct for fresh air, long enough to reach through the outside wall of the building. In contrast to a central-fan system, however, a unit ventilator normally has, at most, a few feet of duct; and each room or zone has its own small conditioning plant, complete except for the remote, central heat-generating plant.

Typical Application of Unit Heater. A typical application of unit heaters might be a factory where one or more units in each bay, suspended above the workers' heads, provide the required additional heat for comfortable working conditions beyond what is supplied by machines and occupants. The occupancy load in heat and humidity gain is usually not very large.

Typical Application of Unit Ventilator. The typical application of a unit ventilator is a school classroom or auditorium. Here the large number of occupants in proportion to space may, even in cold weather, change the heating load to a cooling load early in the day and, at almost all times, makes important the introduction of a considerable volume of outside air for ventilation.

UNIT-HEATER CONTROL METHODS

Examination of the common methods of control applied to unit heaters provides a background for review of the general problems of unit-ventilator control.

Modulating Control Common. Since the typical applications of unit heaters generally utilize the chief feature of this method of heating, which is flexibility in occupancy zoning, both in the sense of use and in the sense of time, it is normal to employ direct control of the steam supply to each unit, where steam unit heaters are used. For this reason, modulating control may almost be considered normal for unit heaters. The controls may be either pneumatic or electric, depending on which is the more convenient and economical in any given installation.

Pneumatic Control. Figure 13-13 is a schematic representation of a typical pneumatic-control system. A room thermostat positions the normally open pneumatic valve in the steam line to the unit coil, according to the heating requirement of the room. A pneumatic-electric relay completes the electric-power supply to the fan motor except when the thermostat branch pressure rises to the point at which the valve is closed. The dashed lines indicate how a reverse-acting pressure control is often wired into the fan-motor circuit to prevent fan operation if the steam pressure falls below the minimum for effective heating. This low-limit feature is discussed later. A three-position manual switch is commonly included, as shown, to permit manual selection of automatic operation (heating), shutdown of the fan when heating is not required in the particular room or zone, or continuous operation of the fan for air circulation in warm

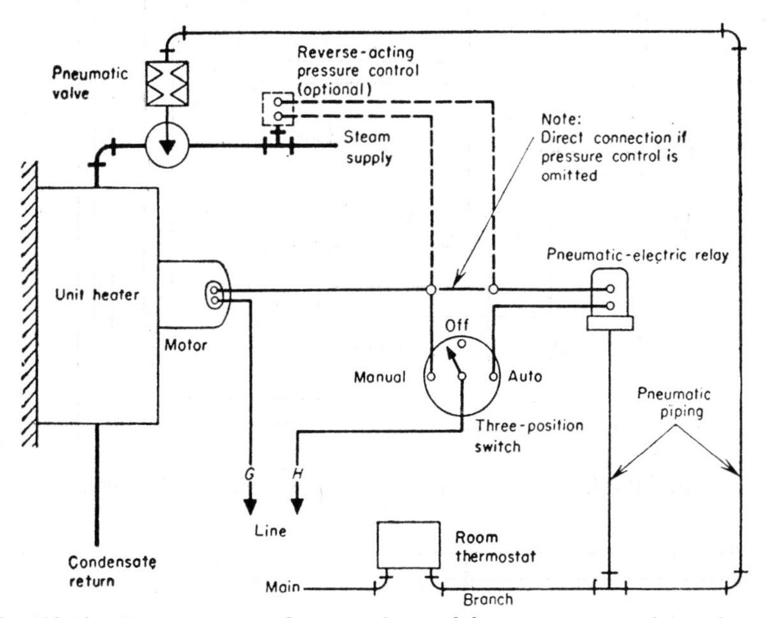

FIG. 13-13. Pneumatic-control system for modulating operation of unit heater.

weather. One thermostat may, of course, control several units in the same zone.

Electric Modulating Control. An electric-control system, identical in function with the pneumatic system just described, is illustrated in Fig. 13-14. The modulating thermostat, potentiometer type, controls a motorized modulating steam valve, and an auxiliary switch on the valve motor stops the fan when steam is shut off. Again, a reverse-acting pressure control may be used to prevent cold drafts, and a three-position selector switch for change-over.

Two-position Control. Two-position control is often used with unit heaters, particularly where some convective heating from the unit coil during off periods is not objectionable. As shown in Fig. 13-15, the coil runs wild (steam valve omitted), and a line-voltage thermostat, usually in conjunction with a low-limit control such as a reverse-acting surface-mounted thermostat on the condensate line below the coil, controls the operation of the unit fan. A manual switch for positive control of the

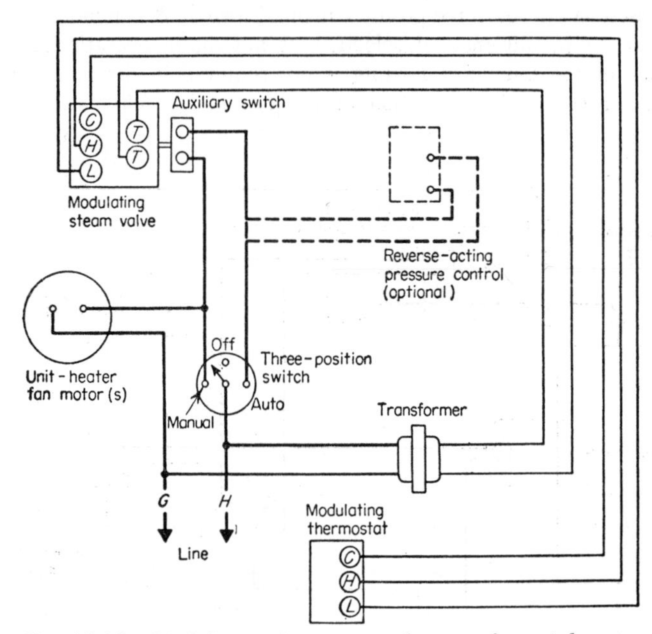

FIG. 13-14. Modulating electric-control system for unit heater.

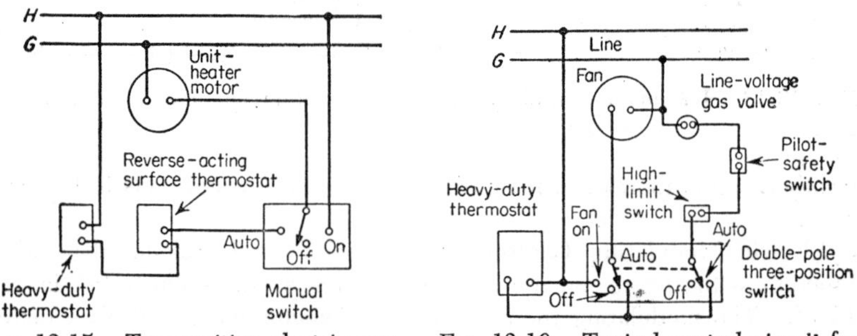

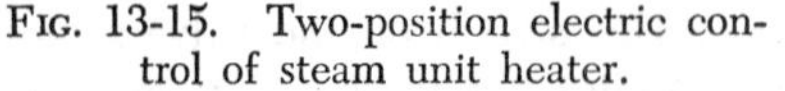

FIG. 13-15. Two-position electric control of steam unit heater.

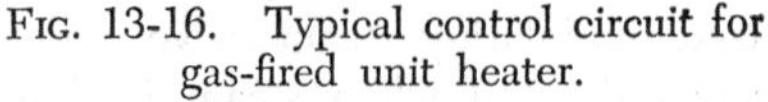

FIG. 13-16. Typical control circuit for gas-fired unit heater.

fan is required, and for maximum utilization of the flexibility of the unit, this is commonly a three-position switch, as indicated. A large unit motor, beyond the rating of the thermostat or one operating on three-phase current, requires addition of a starter relay.

Control of Gas-fired Units. The control systems so far discussed have all been designed for steam-coil unit heaters, since these are probably the most widely used. A good many installations utilizing gas-fired units can be found, particularly where gas for heating is readily available. Figure 13-16 illustrates a typical method of control for gas-fired units. A line-voltage thermostat controls the fan motor, directly or through a relay, and also operates the solenoid gas valve. A switch-model pilot safety control guards against opening of the main valve in the event of extinction of the pilot flame, and a high-limit switch is often included to prevent overheating of the heater surfaces. The manual switch for auto-off–fan-on control must be a double-pole type to separate the fan-motor circuit from the gas valve circuit.

Lowered Night Temperature. During the night hours, unoccupied portions of a building may be maintained at lowered temperature in any of

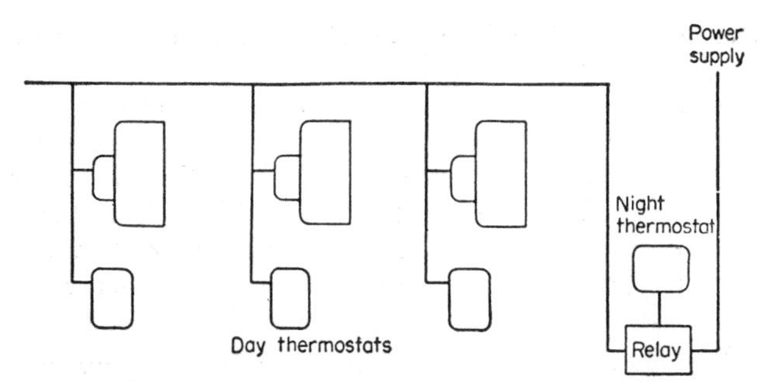

FIG. 13-17. Night-temperature control—relay method.

several ways. If there is no danger of the temperature falling too low, the unit heaters may be shut off entirely, by manual means or by a time control. Where a minimum temperature or low limit must be provided, a night thermostat may be placed in command of the units in a section of the building or of the whole building. Change-over may be instituted manually, or by a separate time switch, or by use of a clock-type night thermostat.

Night-thermostat-relay Method. Figure 13-17 illustrates one method of controlling all the units in the zone or building from one thermostat at night. If the power supply to the units can be carried through a relay, the night thermostat, or a time switch, may trip the relay to cut off the power. All the day thermostats soon close their circuits, since the temperature falls below that called for by the day settings. When the night control is a thermostat, a drop in temperature below the desired minimum causes the thermostat to energize the relay, and this in turn supplies power to the unit-heater motors, but only until the night thermostat is satisfied.

With time control of the change-over, the night-thermostat setting is automatically raised, at some predetermined time in the morning, above the setting of the day thermostats; or a switch is closed to shunt out the night thermostat, so that the relay remains continuously closed and the day thermostats may resume control.

Night-thermostat-valve Method. Another method, illustrated in Fig.

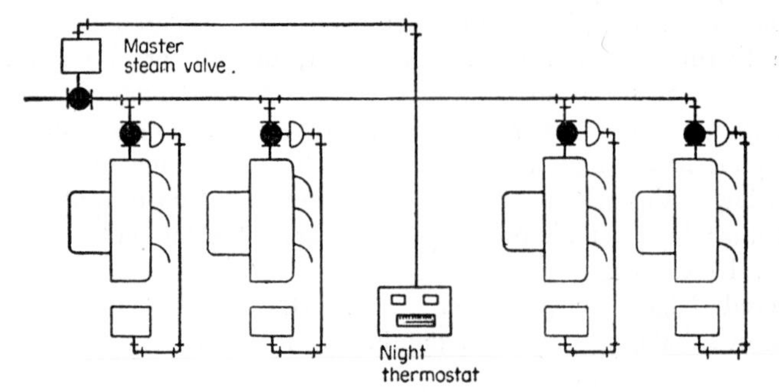

FIG. 13-18. Master-valve method of night-temperature control.

13-18, makes use of a two-position master valve in the steam main serving a particular zone or wing. During the night-control cycle, the night thermostat opens the master valve, if necessary, to keep the temperature above the low limit. This method requires fan-low-limit controls on the individual units, as explained later, in order that the individual thermostats may operate their respective fans only when steam is available at the unit coils.

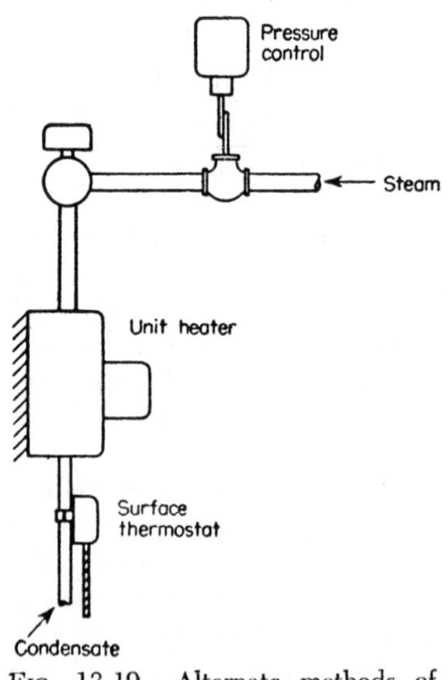

FIG. 13-19. Alternate methods of low-limit control.

Fan-low-limit Control. It is usually desirable and often necessary to provide a *low-limit* feature in the control system, to prevent the thermostat from running the fan when there is no steam in the coil. Without this precaution, uncomfortably cool air may be circulated by the fan. In modulating-control systems, this requirement is, as a rule, adequately met by the pneumatic electric relay (Fig. 13-13) or auxiliary switch (Fig. 13-14), which interrupts the power circuit to the fan motor when the unit valve is throttled down to the closed position under command of the thermostat. In some installations, however, it may be possible for the thermostat to open the valve and thus to start the fan, even though no steam is available. This is particularly true when the master-valve method is used for lowered night temperatures, and, although there may be no question of discomfort, the economic argument against useless operation of the fan motors is obvious.

In two-position control of the unit heater, the use of a low-limit control is practically essential, since there is no other assurance that the fan will operate only when there is steam in the coil.

Figure 13-19 simultaneously illustrates the two types of fan-low-limit controls commonly used on unit heaters. Whether a pressure or temperature control is used, it is reverse-acting; that is, it closes the circuit between thermostat and fan motor when a rise in pressure or temperature signals the presence of steam. The pressure control is the type most common where a steam valve is used, since it responds to the availability of steam even though the valve may be closed or so nearly closed that the return-line piping may be cool. The surface thermostat, when used, is normally clamped to the return line between the unit coil and the trap, where it quickly responds to the rise in pipe temperature that indicates the presence of steam or hot condensate in the return line ahead of the trap. A special type of surface thermostat, designed for the application, is sometimes clamped to the face of the unit-heater core to indicate the presence of steam in the coil itself.

UNIT VENTILATORS

Turning now to the control of unit ventilators, first examine briefly the basic construction and essential operating elements of a typical unit. Figure 13-20 illustrates, schematically rather than pictorially, the most common type. The other types in ordinary use are functionally the same, although they may differ slightly in the arrangement of dampers, coils, blower units, and the like. For instance, a bypass damper may be used to bypass air around the steam coil when no heat is required.

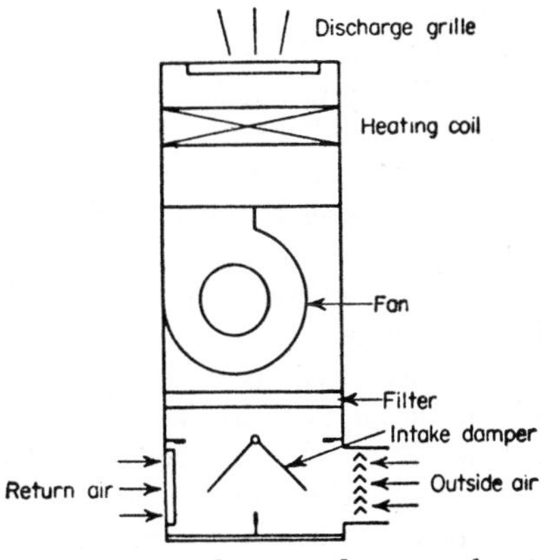

Fig. 13-20. Schematic diagram of unit ventilator.

Referring to Fig. 13-20, note that the unit ventilator has all the necessary elements to deliver air to the room at any temperature desired. At the bottom of the unit is found a dual damper, arranged so as to vary the relative volume of outside air and recirculated air admitted to the unit. As the outside-air damper opens, the return-air damper closes proportionately. By positioning the dampers, it is possible to admit any proportion from 0 to 100 per cent of outdoor air to the unit. A blower unit, consisting of two or more centrifugal blowers, draws a constant volume of air through the unit. A heating coil, supplied with either steam or hot water and equipped with a valve, heats the air leaving the unit.

Stand-by Operation. In a typical unit-ventilator application, the conditioned space is usually unoccupied during a considerable part of each 24-hr period. During this time, only enough heat is required to guard against freeze-up or to avoid the necessity of an excessively long warm-up period in the morning. At the same time, no ventilation is required. Under these conditions, any type of unit ventilator is expected to function as a convector, with the fan shut down, the coil-supply valve open, and the dampers positioned for 100 per cent recirculated air. When auxiliary radiation is used to carry part of the heating load, the radiator valves are also open during *stand-by* operation, and the room thermostat has control of all the steam valves, so that it can throttle them down if the heat available under these conditions is sufficient to raise the room temperature above the desired stand-by level.

Warm-up Cycle. During the warm-up stage also, all types of applications employ a standard control sequence. The fan is started, manually or by an automatic time switch; the valves remain open, for full heat input; and the dampers, as a rule, remain positioned for recirculation of all the air handled by the unit. This continues as long as the room temperature is below the throttling range of the thermostat.

Operating Cycle. As the temperature enters the operating range of the room thermostat, the control-sequence characteristic of the particular application begins. The control sequence to be used is often dictated in large part by legal requirements, particularly as to the amount of outdoor air to be supplied in schools and similar institutions.

Chief Control Sequences. Although a number of control sequences are used, they generally approximate one of the three basic sequences to be described here. In this discussion, it is assumed that the heating load is decreasing, or the occupancy gain increasing, or both, so that the space temperature rises through the full modulating range of the thermostat. In actual practice, of course, the temperature may rise and fall; but it need only be remembered that, on a drop in temperature, the sequence described here is reversed.

The three basic control sequences are distinguished chiefly by the ratio of outside air to recirculated air that is to be maintained during the cycle as a whole, or some significant portion of it. They may be listed as follows:

1. *The fixed-percentage method,* in which the intake dampers are positioned to handle a constant amount of outdoor air whenever the room temperature is above some minimum point. The quantity of outdoor air may be 100 per cent, or some lesser percentage, as specified.
2. *The fixed-minimum method,* in which the intake dampers are positioned to take a fixed minimum percentage of outdoor air as soon as the room temperature enters the throttling range of the thermostat. This percentage is maintained throughout the lower part of the throttling range, but, as the temperature rises through the upper part, the percentage of outdoor air is gradually increased to 100 per cent.

3. *The variable-outdoor-air method,* in which an auxiliary insertion thermostat takes increasing command of the intake dampers as the room temperature rises through the throttling range of the room thermostat and drives the intake dampers to any position, within certain limits, that may be required to maintain a predetermined temperature of the mixed air entering the heating coil.

Any of the three methods may be used in a given application, except as legal codes may limit the choice. The following diagrams and descriptions show how automatic controls accomplish the desired cycles. The temperatures mentioned are illustrative only. Actual temperatures at

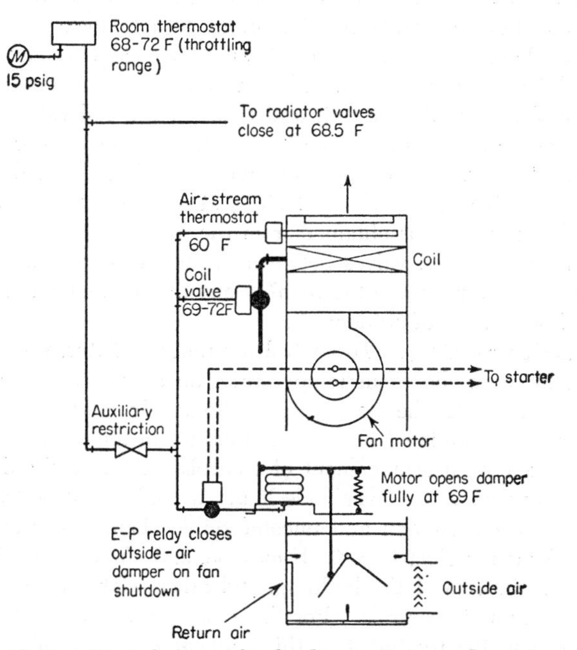

FIG. 13-21. Control system for fixed-percentage outdoor-air cycle.

which the various stages of control action occur may vary somewhat, according to specifications for the individual installation, and according to the characteristics of the equipment and of the building, including occupancy, which affect performance of the heating system. For example, although the illustrations are based on an assumed throttling range of 4°, it may in some cases be possible as well as desirable to adjust the control units for a narrower range; in other cases, however, a somewhat wider throttling range may be necessary to ensure stability of control, that is, to prevent *hunting*.

Fixed-percentage Method. Figure 13-21 schematically illustrates an arrangement of pneumatic controls to provide a *100 per cent outside-air* sequence. Before examining the diagram in detail, remember that valve and damper motors may be provided with springs of the desired rating to cause them to begin moving at a specified branch pressure and to complete the full stroke in a specified range of branch pressure. The throttling range of the thermostat, taken in the example as 68 to 72°F, or 4°, is usually considered to be the change in temperature required to make the thermostat raise the branch pressure from 3 to 13 psig.

The room thermostat, in all the systems illustrated, is a graduate-acting type. Nonbleed thermostats are well adapted to applications such as this. Likewise, the positive-positioning feature in valves and damper motors is highly desirable for exact sequencing, where the position of each at a given temperature is important not only for its own direct effect but also for the combined effect of the relative positions of all the actuators.

Note the effect of positioning the dampers and opening and closing the valve supplying the heating coil. If we position the dampers to admit 100 per cent outdoor air into the unit and close off the coil valve, the air discharged from the unit has the same temperature as the outdoor air. If we arrange the dampers for 100 per cent recirculated air from the room and open the coil valve, the discharged air is at a high temperature, and the maximum heating effect is obtained from the unit. If we control the position of the dampers and valves between the two extremes, we can, within certain limits, deliver air into the room at any desired temperature and with any desired percentage of outdoor air.

Required Control Units. To control the unit ventilator, we need a room thermostat to sense the room temperature, an automatic valve to control the flow of steam to the heating coil, a damper motor to position the dampers, and an additional insertion thermostat in the air stream. Figure 13-21 shows a typical arrangement of controls. Because the discharge-air temperature can be varied over wide limits, a modulating control system is required to prevent sudden and uncomfortable changes in the room temperature. Although electric controls may be, and often are, used, sequence requirements, the size of the building, and other related factors may favor the use of pneumatic controls. For this reason, pneumatic-control systems are illustrated in the treatment of this application. It should be borne in mind that electric controls can be applied to any of the control sequences described.

Typical Temperature Values. Observe, in Fig. 13-21, what happens as the room temperature rises through the modulating range of the room thermostat. At 67°F the outside-air damper would be closed, and the return-air damper and the coil valve would be open. At 68°F, the auxiliary radiation valve starts to close; it is fully closed at 68.5°F. At 69°F, the outdoor-air damper opens to admit the specified percentage of outdoor air, and the coil valve begins to close. At 72°F, the coil valve is completely closed; the outdoor-air damper remains open.

Low-limit Control. The air-stream thermostat shown is a low-limit control commonly provided to prevent discharge of uncomfortably cool air, which would cause a sensation of drafts. The auxiliary restriction is sometimes located in the air line supplying the insertion thermostat and the coil valve. Then, even though the room thermostat may be maintaining full supply pressure in its branch line, the low-limit control can bleed down the pressure in the valve branch and open the valve wide, if necessary, to keep the discharge temperature from falling too low. In the system shown, the location of the restriction enables the low-limit control to close the outdoor-air damper, if that should be necessary.

Electric-pneumatic Relay. In any pneumatic-control system for a unit ventilator, an electric-pneumatic relay is usually used, as shown in Figs. 13-21 to 13-23, to ensure that the outdoor-air damper is closed whenever

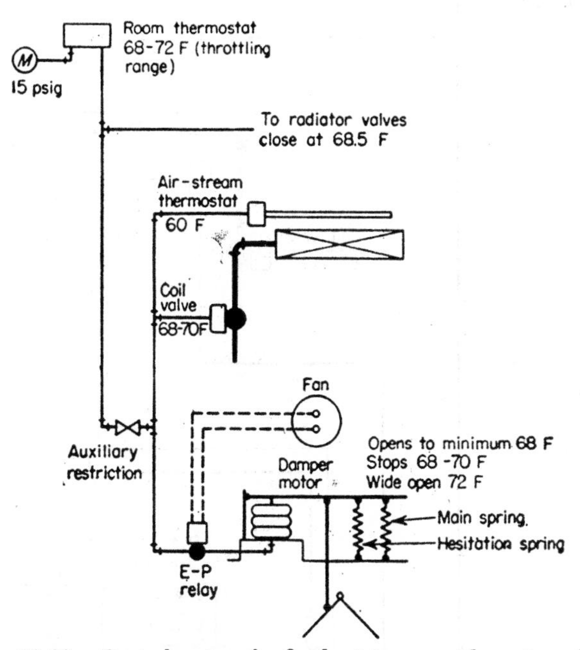

FIG. 13-22. Control system for fixed-minimum outdoor-air cycle.

the fan is shut down. The relay coil is connected to the fan-power circuit, either at the fan as shown or at terminals provided for this purpose in the fan-starter box, so that the coil is energized whenever the fan is running. The electric-pneumatic (E-P) relay is simply a two-position electric valve designed specifically for pneumatic-control applications. When the coil is deenergized, the valve stops the air line from the thermostat and other controls, and exhausts the air from the line running to the damper motor. The damper therefore moves to the 100 per cent recirculated-air position, the thermostat is left in command of the coil valve and radiator valves, and the unit ventilator operates as a convector.

Fixed-minimum Method. Figure 13-22 illustrates a typical control setup for the *fixed-minimum* method. Refer to the temperature notations for the operating sequence. Note first that the damper motor has an auxiliary or *hesitation* spring so adjusted that its tension is applied to the motor arm at the minimum-percentage outside-air position. Now we have rapid closure of the radiator valves, following even more rapid movement of the intake dampers to the *fixed-minimum* position. The unit-coil valve throttles down during the lower half of the thermostat throttling range.

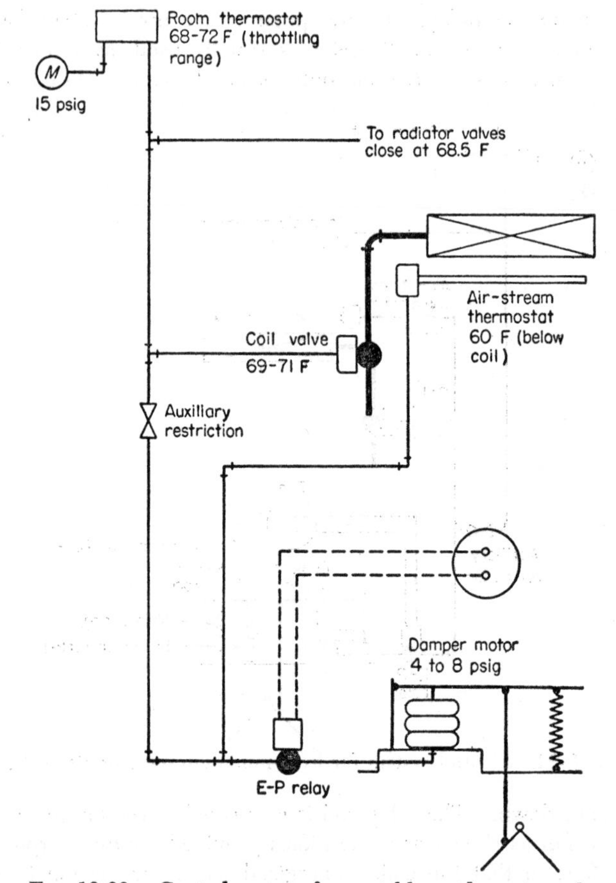

FIG. 13-23. Control system for variable-outdoor-air cycle.

Thereafter, the intake dampers modulate to the full-open position for outside air. The auxiliary restriction permits the low-limit thermostat to close down the outdoor-air damper to the minimum position, and then to open the coil valve, if necessary, to keep the discharge temperature from falling too low.

Variable-outside-air Control. At first glance, the system in Fig. 13-23 appears identical in arrangement with that in Fig. 13-22. Note, however, that the insertion thermostat is located below the coil, where it can measure the temperature of the mixed air entering the coil, and that only the *damper motor* and *insertion thermostat* are supplied with air through the auxiliary restriction. Consequently, the room thermostat has full command of the coil valve, as well as of the radiator valves. The radiators are again shut off immediately after the room temperature enters the throttling range. Room temperature is controlled primarily by throttling the coil steam supply over the middle half of the thermostat throttling range. The damper motor, however, is primarily controlled by the insertion thermostat; regardless of room temperature, except to the extent that the room thermostat determines the *maximum air pressure* available to the motor, the intake dampers are positioned as necessary to maintain the temperature of the air entering the coil within the throttling range of the insertion thermostat. The operating range of the motor on a typical job may be as indicated, 4 to 8 psig. What this range may be in terms of temperature depends on the adjustment of the insertion thermostat, but is typically within a few degrees of 60°F.

Function of Insertion Thermostat. The insertion thermostat, in a *variable-outdoor-air* sequence, is often referred to as a *mixed-air controller.* The designation is appropriate, in that the insertion thermostat can open or close the outdoor-air damper so as to maintain the normal mixed-air temperature. But this is true only when the room thermostat is supplying air pressure of 8 psig or more to the insertion thermostat and damper motor. In addition, the room thermostat can to some extent overrule the insertion thermostat. It might appear that the upper 25 per cent of the room thermostat's throttling range is unused, if we observe only the operating range noted for the valve, as compared with the range noted for the room thermostat. It should be observed, however, that an increase in thermostat branch pressure above the value for valve closure, or above the midpoint value of 8 psig, for that matter, requires a lower temperature in the mixed air to permit the insertion thermostat to reduce the percentage of outside air admitted. Thus, strictly speaking, only the room thermostat can open the outside-air damper; the insertion thermostat can merely reduce the amount of outside air more or less below the maximum permitted by the room thermostat.

Freeze-up Protection. In unit ventilators using hot-water coils, special provisions are often made to prevent freeze-up of the coils. One method is to use an auxiliary immersion thermostat in the hot-water coil to close the outdoor-air damper if the coil temperature drops below a safe point.

CONTROL OF COMMERCIAL CENTRAL-FAN HEATING SYSTEMS

Central-fan systems for heating and air conditioning consist of many separate pieces of apparatus, designed to provide any or all of the following functions:

1. Heating
2. Humidification

3. Ventilation
4. Distribution
5. Air cleaning
6. Cooling by use of cool outdoor air whenever necessary

There are many variations in the physical arrangement of equipment used to accomplish the functions outlined above. Selection of a method which will prove most satisfactory for a specific installation depends on local practice and local climatic and economic factors.

When more than one function is to be accomplished by a single system, particular care must be exercised in arranging and interlocking the equipment, so as to provide a completely coordinated sequence of operation. The use of automatic temperature control provides a satisfactory and economical means of accomplishing this coordination.

In the following sections the factors affecting heating, humidification, ventilation, and atmospheric cooling are discussed separately; and the application of automatic controls to the equipment used for these purposes is analyzed. The diagrams illustrate only the portions of the conditioning system under discussion. Other elements may be combined with those illustrated, in various ways, in making up a complete air-conditioning system.

The control functions discussed may be provided by either electric or pneumatic controls, whichever is indicated by the conditions affecting the individual installation.

Winter-conditioning systems may be divided into two general types:

1. *Tempering systems for ventilation only.* In systems of this type the mixture of outdoor and return air is delivered at space temperature for ventilation only. The actual heat losses of the space are then provided for, by means of direct radiation.

2. *Blast systems for heating and ventilating.* In this type of system, the heating surface in the fan system is made large enough not only to temper the mixture of outdoor and return air, but also to provide for the heat losses from the conditioned space.

TEMPERING SYSTEMS WITH PULL-THROUGH FANS

Coil-valve Control. Figure 13-24 illustrates a *pull-through fan* system utilizing a mixture of outdoor and return air. The mixed air passes through the heating coil to the fan, from which it is discharged to the ventilated space. Since this is a tempering system only, it is desirable to maintain a constant discharge temperature.

A modulating controller T_1 in the fan discharge operates a modulating steam valve V_1 on the heating coil, to provide just enough steam to maintain the desired discharge conditions.

A two-position type of control should not be used, since it would cause alternate high and low discharge temperatures, as the valve opened and closed.

Sizing Equipment. In choosing equipment for this application, care should be taken to size the modulating valve correctly. Because of the rapidity of air movement through the heating coil, a discharge-control system must be accurate if drastic conditions are to be avoided. If, for example, the valve used is twice as large as necessary, then the valve positions from *one-half* to *open* would not change the temperature of the discharge air, since by the time the valve is half open the heating coil would

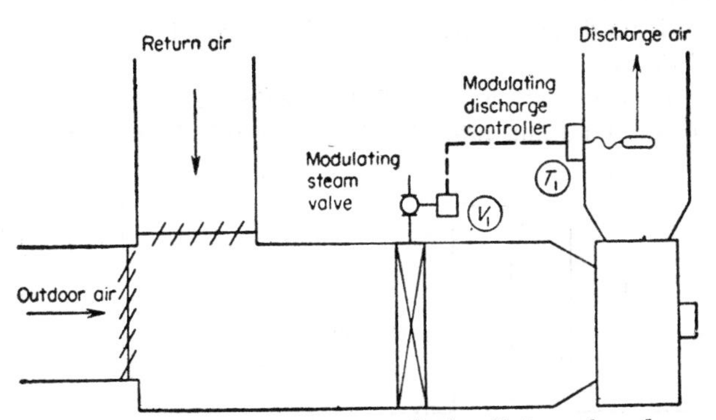

FIG. 13-24. Thermostat control of pull-through coil

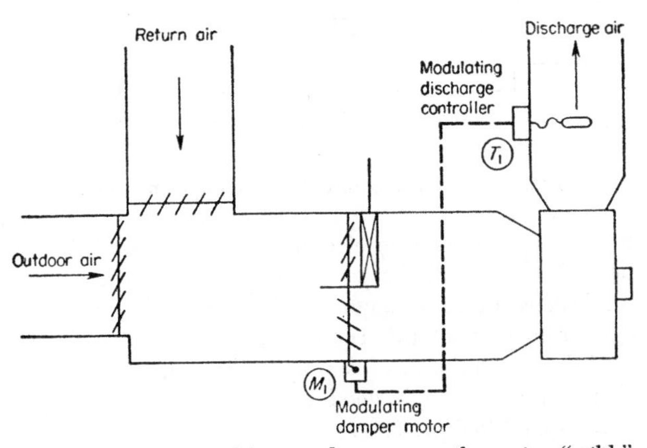

FIG. 13-25. Face and bypass dampers—coil running "wild."

be completely filled with steam. Likewise, if the maximum necessary pickup through the heating coil is 30° and if the heating coil is large enough to heat the air 60°, it is evident that, when the coil itself is operating at more than 50 per cent of capacity, the air discharged is at a temperature higher than that desired.

Bypass System. As shown in Fig. 13-25, installations are often made where the heating coil does not extend from the top to the bottom of the ductwork. When this practice is followed, the air may pass either through or around the heating coil, provided that suitable dampers are installed to cause the proper deflection of air flow. The advantages of this particular arrangement are:

1. The modulating motorized face and bypass dampers move in opposite directions at the command of a modulating discharge controller T_1, and thereby proportion the relative amounts of air passed through and around the heating coil, so as to maintain a constant discharge temperature.
2. There is no danger of freezing the heating coil since it is full of steam at all times.

Damper Leakage. It is difficult to obtain commercially manufactured dampers, of the type illustrated in Fig. 13-25, which are 100 per cent tight.

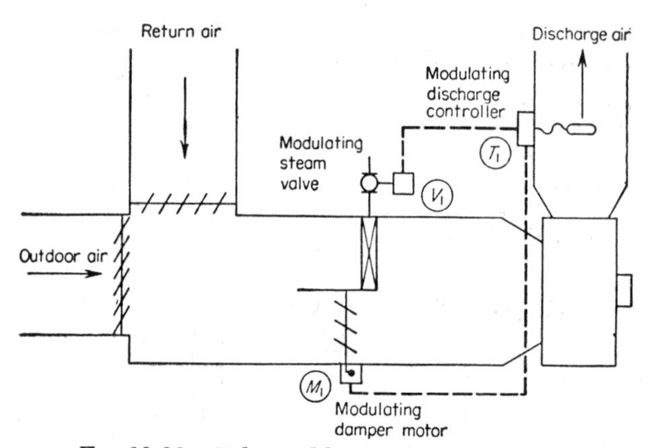

Fig. 13-26. Valve and bypass damper method.

Consequently, when the mixture of outdoor and return air is near the temperature desired for the discharge air, overheating is likely to occur because of the leakage of air through the dampers and because of eddy currents. In order to prevent this condition, it is desirable to provide some means of final shutoff on the steam supply to the coil.

The leakage of air through good commercially built dampers may be expected to be from 2 to 6 per cent.

Valve and Bypass Method. Figure 13-26 illustrates a system similar to that in Fig. 13-25, with the addition of a valve in the steam-supply line to the coil. No face damper is used. The system can be used where danger of coil freezing is not present. The advantages of this system are:

1. The steam valve and bypass damper are modulated together at the command of a modulating discharge controller. As the steam valve throttles, the bypass damper opens, thus reducing heat delivery.
2. There is no danger of overheating in mild weather, as the steam supply is completely shut off when the mixed outdoor-air and return-air temperature rises to the setting of the discharge controller.

This system is an improvement over that shown in Fig. 13-25, inasmuch as the possibilities of overheating are eliminated.

Valve with Face and Bypass Dampers. The system illustrated in Fig. 13-27 is used on applications where there is danger of admitting subfreezing air across the face of the heating coil.

In this instance the modulating discharge controller positions the face and bypass dampers in order to maintain a constant discharge temperature.

The system is so arranged that the action of the damper motor causes throttling of the steam valve as the face damper approaches the closed position. That is, whenever the face damper is closed, the steam valve is also closed; and, whenever the face damper opens by a predetermined amount of its total travel, the steam valve has modulated to the full-open position.

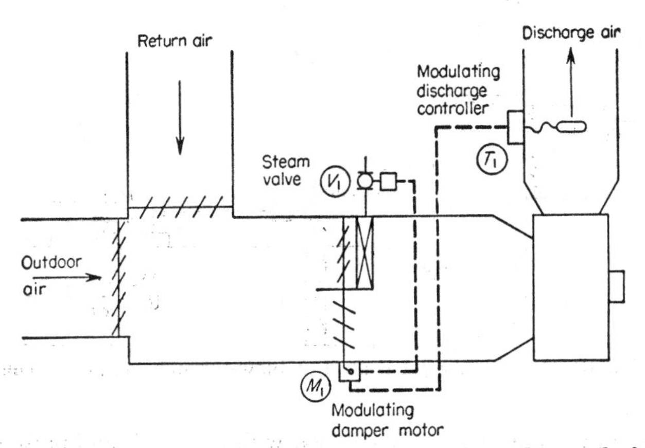

FIG. 13-27. Face and bypass damper control with limited control of valve.

It should be noted that two-position steam valves should not be used in applications of this type because the large, rapid changes in temperature of the air leaking through the closed face damper may cause rapid cycling of the valve on and off under conditions of extremely light load. If a face damper is not used and the valve is a two-position type, such cycling is very pronounced.

In the system shown in Fig. 13-27, the face damper is not always necessary when the bypass damper and modulating valve are used, inasmuch as throttling action of the steam valve is obtained over the final portion of the control range. When the bypass damper is wide open, the valve is fully closed.

Since the valve is wide open at times when the bypass damper around the coil is near the closed position because of a severe load condition, the coil is full of steam when there is danger of subfreezing air coming in contact with it.

Note that in all discharge-control applications it is desirable to use motors on both valves and dampers which have relatively fast timings. Discharge controllers should have adjustable modulating range to permit matching to the job. The wider the range of entering-air temperatures, the wider the modulating range must be for stable control.

TEMPERING SYSTEM WITH BLOW-THROUGH FAN

Figure 13-28 illustrates a simple blow-through tempering system. The steam-distributing type of heating surface should always be used with this type of system because of the possibility of the cold outside air dropping to the bottom of the conditioner and freezing a nondistributing type of coil,

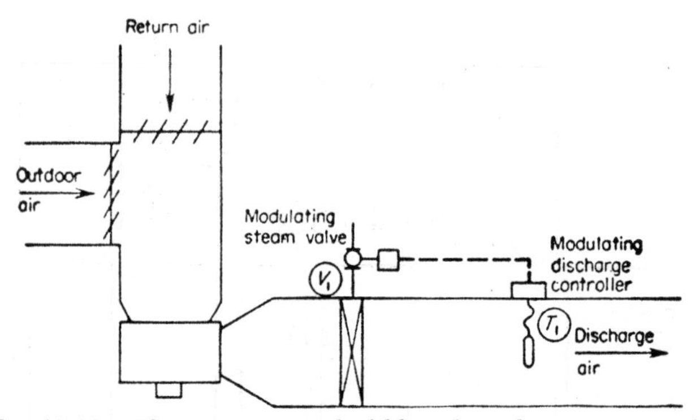

FIG. 13-28. Thermostat control of blow-through tempering coil.

even though the outside air is combined with return air and passed through the fan.

As with pull-through fan systems, it is necessary to exercise extreme care in selecting a valve of the proper size, if the discharge temperatures are to be properly maintained.

BLAST HEATING SYSTEMS

The systems shown in Figs. 13-24 through 13-28 may be increased in capacity to the point where they will provide sufficient heat not only to temper the ventilation air but also to offset external heat losses.

When a system is designed to perform this extra heating function, it is necessary to

1. Select a heating coil of such size that the discharge temperature may be raised enough to allow actual heating within the conditioned space.
2. Use an additional temperature controller of the return-air or room type as a pilot instrument.

Control Requirements. In a typical blast heating system, the automatic-control sequence should provide the following functions:

1. Should the *space temperature fall* below the setting of the return-air or room thermostat, the position of the steam valve and dampers should be changed to provide additional heat. As the *space temperature rises toward* the control setting, the return-air or room instrument should gradually move the steam valve or face damper toward its closed position to reduce the amount of heat supplied.
2. Should the temperature of the *space* rise above the control setting, the return-air or room-type instrument should close the steam valve completely or close the face damper and open the bypass damper.
3. If the *discharge-air temperature* should drop below a predetermined minimum, the discharge-air controller should open the steam valve and face damper sufficiently to raise the discharge air to its minimum temperature level.
4. Protection against the possibility of freezing the heating surface should be provided.
5. Overheating during mild weather should be eliminated.

THERMOSTAT LOCATION

It is frequently difficult to decide between a room-mounted thermostat and a thermostat in the return air for use as a pilot instrument on a blast heating system.

Return-air controllers are frequently chosen in preference to room thermostats because of reduced installation cost. Since the return-air controller is generally located closer to the other control equipment, the cost of installation is reduced, and this point is often the determining factor in the selection.

In general, the return-air control gives a better average measurement of the temperature throughout the space because the return air is commonly drawn from several different grilles throughout the system.

There are, however, other important factors which should be given full consideration in making the choice. These factors are:

1. Air distribution
2. Return-air ducts passing through cold or warm spaces
3. Available location for room thermostat

OUTDOOR-AIR CONTROL

Outdoor air may be used in conditioning systems (1) to provide ventilation, (2) to eliminate odors, (3) to raise indoor-air pressure and counteract infiltration, and (4) to provide atmospheric cooling.

Many systems must provide cooling during the winter months, and, wherever practical, cool outdoor air should be used for this purpose rather than mechanical cooling.

The control of outdoor-air dampers may be made to provide any or all of the following functions:

1. *Closure of outdoor-air damper on fan shutdown.* This is advantageous as it prevents unnecessary cooling of the building. It also reduces the possibility of uncomfortable drafts or freezing of mechanical apparatus.
2. *Manually adjustable quantity of outdoor air.* This may be required to allow for variations in ventilation requirements.
3. *Atmospheric cooling.* Controls may be arranged to increase automatically the percentage of outdoor air taken into the system to provide atmospheric cooling, when the conditioned space becomes overheated.
4. *Summer control of outdoor air.* When outdoor-air temperatures rise too high to be of any use for cooling purposes, only the minimum quantity needed for ventilation should be taken into the system.

Constant Ventilation. Figure 13-29 illustrates a system wherein a fixed percentage of outdoor air is provided whenever the system is in operation.

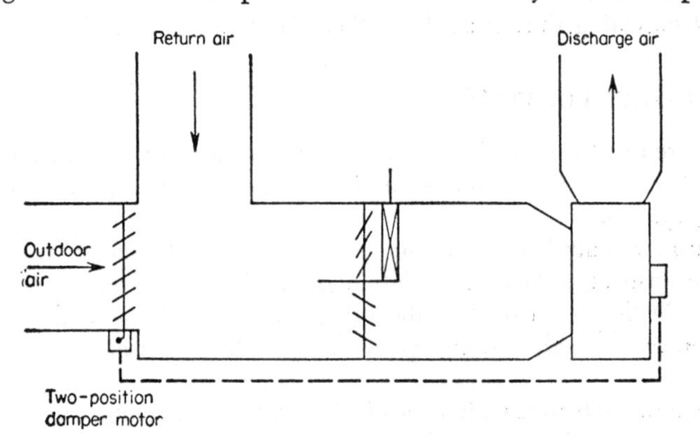

Fig. 13-29. Fixed percentage of outdoor air.

A two-position damper motor is so interconnected with the fan-motor circuit that it opens the outdoor-air damper whenever the fan is operating and closes it whenever the fan is stopped.

If the percentage of outdoor air is properly calculated, this system provides protection against unnecessary cooling of the building, drafts, and freezing of equipment.

Manual Control of Ventilation. Figure 13-30 illustrates an outdoor-air control system which provides

1. Damper closure on fan shutdown
2. Manually adjustable quantity of outdoor air

Modulating damper motor M_1 is so interconnected with the fan-motor circuit that it closes the damper whenever the fan stops, and opens it to the position determined by the manual-positioning switch whenever the fan is started.

The manual-position switch may be set as desired for any percentage of outdoor air and may be located at a remote point for convenient operation.

This type of system is often utilized for spaces with a variable occupancy factor, in order that the amount of air for ventilation may be changed with occupancy.

Mixed-air Control of Outdoor-air Dampers. Figure 13-31 illustrates a system of outdoor-air control wherein the outdoor air is maintained at a minimum except during the periods when overheating exists. At such

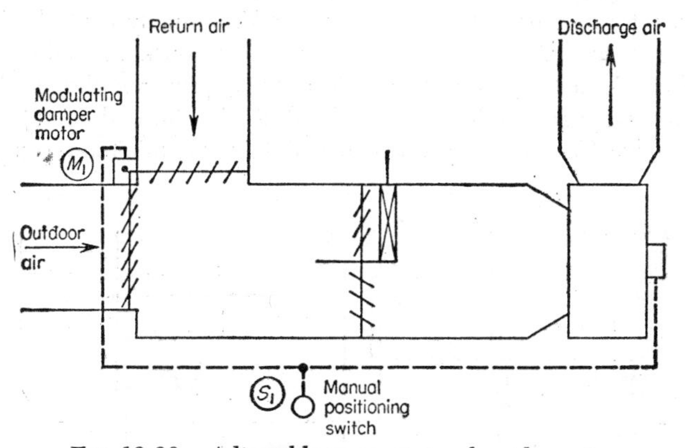

FIG. 13-30. Adjustable percentage of outdoor air.

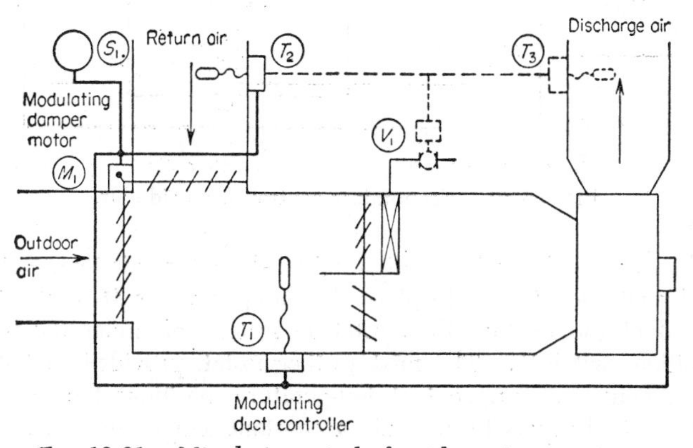

FIG. 13-31. Mixed-air control of outdoor-air percentage.

times the outdoor air is controlled by a thermostat located in the mixture of outdoor and return air, so as to provide an approximately constant temperature in the air entering the heating coil.

In a system of this type, it is extremely difficult, because of air stratification, to find a thermostat-element location which measures the average temperature of mixed outdoor and return air. Normally, the outdoor air follows one section of the duct, and the return air another. Also, as the dampers close, the condition of stratification changes. It is thus almost impossible, because of the constantly changing conditions, to get an average temperature reading through any means. Sometimes it can be done by the use of properly placed baffles. Unless there is definite assurance that stratification will not occur, and that a representative temperature measurement can be made, the use of this system is not recommended.

Sequence of Operation. When mixed-air control is successfully employed, the sequence of operation is as follows:

1. The return-air or room-type controller T_2 normally controls the steam valve V_1 to maintain the space temperature constant. Discharge controller T_3, acting as a low-limit control, prevents the discharge-air temperature from falling below the desired point and operates to keep the valve V_1 sufficiently open to maintain this minimum temperature.

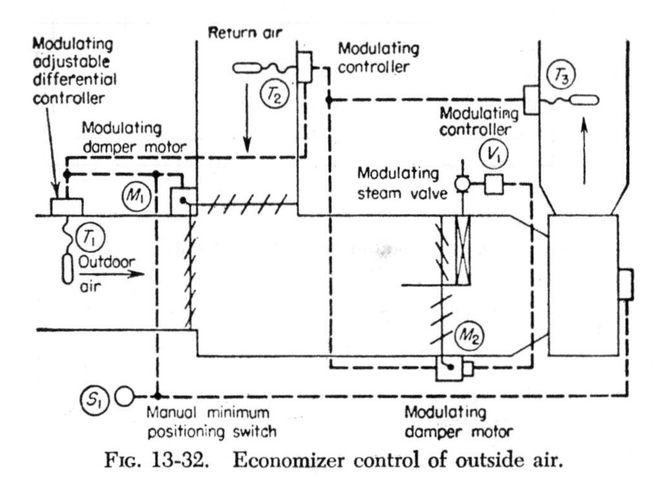

FIG. 13-32. Economizer control of outside air.

2. When an overheated condition exists, the return-air temperature rises and controller T_2 acts to place the damper motor M_1 under the command of the mixture thermostat T_1 instead of under the minimum-position switch S_1. The mixture thermostat, provided that it obtains an average measurement of temperature, positions the outdoor- and return-air dampers to maintain a constant temperature of air entering the coil.
3. Whenever the fan shuts down, the damper motor M_1 drives the outdoor-air damper to a fully closed position.

Atmospheric Cooling. Figure 13-32 illustrates a system for winter control of outdoor air, with notable features of economy. Provisions have been made for

1. Closing the outdoor-air damper when the fan stops
2. A manually adjustable minimum quantity of outdoor air when the fan is in operation
3. Use of additional outdoor air for atmospheric cooling when required

For purposes of illustration, a heating coil with face and bypass dampers has been shown. This method of control may be used with any other method of providing heat.

The *sequence of operation* is as follows:

1. When the fan is inoperative, the outdoor-air damper is automatically kept in the closed position.
2. During the pickup period, with the fan running, the outdoor-air damper opens to a minimum position which may be manually adjusted by a minimum-position switch S_1. This switch may be located at any remote point desired for easy adjustment.
3. The steam valve and the face and bypass dampers are modulated to provide the necessary amounts of heat called for by the modulating return-air controller T_2.
4. As the return-air temperature rises, the return-air controller T_2 gradually throttles the steam valve and opens the bypass damper to provide less heat to the space.
5. Should the return-air thermostat become satisfied and thus call for a closed steam valve, the discharge controller T_3 modulates the valve and the face and bypass dampers to provide just enough heat to maintain a minimum discharge temperature of, say, 65°F. This temperature should be sufficiently low to provide cooling when required. If, however, the temperature of mixed outdoor and return air should be 65°F or more, so that no heat is required to maintain a discharge temperature of 65°F or more, the modulating controller in the fan discharge T_3 allows the steam valve to close tight.
6. If the return-air temperature rises no further, the system then operates with the outdoor-air damper at its minimum position and with all steam shut off.
7. If the internal heat in the space increases sufficiently to cause overheating, the outdoor-air damper motor M_1 is operated at the command of the adjustable-differential modulating outdoor-air controller T_1. This controller is adjusted to open the outdoor-air damper gradually on a rise in outdoor-air temperature, and to return the damper to minimum position as the outdoor-air temperature falls to a point where the minimum percentage of outdoor air would be sufficient to maintain a discharge temperature of 65°F.

For example, in a system designed for a minimum of 20 per cent outdoor air, and with return air at 74°F, the mixture of outdoor and return air would be at 65°F when the outdoor-air temperature reached 29°F. However, as the temperature rises above 29°F, it is necessary to use more than a minimum amount of outside air if a 65° discharge is to be maintained.

For this condition, the adjustable-differential outdoor-air controller T_1 would be set to have the outdoor-air damper wide open at 65°F, and to close it gradually to the minimum position as the outdoor-air temperature falls from 65 to 29°F.

8. As long as the return-air temperature does not indicate an overheated

condition, the outdoor-air damper is held at the minimum opening as dictated by the manual minimum-position switch S_1.

9. If the space temperature should fall because of the temperature of the discharge air provided, the latter containing more than minimum outdoor air, the return-air controller would sense this change and transfer command of the outdoor-air damper from the outdoor-air temperature controller T_1 to the minimum-position switch S_1.

BLOW-THROUGH SYSTEMS

The discussion so far has covered the controlled introduction of outdoor air into systems of the pull-through fan variety. Because of the nature of the blow-through system, the problems accompanying the admission of outdoor air are much more simple, since the outdoor and return air tend to be

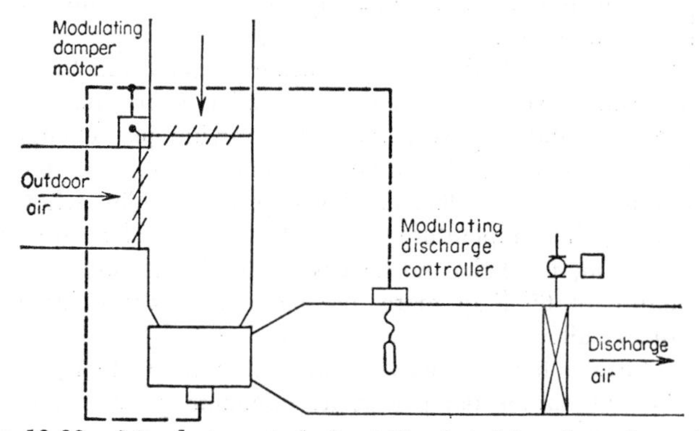

FIG. 13-33. Mixed-air control of outside air in blow-through system.

thoroughly mixed in passing through the fan. It is, therefore, usually possible to get a good average reading of the temperature of the mixed air by locating a duct thermostat downstream from the fan, and using this thermostat for positioning the outdoor-air damper according to the principles outlined in the systems previously described.

Note, however, that single-inlet centrifugal fans mix the air more effectively than double-inlet fans. Axial-flow fans usually provide better mixing than either.

Mixed-air Control without Minimum. Figure 13-33 illustrates outdoor-air control with a blow-through fan system to provide

1. Outdoor-air damper closure on fan shutdown
2. Constant temperature of mixed outdoor and return air
3. No fixed minimum amount of outdoor air

The modulating damper motor is so interconnected with the fan circuit that it closes tight whenever the fan is stopped. When the fan is running, the modulating discharge controller operates the motor to position the outdoor- and return-air dampers so as to give a constant discharge-air temperature of 60 or 65°F. Thus air at a temperature low enough to provide atmospheric cooling is always *available if needed.*

Though no fixed minimum of outdoor air is provided, if the discharge controller is set for 60°F, it requires approximately 14 per cent of outdoor air at 0°F mixed with return air at 70°F to maintain the 60° discharge-air temperature. Fourteen per cent outdoor air is not sufficient for most air-conditioning or ventilating applications. The customary minimum for air conditioning is 25 per cent, and up to 50 per cent is considered the range between good practice and code requirements for school ventilation.

Warm-up Control. The system shown in Fig. 13-33 does not provide full economy, since it takes more than the quantity of outdoor air required for ventilation in all but the coldest weather. Figure 13-34 illustrates a system similar to that shown in Fig. 13-33, except that provisions have been made to keep the outdoor-air damper closed until the return-air temperature

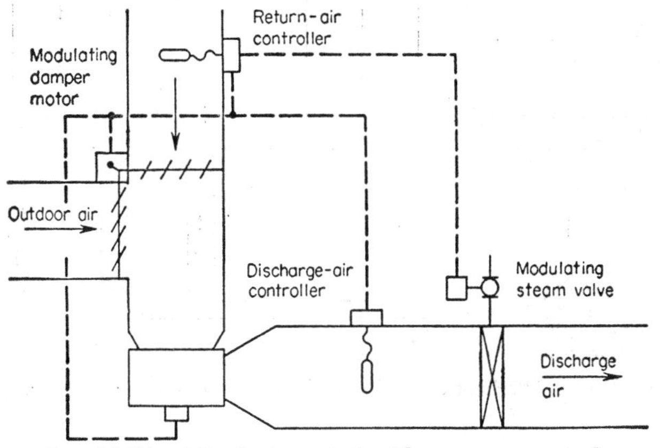

FIG. 13-34. Mixed-air control with warm-up control.

has risen to the value desired in the space being controlled. This provides additional economy of operation.

The foregoing may be accomplished either by an additional mechanism within the return-air thermostat or by means of a connection to the steam valve which does not allow the outdoor-air damper to operate at the command of the discharge-air controller until the steam valve has been throttled.

PREHEATER OUTDOOR-AIR CONTROL

Where it is necessary to use preheat coils, there are two factors which should always be considered from the control standpoint:

1. *Coil capacity.* It is desirable that the control of the preheat coils be so arranged that it is always possible to obtain low-temperature discharge air for cooling. In some cases the operation of a full-size preheat coil may raise the mixed-air temperature so high that cooling cannot be obtained.

2. *Freeze-up protection.* Control of the preheat coil should always be arranged so that there is no danger of freezing the coils.

Simple Preheat. Figure 13-35 illustrates a system using a single preheat coil controlled from outdoor-air temperatures. It may be used for any fixed percentage of outdoor air up to 100 per cent, in which case there is no return-air connection. The system provides

1. Outdoor-air damper closure on fan shutdown
2. Gradual opening of the preheat valve, which would be wide open at 35°F outside temperature

The modulating temperature controller T_1 in the outdoor air is usually set so as to start opening the modulating preheat valve V_1 at about 38°F and to have it wide open by 33 or 35°F. The use of modulating control

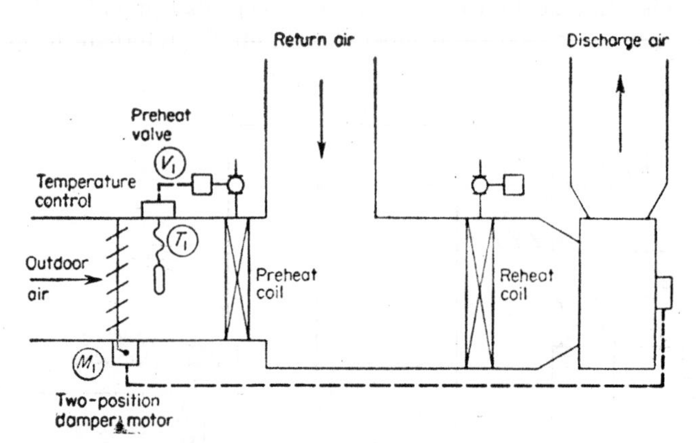

FIG. 13-35. Preheat control with fixed outside-air percentage.

on the preheat coil eliminates the sudden changes in discharge temperature which would result from quickly opening or closing the steam valve completely.

If the preheat coil has a pickup capacity of more than 35°, it is impossible to obtain discharge air for cooling purposes when outdoor-air temperatures are higher than 25 or 30°F. This may result in overheating the conditioned space.

When a steam-distributing type of heating coil is used, it is usually desirable to control the preheat coil from a temperature controller located downstream from the coil. This system prevents overheating in mild weather.

Coil and Damper Combination. Figure 13-36 illustrates a preheat coil with face and bypass dampers arranged to provide:

1. Closing of face and bypass dampers in the outdoor-air duct on fan shutdown
2. Manually adjustable minimum percentage of outdoor air
3. A minimum temperature of mixed outdoor and return air

The sequence of operation is as follows:

1. When the fan is started, the motor M_2 on the outdoor-air damper in front of the preheat coil opens to a minimum position, which is manually adjusted by positioning switch S_1.
2. Two-position temperature controller T_1 in the outdoor air opens valve V_1 on the preheat coil at temperatures of 35°F or lower.
3. Modulating temperature controller T_2 operates damper motor M_1 to position the outdoor-air bypass damper around the preheat coil in step with the return-air damper, to maintain a mixed-air temperature of 65°F.
4. When the fan is shut down, motor M_2 closes the outdoor-air face damper, and an auxiliary switch on M_2 operates motor M_1 to close the outdoor-air bypass damper.

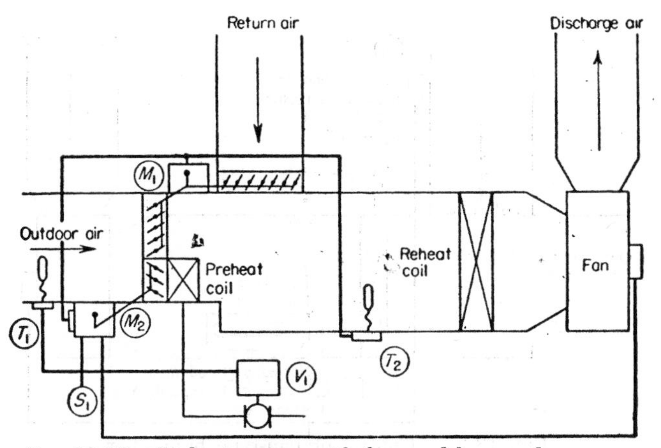

Fig. 13-36. Preheat system with face and bypass dampers.

As explained previously in the paragraph headed "Mixed-air Control of Outdoor-air Dampers," it is difficult to obtain a representative temperature of mixed air without properly baffling the ductwork. Usually, stratification of preheated outdoor air, natural outdoor air, and return air exists, thus making it difficult to measure true temperature.

When using a system such as the one just described, proper consideration should be given to the location of the control bulb T_2. If this system is used, it may not be necessary to have a low-limit controller on the reheat coil, inasmuch as the mixed-air temperature may be warm enough to prevent drafts.

WINTER HUMIDIFICATION

The following types of equipment are in general use for humidification:

1. Water-spray humidifier
2. Pan-type humidifier

3. Steam-jet humidifier
4. Air-washer system

Pan-type humidifiers are somewhat limited in capacity and generally are used only where moisture requirements are low. *Steam-jet humidifiers* are more commonly used for industrial applications. When they are employed for comfort, care must be taken to prevent objectionable odors produced by liberation of boiler compound. *Water-spray humidifiers* are probably the most generally used for comfort-conditioning work. Where there is considerable humidification load, *air washers* may be used because of their greater capacity.

The water used in the washer or spray, or the air entering the spray, must be warm enough for vaporization of moisture to occur. Control valves handling steam or water for humidifiers should always be of the

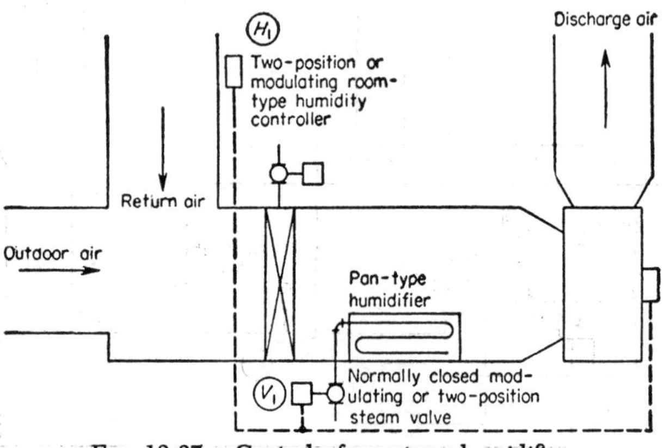

FIG. 13-37. Control of pan-type humidifier.

normally closed type, and arranged to close when the fan stops. The two common methods of heating the washer or spray water are:

1. The use of an air-tempering coil before the washer
2. A heat exchanger to heat the water before it enters the washer

Pan-humidifier System. Figure 13-37 illustrates a typical control application for a pan-type humidifier. A two-position or modulating humidity controller H_1 operates a two-position or modulating steam valve V_1 in the steam line to the humidifier. As the humidity falls, the steam valve is opened to vaporize water by heating it, and thus to supply more moisture to the air.

Control of Steam Jet. Figure 13-38 illustrates a typical application for a steam-jet humidifier. A modulating humidity controller H_1 operates a modulating steam valve V_1 to control the steam flow for humidification.

The valve is so interconnected with the fan-motor circuit that it is tightly closed when the fan is stopped, to prevent condensation of moisture in the ducts.

Water-spray Humidification. Figure 13-39 illustrates an application for humidification using a water spray. The two-position humidity controller H_1 operates a two-position water valve V_1. The water valve is so interconnected with the fan motor that it tightly closes whenever the fan is stopped, to prevent the possibility of moisture condensation in the ductwork.

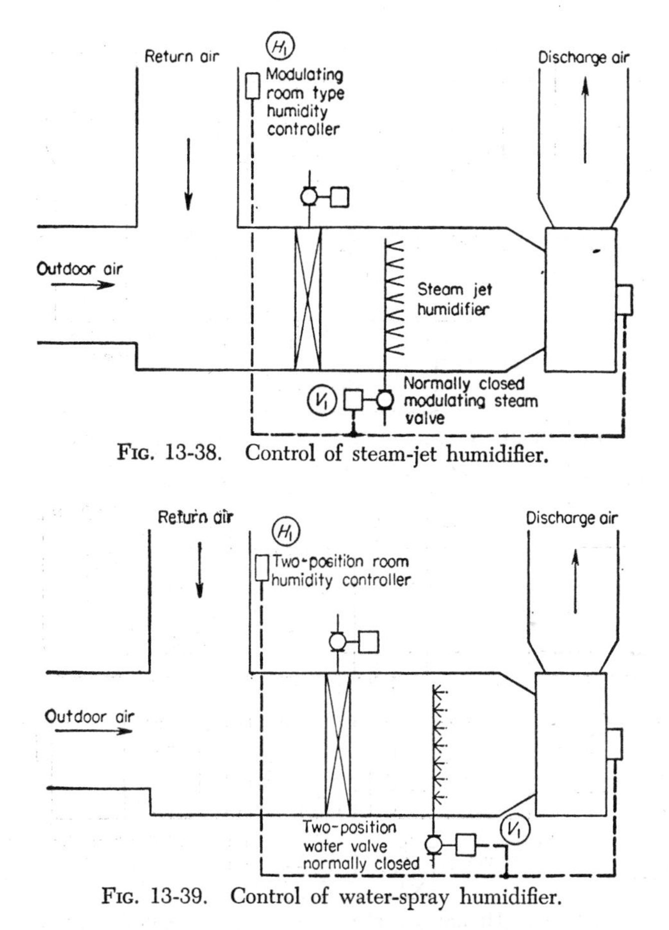

FIG. 13-38. Control of steam-jet humidifier.

FIG. 13-39. Control of water-spray humidifier.

Dew-point Control of Washer. Figure 13-40 illustrates a typical application utilizing an air washer for winter humidification, with dew-point control. The modulating controller T_1 located in the washer water near the suction-line intake operates the modulating steam valve V_1 on the tempering coil. The controller T_1 is set to maintain a washer-water temperature which provides a substantially constant dew point in the discharge air.

It is desirable to use a temperature controller to limit the washer-water temperature when a humidity controller is used to control the tempering coil, which is sometimes done. The washer-water temperature is held within fixed limits, thus preventing the possibility of excessive moisture delivery for short periods or freezing of the washer water when the humidity control is satisfied.

Dew-point plus Relative-humidity Control. Figure 13-41 illustrates a washer system similar to that shown in Fig. 13-40, except that a humidity

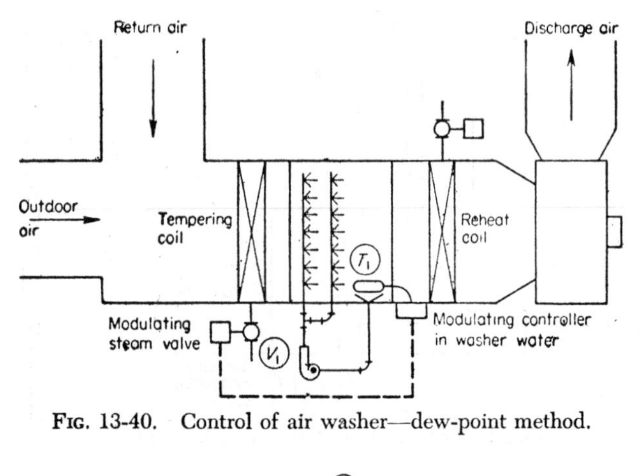

FIG. 13-40. Control of air washer—dew-point method.

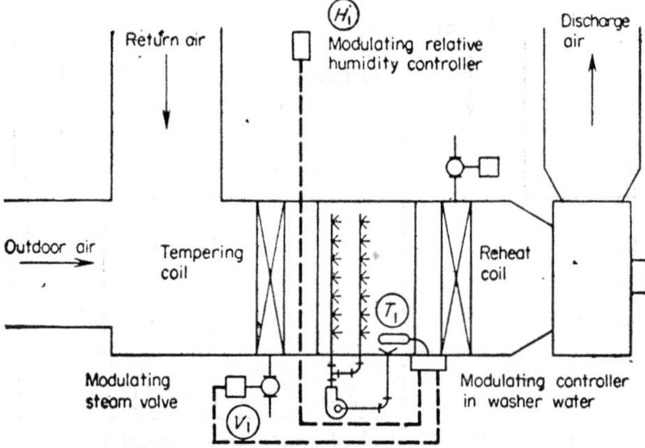

FIG. 13-41. Humidity controller added to dew-point control.

controller has been added to maintain even closer limits on relative-humidity conditions in the space. As in Fig. 13-40, the modulating controller T_1 operates the modulating valve V_1 on the tempering coil.

The relative-humidity controller is of the compensating type and is so interconnected with the water-temperature controller T_1 that it may *reset the control point* of T_1 between any desired limits, for example, from 50 to 55°F. As the relative humidity of the space falls, H_1 raises the control point of T_1 to add more moisture to the air. As the relative humidity rises, H_1 lowers the control point of T_1.

Sometimes water heaters are used in place of air-tempering coils to provide the necessary *heat of vaporization.*

Frosting of Windows. Figure 13-42 graphically illustrates the maximum relative humidity that may be carried without frosting, for both single- and double-paned windows. The curves are based on inside temperatures of 70°F and a wind velocity of 15 mph. These curves are subject to slight corrections for different types of construction, but serve to illustrate that the relative humidity maintained within a space should be definitely lowered in colder weather if frosting is to be prevented.

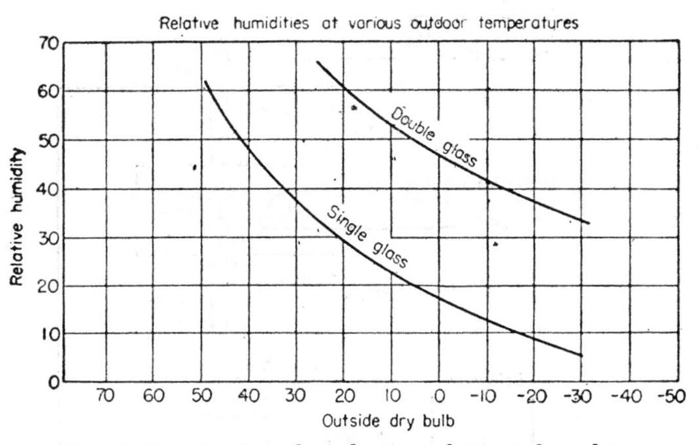

FIG. 13-42. Frosting of windows in relation to humidity.

While these values of relative humidity may prevent window condensation, they may not prevent condensation on doors or within the walls.

DISTRIBUTION AND ZONE CONTROL

The distribution of air in a conditioning system plays an extremely important part in the results obtained. If an installation is to give complete satisfaction, the various parts of the space must be supplied with air in proportion to the existing load.

The load varies with the following factors:

1. Occupancy
2. Exposure
3. Internal loads such as lights, mechanical equipment, and stoves

Thus a room or space with northern exposure requires more heat than a similar space facing south. Likewise, a general office space where occupants are physically inactive requires more heat than a factory space where heavy physical labor is performed.

These conditions of varying load, provided that the fluctuations are not too wide, may usually be satisfactorily handled by proper design of the ductwork and outlet grilles.

When conditions vary widely, it is usually necessary to use zone control in some form so that the amount of heating or cooling for each space may vary with individual requirements.

Zoning for Central-fan Systems. There are three methods for controlling the heat delivery from a central-fan system:

1. Constant air volume delivered at varying temperature
2. Varying air volume delivered at a fixed discharge temperature
3. A combination, varying both volume and temperature

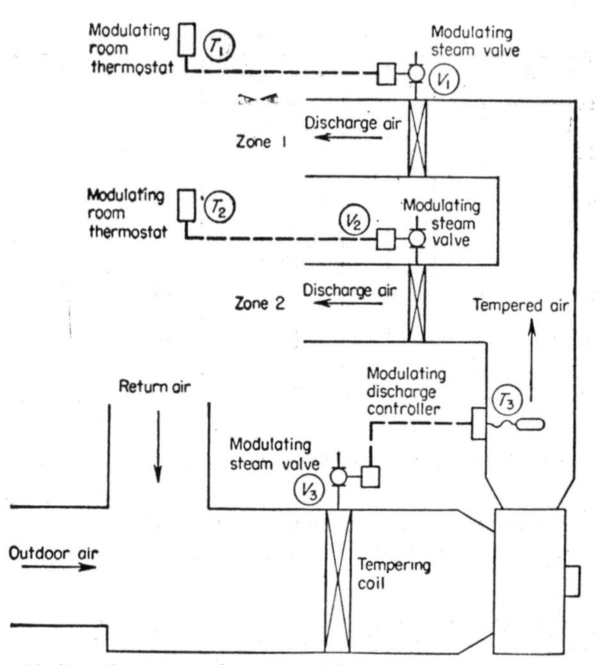

FIG. 13-43. Constant-volume variable-temperature zone system.

Constant Air Volume. A constant-volume system has the following advantages:

1. A means of cooling, if needed, during the winter season
2. A constant volume of air with full ventilation
3. A constant grille velocity providing even distribution

Figure 13-43 illustrates a two-zone system arranged for constant air volume and varying discharge temperature. A modulating controller in the fan discharge operates a modulating steam valve on the tempering coil to maintain a 65°F discharge-air temperature. Room thermostats in each of the zones operate modulating steam valves on booster-heater coils in accordance with heat demands in the zones. When the zone temperature rises above normal, the booster-coil valve is closed, and 65° air is delivered for cooling purposes.

Hot- and Cold-deck System. Figure 13-44 illustrates another type of constant-volume, varying-temperature control. This system makes use of one tempering coil and one reheat coil for all zones, with the ductwork so arranged as to take the necessary proportions of *tempered* and *heated* air for each zone individually.

The modulating discharge controller in the fan outlet maintains a discharge temperature of, say, 65°F by operation of the modulating valve on the tempering coil, in sequence with the bypass damper.

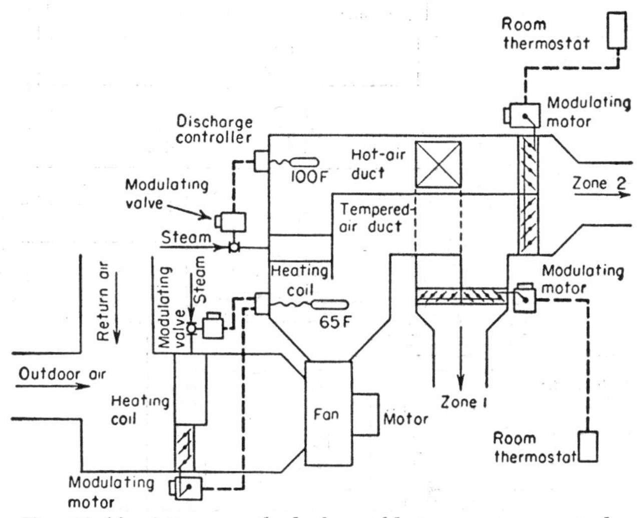

FIG. 13-44. Mixing method of variable-temperature control.

The modulating controller, located beyond the heating coil in the hot-air duct, operates the steam valve to maintain a hot-air-duct temperature of, say, 100°F.

Dampers are placed to control the proportions of hot and tempered air supplied to each zone duct, as shown for zone 2. A modulating room thermostat operates a modulating motor to position these dampers. Should the space tend to overheat, tempered air alone is supplied to the zone to provide cooling.

While this system has been shown as applied to a fan system utilizing heating coils, a similar arrangement is common when warm-air furnaces are used. Dampers and ductwork are so arranged that the warm air from the bonnet may be mixed in varying proportions with tempered air bypassed around the furnace, so as to provide the same results as those of the system shown in Fig. 13-44.

Variable-volume Systems. Though systems providing a varying volume of constant-temperature air, similar to that shown in Fig. 13-45, are sometimes used, it is difficult to control them unless very good air distribution is obtained.

A controller in the fan discharge operates a modulating steam valve and a bypass damper to maintain constant discharge temperatures. The discharge controller must be set high enough to provide for the maximum heat loss.

In each of the zones a room thermostat operates a modulating damper motor to vary the volume of air delivered, according to heat-loss requirements.

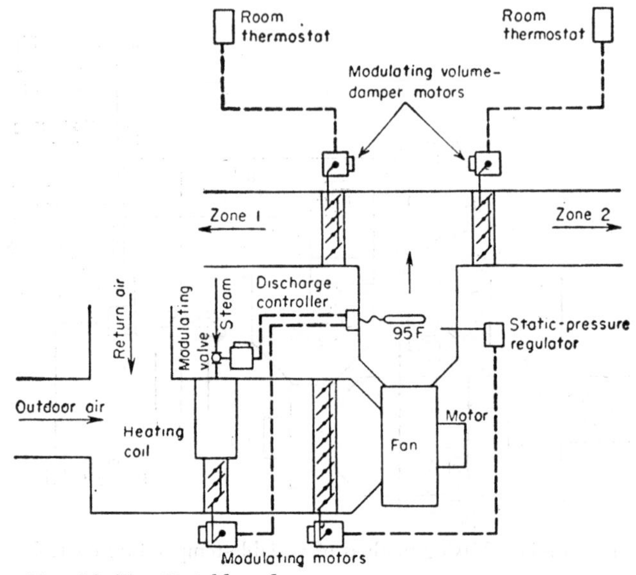

FIG. 13-45. Variable-volume constant-temperature system.

A static-pressure regulator operates a damper at the fan inlet to maintain a constant-discharge-duct pressure, in order that the same volume of air may always be discharged for any given position of a zone-volume damper, and that the noise resulting from high pressure may be avoided.

There are certain inherent disadvantages in a system of this sort that make it impractical for many applications:

1. Since the discharge-air temperature is fixed, it is not possible to provide cooling if it should be required.
2. In mild weather, as the heating load becomes less, the volume of air delivered is progressively reduced to balance heat losses. This has the effect of impairing the ventilation.

Combination Systems. If a varying-volume system is used, its operation may sometimes be improved to some extent by the addition of a return-air controller for varying temperature, as shown in Fig. 13-46. It is not always possible to equal the results obtained from a constant-volume system, but results may be improved by proper control and adequate distribution.

1. The room thermostats control their respective volume dampers, just as in Fig. 13-45.
2. The static-pressure regulator operates as in Fig. 13-45.
3. The return-air controller is so adjusted that it attempts to maintain an actual return-air temperature equal to the average return-air temperature that *would be obtained* if all zones were at their desired

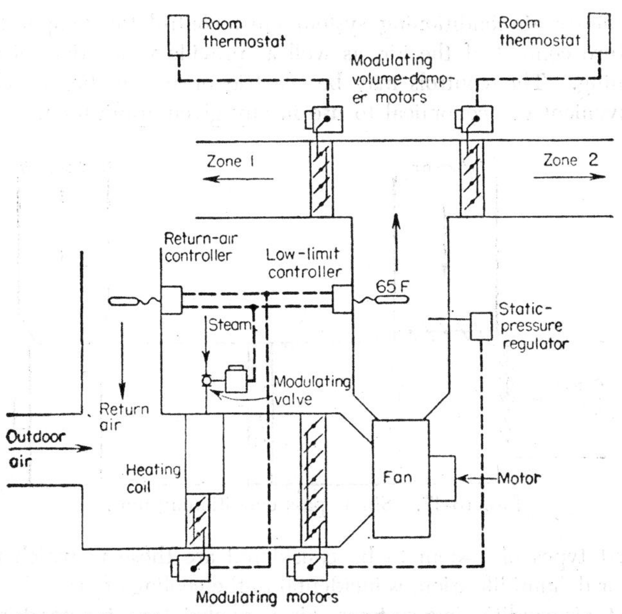

FIG. 13-46. Combination system—variable volume and temperature.

condition. It does this by throttling the modulating steam valve and bypass damper to reduce the discharge-air temperature, when it tends to rise.

4. The modulating controller in the fan discharge acts as a low limit to prevent the discharge of air below, say, 65°F, should the return-air controller become completely satisfied.

Result of Combined Control. Assume that the system is started with both zones below temperature.

1. The volume dampers are wide open, and the maximum temperature of discharge air is delivered.
2. As the zones rise in temperature, their respective volume dampers throttle at the command of the room thermostats.

3. Because of the rise in zone temperature, the return-air temperature also rises. Therefore, the return-air controller throttles the steam valve and opens the bypass damper, so as to reduce the discharge-air temperature.

Thus the heat delivery to the space is reduced, not only by reducing the *volume* of air supplied, but also by reducing the *temperature* of air supplied. Consequently, it is not necessary to reduce the air *volume* quite so far as would be necessary for such a system as that shown in Fig. 13-45.

CONTROL OF COMMERCIAL CENTRAL-FAN COOLING SYSTEMS

The summer air-conditioning system must control the temperature and the moisture content of the air, as well as provide ventilation, circulation, and cleaning. The controls may be electric or pneumatic, whichever is more convenient or economical to use in any given application.

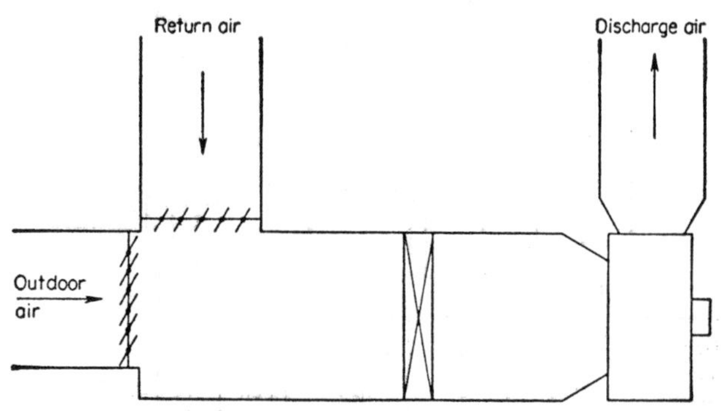

FIG. 13-47. Single-pass conditioning unit.

The first types of system to be considered are those in which moisture removal, or dehumidification, is incidental to the cooling of air.

In most air-conditioning systems, air is cooled to a temperature below the dew point of air in the space being conditioned. Under these circumstances, the air-conditioning plant is said to have a capacity for absorbing latent heat or moisture. The extent of this capacity for dehumidification depends on the type of refrigerant utilized and the cooling surface over which the air passes.

Types of Cooling Equipment. From the standpoint of heat-exchange surface used, central-fan cooling systems usually fall into these classes:

1. Systems which provide cooling through the use of finned coils, with cold water or brine serving as the refrigerant.
2. Direct-expansion systems, in which cooling is provided by the evaporation of liquid refrigerant in a finned coil.
3. Washer systems, in which there is usually no cooling *surface* in the ordinary sense, but the air is cooled in passing through a chamber in which cold water is sprayed in a finely atomized condition. In this

type of system, the air is cooled down to a nearly saturated condition, at a dew point determined by the water temperature used. If the water is cold enough, the air may be cooled below its original dew point, so that moisture is condensed and precipitated out of the air.

A further classification of cooling systems may be made with reference to the arrangement of the plenum chamber. Most of the systems illustrated in this section are shown with the fan drawing air through a conditioning chamber and discharging it to the duct system (*pull-through* systems).

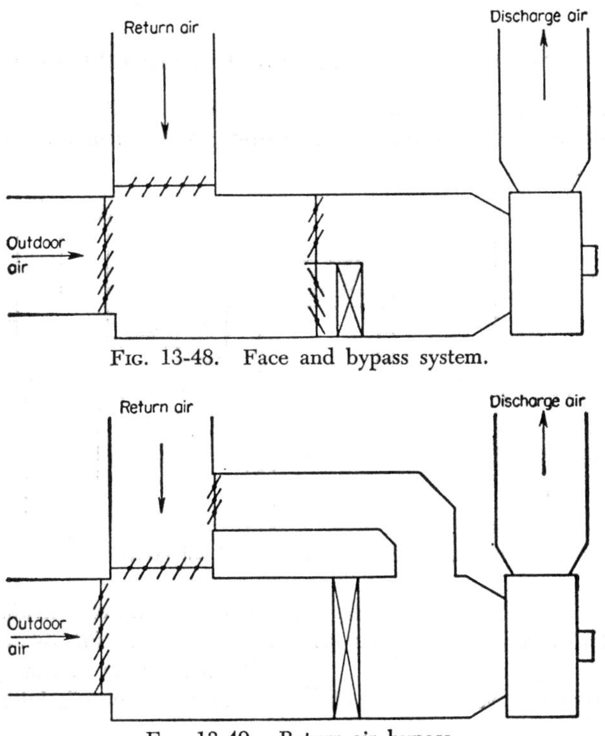

FIG. 13-48. Face and bypass system.

FIG. 13-49. Return-air bypass.

Actually, this arrangement is frequently reversed so that the fan forces the mixture of return and outside air through the cooling surface, and thence to the duct system (blow-through systems). The plenum chamber, or conditioning unit, usually takes one of the forms shown in Figs. 13-47 to 13-49.

Single-pass System. In a single-pass system (Fig. 13-47), all the conditioned air passes through the cooling coil or washer.

Face and Bypass System. In systems of this type (Fig. 13-48), dampers are usually placed across the surface of the cooling coil or washer chamber, and a bypass duct with damper permits part of the air to pass completely around the cooling coil or washer. These dampers are usually operated together, so that as one closes, the other opens. In this type of system, all the air does not pass through the coil; instead, some of it is bypassed, and the air which is bypassed may be a mixture of outside and return air.

Return-air Bypass. This type of system (Fig. 13-49) is similar to the one just described. Instead of the bypass handling a mixture of outside and return air, however, the bypassed air comes directly from the return-air duct, and dampers are installed at that point. The air which is bypassed and mixed with conditioned air is return air from the rooms being served.

CONTROL OF SYSTEMS USING COLD-WATER COOLING COILS

Cold water, when used as a refrigerant in an air-conditioning system, may be obtained from any of three sources—*deep wells, water coolers,* or an *indirect ice system.*

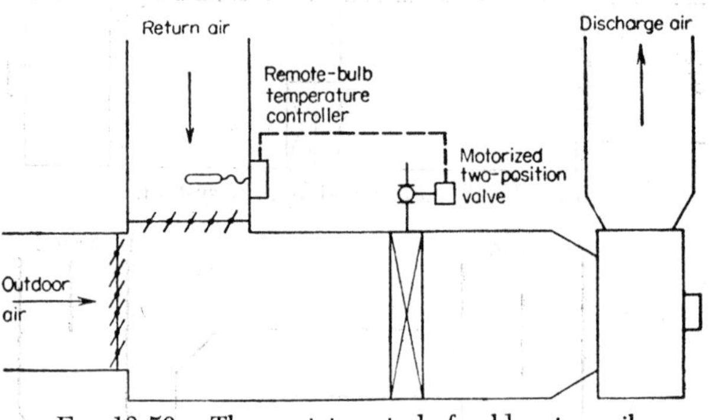

FIG. 13-50. Thermostat control of cold-water coil.

Deep Wells. In some localities, low-temperature water is available from wells, and this may be used as a low-cost refrigerant. Pumps are used to provide the required circulation. Generally, some type of storage is provided, so that the water can be circulated through the coils intermittently as required.

Water Coolers. Where low-temperature well water is not available, mechanical refrigeration is used to reduce the temperature of water circulated through indirect coolers. This type of system is usually set up as a closed circuit, and water leaving the conditioner coil is returned to the indirect cooler for further refrigeration. This prevents the wasting of partly cooled water and increases the economy of operation.

In this type of system, automatic control is usually applied to the pump which provides circulation of the water or to a motorized valve which controls the flow of water to the coil.

In either case, the mechanical compressor used in the refrigerating cycle is controlled from a thermostat set to maintain a constant temperature of the water leaving the cooler.

Indirect Ice Systems. Where a system of this type is used, water returning from the coils is sprayed over ice stored in bunkers. The chilled water is then circulated through the coils by means of a pump.

It is obvious that the capacity of a system for removing latent heat in relation to sensible heat is something inherent in the system and determined by the coil temperature available. It is impossible for the control system to alter such an inherent characteristic. When systems are carefully analyzed to determine their limitations, it is possible to select intelligently a system of control which ensures the best possible results from the equipment in question.

Valve Control of Coil. Figure 13-50 illustrates a common control system for cold-water cooling-coil installations. A return-air thermostat controls the position of the motorized valve admitting cold water to the coil. A room thermostat may be used in place of a return controller.

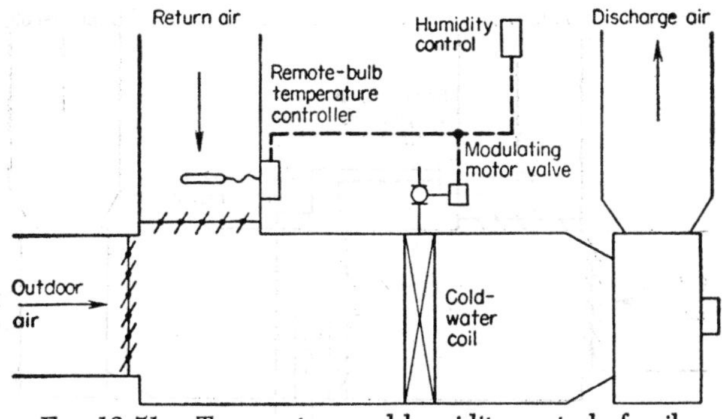

Fig. 13-51. Temperature and humidity control of coil.

A system of this type is frequently installed in such a manner that the valve operates with a modulating or throttling effect. During periods of light sensible-heat load, the valve is throttled to prevent too much cooling. Actually, under the light sensible-load conditions, it is probable that the latent-heat load is proportionately very large. In other words, the S/L ratio, or ratio of sensible- to latent-heat gain in the space, becomes small. This results in a high relative humidity, since the coil temperature rises as the volume of water delivered to the coil is reduced, and little or no moisture is condensed out of the air.

Where a control system like that illustrated in Fig. 13-50 is used, it is preferable to employ two-position valve operation. With this arrangement the valve can be either open or closed, maximum dehumidification occurring whenever the valve is open.

It is frequently possible to control the operation of a pump directly, without the use of a motorized valve. This is particularly true on systems of the indirect type where a supply of cold water is available at all times.

Humidity High-limit Control. Figure 13-51 illustrates a system of control for cold-water cooling which overcomes some of the disadvantages discussed in the previous example. In this system, the thermostat modulates the position of the valve in the cold-water line, thereby causing a variation in coil temperature to give uniform delivery temperatures. In addition, however, the humidity control can cause the thermostat to operate the valve in a two-position manner if the relative humidity becomes too high.

Therefore, when the latent load is high, the coil is operated intermittently at its lowest possible surface temperature, with the result that maximum dehumidification is obtained at intervals.

Face and Bypass Control. Figure 13-52 illustrates a face and bypass control system used with cold-water cooling coils. When the conditioning equipment is arranged in this manner, it is possible to regulate the temperature of the delivered air without sacrificing the dehumidifying effect

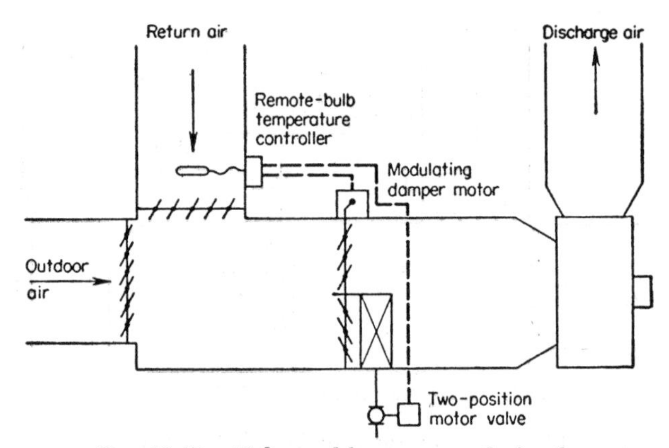

FIG. 13-52. Valve and bypass control of coil.

of low coil temperatures. The coil is maintained at its minimum temperature level at all times, and a throttling effect is obtained by varying the proportions of air passing through and around it.

Under light sensible-load conditions, slightly lower coil temperatures may result with this type of system because there is less load on the coil with a smaller volume of air passing through it, and, therefore, a lesser temperature rise in the water takes place.

The sequence of operation provides that a thermostat in the return air or in the space, if preferred, modulates a control motor which determines the relative positions of the face and bypass dampers. These dampers work in opposition to each other, the face damper closing as the bypass damper opens. It is also possible to make a further provision in this system, as shown, by which a valve controlling the supply of water to the coil can be closed when the face damper reaches its fully closed position, so as to conserve the water supply.

While a face and bypass damper arrangement has been shown here, it is also possible to use a return-air bypass to obtain the same result. The control sequence used would be similar to the one shown here.

Valve and Damper Control. The system shown in Fig. 13-52 may be arranged to provide added economy by varying the coil temperature and permitting it to rise when no dehumidification is required. Figure 13-53 illustrates a system in which the thermostat T_1 positions the face and bypass dampers and also throttles the water valve V_1. A relative-humidity control

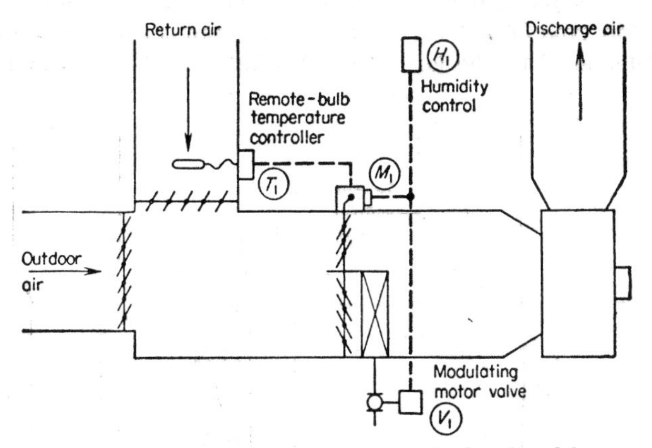

FIG. 13-53. Temperature-humidity control of coil and bypass.

H_1 can open wide the motorized valve on the coil when the humidity is high.

A system of this type may be used to advantage where water rates are high and where the installation of the additional controls involved is warranted on the basis of the economy attained.

CONTROL OF DIRECT-EXPANSION COOLING SYSTEMS

The problem with the direct-expansion cooling system is to balance the control of latent-heat removal against the control of sensible-heat removal. The direct-expansion system provides many possibilities not found with the cold-water cooling just discussed, and a clear understanding of the principles involved makes it possible to choose controls that take full advantage of the inherent features of the system.

Coil-temperature Limitations. It has been shown that low coil temperatures produce maximum dehumidification. If this is so, why not design a system to operate at the lowest possible coil temperature at all times? Why attempt to control the dehumidifying effect? The answer is found by considering economy of operation.

1. It is costly to run a system that is designed to operate on a low coil temperature. In the mechanical-refrigeration system, the quantity

of heat removed per unit of mechanical energy used (Btu per horsepower) decreases very sharply as the suction pressure drops. The operating efficiency thus becomes less at low coil temperatures.

2. It is uneconomical to remove latent heat in excess of comfort requirements.

Control of Refrigerant Flow. Figure 13-54 illustrates a simple type of direct-expansion cooling installation utilizing a single-pass system with a single coil. The flow of refrigerant to the coil is controlled by a solenoid valve V_1. The solenoid is opened and closed upon demand of thermostat T_1.

With this type of system, the suction pressure is allowed to assume any value determined by the load at a given instant. It can be seen that coil

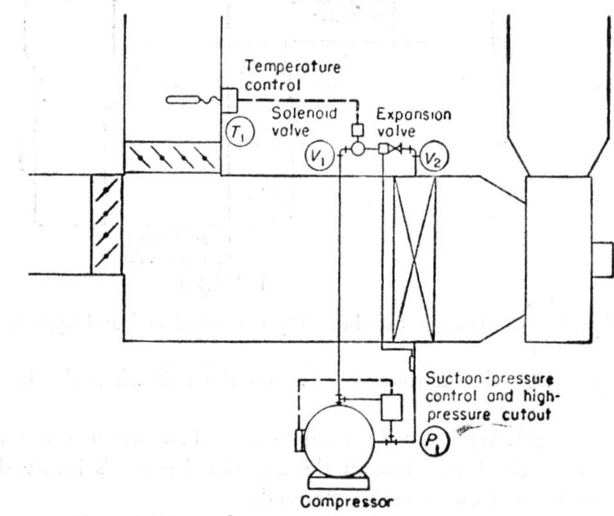

Fig. 13-54. Valve control of direct-expansion coil.

temperatures automatically rise as the load increases. The only variation in instantaneous load on this type of system is that which comes from a reduction in the entering wet-bulb temperature.

Although the control sequence as illustrated shows only the solenoid valve under thermostatic control, it is common practice to deenergize the compressor circuit simultaneously with closure of the valve. If this is not provided for, the compressor is operated by a combination suction-pressure controller and high-pressure cutout P_1. The suction-pressure controller is generally set to cut out at a point which will prevent frost from forming on the coil, and to cut in at a still higher point.

Face and Bypass Control. Figure 13-55 illustrates a face and bypass damper installation in which the temperatures are maintained by positioning dampers to regulate the amount of air flow through the coil. This system provides for gradually changing the temperature of the air discharged, and is therefore a considerable improvement over the simple system previously shown. Discharge-air temperatures remain constant at the proper point for meeting the load requirements, whereas the on-off system results in discharge of alternate cold and warm air.

It is also true that, under light sensible-load conditions, the coil temperature falls. Therefore, the system should permit lower relative humidities than the simple single-coil system. The minimum coil temperature is limited by the point at which frost forms on the coil, and therefore it is necessary to cut off the coil on the refrigerating machine as the face damper approaches the closed position. This is frequently done by the use of an auxiliary switch, operated by the damper motor, which acts to close a solenoid valve. A suction-pressure control on the refrigerating machine cuts off the compressor when the valve closes. This control is set so as to

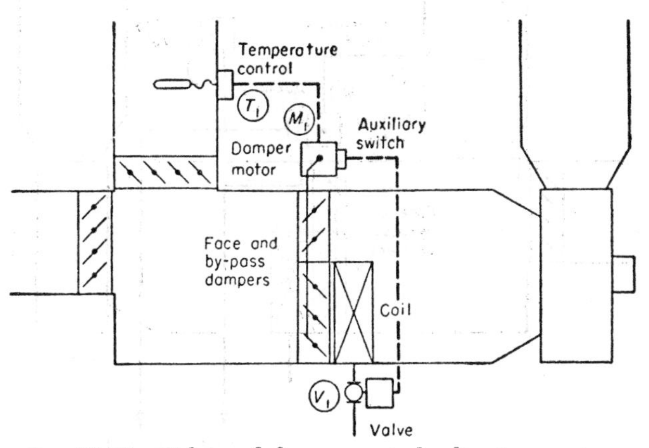

FIG. 13-55. Valve and damper control—direct expansion.

stop the machine whenever the suction pressure reaches a point which would indicate a frosting coil. It should be noted that the reduction in air flow through the coil also tends to limit the point at which frost forms on the coil.

With this arrangement of equipment, substantially constant temperatures are maintained in the discharge from the system, and the noticeable ups and downs of on-off control are eliminated. It is also true that this system provides for more latent cooling under light sensible-load conditions because of the slight reduction in coil temperature that takes place with reduced air flow.

Damper Method with Supplementary Control of Compressor. Figure 13-56 illustrates a system in which the coil usually operates continuously and the delivered-air temperature is determined by the position of face and bypass dampers. Means are provided to shut the machine off completely when the face damper reaches a closed position.

When the refrigerating equipment is operating, its capacity is varied according to relative-humidity conditions. With this system, the compressor equipment is operated at a higher suction pressure under light load conditions, and at the same time an accurate regulation of delivered-air temperatures may be maintained through the throttling effect of the face and bypass dampers. On a rise in relative humidity, representing an increase in latent load, the second-stage compressor is started, and reduced suction pressure, with lower coil temperature, ensures effective dehumidification.

There is no essential difference in the control sequence if a return-air bypass is utilized in place of the face and bypass dampers shown.

Limitations of Two-stage Thermostats. It has been common practice to install a simple two-stage thermostat to control two-speed compressors.

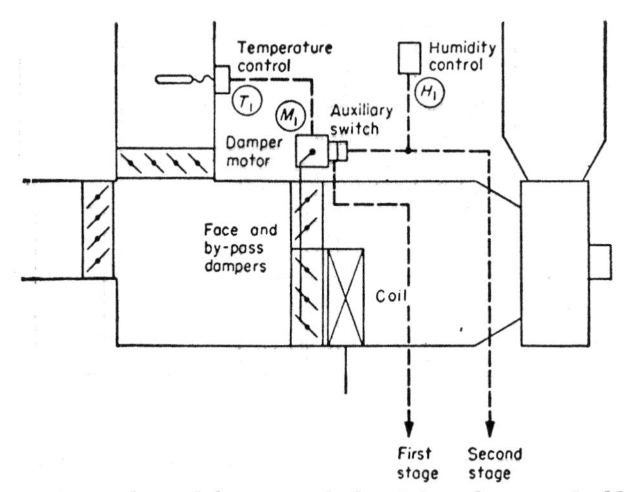

Fig. 13-56. Valve and damper method with humidity control added.

Without the use of a humidity control in the circuit, this system may result in unsatisfactory conditions.

As the load becomes lighter, the sensible load measured by the thermostat decreases, and the machine tends to operate most of the time on low speed or half capacity. Under these conditions, the coil temperatures are high. Actually, under light sensible-load conditions, the latent-heat load often increases as compared with the sensible load; that is, the S/L ratio is low. This results in high relative humidities. In general, therefore, it is better to operate at a low than a high coil temperature during light sensible-load conditions. Obviously, the two-stage thermostat system would operate in just the reverse of the recommended practice.

A similar type of system employs a multistage controller, or sequence switch, piloted by a dry-bulb thermostat, to throttle the capacity of the machine with changes in dry-bulb temperature. Again this permits higher coil temperatures under light sensible-load conditions, which is undesirable if a humidity control is not used as a limiting device.

If a two-stage thermostat is used to control solenoid valves on split direct-expansion coils, it is obvious that under light sensible-load conditions half the coil is used, and therefore the load on the compressor is less, resulting in lower suction pressures. This type of system, therefore, gives low coil temperatures under light sensible-load conditions, and does not result in an upward drift of relative humidities if the S/L ratio decreases.

This statement is also true of a thermostat regulating a multistage controller, which is used to control solenoid valves on several banks of direct-expansion cooling coils. When fewer coils are operating under light sensible load, the suction pressure decreases, as is desirable.

Multiple-compressor Control. On large multiple systems, a battery of individual compressors is very frequently used in preference to one large central machine. The reason for using a number of smaller units instead of one large unit is found in the flexibility of such an arrangement, its adaptability to varying loads.

In controlling a battery of compressors of his type, primary control is very frequently accomplished from the suction pressure. In other words, several compressors may be connected together with common suction and liquid headers. It then becomes desirable to measure the suction pressure and control the number of machines operating in such a way as to maintain a relatively constant suction pressure. Care must be taken in choosing the cutin and cutout pressure of each compressor to prevent short cycling of any machine.

On indirect systems where refrigeration is used to control the temperature of water supplied to the air-conditioning system, the control of the compressors usually takes place from the temperature of the water storage. In direct-expansion systems, the sequence of operation of the compressors may be controlled from room conditions by temperature or humidity controls.

In either event, the control sequences required are similar, and the multistage controller provides a simple means of controlling multiple-compressor operation.

Sequence Control. The sequence switch is a control unit which provides for successive opening and closing of electric circuits in response to the demand of a modulating-type controller. With this type of control, it is possible to operate multiple-power units in a definite sequence so that they may be energized and deenergized in direct proportion to load variations.

Definite precautions must be taken to prevent the possibility of all equipment being energized simultaneously after a failure in power which may occur at a time when the system has been brought up to maximum capacity.

If no particular means of protection were provided and a power failure should occur when all motor units were in operation, these motors would be thrown across the line simultaneously upon a resumption of power, and the electric supply circuit would be dangerously overloaded.

A system which provides protection against such an occurrence is available and should always be installed where the sequence switch is used in the control of multiple-compressor installations. With this system, a modulating-type pressure control or thermostat positions the sequence switch to cause any required number of compressors to operate. However, in the event that a power failure occurs, on the resumption of power service no current is allowed to pass through the starter coils, and therefore none of the compressors starts to run immediately. A relay included in the circuit causes the motor to return to the no-load position, cutting off the switches one by one. When it has reached the starting position, the relay provides power to the starter-coil circuit and the motor is then positioned under command of the modulating control so as to cut in successive compressor motors, one at a time, until it reaches the stage demanded by the controller.

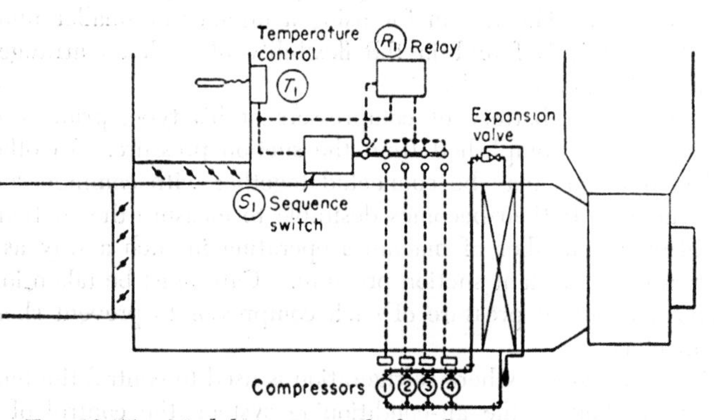

FIG. 13-57. Modulating control of multistage compressors.

Figure 13-57 shows the schematic arrangement of control equipment for sequence control of multiple compressors.

It is not always necessary to have a large number of compressors in order to obtain numerous steps of capacity and a high degree of flexibility in adapting the capacity to the load. It is possible, with the use of the multistage sequence switch, to obtain additional stages of operation by reconnecting the compressors in progressive stages or by reconnecting two-speed compressors. For instance, if three compressors of 5-hp, 7½-hp, and 10-hp capacity are available, a total of seven stages of operation can be obtained, as shown in the following list:

Stage 1: 5-hp compressor
Stage 2: 7½ hp
Stage 3: 10 hp
Stage 4: 5 hp + 7½ hp = 12½ hp
Stage 5: 5 hp + 10 hp = 15 hp
Stage 6: 7½ hp + 10 hp = 17½ hp
Stage 7: 5 hp + 7½ hp + 10 hp = 22½ hp

CONTROL OF AIR-WASHER SYSTEMS

There is a fundamental difference in operation between air washers and closed-coil systems: Air leaving the conditioning chamber of an air washer is always very nearly or completely saturated. Air leaving a closed coil, however, may be saturated to a far less extent, the degree depending on coil temperatures, moisture conditions of air entering the system, and coil efficiency.

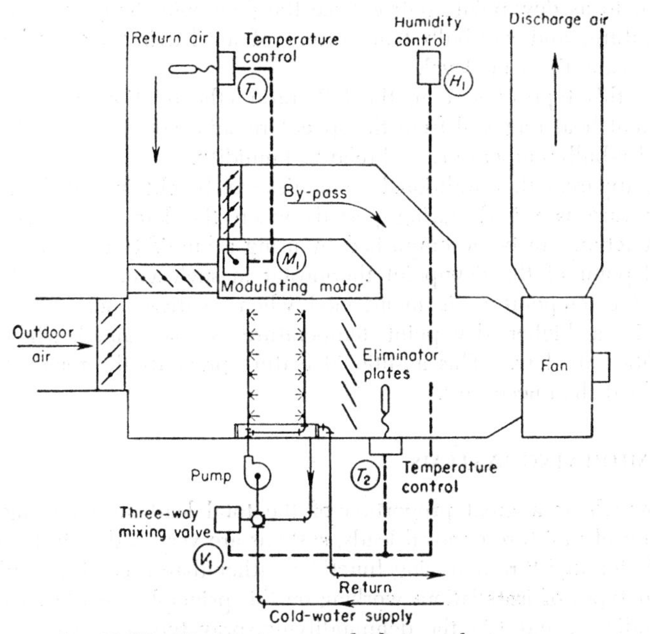

FIG. 13-58. Dew-point control of air-washer system.

In the washer-type system, the temperature of the water in the sprays determines the dew-point temperature of the air leaving the washer chamber.

Dew-point Control. In the closed-coil system, during periods when there is a light latent-heat load and when the dew point of the air passing through the coil is actually below the coil-surface temperature, no latent heat is removed from the air in the rooms. In a washer, since the air always leaves at a dew point determined by the water temperature, it is possible to maintain approximately constant dew point at all times. The effect of this condition is the maintenance of constant relative humidity under a given S/L load ratio. The application of dew-point control is illustrated in Fig. 13-58. The air washer is in continuous operation, and a bypass around the conditioning chamber provides a means of maintaining discharge dry-bulb temperatures at the proper level. A modulating thermostat measures the temperature of air leaving the washer chamber and controls a modulating three-way mixing valve to proportion the relative amounts of refrigerated and bypassed water in accordance with changes in the temperature of the air leaving the chamber. The pump, as illustrated, provides a means for circulating water through the system.

An increase in the temperature of the air leaving the washer causes the valve to reposition in such a manner that a greater proportion of cold water is admitted to the sprays, thus lowering the resultant dew point. This is referred to as dew-point control since the dew-point temperature, dry-bulb temperature, and wet-bulb temperature of air leaving the washer are all at practically the same level.

With this type of system, the S/L ratio (the relation between sensible and latent heat removed from the space) remains constant at all times for a given dry-bulb temperature and relative humidity.

It is apparent that additional economies can be obtained if the dew-point temperature is raised during periods when the latent-heat load is low. This function can be accomplished by using a humidity control to reset the control point of the dew-point thermostat in such a manner that (1) low dew-point temperature is maintained when relative humidities are high; and (2) a higher dew-point temperature is maintained when relative humidities are low. This additional feature prevents the removal of more latent heat than necessary.

DEHUMIDIFYING SYSTEMS

Inasmuch as a great proportion of the total load on a cooling cycle is made up of moisture-removal loads, systems are very widely installed which provide for moisture-removing functions rather than a cooling result. Two popular types of installations working on this principle are (1) solid-sorbent dehumidifiers, and (2) the dehumidifying-spray type of system.

Solid Sorbents. With the solid-sorbent system, air is passed through sorbent beds which absorb or adsorb the moisture from the air. The air is then usually passed through a cooling coil which removes at least the sensible heat released as the moisture is absorbed from the air, even though no sensible cooling is required in the space. The sorbent beds must be periodically reactivated. This is commonly done by using two sorbent beds alternately, reactivating one while the other is in use in the conditioning system.

Spray Dehumidifier. The dehumidifying-spray type of installation employs a solution of halogen salts, glycols, or other chemical liquid sorbents having an affinity for moisture. The solution is concentrated to such an extent that it has an affinity for moisture in the air; then, when the solution is sprayed in the path of the air, it absorbs moisture. In this system, too, the air is usually passed through cooling coils which provide for some sensible cooling; otherwise, the spray itself may be cooled. The solution, when it leaves the spray, is passed at least in part through a concentrator which boils off some of the moisture collected and restores the solution to the required concentration.

Typical Control System for Sorbent Dehumidifier. Figure 13-59 shows typical controls for a dehumidifying system employing a solid sorbent. The sorbent material is arranged in multiple beds in two chambers, and a system of dampers or valves, indicated schematically in the diagram, is positioned so that the moist air to be conditioned passes through one chamber while air heated to perhaps 300°F passes through the other. Until it is saturated, the sorbent, such as silica gel or activated alumina, is capable of taking up moisture, usually by *adsorption* or surface adherence. (Some sorbents act by *absorption,* which involves a physical or chemical change

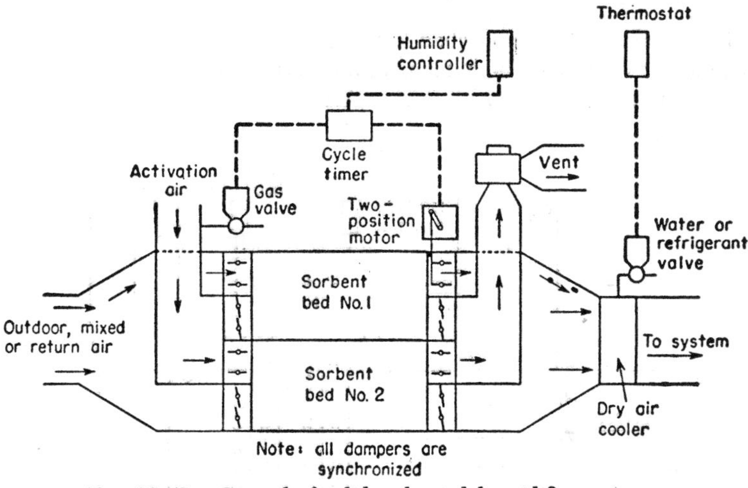

FIG. 13-59. Control of solid-sorbent dehumidifier system.

in the solid material, whereas *adsorption* involves no change in the solid material. An adsorbent has a highly porous structure with very large areas of internal surface.)

The saturated sorbent is dried by the passage of hot air in the regeneration or activation cycle, and then cooled by a short period of ventilation with unheated air.

The humidity controller puts the system into operation, and the cycle timer actuates the gas valve or other heat source and the damper motor to provide alternate periods of use and reactivation for each bed. When the humidity controller is satisfied, the timer halts the alternation of service and activation, at the completion of a cycle. The saturated bed, remaining "in service," is no longer capable of adsorbing moisture, and dehumidification ceases.

The highly dehumidified air coming off the sorbent bed is too hot for discharge into the space, because of the conversion of latent heat into sensible heat in the dehumidifying process. It is therefore cooled by a water, brine, or direct-expansion coil. Water is commonly used for economy. The coil is usually controlled by a space or return-air thermostat.

In an installation of this kind, the humidity controller is usually set to a lower value than could be used on a strictly cooling installation. The reduction in relative humidity permits the maintenance of higher dry-bulb temperatures to get the same feeling of comfort. Maximum dehumidification can be used without excessive cooling or the necessity of reheat. Additional cooling may be provided, after the dehumidifier, to handle the sensible load on the space.

APPLICATIONS REQUIRING REHEAT

Many of the systems described in this chapter have been methods for the independent control of both temperature and humidity within the limitations of the air-conditioning system. Actually, however, the physical laws

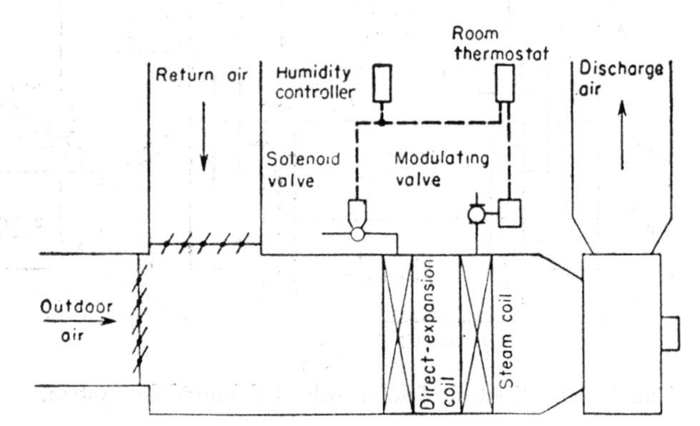

Fig. 13-60. Temperature-humidity control of reheat system.

involved prevent the maintenance of definite temperatures and definite humidities under all load conditions, unless some form of reheat is employed.

Figure 13-60 illustrates a simple system for providing reheat. The thermostat serves two functions:

1. To operate the solenoid valve on the direct-expansion coil to provide cooling in response to a temperature rise
2. To modulate the steam valve to an open position to provide reheat when necessary

The humidity control is also connected to the solenoid valve in such a manner that cooling for dehumidification is always provided when relative humidity is high. Under conditions where the sensible load is light, this cooling for purposes of dehumidification would normally depress the dry-bulb temperature below a comfortable level, but the operation of the thermostat and steam valve as described, to provide reheat, prevents this condition.

Sources of reheat other than steam are hot-water coils, electric-strip heaters, and the heat given off by the condenser in the refrigerating system.

The system as described in Fig. 13-60 definitely provides the means for maintaining the temperature at the setting of the thermostat; and at the same time it maintains the relative humidity at or below the setting of the humidity control. This conclusion is based on the assumption that the refrigeration equipment and steam coil are of sufficient size to handle the maximum loads encountered.

Low Relative Humidity Possible. Since this system does not provide a means of adding moisture to the air, it becomes impossible to prevent the relative humidity from falling *below* the desired level when the primary moisture content of the air is extremely low. If the direct-expansion coil is being operated under command of the thermostat alone, it is obvious that a certain amount of moisture is removed from the air even though the humidity may be at the desired point. This result would, of course, be acceptable in comfort-conditioning applications; but, for certain laboratory and industrial installations, a separate means of humidification would be necessary to prevent the relative humidity from occasionally dropping too low.

Humidification with Cooling. Some additional control over relative humidity may be obtained in this system by adding a water spray in the air stream before the air enters the cooling coil. This spray would be operated in the event of a drop in relative humidity. The water spray performs some sensible cooling of the air and raises its humidity before it enters the cooling coil. Thus the over-all S/L ratio of the system is increased. The degree of control which may be obtained by this method is limited by the ability of the air entering the water spray to absorb moisture.

Reheat-control Requirements. Any of the cooling-cycle air-conditioning systems described earlier can be arranged in such a manner that the function of reheat is provided. In providing a system of reheat the following sequence of operation is necessary:

1. Modulating control of reheat medium to produce heat on a drop in temperature
2. Modulating or two-position control of the cooling medium to offer additional cooling on a rise in temperature
3. Final control of the cooling medium from a relative-humidity controller

When operating refrigerating coils from a relative-humidity controller on a reheat system, it is usually preferable to arrange the equipment in such a manner that the lowest possible coil temperature is obtained when there is a demand for dehumidification and where reheat will be necessary. The reason for this is that the greatest proportion of latent heat is removed from the air at this low coil temperature. Since the proportion of latent heat to sensible heat removed increases with a decrease in coil temperature, less total reheat is required when the system is operated in this manner.

Reheat with Two-stage Cooling. Figure 13-61 illustrates a method of reheat control applied to an installation on which two stages of refrigeration are available. On a rise in temperature, the two-stage thermostat can bring on the two stages of cooling in sequence. Likewise, when the relative humidity is high, the humidity controller operates the cooling equipment at maximum capacity, regardless of the cooling thermostat, to provide positive dehumidification.

Although the diagram indicates use of a two-speed motor on the compressor to provide two-stage cooling, the system could obviously utilize

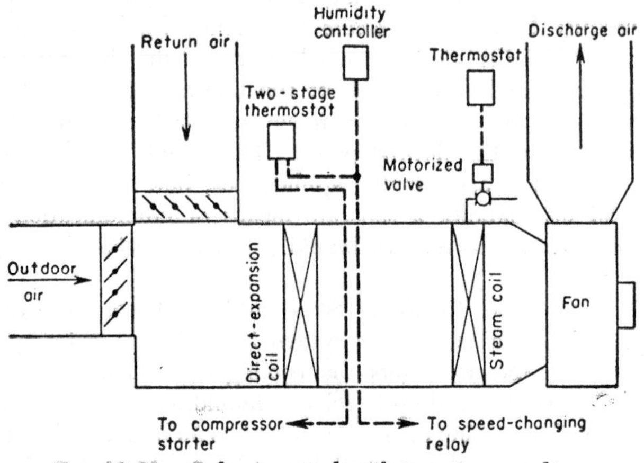

Fig. 13-61. Reheat control with two-stage cooling.

other methods such as two separate compressors or a cylinder bypass on the compressor.

Should the room temperature drop as a result of this demand for dehumidification, the reheat thermostat, set lower than the cooling thermostat, operates the steam valve to provide the necessary degree of reheat. This system is similar to that shown in Fig. 13-60.

This system should maintain uniform temperature and humidity conditions as long as the load is strictly one of cooling and as long as the system is designed in such a manner that the amount of air handled and the coil temperatures maintained bear a proper relation to the maximum sensible and latent loads as well as to the extreme ratios between these loads.

CONTROL OF OUTDOOR-AIR DAMPERS

Ventilation is one of the important functions of the air-conditioning system, and in the usual installation it is required that a certain minimum amount of outdoor air be taken into the system at all times to provide this ventilation.

In addition, however, it can be seen that under some conditions it is much more economical to take outdoor air and condition it instead of recirculating air from the spaces being conditioned.

The control of the outdoor-air dampers, therefore, should bear a relation to the outdoor-air temperature. When the outdoor-air temperature is lower than the temperature in the room, the system should take 100 per cent outdoor air. The economy resulting from this type of control is obvious.

Outdoor-air Control of Dampers. The outdoor-air damper is usually controlled as shown in Fig. 13-62. A modulating damper motor is used to position the return-air and outdoor-air dampers. This motor is controlled by a manual switch that permits setting any minimum required on this

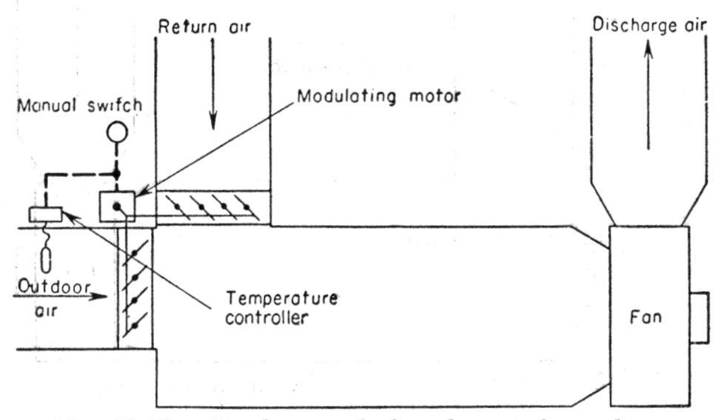

FIG. 13-62. Simple control of outdoor air for cooling.

motor. In the same circuit is a remote-bulb controller measuring the temperature of the outdoor air. This is set at the control point of the inside thermostat. Then, whenever the outdoor air reaches a temperature below the inside control point, the dampers are opened to the 100 per cent outdoor-air position, regardless of the position of the manual positioner. The manual positioner can be located at any convenient point, and it is generally the practice, in theaters and auditoriums, to have the switch located in the manager's office or at some point where the minimum volume of outdoor air can conveniently be adjusted to the number of occupants.

It is obvious that the additional control equipment required for outdoor-air control is well justified in terms of the savings effected by using the cool outdoor air during times when conditions are favorable. Very frequently the outdoor temperature, although quite high during the day, may drop sharply at night. If a building has been thoroughly heated, and has a fairly high internal-heat gain, the temperature can be much higher inside than it is outside. Under that condition, this type of control system takes 100 per cent outdoor air.

Exhaust-damper Control. In tightly constructed buildings, where there is very little leakage, closing off the recirculated-air damper and opening the outdoor-air damper may cause a pressure to be built up which reduces the air volume handled by the fan. On an installation of this type where there is not sufficient leakage, it is necessary to control an exhaust damper simultaneously with the control of the return-air and outdoor-air dampers. This type of control is shown in Fig. 13-63. It consists simply of a second motor operating an exhaust damper and positioned by a dual-control connection with the first motor on the outdoor- and return-air dampers. Then all

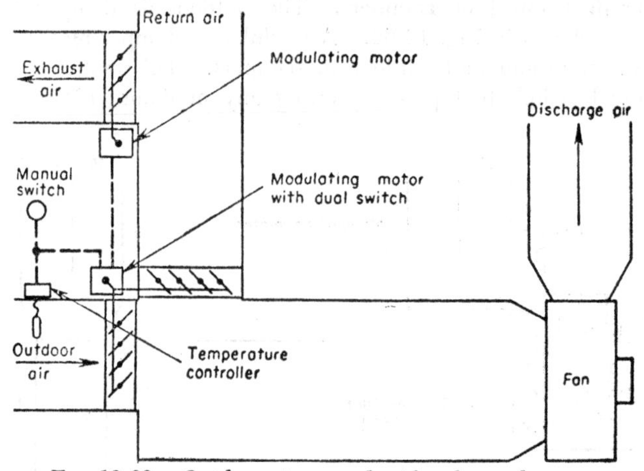

FIG. 13-63. Outdoor-air control with exhaust damper.

three dampers move together, the exhaust and outdoor-air dampers opening as the return-air damper closes.

SUMMER-WINTER SYSTEMS

Where air-conditioning equipment is provided for year-round air conditioning, it is desirable to exercise particular care in the selection of automatic-control equipment, and to pay some additional attention to the automatic-control aspect of the equipment-design problem.

With few exceptions, any of the central-fan heating equipment and control systems described may be combined with any of the central-fan cooling equipment and control systems described. The number of possible combinations is too large to illustrate here. In general, however, a control system suitable for a winter-heating installation, for example, is likely to be suitable for the heating functions of a summer-winter installation.

Change-over. The combination of winter and summer conditioning equipment into one installation, however, introduces an additional control problem. If all elements of the air-conditioning system perform single functions—for example, if separate heating and cooling coils are used—the change-over from heating to cooling or from cooling to heating might conceivably be accomplished by allowing the separate heating and cooling controls to function independently. On a rise in temperature, say, the heating thermostat would shut off the flow of steam to the heating coil, and on a further rise the cooling thermostat would start the operation of the cooling equipment.

But it is not quite as simple as that. For one thing, it is often necessary or desirable to utilize one piece of equipment in both the heating and the cooling cycles. A coil is sometimes used both for heating, with hot water as the medium, and for cooling, with chilled water. Means must then be provided for changing over from one source to the other. Again, the outdoor- and return-air dampers are normally used in both cycles, but the damper motor must be positioned according to different schedules during the heating and cooling cycles. Provision must then be made for transferring command of the damper motor from one controller schedule to the other.

In addition, even the single-purpose equipment elements may require special consideration of change-over means. Even if it is practical to have both heating and cooling capacity available at all times, suitable schedules of control sequence for heating and for cooling may not permit leaving the choice of heating or cooling operation to the independent operation of separate heating and cooling controllers.

It is usually necessary, therefore, to provide for definite change-over, by means of suitable valves or switches, from one mode of operation to the other.

Manual versus Automatic Change-over. If it were always possible to determine that today is the last day of the heating season and tomorrow the cooling season begins, manual means of change-over would be entirely satisfactory. There are, however, during the late spring and early fall periods, times when the demands made on the conditioning system may change, from day to day or even during one day, from a heating to a cooling load and vice versa. During these mild seasons it is often necessary that heat be supplied during the morning and evening hours, and at night if the building is occupied then; whereas temperature and humidity conditions during the day require cooling operation, or at least dehumidification. In some applications, in fact, where very large internal heat gains may occur intermittently, this condition may occur during mild-weather periods in the winter.

Even though a manual change-over can be accomplished easily, if it must be left to the judgment and alertness of the building engineer, some inefficiency and discomfort will often result.

For these reasons, it is generally desirable to provide for automatic change-over. Electric or pneumatic switches and valves may be operated automatically by thermostats or humidity controllers, located within the conditioned space or outdoors, according to the requirements of the specific application. These change-over controls may reverse the action of motors or valves, transfer command of motors or valves from one set of controllers to another, or divert the flow of a medium such as water, in order to convert the mode of operation from heating to cooling or vice versa.

If both the air-conditioning system and the control system are designed with automatic change-over in view, the inclusion of this feature involves only slight additional investment. In view of the gains in operating economy and in comfort of occupants, this investment may be considered self-liquidating.

CONTROL OF REFRIGERATION

Air conditioning makes extensive use of mechanical refrigeration equipment, not only for cooling in the obvious sense, as the removal of sensible heat, but also for dehumidification, or the removal of latent heat. In either of these applications, refrigeration is exactly comparable to heating, except that the process is reversed, that is, removing, not replacing, heat that leaks *into,* not *out of,* the conditioned space.

Means of Dehumidification. For the purpose of dehumidification, other methods are used; for example, *absorption* or *adsorption* of moisture by agents such as silica gel. But the most popular single method is cooling the air below its dew point to precipitate some of its vapor content. And, although this cooling process often makes use of ice, cold water, or absorption-cycle refrigeration, probably the most extensively used method is mechanical refrigeration.

Basic Principles. A brief review of the principles embodied in all varieties of compression-refrigeration machines reveals the essential clues to the possible means of automatic control. And, to begin with, it is important to remember these general rules:

1. *Heat flows from the warmer to the cooler of two adjacent bodies until their temperatures become equal.*
2. *The greater the difference in temperature, the more rapid the transfer of heat.*
3. *At its characteristic boiling point, a liquid changes to vapor if it can absorb heat from its surroundings, or the vapor condenses to liquid if it can give off heat to its surroundings. Which way the change runs depends on whether the surroundings are warmer or cooler than the fluid in question, and in which state the fluid is to begin with.*
4. *Pressure determines the boiling point; the higher the pressure, the higher the boiling point and the condensation temperature of a given substance.*

Note that the quantity of heat that must be gained or lost in the change of state, called *latent heat of vaporization,* is, in general, much greater than the *specific heat,* that is, the heat gained or lost during a unit change in temperature. The latent heat decreases as the boiling point rises with an increase in pressure to the *critical pressure,* above which the fluid cannot exist in the vapor state. But under normal conditions, near standard atmospheric pressure, water, for example, absorbs nearly a thousand times as much heat in vaporizing as in a 1° rise in temperature.

Mechanical-refrigeration Cycle. The way these rules are embodied in the basic operating cycle of any compression-refrigeration machine is illustrated in Fig. 13-64, which schematically represents the essential elements of such a machine.

The drum A, or *receiver*, contains a supply of suitable liquid, the *refrigerant,* under pressure sufficient to ensure that it remains in the liquid state at the room temperatures to which the receiver or the liquid line is exposed. For illustration, assume that the pressure is 150 psig, and that the boiling point of the refrigerant at that pressure is 115°F. As indicated by the

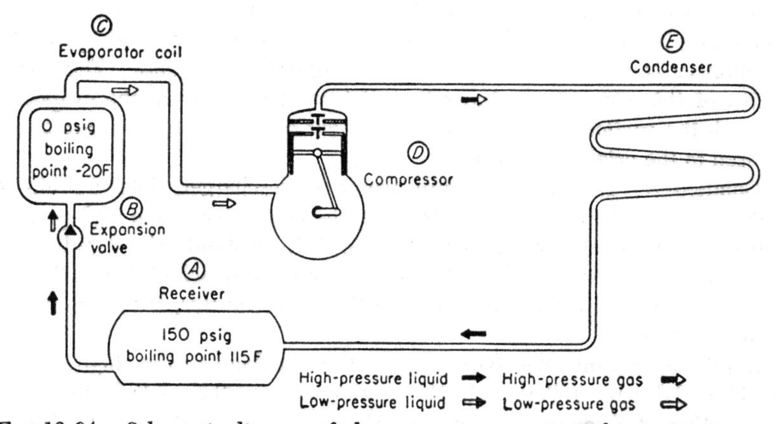

FIG. 13-64. Schematic diagram of elementary compression-refrigeration system.

arrows, liquid refrigerant flows from the receiver through the orifice B, or *expansion valve,* into the *evaporator coil C.* The expansion valve is adjusted to maintain the proper rate of flow and to *step down* the pressure to that maintained in the evaporator.

Assume for illustration, again, that the pressure in the evaporator is 0 psig, that is, atmospheric pressure, equal under standard conditions to approximately 14.7 psia (pounds per square inch *absolute*). Assume, further, that at this pressure the boiling point of the refrigerant is reduced to −20°F, and that the air surrounding the evaporator is at 90°F.

Under these conditions, the refrigerant liquid rapidly vaporizes. This process takes place in two partly overlapping steps. As small particles of the liquid first start to evaporate, they absorb heat from the main body of the liquid refrigerant itself. This process continues until the mixture of vapor and liquid reaches a stable temperature of −20°F. When this condition is reached, additional heat is absorbed from the air surrounding the evaporator.

This second step actually begins as soon as the evaporator temperature falls below the air temperature. It continues as long as the refrigerant exists in the evaporator in liquid form, provided that air at a temperature higher than −20°F is still in contact with the evaporator.

Refrigerated-space Temperatures. In a perfectly insulated refrigerator, the air temperature would fall to −20°F, and no further heat transfer, or evaporation, could occur. But some heat always leaks into the refrigerator, to be absorbed and carried away by the refrigerant. Note also that when the air cools to its *dew point,* for example, 40°F, water vapor in the air begins to condense, giving up its latent heat to the refrigerant. It is this moisture that forms frost on low-temperature evaporator surfaces.

Refrigerant Reused. Since refrigerants are expensive, it would be impractical to waste them. To avoid such waste, the compression-refrigeration system provides a means of reclaiming the vaporized refrigerant and converting it to its original liquid state. This is accomplished by using a compressor *D*, which may be a reciprocating or rotary pump. The suction action of this compressor draws the refrigerant vapor from the evaporator through the *suction* line and into the cylinder of the pump, where it is compressed to the original 150 psig.

Condenser. This high-pressure vapor is discharged into an air- or water-cooled coil called a condenser. In the condenser *E*, the high-pressure vapor gives up heat to the air or water which is constantly circulated around it. The gas is first cooled to its boiling temperature which is also its condensing temperature and then, with further heat transfer, is condensed into its liquid state.

Liquid-refrigerant Receiver. From the condenser, the high-pressure liquid (solid arrow) flows into the receiver, where the cycle begins again.

Function of Expansion Valve. It should be observed, before refrigeration controls are discussed, that the expansion valve need not be considered a control element, even though it does regulate the flow of liquid refrigerant into the evaporator. It corresponds rather to the pressure-reducing valve used in the gas-supply line to a gas burner. The function of the expansion valve is to regulate the flow of refrigerant into the evaporator coil so that the maximum amount of refrigeration can be provided without waste.

If too much refrigerant is allowed to enter the coil in liquid form, it passes through the system so rapidly that complete evaporation cannot take place within the coil. If too little refrigerant is supplied, the full heat-absorbing capacity of the system is not utilized.

Liquid in Suction Line. If so much refrigerant is supplied to the evaporator that a portion of it leaves the coil in liquid form, it is obvious that some additional evaporation takes place in the suction lines leading back to the compressor. This is undesirable, since the heat necessary for this effect is drawn from the air surrounding the suction lines, in spaces where cooling is of no value. Liquid in the suction lines is undesirable for another reason, namely, that it may be drawn into the cylinder of the compressor. Since liquid is not compressible, severe mechanical damage to the compressor valves may result.

Types of Expansion Valves. The expansion valve may take any one of several mechanical forms, such as a fixed orifice, a float-operated liquid-level valve, a pressure-operated valve, or a thermal-expansion unit. In any case, it serves as a high-resistance passageway which performs the dual purpose of metering refrigerant to the coil and maintaining a pressure difference between the *high side* and the *low side* of the system.

Thermal-expansion Valve. The thermal-expansion valve is very popular because of its effectiveness in promoting both evaporator efficiency and compressor safety; it permits filling the evaporator with liquid while guarding against carry-over of liquid into the suction lines. A self-contained thermostatic valve with its bulb fixed firmly in contact with the suction end of the evaporator coil closes down its orifice and reduces the inflow of liquid when a drop in temperature at the bulb indicates that unevaporated liquid has reached the farthest point in the coil compatible with safety from *slugs,* that is, liquid in the suction line.

CONTROL REQUIREMENTS

Various substances are used as refrigerants, the selection being determined by the characteristics of the refrigerant, particularly the temperature-pressure characteristics, in relation to the conditions and requirements of the application. Whatever the refrigerant used, and whatever the variations and refinements in construction of the refrigeration equipment, the basic requirements for automatic control of the refrigeration process are implied in the elementary refrigeration cycle outlined in the preceding paragraphs.

These requirements are fundamentally similar to those governing the application of automatic control to a heating system. In the conventional heating system, controls are provided to accomplish two coordinated but separate functions:

1. The safe and efficient *production* of an adequate supply of heating medium, such as steam, hot water, or warm air
2. The *distribution* of that heating medium in accordance with demands from the conditioned space

The same two functions must be performed when the process of refrigeration is involved, and controls may therefore be divided into (1) those directly related to the operation of the compressor, condenser, and evaporator; and (2) those which are used for regulating the distribution of the cooling medium to various refrigerated areas in accordance with load requirements.

In the smaller and simpler installations, for example, a domestic refrigerator or a cold-storage cabinet, the control of the compressor may also serve to control the delivery of cooling effect. Therefore, this aspect of refrigeration control is reviewed first, after which the commoner methods of regulating distribution are examined.

CAPACITY CONTROL

The first, and in many ways the simplest and most direct, method of controlling the operation of the refrigeration machine is suction-pressure control. *Suction pressure,* incidentally, must *not* be taken in the popular sense of *the opposite of pressure.* Although in some cases, such as steam-boiler operation, pressures below atmospheric may conveniently be measured by a gage reading inches of vacuum on a scale running from *zero gage* toward *zero absolute,* that is, in the direction opposite to that of the *pounds-per-square-inch* scale, in refrigeration calculations all pressures must be taken on the *absolute* scale. Thus the suction pressure, or *pressure in the suction line,* may be in the vacuum range or in the range above atmospheric pressure, depending entirely on the characteristics of the refrigerant used and the evaporator temperatures required. Therefore, a reference to an increase in suction pressure means simply an increase in *absolute* pressure in the suction line.

Pressure an Index of Temperature. Referring again to the discussion of Fig. 13-64 and to the *rules* at the beginning of this chapter, it is evident that

1. *The cooling rate at the evaporator depends on the temperature of the boiling refrigerant, for any temperature of the air coming in contact with the evaporator.*
2. *The boiling point rises and falls with the pressure in the evaporator.*

Consequently, the suction pressure, which is also the pressure in the evaporator, serves as an accurate indication of the evaporator-coil temperature. On many applications where the cooling load is a constant factor, the maintenance of a fixed coil temperature is all that is required for accurate control.

This condition can be compared to a heating system utilizing hot water as the heating medium, in a structure which is subject to absolutely constant heat loss. Under such conditions, the exact temperature of water required to offset those heat losses can be determined and precise room temperature control maintained as long as water is supplied at the temperature calculated.

Thus, in the refrigeration system, if the evaporator, corresponding to the radiator in the heating system, is held at a constant temperature in an area where the heat gain, corresponding to the heat loss in the heating system, is constant, accurate temperature control is obtained.

Suction-pressure Control. Figure 13-65 schematically illustrates the operation of a suction-pressure controller. This is simply a pressure control of suitable scale range, for example, 11 in. vacuum to 35 psig, and suitable differential. It is reverse-acting by heating-control standards; that is, it closes the circuit on a rise in pressure, so as to start the compressor motor.

When enough gaseous refrigerant has been drawn from the suction line and evaporator and driven into the condenser and receiver to lower the suction-line pressure by the desired amount, the control opens the circuit and stops the motor. This method of control by itself serves satisfactorily where the cooling load is constant, or varies only a little. It is inadequate for handling considerable changes in load, because it operates to maintain the same evaporator temperature, regardless of whether that evaporator temperature is just right, too high, or too low to maintain the desired temperature in the fixture or the space.

The dotted lines in Fig. 13-65 show how a *high-pressure cutout* may be, and usually is, wired into the circuit in order that it may be able to stop the motor on an abnormal rise in discharge pressure. This is a safety control,

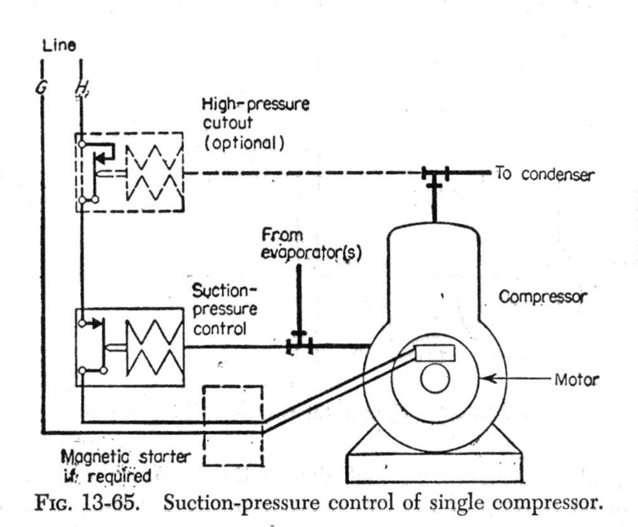

FIG. 13-65. Suction-pressure control of single compressor.

analogous to the high-limit control normally used on a heating plant. It is shown in the diagram as a separate unit, but is often combined with the suction-pressure control, a separate bellows being connected to the discharge line and arranged so that it can override the low-pressure bellows and open the switch. On small machines, such as domestic refrigerators, the high-pressure cutout is sometimes omitted, and reliance is placed on the thermal-overload switch in the motor for shutdown in event of excessive head pressure.

When the suction-pressure control is the operating control, adjustment of the temperature to be maintained is provided by adjusting the pressure setting of the control. Raising the setting permits a corresponding rise in the temperature of the evaporator and thus, indirectly, of the conditioned space.

Temperature Control of Compressor. Another method of controlling the operation of the compressor is illustrated in Fig. 13-66. Here a remote-bulb thermostat is substituted for the suction-pressure control. The bulb is placed where it is exposed to temperatures representative of the cooling load; this may be in the storage space on cold-storage applications, for example, or in the return-air stream on a unit room cooler. A room-type thermostat may be used in place of the remote-bulb thermostat. Thermostatic control has the advantage of responding more directly and more fully to variations in the cooling load than can the suction-pressure method.

Temperature-pressure Control Combined. A temperature control and a suction-pressure control are sometimes wired in series, to provide thermostatic operation under normal conditions and a *limit control* shutdown under abnormal conditions, such as excessive frosting of the coils, reflected in an excessively low suction pressure. Once the suction pressure has fallen to the cutout setting of the pressure control, both the fixture (conditioner

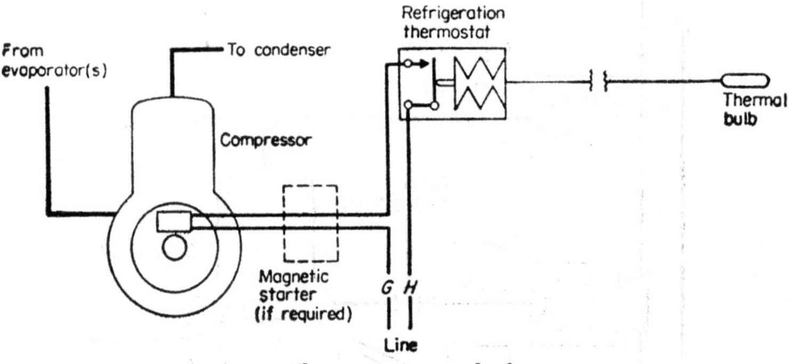

Fig. 13-66. Thermostat control of compressor.

or refrigerator) temperature and the suction pressure must rise to the respective on settings of the two controls before the compressor can start again. In such cases, the suction pressure on setting is selected so that it corresponds to a coil temperature above 32°F. Then the compressor cannot start until the coil temperature has risen above the freezing point far enough to assure defrosting.

On this application it is customary to set the pressure control for a relatively low cutout point, to prevent cycling from pressure-control action alone.

Various combination temperature-pressure controls are in use to provide automatic defrosting on normal cycles, in *above-freezing* applications. In one type, a temperature-actuated bellows engages the switch linkage to trip out the switch on a drop in temperature, but disengages the linkage on a temperature rise; the suction pressure, therefore, must rise to the pressure cutin setting before the switch trips in to start the compressor.

Stage Control of Multiple Compressors. Shown in Fig. 13-67 is one method of control for greater range of load than can be handled satisfactorily by one compressor. Two compressors are connected to the same suction and discharge lines; these may lead to one or more evaporators, and additional controls may be provided at the fixtures for regulating delivery of the cooling effect. As far as the compressors are concerned, the control problem is to maintain a relatively constant *potential,* or ratio of cooling capacity to cooling demand, at the source. It must be remembered that the capacity is limited by the maximum quantity of refrigerant that the compressor can circulate. The diagram shows two suction-pressure controls connected into the common suction line and wired into the circuit so that each controls one of the compressors. Their scale settings are chosen so that, whenever the suction pressure rises to a level indicating that the first

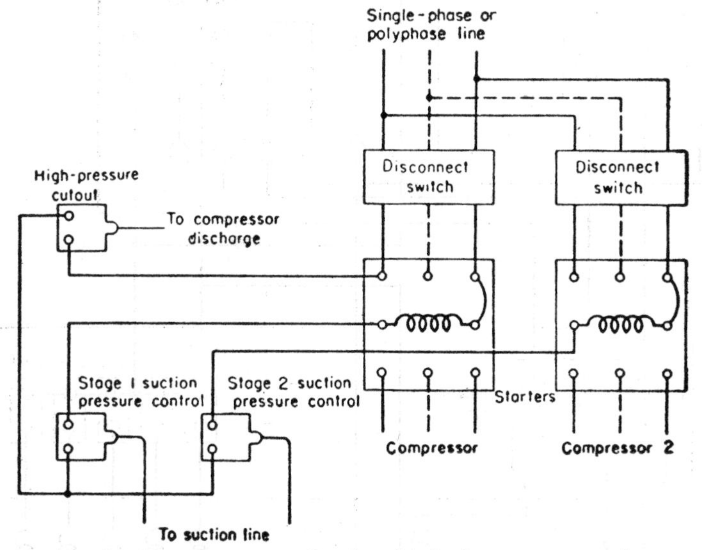

FIG. 13-67. Stage control with individual pressure controls.

compressor cannot handle the load, the second one cuts in. The high-pressure cutout is connected in series with both operating controls so as to stop both compressors on an excessive rise in discharge pressure. One can be used for both, since it acts merely as a safety or limit control.

Control Settings Critical. The method shown in Fig. 13-67 might be extended to provide three, four, or almost any number of stages of compressor capacity, with a suction-pressure control in command of each stage. The more stages combined, however, the more difficult it is to make sure that the successive pressure-control settings remain in step when adjusted for full range of capacity over a relatively narrow range of pressures. Each control is independent of the others, each is bound to show some variation from its nominal calibration, and these variations are almost never consistent.

Sequence-switch Control of Multiple Compressors. A simpler way of obtaining dependable sequence control is shown in Fig. 13-68. This is one that finds considerable use in air-conditioning work, particularly, where the demand for a wide range of refrigeration capacity, to match the variations in *sensible* and *latent* cooling loads, is probably as great as in any refrigeration application. A single suction-pressure control of the modulating type is connected to the suction line, and positions a modulating motor which actuates a sequence-switching mechanism. As the suction pressure rises,

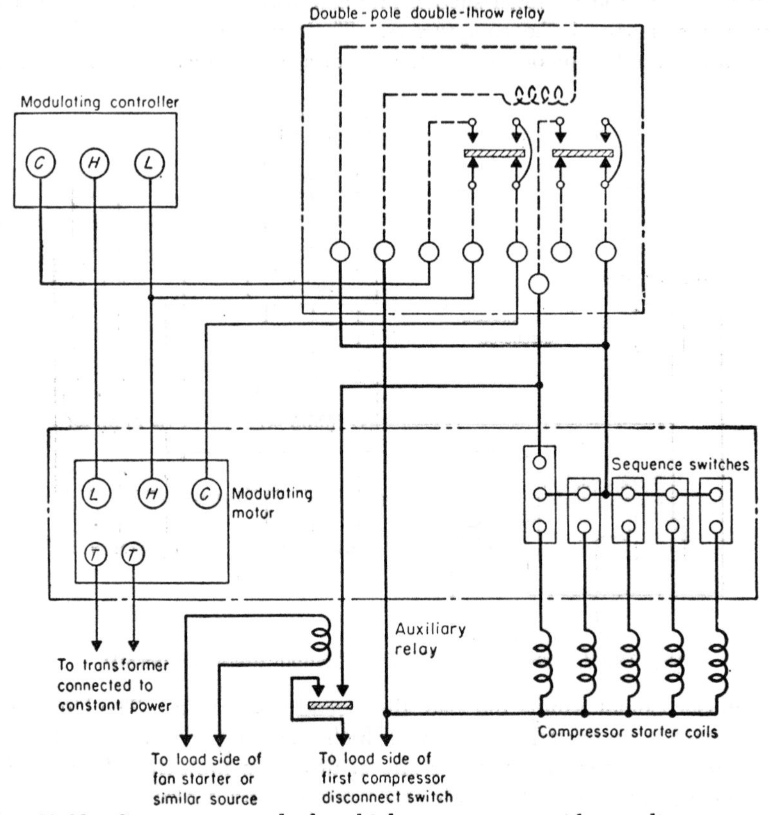

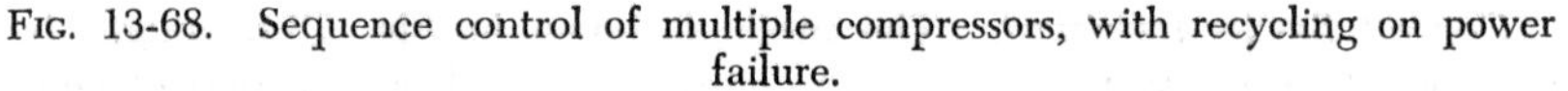

Fig. 13-68. Sequence control of multiple compressors, with recycling on power failure.

for example, the motor is positioned so as to close one switch after another, to energize successive motor starters, and to put successive compressors into operation.

The relay included in the diagram serves a safety purpose. A power interruption may occur while the modulating motor is in a position calling for operation of several compressors. If the corresponding starter circuits remain closed, then, on restoration of power the *inrush* current to several compressor motors under the heavy starting load may overload and damage the electrical wiring. Study of the circuit diagram shows that the relay drops out upon interruption of power, even a momentary one, and breaks the common power lead to the sequence switches. It also switches the control circuit so as to drive the modulating motor to the off position on resumption of power, whereupon a pull-in circuit is completed by the out contacts in the first-stage switch. Once the relay pulls in again, it provides its own holding circuit, reestablishes the power supply to the sequence switches, and restores the normal authority of the pressure control over the modulating motor. Thus the compressors are started, one at a time, until the capacity called for by the pressure control is provided.

Economic Factors. At this point it should perhaps be noted that the use of several compressors to provide a wide range of capacity is dictated chiefly by economic considerations. It would be possible, of course, to use

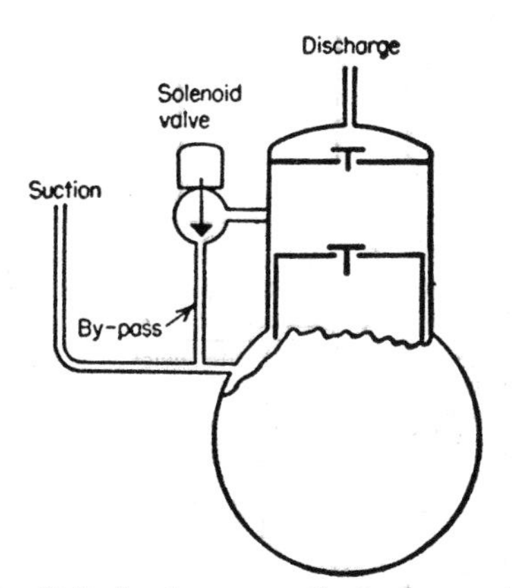

FIG. 13-69. Cylinder bypass method of capacity control.

a single compressor large enough for the heaviest cooling load, and to reduce the length and frequency of operating cycles as necessary for matching lighter loads. Conversely, in heating and power installations, it would be possible to use several boilers in a multistage system for adjusting the heating capacity to varying demands for steam. Except in very large installations with unusually wide ranges of demand, however, it is more economical to use a single boiler operating at various percentages of maximum capacity, since a boiler is reasonably efficient even when operating at a fraction of full capacity. With compressors, however, it is generally more economical to use several smaller ones in varying combinations than to use a single large one, mainly because compressor efficiency falls off seriously, and operating cost rises, at considerably less than full-capacity operation.

Cylinder Bypass or Cutout Control. Several methods of reducing the volume of gas compressed have been used, for adjusting compressor capacity to load or for reducing starting effort, or both. In Fig. 13-69, it is evident that opening a solenoid valve in a bypass line from the suction line to a port in the side of the cylinder allows some of the gas drawn in during the suction stroke to be driven back into the suction line before compression begins. A method frequently used with multicylinder compressors provides means of holding open the intake valves (in the cylinder head) on one or more cylinders, during a start or during operation under light refrigeration load. Either method, like others in use, reduces the volume of refrigerant compressed and therefore reduces the effective capacity of the compressor. Thus, multistage operation can be obtained with a minimum number of compressors.

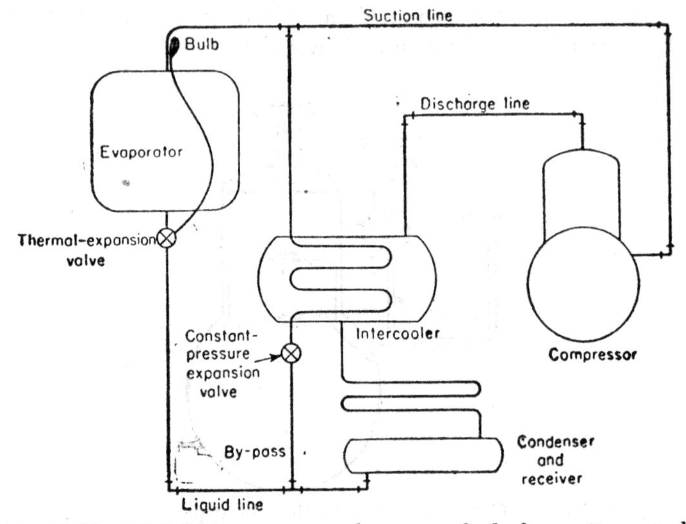

FIG. 13-70. Modulating evaporator-bypass method of capacity control.

Modulating Control of Capacity. The control systems so far discussed are *two-position* or on-off systems. A patented control system for small compressors is described as giving modulation from maximum capacity of the compressor to zero, or over the full range of each step in a multistage system. Figure 13-70 shows a bypass leading from the liquid line, ahead of the evaporator, to the suction line. When the valve in the bypass is open, liquid refrigerant flows through a coil surrounded by hot gas from the compressor discharge. Liquid refrigerant evaporates and precools the compressed gas before it enters the condenser and receiver. Thus heat-absorbing capacity not needed in the main evaporator is utilized in an auxiliary condenser, and the flow of gas into the suction line is maintained at a rate to prevent excessive decrease of suction pressure and evaporator temperature.

What makes this system function as a modulating control is, of course, the bypass valve. This is a constant-pressure type of expansion valve, which admits varying amounts of refrigerant to the bypass in response to small variations in suction pressure. Since an increase in the flow through the bypass counteracts the fall in suction pressure which caused the valve to open more widely, correct design and adjustment of the system can evidently permit a wide variation in the rate of refrigerant flow through the main evaporator, with no more than a normal swing of suction pressure.

CONTROL OF DISTRIBUTION

Almost any of the foregoing methods for regulating the rate at which *cooling effect* is generated may be used in combination with any of several means of regulating the delivery or distribution of the cooling effect. With a single *fixture,* that is, walk-in cooler, storage cabinet, or the like, or with several fixtures nearly identical in cooling requirements, the method

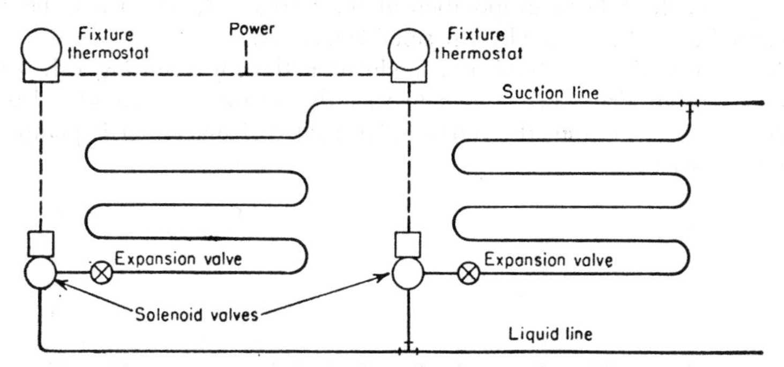

FIG. 13-71. Thermostat control of solenoid valves in liquid lines to evaporators.

shown in Fig. 13-66, with a thermostat in one of the fixtures, provides control of delivery directly by control of capacity from fixture temperature. Similarly, the sequence-switching method shown in Fig. 13-68 may be used on a refrigeration system serving a cooling or dehumidifying coil in an air-conditioning system; and a modulating thermostat in the duct, for example, may be substituted for the pressure control, to regulate the refrigeration rate in accordance with the demand for cooling effect at the coil. But various methods of distribution control, separate from capacity control, are in use.

Solenoid-valve Control of Liquid Refrigerant. Figure 13-71 illustrates one of the most popular methods of distribution control. It is obviously analogous to the valve method of zone control in heating. A thermostat in each *fixture,* which may be a zone-cooling unit, for example, opens and closes a solenoid valve in the liquid refrigerant line to the evaporator in that fixture, so as to maintain the evaporator temperature within the limits necessary for keeping the air temperature between the on and off points of the thermostat. Closing one or more valves, of course, tends to reduce the pressure in the common suction line, and any of the various forms of suction-pressure control may be used for adjusting the capacity at the source, according to the demand.

Similarly, in central-fan conditioning systems, valves may be used to cut off the supply of liquid refrigerant, or of chilled water in indirect refrigeration systems, from one or more sections of a sectional coil in the primary conditioning system.

Modulating Control of Coolant Flow. A variant on the methods of using solenoid valves consists of substituting throttling or modulating valves. This also implies the use of modulating controllers instead of the positive-action controllers to position the valves.

Modulating Control of Air Flow. Another method that may be used, particularly in air-conditioning systems, consists of face and bypass dampers which vary the percentage of circulated air passing through the cooling or dehumidification coils. Changes in the volume of air handled cause changes in the rate of evaporation of the refrigerant, which may be compensated for at the source by any suitable method.

Three-way Valves. Similarly, modulating three-way mixing valves may be used in an air-washer system to vary the mixture of recirculated water and chilled water from the refrigeration heat exchanger that is pumped to the washer jets.

INDEX